BASIC OP AMP MODULES

CIRCUIT	BLOCK DIAGRAM	GAINS

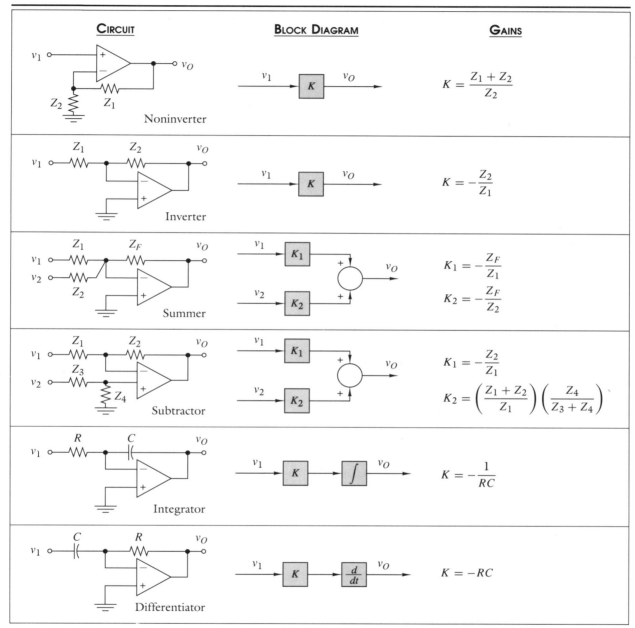

$$K = \frac{Z_1 + Z_2}{Z_2}$$

Noninverter

$$K = -\frac{Z_2}{Z_1}$$

Inverter

$$K_1 = -\frac{Z_F}{Z_1}$$

$$K_2 = -\frac{Z_F}{Z_2}$$

Summer

$$K_1 = -\frac{Z_2}{Z_1}$$

$$K_2 = \left(\frac{Z_1 + Z_2}{Z_1}\right)\left(\frac{Z_4}{Z_3 + Z_4}\right)$$

Subtractor

$$K = -\frac{1}{RC}$$

Integrator

$$K = -RC$$

Differentiator

The Analysis and Design of Linear Circuits

THIRD EDITION

THE ANALYSIS AND DESIGN OF LINEAR CIRCUITS

ROLAND E. THOMAS
Professor Emeritus
United States Air Force Academy

ALBERT J. ROSA
Professor and Chairman of Engineering
University of Denver

JOHN WILEY & SONS, INC.
New York • Chichester • Weinheim • Brisbane • Singapore • Toronto

Acquisitions Editor	Bill Zobrist
Marketing Manager	Katherine Hepburn
Production Editor	Leslie Surovick
Production Management Services	Ingrao Associates
Cover Photograph	Dr. Jeremy Burgess/Photo Researchers

This book was typeset by Pine Tree Composition and printed and bound by Von Hoffmann Press. The cover was printed by Phoenix Color Corporation.

This book is printed on acid-free paper.

Library of Congress Cataloging-in-Publication Data

Thomas, Roland E., 1930–
    The analysis and design of linear circuits / Roland E. Thomas, Albert J. Rosa.—3rd ed.
        p. cm.
    Includes index.
    ISBN 0-471-38679-0 (cloth : alk. paper)
    1. Electric circuit analysis. 2. Electric circuits, Linear—Design and construction. I. Rosa, Albert J., 1942– II. Title.
    TK454 .T466 2000
        621.319′2—dc21   00-023753

Printed in the United States of America

10 9 8 7 6 5 4 3 2 1

*To our wives
Juanita and Kathleen*

PREFACE

The Analysis and Design of Linear Circuits, Third Edition, continues the authors' commitment to three guiding tenets. First, we believe that circuit courses should emphasize topics and techniques used in subsequent upper-division courses. Since these courses have evolved over time, the third edition places more emphasis on systems and signal processing through the use of block diagrams, input–output relationships, and transform methods of describing signals. Second, we believe that in addition to providing early motivation, design problems are a valid method of evaluating a student's comprehension of basic circuit analysis principles. Thus, in addition to teaching basic principles, the book illustrates the application of these principles to real-world design and evaluation problems. Third, we believe that explicitly stated learning objectives make a valuable contribution to the development of problem-solving skills. Consequently, the book provides a set of carefully structured intermediate steps called en route objectives (EROs) that define the basic knowledge and understanding needed to achieve the course mastery.

WHAT'S NEW IN THE THIRD EDITION

- The integrated treatment of computer tools is expanded to five programs: Excel®, MATLAB®, Mathcad®, Electronics Workbench®, and PSpice®. The spreadsheets, math solvers, and circuit simulators are all intrinsic parts of today's engineering environment. It is important for students to begin to develop an analysis style that includes the intelligent use of such tools.

- The integrating problem concept used in previous editions is streamlined by placing an expanded number of such problems at the end of each chapter, thereby enhancing text modularity. Integrating problems require the students to use knowledge gained in several chapters and drive home the point that real-world problems require that they learn how to draw upon their previous experience in new situations.

- The Fourier series chapter has been moved forward so it immediately follows the discussion of frequency response in Chapter 12. This change continues the composite signal representation begun in Chapter 5 (signal waveforms) and enhances the signal processing emphasis in the third edition.

- The treatment of mutual inductance is consolidated in a new Chapter 15 and placed immediately before the treatment of ac power in Chapter 16. For schools that emphasize classical topics, these chapters can be treated immediately after the introduction to phasor analysis in Chapter 8. Schools employing the first edition's "Laplace first" approach can still do so by delaying Chapter 8 until after Chapter 12 (frequency response).

TOPICAL COVERAGE

Chapters 1–4 use resistance circuits to introduce traditional circuit analysis topics including device i–v characteristics, Kirchhoff's laws, equivalent circuits, node and mesh analysis, Thévenin theorem, and superposition. In addition, Chapter 4 introduces active elements including the ideal OP AMP. This model is then used to develop a basic set of the signal processing building blocks.

Chapters 5–8 begin the study of dynamic circuits whose responses depend on both current and previous inputs. Chapter 5 (Signal Waveforms) develops the signal models used to describe dynamic responses while Chapter 6 (Capacitance and Inductance) introduces the circuit elements needed to account for dynamic responses. This background allows the student to gain confidence prior to launching into the complexities of the dynamic circuit analysis topics presented in Chapter 7 (First- and Second-Order Circuits) and Chapter 8 (Sinusoidal Steady-state Response).

Chapters 9–12 use Laplace transforms to develop the s-domain approach to the analysis and design of dynamic circuits. Topics covered include pole-zero diagrams, zero-state and zero-input responses, circuit theorem, network functions, impulse response, step response, frequency response, and Bode diagrams. Computer tools and design examples are used extensively in this block to illustrate practical applications of s-domain concepts.

Chapters 13–16 extend and apply the analysis and design concepts developed in the first 12 chapters. Chapter 13 uses the Fourier series to combine signal representation with frequency response to introduce the concept of a signal spectrum. Chapter 14 then builds on this concept to develop methods of designing frequency selective analog filters. The concept of self-inductance is extended to mutual inductance in Chapter 15 where the applications of transformers are discussed. Finally, Chapter 16 extends the concept of average power from Chapter 8 to include complex power and power flow in single- and three-phase circuits.

KEY FEATURES

Each chapter includes the following features:

1. The **Chapter Opener** provides a brief discussion of the historical context as well as an overview of the topical content of the chapter.

2. **Examples and Exercises**   The third edition includes nearly 300 worked-out examples that apply basic chapter concepts to specific problems. There are over 200 in-text exercises that allow students to practice the analysis or design techniques discussed in the preceding examples.

3. **Design Examples** The third edition includes 40 worked-out design examples. Students are first introduced to circuit design in Chapter 3. Important design coverage also appears in Chapters 4, 6, 7, 8, 11, 12, and 14. Several of the design examples involve evaluating alternate designs, while others emphasize a building block approach to circuit realization.

4. **Computer Examples** The worked examples that involve computer tools break down as follows: Excel (4), Mathcad (7), MATLAB (12), PSpice (6), and Electronics Workbench (6). These examples usually require some preliminary analysis or design to set up the problem in a form amenable to the use of computer tools.

5. A **Chapter Summary** appears at the end of each chapter narrative. The important concepts are summarized in concise statements that help the student review before tackling the homework problems that follow.

6. **Homework Problems** The book includes nearly 900 homework problems, most of them new to the third edition. Most of the homework problems are directly related to and grouped with the En Route Objectives. This arrangement provides clear statements of the intended learning and identifies which problems test which problem-solving skills. About thirty percent of the homework problems have answers in the back of the book. Problems involving design are labeled with a **D** icon.

7. **Integrating Problems** These problems appear at the very end of a chapter and require proficiency across several En Route Objectives. With these problems the student must combine previously mastered skills and choose a problem solving approach. Integrating problems are labeled with **A**, **D**, and **E** icons to indicate whether analysis, design, and evaluation methods are required.

8. **Computer Problems** Those problems for which computer tools offer a significant advantage are marked with a computer icon . As with integrating problems, the students must choose which tool to use. To assist the student in developing this ability, about 100 of the homework problems are solved using computer tools in a separately bound **Student Solutions Manual**. These problems are labeled with an icon.

COURSE DESIGN OPTIONS

The primary goal of the third edition is to provide greater flexibility in course design. The book contains enough material for a full-year sequence in circuit analysis. Assuming three lectures per week, a two-semester sequence can cover Chapters 1–8 in the first semester, leaving Chapters 9–16 for the second semester. Chapter 8 (Sinusoidal Steady-State Response) can be delayed to the second semester and replaced by Chapter 9 (Laplace Transforms) for schools preferring the Laplace before phasors approach. For a three-quarter sequence the logical break points are Chapters 1–6, Chapters 7–11, and Chapters 12–16. Again, schools preferring the Laplace first approach can slide Chapter 12 (Frequency Response) into the second quarter and delay Chapter 8 to the third quarter.

A two-quarter sequence can cover Chapters 1–6 in the first quarter. The instructor has several options for the second quarter. For example, covering Chapters 7, 8, 13, 15, and 16 produces a strong ac circuit emphasis, while treating Chapters 7, 9, 10, 11, and 12 produces a transform/systems emphasis. Other areas of emphasis are possible.

A one-semester course obviously requires that some material in the first eight chapters be deferred to other courses in order to make room for application topics. With careful selection of subsequent topics and homework problems, the following sections in the first eight chapters can be deferred: Section 3–6 (Interface Circuit Design), Section 4–3 (The Transistor), Section 4–6 (OP AMP Circuit Design), Section 4–7 (The Comparator), Section 6–3 (Dynamic OP AMP Circuits), and the treatment of second-order circuits in Chapter 7 (Sections 7–5 through 7–8). With these omissions, a cohesive one-semester course can be designed by covering Chapters 1–8 and then adding one or more areas of emphasis such as:

- Electronics and Instrumentation—Sections 3–6, 4–3, 4–6, and 4–7.
- Transformers and ac Power—Chapters 15 and 16.
- Frequency Response—Deleting Chapter 8 and covering Chapters 9, 10, 11, and 12.
- Signal Processing—Deleting Chapter 8 and covering Chapters 9, 10, 12, 13, and 14.

TEXTBOOK SUPPLEMENTS

1. Student Solutions Manual

Select end-of-chapter problems have been solved using at least one of the following commercial computer tools: MATLAB, Mathcad, Microsoft Excel, Electronics Workbench, or PSpice. These worked out solutions are available in a bound manual that is sold separately. Faculty can order a student bundle containing both the text and the Student Solutions Manual at a special reduced cost.

2. Instructor's Manual

A complete set of detailed solutions to all of the end-of-chapter problems was prepared and documented using Mathcad. These solutions are available in the printed Instructor's Manual and are available free to all adopting faculty. The solutions were given two independent accuracy checks.

3. *Analysis & Design of Linear Circuits, 3rd edition* Website

A title-specific website is available at www.wiley.com/college/thomas. Updates to the existing package and additional resources for both students and instructors will be made available via this site. Check here periodically for updates.

4. **Circuits Extra**

   In an effort to help teachers teach and students learn, John Wiley & Sons has built a circuit website centered on your needs. Circuits Extra provides a variety of resources for both students and instructors. You can access this ever-growing library of material at www.wiley.com/college/circuits.

ACKNOWLEDGMENTS

The authors acknowledge their indebtedness to the many colleagues who have given valuable assistance with one or more editions of this book: Robert M. Anderson, Iowa State University; Doran J. Baker, Utah State University; William E. Bennett, United States Naval Academy; Maqsood A. Chaudhry, California State University at Fullerton; Don E. Cottrell, University of Denver; Michael L. Daley, University of Memphis; James G. Gottling, Ohio State University; Robert Kotiuga, Boston University; Hans H. Kuehl, University of Southern California; K.S.P. Kumar, University of Minnesota; Michael Lightner, University of Colorado at Boulder; Jerry I. Lubell, Jaycor; Reinhold Ludwig, Worcester Polytechnic Institute; Lloyd W. Massengill, Vanderbilt University; Frank L. Merat, Case Western Reserve University; Richard L. Moat, Motorola; William Rison, New Mexico Institute of Mining and Technology; Martin S. Roden, California State University at Los Angeles; Pat Sannuti, the State University of New Jersey; Jacob Shekel, Northeastern University; Kadagattur Srinidhi, Northeastern University; Peter J. Tabolt, University of Massachusetts at Boston; Len Trombetta, University of Houston; David Voltmer, Rose-Hulman Institute of Technology; and Bruce F. Wollenberg, University of Minnesota. We are also indebted to John C. Getty of the University of Denver for his assistance in developing an associated laboratory manual for the first and second editions, and to Ronald R. DeLyser for developing the Student Solutions Manual for the third edition. The editorial assistance of Sondra Chavez, Alan Apt, and Bill Zobrist is also acknowledged. The copy editing provided by Suzanne Ingrao made our task immeasurably easier. Finally, we again thank our wives for their patience and understanding.

CONTENTS

CHAPTER 1 • INTRODUCTION   1
  1–1   About This Book   2
  1–2   Symbols and Units   3
  1–3   Circuit Variables   4
        SUMMARY   11
        PROBLEMS   12
        INTEGRATING PROBLEMS   15

CHAPTER 2 • BASIC CIRCUIT ANALYSIS   16
  2–1   Element Constraints   17
  2–2   Connection Constraints   22
  2–3   Combined Constraints   30
  2–4   Equivalent Circuits   35
  2–5   Voltage and Current Division   41
  2–6   Circuit Reduction   48
  2–7   Computer-aided Circuit Analysis   54
        SUMMARY   58
        PROBLEMS   58
        INTEGRATING PROBLEMS   68

CHAPTER 3 • CIRCUIT ANALYSIS TECHNIQUES   72
  3–1   Node-voltage Analysis   73
  3–2   Mesh-current Analysis   88
  3–3   Linearity Properties   97
  3–4   Thévenin and Norton Equivalent Circuits   105
  3–5   Maximum Signal Transfer   118
  3–6   Interface Circuit Design   121
        SUMMARY   131
        PROBLEMS   132
        INTEGRATING PROBLEMS   144

CHAPTER 4 • ACTIVE CIRCUITS   148
  4–1   Linear Dependent Sources   149
  4–2   Analysis of Circuits with Dependent Sources   151

4–3   The Transistor   164
4–4   The Operational Amplifier   169
4–5   OP AMP Circuit Analysis   176
4–6   OP AMP Circuit Design   191
4–7   The Comparator   203
        SUMMARY   206
        PROBLEMS   207
        INTEGRATING PROBLEMS   218

CHAPTER 5 • SIGNAL WAVEFORMS   224
5–1   Introduction   225
5–2   The Step Waveform   227
5–3   The Exponential Waveform   231
5–4   The Sinusoidal Waveform   235
5–5   Composite Waveforms   243
5–6   Waveform Partial Descriptors   248
        SUMMARY   255
        PROBLEMS   256
        INTEGRATING PROBLEMS   261

CHAPTER 6 • CAPACITANCE AND INDUCTANCE   264
6–1   The Capacitor   265
6–2   The Inductor   272
6–3   Dynamic OP AMP Circuits   278
6–4   Equivalent Capacitance and Inductance   285
        SUMMARY   288
        PROBLEMS   289
        INTEGRATING PROBLEMS   293

CHAPTER 7 • FIRST- AND SECOND-ORDER CIRCUITS   298
7–1   *RC* and *RL* Circuits   299
7–2   First-order Circuit Step Response   309
7–3   Initial and Final Conditions   318
7–4   First-order Circuit Sinusoidal Response   323
7–5   The Series *RLC* Circuit   330
7–6   The Parallel *RLC* Circuit   340
7–7   Second-order Circuit Step Response   345
7–8   Some Other Second-order Circuits   352
        SUMMARY   356
        PROBLEMS   357
        INTEGRATING PROBLEMS   366

CHAPTER 8 • SINUSOIDAL STEADY-STATE RESPONSE   370

8–1   Sinusoids and Phasors   371
8–2   Phasor Circuit Analysis   377
8–3   Basic Circuit Analysis with Phasors   382
8–4   Circuit Theorems with Phasors   393
8–5   General Circuit Analysis with Phasors   403
8–6   Energy and Power   413
      SUMMARY   418
      PROBLEMS   419
      INTEGRATING PROBLEMS   428

CHAPTER 9 • LAPLACE TRANSFORMS   430

9–1   Signal Waveforms and Transforms   431
9–2   Basic Properties and Pairs   435
9–3   Pole-zero Diagrams   444
9–4   Inverse Laplace Transforms   447
9–5   Some Special Cases   453
9–6   Circuit Response Using Laplace Transforms   458
9–7   Initial Value and Final Value Properties   468
      SUMMARY   471
      PROBLEMS   471
      INTEGRATING PROBLEMS   477

CHAPTER 10 • s-DOMAIN CIRCUIT ANALYSIS   480

10–1   Transformed Circuits   481
10–2   Basic Circuit Analysis in the s Domain   490
10–3   Circuit Theorems in the s Domain   495
10–4   Node-voltage Analysis in the s Domain   506
10–5   Mesh-current Analysis in the s Domain   519
10–6   Summary of s-Domain Circuit Analysis   525
       SUMMARY   528
       PROBLEMS   528
       INTEGRATING PROBLEMS   536

CHAPTER 11 • NETWORK FUNCTIONS   538

11–1   Definition of a Network Function   539
11–2   Network Functions of One- and Two-port Circuits   542
11–3   Network Functions and Impulse Response   554
11–4   Network Functions and Step Response   556
11–5   Network Functions and Sinusoidal Steady-state
       Response   561

11–6   Network Function Design   566
        SUMMARY   581
        PROBLEMS   582
        INTEGRATING PROBLEMS   588

CHAPTER 12 • FREQUENCY RESPONSE   591

12–1   Frequency-response Descriptors   592
12–2   First-order Circuit Frequency Response   597
12–3   Second-order Circuit Frequency Response   613
12–4   The Frequency Response of *RLC* Circuits   627
12–5   Bode Diagrams   634
12–6   Bode Diagrams with Complex Critical Frequencies   644
12–7   Frequency Response and Step Response   651
        SUMMARY   656
        PROBLEMS   657
        INTEGRATING PROBLEMS   666

CHAPTER 13 • FOURIER SERIES   669

13–1   Overview of Fourier Analysis   670
13–2   Fourier Coefficients   672
13–3   Waveform Symmetries   681
13–4   Circuit Analysis Using Fourier Series   684
13–5   RMS Value and Average Power   689
        SUMMARY   694
        PROBLEMS   695
        INTEGRATING PROBLEMS   701

CHAPTER 14 • ANALOG FILTER DESIGN   703

14–1   Frequency-domain Signal Processing   704
14–2   Design with First-order Circuits   705
14–3   Design with Second-order Circuits   714
14–4   Low-pass Filter Design   722
14–5   High-pass Filter Design   740
14–6   Bandpass and Bandstop Filter Design   744
        SUMMARY   749
        PROBLEMS   749
        INTEGRATING PROBLEMS   757

CHAPTER 15 • MUTUAL INDUCTANCE 760

15-1   Coupled Inductors 761
15-2   The Dot Convention 764
15-3   Energy Analysis 767
15-4   The Ideal Transformer 769
15-5   Transformers in the Sinusoidal Steady State 775
15-6   Transformer Equivalent Circuits 780
       SUMMARY 785
       PROBLEMS 785
       INTEGRATING PROBLEMS 790

CHAPTER 16 • POWER IN THE SINUSOIDAL STEADY STATE 792

16-1   Average and Reactive Power 793
16-2   Complex Power 796
16-3   AC Power Analysis 800
16-4   Load-flow Analysis 803
16-5   Three-phase Circuits 810
16-6   Three-phase AC Power Analysis 814
       SUMMARY 824
       PROBLEMS 824
       INTEGRATING PROBLEMS 831

APPENDIX A—Standard Values A–1
APPENDIX B—Solution of Linear Equations A–2
APPENDIX C—Complex Numbers A–12
APPENDIX D—References A–16

ANSWERS TO SELECTED PROBLEMS A–19
INDEX I–1

The Analysis and Design of Linear Circuits

CHAPTER 1

INTRODUCTION

The electromotive action manifests itself in the form of two effects which I believe must be distinguished from the beginning by a precise definition. I will call the first of these "electric tension," the second "electric current."

André-Marie Ampère, 1820,
French Mathematician/Physicist

1-1 About This Book

1-2 Symbols and Units

1-3 Circuit Variables

Summary

Problems

Integrating Problems

This book deals with the analysis and design of linear electric circuits. A circuit is an interconnection of electric devices that processes energy or information. Understanding circuits is important because energy and information are the underlying technological commodities in electrical engineering. The study of circuits provides a foundation for areas of electrical engineering such as electronics, power systems, communication systems, and control systems.

This chapter describes the structure of this book, introduces basic notation, and defines the primary physical variables in electric circuits—voltage

1

and current. André Ampère (1775–1836) was the first to recognize the importance of distinguishing between the electrical effects now called voltage and current. He also invented the galvanometer, the forerunner of today's voltmeter and ammeter. A natural genius who had mastered all the then-known mathematics by age 12, he is best known for defining the mathematical relationship between electric current and magnetism. This relationship, now known as Ampère's law, is one of the basic concepts of modern electromagnetics.

The first section of this chapter describes the pedagogical framework and terminology that must be understood to use this book effectively. It describes how the learning objectives are structured to help the student develop the problem-solving abilities needed to analyze and design circuits. The second section provides some of the standard scientific notation and conventions used throughout the book. The last section introduces electric voltage, current, and power—the physical variables used throughout the book to describe the signal-processing and energy-transfer capabilities of linear circuits.

1–1 ABOUT THIS BOOK

The basic purpose of this book is to introduce the analysis and design of linear circuits. Circuits are important in electrical engineering because they process electrical signals that carry energy and information. For the present we can define a **circuit** as an interconnection of electrical devices, and a **signal** as a time-varying electrical entity. For example, the information stored on a compact disk is recovered in the CD-ROM player as electronic signals that are processed by circuits to generate audio and video outputs. In an electrical power system some form of stored energy is converted to electrical form and transferred to loads, where the energy is converted into the form required by the customer. The CD-ROM player and the electrical power system both involve circuits that process and transfer electrical signals carrying energy and information.

In this text we are primarily interested in **linear circuits**. An important feature of a linear circuit is that the amplitude of the output signal is proportional to the input signal amplitude. The proportionality property of linear circuits greatly simplifies the process of circuit analysis and design. Most circuits are only linear within a restricted range of signal levels. When driven outside this range they become nonlinear, and proportionality no longer applies. Although we will treat a few examples of nonlinear circuits, our attention is focused on circuits operating within their linear range.

Our study also deals with interface circuits. For the purposes of this book, we define an **interface** as a pair of accessible terminals at which signals may be observed or specified. The interface idea is particularly important with integrated circuit (IC) technology. Integrated circuits involve many thousands of interconnections, but only a small number are accessible to the user. Designing systems using integrated circuits involves interconnecting complex circuits that have only a few accessible terminals. This often involves relatively simple circuits whose purpose is to change signal levels or formats. Such interface circuits are intentionally introduced to ensure that the appropriate signal conditions exist at the connections between complex integrated circuits.

COURSE OBJECTIVES

This book is designed to help you develop the knowledge and application skills needed to solve three types of circuit problems: analysis, design, and evaluation. The **analysis** level involves determining the output signals of a given circuit with known input signals. Analysis has the important feature that a unique solution exists in linear circuits. Circuit analysis will occupy the bulk of our attention, because it provides the foundation for understanding the interaction of signals and circuits. The **design** level involves devising circuits that perform a prescribed signal-processing function. In contrast to analysis, a design problem may have several solutions or even no solution. The former possibility leads to the **evaluation** level: namely, the rank-ordering of circuits that perform the same function using factors such as cost, power consumption, and part counts. In real life the engineer's role is a blend of analysis, design, and evaluation, and the boundaries between these levels are often blurred.

EN ROUTE OBJECTIVES

To help you develop your problem-solving skills, we have provided a set of carefully structured intermediate steps called en route objectives (EROs). The EROs are listed at the end of each chapter together with homework problems designed to test your understanding of basic concepts and techniques. Although they are listed at the end of each chapter, the EROs are an integral part of the chapter and are keyed to one or more sections of the text. Collectively, the en route objectives define the basic knowledge and understanding needed to achieve the course goals. At the very end of each chapter, there are a number of integrating problems. These problems require the mastery of several en route objectives and offer you an opportunity to test your ability to deal with a broader range of problems involving analysis, design, and evaluation.

This book contains many worked examples to help you master the en route objectives. These examples describe the steps needed to obtain the final answer. They usually treat analysis problems, although design and evaluation examples and application notes are also included. There are in-text exercises that include only the problem statement and the final answer. These exercises provide immediate feedback on your ability to apply the ideas discussed in a particular section. You should work out the solution in detail, and then check your results with the answers given in the exercise. In addition, there is a student solutions manual available in which 100 problems are worked using appropriate computer tools. These problems are marked with a 📖 icon.

1–2 SYMBOLS AND UNITS

Throughout this text we will use the international system (SI) of units. The SI system includes six fundamental units: meter (m), kilogram (kg), second (s), ampere (A), kelvin (K), and candela (cd). All the other units of measure can be derived from these six.

Like all disciplines, electrical engineering has its own terminology and symbology. The symbols used to represent some of the more important physical quantities and their units are listed in Table 1–1. It is not our purpose to define these quantities here, nor to offer this list as an item for memorization. Rather, the purpose of this table is merely to list in one place all the electrical quantities used in this book.

Numerical values in engineering range over many orders of magnitude. Consequently, the system of standard decimal prefixes in Table 1–2 is used. These prefixes on a unit abbreviation symbol indicate the power of 10 that is applied to the numerical value of the quantity.

Exercise 1–1

Given the pattern in the statement 1 kΩ = 1 kilohm = 1 $\times$ 10^3 ohms, fill in the blanks in the following statements using the standard decimal prefixes.

(a) _____ = _____ = 5 $\times$ 10^{-3} watts
(b) 10.0 dB = _____ = _____
(c) 3.6 ps = _____ = _____
(d) _____ = 0.03 microfarads = _____
(e) _____ = _____ gigahertz = 6.6 $\times$ 10^9 Hertz

Answers:
(a) 5.0 mW = 5 milliwatts
(b) 10.0 decibels = 1.0 bel
(c) 3.6 picoseconds = 3.6 $\times$ 10^{-12} seconds
(d) 30 nF or 0.03 μF = 30.0 $\times$ 10^{-9} Farads
(e) 6.6 GHz = 6.6 gigahertz

1–3 CIRCUIT VARIABLES

The underlying physical variables in the study of electronic systems are **charge** and **energy**. The idea of electrical charge explains the very strong electrical forces that occur in nature. To explain both attraction and repulsion, we say that there are two kinds of charge—positive and negative. Like charges repel, while unlike charges attract one another. The symbol q is used to represent charge. If the amount of charge is varying with time, we emphasize the fact by writing $q(t)$. In the international system (SI), charge is measured in **coulombs** (abbreviated C). The smallest quantity of charge in nature is an electron's charge ($q_E = 1.6 \times 10^{-19}$C). Thus, there are $1/q_E = 6.25 \times 10^{18}$ electrons in 1 coulomb of charge.

Electrical charge is a rather cumbersome variable to measure in practice. Moreover, in most situations the charges are moving, so we find it more convenient to measure the amount of charge passing a given point per unit time. If $q(t)$ is the cumulative charge passing through a point, we define a signal variable i called **current** as follows:

$$i = \frac{dq}{dt} \tag{1–1}$$

Current is a measure of the flow of electrical charge. It is the time rate of change of charge passing a given point in a circuit. The physical dimen-

TABLE 1-1   SOME IMPORTANT QUANTITIES, THEIR SYMBOLS,
                  AND UNIT ABBREVIATIONS

QUANTITY	SYMBOL	UNIT	UNIT ABBREVIATION
Time	t	second	s
Frequency	f	hertz	Hz
Radian frequency	ω	radian/second	rad/s
Phase angle	θ, ϕ	degree or radian	° or rad
Energy	w	joule	J
Power	p	watt	W
Charge	q	coulomb	C
Current	i	ampere	A
Electric field	$\mathcal{E}$	volt/meter	V/m
Voltage	v	volt	V
Impedance	Z	ohm	Ω
Admittance	Y	siemens	S
Resistance	R	ohm	Ω
Conductance	G	siemens	S
Reactance	X	ohm	Ω
Susceptance	B	siemens	S
Inductance, self	L	henry	H
Inductance, mutual	M	henry	H
Capacitance	C	farad	F
Magnetic flux	ϕ	weber	wb
Flux linkages	λ	weber-turns	wb-t
Power ratio	$\log_{10}(p_2/p_1)$	bel	B

sions of current are coulombs per second. In the SI system, the unit of current is the **ampere** (abbreviated A). That is,

$$1 \text{ coulomb/second} = 1 \text{ ampere}$$

Since there are two types of electrical charge, there is a bookkeeping problem associated with the direction assigned to the current. In engineering it is customary to define the direction of current as the direction of the net flow of positive charge.

A second signal variable called **voltage** is related to the change in energy that would be experienced by a charge as it passes through a circuit. The symbol w is commonly used to represent energy. In the SI system of units, energy carries the units of **joules** (abbreviated J). If a small charge dq was to experience a change in energy dw in passing from point A to point B in a circuit, then the voltage v between A and B is defined as the change in energy per unit charge. We can express this definition in differential form as

$$v = \frac{dw}{dq} \qquad\qquad (1-2)$$

T A B L E 1–2 STANDARD DECIMAL PREFIXES

MULTIPLIER	PREFIX	ABBREVIATION
10^{18}	exa	E
10^{15}	peta	P
10^{12}	tera	T
10^{9}	giga	G
10^{6}	mega	M
10^{3}	kilo	k
10^{-1}	deci	d
10^{-2}	centi	c
10^{-3}	milli	m
10^{-6}	micro	μ
10^{-9}	nano	n
10^{-12}	pico	p
10^{-15}	femto	f
10^{-18}	atto	a

Voltage does not depend on the path followed by the charge dq in moving from point A to point B. Furthermore, there can be a voltage between two points even if there is no charge motion, since voltage is a measure of how much energy dw would be involved if a charge dq was moved. The dimensions of voltage are joules per coulomb. The unit of voltage in the SI system is the **volt** (abbreviated V). That is,

$$1 \text{ joule/coulomb} = 1 \text{ volt}$$

The general definition of physical variable called **power** is the time rate of change of energy:

$$p = \frac{dw}{dt} \tag{1–3}$$

The dimensions of power are joules per second, which in the SI system is called a **watt** (abbreviated W). In electrical circuits it is useful to relate power to the signal variables current and voltage. Using the chain rule, Eq. (1–3) can be written as

$$p = \left(\frac{dw}{dq}\right)\left(\frac{dq}{dt}\right) \tag{1–4}$$

Now using Eqs. (1–1) and (1–2), we obtain

$$p = vi \tag{1–5}$$

The electrical power associated with a situation is determined by the product of voltage and current. The total energy transferred during the period from t_1 to t_2 is found by solving for dw in Eq. (1–3) and then integrating

$$w_{\text{T}} = \int_{w_1}^{w_2} dw = \int_{t_1}^{t_2} p \, dt \tag{1–6}$$

EXAMPLE 1-1

The electron beam in the cathode-ray tube shown in Figure 1–1 carries 10^{14} electrons per second and is accelerated by a voltage of 50 kV. Find the power in the electron beam.

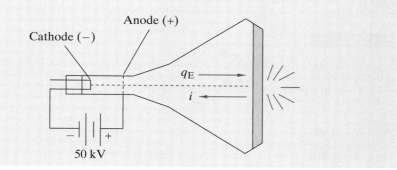

FIGURE   1 - 1

SOLUTION:
Since current is the rate of charge flow, we can find the net current by multiplying the charge of the electron q_E by the rate of electron flow dn_E/dt.

$$i = q_E \frac{dn_E}{dt} = (1.6 \times 10^{-19})(10^{14}) = 1.6 \times 10^{-5} \text{ A}$$

Therefore, the beam power is

$$p = vi = (50 \times 10^3)(1.6 \times 10^{-5}) = 0.8 \text{ W} \qquad \blacksquare$$

EXAMPLE 1-2

The current through a circuit element is 50 mA. Find the total charge and the number of electrons transferred during a period of 100 ns.

SOLUTION:
The relationship between current and charge is given in Eq. (1–1) as

$$i = \frac{dq}{dt}$$

Since the current i is given, we calculate the charge transferred by solving this equation for dq and then integrating

$$q_T = \int_{q_1}^{q_2} dq = \int_0^{10^{-7}} i \, dt$$

$$= \int_0^{10^{-7}} 50 \times 10^{-3} \, dt = 50 \times 10^{-10} \text{C} = 5 \text{ nC}$$

There are $1/q_E = 6.25 \times 10^{18}$ electrons/coulomb, so the number of electrons transferred is

$$n_E = (5 \times 10^{-9} \text{ C})(6.25 \times 10^{18} \text{ electrons/C}) = 31.2 \times 10^9 \text{ electrons} \qquad \blacksquare$$

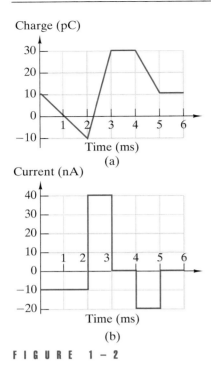

Charge (pC)

(a)

Current (nA)

(b)

FIGURE 1-2

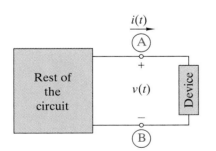

FIGURE 1-3 *Voltage and current reference marks for a two-terminal device.*

A device dissipates 100 W of power. How much energy is delivered to it in 10 seconds?

Answer: 1 kJ

The graph in Figure 1–2(a) shows the charge $q(t)$ flowing past a point in a wire as a function of time.
(a) Find the current $i(t)$ at t = 1, 2.5, 3.5, 4.5, and 5.5 ms.
(b) Sketch the variation of $i(t)$ versus time.

Answers:
(a) –10 nA, +40 nA, 0 nA, –20 nA, 0 nA.
(b) The variations in $i(t)$ are shown in Figure 1–2(b).

THE PASSIVE SIGN CONVENTION

We have defined three circuit variables (current, voltage, and power) using two basic variables (charge and energy). Charge and energy, like mass, length, and time, are basic concepts of physics that provide the scientific foundation for electrical engineering. However, engineering problems rarely involve charge and energy directly, but are usually stated in terms of voltage, current, and power. The reason for this is simple: The circuit variables are much easier to measure and therefore are the most useful working variables in engineering practice.

At this point, it is important to stress the physical differences between current and voltage variables. Current is a measure of the time rate of charge passing a point in a circuit. We think of current as a *through variable*, since it describes the flow of electrical charge through a point in a circuit. On the other hand, voltage is not measured at a single point, but rather between two points or across an electrical device. Consequently, we think of voltage as an *across variable* that inherently involves two points.

The arrow and the plus and minus symbols in Figure 1–3 are *reference marks* that define the positive directions for the current and voltage associated with an electrical device. These reference marks do not represent an assertion about what is happening physically in the circuit. The response of an electrical circuit is determined by physical laws, and not by the reference marks assigned to the circuit variables.

The reference marks are benchmarks assigned at the beginning of the analysis. When the actual direction and reference direction agree, the answers found by circuit analysis will have positive algebraic signs. When they disagree, the algebraic signs of the answers will be negative. For example, if circuit analysis reveals that the current variable in Figure 1–3 is positive [i.e., $i(t) > 0$], then the sign of this answer, together with the assigned reference direction, indicates that the current passes through point A in Figure 1–3 from left to right. Conversely, when analysis reveals that the current variable is negative, then this result, combined with the assigned reference direction, tells us that the current passes through point A from right to left. In summary, the algebraic sign of the answer together with arbitrarily assigned reference marks tell us the actual directions of a voltage or current variable.

In Figure 1–3, the current reference arrow enters the device at the terminal marked with the plus voltage reference mark. This orientation is called the **passive sign convention**. Under this convention, the power $p(t)$ is positive when the device absorbs power and is negative when it delivers power to the rest of the circuit. Since $p(t) = v(t) \times i(t)$, a device absorbs power when the voltage and current variables have the same algebraic sign and delivers power when they have opposite signs. Certain devices, such as heaters (a toaster, for example), can only absorb power. The voltage and current variables associated with these devices must always have the same algebraic sign. On the other hand, a battery absorbs power [$p(t) > 0$] when it is being charged and delivers power [$p(t) < 0$] when it is discharging. Thus, the voltage and current variables for a battery can have the same or the opposite algebraic signs.

In a circuit some devices absorb power and others deliver power, but the sum of the power in all of the devices in the circuit is zero. This is more than just a conservation-of-energy concept. When electrical devices are interconnected to form a circuit, the only way that power can enter or leave the circuit is via the currents and voltages at device terminals. The existence of a power balance in a circuit is one method of checking calculations.

The passive sign convention is used throughout this book. It is also the convention used by circuit simulation computer programs.[1] To interpret correctly the results of circuit analysis, it is important to remember that the reference marks (arrows and plus/minus signs) are reference directions, not indications of the circuit response. The actual direction of a response is determined by comparing its reference direction with the algebraic signs of the result predicted by circuit analysis based on physical laws.

GROUND

Since voltage is defined between two points, it is often useful to define a common voltage reference point called **ground**. The voltages at all other points in a circuit are then defined with respect to this common reference point. We indicate the voltage reference point using the ground symbol shown in Figure 1–4. Under this convention we sometimes refer to the variables $v_A(t)$, $v_B(t)$, and $v_C(t)$ as the voltages at points A, B, and C, respectively. This terminology appears to contradict the fact that voltage is an across variable that involves two points. However, the terminology means that the variables $v_A(t)$, $v_B(t)$, and $v_C(t)$ are the voltages defined between points A, B, and C, and the common voltage reference point at point G.

Using a common reference point for across variables is not an idea unique to electrical circuits. For example, the elevation of a mountain is the number of feet or meters between the top of the mountain and a

F I G U R E   1 – 4 *Ground symbol indicates a common voltage reference point.*

---

1 We discuss computer-aided circuit analysis in subsequent chapters.

common reference point at sea level. If a geographic point lies below sea level, then its elevation is assigned a negative algebraic sign. So it is with voltages. If circuit analysis reveals that the voltage variable at point A is negative [i.e., $v_A(t) < 0$], then this fact together with the reference marks in Figure 1–4 indicate that the potential at point A is less than the ground potential.

EXAMPLE 1–3

Figure 1–5 shows a circuit formed by interconnecting five devices, each of which has two terminals. A voltage and current variable has been assigned to each device using the passive sign convention. The working variables for each device are observed to be as follows:

	DEVICE 1	DEVICE 2	DEVICE 3	DEVICE 4	DEVICE 5
v	+100 V	?	+25 V	+75 V	−75 V
i	?	+5 mA	+5 mA	?	+5 mA
p	−1 W	+0.5 W	?	0.75 W	?

(a) Find the missing variable for each device and state whether the device is absorbing or delivering power.
(b) Check your work by showing that the sum of the device powers is zero.

FIGURE   1 – 5

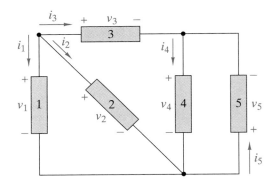

SOLUTION:
(a) We use $p = vi$ to solve for the missing variable since two of the three circuit variables are given for each device.

Device 1: $i_1 = p_1/v_1 = -1/100 = -10$ mA [$p(t) < 0$, delivering power]

Device 2: $v_2 = p_2/i_2 = 0.5/0.005 = 100$ V [$p(t) > 0$, absorbing power]

Device 3: $p_3 = v_3 i_3 = 25 \times 0.005 = 0.125$ W [$p(t) > 0$, absorbing power]

Device 4: $i_4 = p_4/v_4 = 0.75/75 = 10$ mA [$p(t) > 0$, absorbing power]

Device 5: $p_5 = v_5 i_5 = -75 \times 0.005 = -0.375$ W [$p(t) < 0$, delivering power]

(b) Summing the device powers yields

$$p_1 + p_2 + p_3 + p_4 + p_5 = -1 + 0.5 + 0.125 + 0.75 - 0.375$$

$$= +1.375 - 1.375 = 0$$

This example shows that the sum of the power absorbed by devices is equal in magnitude to the sum of the power supplied by devices. A power balance always exists in the types of circuits treated in this book and can be used as an overall check of circuit analysis calculations. ■

Exercise 1–4

The working variables of a set of two-terminal electrical devices are observed to be as follows:

	DEVICE 1	DEVICE 2	DEVICE 3	DEVICE 4	DEVICE 5
v	+10 V	?	−15 V	+5 V	?
i	−3 A	−3 A	+10 mA	?	−12 mA
p	?	+40 W	?	+10 mW	−120 mW

Using the passive sign convention, find the magnitude and sign of the unknown variable and state whether the device is absorbing or delivering power.

Answers:
Device 1: $p = -30$ W (delivering power)
Device 2: $v = -13.3$ V (absorbing power)
Device 3: $p = -150$ mW (delivering power)
Device 4: $i = +2$ mA (absorbing power)
Device 5: $v = +10$ V (delivering power)

SUMMARY

• Circuits are important in electrical engineering because they process signals that carry energy and information. A **circuit** is an interconnection of electrical devices. A **signal** is an electrical current or voltage that carries energy or information. An **interface** is a pair of accessible terminals at which signals may be observed or specified.

• This book defines overall course objectives at the analysis, design, and evaluation levels. In **circuit analysis** the circuit and input signals are given and the object is to find the output signals. The object of **circuit design** is to devise one or more circuits that produce prescribed output signals for given input signals. The **evaluation** problem involves appraising alternative circuit designs using criteria such as cost, power consumption, and parts count.

• Charge (q) and energy (w) are the basic physical variables involved in electrical phenomena. Current (i), voltage (v), and power (p) are the derived variables used in circuit analysis and design. In the SI system, charge is measured in coulombs (C), energy in joules (J), current in amperes (A), voltage in volts (V), and power in watts (W).

- **Current** is defined as dq/dt and is a measure of the flow of electrical charge. **Voltage** is defined as dw/dq and is a measure of the energy required to move a small charge from one point to another. **Power** is defined as dw/dt and is a measure of the rate at which energy is being transferred. Power is related to current and voltage as $p = vi$.

- The **reference marks** (arrows and plus/minus signs) assigned to a device are reference directions, not indications of the way a circuit responds. The actual direction of the response is determined by comparing the reference direction and the algebraic sign of the answer found by circuit analysis using physical laws.

- Under the **passive sign convention**, the current reference arrow is directed toward the terminal with the positive voltage reference mark. Under this convention, the device power is positive when it absorbs power and is negative when it delivers power. When current and voltage have the same (opposite) algebraic signs, the device is absorbing (delivering) power.

PROBLEMS

ERO 1–1 ELECTRICAL SYMBOLS AND UNITS (SECT. 1–2)

Given an electrical quantity described in terms of words, scientific notation, or decimal prefix notation, convert the quantity to an alternate description.

1–1 Write the following statements in symbolic form:
    (a) fifty two kilowatts
    (b) ninety one point five megahertz
    (c) two hundred nanoseconds
    (d) two thousand five hundred volts

1–2 Write the meaning of the following symbols in words:
    (a) 110 kHz
    (b) 56 krad/s
    (c) 200 MW
    (d) 2.02 mJ
    (e) 3.2 MΩ

1–3 Express the following quantities using appropriate engineering prefixes (i.e., state the numeric to the nearest standard prefix).
    (a) 2,200,000 volts
    (b) 2300×10^{-9} farads
    (c) 6200 ohms
    (d) 7.52×10^{5} joules
    (e) 0.000235 henrys

1–4 Commercial electrical power companies measure energy consumption in kilowatt-hours, denoted kWh. One kilowatt-hour is the amount of energy transferred by 1 kW of power in a time period of 1 hour. During a one-month time period, a power company billing statement reports a user's total energy usage to be 856 kWh. Determine the number of joules used during the billing period.

1–5 Fill in the blanks in the following statements:
   (a) To convert capacitance from microfarads to picofarads, multiply by _____.
   (b) To convert resistance from kilohms to megohms, multiply by _____.
   (c) To convert current from amperes to milliamperes, multiply by _____.
   (d) To convert power from watts to megawatts, multiply by _____.

ERO 1–2 CIRCUIT VARIABLES (SECT. 1–3)

Given any two of the three signal variables (i, v, p) or the two basic variables (q, w), find the magnitude and direction (sign) of the unspecified variables.

1–6 A wire carries a constant current of 120 mA. How many coulombs flow past a given point in the wire in 3 s?

1–7 An ampere-hour (Ah) meter measures the integral, with respect to time, of the current in a conductor. During a 12-hour period a certain meter records 3000 Ah.
   (a) Find the number of coulombs that flowed through the meter during the recording period.
   (b) Find the average current during the recording period.

1–8 Figure P1–8 shows a plot of the net positive charge flowing in a wire versus time. Sketch the corresponding current during the same time period.

1–9 The net positive charge flowing through a device varies as $q(t) = 3t^2$ C. Find the current through the device at $t = 0$ s, $t = 1$ s, and $t = 3$ s.

1–10 The current through a device is given by $i(t) = 0.05t$ A. How many coulombs enter the device between time $t = 0$ s and $t = 5$ s?

1–11 For $t \le 5$ s the current through a device is $i(t) = 10$ A. For $t > 5$ s it is $i(t) = 20 - 2t$ A. Sketch $i(t)$ versus time and find the total charge entering the device between $t = 0$ s and $t = 10$ s.

1–12 An incandescent lamp absorbs 40 W when connected to a 120-V source.
   (a) Find the current through the lamp.
   (b) Find the cost of operating the lamp for 4 hours when electricity costs 6.8 cents/kW-hr.

1–13 The 12-V automobile battery in Figure P1–13 has an output capacity of 200 ampere-hours (Ah) when connected to a headlamp that absorbs 50 watts of power. Assume the battery voltage is constant.
   (a) Find the current supplied by the battery.
   (b) How long can the battery power the headlight?

1–14 The net positive charge flowing through a device is $q(t) = e^{-1000t}$ μC. Sketch the current through the device versus time for $t > 0$.

1–15 A total charge of 50 ampere-hours is supplied to a 12-V battery during recharging. Determine the number of joules supplied to the battery. Assume battery voltage is constant.

1–16 The voltage between two points is 250 V. How much energy is required to move a charge of 3 mC from one point to the other?

1–17 When illuminated the i–v relationship for a photocell is $i = e^v - 10$ A.

FIGURE P1–8

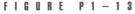

FIGURE P1–13

Calculate the device power and state whether the device is absorbing or delivering power when $v = -3$ V. Repeat for $v = 1.5$ V and 3 V.

1–18 Under the passive sign convention, the current through and voltage across a two-terminal device are $i = 5$ mA and $v = 25$ V.
  (a) Find the device power and state whether the device is absorbing or delivering power.
  (b) Repeat (a) when $i = -20$ mA.

1–19 The voltage across a two-terminal device is always 15 V. The maximum power the device can dissipate is 0.5 W. Determine the maximum current magnitude allowed by the device power rating.

1–20 A laser produces 10-kW bursts of power that last 20 ns. If the burst rate is 50 bursts per second, what is the average power in the laser's output?

1–21 Two electrical devices are connected as shown in Figure P1–21. Using the reference marks shown in the figure, find the power transferred and state whether the power is transferred from A to B or B to A when
  (a) $v = +33$ V and   **(b)** $v = -12$ V and
       $i = -2.2$ A $i = -1.2$ mA
  (c) $v = +37.5$ V and   **(d)** $v = -15$ V and
       $i = +40$ mA $i = -43$ mA

1–22 Figure P1–22 shows an electric circuit with a voltage and a current variable assigned to each of the six devices. The device signal variables are observed to be as follows:
Device 1, $v = 15$ V, $i = -1$ A, and $p = ?$
Device 2, $v = 5$ V, $i = ?$, and $p = 5$ W
Device 3, $v = ?$ V, $i = 0.5$ A, and $p = 5$ W
Device 4, $v = 4$ V, $i = 0.5$ A, and $p = ?$
Device 5, $v = ?$ V, $i = 3$ A, and $p = 18$ W
Device 6, $v = ?$ V, $i = -2.5$ A, and $p = -15$ W
Find the unknown signal variable associated with each device and state whether the device is absorbing or delivering power. Use the power balance to check your work.

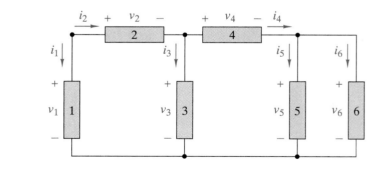

1–23 Figure P1–22 shows an electric circuit with a voltage and a current variable assigned to each of the six devices. The device signal variables are observed to be as follows:

Device 1, $v = 30$ V, $i = -2$ A, and $p = ?$
Device 2, $v = 10$ V, $i = ?$, and $p = 20$ W
Device 3, $v = 20$ V, $i = ?$, and $p = 20$ W
Device 4, $v = 8$ V, $i = 1$ A, and $p = ?$
Device 5, $v = ?$ V, $i = -5$ A, and $p = -60$ W
Device 6, $v = 12$ V, $i = 6$ A, and $p = ?$ W
Find the unknown signal variable associated with each device and state whether the device is absorbing or delivering power. Use the power balance to check your work.

1–24 An integrated circuit package contains 10 devices that are nominally identical. The manufacturer's data sheet states that each device draws between 45 mA and 65 mA of current from a 5-V power supply. What is the power dissipation range for the package?

1–25 Using the passive sign convention the voltage across a device is found to be $v(t) = 20e^{-50t}$ V and the current is $i(t) = 0.5e^{-50t}$ A. Find the energy delivered to the device between $t = 0$ and $t = 10$ ms.

INTEGRATING PROBLEMS

1–26 POWER RATIO IN dB

In complete darkness the voltage across and current through a two-terminal light detector are +5.6 V and +8 nA. In full sunlight the voltage and current are +0.9 V and +4 mA. Express the light/dark power ratio of the device in decibels (dB), where the power ratio in dB is

$$\text{PR}_{db} = 10 \, \text{Log}_{10} \, (p_2/p_1).$$

1–27 EQUIPMENT DATA SHEET

A manufacturer's data sheet for a computer lists the power supply requirements as 26 A @ 5 V, 9 A @ 12 V, 0.5 A @ –5 V, and 0.5 A @ –12 V. The data sheet also states that the total power dissipation is 250 W. Are these data consistent? Explain.

1–28 STORAGE BATTERY EFFICIENCY

The ampere-hour efficiency of a storage battery is the ratio of its ampere-hour output to the ampere-hour input required to recharge the battery. A certain 24-V battery has a rated output of 500 ampere-hours. When the battery is completely drained, a battery charger must deliver 65 A for 8 hours to recharge the battery. Assume the battery voltage is constant.
(a) Determine the ampere-hour efficiency of the battery.
(b) Determine the number of joules required to recharge the battery.

CHAPTER 2

BASIC CIRCUIT ANALYSIS

The equation S = A/L shows that the current of a voltaic circuit is subject to a change, by each variation originating either in the magnitude of a tension or in the reduced length of a part, which latter is itself again determined, both by the actual length of the part as well as its conductivity and its section.

Georg Simon Ohm, 1827,
German Mathematician/Physicist

2–1 Element Constraints

2–2 Connection Constraints

2–3 Combined Constraints

2–4 Equivalent Circuits

2–5 Voltage and Current Division

2–6 Circuit Reduction

2–7 Computer-aided Circuit Analysis

Summary

Problems

Integrating Problems

This chapter begins our study of the ideal models used to describe the physical devices in electrical circuits. Foremost among these is the renowned Ohm's law, defining the model of a linear resistor. Georg Simon Ohm (1789–1854) originally discovered the law that now bears his name in 1827. His results drew heavy criticism and were not generally accepted for many years. Fortunately, the importance of his contribution was eventually recognized during his lifetime. He was honored by the Royal Society of England in 1841 and appointed a Professor of Physics at the University of Munich in 1849.

The first section of this chapter deals with the element constraints derived from the ideal models of resistors, voltage sources, and current sources. The models for other linear devices, such as amplifiers, inductors, capacitors, and transformers, are introduced in later chapters. When devices are interconnected to form a circuit, they are subject to connection constraints based on fundamental conservation principles known as Kirchhoff's laws. The remainder of the chapter uses the combined element and connection constraints to develop a basic set of circuit analysis tools that include equivalent circuits, voltage and current division, and circuit reduction. The basic circuit analysis tools developed in this chapter are used frequently in the rest of the book and by practicing engineers. These analysis tools promote basic understanding because they involve working directly with the circuit model.

2–1 ELEMENT CONSTRAINTS

A **circuit** is a collection of interconnected electrical devices. An electrical **device** is a component that is treated as a separate entity. The rectangular box in Figure 2–1 is used to represent any one of the two-terminal devices used to form circuits. A two-terminal device is described by its *i–v* **characteristic**; that is, by the relationship between the voltage across and current through the device. In most cases the relationship is complicated and nonlinear, so we use a linear model that approximates the dominant features of a device.

To distinguish between a device (the real thing) and its model (an approximate stand-in), we call the model a circuit **element**. Thus, a device is an article of hardware described in manufacturers' catalogs and parts specifications. An element is a model described in textbooks on circuit analysis. This book is no exception, and a catalog of circuit elements will be introduced as we go on. A discussion of real devices and their models is contained in Appendix A.

THE LINEAR RESISTOR

The first element in our catalog is a linear model of the device described in Figure 2–2. The actual *i–v* characteristic of this device is shown in Figure 2–2(b). To model this curve accurately across the full operating range shown in the figure would require at least a cubic equation. However, the graph in Figure 2–2(b) shows that a straight line is a good approximation to the *i–v* characteristic if we operate the device within its linear range. The power rating of the device limits the range over which the *i–v* characteristics can be represented by a straight line through the origin.

FIGURE 2–1 *Voltage and current reference marks for a two-terminal device.*

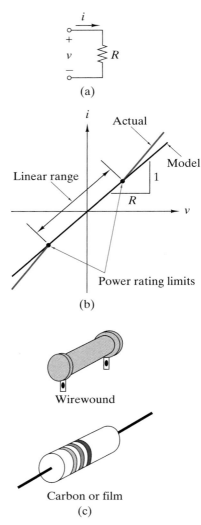

FIGURE 2-2 *The resistor: (a) Circuit symbol. (b) i–v characteristics. (c) Some actual devices.*

For the passive sign convention used in Figure 2–2(a), the equations describing the *linear resistor* element are

$$v = Ri \quad \text{or} \quad i = Gv \tag{2-1}$$

where R and G are positive constants that are reciprocally related.

$$G = \frac{1}{R} \tag{2-2}$$

The relationships in Eq. (2–1) are collectively known as **Ohm's law**. The parameter R is called **resistance** and has the unit **ohms**, Ω. The parameter G is called **conductance**, with the unit **siemens**, S. In earlier times the unit of conductance was cleverly called the mho, with a unit abbreviation symbol ℧ (ohm spelled backward and the ohm symbol upside down). Note that Ohm's law presumes that the passive sign convention is used to assign the reference marks to voltage and current.

The Ohm's law model is represented graphically by the black straight line in Figure 2–2(b). The *i–v* characteristic for the Ohm's law model defines a circuit element that is said to be linear and bilateral. **Linear** means that the defining characteristic is a straight line through the origin. Elements whose characteristics do not pass through the origin or are not a straight line are said to be **nonlinear**. **Bilateral** means that the *i–v* characteristic curve has odd symmetry about the origin.[1] With a bilateral resistor, reversing the polarity of the applied voltage reverses the direction but not the magnitude of the current, and vice versa. The net result is that we can connect a bilateral resistor into a circuit without regard to which terminal is which. This is important because devices such as diodes and batteries are not bilateral, and we must carefully identify each terminal.

Figure 2–2(c) shows sketches of discrete resistor devices. Detailed device characteristics and fabrication techniques are discussed in Appendix A.

The power associated with the resistor can be found from $p = vi$. Using Eq. (2–1) to eliminate v from this relationship yields

$$p = i^2R \tag{2-3}$$

or using the same equations to eliminate i yields

$$p = v^2G = \frac{v^2}{R} \tag{2-4}$$

Since the parameter R is positive, these equations tell us that the power is always nonnegative. Under the passive sign convention, this means that the resistor **always absorbs power**.

EXAMPLE 2-1

A resistor operates as a linear element as long as the voltage and current are within the limits defined by its power rating. Suppose we have a 47-kΩ resistor with a power rating of 0.25 W. Determine the maximum current and voltage that can be applied to the resistor and remain within its linear operating range.

---

1 A curve $i = f(v)$ has odd symmetry if $f(-v) = -f(v)$.

SOLUTION:

Using Eq. (2–3) to relate power and current, we obtain

$$I_{MAX} = \sqrt{\frac{P_{MAX}}{R}} = \sqrt{\frac{0.25}{47 \times 10^3}} = 2.31 \text{ mA}$$

Similarly, using Eq. (2–4) to relate power and voltage, we obtain

$$V_{MAX} = \sqrt{RP_{MAX}} = \sqrt{47 \times 10^3 \times 0.25} = 108 \text{ V} \qquad ■$$

OPEN AND SHORT CIRCUITS

The next two circuit elements can be thought of as limiting cases of the linear resistor. Consider a resistor R with a voltage v applied across it. Let's calculate the current i through the resistor for different values of resistance. If $v = 10$ V and $R = 1$ Ω, using Ohm's law we readily find that $i = 10$ A. If we increase the resistance to 100 Ω, we find i has decreased to 0.1 A or 100 mA. If we continue to increase R to 1 MΩ, i becomes a very small 10 μA. Continuing this process, we arrive at a condition where R is very nearly infinite and i just about zero. When the current $i = 0$, we call the special value of resistance (i.e., $R = \infty$ Ω), an **open circuit**. Similarly, if we reduce R until it approaches zero, we find that the voltage is very nearly zero. When $v = 0$, we call the special value of resistance (i.e., $R = 0$ Ω), a **short circuit**. The circuit symbols for these two elements are shown in Figure 2–3. In circuit analysis the elements in a circuit model are assumed to be interconnected by zero-resistance wire (that is, by short circuits).

FIGURE 2-3 *Circuit symbols: (a) Open-circuit symbol. (b) Short-circuit symbol.*

THE IDEAL SWITCH

A switch is a familiar device with many applications in electrical engineering. The **ideal switch** can be modeled as a combination of an open- and a short-circuit element. Figure 2–4 shows the circuit symbol and the i–v characteristic of an ideal switch. When the switch is closed,

$$v = 0 \quad \text{and} \quad i = \text{ any value} \qquad (2\text{–}5a)$$

and when it is open,

$$i = 0 \quad \text{and} \quad v = \text{ any value} \qquad (2\text{–}5b)$$

When the switch is closed, the voltage across the element is zero and the element will pass any current that may result. When open, the current is zero and the element will withstand any voltage across its terminals. The power is always zero for the ideal switch, since the product $vi = 0$ when the switch is either open ($i = 0$) or closed ($v = 0$). Actual switch devices have limitations, such as the maximum current they can safely carry when closed and the maximum voltage they can withstand when open. The switch is operated (opened or closed) by some external influence, such as a mechanical motion, temperature, pressure, or an electrical signal.

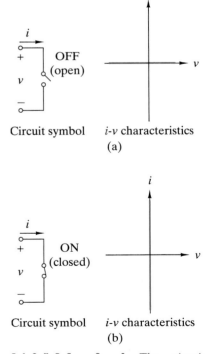

FIGURE 2-4 *The circuit symbol and i–v characteristics of an ideal switch: (a) Switch OFF. (b) Switch ON.*

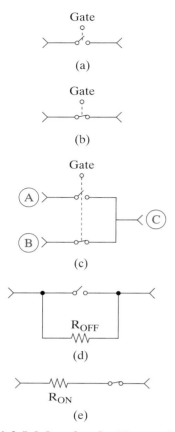

FIGURE 2–5 *The analog switch: (a) Normally open model. (b) Normally closed model. (c) Double throw model. (d) Model with finite OFF resistance. (e) Model with finite ON resistance.*

The **analog switch** is an important device found in analog-to-digital interfaces. Figures 2–5(a) and 2–5(b) show the two basic versions of the device. In either type the switch is actuated by applying a voltage to the terminal labeled "gate." The switch in Figure 2–5(a) is said to be *normally open* because it is open when no voltage is applied to the gate terminal and closes when voltage is applied. The switch in Figure 2–5(b) is said to be *normally closed* because it is closed when no voltage is applied to the controlling gate and opens when voltage is applied.

Figure 2–5(c) shows an application in which complementary analog switches are controlled by the same gate. When gate voltage is applied the upper switch closes and the lower opens so that point A is connected to point C. Conversely, when no gate voltage is applied the upper switch opens and the lower switch closes to connect point B to point C. In the analog world this arrangement is called a double throw switch since point C can be connected to two other points. In the digital world it is called a two-to-one multiplexer (or MUX) because it allows you to select the analog input at point A or point B under control of the digital signal applied to the gate.

In many applications an analog switch can be treated as an ideal switch. In other cases, it may be necessary to account for their nonideal characteristics. When the switch is open an analog switch acts like a very large resistance (R_{OFF}), as suggested in Figure 2–5(d). This resistance is usually negligible because it ranges from perhaps 10^9 to 10^{11} Ω. When the switch is closed it acts like a small resistor (R_{ON}), as suggested in Figure 2–5(e). Depending on other circuit resistances, it may be necessary to account for R_{ON} because it ranges from perhaps 20 to 200 Ω.

This example illustrates how ideal switches and resistors can be combined to model another electrical device. It also suggests that no single model can serve in all applications. It is up to the engineer to select a model that adequately represents the actual device in each application.

IDEAL SOURCES

The signal and power sources required to operate electronic circuits are modeled using two elements: **voltage sources** and **current sources**. These sources can produce either constant or time-varying signals. The circuit symbols and the *i–v* characteristic of an ideal voltage source are shown in Figure 2–6, while the circuit symbol and *i–v* characteristic of an ideal current source are shown in Figure 2–7. The symbol in Figure 2–6(a) represents either a time-varying or constant voltage source. The battery symbol in Figure 2–6(b) is used exclusively for a constant voltage source. There is no separate symbol for a constant current source.

The *i–v* characteristic of an **ideal voltage source** in Figure 2–6(c) is described by the following element equations:

$$v = v_S \quad \text{and} \quad i = \text{any value} \tag{2–6}$$

The element equations mean that the ideal voltage source produces v_S volts across its terminals and will supply whatever current may be required by the circuit to which it is connected.

The $i–v$ characteristic of an **ideal current source** in Figure 2–7(b) is described by the following element equations:

$$i = i_S \quad \text{and} \quad v = \text{any value} \qquad (2-7)$$

The ideal current source produces i_S amperes in the direction of its arrow symbol and will furnish whatever voltage is required by the circuit to which it is connected.

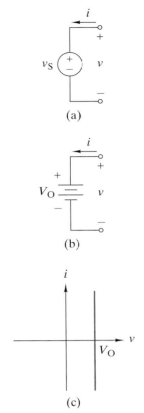

FIGURE 2-7 *Circuit symbols and i–v characteristic of an ideal independent current source: (a) Time-varying or constant source. (b) Constant source i–v characteristics.*

The voltage or current produced by these ideal sources is called a **forcing function** or a **driving function** because it represents an input that causes a circuit response.

FIGURE 2-6 *Circuit symbols and i–v characteristic of an ideal independent voltage source: (a) Time-varying. (b) Constant (Battery). (c) Constant source i–v characteristics.*

EXAMPLE 2-3

Given an ideal voltage source with the time-varying voltage shown in Figure 2–8(a), sketch its $i–v$ characteristic at the times $t = 0, 1,$ and 2 ms.

FIGURE 2-8

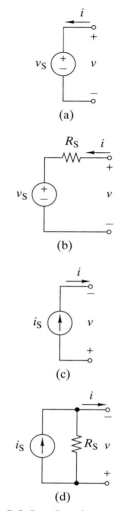

FIGURE 2–9 *Circuit symbols for ideal and practical independent sources: (a) Ideal voltage source. (b) Practical voltage source. (c) Ideal current source. (d) Practical current source.*

SOLUTION:

At any instant of time, the time-varying source voltage has only one value. We can treat the voltage and current at each instant of time as constants representing a snapshot of the source i–v characteristic. For example, at $t = 0$, the equations defining the i–v characteristic are $v_S = 5$ V and $i =$ any value. Figure 2–8(b) shows the i–v relationship at the other instants of time. Curiously, the voltage source i–v characteristic at $t = 1$ ms ($v_S = 0$ and $i =$ any value) is the same as that of a short circuit [see Eq. (2–5a) or Figure 2–3(b)]. ∎

PRACTICAL SOURCES

The practical models for voltage and sources in Figure 2–9 may be more appropriate in some situations than the ideal models used up to this point. These circuits are called practical models because they more accurately represent the characteristics of real-world sources than do the ideal models. It is important to remember that models are interconnections of elements, not devices. For example, the resistance in a model does not always represent an actual resistor. As a case in point, the resistances R_S in the practical source models in Figure 2–9 do not represent physical resistors but are circuit elements used to account for resistive effects within the devices being modeled.

The linear resistor, open circuit, short circuit, ideal switch, ideal voltage source, and ideal current source are the initial entries in our catalog of circuit elements. In Chapter 4 we will develop models for active devices like the transistor and OP AMP. Models for dynamic elements like capacitors and inductors are introduced in Chapter 6.

2–2 CONNECTION CONSTRAINTS

In the previous section, we considered individual devices and models. In this section, we turn our attention to the constraints introduced by interconnections of devices to form circuits. The laws governing circuit behavior are based on the meticulous work of the German scientist Gustav Kirchhoff (1824–1887). **Kirchhoff's laws** are derived from conservation laws as applied to circuits. They tell us that element voltages and currents are forced to behave in certain ways when the devices are interconnected to form a circuit. These conditions are called **connection constraints** because they are based only on the circuit connections and not on the specific devices in the circuit.

In this book, we will indicate that crossing wires are connected (electrically tied together) using the dot symbol, as in Figure 2–10(a). Sometimes crossing wires are not connected (electrically insulated) but pass over or under each other. Since we are restricted to drawing wires on a planar surface, we will indicate unconnected crossovers by *not* placing a dot at their intersection, as indicated in the left of Figure 2–10(b). Other books sometimes show unconnected crossovers using the semicircular "hopover" shown on the right of Figure 2–10(b). In engineering systems two or more separate circuits are often tied together to form a larger circuit (for example, the interconnection of two integrated circuit packages). Interconnecting different circuits forms an interface between the circuits. The special jack or interface symbol in Figure 2–10(c) is used in this book because interface connections represent important points at which the interaction between two circuits can be observed or specified. On certain occasions a control line is required to show a mechanical or other nonelectrical dependency. Figure 2–10(d) shows how this dependency is indicated in this book. Figure 2–10(e) shows how power supply connections are often shown in electronic circuit diagrams. The implied power supply connection is indicated by an arrow pointing to the supply voltage, which may be given in numerical (+15 V) or symbolic form ($+V_{CC}$).

The treatment of Kirchhoff's laws uses the following definitions:

- A **circuit** is an interconnection of electrical devices.
- A **node** is an electrical juncture of two or more devices.
- A **loop** is a closed path formed by tracing through an ordered sequence of nodes without passing through any node more than once.

While it is customary to designate the juncture of two or more elements as a node, it is important to realize that a node is not confined to a point but includes all the zero-resistance wire from the point to each element. In the circuit of Figure 2–11, there are only three different nodes: A, B, and C. The points 2, 3, and 4, for example, are part of node B, while the points 5, 6, 7, and 8 are all part of node C.

(a)   (b)

(c)   (d)

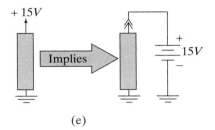

(e)

FIGURE   2 – 1 0 *Symbols used in circuit diagrams: (a) Electrical connection. (b) Crossover with no connection. (c) Jack connection. (d) Control line. (e) Power supply connection.*

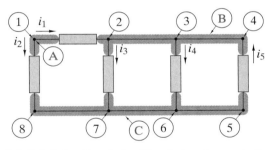

FIGURE 2 – 1 1 *Circuit for demonstrating Kirchhoff's current law.*

KIRCHHOFF'S CURRENT LAW

Kirchhoff's first law is based on the principle of conservation of charge. **Kirchhoff's current law (KCL)** states that

> *the algebraic sum of the currents entering a node is zero at every instant.*

In forming the algebraic sum of currents, we must take into account the current reference direction associated with each device. If the current reference direction is into the node, then we assign a positive sign to the corresponding current in the algebraic sum. If the reference direction is away from the node, we assign a negative sign. Applying this convention to the nodes in Figure 2–11, we obtain the following set of KCL connection equations:

$$\text{Node A: } -i_1 - i_2 \qquad\qquad = 0$$
$$\text{Node B: } i_1 - i_3 - i_4 + i_5 = 0 \qquad (2\text{--}8)$$
$$\text{Node C: } i_2 + i_3 + i_4 - i_5 = 0$$

The KCL equation at node A does not mean that the currents i_1 and i_2 are both negative. The minus signs in this equation simply mean that the reference direction for each current is directed away from node A. Likewise, the equation at node B could be written as

$$i_3 + i_4 = i_1 + i_5 \qquad (2\text{--}9)$$

This form illustrates an alternate statement of KCL:

> *The sum of the currents entering a node equals the sum of the currents leaving the node.*

There are two algebraic signs associated with each current in the application of KCL. First is the sign given to a current in writing a KCL connection equation. This sign is determined by the orientation of the current reference direction relative to a node. The second sign is determined by the actual direction of the current relative to the reference direction. The actual direction is found by solving the set of KCL equations, as illustrated in the following example.

EXAMPLE 2–4

Given $i_1 = +4$ A, $i_3 = +1$ A, $i_4 = +2$ A in the circuit shown in Figure 2–11, find i_2 and i_5.

SOLUTION:
Using the node A constraint in Eq. (2–8) yields

$$-i_1 - i_2 = -(+4) - i_2 = 0$$

The sign outside the parentheses comes from the node A KCL connection constraint in Eq. (2–8). The sign inside the parentheses comes from the actual direction of the current. Solving this equation for the unknown current, we find that $i_2 = -4$ A. In this case, the minus sign indicates that the actual direction of the current i_2 is directed upward in Figure 2–11, which is opposite to the reference direction assigned. Using this current in the second KCL equation in Eq. (2–8), we can write

$$i_1 - i_3 - i_4 + i_5 = (+4) - (+1) - (+2) + i_5$$

which yields the result $i_5 = -1$ A.

Again, the signs inside the parentheses are associated with the actual direction of the current, and the signs outside come from the node B KCL connection constraint in Eq. (2–8). The minus sign in the final answer means that the current i_5 is directed in the opposite direction from its assigned reference direction. We can check our work by substituting the values found into the node C constraint in Eq. (2–8). These substitutions yield

$$+i_2 + i_3 + i_4 - i_5 = (-4) + (+1) + (+2) - (-1) = 0$$

as required by KCL. Given three currents, we determined all the remaining currents in the circuit using only KCL without knowing the element constraints.   ■

In Example 2–4, the unknown currents were found using only the KCL constraints at nodes A and B. The node C equation was shown to be valid, but it did not add any new information. If we look back at Eq. (2–8), we see that the node C equation is the negative of the sum of the node A and B equations. In other words, the KCL connection constraint at node C is not independent of the two previous equations. This example illustrates the following general principle:

> *In a circuit containing a total of* N *nodes there are only* $N – 1$ *independent KCL connection equations.*

Current equations written at $N – 1$ nodes contain all the independent connection constraints that can be derived from KCL. To write these equations, we select one node as the reference or ground node and then write KCL equations at the remaining $N – 1$ nonreference nodes.

Exercise 2–1

Refer to Figure 2–12.

(a) Write KCL equations at nodes A, B, C, and D.
(b) Given $i_1 = -1$ mA, $i_3 = 0.5$ mA, $i_6 = 0.2$ mA, find i_2, i_4, and i_5.

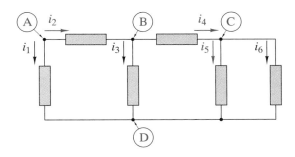

Answers:
(a) Node A: $-i_1 - i_2 = 0$; node B: $i_2 - i_3 - i_4 = 0$; node C: $i_4 - i_5 - i_6 = 0$; node D: $i_1 + i_3 + i_5 + i_6 = 0$
(b) $i_2 = 1$ mA, $i_4 = 0.5$ mA, $i_5 = 0.3$ mA

KIRCHHOFF'S VOLTAGE LAW

The second of Kirchhoff's circuit laws is based on the principle of conservation of energy. **Kirchhoff's voltage law** (abbreviated **KVL**) states that

> *the algebraic sum of all the voltages around a loop is zero at every instant.*

For example, three loops are shown in the circuit of Figure 2–13. In writing the algebraic sum of voltages, we must account for the assigned reference marks. As a loop is traversed, a positive sign is assigned to a voltage when we go from a "+" to "−" reference mark. When we go from "−" to "+", we use a minus sign. Traversing the three loops in Figure 2–13 in the indicated clockwise direction, yields the following set of KVL connection equations:

$$\text{Loop 1: } -v_1 + v_2 + v_3 \qquad = 0$$

$$\text{Loop 2: } -v_3 + v_4 + v_5 \qquad = 0 \qquad\qquad (2\text{–}10)$$

$$\text{Loop 3: } -v_1 + v_2 + v_4 + v_5 = 0$$

FIGURE 2 – 1 3 *Circuit for demonstrating Kirchhoff's voltage law.*

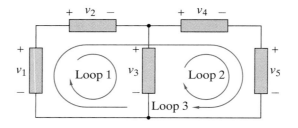

There are two signs associated with each voltage. The first is the sign given the voltage when writing the KVL connection equation. The second is the sign determined by the actual polarity of a voltage relative to its assigned reference polarity. The actual polarities are found by solving the set of KVL equations, as illustrated in the following example.

EXAMPLE 2–5

Given $v_1 = 5$ V, $v_2 = -3$ V, and $v_4 = 10$ V in the circuit shown in Figure 2–13, find v_3 and v_5.

SOLUTION:

Inserting the given numerical values into Eq. (2–10) yields the following KVL equation for loop 1:

$$-v_1 + v_2 + v_3 = -(+5) + (-3) + (v_3) = 0$$

The sign outside the parentheses comes from the loop 1 KVL constraint in Eq. (2–10). The sign inside comes from the actual polarity of the voltage. This equation yields $v_3 = +8$ V. Using this value in the loop 2, KVL constraint in Eq. (2–10) produces

$$-v_3 + v_4 + v_5 = -(+8) + (+10) + v_5 = 0$$

The result is $v_5 = -2$ V. The minus sign here means that the actual polarity of v_5 is the opposite of the assigned reference polarity indicated in Figure 2–13. The results can be checked by substituting all the aforementioned values into the loop 3 KVL constraint in Eq. (2–10). These substitutions yield

$$-(+5) + (-3) + (+10) + (-2) = 0$$

as required by KVL.   ∎

In Example 2–5, the unknown voltages were found using only the KVL constraints for loops 1 and 2. The loop 3 equation was shown to be valid, but it did not add any new information. If we look back at Eq. (2–10), we see that the loop 3 equation is equal to the sum of the loop 1 and 2 equations. In other words, the KVL connection constraint around loop 3 is not independent of the previous two equations. This example illustrates the following general principle:

> *In a circuit containing a total of* **E** *two-terminal elements and* **N** *nodes, there are only* **E – N +1** *independent KVL connection equations.*

Writing voltage summations around a total of $E - N + 1$ *different* loops produces all the independent connection constraints that can be derived from KVL. A *sufficient condition* for loops to be different is that each contains at least one element that is not contained in any other loop. In simple circuits with no crossovers the open space between elements produces $E - N + 1$ independent loops. However, finding all the loops in a more complicated circuit can be a nontrivial problem.

Exercise 2–2

Find the voltages v_x and v_y in Figure 2–14.

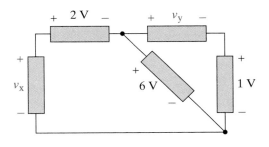

FIGURE 2 – 1 4

Answers: $v_x = +8$ V, $v_y = +5$ V

Exercise 2–3

Find the voltages v_x, v_y, and v_z in Figure 2–15.

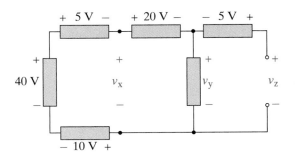

FIGURE 2 – 1 5

Answers: $v_x = +25$ V; $v_y = +5$ V; $v_z = +10$ V. *Note:* KVL yields the voltage v_z even though it appears across an open circuit.

PARALLEL AND SERIES CONNECTIONS

Two types of connections occur so frequently in circuit analysis that they deserve special attention. Elements are said to be connected in **parallel** when they form a loop containing no other elements. For example, loop A in Figure 2–16 contains only elements 1 and 2. As a result, the KVL connection constraint around loop A is

$$- v_1 + v_2 = 0 \tag{2–11}$$

which yields $v_1 = v_2$. In other words, in a parallel connection KVL requires equal voltages across the elements. The parallel connection is not restricted to two elements. For example, loop B in Figure 2–16 contains only elements 2 and 3; hence, by KVL $v_2 = v_3$. As a result, in this circuit we have $v_1 = v_2 = v_3$, and we say that elements 1, 2, and 3 are connected in parallel. In general, then, any number of elements connected between two common nodes are in parallel, and as a result, the same voltage appears across each of them. Existence of a parallel connection does not depend on the graphical position of the elements. For example, the position of elements 1 and 3 could be switched, and the three elements are still connected in parallel.

F I G U R E   2 – 1 6   *A parallel connection.*

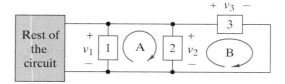

Two elements are said to be connected in **series** when they have one common node to which no other element is connected. In Figure 2–17 elements 1 and 2 are connected in series, since only these two elements are connected at node A. Applying KCL at node A yields

$$i_1 - i_2 = 0 \quad \text{or} \quad i_1 = i_2 \tag{2–12}$$

In a series connection, KCL requires equal current through each element. Any number of elements can be connected in series. For example, element 3 in Figure 2–17 is connected in series with element 2 at node B, and KCL requires $i_2 = i_3$. Therefore, in this circuit $i_1 = i_2 = i_3$, we say that elements 1, 2, and 3 are connected in series, and the same current exists in each of the elements. In general, elements are connected in series when they form a single path between two nodes such that only elements in the path are connected to the intermediate nodes along the path.

F I G U R E   2 – 1 7   *A series connection.*

EXAMPLE 2–6

Identify the elements connected in parallel and in series in each of the circuits in Figure 2–18.

SOLUTION:

In Figure 2–18(a) elements 1 and 2 are connected in series at node A and elements 3 and 4 are connected in parallel between nodes B and C. In Figure 2–18(b) elements 1 and 2 are connected in series at node A, as are elements 4 and 5 at node D. There are no single elements connected in parallel in this circuit. In Figure 2–18(c) there are no elements connected in either series or parallel. It is important to realize that in some circuits there are elements that are not connected in either series or in parallel. ■

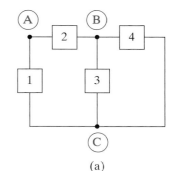

Exercise 2–4

Identify the elements connected in series or parallel when a short circuit is connected between nodes A and B in each of the circuits of Figure 2–18.

Answers:
Circuit in Figure 2–18(a): Elements 1, 3, and 4 are all in parallel.
Circuit in Figure 2–18(b): Elements 1 and 3 are in parallel; elements 4 and 5 are in series.
Circuit in Figure 2–18(c): Elements 1 and 3 are in parallel; elements 4 and 6 are in parallel.

Exercise 2–5

Identify the elements in Figure 2–19 that are connected in (a) parallel, (b) series, or (c) neither.

Answers:
(a) The following elements are in parallel: 1, 8, and 11; 3, 4, and 5.
(b) The following elements are in series: 9 and 10; 6 and 7.
(c) Only element 2 is not in series or parallel with any other element.

DISCUSSION: *The ground symbol indicates the reference node. When ground symbols are shown at several nodes, the nodes are effectively connected together by a short circuit to form a single node.*

(a)

(b)

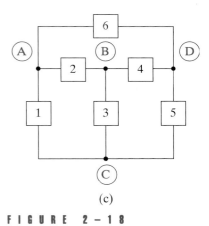

(c)

FIGURE 2 – 18

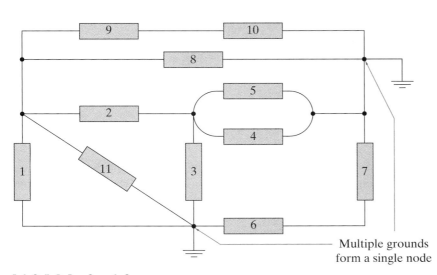

Multiple grounds
form a single node

FIGURE 2 – 19

2–3 COMBINED CONSTRAINTS

The usual goal of circuit analysis is to determine the currents or voltages at various places in a circuit. This analysis is based on constraints of two distinctly different types. The element constraints are based on the models of the specific devices connected in the circuit. The connection constraints are based on Kirchhoff's laws and the circuit connections. The element equations are independent of the circuit connections. Likewise, the connection equations are independent of the devices in the circuit. Taken together, however, the combination of the element and connection constraints supply the equations needed to describe a circuit.

Our study of the combined constraints begins by considering the simple but important example in Figure 2–20. This circuit is driven by a current source i_S and the resulting responses are current/voltage pairs (i_x, v_x) and (i_O, v_O). The reference marks for the response pairs have been assigned using the passive sign convention.

To solve for all four responses, we must write four equations. The first two are the element equations

$$i_x = i_S$$

$$v_O = Ri_O \qquad (2\text{--}13)$$

The first element equation states that the response current i_x and the input driving force i_S are equal in magnitude and direction. The second element equation is Ohm's law relating v_O and i_O under the passive sign convention.

The connection equations are obtained by applying Kirchhoff's laws. The circuit in Figure 2–20 has two elements ($E = 2$) and two nodes ($N = 2$), so we need $E - N + 1 = 1$ KVL equation and $N - 1 = 1$ KCL equation. Selecting node B as the reference node, we apply KCL at node A and apply KVL around the loop to write

$$\text{KCL: } -i_x - i_O = 0$$

$$\text{KVL: } -v_x + v_O = 0 \qquad (2\text{--}14)$$

We now have two element constraints in Eq. (2–13) and two connection constraints in Eq. (2–14), so we can solve for all four responses in terms of the input driving force i_S. Combining the KCL connection equation and the first element equation yields $i_O = -i_x = -i_S$. Substituting this result into the second element equation (Ohm's law) produces

$$v_O = -Ri_S \qquad (2\text{--}15)$$

The minus sign in this equation does not mean that v_O is always negative. Nor does it mean the resistance is negative. It means that when the input driving force i_S is positive, then the response v_O is negative, and vice versa. This sign reversal is a result of the way we assigned reference marks at the beginning of our analysis. The reference marks defined the circuit input and outputs in such a way that i_S and v_O always have opposite algebraic signs. Put differently, Eq. (2–15) is an input-output relationship, not an element i–v relationship.

F I G U R E   2 – 2 0 *Circuit used to demonstrate combined constraints.*

EXAMPLE 2–7

(a) Find the responses i_x, v_x, i_O, and v_O in the circuit in Figure 2–20 when $i_S = +2$ mA and $R = 2$ kΩ.

(b) Repeat for $i_S = -2$ mA.

SOLUTION:

(a) From Eq. (2–13) we have $i_x = i_S = +2$ mA and $v_O = 2000\,i_O$. From Eq. (2–14) we have $i_O = -i_x = -2$ mA and $v_x = v_O$. Combining these results, we obtain

$$v_x = v_O = 2000\,i_O = 2000(-0.002) = -4 \text{ V}$$

(b) In this case $i_x = i_S = -2$ mA, $i_O = -i_x = -(-0.002) = +2$ mA, and

$$v_x = v_O = 2000\,i_O = 2000(+0.002) = +4 \text{ V}$$

This example confirms that the algebraic signs of the outputs v_x, v_O, and i_O are always the opposite sign from that of the input driving force i_S. ∎

Analyzing the circuit in Figure 2–21 illustrates the formulation of combined constraints. We first assign reference marks for all the voltages and currents using the passive sign convention. Then, using these definitions we can write the element constraints as

$$v_A = V_O$$
$$v_1 = R_1 i_1 \tag{2–16}$$
$$v_2 = R_2 i_2$$

FIGURE 2–21 *Circuit used to demonstrate combined constraints.*

These equations describe the three devices and do not depend on how the devices are connected in the circuit.

The connection equations are obtained from Kirchhoff's laws. To apply these laws, we must first label the different loops and nodes. The circuit contains $E = 3$ elements and $N = 3$ nodes, so there are $E - N + 1 = 1$ independent KVL constraints and $N - 1 = 2$ independent KCL constraints. There is only one loop, but there are three nodes in this circuit. We will select one node as the reference point and write KCL equations at the other two nodes. Any node can be chosen as the reference, so we select node C as the reference node and indicate this choice by drawing the ground symbol there. The connection constraints are

$$\text{KCL: Node A } -i_A - i_1 = 0$$
$$\text{KCL: Node B } i_1 - i_2 = 0 \tag{2–17}$$
$$\text{KVL: Loop } -v_A + v_1 + v_2 = 0$$

These equations are independent of the specific devices in the circuit. They depend only on Kirchhoff's laws and the circuit connections.

This circuit has six unknowns: three element currents and three element voltages. Taken together, the element and connection equations give us six independent equations. For a network with (N) nodes and (E) two-terminal elements, we can write ($N - 1$) independent KCL connection equations, $(E - N + 1)$ independent KVL connection equations, and (E) element equations. The total number of equations generated is

Element equations	E
KCL equations	$N - 1$
KVL equations	$\underline{E - N + 1}$
Total	$2E$

The grand total is then $(2E)$ combined connection and element equations, which is exactly the number of equations needed to solve for the voltage across and current through every element—a total of $(2E)$ unknowns.

EXAMPLE 2–8

Find all of the element currents and voltages in Figure 2–21 for $V_O =$ 10 V, $R_1 = 2000\ \Omega$, and $R_2 = 3000\ \Omega$.

SOLUTION:
Substituting the element constraints from Eq. (2–16) into the KVL connection constraint in Eq. (2–17) produces

$$-V_O + R_1 i_1 + R_2 i_2 = 0$$

This equation can be used to solve for i_1 since the second KCL connection equation requires that $i_2 = i_1$.

$$i_1 = \frac{V_O}{R_1 + R_2} = \frac{10}{2000 + 3000} = 2 \text{ mA}$$

In effect, we have found all of the element currents since the elements are connected in series. Hence, collectively the KCL connection equations require that

$$-i_A = i_1 = i_2$$

Substituting all of the known values into the element equations gives

$$v_A = 10 \text{ V} \quad v_1 = R_1 i_1 = 4 \text{ V} \quad v_2 = R_2 i_2 = 6 \text{ V}$$

Every element voltage and current has been found. Note the analysis strategy used. We first found all the element currents and then used these values to find the element voltages. ∎

EXAMPLE 2–9

Consider the source-resistor-switch circuit of Figure 2–22 with $V_O =$ 10 V and $R_1 = 2000\ \Omega$. Find all the element voltages and currents with the switch open and again with the switch closed.

SOLUTION:
The connection equations for the circuit are

$$\text{KCL: Node A} \quad -i_A - i_1 \quad = 0$$
$$\text{KCL: Node B} \quad i_1 - i_2 \quad = 0$$
$$\text{KVL: Loop} \quad -v_A + v_1 + v_2 = 0$$

FIGURE 2 – 2 2

These connection equations are the same as those in Eq. (2–17) for the circuit in Figure 2–21. The two circuits have the same connections but different devices. When the switch is open, the current through the switch is zero, so $i_2 = 0$. The KCL connection equation at node B requires $i_1 = i_2$, and therefore $i_1 = 0$. The element equation for R_1 yields $v_1 = R_1 i_1 = 0$, and the KVL connection equation yields $v_1 + v_2 = v_2 = v_A = 10$ V. Thus, when the switch is open, the current is zero and all the source voltage appears across the switch. When the switch is closed, its element equations require $v_2 = 0$. The KVL connection equation gives $v_1 + v_2 = v_1 = v_A = 10$ V. When the switch is closed, all the source voltage appears across R_1 rather than across the switch. So the current through the circuit is

$$i_1 = \frac{v_1}{R_1} = \frac{10}{2000} = 5 \text{ mA} \qquad \blacksquare$$

Exercise 2–6

For the circuit of Figure 2–23,

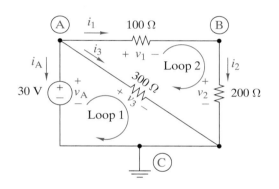

FIGURE 2 – 2 3

(a) Write a complete set of element equations.
(b) Write a complete set of connection equations.
(c) Solve the equations in (a) and (b) for all element currents and voltages.

Answers:
(a) $v_A = 30$ V; $v_3 = 300\, i_3$; $v_1 = 100\, i_1$; $v_2 = 200\, i_2$
(b) $-i_A - i_1 - i_3 = 0$; $+ i_1 - i_2 = 0$; $-30 + v_3 = 0$; $v_1 + v_2 - v_3 = 0$
(c) $v_A = 30$ V; $v_1 = 10$ V; $v_2 = 20$ V; $v_3 = 30$ V; $i_A = -200$ mA; $i_1 = i_2 = 100$ mA; $i_3 = 100$ mA

ASSIGNING REFERENCE MARKS

In all of our previous examples and exercises, the reference marks for the element currents (arrows) and voltages (+ and –) were given. When reference marks are not shown on a circuit diagram, they must be assigned by the person solving the problem. Beginners sometimes wonder how to assign reference marks when the actual voltage polarities and current directions are unknown. It is important to remember that the reference marks do not indicate what is actually happening in the circuit. They are benchmarks assigned at the beginning of the analysis. If it turns out that the actual direction and reference direction agree, then the algebraic sign of the response will be positive. If they disagree, the algebraic sign will be negative. In other words, the sign of the answer together with assigned reference marks tell us the actual voltage polarity or current direction.

In this book the reference marks always follow the passive sign convention. This means that for any given two-terminal element we can arbitrarily assign either the + voltage reference mark or the current reference arrow, but not both. For example, we can arbitrarily assign the voltage reference marks to the terminals of a two-terminal device. Once the voltage reference is assigned, however, the passive sign convention requires that the current reference arrow be directed into the element at the terminal with the + mark. On the other hand, we could start by arbitrarily selecting the terminal at which the current reference arrow is directed into the device. Once the current reference is assigned, however, the passive sign convention requires that the + voltage reference be assigned to the selected terminal.

Following the passive sign convention avoids confusion about the direction of power flow in a device. In addition, the element constraints, such as Ohm's law, assume that the passive sign convention is used to assign the voltage and current reference marks to a device.

The next example illustrates the assignment of reference marks.

EXAMPLE 2–10

Find all of the element voltages and currents in Figure 2–24(a).

FIGURE 2–24

(a)

(b)

(c)

SOLUTION:
Since no reference marks are shown in Figure 2–24(a), we assign references to two voltages and one current, as shown in Figure 2–24(b). Other choices are possible, of course, but once these marks are selected, the pas-

sive sign convention dictates the reference marks for the remaining voltage and currents, as shown in Figure 2–24(c). Using all of these reference marks, we write the element equations as

$$v_1 = 500i_1$$

$$v_2 = 1000i_2$$

$$v_S = -1.5 \text{ V}$$

Using the indicated reference node, we write two KCL and one KVL equations:

KCL: Node A $+ i_S - i_1 \quad = 0$

KCL: Node B $+ i_1 + i_2 \quad = 0$

KVL: Loop $+ v_S + v_1 - v_2 = 0$

Solving the combined element and connection equations yields

$$v_S = -1.5 \text{ V}, \quad i_S = +1.0 \text{ mA}$$

$$v_1 = +0.5 \text{ V}, \quad i_1 = +1.0 \text{ mA}$$

$$v_2 = -1.0 \text{ V}, \quad i_2 = -1.0 \text{ mA}$$

These results show that the reference marks for v_1, i_S, and i_1 agree with the actual voltage polarities and current directions, while the minus signs on the other responses indicate disagreement. It is important to realize that this disagreement does not mean that assigned reference marks are wrong. ∎

2–4 EQUIVALENT CIRCUITS

The analysis of a circuit can often be made easier by replacing part of the circuit with one that is equivalent but simpler. The underlying basis for two circuits to be equivalent is contained in their i–v relationships.

> *Two circuits are said to be equivalent if they have identical* i–v *characteristics at a specified pair of terminals.*

In other words, when two circuits are equivalent the voltage and current at an interface do not depend on which circuit is connected to the interface.

EQUIVALENT RESISTANCE

The two resistors in Figure 2–25(a) are connected in series between a pair of terminals A and B. The objective is to simplify the circuit without altering the electrical behavior of the rest of the circuit.

The KVL equation around the loop from A to B is

$$v = v_1 + v_2 \qquad (2\text{–}18)$$

Since the two resistors are connected in series, the same current i exists in both. Applying Ohm's law, we get $v_1 = R_1 i$ and $v_2 = R_2 i$. Substituting these relationships into Eq. (2–18) and then simplifying yields

$$v = R_1 i + R_2 i = i(R_1 + R_2)$$

(a)

(b)

FIGURE 2 – 2 5 *A series resistance circuit: (a) Original circuit. (b) Equivalent circuit.*

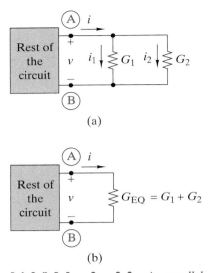

FIGURE 2–26 *A parallel resistance circuit: (a) Original circuit. (b) Equivalent circuit.*

We can write this equation in terms of an equivalent resistance R_{EQ} as

$$v = iR_{EQ} \quad \text{where} \quad R_{EQ} = R_1 + R_2 \qquad (2\text{–}19)$$

This result means the circuits in Figs. 2–25(a) and 2–25(b) have the same i–v characteristic at terminals A and B. As a result, the response of the rest of the circuit is unchanged when the series connection of R_1 and R_2 is replaced by a resistance R_{EQ}.

The parallel connection of two conductances in Figure 2–26(a) is the dual[2] of the series circuit in Figure 2–25(a). Again the objective is to replace the parallel connection by a simpler equivalent circuit without altering the response of the rest of the circuit.

A KCL equation at node A produces

$$i = i_1 + i_2 \qquad (2\text{–}20)$$

Since the conductances are connected in parallel, the voltage v appears across both. Applying Ohm's law, we obtain $i_1 = G_1 v$, and $i_2 = G_2 v$. Substituting these relationships into Eq. (2–20) and then simplifying yields

$$i = vG_1 + vG_2 = v(G_1 + G_2)$$

This result can be written in terms of an equivalent conductance G_{EQ} as follows:

$$i = vG_{EQ}, \quad \text{where} \quad G_{EQ} = G_1 + G_2 \qquad (2\text{–}21)$$

This result means the circuits in Figures 2–26(a) and 2–26(b) have the same i–v characteristic at terminals A and B. As a result, the response of the rest of the circuit is unchanged when the parallel connection of G_1 and G_2 is replaced by a conductance G_{EQ}.

Since conductance is not normally used to describe a resistor, it is sometimes useful to rewrite Eq. (2–21) as an equivalent resistance $R_{EQ} = 1/G_{EQ}$. That is,

$$R_1 \| R_2 = R_{EQ} = \frac{1}{G_{EQ}} = \frac{1}{G_1 + G_2} = \frac{1}{\dfrac{1}{R_1} + \dfrac{1}{R_2}} = \frac{R_1 R_2}{R_1 + R_2} \qquad (2\text{–}22)$$

where the symbol "$\|$" is shorthand for "in parallel." The expression on the far right in Eq. (2–22) is called the product over the sum rule for two resistors in parallel. This rule is useful in derivations in which the resistances are left in symbolic form (see Example 2–11), but it is not an efficient algorithm for calculating numerical values of equivalent resistance.

Caution: The product over sum rule only applies to two resistors connected in parallel. When more than two resistors are in parallel, we must use the following general result to obtain the equivalent resistance:

$$R_{EQ} = \frac{1}{G_{EQ}} = \frac{1}{\dfrac{1}{R_1} + \dfrac{1}{R_2} + \dfrac{1}{R_3} + \cdots} \qquad (2\text{–}23)$$

2 Dual circuits have identical behavior patterns when we interchange the roles of the following parameters: (1) voltage and current, (2) series and parallel, and (3) resistance and conductance. In later chapters we will see duality exhibited by other circuit parameters as well.

EXAMPLE 2 – 11

Given the circuit in Figure 2–27(a),
(a) Find the equivalent resistance R_{EQ1} connected between terminals A and B.
(b) Find the equivalent resistance R_{EQ2} connected between terminals C and D.

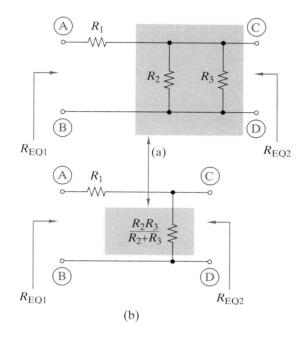

FIGURE   2 – 27

SOLUTION:

First we note that resistors R_2 and R_3 are connected in parallel. Applying the product over sum rule [Eq. (2–22)], we obtain

$$R_2 \| R_3 = \frac{R_2 R_3}{R_2 + R_3}$$

As an interim step, we redraw the circuit, as shown in Figure 2–27(b).
(a) To find the equivalent resistance between terminals A and B, we note that R_1 and the equivalent resistance $R_2 \| R_3$ are connected in series. The total equivalent resistance R_{EQ1} between terminals A and B is

$$R_{EQ1} = R_1 + (R_2 \| R_3)$$

$$R_{EQ1} = R_1 + \frac{R_2 R_3}{R_2 + R_3}$$

$$R_{EQ1} = \frac{R_1 R_2 + R_1 R_3 + R_2 R_3}{R_2 + R_3}$$

(b) Looking between terminals C and D yields a different result. In this case R_1 is not involved, since there is an open circuit (an infinite resistance) between terminals A and B. Therefore, only $R_2 \parallel R_3$ affect the resistance between terminals C and D. As a result, R_{EQ2} is simply

$$R_{EQ2} = R_2\|R_3 = \frac{R_2 R_3}{R_2 + R_3}$$

This example shows that equivalent resistance depends on the pair of terminals involved.   ∎

One final note on checking numerical calculations of equivalent resistance. When several resistances are connected in parallel, the equivalent resistance must be smaller than the smallest resistance in the connection. Conversely, when several resistances are connected in series, the equivalent resistance must be larger than the largest resistance in the connection.

Exercise 2–7

Find the equivalent resistance between terminals A–C, B–D, A–D, and B–C in the circuit in Figure 2–27.

  Answers: $R_{A-C} = R_1$, $R_{B-D} = 0\ \Omega$ (a short circuit), $R_{A-D} = R_1 + R_2 \parallel R_3$, and $R_{B-C} = R_2 \parallel R_3$.

Exercise 2–8

Find the equivalent resistance between terminals A–B, A–C, A–D, B–C, B–D, and C–D in the circuit of Figure 2–28.

  Answers: $R_{A-B} = 100\ \Omega$, $R_{A-C} = 70\ \Omega$, $R_{A-D} = 65\ \Omega$, $R_{B-C} = 90\ \Omega$, $R_{B-D} = 85\ \Omega$, and $R_{C-D} = 55\ \Omega$.

FIGURE 2 – 2 8

EQUIVALENT SOURCES

The practical source models introduced previously are shown in Figure 2–29. These models consist of an ideal voltage source in series with a resistance and an ideal current source in parallel with a resistance. We now determine the conditions under which the practical voltage source and the practical current sources are equivalent.

Figure 2–29 shows the two practical sources connected between terminals labeled A and B. A parallel analysis of these circuits yields the conditions for equivalency at terminals A and B. First, Kirchhoff's laws are applied as

Circuit A

Circuit B

FIGURE   2 – 2 9 *Practical source models that are equivalent when Eq. (2–24) is statisfied.*

Circuit A	Circuit B
KVL	KCL
$v_S = v_R + v$	$i_S = i_R + i$

Next, Ohm's law is used to obtain

Circuit A	Circuit B
$v_R = R_1 i$	$i_R = \dfrac{v}{R_2}$

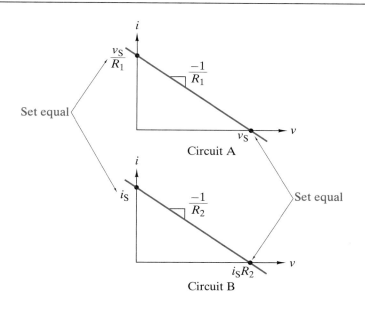

Combining these results, we find that the *i–v* relationships of each of the circuits at terminals A and B are

Circuit A   Circuit B

$$i = -\frac{v}{R_1} + \frac{v_S}{R_1} \qquad i = -\frac{v}{R_2} + i_S$$

These *i–v* characteristics take the form of the straight lines shown in Figure 2–30. The two lines are identical when the intercepts are equal. This requires that $v_S/R_1 = i_S$ and $v_S = i_S R_2$, which, in turn, requires that

$$R_1 = R_2 = R \quad \text{and} \quad v_S = i_S R \qquad (2\text{–}24)$$

When conditions in Eq. (2–24) are met, the response of the rest of the circuit is unaffected when we replace a practical voltage source by an equivalent practical current source, or vice versa. Exchanging one practical source model for an equivalent model is called *source transformation*.

Source transformation means that either model will deliver the same voltage and current to the rest of the circuit. It does not mean the two models are identical in every way. For example, when the practical voltage source drives an open circuit, there is no i^2R power loss since the current in the series resistance is zero. However, the current in the parallel resistance of a practical current source is not zero when the load is an open circuit. Thus, equivalent sources do not have the same internal power loss even though they deliver the same current and voltage to the rest of the circuit.

EXAMPLE 2–12

Convert the practical voltage source in Figure 2–31(a) into an equivalent current source.

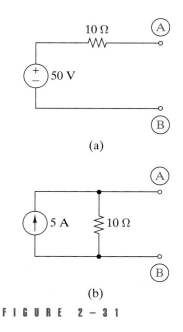

FIGURE 2–31

SOLUTION:
Using Eq. (2–24), we have

$$R_1 = R_2 = R = 10 \ \Omega$$

$$i_S = v_S/R = 5 \text{ A}$$

The equivalent practical current source is shown in Figure 2–31(b). ■

Exercise 2–9

A practical current source consists of a 2-mA ideal current source in parallel with a 0.002-S conductance. Find the equivalent practical voltage source.

Answer: The equivalent is a 1-V ideal voltage source in series with a 500-Ω resistance.

Figure 2–32 shows another source transformation in which a voltage source and resistor in parallel is replaced by a voltage source acting alone. The two circuits are equivalent because i–v constraint at the input to the rest of the circuit is $v = v_S$ in both circuits. In other words, the response of the rest of the circuit is unchanged if a resistor in parallel with a voltage source is removed from the circuit. However, removing the resistor does reduce the total current supplied by the voltage source by v_S/R. While the resistor does not affect the current and voltage delivered to the rest of the circuit, it does dissipate power that must be supplied by the source.

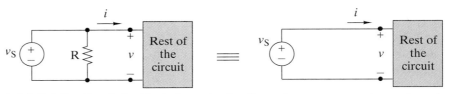

F I G U R E   2 – 3 2 *Equivalent circuit of a voltage source and a resistor in parallel.*

The dual situation is shown in Figure 2–33. In this case a current source connected in series with a resistor can be replaced by a current source acting alone because the i–v constraint at the input to the rest of the circuit is $i = i_S$ for both circuits. In other words, the response of the rest of the circuit is unchanged if a resistor in series with a current source is removed from the circuit.

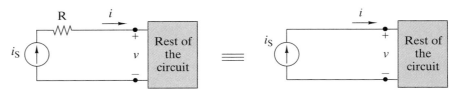

F I G U R E   2 – 3 3 *Equivalent circuit of a current source and a resistor in series.*

FIGURE 2–34 *Summary of two-terminal equivalent circuits.*

SUMMARY OF EQUIVALENT CIRCUITS

Figure 2–34 is a summary of two-terminal equivalent circuits involving resistors and sources connected in series or parallel. The series and parallel equivalences in the first row and the source transformations in the second row are used regularly in subsequent discussions. The last row in Figure 2–34 presents additional source transformations that reduce series or parallel connections to a single ideal current or voltage source. Proof of these equivalences involves showing that the final single-source circuits have the same i–v characteristics as the original connections. The details of such a derivation are left as an exercise for the reader.

2–5 VOLTAGE AND CURRENT DIVISION

We complete our treatment of series and parallel circuits with a discussion of voltage and current division. These two analysis tools find wide application in circuit analysis and design.

FIGURE   2 – 3 5 *A   voltage divider circuit.*

VOLTAGE DIVISION

Voltage division provides a simple way to find the voltage across each element in a series circuit. Figure 2–35 shows a circuit that lends itself to solution by voltage division. Applying KVL around the loop in Figure 2–35 yields

$$v_S = v_1 + v_2 + v_3 \tag{2–25}$$

The elements in Figure 2–35 are connected in series, so the same current i exists in each of the resistors. Using Ohm's law, we find that

$$v_S = R_1 i + R_2 i + R_3 i \tag{2–26}$$

Solving for i yields

$$i = \frac{v_S}{R_1 + R_2 + R_3} \tag{2–27}$$

Once the current in the series circuit is found, the voltage across each resistor is computed using Ohm's law:

$$v_1 = R_1 i = \left(\frac{R_1}{R_1 + R_2 + R_3}\right) v_S \tag{2–28}$$

$$v_2 = R_2 i = \left(\frac{R_2}{R_1 + R_2 + R_3}\right) v_S \tag{2–29}$$

$$v_3 = R_3 i = \left(\frac{R_3}{R_1 + R_2 + R_3}\right) v_S \tag{2–30}$$

Looking over these results, we see an interesting pattern. In a series connection, the voltage across each resistor is equal to its resistance divided by the equivalent series resistance of the connection times the voltage across the series circuit. Thus, the general expression of the *voltage division rule* is

$$v_k = \left(\frac{R_k}{R_{EQ}}\right) v_{TOTAL} \tag{2–31}$$

In other words, the total voltage divides among the series resistors in proportion to their resistance over the equivalent resistance of the series connection. The following examples show several applications of this rule.

FIGURE   2 – 3 6

EXAMPLE 2–13

Find the voltage across the 330-Ω resistor in the circuit of Figure 2–36.

SOLUTION:

Applying the voltage division rule, we find that

$$v_O = \left(\frac{330}{100 + 560 + 330 + 220}\right) 24 = 6.55 \text{ V} \qquad ■$$

Exercise 2–10
Find the voltages v_x and v_y in Figure 2–36.

Answers: $v_x = 11.1$ V, $v_y = 4.36$ V

EXAMPLE 2–14

Select a value for the resistor R_x in Figure 2–37 so $v_O = 8$ V.

SOLUTION:
The unknown resistor is in parallel with the 10-kΩ resistor. Since voltages across parallel elements are equal, the voltage $v_O = 8$ V appears across both. We first define an equivalent resistance $R_{EQ} = R_x \parallel 10$ kΩ as

$$R_{EQ} = \frac{R_x \times 10000}{R_x + 10000}$$

We write the voltage division rule in terms of R_{EQ} as

$$v_O = 8 = \left(\frac{R_{EQ}}{R_{EQ} + 2000}\right)10$$

which yields $R_{EQ} = 8$ kΩ. Finally, we substitute this value into the equation defining R_{EQ} and solve for R_x to obtain $R_x = 40$ kΩ. ■

FIGURE 2–37

EXAMPLE 2–15

Use the voltage division rule to find the output voltage v_O of the circuit in Figure 2–38.

FIGURE 2–38

SOLUTION:
At first glance it appears that the voltage division rule does not apply, since the resistors are not connected in series. However, the current through R_3 is zero since the output of the circuit is an open circuit. Therefore, Ohm's law shows that $v_3 = R_3 i_3 = 0$. Applying KCL at node A shows that the same current exists in R_1 and R_2, since the current through R_3 is zero. Applying KVL around the output loop shows that the voltage across R_2 must be equal to v_O since the voltage across R_3 is zero. In essence, it is as if R_1 and R_2 were connected in series. Therefore, voltage division can be used and yields the output voltage as

$$v_O = \left(\frac{R_2}{R_1 + R_2}\right)v_S$$

The reader should carefully review the logic leading to this result because voltage division applications of this type occur frequently. ■

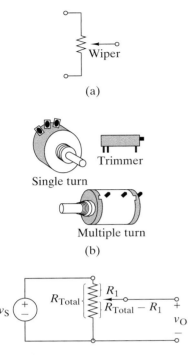

(a)

(b)

(c)

FIGURE 2-39 *The potentiometer: (a) Circuit symbol. (b) Actual device. (c) An application.*

FIGURE 2-40

EXAMPLE 2-16

The operation of a potentiometer is based on the voltage division rule. The device is a three-terminal element that uses voltage (potential) division to meter out a fraction of the applied voltage. Simply stated, a **potentiometer** is an adjustable voltage divider. Figure 2–39 shows the circuit symbol of a potentiometer, sketches of three different types of actual potentiometers, and a typical application.

The voltage v_O in Figure 2–39(c) can be adjusted by turning the shaft on the potentiometer to move the wiper arm contact. Using the voltage division rule, the voltage v_O is found as

$$v_O = \left(\frac{R_{TOTAL} - R_1}{R_{TOTAL}} \right) v_S \qquad (2\text{--}32)$$

Adjusting the movable wiper arm all the way to the top makes R_1 zero, and voltage division yields

$$v_O = \left(\frac{R_{TOTAL} - 0}{R_{TOTAL}} \right) v_S = v_S \qquad (2\text{--}33)$$

In other words, 100% of the applied voltage is delivered to the rest of the circuit. Moving the wiper all the way to the bottom makes R_1 equal to R_{TOTAL}, and voltage division yields

$$v_O = \left(\frac{R_{TOTAL} - R_{TOTAL}}{R_{TOTAL}} \right) v_S = 0 \qquad (2\text{--}34)$$

This opposite extreme delivers zero voltage. By adjusting the wiper arm position, we can obtain an output voltage anywhere between zero and the applied voltage v_S. When the wiper is positioned halfway between the top and bottom, we naturally expect to obtain half of the applied voltage. Setting $R_1 = \frac{1}{2} R_{TOTAL}$ yields

$$v_O = \left(\frac{R_{TOTAL} - \frac{1}{2} R_{TOTAL}}{R_{TOTAL}} \right) v_S = \frac{v_S}{2} \qquad (2\text{--}35)$$

as expected. The many applications of the potentiometer include volume controls, voltage balancing, and fine-tuning adjustment.

EXAMPLE 2-17

Figure 2–40 shows a programmable voltage divider in which the digital signals D_1 and D_0 control the divider resistance R_1 and R_2 by opening and closing the analog switches shown. Determine the output v_O when $(D_1, D_0) = (0, 0), (0, 1), (1, 0)$ and $(1, 1)$. Assume that the analog switches are ideal switches.

SOLUTION:
When $(D_1, D_0) = (0, 0)$ the upper switch is closed and the lower switch is open. In this configuration the divider resistances are

$$R_1 = 3.5 \text{ k}\Omega \| 21 \text{ k}\Omega = 3 \text{ k}\Omega \qquad \text{and} \qquad R_2 = 2 \text{ k}\Omega + 1 \text{ k}\Omega = 3 \text{ k}\Omega$$

and the output voltage is

$$v_O = \frac{R_2}{R_1 + R_2} \times 15 = \frac{1}{2} \times 15 = 7.5 \text{ V}$$

When $(D_1, D_0) = (0, 1)$ both switches are closed and the divider resistances are

$$R_1 = 3.5 \text{ k}\Omega \| 21 \text{ k}\Omega = 3 \text{ k}\Omega \qquad \text{and} \qquad R_2 = 1 \text{ k}\Omega$$

and the output voltage is

$$v_O = \frac{R_2}{R_1 + R_2} \times 15 = \frac{1}{4} \times 15 = 3.75 \text{ V}$$

When $(D_1, D_0) = (1, 0)$ both switches are open and the divider resistances are

$$R_1 = 21 \text{ k}\Omega \qquad \text{and} \qquad R_2 = 2 \text{ k}\Omega + 1 \text{ k}\Omega = 3 \text{ k}\Omega$$

and the output voltage is

$$v_O = \frac{R_2}{R_1 + R_2} \times 15 = \frac{1}{8} \times 15 = 1.875 \text{ V}$$

Finally, when $(D_1, D_0) = (1, 1)$ the upper switch is open, the lower switch is closed, and the divider resistances are

$$R_1 = 21 \text{ k}\Omega \qquad \text{and} \qquad R_2 = 1 \text{ k}\Omega$$

and the output voltage is

$$v_O = \frac{R_2}{R_1 + R_2} \times 15 = \frac{1}{22} \times 15 = 0.682 \text{ V}$$

This example illustrates using digital signals to control an analog signal processor (the divider).   ■

CURRENT DIVISION

Current division provides a simple way to find the current through each element in a parallel circuit. Figure 2–41 shows a parallel circuit that lends itself to solution by current division. Applying KCL at node A yields

$$i_S = i_1 + i_2 + i_3$$

The voltage v appears across all three conductances since they are connected in parallel. Using Ohm's law, we can write

$$i_S = vG_1 + vG_2 + vG_3$$

Solving for v yields

$$v = \frac{i_S}{G_1 + G_2 + G_3}$$

Given the voltage v, the current through any element is found using Ohm's law as

FIGURE   2 – 4 1   *A   current divider circuit.*

$$i_1 = vG_1 = \left(\frac{G_1}{G_1 + G_2 + G_3}\right) i_S \qquad (2\text{--}36)$$

$$i_2 = vG_2 = \left(\frac{G_2}{G_1 + G_2 + G_3}\right) i_S \qquad (2\text{--}37)$$

$$i_3 = vG_3 = \left(\frac{G_3}{G_1 + G_2 + G_3}\right) i_S \qquad (2\text{--}38)$$

These results show that the source current divides among the parallel resistors in proportion to their conductances divided by the equivalent conductances in the parallel connection. Thus, the general expression for the *current division rule* is

$$i_k = \left(\frac{G_k}{G_{EQ}}\right) i_{TOTAL} \qquad (2\text{--}39)$$

Sometimes it is useful to express the current division rule in terms of resistance rather than conductance. For the two-resistor case in Figure 2–42, the current i_1 is found using current division as

$$i_1 = \left(\frac{G_1}{G_1 + G_2}\right) i_S = \frac{\frac{1}{R_1}}{\frac{1}{R_1} + \frac{1}{R_2}} i_S = \left(\frac{R_2}{R_1 + R_2}\right) i_S \qquad (2\text{--}40)$$

FIGURE 2 – 4 2 *Two-path current divider circuit.*

Similarly, the current i_2 in Figure 2–42 is found to be

$$i_2 = \left(\frac{G_2}{G_1 + G_2}\right) i_S = \frac{\frac{1}{R_2}}{\frac{1}{R_1} + \frac{1}{R_2}} i_S = \left(\frac{R_1}{R_1 + R_2}\right) i_S \qquad (2\text{--}41)$$

These two results lead to the following *two-path current division rule:* When a circuit can be reduced to two equivalent resistances in parallel, the current through one resistance is equal to the other resistance divided by the sum of the two resistances times the total current entering the parallel combination.

Caution: Equations (2–40) and (2–41) only apply when the circuit is reduced to two parallel paths in which one path contains the desired current and the other path is the equivalent resistance of all other paths.

EXAMPLE 2–18

Find the current i_x in Figure 2–43(a).

SOLUTION:

To find i_x, we reduce the circuit to two paths, a path containing i_x and a path equivalent to all other paths, as shown in Figure 2–43(b). Now we can use the two-path current divider rule as

$$i_x = \frac{6.67}{20 + 6.67} \times 5 = 1.25 \text{ A} \qquad \blacksquare$$

FIGURE 2-43

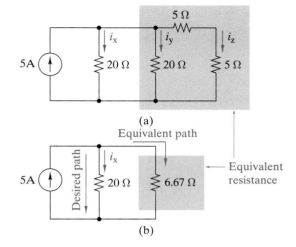

(a)

(b)

Exercise 2-11

(a) Find i_y and i_z in the circuit of Figure 2–43(a).
(b) Show that the sum of i_x, i_y, and i_z equals the source current.

Answers:
(a) $i_y = 1.25$ A, $i_z = 2.5$ A
(b) $i_x + i_y + i_z = 5$ A

Exercise 2-12

The circuit in Figure 2–44 shows a delicate device that is modeled by a 90-Ω equivalent resistance. The device requires a current of 1 mA to operate properly. A 1.5-mA fuse is inserted in series with the device to protect it from overheating. The resistance of the fuse is 10 Ω. Without the shunt resistance R_x, the source would deliver 5 mA to the device, causing the fuse to blow. Inserting a shunt resistor R_x diverts a portion of the available source current around the fuse and device. Select a value of R_x so only 1 mA is delivered to the device.

FIGURE 2-44

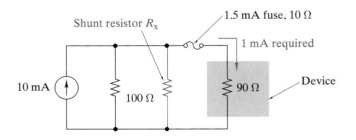

Answer: $R_x = 12.5$ Ω.

(a)

(b)

F I G U R E   2 – 4 5 *The d'Arsonval meter: (a) Meter movement. (b) Equivalent circuit.*

APPLICATION NOTE: **E X A M P L E   2 – 1 9**

The **D'Arsonval meter** shown in Figure 2–45(a) is a device used to measure electrical currents or voltages. In simple terms, a coil of wire is mounted between the poles of a permanent magnet so it is free to rotate. The magnet produces a magnetic field that interacts with the coil current to produce a torque, which causes the coil to turn. The deflection of the pointer attached to the coil is linearly proportional to the current.

D'Arsonval movements are rated by a parameter I_{FS}, which is the electrical current required to produce full-scale deflection of the pointer. Ratings range from a few microamperes to several milliamperes depending on the structure of the device. To measure larger currents, a very precise shunt resistance R_S is connected in parallel with the meter movement, as shown in Figure 2–45(b). The current i_x divides between the shunt resistance and the meter. The shunt resistance diverts a precisely known fraction of i_x around the coil to keep the meter current i_M within the meter's full-scale deflection rating. Different current ranges can be measured with the same D'Arsonval movement by using different values of R_S.

For example, suppose currents up to 10 A are to be measured using a D'Arsonval movement with a full-scale rating of $I_{FS} = 10$ mA and a coil resistance of $R_M = 20\ \Omega$. Using the circuit model in Figure 2–45(b), the problem is to find the value of R_S that will shunt current around the meter so that $i_M = I_{FS} = 10$ mA when $i_x = 10$ A. From the model it is evident the problem can be solved using the two-path current division rule:

$$i_M = \frac{R_S i_x}{R_M + R_S}$$

$$10^{-2} = \frac{R_S(10)}{20 + R_S}$$

Solving the last equation yields $R_S = 20.02$ mΩ.

2–6   CIRCUIT REDUCTION

The concepts of series/parallel equivalence, voltage/current division, and source transformations can be used to analyze **ladder circuits** of the type shown in Figure 2–46. The basic analysis strategy is to reduce the circuit to a simpler equivalent in which the output is easily found by voltage or current division or Ohm's law. There is no fixed pattern to the reduction process, and much depends on the insight of the analyst. In any case, with

F I G U R E   2 – 4 6 *A ladder circuit.*

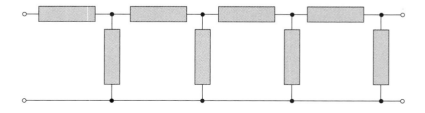

circuit reduction we work directly with the circuit model, and so the process gives us insight into circuit behavior.

With circuit reduction the desired unknowns are found by simplifying the circuit and, in the process, eliminating certain nodes and elements. However, we must be careful not to eliminate a node or element that includes the desired unknown voltage or current. The next three examples illustrate circuit reduction. The final example shows that rearranging the circuit can simplify the analysis.

EXAMPLE 2–20

Use series and parallel equivalence to find the output voltage v_O and the input current i_S in the ladder circuit shown in Figure 2–47(a).

SOLUTION:

One approach is to combine parallel resistors and use voltage division to find v_O, and then combine all resistances into a single equivalent to find the input current i_S. Figure 2–47(b) shows the step required to determine the equivalent resistance between the terminals B and ground. The equivalent resistance of the parallel $2R$ and R resistors is

$$R_{EQ1} = \frac{R \times 2R}{R + 2R} = \frac{2}{3}R$$

The reduced circuit in Figure 2–47(b) is a voltage divider. Notice that the two nodes needed to find the voltage v_O have been retained. The unknown voltage is found in terms of the source voltage as

$$v_O = \frac{\frac{2}{3}R}{\frac{2}{3}R + R}\, v_S = \frac{2}{5}v_S$$

The input current is found by combining the equivalent resistance found previously with the remaining resistor R to obtain

$$R_{EQ2} = R + R_{EQ1}$$

$$= R + \frac{2}{3}R = \frac{5}{3}R$$

Application of series/parallel equivalence has reduced the ladder circuit to the single equivalent resistance shown in Figure 2–47(c). Using Ohm's law, the input current is

$$i_S = \frac{v_S}{R_{EQ2}} = \frac{3}{5}\frac{v_S}{R}$$

Notice that the reduction step between Figures 2–47(b) and 2–47(c) eliminates node B, so the output voltage v_O must be calculated before this reduction step is taken.

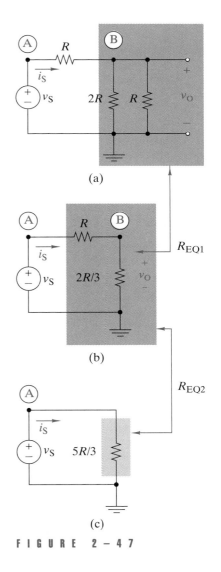

(a)

(b)

(c)

FIGURE   2 – 4 7

EXAMPLE 2–21

Use source transformations to find the output voltage v_O and the input current i_S in the ladder circuit shown in Figure 2–48(a).

FIGURE 2 – 48

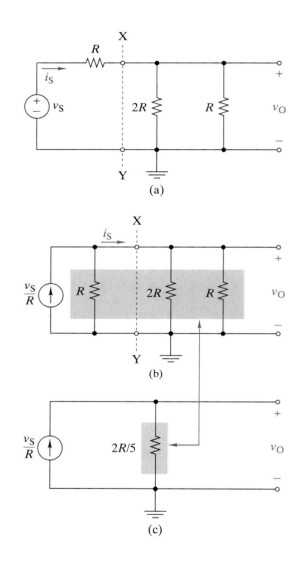

(a)

(b)

(c)

SOLUTION:

Figure 2–48 shows another way to reduce the circuit analyzed in Example 2–20. Breaking the circuit at points X and Y in Figure 2–48(a) produces a voltage source v_S in series with a resistor R. Using source transformation, this combination can be replaced by an equivalent current source in parallel with the same resistor, as shown in Figure 2–48(b).

Caution: The current source v_S/R is *not* the input current i_S, as is indicated in Figure 2–48(b). Applying the two-path current division rule to the circuit in Figure 2–48(b) yields the input current i_S as

$$i_S = \frac{R}{\frac{2}{3}R + R} \times \frac{v_S}{R} = \frac{v_S}{\frac{5}{3}R} = \frac{3}{5}\frac{v_S}{R}$$

The three parallel resistances in Figure 2–48(b) can be combined into a single equivalent conductance without eliminating the node pair used to define the output voltage v_O. Using parallel equivalence, we obtain

$$G_{EQ} = G_1 + G_2 + G_3$$

$$= \frac{1}{R} + \frac{1}{2R} + \frac{1}{R} = \frac{5}{2R}$$

which yields the equivalent circuit in Figure 2–48(c). The current source v_S/R determines the current through the equivalent resistance in Figure 2–48(c). The output voltage is found using Ohm's law.

$$v_O = \left(\frac{v_S}{R}\right) \times \left(\frac{2R}{5}\right) = \frac{2}{5}v_S$$

Of course, these results are the same result obtained in Example 2–20, except that here they were obtained using a different sequence of circuit reduction steps. ∎

EXAMPLE 2−22

Find v_x in the circuit shown in Figure 2–49(a).

SOLUTION:
In the two previous examples the unknown responses were defined at the circuit input and output. In this example the unknown voltage appears across a 10-Ω resistor in the center of the network. The approach is to reduce the circuit at both ends while retaining the 10-Ω resistor defining v_x. Applying a source transformation to the left of terminals X–Y and a series reduction to the two 10-Ω resistors on the far right yields the reduced circuit shown in Figure 2–49(b). The two pairs of 20-Ω resistors connected in parallel can be combined to produce the circuit in Figure 2–49(c). At this point there are several ways to proceed. For example, a source transformation at the points W–Z in Figure 2–49(c) produces the circuit in Figure 2–49(d). Using voltage division in Figure 2–49(d) yields v_x.

$$v_x = \frac{10}{10 + 10 + 10} \times 7.5 = 2.5 \text{ V}$$

Yet another approach is to use the two-path current division rule in Figure 2–49(c) to find the current i_x.

$$i_x = \frac{10}{10 + 10 + 10} \times \frac{3}{4} = \frac{1}{4} \text{ A}$$

FIGURE 2–49

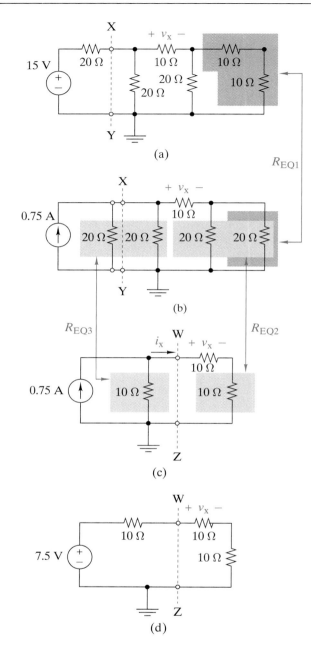

(a)

(b)

(c)

(d)

Then, applying Ohm's law to obtain v_x,

$$v_x = 10 \times i_x = 2.5 \text{ V}$$

◼

Exercise 2–13

Find v_x and i_x using circuit reduction on the circuit in Figure 2–50.

Answers: $v_x = 3.33$ V, $i_x = 0.444$ A

FIGURE 2-50

Exercise 2-14

Find v_x and v_y using circuit reduction on the circuit in Figure 2–51.

FIGURE 2-51

Answers: $v_x = 3.09$ V, $v_y = 9.21$ V

EXAMPLE 2-23

Using circuit reduction, find v_O in Figure 2–52(a).

SOLUTION:

One way to solve this problem is to notice that the source branch and the leftmost two-resistor branch are connected in parallel between node A and ground. Switching the order of these branches and replacing the two resistors by their series equivalent yields the circuit of Figure 2–52(b). A source transformation yields the circuit in Figure 2–52(c). This circuit contains a current source $v_S/2R$ in parallel with two $2R$ resistances whose equivalent resistance is

$$R_{EQ} = 2R\|2R = \frac{2R \times 2R}{2R + 2R} = R$$

Applying a source transformation to the current source $v_S/2R$ in parallel with R_{EQ} results in the circuit of Figure 2–52(d), where

$$V_{EQ} = \left(\frac{v_S}{2R}\right) \times R_{EQ} = \left(\frac{v_S}{2R}\right) R = \frac{v_S}{2}$$

Finally, applying voltage division in Figure 2–52(d) yields

$$v_O = \left(\frac{2R}{R + R + 2R}\right)\frac{v_S}{2} = \frac{v_S}{4}$$

∎

(a)

(b)

(c)

(d)

FIGURE 2-52

FIGURE 2 – 53

Exercise 2–15

Find the voltage across the current source in Figure 2–53.

Answer: $v_S = -0.225$ V

2–7 COMPUTER-AIDED CIRCUIT ANALYSIS

In this book we use three types of computer programs to illustrate computer-aided circuit analysis, namely spreadsheets, math solvers, and circuit simulators. Practicing engineers routinely apply these computer tools to analysis and design problems, so you must learn when and how to use them effectively. The purpose of having computer examples in this book is to help you develop an analysis style that includes the intelligent use of computer tools. As you develop your style, you must always keep in mind that computer tools are not problem solvers. *You* are the problem solver. Computer tools can be very useful once the problem is defined, but they do not substitute for an understanding of the fundamentals needed to formulate the problem.

The spreadsheet examples in this book use Microsoft Excel. The math solver examples use Mathcad by MathSoft and MATLAB by The Math-Works, Inc. The circuit simulation examples use two SPICE-based programs. The first is a program commonly called PSpice that was originally developed by MicroSim Corporation and is now available through OrCAD, Inc. The other is a program called Electronics WorkBench by Interactive Image Technologies, Inc. A student solutions manual supplement is available in which 100 homework problems are worked using these five computer tools. These homework problems are marked with a 📖 icon.

However, keep in mind that our objective is to illustrate the effective use of computer tools rather than to demonstrate the capabilities of these specific software programs. Accordingly this book does not treat the details of how to operate any of these software tools. We assume that you learned how to operate computer tools in previous courses or have enough familiarity with the WINDOWS operating environment to learn how to do so using on-line tutorials or the references given in Appendix D.

The following discussion gives a brief overview of circuit simulation, since you may be less familiar with this tool.

CIRCUIT SIMULATION

Most circuit simulation programs are based on a circuit analysis package called SPICE. The term *SPICE* is an acronym for *S*imulation *P*rogram with *I*ntegrated *C*ircuit *E*mphasis. The original SPICE program was developed in the 1970s at the University of California at Berkeley. Since that time, various companies have added proprietary features to the basic SPICE program to produce an array of SPICE-based commercial products for personal computer and workstation platforms.

Figure 2–54 is a block diagram summarizing the major features of a SPICE-based circuit simulation program. The inputs are a circuit diagram and the type of analysis to be carried out. In contemporary programs, the circuit diagram is drawn on the monitor screen using a graphical schematic editor. When the circuit diagram is complete, the input processor performs *schematic capture,* a process that documents the circuit with what is called a netlist. To initiate circuit simulation, the input processor sends the simulation processor the netlist and analysis commands by means of a circuit file. The simulation processor uses the circuit file together with data from the device library to formulate a set of equations that describe the circuit. The simulation processor then solves the equations, writes a dc analysis summary to a standard SPICE output file, and writes the other analysis results to a response data file. For simple dc analysis, the desired response data are easily accessible by simply examining the SPICE output file. For other types of analysis, the output processor can be used to generate graphical plots of the data in the response data file.

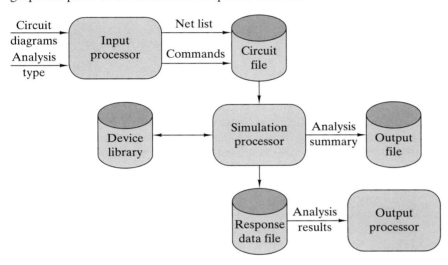

F I G U R E   2 – 5 4 *Flow diagram for circuit simulation programs.*

The following example illustrates circuit simulation using PSpice.

EXAMPLE 2–24

Use PSpice to find the voltage v_x across the 50-Ω resistor in the circuit of Figure 2–55.

F I G U R E   2 – 5 5

SOLUTION:

The purpose of this example is to illustrate the steps involved in using PSpice to analyze a circuit. Figure 2–56 shows the circuit diagram as it might appear in the PSpice schematic editor. The circuit diagram is entered by selecting elements using the **draw/get new part** commands, and then placing the selected parts on the screen. Once all the parts are in place, the interconnecting wires are entered using the **draw/wire** commands. The circuit diagram in Figure 2–56 differs from the diagram in Figure 2–55 in two ways. First, Figure 2–56 includes both the element values and the reference designators (V_1, R_1, R_2, R_3, R_4, R_5). Second, the nodes are labeled A, B, C, and 0, and the reference node is indicated by the ground symbol. In the PSpice syntax the node names can be numbers or letters, except for the ground node, which is automatically labeled with a zero (0). ∎

FIGURE 2-56

When the circuit diagram is completed, it is documented using the **analysis/create netlist** command. If there are mistakes, the program will issue error messages. Beginning users are likely to receive an error message reporting that one or more nodes are "floating." This message generally indicates a violation of one of the following SPICE rules: (1) There must be at least two elements connected to every node, or (2) the circuit must include a reference (ground) node.

Using the **analysis/examine netlist** command produces the netlist display shown in Figure 2–57. A netlist is a sequence of statements that define the circuit elements and connections. The first entry in each statement is the element name. The next two entries are the nodes to which the element is connected. The remaining entries define the element value. For example, the third statement in netlist says that the circuit contains a resistor named R_3, connected between node B and node C, and whose value is 50 Ω. All SPICE-based programs assign the positive reference mark for the element voltage to the first node in the element statement. For example, the plus reference mark for resistor R_2 is at node B and the minus mark is at the ground node. SPICE-based programs use the passive sign convention, so the reference direction for element current is from the first node toward the second node. For example, the reference direction for the current in resistor R_4 is from node A toward node C.

The **analysis/setup** command produces the menu in Figure 2–58, which shows the various types of analysis that PSpice performs. In later chapters,

FIGURE 2-57

FIGURE 2-58

we will use the AC Sweep, DC Sweep, Transfer Function, and Transient analysis options on this menu. The Bias Point Detail option produces a dc analysis of the circuit and writes the results to the output file. This analysis type is the default option, so PSpice always performs a dc analysis even when another analysis type is selected.

The **analysis/simulate** command causes PSpice to analyze the circuit defined by the netlist and to write the calculated dc responses in the output file. When the analysis is completed, the **analysis/examine output** command causes the output file to appear on the screen. Paging down through the file, we find the data shown in Figure 2–59. The output file lists the voltage between the lettered nodes and ground; that is, $v_{A0} = 15$ V, $v_{B0} = 8.1148$ V, and $v_{C0} = 6.8852$ V. The problem statement asks for the voltage v_x across resistor R_3 in Figure 2–56 . To find this voltage, we apply KVL to the loop formed by R_2, R_3, and R_5.

$$- v_{A0} + v_x + v_{C0} = 0$$

Hence,

$$v_x = 8.1148 - 6.8852 = 1.2296 \text{ V}$$

Figure 2–59 also shows that the total power dissipated in the circuit is 3.07 W and that the current through the voltage source V_1 is $-2.049E - 01$, or -0.2049 A. The minus sign means that the actual direction of the current through the voltage source is directed in at the minus terminal and out at the plus terminal, as you would expect. ■

FIGURE 2-59

```
┌─┐                    Notepad - EX2-24.OUT                    ▼ ▲
 File  Edit  Search  Help
 ****        SMALL SIGNAL BIAS SOLUTION        TEMPERATURE =    27.000  ↑
 *************************************************************************

 NODE   VOLTAGE      NODE    VOLTAGE      NODE    VOLTAGE      NODE   V

 (   A)  15.0000  (    B)    8.1148  (     C)    6.8852

      VOLTAGE SOURCE CURRENTS
      NAME            CURRENT

      V_V1          -2.049E-01

      TOTAL POWER DISSIPATION    3.07E+00  WATTS                      ↓
←                                                                    →
```

SUMMARY

- An **electrical device** is a real physical entity, while a **circuit element** is a mathematical or graphical model that approximates major features of the device.

- Two-terminal circuit elements are represented by a circuit symbol and are characterized by a single constraint imposed on the associated current and voltage variables.

- An **electrical circuit** is an interconnection of electrical devices. The interconnections form nodes and loops.

- A **node** is an electrical juncture of the terminals of two or more devices. A **loop** is a closed path formed by tracing through a sequence of devices without passing through any node more than once.

- Device interconnections in a circuit lead to two connection constraints: **Kirchhoff's current law (KCL)** states that the algebraic sum of currents at a node is zero at every instant; and **Kirchhoff's voltage law (KVL)** states that the algebraic sum of voltages around any loop is zero at every instant.

- A pair of two-terminal elements are connected in **parallel** if they form a loop containing no other elements. The same voltage appears across any two elements connected in parallel.

- A pair of two-terminal elements are connected in **series** if they are connected at a node to which no other elements are connected. The same current exists in any two elements connected in series.

- Two circuits are said to be **equivalent** if they each have the same i–v constraints at a specified pair of terminals.

- Series and parallel equivalence and voltage and current division are important tools in circuit analysis and design.

- **Source transformation** changes a voltage source in series with a resistor into an equivalent current source in parallel with a resistor, or vice versa.

- **Circuit reduction** is a method of solving for selected signal variables in ladder circuits. The method involves sequential application of the series/parallel equivalence rules, source transformations, and the voltage/current division rules. The reduction sequence used depends on the variables to be determined and the structure of the circuit and is not unique.

PROBLEMS

ERO 2–1 ELEMENT CONSTRAINTS (SECT. 2–1)

Given a two-terminal element with one or more electrical variables specified, use the element i–v constraint to find the magnitude and direction of the unknown variables.

2–1   Figure P2–1 shows a general two-terminal element with voltage and current reference marks assigned. Find the unknown electrical variables when the element is

FIGURE P2–1

(a) A linear 10-kΩ resistor with $v = 12$ V.

(b) An ideal 10-mA current source with the arrow directed upward and $p = 20$ mW.

(c) An ideal 15-V voltage source with the plus terminal at the top and $i = -8$ mA.

(d) An ideal switch with $i = -20$ A.

2–2 Figure P2–1 shows a general two-terminal element. Determine as much as you can about the element value and signal variables when the element is

(a) A linear resistor with $i = 10$ mA and $p = 50$ mW.

(b) An ideal current source with $p = -250$ mW and $v = 15$ V.

(c) An ideal voltage source with $i = -4$ mA and $v = -15$ V.

(d) An ideal switch with $v = 0$ V.

2–3 Find the current through, the voltage across, and the power dissipated in each of the elements in Figure P2–3.

2–4 The design guidelines for a circuit call for using only ¼-W resistors. The maximum voltage level in the circuit is known to be 12 V. Determine the smallest allowable value of resistance.

2–5 The i–v measurements for an unknown element are as follows, $i = 15$ mA @ $v = 15$ V, $i = 11.7$ mA @ $v = 10$ V, $i = -0.333$ mA @ $v = -8$ V. Use these measurements to estimate the current through the element when $v = -15$ V. Is the device linear?

2–6 The i–v characteristic of a nonlinear resistor is $v = 75i + 0.2i^3$.

(a) Calculate v and p for $i = \pm0.5, \pm1, \pm2, \pm5,$ and ±10 A.

(b) If the operating range of the device is limited to $|i| < 0.5$ A, what is the maximum error in v when the device is approximated by a 75-Ω linear resistance?

2–7 Figure P2–7 shows the circuit symbol for an electronic switch controlled by the voltage between gate G and ground. The manufacturer's data sheet for this device indicates that the switch is closed ($R_{ON} = 200$ Ω) when $v_G > 2$ V and open ($R_{OFF} = 200$ MΩ) when $v_G < 0.8$ V. Use these characteristics to predict the current through the device when $v_G = 5$ V and $v = 0.4$ V, $v_G = 3$ V and $v = -15$ V, $v_G = 0.5$ V and $v = 8$ V, and $v_G = -5$ V and $v = 4$ V.

2–8 A certain type of film resistor is available with resistance values between 10 Ω and 100 MΩ. The maximum ratings for all resistors of this type are 400 V and ½ W. Show that the voltage rating is the controlling limit when $R > 320$ kΩ, and that the power rating is the controlling limit when $R < 320$ kΩ.

2–9 Figure P2–9 shows the circuit symbol for a class of two-terminal devices called diodes. The i–v relationship for a p–n junction diode is

$$i = 2 \times 10^{-16}(e^{40v} - 1)$$

(a) Use this equation to find i and p for $v = 0, \pm0.1, \pm0.2, \pm0.4,$ and ±0.8 V. Use these data to plot the i–v characteristic of the element.

(b) Is the diode linear or nonlinear, bilateral or nonbilateral, and active or passive?

(c) Use the diode model to predict i and p for $v = 5$ V. Do you think the model applies to voltages in this range? Explain.

(d) Repeat (c) for $v = -5$ V.

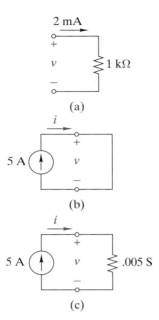

(a)

(b)

(c)

FIGURE P2–3

FIGURE P2–7

FIGURE P2–9

2–10 In a field effect transistor (FET) the current i_S through the device is related to the voltage v_{DS} across the device as $i_S = Gv_{DS}$, where $G = k(v_G - V_{TR})$. The conductance G is controlled by the excess gate voltage $(v_G - V_{TR})$, the amount by which the gate voltage v_G exceeds the threshold voltage V_{TR}. For $k = 1$ mA/V^2, $V_{TR} = 2$ V, and $v_G = 4$ V, 6 V, and 8 V, plot the i_S–v_{DS} characteristics of the device on the range $0 < v_{DS} < 5$ V.

ERO 2–2 CONNECTION CONSTRAINTS (SECT. 2–2)

Given a circuit composed of two-terminal elements,
(a) Identify nodes and loops in the circuit.
(b) Identify elements connected in series and in parallel.
(c) Use Kirchhoff's laws (KCL and KVL) to find selected signal variables.

2–11 For the circuit in Figure P2–11,
    (a) Identify the nodes and at least two loops.
    (b) Identify any elements connected in series or in parallel.
    (c) Write KCL and KVL connection equations for the circuit.
    (d) If $i_1 = 6$ mA and $i_2 = -4$ mA, find the other element currents.

2–12 The currents in Figure P2–11 are observed to be $i_1 = -40$ mA and $i_3 = 20$ mA. Find the other element currents in the circuit.

2–13 For the circuit in Figure P2–13,
    (a) Identify the nodes and at least three loops in the circuit.
    (b) Identify any elements connected in series or in parallel.
    (c) Write KCL and KVL connection equations for the circuit.
    (d) If $v_3 = -8$ V, $v_4 = -8$ V, and $v_5 = 9$ V, find the other element voltages.

2–14 Three of the voltages in Figure P2–13 are $v_1 = 8$ V, $v_4 = -8$ V, and $v_6 = 12$ V. Find the other element voltages in the circuit.

2–15 The circuit in Figure P2–15 is organized around the three signal lines A, B, and C.
    (a) Identify the nodes and at least three loops in the circuit.
    (b) Write KCL connection equations for the circuit.
    (c) If $i_3 = 15$ mA, $i_4 = -12$ mA, and $i_5 = 5$ mA, find the other element currents.

FIGURE P2–11

FIGURE P2–13

FIGURE P2–15

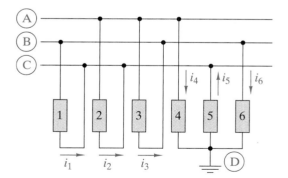

2–16 Repeat Prob. 2–15 when an additional element E_7 is connected between line B and ground. The current through the new element is −8 mA directed from line B to ground.

2–17   The circuit in Figure P2–17 is called a multitap voltage divider. Use KVL to find the output voltage when the switch is in each of the positions A, B, C, D, and E. *Hint:* The output voltage is defined relative to ground.

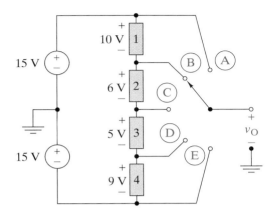

FIGURE   P2–17

2–18   The connection equations for a certain circuit are

$$i_1 - i_2 - i_3 = 0 \qquad v_1 + v_2 = 0$$
$$i_3 + i_4 + i_5 = 0 \qquad -v_2 + v_3 - v_4 = 0$$
$$v_4 - v_5 = 0$$

Draw the circuit diagram and indicate the reference marks for the element voltages and currents.

2–19   The incident matrix $\mathbf{A} = [a_{ij}]$ of a circuit has one row for each node and one column for each element. If the jth element is not connected to the ith node then $a_{ij} = 0$. If the jth element is connected to the ith node then $a_{ij} = \pm 1$, where the plus sign applies if the current reference direction is into the node and the minus sign applies if it is away from the node. The incident matrix for a circuit is

$$\mathbf{A} = \begin{bmatrix} -1 & -1 & -1 & 0 \\ 0 & 0 & +1 & -1 \\ +1 & +1 & 0 & +1 \end{bmatrix}$$

Draw the circuit diagram and indicate the reference directions for the element currents and voltages.

ERO 2–3 COMBINED CONSTRAINTS (SECT. 2–3)

Given a circuit consisting of independent sources and linear resistors, use the element constraints and connection constraints to find selected signal variables.

2–20   Find v_x and i_x in circuits C1 and C2 of Figure P2–20.
2–21   Find v_x and i_x in circuits C3 and C4 of Figure P2–20.
2–22   First use KVL to find the voltage across each resistor in Figure P2–22. Then use these voltages and KCL to find the current through every element, including the voltage sources.

FIGURE   P2–20

FIGURE   P2–22

FIGURE P2 – 23

2–23 Find v_x in Figure P2–23.

2–24 Figure P2–24 shows a subcircuit connected to the rest of the circuit at four points.
(a) Use element and connection constraints to find v_x and i_x.
(b) Show that the sum of the currents into the rest of the circuit is zero.

FIGURE P2 – 24

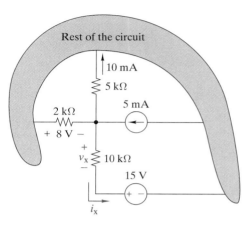

2–25 The circuit in Figure P2–25 is a model of the feedback path in an electronic amplifier circuit. The current i_x is known to be –0.6 mA.
(a) Find the value of R.
(b) Show that the sum of currents into the rest of the circuit is zero.

FIGURE P2 – 25

2–26 The circuit in Figure P2–26 is a model of a dc transmission line connecting a source and load. The current i_x at the line input is known to be 0.4 A.

FIGURE P2 – 26

(a) Find the line input voltage v_x and output voltage v_y.

(b) Find the power delivered at the line input and the power delivered to the load.

(c) Calculate the line efficiency defined as $\eta = (p_{OUT}/p_{IN}) \times 100$.

2-27 The circuit in Figure P2–27 is called a ladder because it is a cascade of alternating series (horizontal) and shunt (vertical) arms. The voltage v_O is known to be 16 V. Find the voltage v_x and the current i_x by successive application of element constraints, KCL, and KVL.

FIGURE P2-27

2-28 Find v_x and i_x in Figure P2–28 when switch S_1 is open and switch S_2 is closed. Repeat when switch S_1 is closed and switch S_2 is open.

FIGURE P2-28

2-29 Find v_x and i_x in Figure P2–28 when both switches are open. Repeat when both switches are closed.

ERO 2–4 EQUIVALENT CIRCUITS (SECT. 2–4)

(a) Given a circuit consisting of linear resistors, find the equivalent resistance between a specified pair of terminals.

(b) Given a circuit consisting of a source-resistor combination, find an equivalent source-resistor circuit.

2-30 Find the equivalent resistance between terminals A–B, B–C, A–C, C–D, B–D, and A–D in the circuit of Figure P2–30.

2-31 Find the equivalent resistance between terminals A–B, A–C, A–D, B–C, B–D, and C–D in the circuit of Figure P2–31.

2-32 Find the equivalent resistance R_{EQ} in Figure P2–32 when $R_L = 10$ kΩ. Repeat when $R_L = 0$. Select the value of R_L so that $R_{EQ} = 23$ kΩ.

2-33 (a) A 10-V voltage source is connected in series with a 50-Ω resistor, and the combination is connected in parallel with a 200-Ω resistor. Use source transformations to obtain an equivalent circuit with one voltage source and one resistor.

(b) A 200-mA current source is connected in parallel with a 100-Ω resistor, and the combination is connected in series with a 33-Ω resistor. Use source transformations to obtain an equivalent circuit with one current source and one resistor.

FIGURE P2-30

FIGURE P2-31

FIGURE P2-32

(c) The equivalent sources found in (a) and (b) above are connected in parallel. Use source transformations to obtain an equivalent circuit with one source and one resistor.

2–34 For the circuit in Figure P2–34, find an equivalent circuit between terminals A and B consisting of one resistor and one current source.

2–35 The circuit of Figure P2–35 is an R-$2R$ resistance array package. All of the following equivalent resistances can be obtained by making proper connections of the array *except for one*: $R/2$, $2R/3$, R, $8R/5$, $2R$, $3R$, and $4R$. Show how to interconnect the terminals of the array to produce the equivalent resistances, and identify the one that cannot be obtained using this array.

F I G U R E   P 2 – 3 4

F I G U R E   P 2 – 3 5

F I G U R E   P 2 – 3 6

2–36 Select the value of R_x in Figure P2–36 so that $R_{EQ} = 58$ kΩ.

2–37 **D** Using no more than a total of 14 resistors, show how to interconnect standard 4.3-kΩ resistors to obtain equivalent resistances of 1 kΩ $\pm 10\%$, 5 kΩ $\pm 10\%$, and 10 kΩ $\pm 10\%$.

2–38 Select the value of R in Figure P2–38 so that $R_{AB} = R_L$.

F I G U R E   P 2 – 4 0

F I G U R E   P 2 – 3 8

F I G U R E   P 2 – 3 9

2–39 What is the range of R_{EQ} in Figure P2–39?

2–40 Each resistor in Figure P2–40 has the same resistance R. The equivalent resistance measured between terminals A and B is 72.9 Ω. Determine the value of R.

ERO 2–5 VOLTAGE AND CURRENT DIVISION (SECT. 2–5)

(a) Given a circuit with elements connected in series or parallel, use voltage or current division to find specified voltages or currents.
(b) Design a voltage or current divider that delivers specified output signals within stated constraints.

2–41 Use voltage division in Figure P2–41 to obtain an expression for v_L in terms of R, R_L, and v_S.

2–42 Find v_x in each of the circuits of Figure P2–42.

F I G U R E   P 2 – 4 1

F I G U R E   P 2 – 4 2

C1

C2

C3

C4

F I G U R E   P 2 – 4 3

2–43 Use voltage or current division in Figure P2–43 to find the indicated unknown signal in circuits C1 and C2.

2–44 Use voltage or current division in Figure P2–43 to find the indicated unknown signal in circuits C3 and C4.

2–45 The electronic switch in Figure P2–45 is controlled by the voltage between gate G and ground. The switch is closed ($R_{ON} = 100\ \Omega$) when $v_G > 2$ V and open ($R_{OFF} = 500$ MΩ) when $v_G < 0.8$ V. Use voltage division to predict v_O when $v_G = 5$ V and $v_G = 0.5$ V.

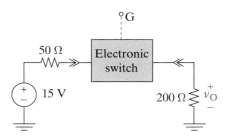

F I G U R E   P 2 – 4 5

2–46 Figure P2–46 shows a resistance divider connected in a general circuit.

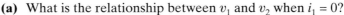

(a) What is the relationship between v_1 and v_2 when $i_1 = 0$?
(b) What is the relationship between v_1 and v_2 when $i_2 = 0$?
(a) What is the relationship between i_1 and i_2 when $v_1 = 0$?
(b) What is the relationship between i_1 and i_2 when $v_2 = 0$?

2–47 ⓓ Figure P2–47 shows an ammeter circuit consisting of a D'Arsonval meter, a two-position selector switch, and two shunt resistors. A current of 0.5 mA produces full-scale deflection of the D'Arsonval meter, whose internal resistance is $R_M = 75\ \Omega$. Select the shunt resistance R_1 and R_2 so that $i_x = 10$ mA produces full-scale deflection when the switch is in position A and $i_x = 50$ mA produces full-scale deflection when the switch is in position B.

2–48 ⓓ Figure P2–48 shows the R-$2R$ integrated circuit package connected as a voltage divider and across a 12-V source. The divider output is 6 V for the connections shown in the figure. Show how to interconnect the source and the R-$2R$ package to obtain outputs of 3 V, 4 V, 8 V, and 9 V.

2–49 ⓓ The circuit in Figure P2–49 delivers $v_L = 4$ V when $R_x = 0$. The circuit is to be modified to deliver $v_L = 3$ V by changing the series resistor R_x. What value of R_x is needed?

2–50 The 1-MΩ potentiometer in Figure P2–50 is adjusted until the voltage across the 1-MΩ load is 5 V. What is the position of the wiper arm in percent?

ERO 2–6 CIRCUIT REDUCTION (SECT. 2–6)

Given a circuit consisting of linear resistors and an independent source, find selected signal variables using successive application of series/parallel equivalence, source transformations, and voltage/current division.

2–51 Use circuit reduction to determine v_x and i_x in the circuit shown in Figure P2–51.

2–52 Use circuit reduction to find i_S and i_x in the circuit shown in Figure P2–52.

FIGURE P2–50

FIGURE P2–51

FIGURE P2–52

2–53 Use circuit reduction to find v_x and i_x in the circuit shown in Figure P2–53.

2–54 Use circuit reduction to find v_x and i_x in the circuit shown in Figure P2–54.

FIGURE P2–54

FIGURE P2–53

2–55 The resistance array circuit in Figure P2–55 has external terminals at pads A, B, C, and D. Connect the "+" terminal of a 10-V voltage to terminal A and the "−" to terminal D. Find the voltage v_{AB}.

2–56 The resistance array circuit in Figure P2–55 has external terminals at pads A, B, C, and D. Connect the "+" terminal of a 10-V voltage to terminal A and the "−" to terminal B. Find the voltage v_{CB}.

2–57 The resistance array circuit in Figure P2–55 has external terminals at pads A, B, C, and D. Connect the "+" terminal of a 5-V voltage to terminal C and the "−" to ground. Connect terminal B to ground. Find the voltage v_{AD}.

FIGURE P2–55

2–58 Select the value of R_L in Figure P2–58 so that the power delivered to R_L is at least 50 mW.

2–59 Select the value of R_L in Figure P2–58 so that the voltage delivered to R_L is at least 2.5 V.

2–60 The box in the circuit in Figure P2–60 is a resistor whose value can be anywhere between 8 kΩ and 80 kΩ. Use circuit reduction to find the range of values of the outputs v_x and i_x.

FIGURE P2–58

FIGURE P2–60

FIGURE P2–61

INTEGRATING PROBLEMS

2–61 **A** DEVICE MODELING

The circuit in Figure P2–61 consists of a 50-Ω linear resistor in parallel with a nonlinear varistor whose i–v characteristic is $i_V = 2.6 \times 10^{-5}v^3$.

(a) Plot the i–v characteristic of the parallel combination.

(b) State whether the parallel combination is linear or nonlinear, active or passive, and bilateral or nonbilateral.

(c) Identify a range of voltages over which the parallel combination can be modeled within ±10% by a linear resistor.

(d) Identify a range of voltages over which the parallel combination can be safely operated if both devices are rated at 50 W. Which device limits this range?

(e) The parallel combination is connected in series with a 50-Ω resistor and a 5-V voltage source. In this circuit, how would you model the parallel combination and why?

2–62 **A** WHEATSTONE BRIDGE

The Wheatstone bridge circuit in Figure P2–62(a) is used in instrumentation systems. The resistance R_x is the equivalent resistance of a transducer (a device that converts energy from one form to another). The value of R_x varies in relation to an external physical phenomenon such as temperature, pressure, or light. The resistance R_M is the equivalent resistance of a measuring instrument such as a D'Arsonval meter. Prior to any measurements, one of the other resistors (usually R_3) is adjusted until the current i_M is zero. The resistance of the transducer changes when exposed to the physical phenomenon it is designed to measure. This change causes the bridge to become unbal-

(a)

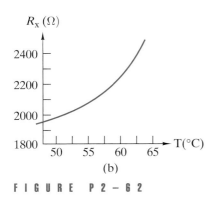

(b)

FIGURE P2–62

anced, and the meter indicates the resulting current through R_M. The deflection of the meter is calibrated to indicate the value of the physical phenomenon measured by the transducer.

(a) **A** Derive the relationship between R_1, R_2, R_3, and R_x when $i_M = 0$ A.

(b) **A** Suppose the transducer resistance R_x varies with temperature, as shown in Figure P2–62(b). With $R_1 = R_2 = 2.2$ kΩ, find the value of R_3 that produces $i_M = 0$ at a temperature of 55° C.

(c) **A** A current $i_M = -1.5$ mA is observed when $R_1 = R_2 = 2.2$ kΩ and R_3 is set to the value found in part (b). Is the temperature higher or lower than 55° C?

(d) **A** For $R_1 = R_2 = R_3 = 2.2$ kΩ, find the temperature at which $i_M = 0$.

2–63 **A** THREE-TERMINAL EQUIVALENCE

Two circuits are said to be equivalent when they have the same i–v characteristics between specified terminal pairs. In this chapter we applied this definition to two-terminal circuits such as resistors in series or parallel and source transformations. The concept can be extended to the three-terminal circuits in Figure P2–63. These three-terminal circuits will be equivalent if the equivalent resistances seen between terminal pairs A and B, B and C, and C and A are the same.

(a) Show that the two circuits are equivalent when

$$R_1 = \frac{R_B R_C}{R_A + R_B + R_C} \quad \text{and} \quad R_2 = \frac{R_C R_A}{R_A + R_B + R_C} \quad \text{and} \quad R_3 = \frac{R_A R_B}{R_A + R_B + R_C}$$

(b) Equivalence is useful because replacing one circuit by an equivalent circuit does not change the response of the rest of the circuit and may simplify an analysis problem. For example, the equations in (a) tell us how to create an equivalent Y-connected subcircuit like Figure P5–63(b) to replace the Δ-connected subcircuit in Figure P5–63(a). Such a replacement is called a Δ-to-Y transformation. Show that a Y-to-Δ transformation is also possible provided

$$G_A = \frac{G_2 G_3}{G_1 + G_2 + G_3} \quad \text{and} \quad G_B = \frac{G_1 G_3}{G_1 + G_2 + G_3} \quad \text{and} \quad G_C = \frac{G_1 G_2}{G_1 + G_2 + G_3}$$

FIGURE P2–63

2–64 **A** DIGITAL-TO-ANALOG CONVERSION

Digital-to-analog (D-to-A) conversion provides a link between the digital and analog worlds. A D-to-A converter produces a single analog output v_O from a multibit digital input $\{b_0, b_1, b_2, \ldots b_{N-1}\}$, where the bits b_j (j = 0, 1, ... $N - 1$) are either 0 or 1. One method is to produce an analog output that is proportional to a fixed reference voltage V_{REF} and related to the digital inputs by the following algorithm:

$$v_O = KV_{REF} \sum_{j=0}^{N-1} b_j 2^j$$

Figure P2–64 shows a programmable voltage divider in which two digital inputs control complementary analog switches connecting a

FIGURE   P2 - 64

(a)

(b)

FIGURE   P2 - 65

FIGURE   P2 - 66

multitap voltage divider to the output terminal. Show that the programmable voltage divider implements this D-to-A algorithm with $K = 0.25$ and $N = 2$.

2–65   A D E   ANALOG VOLTMETER DESIGN

A voltmeter can be made using a series resistor and a D'Arsonval meter. Figure P2–65(a) shows a voltmeter circuit consisting of a D'Arsonval meter, a two-position selector switch, and two series resistors. A current of 400 µA produces full-scale deflection of the D'Arsonval meter, whose internal resistance is $R_M = 25 \, \Omega$.

   (a)   **D**   Select the series resistance R_1 and R_2 so a voltage $v_x = 50$ V produces full-scale deflection when the switch is in position A, and voltage $v_x = 10$ V produces full-scale deflection when the switch is in position B.

   (b)   **A**   By voltage division, the voltage across the 20-kΩ resistor in Figure P2–65(b) is 20 V when the voltmeter is disconnected. What voltage reading is obtained when the voltmeter designed in part (a) is connected across the 20-kΩ resistor? What is the percentage error in the voltmeter reading?

   (c)   **E**   A D'Arsonval meter with an internal resistance of 200 Ω and a full-scale deflection current of 100 µA is available. If the voltmeter in part (a) is redesigned using this D'Arsonval meter, would the error found in part (b) be smaller or larger? Explain.

2–66   E   DC CIRCUIT ANALYSIS EVALUATION

Later in life you become an Assistant Professor of Electrical Engineering teaching (as luck would have it) the introduction to circuit analysis course. You give your students a quiz wherein you ask them to find the power delivered by the 100-V voltage source in Figure P2–66. One of your students performs a source transformation and triumphantly reports that the power is $p = (100 \text{ V})(4 \text{ A}) = 100$ W. Assign an appropriate letter grade to this paper and provide the student with helpful feedback. That is, explain why you awarded the grade you did. Your feedback should include the correct numerical answer.

2–67 ⓘ VOLTAGE DIVIDER DESIGN

A 30-V source with negligible internal resistance is available to supply two constant-resistance loads. The first load requires 250 mW at +15 V (relative to ground) and the second 100 mW at −10 V (relative to ground). Design a voltage divider to supply the two loads from the 30-V source.

CHAPTER 3

CIRCUIT ANALYSIS TECHNIQUES

Assuming any system of linear conductors connected in such a manner that to the extremities of each one of them there is connected at least one other, a system having electromotive forces E_1, E_2 ... E_3, no matter how distributed, we consider two points A and A' belonging to the system and having potentials V and V'. If the points A and A' are connected by a wire ABA', which has a resistance r, with no electromotive forces, the potentials of points A and A' assume different values from V and V', but the current i flowing through this wire is given by $i = (V - V')/(r + R)$ in which R represents the resistance of the original wire, this resistance being measured between the points A and A', which are considered to be electrodes.

Leon Charles Thévenin, 1883,
French Telegraph Engineer

3–1 Node-voltage Analysis

3–2 Mesh-current Analysis

3–3 Linearity Properties

3–4 Thévenin and Norton Equivalent Circuits

3–5 Maximum Signal Transfer

3–6 Interface Circuit Design

72

Summary

Problems

Integrating Problems

Thévenin's theorem is one of the important circuit analysis concepts developed in this chapter. Leon Charles Thévenin (1857–1926), a distinguished French telegraph engineer and teacher, was led to his theorem in 1883 following an extensive study of Kirchhoff's laws. Norton's theorem, which is the dual of Thévenin's theorem, was not proposed until 1926 by Edward L. Norton, an American electrical engineer working on long-distance telephony. Curiously, it turns out that the basic concept had been discovered in 1853 by Herman von Helmholtz. The earlier discovery by Helmholtz is not recognized in engineering terminology possibly because Thévenin and Norton both worked in areas of technology that offered immediate applications for their results, whereas Helmholtz was studying electricity in animal tissue at the time.

The analysis methods developed in the previous chapter offer insight into circuit analysis and design because we work directly with the circuit model to find responses. With practice and experience we learn which tools to use and in what order to avoid going down blind alleys. This ad hoc approach is practical as long as the circuits are fairly simple. As circuit complexity increases, however, a more systematic approach is needed.

Two basic methods of systematic circuit analysis—node-voltage analysis and mesh-current analysis—are developed in the first two sections of this chapter. With these methods the device and connection constraints are used to formulate a set of linear algebraic equations that characterize the circuit. Solving these equations simultaneously then yields the desired circuit responses. General methods of circuit analysis are necessarily more abstract because we manipulate sets of equations, rather than the circuit itself. In doing so we lose some contact with the more intuitive approach developed in the previous chapter. However, systematic methods make it possible for us to treat a wider range of applications and provide a framework for exploiting other properties of linear circuits.

Two key properties of linear circuits are developed in the third section. The superposition and proportionality properties lead to new circuit analysis techniques and are often used to derive other properties of linear circuits. For example, the superposition property is used in the fourth section to derive Thévenin's theorem. This theorem provides a viewpoint for dealing with circuit interfaces, and that leads directly to the maximum signal transfer properties developed in the fifth section. In the final section, circuit design is introduced by showing how the maximum signal transfer principles guide the design of interface circuit.

3–1   NODE-VOLTAGE ANALYSIS

Before describing node-voltage analysis, we first review the foundation for every method of circuit analysis. As noted in Sec. 2–3, circuit behavior is based on constraints of two types: (1) connection constraints (Kirchhoff's

Notation

Interpretation

FIGURE 3 – 1 *Node-voltage definition and notation.*

Case A

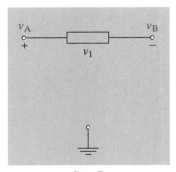

Case B

FIGURE 3 – 2 *Two possible connections of a two-terminal element.*

laws) and (2) device constraints (element *i–v* relationships). As a practical matter, however, using element voltages and currents to express the circuit constraints produces a large number of equations that must be solved simultaneously to find the circuit responses. For example, a circuit with only six devices requires us to treat 12 equations with 12 unknowns. Although this is not an impossible task using software tools like Mathcad, it is highly desirable to reduce the number of equations that must be solved simultaneously.

You should not abandon the concept of element and connection constraints. This method is vital because it provides the foundation for all methods of circuit analysis. In subsequent chapters, we use element and connection constraints many times to develop important ideas in circuit analysis.

Using node voltages instead of element voltages as circuit variables can reduce the number of equations that must be treated simultaneously. To define a set of node voltages we first select a reference node. The **node voltages** are then defined as the voltages between the remaining nodes and the selected reference node. Figure 3–1 shows the notation used to define node-voltage variables. In this figure the reference node indicated by the ground symbol and the node voltages are identified by a voltage symbol next to all the other nodes. This notation means that the positive reference mark for the node voltage is located at the node in question while the negative mark is at the reference node. Obviously, any circuit with N nodes involves $N - 1$ node voltages.

A fundamental property of node voltages needs to be covered at the outset. Suppose we are given a two-terminal element whose element voltage is labeled v_1. Suppose further that the terminal with the plus reference mark is connected to a node, say node A. The two cases shown in Figure 3–2 are the only two possible ways the other element terminal can be connected. In case A, the other terminal is connected to the reference node, in which case KVL requires $v_1 = v_A$. In case B, the other terminal is connected to a non-reference node, say node B, in which case KVL requires $v_1 = v_A - v_B$. This example illustrates the following fundamental property of node voltages:

*If the **K**th two-terminal element is connected between nodes **X** and **Y**, then the element voltage can be expressed in terms of the two node voltages as*

$$v_K = v_X - v_Y \qquad (3\text{–}1)$$

*where **X** is the node connected to the positive reference for element voltage* **v**$_K$.

Equation (3–1) is a KVL constraint at the element level. If node Y is the reference node, then by definition $v_Y = 0$ and Eq. (3–1) reduces to $v_K = v_X$. On the other hand, if node X is the reference node, then $v_X = 0$ and therefore $v_K = -v_Y$. The minus sign occurs here because the positive reference for the element is connected to the reference node. In any case, the important fact is that the voltage across any two-terminal element can be expressed as the difference of two node voltages, one of which may be zero.

The reference node and node voltages in the bridge circuit of Figure 3–3 are $v_A = $ 5 V, $v_B = 10$ V, and $v_C = -3$ V. Find the element voltages.

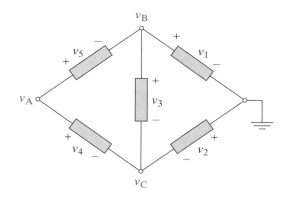

FIGURE   3 – 3

Answers:

$$v_1 = 10 \text{ V}, v_2 = 3 \text{ V}, v_3 = 13 \text{ V}, v_4 = 8 \text{ V, and } v_5 = -5 \text{ V}.$$

FORMULATING NODE-VOLTAGE EQUATIONS

To formulate a circuit description using node voltages, we use element and connection constraints, except that the KVL connection equations are not explicitly written. Instead we use the fundamental property of node analysis to express the element voltages in terms of the node voltages.

The circuit in Figure 3–4 will demonstrate the formulation of node-voltage equations. In Figure 3–4 we have identified a reference node (indicated by the ground symbol), four element currents (i_0, i_1, i_2, and i_3), and two node voltages (v_A and v_B).

The KCL constraints at the two nonreference nodes are

$$\text{Node A: } -i_0 - i_1 - i_2 = 0$$

$$\text{Node B: } i_2 - i_3 = 0 \tag{3–2}$$

FIGURE   3 – 4   *Circuit for demonstrating node-voltage analysis.*

Using the fundamental property of node analysis, we use the element equations to relate the element currents to the node voltages.

$$\text{Resistor } R_1 \text{: } i_1 = G_1 v_A$$

$$\text{Resistor } R_2 \text{: } i_2 = G_2(v_A - v_B)$$

$$\text{Resistor } R_3 \text{: } i_3 = G_3 v_B$$

$$\text{Current Source: } i_0 = -i_S \tag{3–3}$$

We have written six equations in six unknowns—four element currents and two node voltages. The right side of the element equations in Eq. (3–3) involves unknown node voltages and the input signal i_S. Substituting the element constraints in Eq. (3–3) into the KCL connection constraints in Eq. (3–2) yields

$$\text{Node A: } i_S - G_1 v_A - G_2(v_A - v_B) = 0$$

$$\text{Node B: } \quad G_2(v_A - v_B) - G_3 v_B = 0$$

which can be arranged in the following standard form:

$$\text{Node A:} \quad (G_1 + G_2)v_A - G_2 v_B = i_S$$

$$\text{Node B:} \quad -G_2 v_A + (G_2 + G_3)v_B = 0 \tag{3–4}$$

In this standard form all of the unknown node voltages are grouped on one side and the independent sources on the other.

By systematically eliminating the element currents, we have reduced the circuit description to two linear equations in the two unknown node voltages. The coefficients in the equations on the left side ($G_1 + G_2$, G_2, $G_2 + G_3$) depend only on circuit parameters, while the right side contains the known input driving force i_S.

As noted previously, every method of circuit analysis must satisfy KVL, KCL, and the element i–v relationships. In developing the node-voltage equations in Eq. (3–4), it may appear that we have not used KVL. However, KVL is satisfied because the equations $v_1 = v_A$, $v_2 = v_A - v_B$, and $v_3 = v_B$ were used to write the right side of the element equations in Eq. (3–3). The KVL constraints do not appear explicitly in the formulation of node equations, but are implicitly included when the fundamental property of node analysis is used to write the element voltages in terms of the node voltages.

In summary, four steps are needed to develop node-voltage equations.

STEP 1   Select a reference node. Identify a node voltage at each of the remaining $N - 1$ nodes and a current with every element in the circuit.

STEP 2   Write KCL connection constraints in terms of the element currents at the $N - 1$ nonreference nodes.

STEP 3   Use the element i–v relationships and the fundamental property of node analysis to express the element currents in terms of the node voltages.

STEP 4   Substitute the element constraints from step 3 into the KCL connection constraints from step 2 and arrange the resulting $N - 1$ equations in a standard form.

Writing node-voltage equations leads to $N - 1$ equations that must be solved simultaneously. If we write the element and connection constraints in terms of element voltages and currents, we must solve $2E$ simultaneous equations. The reduction from $2E$ to $N - 1$ is particularly impressive in circuits with a large number of elements (large E) connected in parallel (small N).

EXAMPLE 3–1

Formulate node-voltage equations for the bridge circuit in Figure 3–5.

SOLUTION:

Step 1: The reference node, node voltages, and element currents are shown in Figure 3–5.
Step 2: The KCL constraints at the three nonreference nodes are:

Reference node

FIGURE 3 – 5

$$\text{Node A: } i_0 - i_1 - i_2 = 0$$

$$\text{Node B: } i_1 - i_3 + i_5 = 0$$

$$\text{Node C: } i_2 - i_4 - i_5 = 0$$

Step 3: We write the element equations in terms of the node voltages and input signal sources.

$$i_0 = i_{S1} \qquad i_3 = G_3 v_B$$

$$i_1 = G_1(v_A - v_B) \quad i_4 = G_4 v_C$$

$$i_2 = G_2(v_A - v_C) \quad i_5 = i_{S2}$$

Step 4: Substituting the element equations into the KCL constraints and arranging the result in standard form yields three equations in the three unknown node voltages.

$$\text{Node A: } (G_1 + G_2)v_A - G_1 v_B - G_2 v_C = i_{S1}$$

$$\text{Node B: } \qquad - G_1 v_A + (G_1 + G_3)v_B = i_{S2}$$

$$\text{Node C: } \qquad - G_2 v_A + (G_2 + G_4)v_C = -i_{S2} \qquad \blacksquare$$

WRITING NODE-VOLTAGE EQUATIONS BY INSPECTION

The node-voltage equations derived in Example 3–1 have a symmetrical pattern. The coefficient of v_B in the node A equation and the coefficient of v_A in the node B equation are both the negative of the conductance connected between the nodes ($-G_1$). Likewise, the coefficients of v_A in the node C equation and v_C in the node A equation are both $-G_2$. Finally, coefficients of v_A in the node A equation, v_B in the node B equation, and v_C in the node C equation are the sum of the conductances connected to the node in question.

The symmetrical pattern always occurs in circuits containing only resistors and independent current sources. To understand why, consider any general two-terminal conductance G with one terminal connected to, say, node A. According to the fundamental property of node analysis, there are only two possibilities. Either the other terminal of G is connected to the reference node, in which case the current *leaving* node A via conductance G is

$$i = G(v_A - 0) = G v_A$$

or else it is connected to another nonreference node, say, node B, in which case the current *leaving* node A via G is

$$i = G(v_A - v_B)$$

The pattern for node equations follows from these observations. The sum of the currents leaving any node A via conductances involves the following terms:

1. v_A times the sum of conductances connected to node A

2. Minus v_B times the sum of conductances connected between nodes A and B and similar terms for all other nodes connected to node A by conductances.

FIGURE 3–6 *Circuit for demonstrating writing node-voltage equations by inspection.*

Because of KCL, the sum of currents leaving node A via conductances plus the sum of currents directed away from node A by independent current sources must equal zero.

The aforementioned process allows us to write node-voltage equations by inspection without going through the intermediate steps involving the KCL constraints and the element equations. For example, the circuit in Figure 3–6 contains two independent current sources and four resistors. Starting with node A, the sum of conductances connected to node A is $G_1 + G_2$. The conductances between nodes A and B is G_2. The reference direction for the source current i_{S1} is into node A, while the reference direction for i_{S2} is directed away from node A. Pulling all of the observations together, we write the sum of currents directed out of node A as

$$\text{Node A: } (G_1 + G_2)v_A - G_2v_B - i_{S1} + i_{S2} = 0 \qquad (3\text{–}5)$$

Similarly, the sum of conductances connected to node B is $G_2 + G_3 + G_4$, the conductance connected between nodes B and A is again G_2, and the source current i_{S2} is directed away from node B. These observations yield the following node-voltage equation:

$$\text{Node B: } (G_2 + G_3 + G_4)v_B - G_2v_A - i_{S2} = 0 \qquad (3\text{–}6)$$

Rearranging Eqs. (3–5) and (3–6) in standard form yields

$$\text{Node A: } \qquad (G_1 + G_2)v_A - G_2v_B = i_{S1} - i_{S2}$$
$$\text{Node B: } -G_2v_A + (G_2 + G_3 + G_4)v_B = i_{S2} \qquad (3\text{–}7)$$

We have two symmetrical equations in the two unknown node voltages. The equations are symmetrical because the conductance G_2 connected between nodes A and B appears as the cross-coupling term in each equation.

EXAMPLE 3–2

Formulate node-voltage equations for the bridged-T circuit in Figure 3–7.

FIGURE 3–7

SOLUTION:

The total conductance connected to node A is $1/2R + 2/R = 2.5G$, to node B is $1/2R + 1/2R + 2/R = 3G$, and to node C is $1/R + 2/R + 1/2R = 3.5G$. The conductance connected between nodes A and B is $1/2R = 0.5G$, between nodes A and C is $2/R = 2G$, and between nodes B and C is $1/2R = 0.5G$. The independent current source is directed into node A. By inspection, the node-voltage equations are

$$\text{Node A: } \quad 2.5Gv_A - 0.5Gv_B - 2Gv_C = i_S$$
$$\text{Node B: } -0.5Gv_A + 3Gv_B - 0.5Gv_C = 0$$
$$\text{Node C: } -2Gv_A - 0.5Gv_B + 3.5Gv_C = 0$$

Written in matrix form,

$$\begin{bmatrix} 2.5G & -0.5G & -2G \\ -0.5G & 3G & -0.5G \\ -2G & -0.5G & 3.5G \end{bmatrix} \begin{bmatrix} v_A \\ v_B \\ v_C \end{bmatrix} = \begin{bmatrix} i_S \\ 0 \\ 0 \end{bmatrix}$$

This matrix equation is of the form $\mathbf{Ax} = \mathbf{B}$, where $\mathbf{A}$ is a 3×3 square matrix describing the circuit, $\mathbf{x}$ is a 3×1 column matrix of unknown node voltages, and $\mathbf{B}$ is a 3×1 column matrix of known inputs. Note that the $\mathbf{A}$ matrix is symmetrical about its main diagonal.[1] ■

Exercise 3–2

Formulate node-voltage equations for the circuit in Figure 3–8.

FIGURE 3 – 8

Answers:

$$(1.5 \times 10^{-3})v_A - (0.5 \times 10^{-3})v_B = i_{S1}$$

$$-(0.5 \times 10^{-3})v_A + (2.5 \times 10^{-3})v_B = -i_{S2}$$

SOLVING LINEAR ALGEBRAIC EQUATIONS

So far, we have only dealt with the problem of formulating node-voltage equations. To complete a circuit analysis problem, we must solve these linear equations for selected responses. Cramer's rule and Gaussian elimination are standard mathematical tools commonly used for hand solution of circuit equations. These tools are assumed to be part of the reader's mathematical background. Those needing a review of these matters are referred to Appendix B.

Cramer's rule and Gaussian elimination are suitable for hand calculations involving up to three or perhaps four simultaneous equations. Cramer's rule is useful when the circuit parameters are left in symbolic form, while the Gaussian method is more efficient for numerical examples involving four or more equations. However, any problem in which Gaussian methods offer significant advantages over Cramer's rule is probably best handled using computer tools. In other words, the ready availability of computer tools makes hand solution by Gaussian elimination an obsolete skill.

At about four or five simultaneous equations, numerical solutions are best obtained using computer-aided analysis. Many scientific hand-held calculators have a built-in capability to solve five or more linear equations. Virtually all PC-based mathematical software can solve systems of linear equations or, what is equivalent, perform matrix manipulations. In particular, Appendix B illustrates that both MATLAB and Mathcad can solve systems of linear equations written in matrix form.

---

1 See Appendix B for a discussion of matrix algebra, including the definition of a symmetrical matrix.

In this book we often use Cramer's rule to solve simultaneous equations. This does not mean Cramer's rule is the optimum method, but only that it can easily handle and compactly document the solution of the class of problems treated in this book. We will also use MATLAB and Mathcad to demonstrate the use of computer tools to solve circuit equations in the matrix form. The student solutions manual supplement has additional examples of solving matrix equations using computer tools. Readers needing a review of or an introduction to matrix methods are advised to study Appendix B.

Earlier in this section we formulated node-voltage equations for the circuit in Figure 3–4 [See Eq. (3–4).]

$$\text{Node A:} \quad (G_1 + G_2)v_A - G_2v_B = i_S$$

$$\text{Node B:} \quad -G_2v_A + (G_2 + G_3)v_B = 0$$

We use Cramer's rule to solve these equations because it easily handles the case in which the circuit parameters are left in symbolic form:

$$v_A = \frac{\Delta_A}{\Delta} = \frac{\begin{vmatrix} i_S & -G_2 \\ 0 & G_2 + G_3 \end{vmatrix}}{\begin{vmatrix} G_1 + G_2 & -G_2 \\ -G_2 & G_2 + G_3 \end{vmatrix}} = \left(\frac{G_2 + G_3}{G_1G_2 + G_1G_3 + G_2G_3}\right)i_S \qquad (3\text{–}8)$$

$$v_B = \frac{\Delta_B}{\Delta} = \frac{\begin{vmatrix} G_1 + G_2 & i_S \\ -G_2 & 0 \end{vmatrix}}{\Delta} = \left(\frac{G_2}{G_1G_2 + G_1G_3 + G_2G_3}\right)i_S \qquad (3\text{–}9)$$

These results express the two node voltages in terms of the circuit parameters and the input signal. Given the two node voltages v_A and v_B, we can now determine every element voltage and every current using Ohm's law and the fundamental property of node voltages.

$$v_1 = v_A \qquad v_2 = v_A - v_B \qquad v_3 = v_B$$

$$i_1 = G_1v_A \qquad i_2 = G_2(v_A - v_B) \qquad i_3 = G_3v_B$$

In solving the node equations, we left everything in symbolic form to emphasize that responses depend on the values of the circuit parameters (G_1, G_2, G_3) and the input signal (i_S). Even when numerical values are given, it is sometimes useful to leave some parameters in symbolic form to obtain input-output relationships or to reveal the effect of specific parameters on circuit response.

EXAMPLE 3–3

Given the circuit in Figure 3–9, find the input resistance R_{IN} seen by the current source and the output voltage v_O.

FIGURE 3-9

SOLUTION:
In Example 3–2 we formulated node-voltage equations for this circuit as
follows:

$$\text{Node A:} \quad 2.5Gv_A - 0.5Gv_B - 2Gv_C = i_S$$

$$\text{Node B:} \quad -0.5Gv_A + 3Gv_B - 0.5Gv_C = 0$$

$$\text{Node C:} \quad -2Gv_A - 0.5Gv_B + 3.5Gv_C = 0$$

The input resistance is the ratio v_A/i_S, so we first solve for v_A:

$$v_A = \frac{\Delta_A}{\Delta} = \frac{\begin{vmatrix} i_S & -0.5G & -2G \\ 0 & 3G & -0.5G \\ 0 & -0.5G & 3.5G \end{vmatrix}}{\begin{vmatrix} 2.5G & -0.5G & -2G \\ -0.5G & 3G & -0.5G \\ -2G & -0.5G & 3.5G \end{vmatrix}}$$

$$= \frac{i_S \times G^2 \begin{vmatrix} 3 & -0.5 \\ -0.5 & 3.5 \end{vmatrix}}{2.5G^3 \begin{vmatrix} 3 & -0.5 \\ -0.5 & 3.5 \end{vmatrix} + 0.5G^3 \begin{vmatrix} -0.5 & -2 \\ -0.5 & 3.5 \end{vmatrix} - 2G^3 \begin{vmatrix} -0.5 & -2 \\ 3 & -0.5 \end{vmatrix}}$$

$$= \frac{10.25 i_S}{11.75G}$$

Hence the input resistance is

$$R_{IN} = \frac{v_A}{i_S} = \frac{10.25}{11.75G} = 0.872R$$

To find the output voltage, we solve for v_C:

$$v_C = \frac{\Delta_C}{\Delta} = \frac{\begin{vmatrix} 2.5G & -0.5G & i_S \\ -0.5G & 3G & 0 \\ -2G & -0.5G & 0 \end{vmatrix}}{\Delta} = \frac{i_S \times G^2 \begin{vmatrix} -0.5 & 3 \\ -2 & -0.5 \end{vmatrix}}{\Delta}$$

$$= \frac{6.25G^2 i_S}{11.75G^3} = 0.532 i_S R$$

∎

Exercise 3–3

Solve for the node-voltage equations in Exercise 3–2 for v_O in Figure 3–8.

Answer:

$$v_O = 1000(i_{S1} - 3i_{S2})/7$$

Exercise 3–4

Use node-voltage equations to solve for v_1, v_2, and i_3 in Figure 3–10.

FIGURE 3–10

Answers:

$$v_1 = 12 \text{ V}, v_2 = 32 \text{ V}, \text{ and } i_3 = -10 \text{ mA}$$

NODE ANALYSIS WITH VOLTAGE SOURCES

Up to this point we have analyzed circuits containing only resistors and independent current sources. Applying KCL in such circuits is simplified because the sum of currents at a node only involves the output of current sources or resistor currents expressed in terms of the node voltages. Adding voltage sources to circuits modifies node analysis procedures because the current through a voltage source is not directly related to the voltage across it. While initially it may appear that voltage sources complicate the situation, they actually simplify node analysis by reducing the number of equations required.

Figure 3–11 shows three ways to deal with voltage sources in node analysis. Method 1 uses a source transformation to replace the voltage source and series resistance with an equivalent current source and parallel resistance. We can then formulate node equations at the remaining nonreference nodes in the usual way. The source transformation eliminates node C, so there are only $N - 2$ nonreference nodes left in the circuit. Obviously, method 1 only applies when there is a resistance in series with the voltage source.

Method 2 in Figure 3–11 can be used whether or not there is a resistance in series with the voltage source. When node B is selected as the reference node, then by definition $v_B = 0$ and the fundamental property of node voltages says that $v_A = v_S$. We do not need a node-voltage equation at node A because its voltage is known to be equal to the source voltage. We write the node equations at the remaining $N - 2$ nonreference nodes in the usual way. In the final step, we move all terms involving v_A to the right side, since it is a known input and not an unknown response. Method 2 reduces the number of node equations by 1 since no equation is needed at node A.

FIGURE 3 – 11 *Three methods of treating voltage sources in node analysis.*

Method 1

Method 2

Method 3

The third method in Figure 3–11 is needed when neither node A nor node B can be selected as the reference and the source is not connected in series with a resistance. In this case we combine nodes A and B into a **supernode**, indicated by the boundary in Figure 3–11. We use the fact that KCL applies to the currents penetrating this boundary to write a node equation at the supernode. We then write node equations at the remaining $N - 3$ nonreference nodes in the usual way. We now have $N - 3$ node equations plus one supernode equation, leaving us one equation short of the $N - 1$ required. Using the fundamental property of node voltages, we can write

$$v_A - v_B = v_S \qquad (3\text{–}10)$$

The voltage source inside the supernode constrains the difference between the node voltages at nodes A and B. The voltage source constraint provides the additional relationship needed to write $N - 1$ independent equations in $N - 1$ node voltages.

For reference purposes we will call these modified node equations, since we either modify the circuit (method 1), use voltage source constraints to define node voltage at some nodes (method 2), or combine nodes to produce a supernode (method 3). The three methods are not mutually exclusive. We frequently use a combination of methods, as illustrated in the following examples.

EXAMPLE 3–4

Use node-voltage analysis to find v_O in the circuit in Figure 3–12(a).

FIGURE 3 – 12

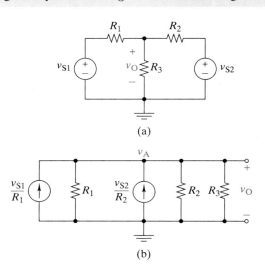

(a)

(b)

SOLUTION:

The given circuit in Figure 3–12(a) has four nodes, so we appear to need $N - 1 = 3$ node-voltage equations. However, applying source transformations to the two voltage sources (method 1) produces the two-node circuit in Figure 3–12(b). For the modified circuit, we need only one node equation.

$$(G_1 + G_2 + G_3)v_A = G_1 v_{S1} + G_2 v_{S2}$$

To find the output voltage, we solve for v_A:

$$v_O = v_A = \frac{G_1 v_{S1} + G_2 v_{S2}}{G_1 + G_2 + G_3}$$

Because of the two voltage sources, we need only one node equation in what appears to be a three-node circuit. The two voltage sources have a common node, so the number of unknown node voltages is reduced from three to one. The general principle illustrated is that the number of independent KCL constraints in a circuit containing N nodes and N_V voltage sources is $N - 1 - N_V$. ∎

EXAMPLE 3–5

Find the input resistance of the circuit in Figure 3–13.

FIGURE 3 – 13

SOLUTION:

Method 1 of handling voltage sources will not work here because the source in Figure 3–13 is not connected in series with a resistor. Method 2 will work in this case because the voltage source is connected to the reference node. As a result, we only need node equations at nodes B and C since the node A voltage is $v_A = v_S$. By inspection, the two required node equations are

$$\text{Node B:} \quad -0.5Gv_A + 3Gv_B - 0.5Gv_C = 0$$

$$\text{Node C:} \quad -2Gv_A - 0.5Gv_B + 3.5Gv_C = 0$$

Since $v_A = v_S$, these equations can be written in standard form as follows:

$$\text{Node B:} \qquad 3Gv_B - 0.5Gv_C = 0.5Gv_S$$

$$\text{Node C:} \quad -0.5Gv_B + 3.5Gv_C = 2Gv_S$$

Solving for the two unknown node voltages yields

$$v_B = \frac{\Delta_B}{\Delta} = \frac{\begin{vmatrix} 0.5Gv_S & -0.5G \\ 2Gv_S & 3.5G \end{vmatrix}}{\begin{vmatrix} 3G & -0.5G \\ -0.5G & 3.5G \end{vmatrix}} = \frac{2.75G^2v_S}{10.25G^2} = \frac{2.75v_S}{10.25}$$

$$v_C = \frac{\Delta_C}{\Delta} = \frac{\begin{vmatrix} 3G & 0.5Gv_S \\ -0.5G & 2Gv_S \end{vmatrix}}{\Delta} = \frac{6.25G^2v_S}{10.25G^2} = \frac{6.25v_S}{10.25}$$

Given the two node voltages, we can now solve for the input current.

$$i_{IN} = \frac{v_S - v_B}{2R} + \frac{v_S - v_C}{R/2} = \frac{3.75v_S}{10.25R} + \frac{8v_S}{10.25R} = \frac{11.75v_S}{10.25R}$$

Hence, the input resistance is

$$R_{IN} = \frac{v_S}{i_{IN}} = \frac{10.25R}{11.75} = 0.872R$$

This is the same answer as in Example 3–3, where the same circuit was driven by a current source rather than a voltage source. Input resistance is an intrinsic property of a circuit that does not depend on how the circuit is driven. ■

EXAMPLE 3-6

For the circuit in Figure 3–14,

(a) Formulate node-voltage equations.
(b) Solve for the output voltage v_O using $R_1 = R_4 = 2$ kΩ and $R_2 = R_3 = 4$ kΩ.

SOLUTION:

(a) The voltage sources in Figure 3–14 do not have a common node, and we cannot select a reference node that includes both sources. Selecting node D as the reference forces the condition $v_B = v_{S2}$ (method 2) but leaves the other source v_{S1} ungrounded. We surround the ungrounded source,

FIGURE 3 – 1 4

and all wires leading to it, by the supernode boundary shown in Figure 3–14 (method 3). KCL applies to the four element currents that penetrate the supernode boundary, and we can write

$$i_1 + i_2 + i_3 + i_4 = 0$$

These currents can easily be expressed in terms of the node voltages.

$$G_1 v_A + G_2(v_A - v_B) + G_3(v_C - v_B) + G_4 v_C = 0$$

Since $v_B = v_{S2}$, the standard form of this equation is

$$(G_1 + G_2)v_A + (G_3 + G_4)v_C = (G_2 + G_3)v_{S2}$$

We have one equation in the two unknown node voltages v_A and v_C. Applying the fundamental property of node voltages inside the supernode, we can write

$$v_A - v_C = v_{S1}$$

That is, the ungrounded voltage source constrains the difference between the two unknown node voltages inside the supernode. It thereby supplies the relationship needed to obtain two equations in two unknowns.

(b) Inserting the given numerical values yields

$$(7.5 \times 10^{-4})v_A + (7.5 \times 10^{-4})v_C = (5 \times 10^{-4})v_{S2}$$

$$v_A - v_C = v_{S1}$$

To find the output v_O, we need to solve these equations for v_C. The second equation yields $v_A = v_C + v_{S1}$, which, when substituted into the first equation, yields the required output:

$$v_O = v_C = \frac{v_{S2}}{3} - \frac{v_{S1}}{2}$$

■

Exercise 3–5

Find v_O in Figure 3–15 when the element E is

(a) A 10-kΩ resistance,
(b) A 4-mA independent current source with reference arrow pointing left.

Answers:

(a) 2.53 V
(b) −17.3 V

FIGURE 3 – 1 5

Exercise 3–6

Find v_O in Figure 3–15 when the element E is

(a) An open circuit
(b) A 10-V independent voltage source with the plus reference on the right.

Answers:

(a) 1.92 V
(b) 12.96 V

SUMMARY OF NODE-VOLTAGE ANALYSIS

We have seen that node-voltage equations are very useful in the analysis of a variety of circuits. These equations can always be formulated using KCL, the element constraints, and the fundamental property of node voltages. When in doubt, always fall back on these principles to formulate node equations in new situations. With practice and experience, however, we eventually develop an analysis approach that allows us to recognize shortcuts in the formulation process. The following guidelines summarize our approach and may help you develop your own analysis style:

1. Simplify the circuit by combining elements in series and parallel wherever possible.

2. If not specified, select a reference node so that as many voltage sources as possible are directly connected to the reference.

3. Node equations are required at supernodes and all other nonreference nodes except those that are directly connected to the reference by voltage sources.

4. Use KCL to write node equations at the nodes identified in step 3. Express element currents in terms of node voltages or the currents produced by independent current sources.

5. Write expressions relating the node voltages to the voltages produced by independent voltage sources.

6. Substitute the expressions from step 5 into the node equations from step 4 and arrange the resulting equations in standard form.

7. Solve the equations from step 6 for the node voltages of interest. Cramer's rule is often useful when circuit parameters are left in symbolic form. Computer tools are useful when there are four or more equations and numerical values are given.

3-2 MESH-CURRENT ANALYSIS

Mesh currents are analysis variables that are useful in circuits containing many elements connected in series. To review terminology, a loop is a closed path formed by passing through an ordered sequence of nodes without passing through any node more than once. A mesh is a special type of loop that does not enclose any elements. For example, loops A and B in Figure 3–16 are meshes, while the loop X is not a mesh because it encloses an element.

Mesh-current analysis is resistricted to planar circuits. A **planar circuit** can be drawn on a flat surface without crossovers in the "window pane" fashion shown in Figure 3–16. To define a set of variables, we associate a **mesh current** (i_A, i_B, i_C, etc.) with each window pane and assign a reference direction. The reference directions for all mesh currents are customarily taken in a clockwise sense. There is no momentous reason for this, except perhaps tradition.

We think of these mesh currents as circulating through the elements in their respective meshes, as suggested by the reference directions shown in Figure 3–16. We should emphasize that this viewpoint is not based on the physics of circuit behavior. There are not red and blue electrons running around that somehow get assigned to mesh currents i_A or i_B. Mesh currents are variables used in circuit analysis. They are only somewhat abstractly related to the physical operation of a circuit and may be impossible to measure directly. For example, there is no way to cut the circuit in Figure 3–16 to insert an ammeter that only measures i_E.

Mesh currents have a unique feature that is the dual of the fundamental property of node voltages. If we examine Figure 3–16, we see the elements around the perimeter are contained in only one mesh, while those in the interior are in two meshes. In a planar circuit any given element is contained in at most two meshes. When an element is in two meshes, the two mesh currents circulate through the element in opposite directions. In such cases KCL declares that the net current through the element is the difference of the two mesh currents.

FIGURE 3 – 1 6   *Meshes in a planar circuit.*

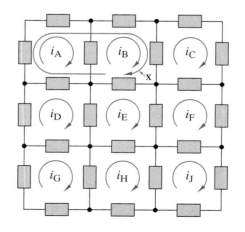

These observations lead us to the fundamental property of mesh currents:

> *If the **K**th two-terminal element is contained in meshes **X** and **Y**, then the element current can be expressed in terms of the two mesh currents as*
>
> $$i_K = i_X - i_Y \qquad (3\text{–}11)$$
>
> *where **X** is the mesh whose reference direction agrees with the reference direction of* i_K.

Equation (3–11) is a KCL constraint at the element level. If the element is contained in only one mesh, then $i_K = i_X$ or $i_K = -i_Y$, depending on whether the reference direction for the element current agrees or disagrees with the reference direction of the mesh current. The key idea is that the current through every two-terminal element in a planar circuit can be expressed in terms of no more than two mesh currents.

F I G U R E   3 – 1 7

Exercise 3–7

In Figure 3–17 the mesh currents are $i_A = 10$ A, $i_B = 5$ A, and $i_C = -3$ A. Find the element currents i_1 through i_6 and show that KCL is satisfied at nodes A, B, and C.

Answers:

$i_1 = -10$ A, $i_2 = 13$ A, $i_3 = 5$ A, $i_4 = 8$ A, $i_5 = 5$ A, and $i_6 = -3$ A.

To use mesh currents to formulate circuit equations, we use elements and connection constraints, except that the KCL constraints are not explicitly written. Instead, we use the fundamental property of mesh currents to express the element voltages in terms of the mesh currents. By doing so we avoid using the element currents and work only with the element voltages and mesh currents.

For example, the planar circuit in Figure 3–18 can be analyzed using the mesh-current method. In the figure we have defined two mesh currents and five element voltages. We write KVL constraints around each mesh using the element voltages.

F I G U R E   3 – 1 8 *Circuit for demonstrating mesh-current analysis.*

Mesh A: $-v_0 + v_1 + v_3 = 0$

Mesh B: $-v_3 + v_2 + v_4 = 0 \qquad (3\text{–}12)$

Using the fundamental property of mesh currents, we write the element voltages in terms of the mesh currents and input voltages:

$$v_1 = R_1 i_A \qquad v_0 = v_{S1}$$
$$v_2 = R_2 i_B \qquad v_4 = v_{S2} \tag{3–13}$$
$$v_3 = R_3(i_A - i_B)$$

We substitute these element equations into the KVL connection equations and arrange the result in standard form.

$$(R_1 + R_3)i_A - R_3 i_B = v_{S1}$$
$$-R_3 i_A + (R_2 + R_3)i_B = -v_{S2} \tag{3–14}$$

We have completed the formulation process with two equations in two unknown mesh currents.

As we have previously noted, every method of circuit analysis must satisfy KCL, KVL, and the element i–v relationships. When formulating mesh equations, it may appear that we have not used KCL. However, writing the element constraints in the form in Eq. (3–13) requires the KCL equations $i_1 = i_A$, $i_2 = i_B$, and $i_3 = i_A - i_B$. Mesh-current analysis implicitly satisfies KCL when the element constraints are expressed in terms of the mesh currents. In effect, the fundamental property of mesh currents ensures that the KCL constraints are satisfied.

We use Cramer's rule to solve for the mesh currents in Eq. (3–14):

$$i_A = \frac{\Delta_A}{\Delta} = \frac{\begin{vmatrix} v_{S1} & -R_3 \\ -v_{S2} & R_2 + R_3 \end{vmatrix}}{\begin{vmatrix} R_1 + R_3 & -R_3 \\ -R_3 & R_2 + R_3 \end{vmatrix}} = \frac{(R_2 + R_3)v_{S1} - R_3 v_{S2}}{R_1 R_2 + R_1 R_3 + R_2 R_3} \tag{3–15}$$

and

$$i_B = \frac{\Delta_B}{\Delta} = \frac{\begin{vmatrix} R_1 + R_3 & v_{S1} \\ -R_3 & -v_{S2} \end{vmatrix}}{\begin{vmatrix} R_1 + R_3 & -R_3 \\ -R_3 & R_2 + R_3 \end{vmatrix}} = \frac{R_3 v_{S1} - (R_1 + R_3)v_{S2}}{R_1 R_2 + R_1 R_3 + R_2 R_3} \tag{3–16}$$

Equations (3–15) and (3–16) can now be substituted into the element constraints in Eq. (3–13) to solve for every voltage in the circuit. For instance, the voltage across R_3 is

$$v_3 = R_3(i_A - i_B) = \frac{R_2 R_3 v_{S1} + R_1 R_3 v_{S2}}{R_1 R_2 + R_1 R_3 + R_2 R_3} \tag{3–17}$$

You are invited to show that the result in Eq. (3–17) agrees with the node analysis result obtained in Example 3–4 for the same circuit.

The mesh-current analysis approach just illustrated can be summarized in four steps:

STEP 1: Identify a mesh current with every mesh and a voltage across every circuit element.

STEP 2: Write KVL connection constraints in terms of the element voltages around every mesh.

STEP 3: Use KCL and the i–v relationships of the elements to express the element voltages in terms of the mesh currents.

STEP 4: Substitute the element constraints from step 3 into the connection constraints from step 2 and arrange the resulting equations in standard form.

The number of mesh-current equations derived in this way equals the number of KVL connection constraints in step 2. When discussing combined constraints in Chapter 2, we noted that there are $E - N + 1$ independent KVL constraints in any circuit. Using the window panes in a planar circuit generates $E - N + 1$ independent mesh currents. Mesh analysis works best when the circuit has many elements (E large) connected in series (N also large).

WRITING MESH-CURRENT EQUATIONS BY INSPECTION

The mesh equations in Eq. (3–14) have a symmetrical pattern that is similar to the coefficient symmetry observed in node equations. The coefficients of i_B in the first equation and i_A in the second equation are the negative of the resistance common to meshes A and B. The coefficients of i_A in the first equation and i_B in the second equation are the sum of the resistances in meshes A and B, respectively.

This pattern will always occur in planar circuits containing resistors and independent voltage sources when the mesh currents are defined in the window panes of a planar circuit, as shown in Figure 3–16. To see why, consider a general resistance R that is contained in, say, mesh A. There are only two possibilities. Either R is not contained in any other mesh, in which case the voltage across it is

$$v = R(i_A - 0) = Ri_A$$

or else it is also contained in only one adjacent mesh, say mesh B, in which case the voltage across it is

$$v = R(i_A - i_B)$$

These observations lead to the following conclusions. The voltages across resistance in mesh A involves the following terms:

1. i_A times the sum of the resistances in mesh A

2. $-i_B$ times the sum of resistances common to mesh A and mesh B, and similar terms for any other mesh adjacent to mesh A.

The sum of the voltages across resistors plus the sum of the independent voltage sources around mesh A must equal zero.

The aforementioned process makes it possible for us to write mesh-current equations by inspection without going through the intermediate steps involving the KVL connection constraints and the element constraints.

EXAMPLE 3–7

For the circuit of Figure 3–19,

(a) Formulate mesh-current equations.
(b) Find the output v_O using $R_1 = R_4 = 2 \text{ k}\Omega$ and $R_2 = R_3 = 4 \text{ k}\Omega$.

FIGURE 3-19

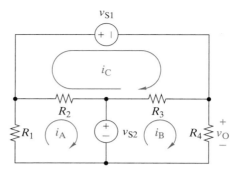

SOLUTION:

(a) To write mesh-current equations by inspection, we note that the total resistances in meshes A, B, and C are $R_1 + R_2$, $R_3 + R_4$, and $R_2 + R_3$, respectively. The resistance common to meshes A and C is R_2. The resistance common to meshes B and C is R_3. There is no resistance common to meshes A and B. Using these observations, we write the mesh equations as

$$\text{Mesh A: } (R_1 + R_2)i_A - 0\,i_B - R_2 i_C + v_{S2} = 0$$

$$\text{Mesh B: } (R_3 + R_4)i_B - 0\,i_A - R_3 i_C - v_{S2} = 0$$

$$\text{Mesh C: } (R_2 + R_3)i_C - R_2 i_A - R_3 i_B + v_{S1} = 0$$

The algebraic signs assigned to voltage source terms follow the passive convention for the mesh current in question. Arranged in standard form, these equations become

$$(R_1 + R_2)i_A - R_2 i_C = -v_{S2}$$

$$+ (R_3 + R_4)i_B - R_3 i_C = +v_{S2}$$

$$- R_2 i_A - R_3 i_B + (R_2 + R_3)i_C = -v_{S1}$$

Coefficient symmetry greatly simplifies the formulation of these equations compared with the more fundamental, but time-consuming, process of writing element and connection constraints. (b) Inserting the numerical values into these equations yields

$$6000\,i_A \qquad\qquad - 4000\,i_C = -v_{S2}$$

$$6000\,i_B - 4000\,i_C = v_{S2}$$

$$- 4000\,i_A - 4000\,i_B + 8000\,i_C = -v_{S1}$$

Placing these three mesh equations in matrix form produces

$$\begin{bmatrix} 6000 & 0 & -4000 \\ 0 & 6000 & -4000 \\ -4000 & -4000 & +8000 \end{bmatrix} \begin{bmatrix} i_A \\ i_B \\ i_C \end{bmatrix} = \begin{bmatrix} -v_{S2} \\ v_{S2} \\ -v_{S1} \end{bmatrix}$$

This is a matrix equation of the form $\mathbf{AX} = \mathbf{B}$, where

$$\mathbf{A} = \begin{bmatrix} 6000 & 0 & -4000 \\ 0 & 6000 & -4000 \\ -4000 & -4000 & +8000 \end{bmatrix} \quad \mathbf{X} = \begin{bmatrix} i_A \\ i_B \\ i_C \end{bmatrix} \quad \mathbf{B} = \begin{bmatrix} -v_{S2} \\ v_{S2} \\ -v_{S1} \end{bmatrix}$$

The matrix $\mathbf{A}$ is a square matrix of the coefficients on the left side of the mesh equations, $\mathbf{X}$ is a column matrix of the unknown mesh currents, and $\mathbf{B}$ is a column matrix of the input voltages on the right side of the mesh equations. To solve this matrix equation we multiply by the inverse of the coefficient matrix $(\mathbf{A}^{-1})$. On the left side the result is $\mathbf{A}^{-1}\mathbf{A}\mathbf{X} = \mathbf{X}$, and the right side becomes $\mathbf{A}^{-1}\mathbf{B}$. In other words, multiplying by $\mathbf{A}^{-1}$ yields the solution for the unknown mesh currents as

$$\mathbf{X} = \mathbf{A}^{-1}\mathbf{B}$$

In sum, to solve a system of linear equations by matrix methods we form the matrix product $\mathbf{A}^{-1}\mathbf{B}$. It is at this point that MATLAB comes into play.

Using MATLAB to solve for the mesh currents, we first enter the $\mathbf{A}$ matrix by the statement

```
A=[6000 0 -4000;0 6000 -4000;-4000 -4000 8000];
```

The elements in $\mathbf{B}$ matrix are the symbolic variables v_{S1} and v_{S2}. These quantities are not unknowns, but symbols that represent all possible values of the input voltages. Thus, we must first write the statement

```
syms VS1 VS2
```

which declares these identifiers to be symbolic rather than numerical quantities. We can now enter the $\mathbf{B}$ matrix as

```
B=[-VS2;VS2;-VS1];
```

and solve for the unknown mesh currents using MATLAB statement

```
X=inv(A)*B
```

which yields

```
X =

[-1/6000*VS2-1/4000*VS1]
[1/6000*VS2-1/4000*VS1]
[          -3/8000*VS1]
```

The elements of the column matrix $\mathbf{X}$ are the three unknown mesh currents expressed in terms of the input voltages. The output voltage in Figure 3–19 is written in terms of the mesh currents as $v_O = R_4 i_B$. In MATLAB notation the required mesh current is

$$i_B = X(2)=1/6000*VS2-1/4000*VS1$$

Since $R_4 = 2000$ we conclude that

$$v_O = \frac{v_{S2}}{3} - \frac{v_{S1}}{2}$$

The mesh current analysis result obtained here is the same as the node voltage result obtained in Example 3–6. Either approach produces the same answer, but which method do you think is easier? ■

MESH EQUATIONS WITH CURRENT SOURCES

In developing mesh analysis, we assumed that circuits contain only voltage sources and resistors. This assumption simplifies the formulation process because the sum of voltages around a mesh is determined by voltage sources and the mesh currents through resistors. A current source complicates the picture because the voltage across it is not directly related to its current. We need to adapt mesh analysis to accommodate current sources just as we revised node analysis to deal with voltage sources.

There are three ways to handle current sources in mesh analysis:

1. If the current source is connected in parallel with a resistor, then it can be converted to an equivalent voltage source by source transformation. Each source conversion eliminates a mesh and reduces the number of equations required by 1. This method is the dual of method 1 for node analysis.

2. If a current source is contained in only one mesh, then that mesh current is determined by the source current and is no longer an unknown. We write mesh equations around the remaining meshes in the usual way and move the known mesh current to the source side of the equations in the final step. The number of equations obtained is one less than the number of meshes. This method is the dual of method 2 for node analysis.

3. Neither of the first two methods will work when a current source is contained in two meshes or is not connected in parallel with a resistance. In this case we create a **supermesh** by excluding the current source and any elements connected in series with it, as shown in Figure 3–20. We write one mesh equation around the

FIGURE 3 – 20   *Example of a supermesh.*

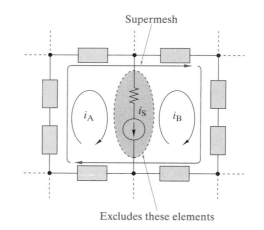

supermesh using the currents i_A and i_B. We then write mesh equations of the remaining meshes in the usual way. This leaves us one equation short because parts of meshes A and B are included in the supermesh. However, the fundamental property of mesh currents relates the currents i_S, i_A, and i_B as

$$i_A - i_B = i_S$$

This equation supplies the one additional relationship needed to get the requisite number of equations in the unknown mesh currents.

The aforementioned three methods are not mutually exclusive. We can use more than one method in a circuit, as the following examples illustrate.

EXAMPLE 3-8

Use mesh-current equations to find i_O in the circuit in Figure 3–21(a).

FIGURE   3 - 2 1

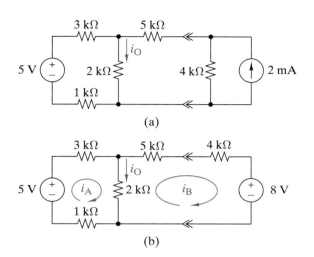

(a)

(b)

SOLUTION:

The current source in this circuit can be handled by a source transformation (method 1). The 2-mA source in parallel with the 4-kΩ resistor in Figure 3–21(a) can be replaced by an equivalent 8-V voltage source in series with the same resistor, as shown in Figure 3–21(b). In this circuit the total resistance in mesh A is 6 kΩ, the total resistance in mesh B is 11 kΩ, and the resistance contained in both meshes is 2 kΩ. By inspection, the mesh equations for this circuit are

$$(6000)i_A - (2000)i_B = 5$$

$$-(2000)i_A + (11000)i_B = -8$$

Solving for the two mesh currents yields $i_A = 0.6290$ mA and $i_B = -0.6129$ mA. By KCL the desired current is $i_O = i_A - i_B = 1.2419$ mA. The given circuit in Figure 3–21(a) has three meshes and one current source. The source transformation leading to Figure 3–21(b) produces a circuit with only two meshes. The general principle illustrated is that the number of independent mesh equations in a circuit containing E elements, N nodes, and N_I current sources is $E - N + 1 - N_I$. ∎

EXAMPLE 3-9

Use mesh-current equations to find the v_O in Figure 3–22.

SOLUTION:

Source transformation (method 1) is not possible here since neither current source is connected in parallel with a resistor. The current source i_{S2} is in both mesh B and mesh C, so we exclude this element and create the

FIGURE 3 – 22

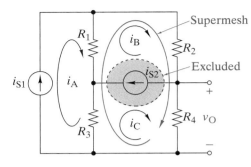

supermesh (method 3) shown in the figure. The sum of voltages around the supermesh is

$$R_1(i_B - i_A) + R_2(i_B) + R_4(i_C) + R_3(i_C - i_A) = 0$$

The supermesh voltage constraint yields one equation in the three unknown mesh currents. Applying KCL to each of the current sources yields

$$i_A = i_{S1}$$

$$i_B - i_C = i_{S2}$$

Because of KCL the two current sources force constraints that supply two more equations. Using these two KCL constraints to eliminate i_A and i_B from the supermesh KVL constraint yields

$$(R_1 + R_2 + R_3 + R_4)i_C = (R_1 + R_3)i_{S1} - (R_1 + R_2)i_{S2}$$

Hence, the required output voltage is

$$v_O = R_4 i_C = R_4 \times \left[\frac{(R_1 + R_3)i_{S1} - (R_1 + R_2)i_{S2}}{R_1 + R_2 + R_3 + R_4} \right] \qquad \blacksquare$$

Exercise 3–8

Use mesh analysis to find the current i_O in Figure 3–23 when the element E is

(a) A 5-V voltage source with the positive reference at the top
(b) A 10-kΩ resistor.

Answers:

(a) –0.136 mA
(b) –0.538 mA

FIGURE 3 – 23

Exercise 3–9

Use mesh analysis to find the current i_O in Figure 3–23 when the element E is

(a) A 1-mA current source with the reference arrow directed down
(b) Two 20-kΩ resistors in parallel.

Answers:

(a) –1 mA
(b) –0.538 mA

FIGURE 3-24

Exercise 3-10

Use mesh-current equations to find v_O in Figure 3-24.

Answer:
$$v_O = (v_1 + v_2)/4$$

SUMMARY OF MESH-CURRENT ANALYSIS

Mesh-current equations can always be formulated from KVL, the element constraints, and the fundamental property of mesh currents. When in doubt, always fall back on these principles to formulate mesh equations in new situations. The following guidelines summarize an approach to formulating mesh equations for resistance circuits:

1. Simplify the circuit by combining elements in series or parallel wherever possible.

2. Mesh equations are required for supermeshes and all other meshes except those where current sources are contained in only one mesh.

3. Use KVL to write mesh equations for the meshes identified in step 2. Express element voltages in terms of mesh currents or the voltage produced by independent voltage sources.

4. Write expressions relating the mesh currents to the currents produced by independent current sources.

5. Substitute the expressions from step 4 into the mesh equations from step 3 and place the result in standard form.

6. Solve the equations from step 5 for the mesh currents of interest. Cramer's rule is often useful when circuit parameters are left in symbolic form. Computer tools are useful when there are four or more equations and numerical values are given.

3-3 LINEARITY PROPERTIES

This book treats the analysis and design of **linear circuits**. A circuit is said to be linear if it can be adequately modeled using only linear elements and independent sources. The hallmark feature of a linear circuit is that outputs are linear functions of the inputs. Circuit **inputs** are the signals produced by external sources, and **outputs** are any other designated signals. Mathematically, a function is said to be linear if it possesses two properties—homogeneity and additivity. In linear circuits, **homogeneity** means that the output is proportional to the input. **Additivity** means that the output due to two or more inputs can be found by adding the outputs obtained when each input is applied separately. Mathematically, these properties are written as follows:

$$f(Kx) = Kf(x) \quad \text{(homogeneity)} \tag{3-18}$$

and

$$f(x_1 + x_2) = f(x_1) + f(x_2) \quad \text{(additivity)} \tag{3-19}$$

where K is a scalar constant. In circuit analysis the homogeneity property is called **proportionality**, while the additivity property is called **superposition**.

THE PROPORTIONALITY PROPERTY

The **proportionality property** applies to linear circuits with one input. For linear resistive circuits, proportionality states that every input-output relationship can be written as

$$y = Kx \qquad \text{(3–20)}$$

x Input → K → $y = Kx$ Output

FIGURE 3 – 2 5 *Block diagram representation of the proportionality property.*

where x is the input current or voltage, y is an output current or voltage, and K is a constant. The block diagram in Figure 3–25 describes this linear input–output relationship. In a block diagram the lines headed by arrows indicate the direction of signal flow. The arrow directed into the block indicates the input, while the output is indicated by the arrow directed out of the block. The variable names written next to these lines identify the input and output signals. The scalar constant K written inside the block indicates that the input signal x is multiplied by K to produce the output signal as $y = Kx$.

Caution: Proportionality only applies when the input and output are current or voltage. It does not apply to output power since power is equal to the product of current and voltage. In other words, output power is not linearly related to the input current or voltage.

We have already seen several examples of proportionality. For instance, using voltage division in Figure 3–26(a) produces

$$v_O = \left(\frac{R_2}{R_1 + R_2} \right) v_S$$

which means

$$x = v_S \qquad y = v_O$$

$$K = \frac{R_2}{R_1 + R_2}$$

(a)

Similarly, applying current division in Figure 3–26(b) yields

$$i_O = \left(\frac{G_2}{G_1 + G_2} \right) i_S$$

so that

$$x = i_S \qquad y = i_O$$

$$K = \frac{G_2}{G_1 + G_2}$$

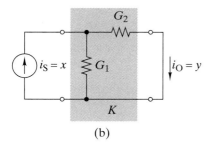

(b)

FIGURE 3 – 2 6 *Examples of circuit exhibiting of proportionality: (a) Voltage divider. (b) Current divider.*

In these two examples the proportionality constant K is dimensionless because the input and output have the same units. In other situations K could have the units of ohms or siemens when the input and output have different units.

The next example illustrates that the proportionality constant K can be positive, negative, or even zero.

EXAMPLE 3–10

Given the bridge circuit of Figure 3–27,

(a) Find the proportionality constant K in the input-output relationship
$v_O = Kv_S$
(b) Find the sign of K when $R_2R_3 > R_1R_4$, $R_2R_3 = R_1R_4$, and $R_2R_3 < R_1R_4$.

FIGURE 3 – 2 7

SOLUTION:

(a) We observe that the circuit consists of two voltage dividers. Applying the voltage division rule to each side of the bridge circuit yields

$$v_A = \frac{R_3}{R_1 + R_3}v_S \quad \text{and} \quad v_B = \frac{R_4}{R_2 + R_4}v_S$$

The fundamental property of node voltages allows us to write

$$v_O = v_A - v_B$$

Substituting the equations for v_A and v_B into this KVL equation yields

$$v_O = \left(\frac{R_3}{R_1 + R_3} - \frac{R_4}{R_2 + R_4}\right)v_S$$

$$= \left(\frac{R_2R_3 - R_1R_4}{(R_1 + R_3)(R_2 + R_4)}\right)v_S$$

$$= \qquad (K) \qquad v_S$$

(b) The proportionality constant K can be positive, negative, or zero. Specifically,

$$\text{If } R_2R_3 > R_1R_4 \text{ then } K > 0$$

$$\text{If } R_2R_3 = R_1R_4 \text{ then } K = 0$$

$$\text{If } R_2R_3 < R_1R_4 \text{ then } K < 0$$

When the products of the resistances in opposite legs of the bridge are equal, then $K = 0$ and the bridge is said to be balanced. ∎

UNIT OUTPUT METHOD

The **unit output method** is an analysis technique based on the proportionality property of linear circuits. The method involves finding the input-output proportionality constant K by assuming an output of one unit and determining the input required to produce that unit output. This technique is most useful when applied to ladder circuits, and it involves the following steps:

1. A unit output is assumed; that is, $v_O = 1$ V or $i_O = 1$ A.

2. The input required to produce the unit output is then found by successive application of KCL, KVL, and Ohm's law.

3. Because the circuit is linear, the proportionality constant relating input and output is

$$K = \frac{\text{Output}}{\text{Input}} = \frac{1}{\text{Input for unit output}}$$

Given the proportionality constant K, we can find the output for any input using Eq. (3–20).

In a way, the unit output method solves the circuit response problem backwards—that is, from output to input—as illustrated by the next example.

EXAMPLE 3–11

Use the unit output method to find v_O in the circuit shown in Figure 3–28(a).

FIGURE 3–28

(a)

(b)

SOLUTION:

We start by assuming $v_O = 1$, as shown in Figure 3–28(b). Then, using Ohm's law, we find i_O.

$$i_O = \frac{v_O}{20} = 0.05 \text{ A}$$

Next, using KCL at node B, we find i_1.

$$i_1 = i_O = 0.05 \text{ A}$$

Again, using Ohm's law, we find v_1.

$$v_1 = 10i_1 = 0.5 \text{ V}$$

Then, writing a KVL equation around loop L2, we find v_2 as

$$v_2 = v_1 + v_O = 0.5 + 1.0 = 1.5 \text{ V}$$

Using Ohm's law once more produces

$$i_2 = \frac{v_2}{15} = \frac{1.5}{15} = 0.1 \text{ A}$$

Next, writing a KCL equation at node A yields

$$i_3 = i_1 + i_2 = 0.05 + 0.1 = 0.15 \text{ A}$$

Using Ohm's law one last time,

$$v_3 = 10i_3 = 1.5 \text{ V}$$

We can now find the source voltage v_S by applying KVL around loop L1:

$$v_S|_{\text{for } v_O = 1\text{V}} = v_3 + v_2 = 1.5 + 1.5 = 3 \text{ V}$$

A 3-V source voltage is required to produce a 1-V output. From this result, we calculate the proportionality constant K to be

$$K = \frac{v_O}{v_S} = \frac{1}{3}$$

Once K is known, the output for the specified 5-V input is $v_O = (\frac{1}{3})5 = 1.667$ V. ∎

Find v_O in the circuit of Figure 3–28(a) when v_S is –5 V, 10 mV, and 3 kV.

Answers:

$$v_O = -1.667 \text{ V}, 3.333 \text{ mV, and } 1 \text{ kV}$$

Exercise 3–12

Use the unit output method to find $K = i_O/i_{\text{IN}}$ for the circuit in Figure 3–29. Then use the proportionality constant K to find i_O for the input current shown in the figure.

FIGURE 3 – 29

Answers:

$$K = \frac{1}{4}; i_O = 0.15 \text{ mA}$$

Exercise 3–13

Use the unit output method to find $K = v_O/i_{\text{IN}}$ for the circuit in Figure 3–29. Then use K to find v_O for the input current shown in the figure.

Answers:

$$K = 750 \text{ }\Omega; v_O = 450 \text{ mV}$$

Note: In this exercise K has the dimensions of ohms because the input is a current and the output a voltage.

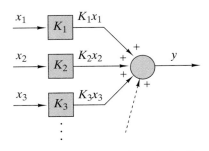

FIGURE 3-30 *Block diagram representation of the additivity property.*

The **additivity property** states that any output current or voltage of a linear resistive circuit with multiple inputs can be expressed as a linear combination of the several inputs:

$$y = K_1 x_1 + K_2 x_2 + K_3 x_3 + \dots \tag{3-21}$$

where $x_1, x_2, x_3, \dots$ are current or voltage inputs, and $K_1, K_2, K_3 \dots$ are constants that depend on the circuit parameters. Figure 3–30 shows how we represent this relationship in block diagram form. Again the arrows indicate the direction of signal flow and the K's within the blocks are scalar multipliers. The circle in Figure 3–30 is a new block diagram element called a summing point that implements the operation $y = \sum K_i x_i$. Although the block diagram in Figure 3–30 is nothing more than a pictorial representation of Eq. (3–21), the diagram often helps us gain a clearer picture of how signals interact in different parts of a circuit.

To illustrate this property, we analyze the two-input circuit in Figure 3–31 using node-voltage analysis. Applying KCL at node A, we obtain

$$\frac{v_A - v_S}{R_1} + i_S + \frac{v_A}{R_2} = 0$$

FIGURE 3-31 *Circuit used to demonstrate superposition.*

Moving the inputs to the right side of this equation yields

$$\left[\frac{1}{R_1} + \frac{1}{R_2}\right] v_A = \frac{v_S}{R_1} + i_S$$

Since $v_O = v_A$, we obtain the input-output relationship in the form

$$v_O = \left[\frac{R_2}{R + R_2}\right] v_S + \left[\frac{R_1 R_2}{R_1 + R_2}\right] i_S$$

$$y = [K_1] x_1 + [K_2] x_2 \tag{3-22}$$

This result shows that the output is a linear combination of the two inputs. Note that K_1 is dimensionless since its input and output are voltages, and that K_2 has the units of ohms since its input is a current and its output is a voltage.

SUPERPOSITION PRINCIPLE

Since the output in Eq. (3–21) is a linear combination, the contribution of each input source is independent of all other inputs. This means that the output can be found by finding the contribution from each source acting alone and then adding the individual response to obtain the total response. This suggests that the output of a multiple-input linear circuit can be found by the following steps:

STEP 1: "Turn off" all independent sources except one and find the output of the circuit due to that source acting alone.

STEP 2: Repeat the process in step 1 until each independent source has been turned on and the output due to that source found.

STEP 3: The total output with all independent sources turned on is the algebraic sum of the outputs caused by each source acting alone.

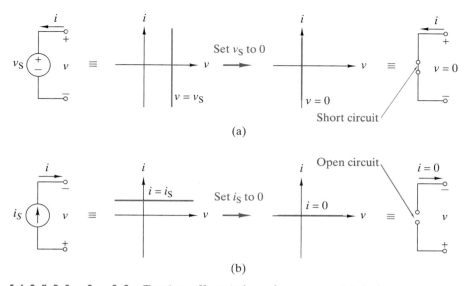

FIGURE 3-32 *Turning off an independent source: (a) Voltage source. (b) Current source.*

These steps describe a circuit analysis technique called the **superposition principle**. Before applying this method, we must discuss what happens when a voltage or current source is "turned off."

The i–v characteristics of voltage and current sources are shown in Figure 3–32. A voltage source is "turned off" by setting its voltage to zero ($v_S = 0$). This step translates the voltage source i–v characteristic to the i–axis, as shown in Figure 3–32(a). In Chapter 2 we found that a vertical line on the i–axis is the i–v characteristic of a short circuit. Similarly, "turning off" a current source ($i_S = 0$) in Figure 3–32(b) translates its i–v characteristic to the v-axis, which is the i–v characteristic of an open circuit. Therefore, when a voltage source is "turned off" we replace it by a short circuit, and when a current source is "turned off" we replace it by an open circuit.

The superposition principle is now applied to the circuit in Figure 3–31 to duplicate the response in Eq. (3–22), which was found by node analysis. Figure 3–33 shows the steps involved in applying superposition to the circuit in Figure 3–31. Figure 3–33(a) shows that the circuit has two input sources. We will first "turn off" i_S and replace it with an open circuit, as shown in Figure 3–33(b). The output of the circuit in Figure 3–33(b) is called v_{O1} and represents that part of the total output caused by the voltage source. Using voltage division in Figure 3–33(b) yields v_{O1} as

$$v_{O1} = \frac{R_2}{R_1 + R_2} v_S$$

Next we "turn off" the voltage source and "turn on" the current source, as shown in Figure 3–33(c). Using Ohm's law, we get $v_{O2} = i_{O2}R_2$. We use current division to express i_{O2} in terms of i_S to obtain v_{O2}:

$$v_{O2} = i_{O2}R_2 = \left[\frac{R_1}{R_1 + R_2} i_S\right] R_2 = \frac{R_1 R_2}{R_1 + R_2} i_S$$

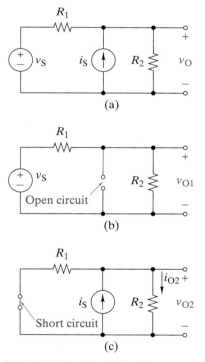

FIGURE 3-33 *Circuit analysis using superposition: (a) Current source off. (b) Voltage source off.*

Applying the superposition principle, we find the response with both sources "turned on" by adding the two responses v_{O1} and v_{O2}.

$$v_O = v_{O1} + v_{O2}$$

$$= \left[\frac{R_2}{R_1 + R_2}\right]v_S + \left[\frac{R_1 R_2}{R_1 + R_2}\right]i_S$$

This superposition result is the same as the circuit reduction result given in Eq. (3–22).

EXAMPLE 3–12

Figure 3–34(a) shows a resistance circuit used to implement a signal-summing function. Use superposition to show that the output v_O is a weighted sum of the inputs v_{S1}, v_{S2}, and v_{S3}.

(a)

Short circuits

(b)

F I G U R E   3 – 3 4

SOLUTION:

To determine v_O using superposition, we first turn off sources 1 and 2 ($v_{S1} = 0$ and $v_{S2} = 0$) to obtain the circuit in Figure 3–34(b). This circuit is a voltage divider in which the output leg consists of two equal resistors in parallel. The equivalent resistance of the output leg is $R/2$, so the voltage division rule yields

$$v_{O3} = \frac{R/2}{R + R/2}v_{S3} = \frac{v_{S3}}{3}$$

Because of the symmetry of the circuit, it can be seen that the same technique applies to all three inputs; therefore

$$v_{O2} = \frac{v_{S2}}{3} \text{ and } v_{O1} = \frac{v_{S1}}{3}$$

Applying the superposition principle, the output with all sources "turned on" is

$$v_O = v_{O1} + v_{O2} + v_{O3}$$

$$= \frac{1}{3}[v_{S1} + v_{S2} + v_{S3}]$$

That is, the output is proportional to the sum of the three input signals with $K_1 = K_2 = K_3 = \frac{1}{3}$. ∎

Exercise 3–14

The circuit of Figure 3–35 contains two of the R-$2R$ modules discussed in Problem 2–35. Use superposition to find v_O.

Answer:

$$v_O = \frac{1}{2}v_{S1} + \frac{1}{4}v_{S2}$$

FIGURE   3 – 35

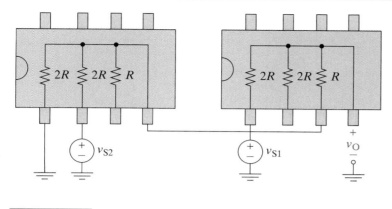

Exercise 3–15

Repeat Exercise 3–14 with the voltage source v_{S2} replaced by a current source i_{S2} with the current reference arrow directed toward ground.

Answer:

$$v_O = 3v_{S1}/5 - 4i_{S2}R/5$$

The preceding examples and exercises illustrate the applications of the superposition principle. You should not conclude that superposition is used primarily to solve for the response of circuits with multiple independent sources. In fact, superposition is not a particularly attractive method of analysis since a circuit with N sources requires N different circuit analyses to obtain the final result. Unless the circuit is relatively simple, superposition does not reduce the analysis effort compared with, say, node-voltage analysis. Superposition is still an important property of linear circuits because it is often used as a conceptual tool to develop other circuit analysis techniques. For example, superposition is used in the next section to prove Thévenin's theorem.

3–4 THÉVENIN AND NORTON EQUIVALENT CIRCUITS

An *interface* is a connection between circuits. Circuit interfaces occur frequently in electrical and electronic systems, so special analysis methods are used to handle them. For the two-terminal interface shown in Figure 3–36, we normally think of one circuit as the source S and the other as the load L. We think of signals as being produced by the source circuit and delivered to the load circuit. The source-load interaction at an interface is one of the central problems of circuit analysis and design.

The Thévenin and Norton equivalent circuits shown in Figure 3–37 are valuable tools for dealing with circuit interfaces. The conditions under which these equivalent circuits exist can be stated as a theorem:

> *If the source circuit in a two-terminal interface is linear, then the interface signals v and i do not change when the source circuit is replaced by its Thévenin or Norton equivalent circuit.*

The equivalence requires the source circuit to be linear, but places no restriction on the linearity of the load circuit. Later in this section we

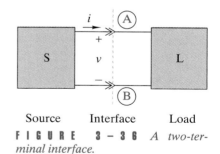

Source   Interface   Load

FIGURE   3 – 36 *A two-terminal interface.*

F I G U R E 3 – 3 7 *Equivalent circuits for the source: (a) Thévenin equivalent. (b) Norton equivalent.*

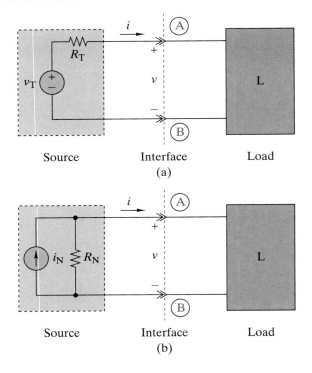

Source   Interface   Load
(a)

Source   Interface   Load
(b)

consider cases in which the load is nonlinear. In subsequent chapters we will study circuits in which the loads are energy storage elements called capacitors and inductors.

The Thévenin equivalent circuit consists of a voltage source (v_T) in series with a resistance (R_T). The Norton equivalent circuit is a current source (i_N) in parallel with a resistance (R_N). Note that the Thévenin and Norton equivalent circuits are practical sources in the sense discussed in Chapter 2.

The two circuits have the same i–v characteristics, since replacing one by the other leaves the interface signals unchanged. To derive the equivalency conditions, we apply KVL and Ohm's law to the Thévenin equivalent in Figure 3–37(a) to obtain its i–v relationship at the terminals A and B:

$$v = v_T - iR_T \qquad (3\text{–}23)$$

Next, applying KCL and Ohm's law to the Norton equivalent in Figure 3–37(b) yields its i–v relationship at terminals A–B:

$$i = i_N - \frac{v}{R_N} \qquad (3\text{–}24)$$

Solving Eq. (3–24) for v yields

$$v = i_N R_N - iR_N \qquad (3\text{–}25)$$

The Thévenin and Norton circuits have identical i–v relationships. Comparing Eqs. (3–23) and (3–25), we conclude that

$$R_N = R_T$$

$$i_N R_N = v_T \qquad (3\text{–}26)$$

In essence, the Thévenin and Norton equivalent circuits are related by the source transformation studied in Chapter 2. We do not need to find both equivalent circuits. Once one of them is found, the other can be determined by a source transformation. The Thévenin and Norton circuits involve four parameters (v_T, R_T, i_N, R_N) and Eq. (3–26) provides two relations between the four parameters. Therefore, only two parameters are needed to specify either equivalent circuit.

In circuit analysis problems it is convenient to use the short-circuit current and open-circuit voltage to specify Thévenin and Norton circuits. The circuits in Figure 3–38(a) show that when the load is an open circuit the interface voltage equals the Thévenin voltage; that is, $v_{OC} = v_T$, since there is no voltage across R_T when $i = 0$. Similarly, the circuits in Figure 3–38(b) show that when the load is a short circuit the interface current equals the Norton current; that is, $i_{SC} = i_N$, since all the source current i_N is diverted through the short-circuit load.

In summary, the parameters of the Thévenin and Norton equivalent circuits at a given interface can be found by determining the open-circuit voltage and the short-circuit current.

$$v_T = v_{OC}$$
$$i_N = i_{SC}$$
$$R_N = R_T = v_{OC}/i_{SC}$$

(3–27)

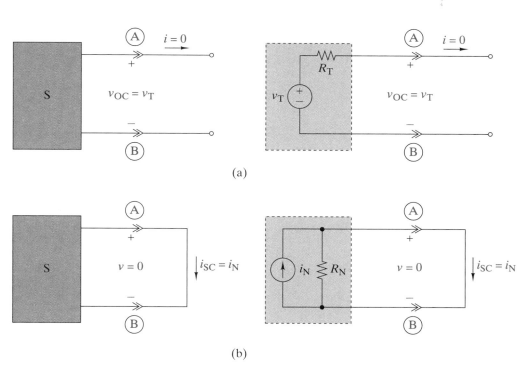

(a)

(b)

FIGURE 3 – 3 8 *Loads used to find Thévenin and Norton equivalent circuits: (a) Open circuit yields the Thévenin voltage. (b) Short-circuit yields the Norton current.*

APPLICATIONS OF THÉVENIN AND NORTON EQUIVALENT CIRCUITS

Replacing a complex circuit by its Thévenin or Norton equivalent can greatly simplify the analysis and design of interface circuits. For example, suppose we need to select a load resistance in Figure 3–39(a) so the source circuit to the left of the interface A–B delivers 4 volts to the load. This task is easily handled once we have the Thévenin or Norton equivalent for the source circuit.

To obtain the Thévenin and Norton equivalents, we need v_{OC} and i_{SC}. The open-circuit voltage v_{OC} is found by disconnecting the load at the terminals A and B, as shown in Figure 3–39(b). The voltage across the 15-Ω resistor is zero because the open circuit causes the current through the resistor to be zero. The open-circuit voltage at the interface is the same as the voltage across the 10-Ω resistor. Using voltage division, this voltage is

$$v_{T} = v_{OC} = \frac{10}{10 + 5} \times 15 = 10 \text{ V}$$

(a)

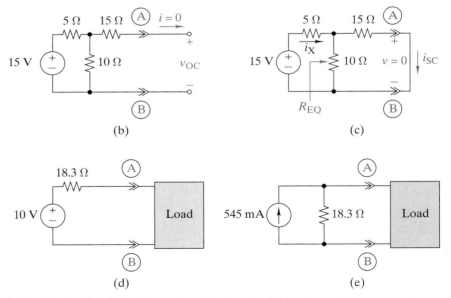

(b)   (c)

(d)   (e)

F I G U R E   3 – 3 9 *Example of finding the Thévenin and Norton equivalent circuits: (a) The given circuit. (b) Open circuit yields the Thévenin voltage. (c) Short circuit yields the Norton current. (d) Thévenin equivalent circuit. (e) Norton equivalent circuit.*

Next we find the short-circuit current i_{SC} using the circuit in Figure 3–39(c). The total current i_X delivered by the 15-V voltage source is

$$i_X = 15/R_{EQ}$$

where R_{EQ} is the equivalent resistance seen by the voltage source with a short circuit at the interface.

$$R_{EQ} = 5 + \cfrac{1}{\cfrac{1}{10} + \cfrac{1}{15}} = 11 \ \Omega$$

We find $i_X = 15/11 = 1.36$ A. Given i_X, we now use current division to obtain the short-circuit current:

$$i_N = i_{SC} = \frac{10}{10 + 15} \times i_X = 0.545 \text{ A}$$

Finally, we compute the Thévenin and Norton resistances:

$$R_T = R_N = \frac{v_{OC}}{i_{SC}} = 18.3 \ \Omega$$

The resulting Thévenin and Norton equivalent circuits are shown in Figures 3–39(d) and 3–39(e).

It now is an easy matter to select a load R_L so 4 V is supplied to the load. Using the Thévenin equivalent circuit, the problem reduces to a voltage divider:

$$\frac{R_L}{R_L + R_T} \times v_T = \frac{R_L}{R_L + 18.3} \times 10 = 4 \text{ V}$$

Solving for R_L yields $R_L = 12.2 \ \Omega$.

The Thévenin or Norton equivalent can always be found from the open-circuit voltage and short-circuit current at the interface. The following examples illustrate other methods of determining these equivalent circuits.

EXAMPLE 3-13

(a) Find the Thévenin equivalent circuit of the source circuit to the left of the interface in Figure 3–40(a).
(b) Use the Thévenin equivalent to find the power delivered to two different loads. The first load is a 10-kΩ resistor and the second is a 5-V voltage source whose positive terminal is connected to the upper interface terminal.

SOLUTION:

(a) To find the Thévenin equivalent, we use the sequence of circuit reductions shown in Figure 3–40. In Figure 3–40(a) the 15-V voltage source in series with the 3-kΩ to the left of terminals A and B is replaced by a 3-kΩ resistor in parallel with an equivalent current source with $i_S = 15/3000 = 5$ mA. In Figure 3–40(b), looking to the left at terminals C and D, we see two resistors in parallel whose equivalent resistance is $(3 \text{ k}\Omega)\|(6 \text{ k}\Omega) = 2 \text{ k}\Omega$. We also see two current sources in parallel whose equivalent is $i_S = 5$ mA $- 2$ mA $= 3$ mA. This equivalent current source is shown in Figure 3–40(c) to the left of terminals C and D. Figure 3–40(d) shows this current source converted to an equivalent voltage source $v_S = (3$ ma$) \times (2$ k$\Omega) = 6$ V in series with 2 kΩ. In Figure 3–40(d) the three resistors are connected in series and can be replaced by an equivalent resistance $R_{EQ} = 2$ k$\Omega + 3$ k$\Omega + 4$ k$\Omega = 9$ kΩ. This step produces the Thévenin equivalent shown in Figure 3–40(e).

Note: The steps leading from Figure 3–40(a) to 3–40(e) involve circuit reduction techniques studied in Chapter 2, so we know that this approach only works on ladder circuits like the one in Figure 3–40(a).

FIGURE 3 – 40

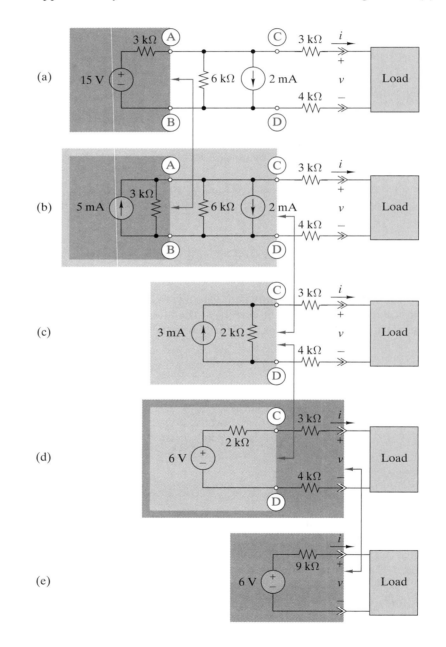

(b) Figure 3–41 shows the Thévenin equivalent found in (a) and the two loads. When the load is a 10-kΩ resistor, the interface current is i = (6)/(9000 + 10,000) = 0.3158 mA, and the power delivered to the load is $i^2 R_L$ = 0.9973 mW. When the load is a 5-V source, the interface voltage and current are v = 5 V and i = (6 − 5)/9000 = 0.1111 mA, and the power to the load is vi = 0.5555 mW. Since $p > 0$ in the latter case, we see that the voltage source load is absorbing rather than delivering power. A practical example of this situation is a battery charger.

Caution: The Thévenin equivalent allows us to calculate the power delivered to a load, but it does not tell us what power is dissipated in the source circuit. For instance, if the load in Figure 3–40(e) is an open circuit, then no power is dissipated in the Thévenin equivalent since i = 0. This does not mean that the power dissipated in the source circuit is zero, as we can easily see by looking back at Figure 3–40(a). The Thévenin equivalent circuit has the same i–v characteristic at the interface, but it does not duplicate the internal characteristics of the source circuit. ■

FIGURE 3 – 41

EXAMPLE 3–14

(a) Find the Norton equivalent of the source circuit to the left of the interface in Figure 3–42.
(b) Find the interface current i when the power delivered to the load is 5 W.

FIGURE 3 – 42

SOLUTION:

(a) The circuit reduction method will not work here since the source circuit is not a ladder. In this example we write mesh-current equations and solve directly for the source circuit i–v relationship. We only need to write equations for meshes A and B since the 2-A current source determines the mesh C current. The voltages sums around these meshes are

$$\text{Mesh A:} \quad -40 + 60(i_A - i_C) + 180(i_A - i_B) = 0$$

$$\text{Mesh B:} \quad -180(i_A - i_B) + 15(i_B - i_C) + v = 0$$

But since $i_B = i$ and the current source forces the condition $i_C = -2$, these equations have the form

$$240i_A - 180i = -80$$

$$-180i_A + 195i = -30 - v$$

Solving for i in terms of v yields

$$i = \frac{\begin{vmatrix} 240 & -80 \\ -180 & -30-v \end{vmatrix}}{\begin{vmatrix} 240 & -180 \\ -180 & 195 \end{vmatrix}} = \frac{-21600 - 240v}{14400}$$

$$= -1.5 - \frac{v}{60}$$

At the interface the i–v relationship of the source circuit is $i = -1.5 - v/60$. Equation (3–24) gives the i–v relationship of the Norton circuit as $i = i_N - v/R_N$. By direct comparison, we conclude that $i_N = -1.5$ A and $R_N = 60\ \Omega$. This equivalent circuit is shown in Figure 3–43.

(b) When 5 W is delivered to the load, we have $vi = 5$ or $v = 5/i$. Substituting $v = 5/i$ into the source i–v relationship $i = -1.5 - v/60$ yields a quadratic equation

$$12i^2 + 18i + 1 = 0$$

whose roots are $i = -0.05778$ A and -1.442 A. Thus, there are two values of interface current that deliver 5 W to the load. ∎

F I G U R E   3 – 4 3

DERIVATION OF THÉVENIN'S THEOREM

The derivation of Thévenin's theorem is based on the superposition principle. We begin with the circuit in Figure 3–44(a), where the source circuit S is linear. Our approach is to use superposition to show that the source circuit and the Thévenin circuit have the same i–v relationship at the interface. To find the source circuit i–v relationship, we first disconnect the load and apply a current source i_{TEST}, as shown in Figure 3–44(b). Using superposition to find v_{TEST}, we first turn i_{TEST} off and leave all the sources inside S on, as shown in Figure 3–44(c). Turning a current source off leaves an open circuit, so

$$v_{TEST1} = v_{OC}$$

Next we turn i_{TEST} back on and turn off all of the sources inside S. Since the source circuit S is linear, it reduces to the equivalent resistance shown in Figure 3–44(d) when all internal sources are turned off. Using Ohm's law, we write

$$v_{TEST2} = (R_{EQ})(-i_{TEST})$$

The minus sign in this equation results from the reference directions originally assigned to i_{TEST} and v_{TEST} in Figure 3–44(b). Using the superposition principle, we find the i–v relationship of the source circuit at the interface to be

$$v_{TEST} = v_{TEST1} + v_{TEST2}$$
$$= v_{OC} - R_{EQ}i_{TEST}$$

This equation has the same form as the i–v relationship of the Thévenin equivalent circuit in Eq. (3–23) when $v_{TEST} = v$, $i_{TEST} = i$, $v_{OC} = v_T$, and $R_T = R_{EQ}$.

The derivation points out another method of finding the Thévenin resis-

(a)

(b)

(c)

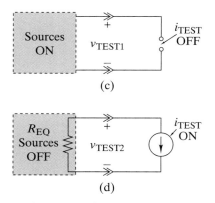

(d)

F I G U R E   3 – 4 4   *Using superposition to prove Thévenin's theorem.*

tance. As indicated in Figure 3–44(d), when all the sources are turned off, the i–v relationship of the source circuit reduces to $v = -iR_{EQ}$. Similarly, the i–v relationship of a Thévenin equivalent circuit reduces to $v = -iR_T$ when $v_T = 0$. We conclude that

$$R_T = R_{EQ} \qquad (3\text{–}28)$$

We can find the value of R_T by determining the resistance seen looking back into the source circuit with all sources turned off. For this reason the Thévenin resistance R_T is sometimes called the **lookback resistance**.

The next example shows how lookback resistance contributes to finding a Thévenin equivalent circuit.

EXAMPLE 3–15

(a) Find the Thévenin equivalent of the source circuit to the left of the interface in Figure 3–45.
(b) Use the Thévenin equivalent to find the voltage delivered to the load.

FIGURE 3–45

SOLUTION:
(a) The source circuit in Figure 3–45 is treated in Example 3–12 by using superposition to calculate the open-circuit voltage between terminals A and B. Using the results from Example 3–12, we have

$$v_T = v_{OC} = \frac{1}{3}(v_{S1} + v_{S2} + v_{S3})$$

Turning all sources off in the Figure 3–45 leads to the resistance circuit in Figure 3–46. Looking back into the source circuit in Figure 3–46, we see three equal resistances connected in parallel whose equivalent resistance is $R/3$. Hence, the Thévenin resistance is

$$R_T = R_{EQ} = \frac{R}{3}$$

(b) Given the Thévenin circuit parameters v_T and R_T, we apply voltage division in Figure 3–45 to find the interface voltage.

FIGURE 3 – 46

$$v = \frac{R_L}{R_L + R_T} v_T = \left(\frac{R_L}{R_L + R/3}\right)\left(\frac{v_{S1} + v_{S2} + v_{S3}}{3}\right)$$

$$= \left(\frac{R_L}{3R_L + R}\right)(v_{S1} + v_{S2} + v_{S3})$$

The interface voltage is proportional to the sum of the three source voltages. The proportionality constant $K = R_L/(3R_L + R)$ depends on both the source and the load since these two circuits are connected at the interface. ∎

Exercise 3–16

(a) Find the Thévenin and Norton equivalent circuits seen by the load in Figure 3–47.
(b) Find the voltage, current, and power delivered to a 50-Ω load resistor.

FIGURE 3 – 47

Answers:
(a) $v_T = -30$ V, $i_N = -417$ mA, $R_N = R_T = 72$ Ω
(b) $v = -12.3$ V, $i = -246$ mA, $p = 3.03$ W

Exercise 3–17

Find the current and power delivered to an unknown load in Figure 3–47 when $v = +6$ V.

Answers:

$$i = -\tfrac{1}{2}\text{A}, \, p = -3 \text{ W}$$

APPLICATION TO NONLINEAR LOADS

Thévenin and Norton equivalent circuits can be used to find the response of a two-terminal nonlinear element (NLE). The method of analysis is a straightforward application of device and interface i–v characteristics. An interface is defined at the terminals of the nonlinear element, and the linear part of the circuit is reduced to the Thévenin equivalent in Figure 3–48(a). The i–v relationship of the Thévenin equivalent can be written with interface current as the dependent variable:

$$ i = \left(-\frac{1}{R_T}\right)v + \left(\frac{v_T}{R_T}\right) \tag{3–29} $$

This is the equation of a straight line in the i–v plane shown in Figure 3–48(b). The line intersects the i-axis ($v = 0$) at $i = v_T/R_T = i_{SC}$ and inter-

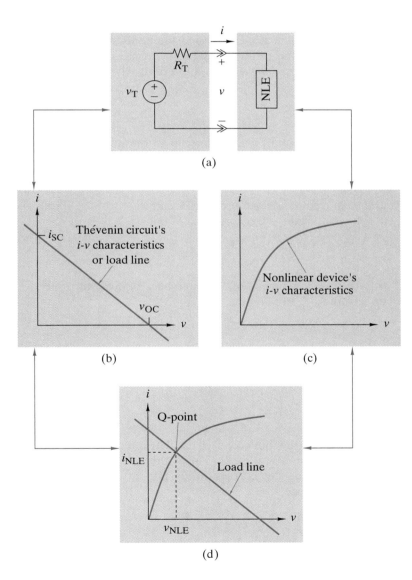

(a)

(b)   (c)

(d)

FIGURE 3 – 4 8 *Graphical analysis of a nonlinear circuit: (a) Given circuit. (b) Load line. (c) Nonlinear device i–v characteristics. (d) Q-point.*

sects the v–axis ($i = 0$) at $v = v_T = v_{OC}$. This line could logically be called the source line since it is determined by the Thévenin parameters of the source circuit. Logic notwithstanding, electrical engineers call this the **load line** for reasons that have blurred with the passage of time.

The nonlinear element has the i–v characteristic shown in Figure 3–48(c). Mathematically, this nonlinear characteristic has the form

$$i = f(v) \qquad (3\text{–}30)$$

To find the circuit response, we must solve Eqs. (3–29) and (3–30) simultaneously. Computer software tools like Mathcad can easily solve this problem when a numerical expression for the function $f(v)$ is known explicitly. However, in practice an approximate graphical solution is often adequate, particularly when $f(v)$ is only given in graphical form.

In Figure 3–48(d) we superimpose the load line on the i–v characteristic curve of the nonlinear element. The two curves intersect at the point $i = i_{NLE}$ and $v = v_{NLE}$, which yields the values of interface variables that satisfy both the source constraints in Eq. (3–29) and the nonlinear element constraints in Eq. (3–30). In the terminology of electronics, the point of intersection is called the operating point or **Q-point**, where Q stands for quiescent.

EXAMPLE 3–16

Find the voltage, current, and power delivered to the diode in Figure 3–49(a).

SOLUTION:

We first find the Thévenin equivalent of the circuit to left of the terminals A and B. By voltage division, the open-circuit voltage is

$$v_T = v_{OC} = \frac{100}{100 + 100} \times 5 = 2.5 \text{ V}$$

When the voltage source is turned off, the lookback equivalent resistance seen between terminals A and B is

$$R_T = 10 + 100 \| 100 = 60 \ \Omega$$

The source circuit load line is given by

$$i = -\frac{1}{60}v + \frac{1}{60} \times 2.5$$

This line intersects the i-axis ($v = 0$) at $i = i_{SC} = 2.5/60 = 41.7$ mA and intersects the v–axis ($i = 0$) at $v = v_{OC} = 2.5$ V. Figure 3–49(b) superimposes the source circuit load line on the diode's i–v curve. The intersection (Q-point) is at $i = i_D = 15$ mA and $v = v_D = 1.6$ V. This is the point (i_D, v_D) at which both the source and diode device constraints are satisfied. Finally, the power delivered to the diode is

$$p_D = i_D v_D = (15 \times 10^{-3})(1.6) = 24 \text{ mW}$$

Because of the nonlinear element, the proportionality and superposition properties do not apply to this circuit. For instance, if the source voltage in

FIGURE 3–49

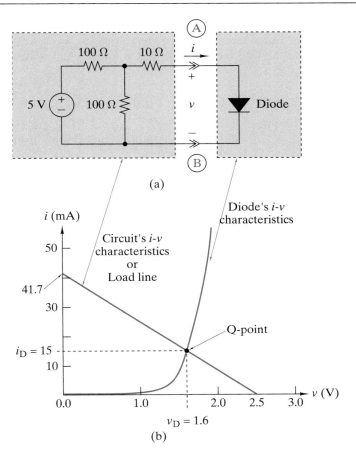

(a)

(b)

Figure 3–49 is decreased from 5 V to 2.5 V, the diode current and voltage do not decrease by one-half. Try it.   ■

Exercise 3–18

Find the voltage, current, and power delivered to the diode in Figure 3–49(a) when the 10-Ω resistor is replaced by a short circuit.

Answer:
$$v_D = 1.7 \text{ V}, \, i_D = 18 \text{ mA, and } p_D = 30.6 \text{ mW}$$

In summary, any two of the following parameters determine the Thévenin or Norton equivalent circuit at a specified interface:

- The open-circuit voltage at the interface
- The short-circuit current at the interface
- The source circuit lookback resistance.

Alternatively, for ladder circuits the Thévenin or Norton equivalent circuit can be found by a sequence of circuit reductions (see Example 3–13). For general circuits they can always be found by directly solving for the i–v relationship of the source circuit using node-voltage or mesh-current equations that include the interface current and voltage as unknowns (see Example 3–14).

3–5 MAXIMUM SIGNAL TRANSFER

An interface is a connection between two circuits at which the signal levels may be observed or specified. In this regard an important consideration is the maximum signal levels that can be transferred across a given interface. This section defines the maximum voltage, current, and power available at an interface between a *fixed source* and an *adjustable load*.

For simplicity we will treat the case in which both the source and load are linear resistance circuits. The source can be represented by a Thévenin equivalent and the load by an equivalent resistance R_L, as shown in Figure 3–50. For a fixed source, the parameters v_T and R_T are given and the interface signal levels are functions of the load resistance R_L.

By voltage division, the interface voltage is

$$v = \frac{R_L}{R_L + R_T} v_T \qquad (3\text{–}31)$$

For a fixed source and a variable load, the voltage will be a maximum if R_L is made very large compared with R_T. Ideally, R_L should be made infinite (an open circuit), in which case

$$v_{\text{MAX}} = v_T = v_{\text{OC}} \qquad (3\text{–}32)$$

Therefore, the maximum voltage available at the interface is the source open-circuit voltage v_{OC}.

The current delivered at the interface is

$$i = \frac{v_T}{R_L + R_T} \qquad (3\text{–}33)$$

For a fixed source and a variable load, the current will be a maximum if R_L is made very small compared with R_T. Ideally, R_L should be zero (a short circuit), in which case

$$i_{\text{MAX}} = \frac{v_T}{R_T} = i_N = i_{\text{SC}} \qquad (3\text{–}34)$$

FIGURE   3 – 5 0 *Two-terminal interface for deriving the maximum signal transfer conditions.*

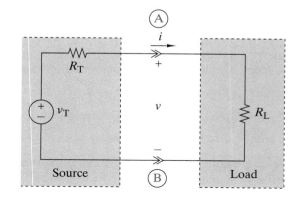

Therefore, the maximum current available at the interface is the source short-circuit current i_{SC}.

The power delivered at the interface is equal to the product $v \times i$. Using Eqs. (3-31) and (3-33), the power is

$$p = v \times i$$

$$= \frac{R_L v_T^2}{(R_L + R_T)^2} \tag{3-35}$$

For a given source, the parameters v_T and R_T are fixed and the delivered power is a function of a single variable R_L. The condition for maximum voltage ($R_L \to \infty$) and the condition for maximum current ($R_L = 0$) both produce zero power. The value of R_L that maximizes the power lies somewhere between these two extremes. To find this value, we differentiate Eq. (3-35) with respect to R_L and solve for the value of R_L for which $dp/dR_L = 0$.

$$\frac{dp}{dR_L} = \frac{[(R_L + R_T)^2 - 2R_L(R_L + R_T)]v_T^2}{(R_L + R_T)^4} = \frac{R_T - R_L}{(R_L + R_T)^3} v_T^2 = 0 \tag{3-36}$$

Clearly, the derivative is zero when $R_L = R_T$. Therefore, **maximum power transfer** occurs when the load resistance equals the Thévenin resistance of the source. When the condition $R_L = R_T$ exists, the source and load are said to be **matched**.

Substituting the condition $R_L = R_T$ back into Eq. (3-35) shows the maximum power to be

$$p_{MAX} = \frac{v_T^2}{4R_T} \tag{3-37}$$

Since $v_T = i_N R_T$, this result can also be written as

$$p_{MAX} = \frac{i_N^2 R_T}{4} \tag{3-38}$$

or

$$p_{MAX} = \frac{v_T i_N}{4} = \left[\frac{v_{OC}}{2}\right]\left[\frac{i_{SC}}{2}\right] \tag{3-39}$$

These equations are consequences of what is known as the **maximum power transfer theorem**:

> *A source with a fixed Thévenin resistance R_T delivers maximum power to an adjustable load R_L when $R_L = R_T$.*

To summarize, at an interface with a fixed source,

1. The maximum available voltage is the open-circuit voltage.

2. The maximum available current is the short-circuit current.

3. The maximum available power is the product of one-half the open-circuit voltage times one-half the short-circuit current.

Figure 3-51 shows plots of the interface voltage, current, and power as functions of R_L/R_T. The plots of v/v_{OC}, i/i_{SC}, and p/p_{MAX} are normalized to

FIGURE 3 – 5 1 *Normalized plots of current, voltage, and power versus R_L/R_T.*

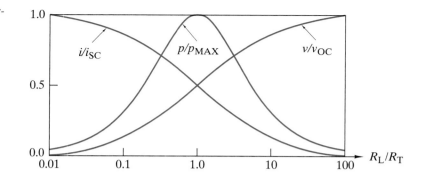

the maximum available signal levels, so the ordinates in Figure 3–51 range from 0 to 1. The plot of the normalized power p/p_{MAX} in the neighborhood of the maximum is not a particularly strong function of R_L/R_T. Changing the ratio R_L/R_T by a factor of 2 in either direction from the maximum reduces p/p_{MAX} by less than 20%. The normalized voltage v/v_{OC} is within 20% of its maximum when $R_L/R_T = 4$. Similarly, the normalized current is within 20% of its maximum when $R_L/R_T = \frac{1}{4}$. In other words, for engineering purposes we can get close to the maximum signal levels with load resistances that only approximate the theoretical requirements.

EXAMPLE 3–17

A source circuit with $v_T = 2.5$ V and $R_T = 60\ \Omega$ drives a load with $R_L = 30\ \Omega$.

(a) Determine the maximum signal levels available from the source circuit.
(b) Determine the actual signal levels delivered to the load.

SOLUTION:
(a) The maximum available voltage and current are

$$v_{MAX} = v_{OC} = v_T = 2.5\text{ V } (R_L \rightarrow \infty)$$

$$i_{MAX} = i_{SC} = \frac{v_T}{R_T} = 41.7\text{ mA } (R_L = 0)$$

The maximum available power is found using Eq. (3–39).

$$p_{MAX} = \left[\frac{v_{OC}}{2}\right]\left[\frac{i_{SC}}{2}\right] = 26.0\text{ mW } (R_L = R_T = 60\ \Omega)$$

(b) The actual signal levels delivered to the 30-Ω load are

$$v_L = \frac{30}{30 + 60}\,2.5 = 0.833\text{ V}$$

$$i_L = \frac{2.5}{30 + 60} = 27.8\text{ mA}$$

$$p_L = v_L i_L = 23.1\text{ mW}$$

Although these levels are less than the maximum available values, the power delivered to the 30-Ω load is nearly 90% of the maximum.   ■

Exercise 3–19

A resistive source circuit delivers 4 V when a 50-Ω resistor is connected across its output and 5 V when a 75-Ω resistor is connected. Find the maximum voltage, current, and power available from the source.

A n s w e r s :

$$10 \text{ V}, 133 \text{ mA}, \text{ and } 333 \text{ mW}$$

Remember that the maximum signal levels just derived are for a fixed source resistance and an adjustable load resistance. This situation often occurs in communication systems where devices such as antennas, transmitters, and signal generators have fixed source resistances such as 50, 75, 300, or 600 ohms. In such cases the load resistance is selected to achieve the desired interface conditions, which often involves matching.

Matching source and load applies when the load resistance R_L in Figure 3–50 is adjustable and the Thévenin source resistance R_T is fixed. When R_L is fixed and R_T is adjustable, then Eqs. (3–31), (3–33), and (3–35) point out that the maximum voltage, current, and power are delivered when the Thévenin source resistance is zero. If the source circuit at an interface is adjustable, then ideally the Thévenin source resistance should be zero. In the next chapter we will see that OP AMP circuits approach this ideal.

3–6   INTERFACE CIRCUIT DESIGN

The maximum signal levels discussed in the previous section place bounds on what is achievable at an interface. However, those bounds are based on a fixed source and an adjustable load. In practice, there are circumstances in which the source or the load, or both, can be adjusted to produce prescribed interface signal levels. Sometimes it is necessary to insert a circuit between the source and the load to achieve the desired results. Figure 3–52 shows the general situations and some examples of resistive interface circuits. By its nature, the inserted circuit has two terminal pairs, or interfaces, at which voltage and current can be observed or specified. These terminal pairs are also called *ports*, and the interface circuit is referred to as a **two-port network**. The port connected to the source is called the input, and the port connected to the load is called the output. The purpose of this two-port network is to make certain that the source and load interact in a prescribed way.

BASIC CIRCUIT DESIGN CONCEPTS

Before we treat examples of different interface situations, you should recognize that we are now discussing a limited form of circuit design, as contrasted with circuit analysis. Although we use circuit analysis tools in design, there are important differences. A linear circuit analysis problem generally has a unique solution. A circuit design problem may have many solutions or even no solution. The maximum available signal levels found in the preceding section provide bounds that help us test for the existence of a solution. Generally there will be several ways to meet the interface

F I G U R E 3 – 5 2 *A general interface circuit and some examples.*

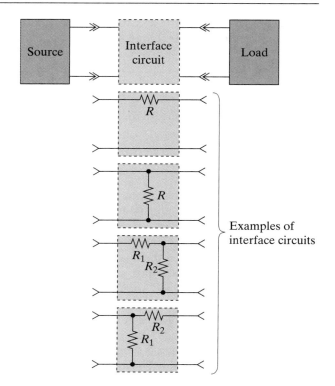

constraints, and it then becomes necessary to evaluate the alternatives using other factors, such as cost, power consumption, or reliability.

At this point in our study, resistors are the only elements we can use to design interface circuits. In subsequent chapters we will introduce other devices, such as OP AMPs (Chapter 4) and capacitors and inductors (Chapter 6). In a design situation the engineer must choose the resistance values in a proposed circuit. This decision is influenced by a host of practical considerations, such as standard values and tolerances, power ratings, temperature sensitivity, cost, and fabrication methods. Some of these issues are discussed in Appendix A. We will occasionally introduce some of these considerations into our design examples. Gaining a full understanding of these practical matters is not one of our en route objectives. Rather, our goal is simply to illustrate how different constraints can influence the design process.

D E S I G N E X A M P L E 3–18

Select the load resistance in Figure 3–53 so that the interface signals are in the range defined by $v \geq 4$ V and $i \geq 30$ mA.

SOLUTION:

In this design problem, the source circuit is given and we are free to select the load. For a fixed source the maximum signal levels available at the interface are

$$v_{\text{MAX}} = v_{\text{T}} = 10 \text{ V}$$

$$i_{\text{MAX}} = \frac{v_{\text{T}}}{R_{\text{T}}} = 100 \text{ mA}$$

The bounds given as design requirements are below the maximum available signal levels, so we should be able to find a suitable resistor. Using voltage division, the interface voltage constraint requires

$$\frac{R_{\text{L}}}{100 + R_{\text{L}}} \times 10 \geqslant 4$$

or

$$10R_{\text{L}} \geqslant 4R_{\text{L}} + 400$$

This condition yields $R_{\text{L}} \geq 400/6 = 66.7 \ \Omega$. The interface current constraint can be written as

$$\frac{10}{100 + R_{\text{L}}} \geqslant 0.03$$

or

$$10 \geqslant 3 + 0.03R_{\text{L}}$$

which requires $R_{\text{L}} \leq 7/0.03 = 233 \ \Omega$. In theory, any value of R_{L} between 66.7 Ω and 233 Ω will work. However, to allow for parameter variations we select $R_{\text{L}} = 150 \ \Omega$ because it lies at the arithmetic midpoint of the allowable range and is a standard values of resistance (see Table A–1, Appendix A). ∎

D DESIGN EXAMPLE 3–19

Select the load resistor in Figure 3–53 so that the voltage delivered to the load is 3 V ±10%. Use only the standard resistance value given in Appendix A for ±10% tolerance.

SOLUTION:
The ±10% voltage tolerance means that the interface voltage must fall between 2.7 V and 3.3 V. The required range can be achieved since the open-circuit voltage of the source circuit is 10 V. Using voltage division, we can write the constraints on the value of R_{L}:

$$2.7 \leqslant \frac{R_{\text{L}}}{100 + R_{\text{L}}} \times 10 \leqslant 3.3$$

The left inequality requires

$$270 + 2.7R_{\text{L}} \leqslant 10R_{\text{L}}$$

or $R_{\text{L}} \geq 270/7.3 = 37.0 \ \Omega$. The right inequality requires

$$10R_{\text{L}} \leqslant 330 + 3.3R_{\text{L}}$$

or $R_{\text{L}} \leq 330/6.7 = 49.2 \ \Omega$. Thus, R_{L} must lie in the range from 37.0 Ω to 49.2 Ω, which has a midpoint at 43.1 Ω.

Commercial resistors are not available in infinitely many values. To limit inventory costs, the electronic industry has agreed on a finite set of standard resistance values. For resistors with ±10% tolerance the standard values are

10 12 15 18 22 27 33 39 47 56 68 82 Ω

and multiples of 10 times these values. The standard values of 39 Ω and 47 Ω fall within the required 37.0 to 49.2 Ω range. However, these nominal values and the ±10% tolerance could carry the actual resistance value outside the range. Various series and parallel combinations of standard ±10% resistors produce values that fall in the desired range even with 10% tolerance. For example, two 22-Ω resistors connected in series produce a nominal equivalent resistance of 44 Ω with a ±10% range from 39.6 Ω to 48.4 Ω. Finally, it turns out that 43 Ω is a standard value for resistors with ±5% tolerance (see Appendix A, Table A–1). This option is attractive since 43 Ω falls almost exactly at the midpoint of the desired resistance range and the 5% tolerance easily meets the interface voltage range requirement.

In summary, either a standard 39-Ω resistor or a standard 47-Ω resistor produces nominal designs that meet the design requirements, but the 10% tolerance on the resistance could cause the voltage to fall outside the specified range. Series and parallel combinations of two 10% resistors meet all requirements but lead to a more complicated and costly design. A single 43-Ω 5% resistor meets all requirements, but its tighter tolerance could mean higher parts costs than two 10% resistors. The final design decision must take into account many factors, particularly the sensitivity of the system to changes in the specified interface voltage. ■

DESIGN EXAMPLE 3–20

The resistance array package shown in Figure 3–54 includes seven nearly identical resistors with nominal resistances of 1 kΩ. Use this resistance array to design a voltage divider that reduces the output of a 10-V dc power supply to 4.3 V.

FIGURE 3–54

$R = 1 \text{ k}\Omega$

SOLUTION:
The required interface voltage is well within the capability of a 10-V source. Using voltage division, we can express the design requirement as

$$\frac{R_2}{R_1 + R_2} \times 10 = 4.3$$

where R_1 and R_2 are the series and shunt legs of the divider, respectively. This design constraint can be rearranged as

$$\frac{R_1}{R_2} = 1.33$$

In other words, the design requirement imposes a constraint on the ratio of R_1 and R_2.

Resistance array packages are particularly useful when the design constraints involve resistance ratios. The nominal resistance can change from one array to the next, but within a given array the resistances are all nearly equal, so their ratios are nearly unity. In effect, the resistance array in Figure 3–54 contains seven identical resistors, although we do not know the exact value of their resistance. Constructing the series leg using three resistors in parallel and the shunt leg using four resistors in parallel produces R_1 = $R/3$ and R_2 = $R/4$, where R is nominally 1 kΩ. But regardless of the actual value of R, the ratio $R_1/R_2 = 1.33$ as required.

The final design shown in Figure 3–55 may seem wasteful of resistors. However, in large-volume production an integrated circuit resistance array can cost less than a discrete resistor design. Resistance array packages are better suited to automated manufacturing than discrete resistors. ■

FIGURE 3 – 5 5

D ◖ DESIGN EXAMPLE 3–21 ◗

Design the two-port interface circuit in Figure 3–56 so the 10-A source delivers 100 V to the 50-Ω load.

FIGURE 3 – 5 6

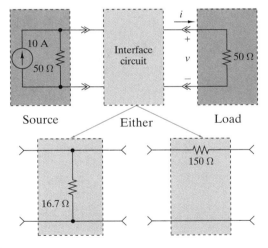

SOLUTION:

The problem requires that the current delivered to the load be $i = 100/50 = 2$ A, which is well within the maximum value of current available from the source. In fact, if the 10-A source is connected directly to the load, the source current divides equally between two 50-Ω resistances, producing 5 A through the load. Therefore, an interface circuit is needed to reduce the load current to the specified 2-A level. Two possible design solutions are shown in Figure 3–56. Applying current division to the parallel resistor case yields the following constraint:

$$i = \frac{1/50}{1/50 + 1/50 + 1/R_{\text{PAR}}} \times 10$$

After inserting the $i = 2$ A design requirement, this equation becomes

$$2 = \frac{10}{2 + 50/R_{\text{PAR}}}$$

Solving for R_{PAR} yields

$$R_{\text{PAR}} = \frac{50}{3} = 16.7 \ \Omega$$

Applying the two-path current division rule to the series resistor case yields the following constraint:

$$i = \frac{50}{50 + (50 + R_{\text{SER}})} \times 10 = 2 \text{ A}$$

Solving for R_{SER} yields

$$R_{\text{SER}} = 150 \ \Omega$$

We have two designs that meet the basic $i = 2$ A requirement. In practice, engineers evaluate alternative designs using additional criteria. One such consideration is the required power ratings of the resistors in each design. The voltage across the parallel resistor is $v = 100$ V, so the power loss is

$$p_{\text{PAR}} = \frac{100^2}{50/3} = 600 \text{ W}$$

The current through the series resistor interface is $i = 2$ A, so the power loss is

$$p_{\text{SER}} = 2^2 \times 150 = 600 \text{ W}$$

In either design the resistors must have a power rating of at least 600 W. Other factors besides power rating determine which design should be selected. ■

D **DESIGN EXAMPLE 3–22**

Figure 3–57 is a simplified circuit for a digital inverter. The basic performance requirement is that the output be low when the input is high and vice versa. In this example the switch closes when $v_{\text{S}} > 2.0$ V ($R_{\text{ON}} = 200$ Ω) and opens when $v_{\text{S}} < 0.8$ V ($R_{\text{OFF}} = 10^{11}$ Ω). Stated in terms of voltages, the design requirements are (1) $v_{\text{O}} < 0.5$ V when $v_{\text{S}} > 2.0$ V and (2) $v_{\text{O}} > 3.5$ V when $v_{\text{S}} < 0.8$ V. In addition, the current through the switch must be less than 1 mA and the total power dissipated in the circuit less than 2.5 mW. For $V_{\text{DD}} = 5$ V, select the value of R_{D} to meet all of these requirements.

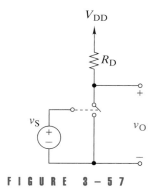

FIGURE 3–57

SOLUTION:

When $v_{\text{S}} > 2.0$ V the switch is closed and the output voltage is found by voltage division as

$$v_{\text{O}} = \frac{R_{\text{ON}}}{R_{\text{ON}} + R_{\text{D}}} \times V_{\text{DD}} = \frac{200}{200 + R_{\text{D}}} \times 5 < 0.5 \text{ V}$$

which yields the requirement $R_D > 1800 \ \Omega$. The switch current is a maximum when the switch is closed, hence

$$I_{SW} = \frac{V_{DD}}{R_{ON} + R_D} = \frac{5}{200 + R_D} < 1 \text{ mA}$$

which yields the requirement $R_D > 4800 \ \Omega$. Likewise, the maximum power is dissipated in the circuit when the switch is closed, hence

$$p_D = \frac{V_{DD}^2}{R_{ON} + R_D} = \frac{25}{200 + R_D} < 2.5 \text{ mW}$$

which yields the requirement $R_D > 9800 \ \Omega$. Clearly the power dissipation requirement places the greatest lower bound on R_D. When $v_S < 0.8$ V the switch is open and the output voltage is found by voltage division as

$$v_O = \frac{R_{OFF}}{R_{OFF} + R_D} \times V_{DD} = \frac{10^{11}}{10^{11} + R_D} \times 5 > 3.5 \text{ V}$$

which yields the requirement $R_D < 4.28 \times 10^{11} \ \Omega$. Thus, all requirements are met for

$$9800 < R_D < 4.28 \times 10^{11} \ \Omega$$

Any value of R_D in this very wide range would meet the design constraints given here. In later chapters we will see that the speed of response improves as we move toward the lower end of this range. ∎

Exercise 3–20

Find the lower bound R_D in Example 3–22 when the allowable power dissipation is 5 mW.

Answer: $4800 \ \Omega$

▶ DESIGN EXAMPLE 3–23

Design the two-port interface circuit in Figure 3–58 so the load "sees" a Thévenin resistance of 50 Ω between terminals C and D, while simultaneously the source "sees" a load resistance of 300 Ω between A and B.

SOLUTION:

We first try a single resistor in the interface circuit. A 60-Ω parallel resistor in the interface circuit would make the load see 50 Ω Thévenin resistance, but then the source would see a load resistance much less than 300 Ω. A 250-Ω series resistor in the interface circuit would make the source see a 300-Ω load, but then the load would see a Thévenin resistance much greater than 50 Ω.

We next try an interface circuit containing two resistors. Since the load must see a smaller resistance than the source, it should "look" into a parallel resistor. On the other hand, since the source must see a larger resistance

FIGURE 3 – 5 8

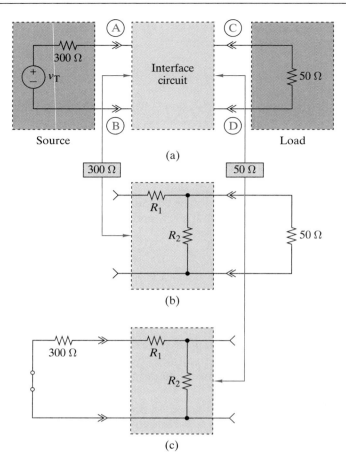

(a)

(b)

(c)

than the load, it should look into a series resistor. A configuration that meets these conditions is the L-circuit shown in Figures 3–58(b) and 3–58(c).

The preceding discussion can be summarized mathematically. Using the L-circuit, the design requirement at terminals C and D is

$$\frac{(R_1 + 300)\, R_2}{R_1 + 300 + R_2} = 50\ \Omega$$

At terminals A and B, the requirement is

$$R_1 + \frac{50\, R_2}{R_2 + 50} = 300\ \Omega$$

The design requirements yield two equations in two unknowns; what could be simpler? It turns out that solving these nonlinear equations by hand analysis is a bit of a chore. These equations can easily be solved using a math solver, as we will shortly demonstrate. However, at this point in your development we encourage you to think about the problem in physical terms.

Given the L-circuits in Figure 3–58(b), one approach goes as follows. If we let $R_2 = 50$ ohms, then the requirement at terminals C and D will be met, at least approximately. Similarly, if $R_1 + R_2 = 300\ \Omega$, the requirements at terminals A and B will be approximately satisfied. In other words, try

$R_1 = 250 \ \Omega$ and $R_2 = 50 \ \Omega$ as a first cut. These values yield equivalent resistances of $R_{CD} = 50\|550 = 45.8 \ \Omega$ and $R_{AB} = 250 + 50\|50 = 275 \ \Omega$. These equivalent resistances are not the exact values specified, but are within ± 10%. Since the tolerance on electrical components may be at least this high, a design using these values could be adequate.

Figure 3–59 shows how Mathcad solves the two design equations to find the exact values of R_1 and R_2. The approximate values derived previously are used as the first estimates required by a Mathcad solve block. Mathcad uses these estimates in an iterative process to solve the design equations listed between the Given and the Find statements. The final results at the bottom in Figure 3–59 show that the approximate values derived from physical reasoning are within 10% of the exact values derived using Mathcad.

In this example we used physical reasoning to find an approximation solution of a problem. Using physical reasoning to find practical solutions is what engineering is all about. ■

FIGURE   3 – 5 9

APPLICATION NOTE: EXAMPLE 3–24

An **attenuation pad** is a two-port resistance circuit that provides a non-adjustable reduction in signal level and provides resistance matching at the input and output ports. An example of an attenuation pad is shown in Figure 3–60. The manufacturer's data sheet states that the pad in Figure 3–60 has the following characteristics:

CHARACTERISTICS	PORT	CONDITION	VALUE	UNITS
Thévenin voltage	Output	600-Ω source connected at the input port	$v_S/4$	V
Thévenin resistance	Output	600-Ω source connected at the input port	600	Ω
Input resistance	Input	600-Ω load connected at the output port	600	Ω

Verify these characteristics.

FIGURE 3 - 60

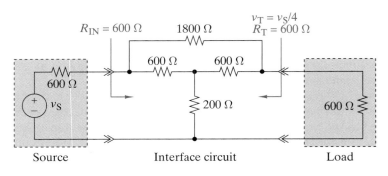

Source   Interface circuit   Load

SOLUTION:

Figure 3–61 shows a PSpice circuit diagram that can be used to calculate the two-port parameters of the attenuation pad. A 600-Ω source is connected at the input port (node A). The 600-Ω load is not connected since we want to find the Thévenin equivalent circuit seen looking back into the output port (node C). Under the **Analysis** menu we select **Analysis Set Up** and activate the **Transfer Function** option. Selecting the Transfer Function option produces the menu at the bottom left in Figure 3–61. Here we identify the output variable as V(C)—the voltage at node C. Note that V(C) is the open-circuit voltage at the output port. We also identify the input as the voltage source V1. With these selections PSpice will calculate three things: (1) the transfer function V(C)/V1, (2) the input resistance seen by V1, and (3) the output resistance at node C.

FIGURE 3 - 61

Returning to the **Analysis** menu, we select the **Simulate** command. PSpice performs a dc analysis of the circuit and writes the results to an output file. Included in the output file are the small signal characteristics shown at the bottom right in Figure 3–61. PSpice reports that the transfer function V(C)/V1 is 2.500E–01, the input resistance is 1.280E+03, and the output resistance is 6.000E+02. That is, the open-circuit (Thévenin) voltage at the output port is exactly one-fourth of the source voltage and the output (Thévenin) resistance is 600 Ω, as advertised. The input resistance (1280 Ω) given in the output file is the resistance seen by the source V1 when the output port is an open circuit. The manufacturer specifies the input resistance looking in at node A when a 600-Ω load is connected at the output port. The input port resistance is 600 Ω since the output resistance is 600 Ω and the two-port circuit is symmetrical. That is, the input resistance looking in at node A with a 600-Ω load connected is the same as the output resistance looking in at node C with the source V1 turned off.

In summary, the pad reduces the signal voltage by one-fourth and has 600-Ω input and output resistances, as stated by the manufacturer. ∎

D DESIGN EXERCISE: EXERCISE 3–21

Using only standard ±5% resistors, design the two-port interface circuit in Figure 3–62 to meet the following requirements:

(a) The power delivered to the 50-Ω load is 0.2 W ±5%.
(b) Repeat (a) for $p = 0.6$ W ±5%.

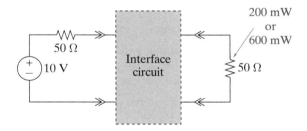

FIGURE 3–62

Answer: There are no unique answers to this problem. Some possibilities are discussed next.

DISCUSSION: *(a) A standard 56-Ω resistor in series (ideally 58.1 Ω) yields 200 mW ±5% into the 50-Ω load. (b) No resistive interface will work since the maximum power available from the source is only 0.5 W. Delivering more power than the signal source can provide requires an active device, such as an OP AMP or transistor (treated in the next chapter).*

SUMMARY

- Node-voltage analysis involves identifying a reference node and the node to datum voltages at the remaining $N - 1$ nodes. The KCL connection constraints at the $N - 1$ nonreference nodes combined with the element constraints written in terms of the node voltages produce $N - 1$ linear equations in the unknown node voltages.

- Mesh-current analysis involves identifying mesh currents that circulate around the perimeter of each mesh in a planar circuit. The KVL connection constraints around $E - N + 1$ meshes combined with the element constraints written in terms of the mesh currents produce $E - N + 1$ linear equations in the unknown mesh currents.

- Node and mesh analysis can be modified to handle both types of independent sources using a combination of three methods: (1) source transformations, (2) selecting circuit variables so independent sources specify the values of some of the unknowns, and (3) using supernodes or supermeshes.

- A circuit is linear if it contains only linear elements and independent sources. For single-input linear circuits, the proportionality property states that any output is proportional to the input. For multiple-input linear circuits, the superposition principle states that any output can be found by summing the output produced when each input acts alone.

- A Thévenin equivalent circuit consists of a voltage source in series with a resistance. A Norton equivalent circuit consists of a current source in parallel with a resistance. The Thévenin and Norton equivalent circuits are related by a source transformation.

- The parameters of the Thévenin and Norton equivalent circuits can be determined using any two of the following: (1) the open-circuit voltage at the interface, (2) the short-circuit current at the interface, and (3) the equivalent resistance of the source circuit with all sources turned off.

- The parameters of the Thévenin and Norton equivalent circuits can also be determined using circuit reduction methods or by directly solving for the source $i–v$ relationship using node-voltage or mesh-current analysis.

- For a fixed source and an adjustable load, the maximum interface signal levels are $v_{MAX} = v_{OC}$ ($R_L = \infty$), $i_{MAX} = i_{SC}$ ($R_L = 0$), and $p_{MAX} = v_{OC}i_{SC}/4$ ($R_L = R_T$). When $R_L = R_T$, the source and load are said to be matched.

- Interface signal transfer conditions are specified in terms of the voltage, current, or power delivered to the load. The design constraints depend on the signal conditions specified and the circuit parameters that are adjustable. Some design requirements may require a two-port interface circuit. An interface design problem may have one, many, or no solutions.

PROBLEMS

ERO 3–1 GENERAL CIRCUIT ANALYSIS (SECTS. 3–1 TO 3–2)

Given a circuit consisting of linear resistors and independent sources,

(a) (Formulation) Write node-voltage or mesh-current equations for the circuit.
(b) (Solution) Solve the equations from (a) for selected signal variables or input–output relationships.

3–1   **(a)** Formulate a set of node-voltage equations for the circuit in Figure P3–1.

       (b) Use the node-voltage equations to find v_x and i_x.

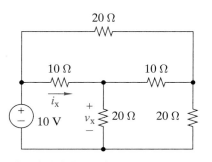

F I G U R E   P 3 – 1

3–2 (a) Formulate a set of mesh-current equations for the circuit in Figure P3–1.
(b) Use the mesh-current equations to find v_x and i_x.

3–3 (a) Formulate a set of node-voltage equations for the circuit in Figure P3–3.
(b) Use the node-voltage equations to find v_x and i_x.

3–4 (a) Formulate a set of mesh-current equations for the circuit in Figure P3–3.
(b) Use the mesh-current equations to find v_x and i_x.

3–5 (a) Formulate node-voltage equations for the circuit in Figure P3–5.
(b) Solve for v_x and i_x when $v_S = 5V$, $i_S = 1$ mA, $R_1 = R_2 = 10$ kΩ, and $R_3 = R_4 = 5$ kΩ.

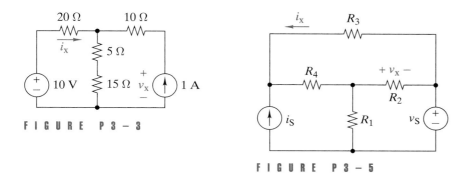

FIGURE P 3 – 3

FIGURE P 3 – 5

3–6 (a) Formulate mesh-current equations for the circuit in Figure P3–5.
(b) Solve for v_x and i_x when $v_S = 25$ V, $i_S = 10$ A, $R_1 = R_2 = 10$ Ω, and $R_3 = R_4 = 15$ Ω.

3–7 (a) Formulate node-voltage equations for the circuit in Figure P3–7.
(b) Solve for v_x and i_x for $R_1 = R_2 = R_3 = R_4 = 1$ kΩ, $R_5 = R_6 = 2$ kΩ, and $i_{S1} = i_{S2} = 100$ mA.
(c) Find the total power delivered to the circuit by the two current sources.

3–8 (a) Formulate node-voltage equations for the circuit in Figure P3–8.
(b) Solve for v_x and i_x using $R_1 = 10$ kΩ, $R_2 = 20$ kΩ, $R_3 = 30$ kΩ, $R_4 = 40$ kΩ, $R_x = 3$ kΩ, and $v_S = 10$ V.
(c) Find the power absorbed by resistor R_2.

FIGURE P 3 – 7

FIGURE P 3 – 8

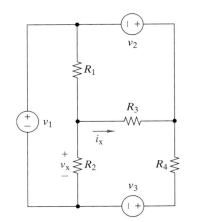

3–9   **(a)** Formulate node-voltage equations for the circuit in Figure P3–9.
  (b) Solve for v_x and i_x using $R_1 = 10$ kΩ, $R_2 = 10$ kΩ, $R_3 = 40$ kΩ, $R_4 = 20$ kΩ, and $v_1 = v_2 = v_3 = 15$ V.
  (c) Find the power delivered to resistor R_1.

3–10   **(a)** Formulate node-voltage equations for the circuit in Figure P3–10.
  (b) Solve for v_x and i_x using $R_1 = 10$ Ω, $R_2 = 20$ Ω, $R_3 = 20$ Ω, $R_4 = 10$ Ω, $R_5 = 5$ Ω, $R_x = 30$ Ω, $i_S = 2$ A, and $v_S = 20$ V.
  (c) Find the power delivered by the voltage source.

FIGURE   P 3 – 1 0

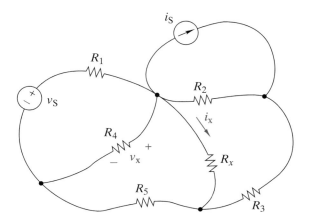

3–11   **(a)** Formulate mesh-current equations for the circuit in Figure P3–10.
  (b) Solve for v_x and i_x using $R_1 = 10$ Ω, $R_2 = 20$ Ω, $R_3 = 10$ Ω, $R_4 = 20$ Ω, $R_5 = 5$ Ω, $R_x = 30$ Ω, $i_S = 2$ A and $v_S = 10$ V.
  (c) Find the power delivered by the voltage source.

3–12   **(a)** Formulate mesh-current equations for the circuit in Figure P3–12.
  (b) Solve for v_x and i_x using $R_1 = 200$ Ω, $R_2 = 500$ Ω, $R_3 = 60$ Ω, $R_4 = 240$ Ω, $R_5 = 200$ Ω, $i_S = 40$ mA, and $v_S = 12$ V.
  (c) Find the total power dissipated in the circuit.

3–13   **(a)** Formulate mesh-current equations for the circuit in Figure P3–13.
  (b) Solve for v_x and i_x using $R_1 = R_2 = 10$ kΩ, $R_3 = 2$ kΩ, $R_4 = 1$ kΩ, $i_S = 2.5$ mA, $v_{S1} = 12$ V, and $v_{S2} = 0.5$ V.
  (c) Find the power supplied by v_{S1}.

3–14   The circuit in Figure P3–14 seems to require two supermeshes because both current sources appear in two meshes. However, a circuit diagram can sometimes be rearranged to eliminate the need for supermesh equations.
  (a) Show that supermeshes in Figure P3–14 can be avoided by connecting resistor R_6 between node A and node D via a different route.
  (b) Formulate mesh-current equations for the modified circuit as redrawn in (a).

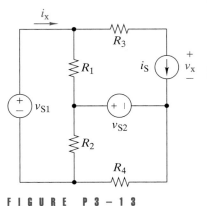

(c) Solve for v_x using $R_1 = R_2 = R_3 = R_4 = 2$ kΩ, $R_5 = R_6 = 1$ kΩ, $i_{S1} = 40$ mA, and $i_{S2} = 20$ mA.

FIGURE   P 3 – 1 4

3–15 Figure P3–15 shows a resistive circuit in which the elements E_1 and E_2 are not yet specified. To define an analysis problem we must specify these elements and state the response to be found. Write the node-voltage or mesh-current equations that offer the best approach to solving the following problems. Do not solve the equations.
- **(a)** Find the voltage across R_L when element E_1 is an open circuit and element E_2 is a resistor R_3.
- **(b)** Find the power delivered by v_S when element E_1 is a resistor R_3 and element E_2 is an independent current source i_S.

3–16 Figure P3–16 shows a resistive circuit in which the elements E_1 and E_2 are not yet specified. To define an analysis problem we must specify these elements and state the response to be found. Write the node-voltage or mesh-current equations that offer the best approach to solving the following problems. Do not solve the equations.
- **(a)** Find the voltage across R_3 when element E_1 is a short circuit and element E_2 is a resistor R_6.
- **(b)** Find the current through R_3 when element E_1 is an open circuit and element E_2 is an independent current source i_S.

3–17 A circuit analysis problem involves a given circuit with known inputs and unknown responses. A design problem involves a mixture in which some of the circuit parameters and responses are given and the unknowns are the unspecified circuit parameters and responses. Write node-voltage or mesh-current equations for the circuit in Figure P3–17 and use these equations to find the value of R that makes $v_O = 5$ V. Then find the power delivered by the 15-V source for this value of R.

3–18 A circuit analysis problem involves a given circuit with known inputs and unknown responses. A design problem involves a mixture in which some of the circuit parameters and responses are given and the unknowns are the unspecified circuit parameters and responses. Write node-voltage or mesh-current equations for the circuit in Figure P3–18 and use these equations to find the value of R that makes $v_O = (v_1 + v_2)/4$.

3–19 A linear resistive circuit has four nodes labeled A, B, C, and D. When node D is selected as the reference, the node-voltage equations for the circuit are

FIGURE   P 3 – 1 5

FIGURE   P 3 – 1 6

FIGURE   P 3 – 1 7

FIGURE   P 3 – 1 8

$$\text{Node A:} \quad 3v_A - v_B \qquad = 2$$
$$\text{Node B:} \quad -v_A + 5v_B - 2v_C = 0$$
$$\text{Node C:} \qquad\qquad -2v_B + 3v_C = 0$$

(a) Draw a diagram of a circuit described by these three node-voltage equations.
(b) Write the mesh-current equations of the circuit found in (a).
(c) How would you modify the node-voltage and mesh-current equations if node C is shorted to ground?

3–20 The mesh-current equations for a linear resistive circuit are

$$\text{Mesh A:} \quad 30i_A - 10i_B \qquad = 15$$
$$\text{Mesh B:} \quad -10i_A + 50i_B - 20i_C = 0$$
$$\text{Mesh C:} \qquad\qquad -20i_B + 30i_C = -10$$

(a) Draw a diagram of a circuit described by these three mesh-current equations.
(b) Write the node-voltage equations for the circuit found in (a).
(c) How would you modify the mesh-current and node-voltage equations if the resistor common to meshes A and B is replaced by an open circuit?

ERO 3–2 LINEARITY PROPERTIES (SECT. 3–3)

(a) Given a circuit containing linear resistors and one independent source, use the proportionality principle to find selected signal variables.
(b) Given a circuit containing linear resistors and two or more independent sources, use the superposition principle to find selected signal variables.

3–21 Find the proportionality constant $K = v_O/i_S$ for the circuit in Figure P3–21.

3–22 Find the proportionality constant $K = i_O/i_S$ for the circuit in Figure P3–22.

3–23 Find the proportionality constant $K = v_O/i_S$ for the circuit in Figure P3–23.

F I G U R E   P 3 – 2 1

F I G U R E   P 3 – 2 2

F I G U R E   P 3 – 2 3

3–24 Use the unit output method to find v_O in the circuit in Figure P3–24.

F I G U R E   P 3 – 2 4

3–25 Use the unit output method to find i_O in the circuit in Figure P3–25.

3–26 Use the unit output method to select R in the circuit in Figure P3–26 so that the proportionality constant $K = v_O/v_S = 1/3$. Find the largest possible value of K for any value of R.

3–27 Input/output test data for a circuit are as follows:

$$\{v_{IN}(V), v_{OUT}(V)\} = \{-2, 3\}, \{-1, 0\}, \{3, 20\}, \{5, 30\}$$

What is the input-output relationship? Is this a linear relationship?

3–28 Use the superposition principle in the circuit of Figure P3–28 to find v_O.

3–29 Use the superposition principle in the circuit of Figure P3–29 to find i_O.

FIGURE P3-30

FIGURE P3-31

3–30 Use the superposition principle in the circuit of Figure P3–30 to find the output v_O in terms of v_1, v_2, and R.

3–31 Use the superposition principle in the circuit of Figure P3–31 to find the output v_O in terms of i_1, i_2, and R.

3–32 A linear circuit containing two sources drives a 50-Ω load resistor. Source number 1 delivers 500 mW to the load when source number 2 is off. Source number 2 delivers 2 W to the load when source number 1 is off. Find the power delivered to the load when both sources are on. *Hint:* The answer is not 2.5 W. Why?

3–33 A linear circuit is driven by an independent voltage source $v_S = 5$ V and an independent current source $i_S = 10$ mA. When the voltage source is on and the current source off, the output voltage is $v_O = 3$ V. When both sources are on, the output is $v_O = 2$ V. Find the output when $v_S = 10$ V and $i_S = -10$ mA.

3–34 The following table lists test data of the output of a linear resistive circuit for different values of its three inputs. Find the input-output relationship for the circuit.

$v_{S1}(V)$	$v_{S2}(V)$	$v_{S3}(V)$	$v_O(V)$
0	4	−6	0
2	0	−3	1.5
2	4	0	2

3–35 **D** This problem involves designing a resistive circuit with two inputs v_{S1} and v_{S2} and a single output voltage v_O. Design the circuit so that $v_O = K(v_{S1} + 3v_{S2})$ is delivered across a 500-Ω load. The value of K is not specified but should be greater than 1/20.

ERO 3–3 THÉVENIN AND NORTON EQUIVALENT CIRCUITS (SECT. 3–4)

Given a circuit containing linear resistors and independent sources,

(a) Find the Thévenin or Norton equivalent at a specified pair of terminals.
(b) Use the Thévenin or Norton equivalent to find the signals delivered to linear or nonlinear loads.

3–36 (a) Find the Thévenin or Norton equivalent circuit seen by R_L in the circuit in Figure P3–36.
 (b) Use the equivalent circuit found in (a) to find i_L in terms of i_S, R_1, R_2, and R_L.
 (c) Check your answer in (b) using current division.

3–37 (a) Find the Thévenin or Norton equivalent seen by R_L in the circuit in Figure P3–37.
 (b) Use the equivalent circuit found in (a) to find load power when $R_L = 50$ Ω, 100 Ω, and 500 Ω.

3–38 (a) Find the Thévenin or Norton equivalent seen by R_L in the circuit in Figure P3–38.
 (b) Use the equivalent circuit found in (a) to find load power when $R_L = 10$ kΩ.
 (c) Repeat (b) for $R_L = 20$ kΩ, 40 kΩ, and 80 kΩ.
 (d) Which value of R_L draws the most output power?

FIGURE P3-36

FIGURE P3-37

3–39 **(a)** Find the Thévenin or Norton equivalent seen by R_L in the circuit in Figure P3–39.
(b) Use the equivalent circuit found in (a) to find i_O for $R_L = 5\ k\Omega$.
(c) Repeat (b) for $R_L = 10\ k\Omega$, $20\ k\Omega$, and $40\ k\Omega$.
(d) Which load draws the most output current?

3–40 **(a)** Find the Thévenin or Norton equivalent at terminals A and B in the circuit in Figure P3–40.
(b) Use the equivalent circuit to find interface power when a 10-Ω load is connected between terminals A and B.
(c) Repeat (b) when a 5-V source is connected between terminals A and B with the plus terminal at terminal A.

3–41 **(a)** Find the Thévenin or Norton equivalent seen to the left of terminals A and B in Figure P3–41. (*Hint:* Use source transformations and circuit reduction.)
(b) Use the equivalent found in (a) to find the output v_O in terms of v_1, v_2, and R.

3–42 The open-circuit voltage of a two-terminal source is 8 V. When a 3-kΩ resistor is connected across the source, the output drops to 6 V.
(a) Find the Thévenin equivalent of the source.
(b) Use the equivalent circuit to find the power delivered to loads of 500 Ω, 1 kΩ, and 2 kΩ.
(c) Which load draws the most output power?

3–43 When a 100-Ω resistor is connected across a two-terminal source the output voltage is 10 V. When a 200-Ω resistor is connected the output voltage is 12 V. How much current would this source deliver to a 6-V battery?

3–44 Some measurements of the i–v characteristic of a two-terminal source are:

v (V)	−10	−5	0	+5	+10	12	13	14
i (mA)	+5	+4	+3	+2	+1	0	−1	−2

(a) Plot the source i–v characteristic using these data.
(b) Develop a Thévenin equivalent circuit valid on the range $|\,v\,| < 10$ V.
(c) Use the equivalent circuit to predict v_{OC} and i_{SC}.

FIGURE P3-46

FIGURE P3-47

FIGURE P3-48

(d) Compare your results in (c) with the given measurements and explain any differences.

3-45 When the logic output of a digital device is high (logic one) the Thévenin equivalent at output terminals is $v_T = 5$ V and $R_T = 150$ Ω. The device drives standard loads connected in parallel across the output terminals. When the logic output is high each load draws 2.2 mA. To maintain noise immunity, the output-high voltage must not drop below 3.5 V under load. How many standard loads can the device drive?

3-46 The two dc power supplies in Figure P3–46 operate in parallel to supply a load R_L.
   (a) Find the smallest value of R_L for which $v_L \geq 5.5$ V.
   (b) Find the power delivered to the load when $v_L = 5.5$ V.
   (c) Find the power delivered by each source when $v_L = 5.5$ V.

3-47 **D** Select values for V_{CC} and R so the circuit in Figure P3–47 approximates (±2.5%) the i–v characteristic of the 25-mA current source for any load resistance from zero to 500 Ω.

3-48 In the Wheatstone bridge in Figure P3–48, the lower right leg is slightly out of balance by a small amount ΔR.
   (a) Show that for $\Delta R << R$, the Thévenin equivalent seen between terminals A and B is approximately $v_T = (\Delta R/R)(V_{CC}/4)$ and $R_T = R$.
   (b) An ammeter with $R_M = 5$ Ω is connected between terminals A and B. Calculate the meter reading for $V_{CC} = 10$ V, $R = 5$ kΩ, and $\Delta R/R = 2\%$.

3-49 A nonlinear resistor is connected across a two-terminal source whose Thévenin equivalent is $v_T = 5$ V and $R_T = 500$ Ω.
   (a) Plot the i–v characteristic of the source in the first quadrant ($i \geq 0$, $v \geq 0$).
   (b) The i–v characteristic of the resistor is $i = 10^{-4}(v + 2v^{3.3})$. Plot this characteristic on the source plot obtained in (a) and graphically determine the voltage across and current through the nonlinear resistor.

3-50 A nonlinear resistor is connected across a two-terminal source whose Thévenin equivalent is $v_T = 10$ V and $R_T = 200$ Ω.
   (a) Plot the i–v characteristic of the source in the first quadrant ($i \geq 0$, $v \geq 0$).
   (b) The i–v characteristic of the resistor is $v = 4000\,i^2$. Plot this characteristic on the source plot obtained in (a) and graphically determine the voltage across and current through the nonlinear resistor.

ERO 3-4 MAXIMUM SIGNAL TRANSFER (SECT. 3-5)

Given a circuit containing linear resistors and independent sources,

(a) Find the maximum voltage, current, and power available at a specified pair of terminals.

(b) Find the resistive loads required to obtain the maximum available signal levels.

3-51 **(a)** The load resistance in Figure P3–51 is adjusted until the maximum power is delivered. Find the power delivered and the value of R_L.
   (b) The load resistance is adjusted until maximum voltage is delivered. Find the voltage delivered and the value of R_L.

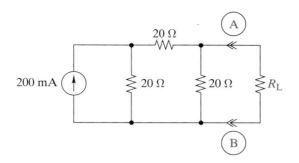

FIGURE P3−51

(c) The load resistance is adjusted until maximum current is delivered. Find the current delivered and the value of R_L.

3–52 **(a)** Find the maximum power available to the load resistance in Figure P3–52. What value of R_L will extract maximum power?

(b) Find the maximum voltage available across the load resistance in Figure P3–52. What value of R_L will produce the maximum voltage?

(c) Find the maximum available current through the load resistance in Figure P3–52. What value of R_L will produce this maximum?

3–53 The circuit in Figure P3–53 has been analyzed, and it is reported that the power delivered to the load is 900 mW. Is this value correct? If so, what is the value of R_L? If not, what is the maximum power available to the load resistance R_L?

3–54 The resistance R in Figure P3–54 is adjusted until maximum power is delivered across the interface to the load consisting of R and the 6-kΩ resistor in parallel. Find the voltage and power delivered to the load and the value of R.

FIGURE P3−52

FIGURE P3−53

FIGURE P3−54

3–55 A current of 0.75 mA is delivered when a 10-kΩ resistor is connected across a two-terminal source. A total current of 1 mA is delivered when a second 10-kΩ resistor is connected in parallel with the first. Find the maximum power available from the source and specify the load resistance required to extract maximum power from the source.

3–56 **(a)** The potentiometer in Figure P3–56 is adjusted until maximum power is delivered to the 2-kΩ load. Find the wiper position x. *Caution:* The load is fixed.

(b) The potentiometer in Figure P3–56 is adjusted until maximum voltage is delivered to the 2-kΩ load. Find the wiper position x.

FIGURE P3−56

3–57 A stereo speaker can be modeled as an 8-Ω linear resistance. A 75-W stereo amplifier with an output resistance of 16 Ω is connected to the speaker via two wires each 75 meters long. The wire resistance is 0.02 Ω/meter.

(a) Find the amplifier's Thévenin voltage when its volume control is set to deliver 50 W to an 8-Ω load.

(b) Find the maximum power available at the speaker when the amplifier volume control is set to the level in part (a).

(c) Find the power delivered to the speaker when the amplifier volume control is set to the level in part (a).

3–58 A number of identical sources are available, each with $v_T = 5$ V and $R_T = 2$ Ω. A compound source is formed by connecting three such sources in series and then connecting two such series strings in parallel. What is the maximum power available from each source individually? What is the maximum power available from this series/parallel combination of six sources?

3–59 A 15-V source with negligible internal resistance is connected in series with a resistor R_S. Select the value of R_S so that outputs of the series combination are bounded by $i < 50$ mA and $p < 200$ mW for any load resistance.

3–60 The maximum power available from a practical source is 250 W. Find the maximum power available from N identical versions of the source when they are connected in parallel. Repeat for N identical versions connected in series.

ERO 3–5 INTERFACE CIRCUIT DESIGN (SECT. 3–6)

Given the signal transfer objectives at a source-load interface, adjust the circuit parameters or design one or more two-port interface circuits to achieve the specified objectives within stated constraints.

3–61 **D** Figure P3–61 shows a two-port interface circuit connecting source and load circuits. In this problem $v_S = 10$ V, $R_S = 50$ Ω, and the load is a 50-Ω resistor. To avoid damaging the source, its output current must be less than 100 mA. Design a resistive interface circuit so that the voltage delivered to the load is 4 V and the source current is less than 100 mA.

FIGURE P 3 – 6 1

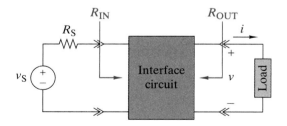

3–62 **D** Figure P3–61 shows a two-port interface circuit connecting source and load circuits. In this problem, $v_S = 10$ V, $R_S = 200$ Ω, and the load is a diode whose i–v characteristic is

$$i = 10^{-15}(e^{40v} - 1)$$

Design an interface circuit that dissipates less than 50 mW and produces an interface voltage of $v = 0.7$ V.

3–63 **D** Figure P3–61 shows a two-port interface circuit connecting source and load circuits. In this problem, $v_S = 15$ V, $R_S = 100$ Ω, and the load is a 1-kΩ resistor. Design a resistive interface circuit so that the voltage delivered to the load is 10 V ±10% using one or more of the following standard resistors: 68 Ω, 100 Ω, 220 Ω, 330 Ω, 470 Ω, and 680 Ω. The resistors all have a tolerance of ± 5%, which you must account for in your design.

3–64 **D** In Figure P3–61 the load is a 500-Ω resistor and $R_S = 75$ Ω. Design an interface circuit so that the input resistance of the two-port is 75 Ω ±10% and the output resistance seen by the load is 500 Ω ±10%.

3–65 **D** Use the resistor array in Figure P3–65 to design a two-port resistance circuit whose voltage gain $K = v_O/v_{IN}$ is at least 0.4 and whose input resistance is as close to 100 Ω as possible.

3–66 **D** Figure P3–61 shows a two-port interface circuit connecting source and load circuits. In this problem $v_S = 12$ V, $R_S = 300$ Ω, and the load is a 50-Ω resistor. Design an interface circuit so that $v < 2$ V and $i > 40$ mA.

3–67 **D** Figure P3–61 shows a two-port interface circuit connecting source and load circuits. In this problem $v_S = 10$ V, $R_S = 50$ Ω, and the load is a 50-Ω resistor. Design an interface so that 200 mW is delivered to the load.

3–68 **D** Figure P3–61 shows a two-port interface circuit connecting a source and load. In this problem the source with $v_S = 5$ V and $R_S = 5$ Ω is to be used in production testing of two-terminal semiconductor devices. The devices are to be connected as the load in Figure P3–61 and have highly nonlinear and variable i–v characteristics. The normal operating range for acceptable devices is {$i > 10$ mA or $v > 0.7$ V} and {$p < 10$ mW}. Design an interface circuit so that the operating point always lies within the specified normal range regardless of the test article's i–v characteristic.

3–69 **D** Figure P3–69 shows an interface circuit with a three-position switch that allows three different signals to be delivered to the load. In position A the interface circuit must deliver 4 V to the 50-Ω load. In position B the interface must deliver 33.3 mA to the load. In position C it must deliver 250 mW to the load. Design the interface circuit using only 100-Ω resistors.

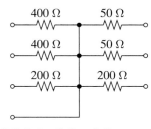

F I G U R E   P 3 – 6 5

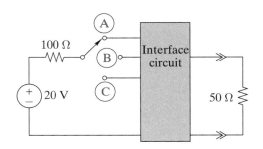

F I G U R E   P 3 – 6 9

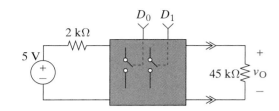

3-70 D The source circuit in Figure P3-70 is $v_S = 5$ V and $R_S = 2$ kΩ and the load is $R_L = 45$ kΩ. The interface circuit contains two identical electronic switches controlled by the digital signals D_0 and D_1. The switches are closed ($R_{ON} = 25$ Ω) when the digital signals are greater than 4 V and open ($R_{OFF} = 500$ MΩ) when they are less than 1 V. Design the interface circuit to produce the following output combinations.

D_1(V)	D_0(V)	v_O(V)
0	0	> 4.5
0	5	< 0.5
5	0	< 0.5
5	5	< 0.5

INTEGRATING PROBLEMS

3-71 A COMPARISON OF ANALYSIS METHODS

The circuit in Figure P3-71 is called an *R-2R* ladder for obvious reasons. The additive property of linear circuits states that the output of the form

$$v_O = k_1 v_1 + k_2 v_2 + k_3 v_3$$

We could use superposition to find the gains k_1, k_2, and k_3. But applying the input sources one at a time may not be the best way to find these gains. The purpose of this problem is to compare two other alternatives.

(a) Write node-voltage equations for the circuit in Figure P3-71 using node B as the reference node. Solve these equations to show that $v_O = v_1/8 + v_2/4 + v_3/2$.

(b) We can also use circuit reduction techniques since the circuit is a ladder. Use successive source transformations and series/parallel equivalence to reduce the circuit between nodes A and B to a single equivalent voltage source connected in series with an equivalent resistor. Use this equivalent circuit to show that $v_O = v_1/8 + v_2/4 + v_3/2$.

(c) Which method do you think is easier and why?

3–72 **A** THREE-TERMINAL DEVICE MODELING

Figure P3–72 shows a three-terminal device with a voltage source v_S is connected at the input and a 50-Ω load resistor connected at the output. The purpose of the problem is to develop an input-output relationship for the device using the following experimental data:

v_S(V)	-12	-9	-6	-3	0	$+3$	$+6$	$+9$	$+12$
v_O(V)	3	2.5	2	1	0	-1	-2	-2	-2

F I G U R E   P 3 – 7 2

The device input resistance is $R_{IN} = 1$ kΩ when the 50-Ω load is connected at the output.

(a) Plot v_O versus v_S and state whether the graph is linear or nonlinear.

(b) Devise a linear model of the form $v_O = Kv_S$ that approximates the data for $|v_S| < 6$ V.

(c) Use your model from part (c) to predict the circuit output for $v_S = \pm10$ V and explain why the predictions do not agree with the experimental results.

(d) For $v_S = 1$ V use your model from part (c) to find the circuit power gain $= P_O/P_{IN}$.

(e) In view of the result in (d), is the device active or passive?

3–73 **A** ADJUSTABLE THÉVENIN EQUIVALENT

Figure P3–73 shows an ideal voltage source in parallel with an adjustable potentiometer. This problem concerns the effect of adjusting the potentiometer on the interface Thévenin equivalent circuit and the signals available at the interface.

(a) Find the parameters v_T and R_T of the Thévenin equivalent circuit at the interface in terms of circuit parameters k, v_S, and R.

(b) Find the value of R_T for $k = 0$, $k = 0.5$, and $k = 1$. Explain your results physically in terms of the position of the movable arm of the potentiometer.

(c) What is the power available at the interface for $k = 0$, $k = 0.5$, and $k = 1$? Justify your answers physically in terms of the position of the movable arm of the potentiometer.

(d) A load resistor $R_L = R/4$ is connected across the interface. Where should the potentiometer be positioned ($k = $?) to transfer the maximum power to this load?

(e) A load resistor $R_L = R/4$ is connected across the interface. Where should the potentiometer be positioned ($k = $?) to deliver an interface voltage $v = v_S/4$ to this load?

F I G U R E   P 3 – 7 3

3–74 ⓓ BATTERY DESIGN

A satellite requires a battery with an open-circuit voltage $v_{OC} = 36$ V and a Thévenin resistance $R_T \leq 10$ Ω. The battery is to be constructed using series and parallel combinations of one of the two types of cells. The first type has $v_{OC} = 9$ V, $R_T = 4$ Ω, and a weight of 40 grams. The second type has $v_{OC} = 4$ V, $R_T = 0.5$ Ω, and a weight of 15 grams. Design a minimum weight battery to meet the open-circuit voltage and Thévenin resistance requirements.

3–75 ⓔ TTL TO ECL CONVERTER

It is claimed that the resistive circuit in Figure P3–75 converts transistor-transistor logic (TTL) input signals into output signals compatible with emitter coupled logic (ECL). Specifically, the claim is that for any output current in the range -0.025 mA $\leq i \leq 0.025$ mA, the circuit converts any input in the TTL low range ($0 \leq v_S \leq 0.4$ V) to an output in the ECL low range (-1.7 V $\leq v < -1.5$ V), and converts any input in the TTL high range ($3.0 \leq v_S \leq 3.8$ V) to an output in the ECL high range (-0.9 V $\leq v \leq -0.6$ V). The purpose of this problem is to verify this claim.

(a) The output voltage can be written in the form

$$v = k_1 v_S + k_2 i + k_3$$

Write a KCL equation at the output with the current i as an unknown and solve for the constants k_1, k_2, and k_3.

(b) Use the relationship found in (a) to verify that output voltage falls in one of the allowed ECL ranges for every allowed combination of TTL inputs and load currents.

FIGURE P3–75

3–76 ⓐ DEGENERATE CIRCUITS

A linear resistive circuit normally has a unique response to a given set of inputs. It is possible to draw circuit diagrams consisting of ideal elements that have no solutions. An example is shown in Figure P3–76. The problem with this circuit is a conflict between the element constraints and the connection constraints. The current sources require that the total current entering node A or B to be 2 A, whereas KCL requires the total to be zero. In reality the response of an insolvable circuit is controlled by parasitic components not included in the models of our ideal circuit elements. These parasitic components resolve the irreconcilable conflict, often in a blinding flash.

(a) Formulate node equations for the circuit and show that these equations have no solution.

(b) Connect a 1-GΩ resistor (almost an open circuit) between node A and ground. Modify the node equations in (a) to account for this element and solve for the two node voltages.

FIGURE   P3–76

3–77   D PHOTODIODE

The circuit symbol of a photodiode is shown in Figure P3–77. When the voltage $v_P > 0$ the diode current is determined by $i_P = \beta L_I$, where L_I is the intensity of the incoming light measured in mW/cm^2 and β is called the photoconductance of the diode. A diode with $\beta = 15$ μA per mW/cm^2 is to be used in an application where the incoming light intensity varies from 0 to 0.1 mW/cm^2. Design a simple photodiode circuit that produces an output voltage of 0 to 5 V for this range of light intensity.

FIGURE   P3–77

CHAPTER 4

ACTIVE CIRCUITS

Then came the morning of Tuesday, August 2, 1927, when the concept of the negative feedback amplifier came to me in a flash while I was crossing the Hudson River on the Lacakawana Ferry, on my way to work.

Harold S. Black, 1927,
American Electrical Engineer

4–1 Linear Dependent Sources

4–2 Analysis of Circuits with Dependent Sources

4–3 The Transistor

4–4 The Operational Amplifier

4–5 OP AMP Circuit Analysis

4–6 OP AMP Circuit Design

4–7 The Comparator

Summary

Problems

Integrating Problems

The integrated circuit operational amplifier (OP AMP) is the workhorse of present-day linear electronic circuits. However, to operate as a linear amplifier the OP AMP must be provided with "negative feedback." The negative feedback amplifier is one of the key inventions of all time. During the 1920s, Harold S. Black (1898–1983) worked for several years without much success on the problem of improving the performance of vacuum tube amplifiers in telephone systems. The feedback amplifier solution came to him suddenly on his way to work. He documented his invention by writing the key concepts of negative feedback on his morning copy of the *New York Times*. His invention paved the way for the development of worldwide communication systems and spawned whole new areas of technology, such as feedback control systems and robotics.

An **active circuit** contains one or more devices that require an external power supply to operate correctly. These active devices control the flow of electrical power from the external supply and can operate in both linear and nonlinear modes. In their linear range active devices can be modeled using the four dependent source elements introduced in the first section. The second section shows how dependent sources change the properties of resistive circuits and how we modify our methods of analysis to cover these new linear elements. Dependent sources and resistors are used to model many active devices, including the transistor and OP AMP devices discussed in the third and fourth sections. The fourth section also develops an ideal OP AMP model that is useful in circuit analysis and design. In the fifth section the ideal OP AMP model is used to analyze circuit realizations of basic analog signal-processing functions, such as amplifiers, summers, subtractors, and inverters. These basic building blocks are then used to design OP AMP circuits that realize specified signal-processing functions. The comparator section treats a case in which the OP AMP intentionally operates as a nonlinear element.

4–1   LINEAR DEPENDENT SOURCES

This chapter treats the analysis and design of circuits containing active devices, such as transistors or operational amplifiers (OP AMPs). An **active device** is a component that requires an external power supply to operate correctly. An **active circuit** is one that contains one or more active devices. An important property of active circuits is that they are capable of providing signal amplification, one of the most important signal-processing functions in electrical engineering. Linear active circuits are governed by the proportionality property, so their input-output relationships are of the form $y = Kx$. The term **signal amplification** means the proportionality factor K is greater than 1 when the input x and output y have the same dimensions. Thus, active circuits can deliver more signal voltage, current, and power at their output than they receive from the input signal. The passive resistance circuits studied thus far cannot produce voltage, current, or power gains greater than unity.

Active devices operating in a linear mode are modeled using resistors and one or more of the dependent sources shown in Figure 4–1. A **dependent source** is a voltage or current source whose output is controlled by a voltage or current in a different part of the circuit. As a result, there are

FIGURE 4–1 *Dependent source circuit symbols: (a) Current-controlled voltage source. (b) Voltage-controlled voltage source. (c) Current-controlled current source. (d) Voltage-controlled current source.*

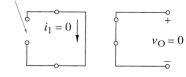

FIGURE 4–2 *Turning off the independent source affects the dependent source.*

four possible types of dependent sources: a current-controlled voltage source (CCVS), a voltage-controlled voltage source (VCVS), a current-controlled current source (CCCS), and a voltage-controlled current source (VCCS). The properties of these dependent sources are very different from those of the independent sources described in Chapter 2. The output voltage (current) of an independent voltage (current) source is a specified value that does not depend on the circuit to which it is connected. To distinguish between the two types of sources, the dependent sources are represented by the diamond symbol in Figure 4–1, in contrast to the circle symbol used for independent sources.

Caution: This book uses the diamond symbol to indicate a dependent source and the circle to show an independent source. However, some books and circuit analysis programs use the circle symbol for both dependent and independent sources.

A **linear dependent source** is one whose output is proportional to the controlling voltage or current. The defining relationship for dependent sources in Figure 4–1 are all of the form $y = Kx$, where x is the controlling variable, y is the source output variable, and K is the proportionality factor. Each type of dependent source is characterized by a proportionality factor, either μ, β, r, or g. These parameters are often called simply the **gain** of the controlled source. Strictly speaking, the parameters μ and β are dimensionless quantities called the **voltage gain** and **current gain**, respectively. The parameter r has the dimensions of ohms and is called the **transresistance**, a contraction of transfer resistance. The parameter g is then called **transconductance** and has the dimensions of siemens.

Although dependent sources are elements used in circuit analysis, they are conceptually different from the other circuit elements we have studied. The linear resistor and ideal switch are models of actual devices called resistors and switches. However, you will not find dependent sources listed in electronic part catalogs. For this reason, dependent sources are more abstract, since they are not models of identifiable physical devices. Dependent sources are used in combination with other resistive elements to create models of active devices.

In Chapter 3 we found that a voltage source acts as a short circuit when it is turned off. Likewise, a current source behaves as an open circuit when it is turned off. The same results apply to dependent sources, with one important difference. Dependent sources cannot be turned on and off individually because they depend on excitation supplied by independent sources.

Some consequences of this dependency are illustrated in Figure 4–2. When the independent current source is turned on, KCL requires that $i_1 = i_S$. Through controlled source action, the current controlled voltage source is on and its output is

$$v_O = ri_1 = ri_S$$

When the independent current source is off ($i_S = 0$), it acts as an open circuit and KCL requires that $i_1 = 0$. The dependent source is now off and its output is

$$v_O = ri_1 = 0$$

When the independent voltage source is off, the dependent source acts as a short circuit.

In other words, turning the independent source on and off turns the dependent source on and off as well. We must be careful when applying the superposition principle and Thévenin's theorem to active circuits, since the state of a dependent source depends on the excitation supplied by independent sources. To account for this possibility, we modify the superposition principle to state that the response due to all *independent* sources acting simultaneously is equal to the sum of the responses due to each *independent* source acting one at a time.

EXAMPLE 4−1

Figure 4–3(a) shows the PSpice circuit symbols for the four dependent sources. These elements are found on the **Analog** symbol library as part names starting with the letters:

     E for a voltage-controlled voltage source

     F for a current-controlled current source

     G for a voltage-controlled current source

     H for a current-controlled voltage source

These letters are the standard first letter in SPICE reference designators for controlled sources. Figure 4–3(b) shows the corresponding Electronics Workbench symbols for the controlled sources. These elements are available in the Sources Group in the Parts Bin toolbar. In either case the input ports are open circuits for the voltage-controlled elements and short circuits for the current-controlled elements. The output ports are voltage sources or current sources depending on the controlled variable. Note that the PSpice controlled sources are indicated by circles rather than diamonds.

The PSpice dependent sources are characterized by a single parameter called GAIN. The units of the PSpice GAIN depend on the units of the signals at the input and output ports. The Electronics Workbench refers to these parameters as voltage gain, current gain, transresistance, and transconductance, respectively, and its screen symbols explicitly show the units of the gain parameters. In both programs the default value is GAIN = 1.

4–2 ANALYSIS OF CIRCUITS WITH DEPENDENT SOURCES

With certain modifications the analysis tools developed for passive circuits apply to active circuits as well. Circuit reduction applies to active circuits, but in so doing we must not eliminate the control variable for a dependent source. As noted previously, when applying the superposition principle or Thévenin's theorem, we must remember that dependent sources cannot be turned on and off independently since their states depend on excitation supplied by one or more independent sources. Applying a source transfor-

FIGURE 4 – 3 *Dependent source representations. (a) PSpice. (b) Electronics Workbench.*

(a)

(b)

mation to a dependent source is sometimes helpful, but again we must not lose the identity of a controlling signal for a dependent source. Methods like node and mesh analysis can be adapted to include dependent sources as well.

However, the main difference is that the properties of active circuits can be significantly different from those of the passive circuits treated in Chapters 2 and 3. Our analysis examples are chosen to highlight these differences.

In our first example the objective is to determine the current, voltage, and power delivered to the 500-Ω output load in Figure 4–4. The control current i_x is found using current division in the input circuit:

$$i_x = \left(\frac{50}{50 + 25}\right) i_S = \frac{2}{3} i_S \qquad (4\text{–}1)$$

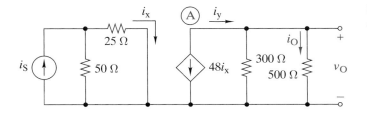

FIGURE 4 – 4 *A circuit with a dependent source.*

Similarly, the output current i_O is found using current division in the output circuit:

$$i_O = \left(\frac{300}{300 + 500}\right) i_y = \frac{3}{8} i_y \qquad (4\text{–}2)$$

At node A, KCL requires that $i_y = -48i_x$. Combining this result with Eqs. (4–1) and (4–2) yields the output current:

$$i_O = \left(\frac{3}{8}\right)(-48)i_x = (-18)\left(\frac{2}{3}i_S\right)$$

$$= -12\, i_S \qquad (4\text{–}3)$$

The output voltage v_O is found using Ohm's law:

$$v_O = i_O\, 500 = -6000\, i_S \qquad (4\text{–}4)$$

The input-output relationships in Eqs. (4–3) and (4–4) are of the form $y = Kx$ with $K < 0$. The proportionality constants are negative because the reference direction for i_O in Figure 4–4 is the opposite of the orientation of the dependent source reference arrow. Active circuits often produce negative values of K, which means that the input and output signals have opposite algebraic signs. Circuits for which $K < 0$ are said to provide **signal inversion**. In the analysis and design of active circuits, it is important to keep track of signal inversions.

Using Eqs. (4–3) and (4–4), the power delivered to the 500-Ω load in Figure 4–4 is

$$p_O = v_O i_O = (-6000\, i_S)(-12\, i_S) = 72{,}000 i_S^2 \qquad (4\text{–}5)$$

The independent source at the input delivers its power to the parallel combination of 50 Ω and 25 Ω. Hence, the input power supplied by the independent source is

$$p_S = (50 \| 25)i_S^2 = \left(\frac{50}{3}\right) i_S^2$$

Given the input power and output power, we find the power gain in the circuit:

$$\text{Power gain} = \frac{p_O}{p_S} = \frac{72{,}000 i_S^2}{(50/3) i_S^2} = 4320$$

A power gain greater than unity means that the circuit delivers more power at its output than it receives from the input source. At first glance this appears to be a violation of energy conservation, until we remember that dependent sources are models of active devices that require an external power supply to operate. Usually the external power supply is not shown in circuit diagrams. When using a dependent source to model an active circuit, we assume that the external supply and the active device itself can handle whatever power is required by the circuit. When designing the actual circuit, the engineer must make certain that the active device and its power supply operate within their power ratings.

Exercise 4–1

Find the output v_O in terms of the input v_S in the circuit in Figure 4–5.

FIGURE 4–5

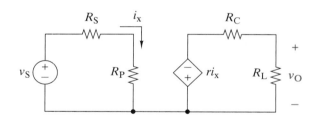

Answer:

$$v_O = \left[\frac{-R_L r}{(R_S + R_P)(R_C + R_L)} \right] v_S$$

NODE-VOLTAGE ANALYSIS WITH DEPENDENT SOURCES

Node analysis of active circuits is much the same as for passive circuits except that we must account for the additional constraints caused by the dependent sources. For example, the circuit in Figure 4–6 has five nodes. With node E as the reference, each independent voltage source has one terminal connected to ground. These connections force the conditions $v_A = v_{S1}$ and $v_B = v_{S2}$. Therefore, we only need to write node equations at nodes C and D because voltages at nodes A and B are already known.

Node analysis involves expressing element currents in terms of the node voltages and applying KCL at each unknown node. The sum of the currents leaving node C is

$$G_1(v_C - v_{S1}) + G_2(v_C - v_{S2}) + G_B v_C + G_P(v_C - v_D) = 0$$

Similarly, the sum of currents leaving node D is

$$G_P(v_D - v_C) + G_E v_D - \beta i_B = 0$$

These two node equations can be rearranged into the form

Node C: $(G_1 + G_2 + G_B + G_P)v_C - G_P v_D = G_1 v_{S1} + G_2 v_{S2}$

Node D: $\qquad - G_P v_C + (G_P + G_E)v_D = \beta i_B$ (4–6)

We could write these two symmetrical node equations by inspection if the dependent current source βi_B had been an independent source. Since it is not independent, we must express its controlling variable i_B in terms of the unknown node voltages. Using the fundamental property of node voltages and Ohm's law, we express the current i_B in terms of the node voltages as

$$i_B = G_P(v_C - v_D)$$

Substituting this expression for i_B into Eq. (4–6) and putting the results in standard form yields

Node C: $\qquad (G_1 + G_2 + G_B + G_P)v_C - G_P v_D = G_1 v_{S1} + G_2 v_{S2}$

Node D: $-(\beta + 1)G_P v_C + [(\beta + 1)G_P + G_E]v_D = 0$ (4–7)

The result in Eq. (4–7) involves two equations in the two unknown node voltages and includes the effect of the dependent source. Note that including the dependent source constraint destroys the coefficient symmetry in Eq. (4–6).

This example illustrates a general approach to writing node-voltage equations for circuits with dependent sources. We start out treating the dependent sources as if they are independent sources and write node equations for the resulting passive circuit using the inspection method developed in Chapter 3. This step produces a set of symmetrical node-voltage equations with the independent and dependent source terms on the right-hand side. Then we express the dependent source terms in terms of the unknown node voltages and move them to the left-hand side of the equations with the other terms involving the unknown node voltages. This step destroys the coefficient symmetry but leads to a set of node-voltage equations that describe the active circuit.

EXAMPLE 4–2

For the circuit in Figure 4–6, use the node-voltage equations in Eq. (4–7) to find the output voltage v_O when $R_1 = 1\ k\Omega$, $R_2 = 3\ k\Omega$, $R_B = 100\ k\Omega$, $R_P = 1.3\ k\Omega$, $R_E = 3.3\ k\Omega$, and $\beta = 50$.

SOLUTION:

Substituting the given numerical values into Eq. (4–7) yields

$$(2.11 \times 10^{-3})v_C - (7.69 \times 10^{-4})v_D = (10^{-3})v_{S1} + (3.33 \times 10^{-4})v_{S2}$$

$$-(3.92 \times 10^{-2})v_C + (3.95 \times 10^{-2})v_D = 0$$

We solve the second equation for $v_C = 1.008v_D$. When this equation is substituted into the first equation, we obtain

$$v_O = v_D = 0.736v_{S1} + 0.245v_{S2}$$

This circuit is a signal summer that does not involve a signal inversion. The fact that the output is a linear combination of the two inputs reminds us that the circuit is linear. ∎

EXAMPLE 4–3

The circuit in Figure 4–7(a) is a model of an inverting OP AMP circuit.

FIGURE   4 – 7

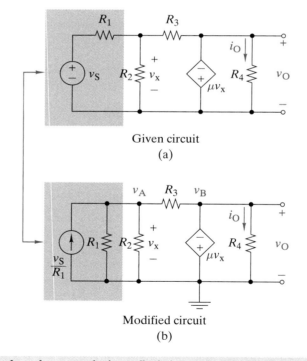

Given circuit
(a)

Modified circuit
(b)

(a) Use node-voltage analysis to find the output v_O in terms of the input v_S.

(b) Evaluate the input-output relationship found in (a) as the gain μ becomes very large.

SOLUTION:

(a) Applying a source transformation to the independent source leads to the modified three-node circuit shown in Figure 4–7(b). With the indicated reference node the dependent voltage source constrains the voltage at node B. The control voltage is $v_x = v_A$, and the controlled source forces the node B voltage to be

$$v_B = -\mu v_x = -\mu v_A$$

Thus, node A is the only independent node in the circuit. We can write the node A equation by inspection as

$$(G_1 + G_2 + G_3)v_A - G_3 v_B = G_1 v_S$$

Substituting in the control source constraint yields the standard form for this equation:

$$[G_1 + G_2 + G_3(1 + \mu)] v_A = G_1 v_S$$

We end up with only one node equation even though at first glance the given circuit appears to need three node equations. The reason is that there are two voltage sources in the original circuit in Figure 4–7(a). Since the two sources share the reference node, the number of unknown node voltages is reduced from three to one. The general principle illustrated is that the number of independent KCL constraints in a circuit containing N nodes and N_V voltage sources (dependent or independent) is $N - 1 - N_V$.

   The one-node equation can easily be solved for the output voltage $v_O = v_B$.

$$v_O = v_B = -\mu v_A = \left(\frac{-\mu G_1}{G_1 + G_2 + G_3(1 + \mu)} \right) v_S$$

The minus sign in the numerator means that the circuit provides signal inversion. The output voltage does not depend on the value of the load resistor R_4, since the load is connected across an ideal (though dependent) voltage source.

(b) For large gains μ we have $(1 + \mu)G_3 \gg G_1 + G_2$ and the input-output relationship reduces to

$$v_O \approx \left[\frac{-\mu G_1}{(1 + \mu) G_3} \right] v_S \approx -\left[\frac{R_3}{R_1} \right] v_S$$

That is, when the active device gain is large, the voltage gain of the active circuit depends on the ratio of two resistances. We will see this situation again with OP AMP circuits. ∎

Exercise 4–2

(a) Formulate node-voltage equations for the circuit in Figure 4–8.
(b) Solve the node-voltage equations for v_O and i_O in terms of i_S.

FIGURE 4 – 8

Answers:

(a)

$$(1.5 \times 10^{-3}) v_A - (0.5 \times 10^{-3}) v_B = i_S$$

$$-(2.5 \times 10^{-3}) v_A + (2.5 \times 10^{-3}) v_B = 0$$

(b) $v_O = 1000 i_S$; $i_O = 2 i_S$.

FIGURE 4 – 9

Exercise 4–3

Use node-voltage analysis to find v_O in Figure 4–9.

Answer:

$$v_O = \frac{G_x + \mu G_2}{G_x + G_L + (\mu + 1)G_2} v_S$$

MESH-CURRENT ANALYSIS WITH DEPENDENT SOURCES

Mesh-current analysis of active circuits follows the same pattern noted for node-voltage analysis. We initially treat the dependent sources as independent sources and write the mesh equations of the resulting passive circuit using the inspection method from Chapter 3. We then account for the dependent sources by expressing their constraints in terms of unknown mesh currents. The following example illustrates the method.

EXAMPLE 4–4

(a) Formulate mesh-current equations for the circuit in Figure 4–10.
(b) Use the mesh equations to find v_O and R_{IN} when $R_1 = 50\ \Omega$, $R_2 = 1\ k\Omega$, $R_3 = 100\ \Omega$, $R_4 = 5\ k\Omega$, and $g = 100$ mS.

SOLUTION:
(a) Applying source transformation to the parallel combination of R_3 and gv_x in Figure 4–10(a) produces the dependent voltage source $R_3gv_x = \mu v_x$ in Figure 4–10(b). In the modified circuit we have identified two mesh currents. Initially treating the dependent source $(gR_3)v_x$ as an independent source leads to two symmetrical mesh equations.

$$\text{Mesh A: } (R_1 + R_2 + R_3)i_A - R_3i_B = v_S - (gR_3)v_x$$

$$\text{Mesh B: } \quad -R_3i_A + (R_3 + R_4)i_B = (gR_3)v_x$$

FIGURE 4–10

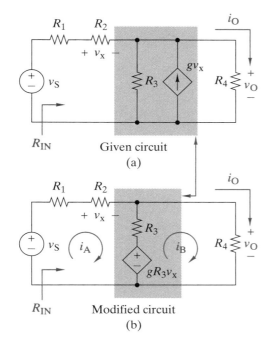

Given circuit
(a)

Modified circuit
(b)

The control voltage v_x can be written in terms of mesh currents.

$$v_x = R_2 i_A$$

Substituting this equation for v_x into the mesh equations and putting the equations in standard form yields

$$(R_1 + R_2 + R_3 + gR_2R_3)i_A - R_3 i_B = v_S$$

$$-(R_3 + gR_2R_3)i_A + (R_3 + R_4)i_B = 0$$

The resulting mesh equations are not symmetrical because of the controlled source.

(b) Substituting the numerical values into the mesh equations gives

$$(1.115 \times 10^4)i_A - (10^2)i_B = v_S$$

$$-(1.01 \times 10^4)i_A + (5.1 \times 10^3)i_B = 0$$

Solving for the two mesh currents yields

$$i_A = \frac{\Delta_A}{\Delta} = \frac{\begin{vmatrix} v_S & -10^2 \\ 0 & 5.1 \times 10^3 \end{vmatrix}}{\begin{vmatrix} 1.115 \times 10^4 & -10^2 \\ -1.01 \times 10^4 & 5.1 \times 10^3 \end{vmatrix}} = \frac{5.1 \times 10^3 v_S}{5.5855 \times 10^7}$$

$$= 0.9131 \times 10^{-4} v_S$$

$$i_B = \frac{\Delta_B}{\Delta} = \frac{\begin{vmatrix} 1.115 \times 10^4 & v_S \\ -1.01 \times 10^4 & 0 \end{vmatrix}}{5.5885 \times 10^7} = 1.808 \times 10^{-4} v_S$$

The output voltage and input resistance are found using Ohm's law.

$$v_O = R_4 i_B = 0.904 v_S$$

$$R_{IN} = \frac{v_S}{i_A} = 10.95 \text{ k}\Omega \qquad \blacksquare$$

EXAMPLE 4–5

The circuit in Figure 4–11 is a model of a bipolar junction transistor operating in the active mode. Use mesh analysis to find the transistor base current i_B.

FIGURE 4 – 11 *Junction transistor circuit model.*

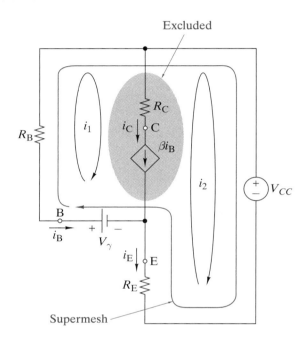

SOLUTION:
The two mesh currents in Figure 4–11 are labeled i_1 and i_2 to avoid possible confusion with the transistor base current i_B. As drawn, the circuit requires a supermesh since the dependent current source βi_B is included in both meshes and is not connected in parallel with a resistance. A supermesh is created by combining meshes 1 and 2 after excluding the series subcircuit consisting of βi_B and R_C. Beginning at the bottom of the circuit, we write a KVL mesh equation around the supermesh using unknowns i_1 and i_2:

$$i_2 R_E - V_\gamma + i_1 R_B + V_{CC} = 0$$

This KVL equation provides one equation in the two unknown mesh currents. Since the two mesh currents have opposite directions through the dependent current source βi_B, the currents i_1, i_2, and βi_B are related by KCL as

$$i_1 - i_2 = \beta i_B$$

This constraint supplies the additional relationship needed to obtain two equations in the two unknown mesh-current variables. Since $i_B = -i_1$, the preceding KCL constraint means that $i_2 = (\beta + 1)i_1$. Substituting $i_2 = (\beta + 1)i_1$ into the supermesh KVL equation and solving for i_B yields

$$i_B = -i_1 = \frac{V_{CC} - V_\gamma}{R_B + (\beta + 1)R_E}$$ ∎

FIGURE   4 – 1 2

Exercise 4–4

Use mesh analysis to find the current i_O in Figure 4–12 when the element E is a dependent current source $2i_x$ with the reference arrow directed down.

Answer:
–0.857 mA

Exercise 4–5

Use mesh analysis to find the current i_O in Figure 4–12 when the element E is a dependent voltage source $2000i_X$ with the plus reference at the top.

Answer:
–0.222 mA

EXAMPLE 4–6

The circuit in Figure 4–13 is a small signal model of a field effect transistor (FET) amplifier with two inputs v_{S1} and v_{S2}. Use PSpice to solve for the input-output relationship of the circuit.

$R_1 = R_2 = 500 \text{ k}\Omega$
$R_3 = R_4 = 8 \text{ k}\Omega$
$r_{ds} = 40 \text{ k}\Omega$
$g = 3 \text{ mS}$

FIGURE   4 – 1 3   *Field effect transistor amplifier.*

SOLUTION:
Since the circuit is linear, the input-output relationship is of the form

$$v_O = K_1 v_{S1} + K_2 v_{S2}$$

We can use the superposition principle to find the gain K_1 by setting $v_{S1} = 1$ and $v_{S2} = 0$ and solving for the node voltage V(4). The gain K_2 is then found by setting $v_{S1} = 0$ and $v_{S2} = 1$ and again solving for V(4). Figure 4–14 shows a PSpice Schematics circuit diagram for the case $v_{S1} = 1$ and $v_{S2} = 0$. The figure also shows the resulting net list and a portion of the output file that includes the dc solution for the five node voltages. For this case the output file reports that V(4) = 10; therefore, gain $K_1 = V(4)/v_{S1} = 10$.

To find K_2, we change the source attributes to $v_{S1} = 0$ and $v_{S2} = 1$ and again simulate the circuit using PSpice. For this case the output file reports V(4) = –10.000, and therefore $K_2 = V(4)/v_{S2} = -10$. The input-output relationship for the circuit is

$$v_O = 10(v_{S1} - v_{S2})$$

The circuit is a differential amplifier of a type often used as the input circuit of an OP AMP.   ∎

F I G U R E   4 – 1 4 *PSpice schematic diagram of the circuit in Figure 4–13.*

THÉVENIN EQUIVALENT CIRCUITS WITH DEPENDENT SOURCES

To find the Thévenin equivalent of an active circuit, we must leave the independent sources on or else supply excitation from an external test source. This means that the Thévenin resistance cannot be found by the lookback method because that method requires that all independent sources be turned off. Turning off the independent sources deactivates the dependent sources as well and can result in a profound change in the input and output characteristics of an active circuit. Thus, there are two ways of finding active circuit Thévenin equivalents. We can either find the open-circuit voltage and short-circuit current at the interface or directly solve for the interface i–v relationship.

EXAMPLE 4-7

Find the input resistance of the circuit in Figure 4–15.

FIGURE 4-15

SOLUTION:

With the independent source turned off ($i_{IN} = i_S = 0$), the resistance seen at the input port is R_E since the dependent current source βi_{IN} is inactive and acts like an open circuit. Applying KCL at node A with the input source turned on yields

$$i_E = i_{IN} + \beta i_{IN} = (\beta + 1)i_{IN}$$

By Ohm's law, the input voltage is

$$v_{IN} = i_E R_E = (\beta + 1)i_{IN}R_E$$

Hence, the active input resistance is

$$R_{IN} = \frac{v_{IN}}{i_{IN}} = (\beta + 1)R_E$$

The circuit in Figure 4–15 is a model of a transistor circuit in which the gain parameter β typically lies between 50 and 100. The input resistance with external excitation is $(\beta + 1)R_E$, which is significantly different from the value of R_E without external excitation. ∎

EXAMPLE 4-8

Find the Thévenin equivalent at the output interface of the circuit in Figure 4–16.

FIGURE 4-16

SOLUTION:

In this circuit the controlled voltage v_x appears across an open circuit between nodes A and B. By the fundamental property of node voltages, $v_x = v_S - v_O$. With the load disconnected and the input source turned off ($v_x = 0$), the dependent voltage source μv_x acts like a short circuit, and the Thévenin resistance looking back into the output port is R_O. With the load connected and the input source turned on, the sum of currents leaving node B is

$$\frac{v_O - \mu v_x}{R_O} + i_O = 0$$

Using the relationship $v_x = v_S - v_O$ to eliminate v_x and then solving for v_O produces the i–v characteristic at the output interface as

$$v_O = \frac{\mu v_S}{\mu + 1} - i_O \left[\frac{R_O}{\mu + 1} \right]$$

The i–v relationship of a Thévenin circuit is $v = v_T - iR_T$. By direct comparison, we find the Thévenin parameters of the active circuit to be

$$v_T = \frac{\mu v_S}{\mu + 1} \quad \text{and} \quad R_T = \frac{R_O}{\mu + 1}$$

The circuit in Figure 4–16 is a model of an OP AMP circuit called a voltage follower. The resistance R_O for a general-purpose OP AMP is around 100 Ω, while the gain μ is about 10^5. Thus, the active Thévenin resistance of the voltage follower is not 100 Ω, as the lookback method suggests, but is only a milliohm.

R_{IN} v_T, R_T

F I G U R E   4 – 1 7

Exercise 4–6

Find the input resistance and output Thévenin equivalent circuit of the circuit in Figure 4–17.

Answers:

$$R_{IN} = (1 + \mu)R_F$$

$$v_T = \frac{\mu}{\mu + 1} v_S$$

$$R_T = R_O$$

4–3   THE TRANSISTOR

In the previous sections we introduced the four controlled sources and showed how to analyze linear circuits containing these elements. In this and the next section, we show how these elements are used to describe electronic devices like transistors and OP AMPS. The *behavioral models* developed here describe the large-scale input-output characteristics at the device terminals and not the underlying physical processes in the device. They are called behavioral models because they allow us to predict circuit behavior without describing the internal device characteristics.

To illustrate the process, we develop a behavioral model for the bipolar junction transistor (BJT)—only one of the many different kinds of transistors. The objective of this section is to show that linear circuit elements, including dependent sources, can duplicate the major nonlinear features of a semiconductor device, the BJT. In this sense, this section lays the circuit analysis groundwork for subsequent study of semiconductor devices in electronics courses.

The circuit symbol of the BJT is shown in Figure 4–18. The device has three terminals called the **emitter (E)**, the **base (B)**, and the **collector (C)**. The voltages v_{BE} and v_{CE} in the figure are called the *base-emitter voltage* and the *collector-emitter voltage,* respectively. The three currents i_E, i_B, and

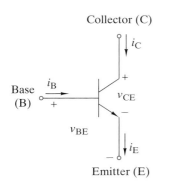

Collector (C)

Base (B)

Emitter (E)

F I G U R E   4 – 1 8   *Circuit symbol for the bipolar junction transistor.*

i_C are called the emitter, base, and collector currents. Applying KCL to the BJT as a whole yields

$$i_E = i_B + i_C \qquad (4-8)$$

This constraint means that only two of the three transistor currents can be independently specified. Normally, we calculate i_B and i_C, and invoke Eq. (4-8) to find i_E if need be.

Our behavioral model for the BJT applies to a region where $i_B \geq 0$ and $v_{CE} \geq 0$. The dominant feature of the interior of this region is that the collector current is proportional to the base current and the base-to-emitter voltage is constant. These conditions define what is called the **active mode** in which the BJT has the following i–v characteristics:

$$\text{Active Mode:} \quad i_C = \beta i_C \quad \text{and} \quad v_{BE} = V_\gamma \qquad (4-9)$$

The proportionality factor β is called the *forward current gain* and typically ranges from 50 to several hundred. The constant V_γ is called the *turn-on* or *threshold* voltage and generally is less than one volt. In this mode the transistor acts like a current-controlled current source between the collector and emitter terminals, as suggested by the circuit diagram in Figure 4–19(a).

Two other operating modes exist at the boundary of the region defined by $i_B \geq 0$ and $v_{CE} \geq 0$. When $i_B = 0$ and $v_{CE} > 0$ the transistor is said to be in the **cutoff mode** with $i_C = 0$. In this mode the transistor acts like an open circuit ($i_C = 0$) between the collector and emitter terminals, as indicated in Figure 4–19(b). Conversely, when $i_B > 0$ and $v_{CE} = 0$ the transistor is said to be in the **saturation mode** with $v_{BE} = V_\gamma$. In this mode the transistor acts like a short circuit ($v_{CE} = 0$) between the collector and emitter terminals, as suggested in Figure 4–19(c). We can summarize these two modes by writing

$$\begin{aligned} \text{Cutoff Mode:} \quad & i_B = 0 \quad \text{and} \quad i_C = 0 \\ \text{Saturation Mode:} \quad & v_{CE} = 0 \quad \text{and} \quad v_{BE} = V_\gamma \end{aligned} \qquad (4-10)$$

Thus, our behavioral model has three operating modes (active, cutoff, saturation), each with its own distinct set of i–v characteristics.

In any given situation the appropriate mode is determined by the circuit in which the transistor is connected. On the other hand, we cannot analyze the circuit until we know the BJT's i–v characteristics, that is, until we know which operating mode is appropriate. We appear to be at an impasse. We cannot analyze the circuit until we know the operating mode, and we cannot find the operating mode until we analyze the circuit. How do we resolve this dilemma?

One way is to assume that the BJT is in the active mode and use the element constraints in Eq. (4-9) together with normal circuit analysis methods to calculate the responses i_B and v_{CE}. We then test our assumption as follows:

1. If $i_B > 0$ and $v_{CE} > 0$, then the active mode assumption is correct.

2. If $i_B < 0$, then the assumption is incorrect and the transistor is in the cutoff mode with $i_B = 0$ and $i_C = 0$.

3. If $v_{CE} < 0$, then the assumption is incorrect and the transistor is in the saturation mode with $v_{CE} = 0$ and $v_{BE} = V_\gamma$.

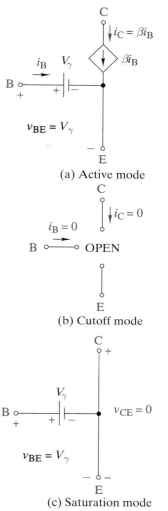

(a) Active mode

(b) Cutoff mode

(c) Saturation mode

F I G U R E   4 - 1 9 *Circuit models for transistor operating modes.*

Before working examples it is important to emphasize that the circuit diagrams in Figure 4–19 do not represent the internal physical structure of a BJT. They merely show how the i–v relations in Eqs. (4–9) and (4–10) could be created using the circuit elements we have previously studied.

EXAMPLE 4–9

The parameters of the transistor circuit of Figure 4–20 are $\beta = 100$, $V_\gamma = 0.7$ V, $R_B = 100$ kΩ, $R_C = 1$ kΩ, and $V_{CC} = 5$ V. The dc source V_{CC} is necessary because the transistor is an active device that requires an external power supply to operate correctly.

(a) Find i_C and v_{CE} when $v_S = 3$ V.
(b) Find the range of the input voltage v_S over which the transistor operates in the active mode.

FIGURE 4–20

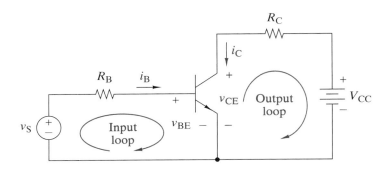

SOLUTION:
(a) First we assume that the transistor is operating in the active mode. Under this assumption $v_{BE} = V_\gamma = 0.7$ V and the transistor base current is

$$i_B = \frac{v_S - V_\gamma}{R_B} = \frac{3 - 0.7}{10^5} = 23 \ \mu\text{A} \qquad (4\text{–}11)$$

With the transistor in the active mode, the collector current $i_C = \beta i_B = 100 \times 23 \times 10^{-6} = 2.3$ mA. Writing a KVL equation around the output loop in Figure 4–20 yields

$$-v_{CE} - i_C R_C + V_{CC} = 0$$

from which we obtain v_{CE} as

$$v_{CE} = V_{CC} - i_C R_C = 5 - 2.3 = 2.7 \text{ V} \qquad (4\text{–}12)$$

This analysis yields $i_B = 23 \ \mu\text{A} > 0$ and $v_{CE} = 2.7$ V > 0. According to Test No. 1 our assumption is correct and the transistor is operating in the active mode.
(b) The transistor operates in the active mode as long as $i_B > 0$ and $v_{CE} > 0$. Looking back at Eq. (4–11), we see that the base current is positive when $v_S > V_\gamma = 0.7$ V. Next, by looking at Eq. (4–12), we see that $v_{CE} > 0$ requires that $i_C < V_{CC}/R_C = 5$ mA. In the active mode, $i_C < 5$ mA means that $i_B < i_C/\beta = 5 \times 10^{-3}/100 = 50 \ \mu\text{A}$. Again invoking Eq. (4–11), we have

$$i_B = \frac{v_S - V_\gamma}{R_B} = \frac{v_S - 0.7}{10^5} < 50 \ \mu A$$

This constraint requires that $v_S - 0.7 < 50 \times 10^{-6} \times 10^5$ or $v_S < 5.7$ V. The net result is that the transistor is in the active mode when the input voltage is in the range $0.7 < v_S < 5.7$ V. ∎

EXAMPLE 4–10

The parameters of the transistor circuit of Figure 4–20 are $\beta = 100$, $V_\gamma = 0.7$ V, $R_B = 100$ kΩ, $R_C = 1$ kΩ, and $V_{CC} = 5$ V. Plot the output voltage v_{CE} versus the input voltage for v_S on the range from 0 to 10 V.

SOLUTION:
The numerical values given here are the same as those in the preceding example, where we found that when $0.7 < v_S < 5.7$, the transistor is in the active mode with $i_C = \beta i_B$. Using this result together with Eqs. (4–11) and (4–12), we find that

$$v_{CE} = V_{CC} - \beta i_B R_C = V_{CC} - \beta \left(\frac{v_S - V_\gamma}{R_B} \right) R_C$$

$$= 5.7 - v_S$$

The result $v_{CE} = 5.7 - v_S$ is the equation of a straight line passing through $v_{CE} = 5$ when $v_S = 0.7$ and $v_{CE} = 0$ when $v_S = 5.7$ V. When $v_S > 5.7$ this relationship predicts that $v_{CE} < 0$, which is not consistent with the active mode and Test No. 3 tells us that the transistor is in the saturation mode with $v_{CE} = 0$ V. When $0 < v_S < 0.7$, Eq. (4–11) points out that $i_B < 0$, which is not consistent with the active mode and Test No. 2 tells us that the transistor is in cutoff with $i_C = 0$, in which case Eq. (4–12) points out that $v_{CE} = V_{CC} = 5$ V. Figure 4–21 shows that the input-output characteristic is nonlinear with changes in slope occurring at points where the transistor changes modes. That is, we are able to model the nonlinear features of the BJT using three operating modes each with its own set of i–v characteristics.

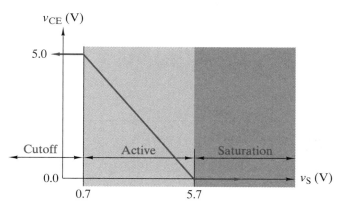

FIGURE   4–21 *Transfer characteristics of the transistor circuit in Figure 4–20.*

Comment: The plot in Figure 4–21 is called the transfer characteristic of the transistor circuit in Figure 4–20. In digital applications the input drives the transistor between the cutoff and saturation modes, passing as quickly as possible through the active mode. In analog applications the transistor remains in the active mode and operates as an amplifier. In the next section we see that the OP AMP also has a similar transfer characteristic. ∎

EXAMPLE 4-11

In the transistor circuit in Figure 4–22, $V_{CC} = 10$ V, $R_C = 2.2$ kΩ, $V_\gamma = 0.7$ V, $\beta = 50$, and $v_S = 5$ V. Select the value of R_B so that $v_{CE} = 5$ V.

FIGURE 4-22

SOLUTION:

In the active mode collector voltage falls in the range $0 < v_{CE} < V_{CC}$. Since the specified value is in this range, the transistor must operate in the active mode. By KVL the voltage across R_C is $V_{CC} - v_{CE} = 10 - 5 = 5$ V, and hence the collector current is $i_C = 5/R_C = 2.27$ mA. In the active mode $i_C = \beta i_B$, and hence the base current is $i_B = i_C/50 = 45.5$ μA. In the active mode $v_{BE} = V_\gamma = 0.7$ V. By KVL, the voltage across R_B is $v_S - v_{BE} = 5 - 0.7 = 4.3$ V. So finally we have

$$R_B = \frac{v_S - v_{BE}}{i_B} = \frac{4.3}{45.5 \times 10^{-6}} = 94.5 \text{ k}\Omega$$

Note that for $R_B = 94.5$ kΩ the transistor is in the active mode since $i_B > 0$ and $v_{CE} > 0$. ∎

Exercise 4–7

In the transistor circuit in Figure 4–22, $V_{CC} = 10$ V, $R_C = 2.2$ kΩ, $V_\gamma = 0.7$ V, $\beta = 50$, and $R_B = 100$ kΩ. Find the i_C and v_{CE} for $v_S = 3$ V.

Answer:
$i_C = 1.15$ mA; $v_{CE} = 7.47$ V.

Exercise 4–8

In the transistor circuit in Figure 4–22, $V_{CC} = 10$ V, $R_C = 2.2$ kΩ, $V_\gamma = 0.7$ V, $\beta = 50$, and $v_S = 5$ V. Select the value of R_B so that the transistor operates in the saturated mode.

Answer:
$R_B < 47.3$ kΩ

4–4 THE OPERATIONAL AMPLIFIER

The integrated circuit operational amplifier is the premier linear active device in present-day analog circuit applications. The term *operational amplifier* was apparently first used in a 1947 paper by John R. Ragazzini and his colleagues, who reported on work carried out for the National Defense Research Council during World War II. The paper described high-gain dc amplifier circuits that perform mathematical operations (addition, subtraction, multiplication, division, integration, etc.), hence the name *operational amplifier*. For more than a decade the most important applications were general- and special-purpose analog computers using vacuum tube amplifiers. In the early 1960s general-purpose, discrete-transistor operational amplifiers became readily available, and by the mid-1960s the first commercial integrated circuit OP AMPs entered the market. The transition from vacuum tubes to integrated circuits decreased the size, power consumption, and cost of OP AMPs by nearly three orders of magnitude. By the early 1970s the integrated circuit version became the dominant active device in analog circuits.

The device itself is a complex array of transistors, resistors, diodes, and capacitors all fabricated and interconnected on a silicon chip less than 0.5 mm^2 in area and about 1 mm thick. Figure 4–23 shows examples of ways OP AMPs are packaged for use in circuits. In spite of its complexity, the device can be modeled by rather simple i–v characteristics. We do not need to concern ourselves with what is going on inside the package; rather, we treat the OP AMP using a behavioral model that constrains the voltages and currents at the external terminals of the device.

OP AMP NOTATION

Certain matters of notation and nomenclature must be discussed before developing a circuit model for the OP AMP. The OP AMP is a five-terminal device, as shown in Figure 4–24(a). The "+" and "−" symbols identify the input terminals and are a shorthand notation for the noninverting and inverting input terminals, respectively. These "+" and "−" symbols identify the two input terminals and have nothing to do with the polarity of the voltages applied. The other terminals are the output and the positive and negative supply voltages, usually labeled $+V_{CC}$ and $-V_{CC}$. While some OP AMPs have more than five terminals, these five are always present and are the only ones we will use in this text. Figure 4–24(b) shows how these terminals are arranged in a common eight-pin integrated circuit package.

The two power supply terminals in Figure 4–24 are not usually shown in circuit diagrams. Be assured that they are always there because the external power supplies are required for the OP AMP to operate as an active device. The power required for signal amplification comes through these terminals from an external power source. The $+V_{CC}$ and $-V_{CC}$ voltages applied to these terminals also determine the upper and lower limits on the OP AMP output voltage.

Figure 4–25(a) shows a complete set of voltage and current variables for the OP AMP, while Figure 4–25(b) shows the abbreviated set of signal variables we will use. All voltages are defined with respect to a common reference node, usually ground. Voltage variables v_P, v_N, and v_O are de-

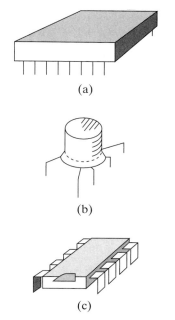

FIGURE 4 – 2 3 *Examples of integrated circuit OP AMP packages: (a) Encapsulated hybrid. (b) TO-5 can. (c) Dual in-line package.*

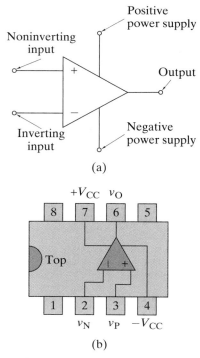

FIGURE 4 – 2 4 *The OP AMP: (a) Circuit symbol. (b) Pin out diagram for an eight-pin package.*

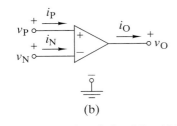

FIGURE 4-25 *OP AMP voltage and current definitions: (a) Complete set. (b) Shorthand set.*

FIGURE 4-26 *OP AMP transfer characteristics.*

fined by writing a voltage symbol beside the corresponding terminals. This notation means the "+" reference mark is at the terminal in question and the "−" reference mark is at the reference or ground terminal. In this book the reference directions for the currents are directed in at input terminals and out at the output. At times the abbreviated set of current variables may appear to violate KCL. For example, a global KCL equation for the complete set of variables in Figure 4–25(a) is

$$i_O = I_{C+} + I_{C-} + i_P + i_N \quad \text{(correct equation)} \quad (4–13)$$

A similar equation using the shorthand set of current variables in Figure 4–25(b) reads

$$i_O = i_N + i_P \quad \text{(incorrect equation)} \quad (4–14)$$

This equation is *not* correct, since it does not include all the currents. What is more important, it implies that the output current comes from the inputs. In fact, this is wrong. The input currents are very small, ideally zero. The output current comes from the supply voltages, as Eq. (4–13) points out, even though these terminals are not shown on the abbreviated circuit diagram.

TRANSFER CHARACTERISTICS

The dominant feature of the OP AMP is the transfer characteristic shown in Figure 4–26. This characteristic provides the relationships between the **noninverting input** v_P, the **inverting input** v_N, and the **output voltage** v_O. The transfer characteristic is divided into three regions or modes called **+saturation**, **−saturation**, and **linear**. In the linear region the OP AMP is a **differential amplifier** because the output is proportional to the difference between the two inputs. The slope of the line in the linear range is called the voltage gain. In this linear region the input-output relation is

$$v_O = A(v_P - v_N) \quad (4–15)$$

The voltage gain of an OP AMP is very large, usually greater than 10^5. As long as the net input $(v_P - v_N)$ is very small, the output will be proportional to the input. However, when $A|v_P - v_N| > V_{CC}$, the OP AMP is saturated and the output voltage is limited by the supply voltages (less some small internal losses).

In the previous section, we found that the transistor has three operating modes. The input-output characteristic in Figure 4–26 points out that the OP AMP also has three operating modes:

1. +Saturation mode when $A(v_P - v_N) > V_{CC}$ and $v_O = +V_{CC}$.

2. −Saturation mode when $A(v_P - v_N) < -V_{CC}$ and $v_O = -V_{CC}$.

3. Linear mode when $A|v_P - v_N| < V_{CC}$ and $v_O = A(v_P - v_N)$.

Usually we analyze and design OP AMP circuits using the model for the linear mode. When the operating mode is not given, we use a self-consistent approach similar to the one used for the transistor. That is, we assume that the OP AMP is in the linear mode and then calculate the output voltage v_O. If it turns out that $-V_{CC} < v_O < +V_{CC}$, then the assumption is correct and the OP AMP is indeed in the linear mode. If $v_O < -V_{CC}$, then the assumption is wrong and the OP AMP is in the −saturation mode with $v_O = -V_{CC}$. If $v_O > +V_{CC}$, then the assumption is wrong and the OP AMP is in the +saturation mode with $v_O = +V_{CC}$.

IDEAL OP AMP MODEL

A dependent-source model of an OP AMP operating in its linear range is shown in Figure 4–27. This model includes an input resistance (R_I), an output resistance (R_O), and a voltage-controlled voltage source whose gain is A. Numerical values of these OP AMP parameters typically fall in the following ranges:

$$10^6 < R_I < 10^{12} \ \Omega$$

$$10 < R_O < 100 \ \Omega$$

$$10^5 < A < 10^8$$

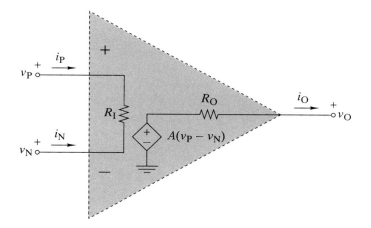

FIGURE 4−27 *Dependent source model of an OP AMP operating in the linear mode.*

Clearly, high input resistance, low output resistances, and high voltage gain are the key attributes of an OP AMP.

The dependent-source model can be used to develop the i–v relationships of the ideal model. For the OP AMP to operate in its linear mode, the output voltage is bounded by

$$- V_{CC} \leqslant v_O \leqslant + V_{CC}$$

Using Eq. (4–15), we can write this bound as

$$-\frac{V_{CC}}{A} \leqslant (v_P - v_N) \leqslant +\frac{V_{CC}}{A}$$

The supply voltage V_{CC} is typically about 15 V, while A is a very large number, usually 10^5 or greater. Consequently, linear operation requires that $v_P \approx v_N$. In the ideal OP AMP model, the voltage gain is assumed to be infinite ($A \to \infty$), in which case linear operation requires $v_P = v_N$. The input resistance R_I of the ideal OP AMP is assumed to be infinite, so the currents entering both input terminals are both zero. In summary, the i–v relationships of the **ideal model** of the OP AMP are

FIGURE 4 – 2 8 *Ideal OP AMP characteristics.*

(a)

(b)

FIGURE 4 – 2 9 *The noninverting amplifier circuit.*

$$v_P = v_N$$
$$i_P = i_N = 0 \qquad (4\text{–}16)$$

The implications of these element equations are illustrated on the OP AMP circuit symbol in Figure 4–28.

At first glance the element constraints of the ideal OP AMP appear to be fairly useless. They look more like connection constraints and are totally silent about the output quantities (v_O and i_O), which are usually the signals of greatest interest. They seem to say that the OP AMP input terminals are simultaneously a short circuit ($v_P = v_N$) and an open circuit ($i_P = i_N = 0$). In practice, however, the ideal model of the OP AMP is very useful because in linear applications feedback is always present. That is, for the OP AMP to operate in a linear mode, it is necessary for there to be feedback paths from the output to one or both of the inputs. These feedback paths ensure that $v_P \approx v_N$ and make it possible for us to analyze OP AMP circuits using the ideal OP AMP element constraints in Eq. (4–16).

NONINVERTING OP AMP

To illustrate the effects of feedback, let us find the input-output characteristics of the circuit in Figure 4–29. In this circuit the voltage divider provided a feedback path from the output to the inverting input.[1] Since the ideal OP AMP draws no current at either input ($i_P = i_N = 0$), we can use voltage division to determine the voltage at the inverting input:

$$v_N = \frac{R_2}{R_1 + R_2} v_O \qquad (4\text{–}17)$$

The input source connection at the noninverting input requires the condition

$$v_P = v_S \qquad (4\text{–}18)$$

The ideal OP AMP element constraints demand that $v_P = v_N$, therefore, we can equate the right sides of Eqs. (4–17) and (4–18) to obtain the input–output relationship of the overall circuit.

$$v_O = \frac{R_1 + R_2}{R_2} v_S \qquad (4\text{–}19)$$

The preceding analysis illustrates a general strategy for analyzing OP AMP circuits. We use normal circuit analysis methods to express the OP AMP input voltages v_P and v_N in terms of circuit parameters. We then use the ideal OP AMP constraint $v_P = v_N$ to solve for the overall circuit input–output relationship.

The circuit in Figure 4–29(a) is called a **noninverting amplifier**. The input-output relationship is of the form $v_O = Kv_S$, which reminds us that the circuit is linear. Figure 4–29(b) shows the functional building block for this circuit, where the proportionality constant K is

---

1 The feedback must always be to the inverting input, otherwise the circuit will be unstable for reasons that cannot be explained by what we have learned up to now.

$$K = \frac{R_1 + R_2}{R_2} \qquad (4\text{–}20)$$

In an OP AMP circuit the proportionality constant K is sometimes called the **closed-loop** gain, because it defines the input-output relationship when the feedback loop is connected (closed).

When discussing OP AMP circuits, it is necessary to distinguish between two types of gains. The first is the large voltage gain provided by the OP AMP device itself. The second is the voltage gain of the OP AMP circuit with a negative feedback path. Note that Eq. (4–20) indicates that the circuit gain is determined by the resistors in the feedback path and not by the value of the OP AMP gain. The gain in Eq. (4–20) is really the voltage division rule upside down. Variation of the value of K depends on the tolerance on the resistors in the feedback path, and not the variation in the value of the OP AMP's gain. In effect, feedback converts the OP AMP's very large but variable gain into a much smaller but well-defined gain.

D DESIGN EXAMPLE 4–12

Design an amplifier with a gain of $K = 10$.

SOLUTION:

Using a noninverting OP AMP circuit, the design problem is to select the values of the resistors in the feedback path. From Eq. (4–20) the design constraint is

$$10 = \frac{(R_1 + R_2)}{R_2}$$

We have one constraint with two unknowns. Arbitrarily selecting $R_2 = 10$ kΩ, we find $R_1 = 90$ kΩ. These resistors would normally have low tolerances ($\pm 1\%$ or less) to produce a precisely controlled closed-loop gain.

Comment: The problem of choosing resistance values in OP AMP circuit design problems deserves some discussion. Although values of resistance from a few ohms to several hundred megohms are commercially available, we generally limit ourselves to the range from about 1 kΩ to perhaps 1 MΩ. The lower limit of 1 kΩ is imposed in part because of power dissipation in the resistors. Typically, we use resistors with ¼-W power ratings or less. The maximum voltage in OP AMP circuits is often around 15 V. The smallest ¼-W resistance we can use is $R_{\text{MIN}} \geq (15)^2/0.25 = 900\ \Omega$, or about 1 k$\Omega$. The upper bound of 1 M$\Omega$ comes about because surface leakage makes it difficult to maintain the tolerance in a high-value resistance. High-value resistors are also noisy, which leads to problems when they are connected in the feedback path. The 1-kΩ to 1-MΩ range should be used as a guideline and not an inviolate design rule. Actual design choices are influenced by system-specific factors and changes in technology. ■

Exercise 4–9

The noninverting amplifier circuit in Figure 4–29(a) is operating with $R_1 = 2R_2$ and $V_{\text{CC}} = \pm 12$ V. Over what range of input voltages v_S is the OP AMP in the linear mode?

EFFECTS OF FINITE OP AMP GAIN

The ideal OP AMP model has an infinite gain. Actual OP AMP devices have very large, but finite voltage gains. We now address the effect of large but finite gain on the input-output relationships of OP AMP circuits.

The circuit in Figure 4–30 shows a finite gain OP AMP circuit model in which the input resistance R_I is infinite. The actual values of OP AMP input resistance range from 10^6 to 10^{12} Ω, so no important effect is left out by ignoring this resistance. Examining the circuit, we see that the noninverting input voltage is determined by the independent voltage source. The inverting input can be found by voltage division, since the current i_N is zero. In other words, Eqs. (4–17) and (4–18) apply to this circuit as well.

FIGURE 4–30 *The noninverting amplifier circuit with the dependent source model.*

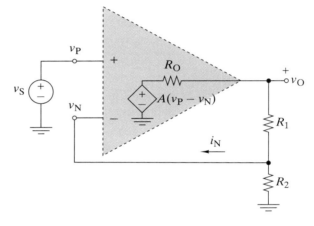

We next determine the output voltage in terms of the controlled source voltage using voltage division on the series connection of the three resistors R_O, R_1, and R_2:

$$v_O = \frac{R_1 + R_2}{R_O + R_1 + R_2} A(v_P - v_N)$$

Substituting v_P and v_N from Eqs. (4–17) and (4–18) yields

$$v_O = \left[\frac{R_1 + R_2}{R_O + R_1 + R_2}\right] A \left[v_S - \frac{R_2}{R_1 + R_2} v_O\right] \tag{4–21}$$

The intermediate result in Eq. (4–21) shows that feedback is present since v_O appears on both sides of the equation. Solving for v_O yields

$$v_O = \frac{A(R_1 + R_2)}{R_O + R_1 + R_2(1 + A)} v_S \tag{4–22}$$

In the limit, as $A \to \infty$, Eq. (4–22) reduces to

$$v_O = \frac{R_1 + R_2}{R_2} v_S = K v_S$$

where K is the closed-loop gain we previously found using the ideal OP AMP model.

To see the effect of a finite A, we ignore R_O in Eq. (4–22) since it is generally quite small compared with $R_1 + R_2$. With this approximation Eq. (4–22) can be written in the following form:

$$v_O = \frac{K}{1 + (K/A)} v_S \qquad (4\text{--}23)$$

When written in this form, we see that the closed-loop gain reduces to K as $A \to \infty$. Moreover, we see that the finite-gain model yields a good approximation to the ideal model results as long as $K \ll A$. In other words, the ideal model yields good results as long as the closed-loop gain is much less than the gain of the OP AMP device. One practical rule of thumb is to limit the closed-loop gain to less than 1% of the OP AMP gain (i.e., $K < A/100$).

The feedback path also affects the active output resistance. To see this, we construct a Thévenin equivalent circuit using the open-circuit voltage and the short-circuit current. Equation (4–23) is the open-circuit voltage, and we need only find the short-circuit current. Connecting a short-circuit at the output in Figure 4–30 forces $v_N = 0$ but leaves $v_P = v_S$. Therefore, the short-circuit current is

$$i_{SC} = A(v_S/R_O)$$

As a result, the Thévenin resistance is

$$R_T = \frac{v_{OC}}{i_{SC}} = \frac{K/A}{1 + K/A} R_O$$

When $K \ll A$, this expression reduces to

$$R_T = \frac{K}{A} R_O \approx 0\ \Omega$$

The OP AMP circuit with feedback has an output Thévenin resistance that is much smaller than the output Thévenin resistance of the OP AMP device itself. In fact, the Thévenin resistance is very small since R_O is typically less than 100 Ω and A is greater than 10^5.

At this point we can summarize our discussion. We introduced the OP AMP as an active five-terminal device including two supply terminals not normally shown on the circuit diagram. We then developed an ideal model of this device that is used to analyze and design circuits that have feedback. Feedback must be present for the device to operate in the linear mode. The most dramatic feature of the ideal model is the assumption of infinite gain. Using a finite-gain model, we found that the ideal model predicts the circuit input-output relationship quite closely as long as the circuit gain K is much smaller than the OP AMP gain A. We also discovered that the Thévenin output resistance of an OP AMP with feedback is essentially zero.

In the rest of this book we use the ideal i–v constraints in Eq. (4–16) to analyze OP AMP circuits. The OP AMP circuits have essentially zero output resistance, which means that the output voltage does not change with different loads. Unless otherwise stated, from now on the term *OP AMP* refers to the ideal model.

4-5 OP AMP CIRCUIT ANALYSIS

This section introduces OP AMP circuit analysis using examples that are building blocks for analog signal-processing systems. We have already introduced one of these circuits—the noninverting amplifier discussed in the preceding section. The other basic circuits are the voltage follower, the inverting amplifier, the summer, and the subtractor. The key to using the building block approach is to recognize the feedback pattern and to isolate the basic circuit as a building block. The first example illustrates this process.

EXAMPLE 4-13

Find the input-output relationship of the circuit in Figure 4–31(a).

FIGURE 4-31

Source⟶ Noninverting amplifier (Active circuit) (a) ⟵ Load

(b)

SOLUTION:

When the circuit is partitioned as shown in Figure 4–31(a), we recognize two building block gains: (1) K_S, the proportionality constant of the source circuit, and (2) K_{AMP}, the gain of the noninverting amplifier. The OP AMP circuit input current $i_P = 0$; hence, we can use voltage division to find K_S as

$$K_S = \frac{v_P}{v_S} = \frac{R_2}{R_1 + R_2}$$

Since the noninverting amplifier has zero output resistance, the load R_L has no effect on the output voltage v_O. Using Eq. (4–19), the gain of the noninverting amplifier circuit is

$$K_{AMP} = \frac{v_O}{v_P} = \frac{R_3 + R_4}{R_4}$$

The overall circuit gain is found as

$$K_{\text{CIRCUIT}} = \frac{v_O}{v_S} = \left[\frac{v_P}{v_S}\right]\left[\frac{v_O}{v_P}\right]$$

$$= K_S \times K_{\text{AMP}}$$

$$= \left[\frac{R_2}{R_1 + R_2}\right]\left[\frac{R_3 + R_4}{R_4}\right]$$

The gain K_{CIRCUIT} is the product of K_S times K_{AMP} because the amplifier circuit does not load the source circuit since $i_P = 0$. ■

VOLTAGE FOLLOWER

The OP AMP in Figure 4–32(a) is connected as a **voltage follower** or **buffer**. In this case, the feedback path is a direct connection from the output to the inverting input. The feedback connection forces the condition $v_N = v_O$. The input current $i_P = 0$, so there is no voltage across the source resistance R_S. Applying KVL, we have the input condition $v_P = v_S$. The ideal OP AMP model requires $v_P = v_N$, so we conclude that $v_O = v_S$. By inspection, the closed-loop gain is $K = 1$. Since the output exactly equals the input, we say that the output follows the input (hence the name *voltage follower*).

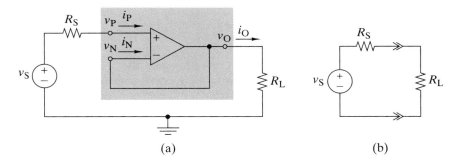

FIGURE 4–32 (a) Source-load interface with a voltage follower. (b) Interface without the voltage follower.

(a)   (b)

The voltage follower is used in interface circuits because it isolates the source from the load. Note that the input-output relationship $v_O = v_S$ does not depend on the source or load resistance. When the source is connected directly to the load, as in Figure 4–32(b), the voltage delivered to the load depends on R_S and R_L. The source and load interaction limits the signals that can be transferred across the interface, as discussed in Chapter 3. When the voltage follower is inserted between the source and load, the signal levels are limited by the capability of the OP AMP.

By Ohm's law, the current delivered to the load is $i_O = v_O/R_L$. But since $v_O = v_S$, the output current can be written in the form

$$i_O = v_S/R_L$$

Applying KCL at the reference node, we discover an apparent dilemma:

$$i_P = i_O$$

For the ideal model $i_P = 0$, but the preceding equations say that i_O cannot be zero unless v_S is zero. It appears that KCL is violated.

The dilemma is resolved by noting that the circuit diagram does not include the supply terminals. The output current comes from the power supply and not from the input. This dilemma arises only at the reference node (the ground terminal). In OP AMP circuits, as in all circuits, KCL must be satisfied. However, we must be alert to the fact that a KCL equation at the reference node could yield misleading results because the power supply terminals are not usually included in circuit diagrams.

Exercise 4–10

The circuits in Figure 4–32 have $v_S = 1.5$ V, $R_S = 2$ kΩ, and $R_L = 1$ kΩ. Compute the maximum power available from the source. Compute the power absorbed by the load resistor in the direct connection in Figure 4–32(b) and in the voltage follower circuit in Figure 4–32(a). Discuss any differences.

Answers:
281 μW; 250 μW; 2250 μW.

DISCUSSION: *With the direct connection, the power delivered to the load is less than the maximum power available. With the voltage follower circuit, the power delivered to the load is greater than the maximum value specified by the maximum power transfer theorem. However, the maximum power transfer theorem does not apply to the voltage follower circuit since the load power comes from the OP AMP power supply rather than the signal source.*

(a)

(b)

F I G U R E   4 – 3 3 *The inverting amplifier circuit.*

THE INVERTING AMPLIFIER

The circuit in Figure 4–33 is called an **inverting amplifier**. The key feature of this circuit is that the input signal and the feedback are both applied at the inverting input. Since the noninverting input is grounded, we have $v_P = 0$, an observation we will use shortly. The sum of currents entering node A can be written as

$$\frac{v_S - v_N}{R_1} + \frac{v_O - v_N}{R_2} - i_N = 0 \qquad (4\text{–}24)$$

The element constraints for the OP AMP are $v_P = v_N$ and $i_P = i_N = 0$. Since $v_P = 0$, it follows that $v_N = 0$. Substituting the OP AMP constraints into Eq. (4–24) and solving for the input-output relationship yields

$$v_O = -\left(\frac{R_2}{R_1}\right) v_S \qquad (4\text{–}25)$$

This result is of the form $v_O = Kv_S$, where K is the closed-loop gain. However, in this case the voltage gain $K = -R_2/R_1$ is negative, indicating a signal inversion (hence the name *inverting amplifier*). We use the block diagram symbol in Figure 4–33(b) to indicate either the inverting or the noninverting OP AMP configuration.

The OP AMP constraints mean that the input current i_1 in Figure 4–33(a) is

$$i_1 = \frac{v_S - v_N}{R_1} = \frac{v_S}{R_1}$$

This, in turn, shows that the input resistance seen by the source v_S is

$$R_{\text{IN}} = \frac{v_S}{i_1} = R_1 \qquad\qquad (4\text{--}26)$$

In other words, the inverting amplifier has a finite input resistance determined by the external resistor R_1.

The next example shows that the finite input resistance must be taken into account when analyzing circuits with OP AMPs in the inverting amplifier configuration.

EXAMPLE 4−14

Find the input-output relationship of the circuit in Figure 4–34(a).

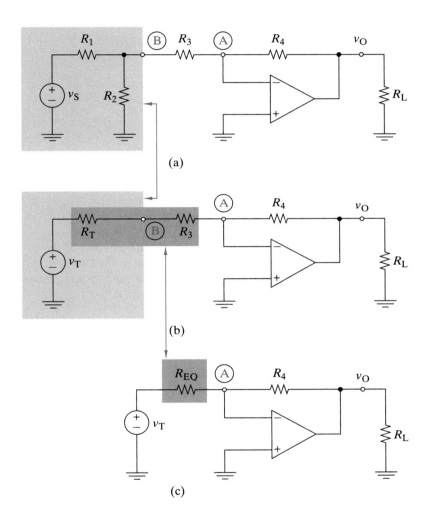

FIGURE 4 − 3 4

(a)

(b)

(c)

SOLUTION:

The circuit to the right of node B is an inverting amplifier. The load resistance R_L has no effect on the circuit transfer characteristics since the OP AMP has zero output resistance. However, the source circuit to the left of node B is influenced by the input resistance of the inverting amplifier circuit. The effect can be seen by constructing a Thévenin equivalent of the circuit to the left of node B, as shown in Figure 4–34(b). By inspection of Figure 4–34(a),

$$v_T = \frac{R_2}{R_1 + R_2} v_S$$

$$R_T = \frac{R_1 R_2}{R_1 + R_2}$$

In Figure 4–34(b) this Thévenin resistance is connected in series with the input resistor R_3, yielding the equivalence resistance $R_{EQ} = R_T + R_3$ shown in Figure 4–34(c). This reduced circuit is in the form of an inverting amplifier, so we can write the input-output relationship relating v_O and v_T as

$$K_1 = \frac{v_O}{v_T} = -\frac{R_4}{R_{EQ}} = -\frac{R_4(R_1 + R_2)}{R_1 R_2 + R_1 R_3 + R_2 R_3}$$

The overall input-output relationship from the input source v_S to the OP AMP output v_O is obtained by writing

$$K_{CIRCUIT} = \frac{v_O}{v_S} = \left[\frac{v_O}{v_T}\right]\left[\frac{v_T}{v_S}\right]$$

$$= -\left[\frac{R_4(R_1 + R_2)}{R_1 R_2 + R_1 R_3 + R_2 R_3}\right]\left[\frac{R_2}{R_1 + R_2}\right]$$

$$= -\left[\frac{R_2 R_4}{R_1 R_2 + R_1 R_3 + R_2 R_3}\right]$$

It is important to note that the overall gain is *not* the product of the source circuit voltage gain $R_2/(R_1 + R_2)$ and the inverting amplifier gain $-R_4/R_3$. In this circuit the two building blocks interact because the input resistance of the inverting amplifier circuit loads the source circuit. ∎

Exercise 4–11

Sketch the transfer characteristic of the OP AMP circuit in Figure 4–35(a) for $-10 \text{ V} < v_S < 10 \text{ V}$.

Answer:
The solution is shown in Figure 4–35(b).

THE SUMMING AMPLIFIER

The **summing amplifier** or **adder** circuit is shown in Figure 4–36(a). This circuit has two inputs connected at node A, which is called the **summing point**. Since the noninverting input is grounded, we have the condition $v_P = 0$. This configuration is similar to the inverting amplifier, so we start by applying KCL to write the sum of currents entering the node A summing point.

$$\frac{v_1 - v_N}{R_1} + \frac{v_2 - v_N}{R_2} + \frac{v_O - v_N}{R_F} - i_N = 0 \qquad (4\text{–}27)$$

With the noninverting input grounded, the OP AMP element constraints

FIGURE 4–35

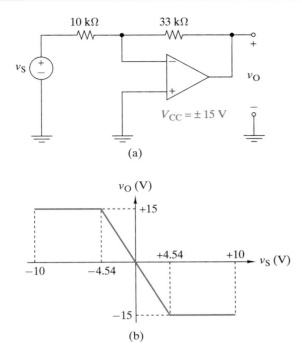

(a)

(b)

are $v_N = v_P = 0$ and $i_N = 0$. Substituting these OP AMP constraints into Eq. (4–27), we can solve for the circuit input-output relationship.

$$v_O = \left(-\frac{R_F}{R_1}\right) v_1 + \left(-\frac{R_F}{R_2}\right) v_2$$

$$= (K_1) v_1 + (K_2) v_2$$

(4–28)

The output is a weighted sum of the two inputs. The scale factors (or gains, as they are called) are determined by the ratio of the feedback resistor R_F to the input resistor for each input: that is, $K_1 = -R_F/R_1$ and $K_2 = -R_F/R_2$. In the special case $R_1 = R_2 = R$, Eq. (4–28) reduces to

$$v_O = -\frac{R_F}{R}(v_1 + v_2)$$

In this special case the output is proportional to the sum of the two inputs (hence the name *summing amplifier* or, more precisely, *inverting summer*). A block diagram representation of this circuit is shown in Figure 4–36(b).

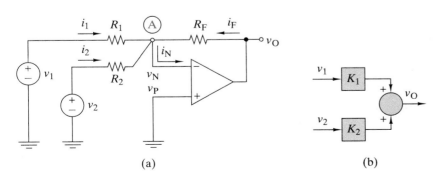

(a)   (b)

FIGURE 4–36 *The inverting summer.*

The summing amplifier in Figure 4–36 has two inputs, so there are two gains to contend with, one for each input. The input-output relationship in Eq. (4–28) is easily generalized to the case of n inputs as

$$v_O = \left(-\frac{R_F}{R_1}\right) v_1 + \left(-\frac{R_F}{R_2}\right) v_2 + \cdots + \left(-\frac{R_F}{R_n}\right) v_n$$

$$= K_1 v_1 + K_2 v_2 + \cdots + K_n v_n$$

(4–29)

where R_F is the feedback resistor and $R_1, R_2, \ldots R_n$ are the input resistors for the n input voltages $v_1, v_2, \ldots v_n$. You can easily verify this result by expanding the KCL sum in Eq. (4–27) to include n inputs, invoking the OP AMP constraints, and then solving for v_O.

▶ DESIGN EXAMPLE 4–15

Design an inverting summer that implements the input-output relationship

$$v_O = -(5 v_1 + 13 v_2)$$

SOLUTION:
The design problem involves selecting the input and feedback resistors so that

$$\frac{R_F}{R_1} = 5 \quad \text{and} \quad \frac{R_F}{R_2} = 13$$

One solution is arbitrarily to select $R_F = 65$ kΩ, which yields $R_1 = 13$ kΩ and $R_2 = 5$ kΩ. The resulting circuit is shown in Figure 4–37(a). The design can be modified to use standard resistance values for resistors with ±5% tolerance (see Appendix A, Table A-1). Selecting the standard value $R_F = 56$ kΩ requires $R_1 = 11.2$ kΩ and $R_2 = 4.31$ kΩ. The nearest standard values are 11 kΩ and 4.3 kΩ. The resulting circuit shown in Figure 4–37(b) incorporates standard value resistors and produces gains of $K_1 = 56/11 = 5.09$ and $K_2 = 56/4.3 = 13.02$. These nominal gains are within 2% of the values in the specified input-output relationship.   ■

FIGURE 4-37

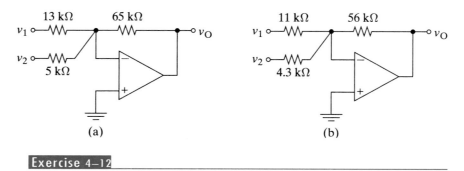

(a)   (b)

Exercise 4–12

(a) Find v_O in Figure 4–37(a) when $v_1 = 2$ V and $v_2 = -0.5$ V.
(b) If $v_1 = 400$ mV and $V_{CC} = \pm 15$ V, what is the maximum value of v_2 for linear mode operation?

(c) If $v_1 = 500$ mV and $V_{CC} = \pm 15$ V, what is the minimum value of v_2 for linear mode operation?

Answers:
(a) –3.5 V; (b) 1 V; (c) –1.346 V

THE DIFFERENTIAL AMPLIFIER

The circuit in Figure 4–38(a) is called a **differential amplifier** or **subtractor**. Like the summer, this circuit has two inputs, one applied at the inverting input and one at the noninverting input of the OP AMP. The input-output relationship can be obtained using the superposition principle.

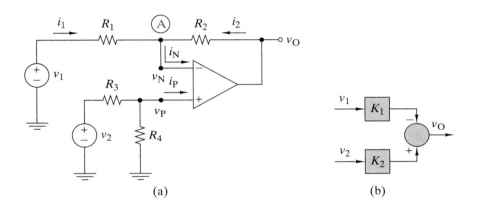

FIGURE 4 – 3 8 *The differential amplifier.*

(a)   (b)

First, we turn off source v_2, in which case there is no excitation at the noninverting input and $v_P = 0$. In effect, the noninverting input is grounded and the circuit acts like an inverting amplifier with the result that

$$v_{O1} = -\frac{R_2}{R_1} v_1 \tag{4–30}$$

Next, turning v_2 back on and turning v_1 off, we see that the circuit looks like a noninverting amplifier with a voltage divider connected at its input. This case was treated in Example 4–13, so we can write

$$v_{O2} = \left[\frac{R_4}{R_3 + R_4}\right]\left[\frac{R_1 + R_2}{R_1}\right] v_2 \tag{4–31}$$

Using superposition, we add outputs in Eqs. (4–30) and (4–31) to obtain the output with both sources on:

$$v_O = v_{O1} + v_{O2}$$
$$= -\left[\frac{R_2}{R_1}\right] v_1 + \left[\frac{R_4}{R_3 + R_4}\right]\left[\frac{R_1 + R_2}{R_1}\right] v_2 \tag{4–32}$$
$$= -[K_1]\, v_1 + [K_2]v_2$$

where K_1 and K_2 are the inverting and noninverting gains. Figure 4–38(b) shows how the differential amplifier is represented in a block diagram.

For the special case of $R_3/R_1 = R_4/R_2$, Eq. (4–32) reduces to

$$v_O = \frac{R_2}{R_1}(v_2 - v_1) \qquad (4-33)$$

In this case the output is proportional to the difference between the two inputs (hence the name *differential amplifier* or *subtractor*).

Exercise 4–13

(a) Find v_O in Figure 4–39 when $v_S = 10$ V.
(b) When $V_{CC} = \pm 15$ V, find the maximum value of v_S for linear mode operation.
(c) When $V_{CC} = \pm 15$ V, find the minimum value of v_S for linear mode operation.

Answers:
(a) –5 V; (b) 30 V; (c) –30 V

F I G U R E   4 - 3 9

BASIC OP AMP BUILDING BLOCKS

The block diagram representations of the basic OP AMP circuit configurations are shown in Figure 4–40. The noninverting and inverting amplifiers are represented as gain blocks. The summing amplifier and differential amplifier require both gain blocks and the summing point symbol. Considerable care must be used when translating from a block diagram to a circuit, or vice versa, since some gain blocks involve negative gains. For example, the gains of the inverting summer are negative. The required minus sign is sometimes moved to the summing point and the value of K within the gain block changed to a positive number. Since there is no standard convention for doing this, it is important to keep track of the signs associated with gain blocks and summing points.

EXAMPLE 4–16

Find the input-output relationship of the circuit in Figure 4–41(a).

SOLUTION:
This circuit is a cascade connection of three OP AMP circuits. The first circuit is an inverting amplifier with $K_1 = -0.33$, the second is a unity-gain inverting summer, and the final circuit is another inverting amplifier with $K_3 = -5/9$. Given these observations, we construct the block diagram shown in Figure 4–41(b). To trace the signal from input to output, we note that the input to the first circuit is 9.7 V and its output is $K_1 \times 9.7 = (-0.33) \times (9.7) = -3.2$ V. This voltage is added to the variable input v_F in the second circuit to produce $(-1) \times (-3.2) + (-1) \times (v_F) = 3.2 - v_F$. This result is then multiplied by the gain of the final circuit ($K_3 = -5/9$) to obtain the overall circuit output.

$$v_C = -\frac{5}{9}(3.2 - v_F) = \frac{5}{9}(v_F - 3.2)$$

This circuit implements a conversion of temperature from degrees Fahrenheit to degrees Celsius. Specifically, when the input voltage is $v_F =$

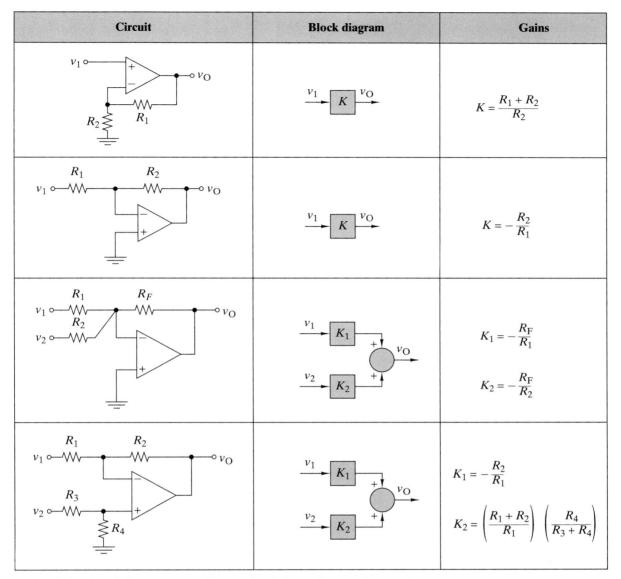

Circuit	Block diagram	Gains
		$K = \dfrac{R_1 + R_2}{R_2}$
		$K = -\dfrac{R_2}{R_1}$
		$K_1 = -\dfrac{R_F}{R_1}$ $K_2 = -\dfrac{R_F}{R_2}$
		$K_1 = -\dfrac{R_2}{R_1}$ $K_2 = \left(\dfrac{R_1 + R_2}{R_1}\right)\left(\dfrac{R_4}{R_3 + R_4}\right)$

FIGURE 4 – 4 0 *Summary of basic OP AMP signal processing circuits.*

$\theta_F/10$, where θ_F is a temperature in degrees Fahrenheit, then the output voltage is $v_C = \theta_C/10$, where θ_C is the equivalent temperature in degrees Celsius.

Each circuit in Figure 4–41 is an inverting amplifier with input resistances of about 10 kΩ. Since $V_{CC} = 15$ V, this means that the maximum current drawn at the input of any stage is about $15 \div 10^4 = 1.5$ mA, which is well within the power capabilities of general-purpose OP AMP devices.

Show that all OP AMPs are in the linear mode when the input temperature is in the range from −128°F to +182°F. ∎

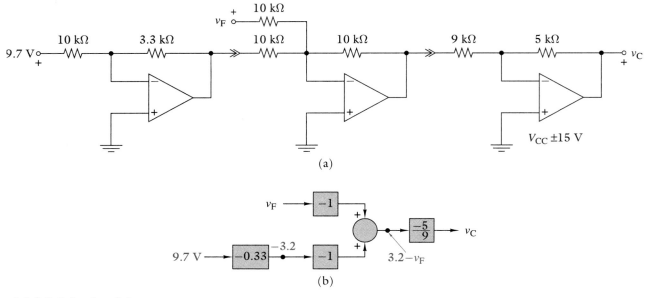

(a)

$v_F \longrightarrow \boxed{-1}$

$9.7\ V \longrightarrow \boxed{-0.33} \xrightarrow{-3.2} \boxed{-1}$

$3.2 - v_F$

$\boxed{\dfrac{-5}{9}} \longrightarrow v_C$

(b)

FIGURE 4-41

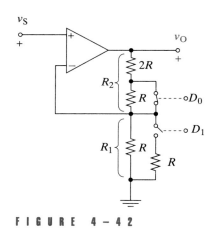

FIGURE 4-42

EXAMPLE 4-17

Figure 4–42 shows a noninverting amplifier in which the digital signals D_1 and D_0 adjust the effective feedback resistance R_1 and R_2 by opening and closing the analog switches shown. Determine the amplifier gain $K = v_O/v_S$ for $(D_1, D_0) = (0, 0)$, $(0, 1)$, $(1, 0)$ and $(1, 1)$. Assume an ideal analog switch with $R_{ON} = 0$ and $R_{OFF} = \infty$.

SOLUTION:

Note that D_1 controls a normally open switch, whereas D_0 controls a normally closed switch. The following truth table describes the switch states for each of the four digital input combinations.

DIGITAL INPUTS		SWITCH STATE	
D_1	D_0	LOWER	UPPER
0	0	open	closed
0	1	open	open
1	0	closed	closed
1	1	closed	open

When $(D_1, D_0) = (0, 0)$ the lower switch is open, the upper switch is closed, the feedback resistances are $R_1 = R$ and $R_2 = 2R$, and the amplifier gain is

$$K = \frac{R_1 + R_2}{R_1} = \frac{3R}{R} = 3$$

When $(D_1, D_0) = (0, 1)$ both switches are open, the feedback resistances are $R_1 = R$ and $R_2 = 3R$, and the amplifier gain is

$$K = \frac{R_1 + R_2}{R_1} = \frac{4R}{R} = 4$$

When $(D_1, D_0) = (1, 0)$, both switches are closed, the feedback resistances are $R_1 = R/2$ and $R_2 = 2R$, and the amplifier gain is

$$K = \frac{R_1 + R_2}{R_1} = \frac{2.5R}{0.5R} = 5$$

Finally, when $(D_1, D_0) = (1, 1)$, the lower switch is closed, the upper switch is open, the feedback resistances are $R_1 = R/2$ and $R_2 = 3R$, and the amplifier gain is

$$K = \frac{R_1 + R_2}{R_1} = \frac{3.5R}{0.5R} = 7$$

This example illustrates how digital signals can control an analog signal processor.   ■

NODE-VOLTAGE ANALYSIS WITH OP AMPS

In many applications we encounter OP AMP circuits that are more complicated than the four basic configurations in Figure 4–40. In such cases we use a modified form of node-voltage analysis that is based on the OP AMP connections in Figure 4–43. The overall circuit contains N nodes, including the three associated with the OP AMP. Normally the objective is to find the OP AMP output voltage (v_O) relative to the reference node (ground). We assign node voltage variables to the $N - 1$ nonreference nodes, including a variable at the OP AMP output. However, an ideal OP AMP acts like a dependent voltage source connected between the output terminal and ground. As a result, the OP AMP output voltage is determined by the other node voltages, so we do not need to write a node equation at the OP AMP output node.

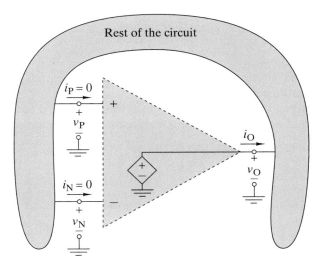

FIGURE   4 – 4 3   *General OP AMP circuit analysis.*

We formulate node equations at the other $N - 2$ nonreference nodes in the usual way. Since there are $N - 1$ node voltages, we seem to have more unknowns than equations. However, the OP AMP forces the condition $v_P = v_N$ in Figure 4–43. This eliminates one unknown node voltage since these two nodes are forced to have identical voltages. Finally, remember that the ideal OP AMP draws no current at its inputs ($i_P = i_N = 0$) in Figure 4–43, so these currents can be ignored when formulating node equations.

The following steps outline an approach to the formulation of node equations for OP AMP circuits:

STEP 1: Identify a node voltage at all nonreference nodes, including OP AMP outputs, but do *not* formulate node equations at the OP AMP output nodes.

STEP 2: Formulate node equations at the remaining nonreference nodes and then use the ideal OP AMP voltage constraint $v_P = v_N$ to reduce the number of unknowns.

EXAMPLE 4–18

Find the input-output relationship of the circuit in Figure 4–44.

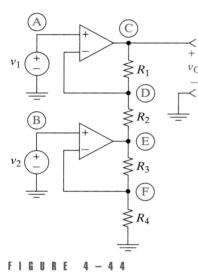

FIGURE 4 – 4 4

SOLUTION:

The circuit contains two OP AMPs and a total of six nonreference nodes. Nodes C and E are connected to OP AMP outputs, and nodes A and B are connected to the grounded independent voltage sources. As a result, we only require node equations at nodes D and F. By inspection,

$$\text{Node D: } (G_1 + G_2)v_D - G_1 v_C - G_2 v_E = 0$$

$$\text{Node F: } \qquad (G_3 + G_4)v_F - G_3 v_E = 0$$

This formulation yields two equations in four unknowns. However, the noninverting inputs are connected to independent voltage sources, so $v_A = v_1$ and $v_B = v_2$. The OP AMP voltage constraint ($v_P = v_N$) means $v_D = v_A = v_1$ and $v_F = v_B = v_2$. Substituting these constraints reduces the two node equations to

$$G_1 v_C + G_2 v_E = (G_1 + G_2) v_1$$

$$G_3 v_E = (G_3 + G_4)v_2$$

Node analysis quickly reduces this rather formidable appearing OP AMP circuit to two equations in two unknowns.

The circuit output voltage is $v_O = v_C$, so we use the second equation to eliminate v_E from the first equation and then solve for the input-output relationship as

$$v_O = v_C = \left[\frac{G_1 + G_2}{G_1}\right]v_1 - \frac{G_2}{G_1}\left[\frac{G_3 + G_4}{G_3}\right]v_2$$

For the special case $R_1 = R_4$ and $R_2 = R_3$, this equation reduces to

$$v_O = \frac{R_1 + R_2}{R_2}[v_1 - v_2]$$

The configuration in Figure 4–44 is a two-OP-AMP subtractor. The advantage of this circuit over the one-OP-AMP configuration in Figure 4–40 is that both inputs are connected to noninverting inputs, so the two-OP-AMP subtractor does not load the input sources. ∎

Exercise 4–14

Find the input-output relationship for the circuit in Figure 4–45.

Answer:

$$v_O = -\left[\frac{R_2R_3 + R_2R_4 + R_3R_4}{R_1R_3}\right]v_S$$

FIGURE 4 – 45

EXAMPLE 4 – 19

Circuit simulation programs like PSpice or Electronics Workbench do not support the theoretical characteristics of the ideal OP AMP model we use in circuit analysis and design. The default OP AMP in the Electronics Workbench library lists its model as "ideal." The purpose of this example is to evaluate this default model as a stand-in for the theoretical ideal OP AMP.

(a) Determine the parameters of the default OP AMP model in Electronics Workbench.
(b) Compare the theoretical voltage gain, input resistance, and output resistance of an inverting OP AMP circuit with $R_1 = 10$ kΩ and $R_2 = 50$ kΩ with the simulation results predicted by Electronics Workbench using its default model.

SOLUTION:

(a) Figure 4–46 shows the circuit symbol for the three-terminal OP AMP component available in the Analog IC's group of the Parts Bin toolbar. Double-clicking on the circuit symbol brings up a menu of OP AMPs including the default OP AMP whose model is listed as "ideal." Clicking on edit brings up the parameter list shown in Figure 4–46. The voltage gain, input resistance, and output resistance of the "ideal" model are listed as $A = 10^6$, $R_I = 10^{10}$, and $R_O = 1$. The corresponding theoretical values for an ideal OP AMP are $A = \infty$, $R_I = \infty$, and $R_O = 0$. Thus, the default OP AMP approximates, but does not duplicate the theoretical ideal. The question is, how much difference does this make?
(b) Figure 4–47 shows different stages of constructing the circuit diagram for an inverting amplifier with $R_1 = 10$ kΩ and $R_2 = 50$ kΩ. The OP AMP is the default model whose parameters were found in (a). The left side of the figure shows the component placements prior to interconnection. The resistors are available in the Basic Group of the Parts Bin toolbar. The input voltage source and ground are found in the Sources Group. To wire components together, we point at a component terminal, click and drag a wire to the terminal of another component, and release the mouse button. The program then automatically routes the wire

at right angles, without overlapping any of the components. The circuit with all components interconnected is shown in the right side of Figure 4–47.

From the Analysis Menu we select the Transfer Function option, which brings up the window shown at the top of Figure 4–48. In this window we identify the output at the voltage between node 3 and node 0 (ground), and the input as the voltage source VS. Clicking on the

FIGURE 4–48

(b) Components Wired Together

simulate button causes the program to perform a dc analysis of the circuit and report the results shown at the bottom of Figure 4–48. Using the default model the program predicts an output resistance of 5.999997 $\mu\Omega$, a voltage gain of –4.99997, and an input resistance of 10.00005 kΩ. An ideal OP AMP in the same inverting configuration would have an output resistance of zero, a voltage gain of $-R_2/R_1 = -5$, and an input resistance of $R_{IN} = R_1 = 10$ kΩ. The results using the default model agree with those of the ideal model to within four significant figures for voltage gain and input resistance. The output resistance is not exactly zero, but the value 6 $\mu\Omega$ is negligible compared with other circuit resistances. In summary, the default model and the ideal model predict essentially identical results in typical OP AMP circuit applications. From this point on we will use the default model in circuit simulations as a stand-in for the ideal OP AMP.   ■

4–6 OP AMP Circuit Design

With OP AMP circuit analysis, we are asked to find the input-output relationship for a given circuit configuration. An OP AMP circuit analysis problem has a unique answer. In OP AMP circuit design, we are given an equation or block diagram representation of a signal-processing function and asked to devise a circuit configuration that implements the desired function. Circuit design can be accomplished by interconnecting the amplifier, summer, and subtractor building blocks shown in Figure 4–40. The design process is greatly simplified by the nearly one-to-one correspondence between the OP AMP circuits and the elements in a block diagram. However, a design problem may not have a unique answer since often there are several OP AMP circuits that meet the design objective. The following example illustrates the design process.

D DESIGN EXAMPLE 4–20

Design an OP AMP circuit that implements the block diagram in Figure 4–49.

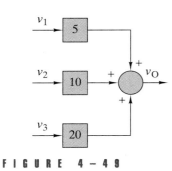

FIGURE 4–49

SOLUTION:

The input-output relationship represented by the block diagram is

$$v_O = 5v_1 + 10v_2 + 20v_3$$

The inverting summer can be used to produce the summation required in this relationship. A three-input adder implements the relationship

$$v_O = -\left[\frac{R_F}{R_1} v_1 + \frac{R_F}{R_2} v_2 + \frac{R_F}{R_3} v_3 \right]$$

The required scale factors are obtained by first selecting $R_F = 100$ kΩ, and then choosing $R_1 = 20$ kΩ, $R_2 = 10$ kΩ, and $R_3 = 5$ kΩ. However, the circuit involves a signal inversion. To implement the block diagram, we must add an inverting amplifier ($K = -R_2/R_1$) with $R_1 = R_2 = 100$ kΩ. The final implementation is shown in Figure 4–50. ∎

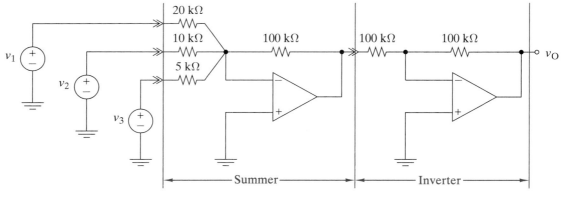

FIGURE 4–50

EVALUATION EXAMPLE 4–21

It is claimed that the OP AMP circuit in Figure 4–51 implements the same input-output function as the circuit in Figure 4–50. Verify this claim and compare the two circuits in terms of element count.

FIGURE 4–51

SOLUTION:
The circuit in Figure 4–51 is a modification of the OP AMP subtractor in Figure 4–40. In the modified circuit, several inputs are connected to the noninverting input and the negative input is grounded. The output voltage is related to the signal v_P by a noninverting amplifier gain; that is,

$$v_O = Kv_P = \frac{100 \times 10^3 + 2.94 \times 10^3}{2.94 \times 10^3} v_P$$

$$= 35v_P$$

Applying KCL, the sum of currents entering the noninverting input is

$$\frac{v_1 - v_P}{2 \times 10^4} + \frac{v_2 - v_P}{10^4} + \frac{v_3 - v_P}{0.5 \times 10^4} = 0$$

Multiplying this equation by 10^4 and solving for v_P yields

$$v_P = \frac{0.5}{3.5} v_1 + \frac{1}{3.5} v_2 + \frac{2}{3.5} v_3$$

Thus the overall input-output relationship is

$$v_O = 35v_P = 5v_1 + 10v_2 + 20v_3$$

and the claim is verified. The circuit in Figure 4–51 has the same input–output relationship as the circuit in Figure 4–50. The circuit in Figure 4–51 requires only one OP AMP since the inverting stage in Figure 4–50 is not required.

The circuit in Figure 4–51 is an example of a **noninverting summer**. The input-output relationship for a general noninverting summer in Figure 4–52 is

FIGURE 4–52

$$v_O = K\left[\left(\frac{R_{EQ}}{R_1}\right)v_1 + \left(\frac{R_{EQ}}{R_2}\right)v_2 + \cdots + \left(\frac{R_{EQ}}{R_n}\right)v_n\right] \qquad (4\text{-}34)$$

where R_{EQ} is the Thévenin resistance looking to the left at point P with all sources turned off (i.e., $R_{EQ} = R_1\|R_2\|R_3 \ldots \|R_n$) and K is the gain of the noninverting amplifier circuit to the right of point P. There are several similarities between this equation and the general inverting summer result in Eq. (4–29). In both cases the gain factor assigned to each input voltage is inversely proportional to the input resistance to which it is connected. The gain factor is directly proportional to the feedback resistor R_F in the inverting summer and is directly proportional to R_{EQ} in the noninverting summer. ∎

DIGITAL-TO-ANALOG CONVERTERS

A **digital-to-analog converter** (DAC) is a signal processor whose input is an n-bit digital word and whose output is an analog signal proportional to the binary value of the digital input. For example, the parallel four-bit digital signal in Figure 4–53 represents the value of a signal. Each bit can only have two values: (1) a high or "1" (e.g., +5 V) and (2) a low or "0" (e.g., 0 V). The bits have binary weights, so v_1 is worth $(2)^3 = 8$ times as much v_4, v_2 is worth $(2)^2 = 4$ times as much as v_4, and v_3 is worth $(2)^1 = 2$ times as much v_4. We call v_4 the **least significant bit** (LSB) and call v_1 the **most significant bit** (MSB). To convert the digital representation of the signal to analog form, we must weight the bits so that the analog output v_O is

$$v_O = \pm K(8v_1 + 4v_2 + 2v_3 + v_4) \qquad (4\text{-}35)$$

where K is an overall scale factor. This result is the input-output relationship of a 4-bit DAC.

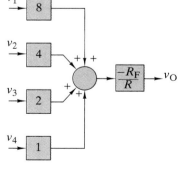

(a)

(b)

FIGURE 4 – 54 *A binary-weighted DAC: (a) Circuit diagram. (b) Block diagram.*

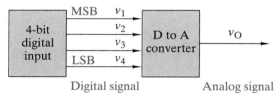

FIGURE 4 – 53 *A digital-to-analog converter (DAC).*

One way to implement Eq. (4–35) is to use an inverting summer with binary-weighted input resistors. Figure 4–54 shows the OP AMP circuit and a block diagram of the circuit input-output relationship. In either form, the output is seen to be a binary-weighted sum of the digital input scaled by $-R_F/R$. That is, the output voltage is

$$v_O = \frac{-R_F}{R}(8v_1 + 4v_2 + 2v_3 + v_4) \qquad (4\text{-}36)$$

The R-2R ladder circuit in Figure 4–55(a) also produces a 4-bit DAC. The resistance seen looking back into the R-2R ladder at point A, with all

FIGURE 4–55 *An R-2R ladder DAC: (a) Circuit diagram. (b) Thévenin equivalent.*

(a)

(b)

sources turned off, is seen to be $R_T = R$. The Thévenin equivalent circuit of the R-2R circuit is shown in Figure 4–55(b), where

$$v_T = \frac{v_1}{2} + \frac{v_2}{4} + \frac{v_3}{8} + \frac{v_4}{16}$$

The output voltage is found using the inverting amplifier gain relationship:

$$v_O = \frac{-R_F}{R} v_T = \frac{-R_F}{R}\left(\frac{v_1}{2} + \frac{v_2}{4} + \frac{v_3}{8} + \frac{v_4}{16}\right) \quad (4\text{–}37)$$

Using $R_F = R$ yields

$$v_O = -\frac{1}{16}(8v_1 + 4v_2 + 2v_3 + v_4) \quad (4\text{–}38)$$

This equation shows that the ladder assigns the correct binary weights to the digital inputs—namely, 8, 4, 2, and 1.

In theory, the circuits in Figures 4–54 and 4–55 perform the same signal-processing function—namely, a 4-bit DAC. However, there are important practical differences between the two circuits. The inverting summer in Figure 4–54 requires precision resistors with four different values spanning an 8:1 range. An 8-bit converter would require eight precision resistors spanning a 256:1 range. Moreover, the digital voltage sources in Figure 4–54 see input resistances that span an 8:1 range; therefore, the source-load interface is not the same for each bit. On the other hand, the resistances in the R-2R ladder converter in Figure 4–55 only span a 2:1 range regardless of the number of digital bits. Another desirable feature of the R-2R ladder is that it presents the same input resistance to each binary input.

Exercise 4–15

Find the output voltage when the following inputs are applied to the *R-2R* ladder D/A converter in Figure 4–55(a) with $R_F = 16R/5$.

(a) $v_1 = 0$ V; $v_2 = 0$ V; $v_3 = 0$ V; $v_4 = 5$ V.
(b) $v_1 = 0$ V; $v_2 = 0$ V; $v_3 = 5$ V; $v_4 = 5$ V.
(c) $v_1 = 0$ V; $v_2 = 5$ V; $v_3 = 0$ V; $v_4 = 5$ V.
(d) $v_1 = 5$ V; $v_2 = 0$ V; $v_3 = 0$ V; $v_4 = 5$ V.

Answers:
(a) $v_O = -1$ V, (b) $v_O = -3$ V, (c) $v_O = -5$ V, (d) $v_O = -9$ V.

D ◗ **DESIGN EXAMPLE 4–22**

Design a noninverting summer that implements a 3-bit D/A converter defined by

$$v_O = \frac{1}{5}[4v_1 + 2v_2 + v_3]$$

SOLUTION:

From Eq. (4–34) the input-output relationship for a three-input noninverting summer is

$$v_O = K\left[\left(\frac{R_{EQ}}{R_1}\right)v_1 + \left(\frac{R_{EQ}}{R_2}\right)v_2 + \left(\frac{R_{EQ}}{R_3}\right)v_3\right]$$

To implement the given relationship, we first select the input resistors so that the correct weight is assigned to each input. One way to accomplish this is to assign $R_1 = 10$ kΩ, $R_2 = 20$ kΩ, and $R_3 = 40$ kΩ. For this assignment $R_{EQ} = R_1\|R_2\|R_3 = 1/0.175$ kΩ, and the input-output relationship is

$$v_O = K\left[\frac{1}{1.75}v_1 + \frac{1}{3.5}v_2 + \frac{1}{7}v_3\right]$$

$$= \frac{K}{7}[4v_1 + 2v_2 + v_3]$$

To achieve the specified scale factor, we need $K = 7/5 = 1.4$. Figure 4–56 shows the circuit diagram for the noninverting summer design.   ■

FIGURE 4 - 5 6

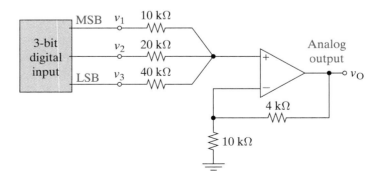

INSTRUMENTATION SYSTEMS

One of the most interesting and useful applications of OP AMP circuits is in instrumentation systems that collect and process data about physical phenomena. Such systems normally involve the signal processing elements shown in Figure 4–57. For our purposes we define a **transducer** as a device that converts a change in some physical quantity, such as temperature or pressure, into a change in an electrical signal. The transducer's electrical output signal is processed by OP AMP circuits that perform signal conditioning and amplification. **Signal conditioning** compensates for transducer shortcomings and produces a signal suitable for further electrical processing. Signal **amplification** is required because the transducer output voltage is quite small, often in the millivolt range or less. In simple systems the amplified analog output goes to a **display**, such as a digital voltmeter. In data acquisition systems with many transducers, the analog signals are converted to digital form and sent to a **computer** for further processing and storage. In some systems the processed data may be used in a feedback system to control the physical process monitored by the transducers.

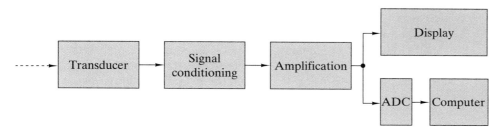

FIGURE 4 – 57   *A block diagram of a basic instrumentation system.*

Our interest here is the design of the OP AMP circuits that perform the signal conditioning and amplification functions in an instrumentation system. The overall objective of these circuits is to deliver an analog output signal that falls within a prescribed range and is directly proportional to the value of the physical variable measured by the input transducer. The input transducer converts a physical variable x into an electrical voltage v_{TR}. For many transducers there is a linear range in which this voltage is of the form

$$v_{TR} = mx + b$$

where m is a calibration constant and b is a constant offset or bias. The transducer voltage is usually quite small and must be amplified by a gain K. Direct amplification of the transducer output would produce both the desired signal component $K(mx)$ and an unwanted bias component $K(b)$. The signal conditioning circuit removes the bias component so that the analog output $K(mx)$ is directly proportional to the physical variable measured.

Let's look at a specific example. In a laboratory experiment the amount (5 to 20 lumens) of incident light is to be measured using a photocell as the input transducer. The system output is to be sent to an analog-to-digital converter whose input range is 0 to 5 V. The transducer characteristics of the photocell are shown in Figure 4–58. The overall design requirements

FIGURE   4 – 58   *Photocell transducer characteristics.*

are that 5 lumens produce an analog output of 0 V, and that 20 lumens produce 5 V. From the transducer characteristics we see that a change in light intensity $\Delta\Phi = (20 - 5) = 15$ lumens will produce a change in transducer output of $\Delta v = (0.6 - 0.2) = 0.4$ mV. This 0.4-mV change must be translated into a 5-V change at the output. To accomplish this, the transducer voltage must be amplified by an overall gain of

$$K = \frac{\text{specified output range}}{\text{available input range}} = \frac{(5 - 0)}{(0.6 - 0.2) \times 10^{-3}} = 1.25 \times 10^4$$

If the transducer's output voltage range (0.2 to 0.6 mV) is multiplied by the gain $K = 1.25 \times 10^4$, the resulting output voltage range is 2.5 to 7.5 V, rather than the required range of 0 to 5 V. The bias in the output range is removed by subtracting a constant voltage at some point in the signal conditioning circuit.

Figure 4–59(a) shows a block diagram of one possible way to distribute the required signal-processing functions. The required gain is divided between two stages because the overall voltage gain of $K = 1.25 \times 10^4$ is too large to achieve with a single-stage OP AMP circuit. The transducer bias is removed by subtracting a constant voltage following the first-stage amplification ($K_1 = 100$). The second-stage amplification ($K_2 = 125$) then produces the required output voltage range. Figure 4–59(b) shows one possible OP AMP circuit that implements this block diagram. The first stage is a subtractor circuit, and the second stage is an inverting amplifier. This arrangement has two signal inversions so that the overall gain in the signal path is

$$K = (-100)(-125) = 1.25 \times 10^4$$

which is positive as required. As always, many other alternative designs are possible.

F I G U R E 4 – 5 9 *Transducer signal conditioning design. (a) Block diagram. (b) Circuit diagram.*

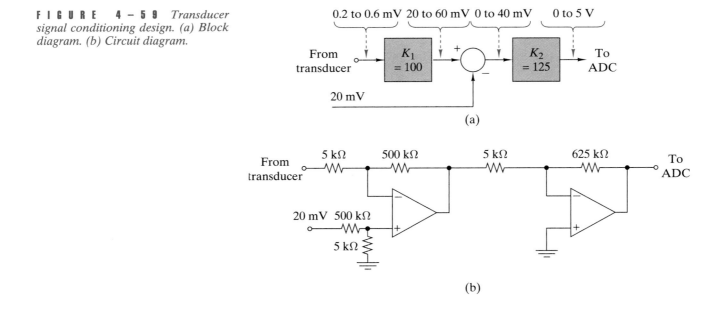

(a)

(b)

⬤ **DESIGN EXAMPLE 4–23**

A strain gauge is a resistive device that measures the elongation (strain) of a solid material caused by applied forces (stress). A typical strain gauge consists of a thin film of conducting material deposited on an insulating substrate. When bonded to a member under stress, the resistance of the gauge changes by an amount

$$\Delta R = 2R_G \frac{\Delta L}{L}$$

where R_G is the resistance of the gauge with no applied stress and $\Delta L/L$ is the elongation of the material expressed as a fraction of the unstressed length L. The change in resistance ΔR is only a few tenths of a milliohm, far too small a value to be measured with a standard ohmmeter. To detect such a small change, the strain gauge is placed in a Wheatstone bridge circuit like the one shown in Figure 4–60. The bridge contains fixed resistors R_A and R_B, matched strain gauges R_{G1} and R_{G2}, and a precisely controlled reference voltage V_{REF}. The values of R_A and R_B are chosen so that the bridge is balanced ($v_1 = v_2$) when no stress is applied. When stress is applied, the resistance of the stressed gauge changes to $R_{G2} + \Delta R$, and the bridge is unbalanced ($v_1 \neq v_2$). The differential signal ($v_2 - v_1$) indicates the strain resulting from the applied stress.

Design an OP AMP circuit to translate strains over the range $0 < \Delta L/L < 0.02\%$ into an output voltage on the range $0 < v_O < 4$ V, for $R_G = 120\ \Omega$ and $V_{REF} = 25$ V.

FIGURE   4 – 6 0 *Strain gauges in a Wheatstone bridge.*

SOLUTION:

When an external stress is applied, the resistance R_{G2} changes to $R_{G2} + \Delta R$. Applying voltage division to each leg of the bridge yields

$$v_2 = \frac{R_{G2} + \Delta R}{R_{G1} + R_{G2}} V_{REF}$$

$$v_1 = \frac{R_B}{R_A + R_B} V_{REF}$$

The differential voltage ($v_2 - v_1$) can be written as

$$v_2 - v_1 = V_{REF} \left[\frac{R_{G2} + \Delta R}{R_{G1} + R_{G2}} - \frac{R_A}{R_A + R_B} \right]$$

To achieve bridge balance in the unstressed state, we select $R_{G1} = R_{G2} = R_A = R_B = R_G$, in which case the differential voltage reduces to

$$v_2 - v_1 = V_{REF} \left[\frac{\Delta R}{2R_G} \right] = V_{REF} \left[\frac{\Delta L}{L} \right]$$

Thus, the differential voltage is directly proportional to the strain $\Delta L/L$. However, for $V_{REF} = 25$ V and $\Delta L/L = 0.02\%$, the differential voltage is only $(V_{REF})(\Delta L/L) = 25 \times 0.0002 = 5$ mV. To obtain the required 4-V output, we need a voltage gain of $K = 4/0.005 = 800$.

The OP AMP subtractor in Figure 4–40 is ideally suited to the task of amplifying differential signals. Selecting $R_1 = R_3 = 10$ kΩ and $R_2 = R_4 = 8$ MΩ produces an input-output relationship for the subtractor circuit of

$$v_O = 800(v_2 - v_1)$$

Figure 4–61 shows the basic design.

FIGURE 4 – 61

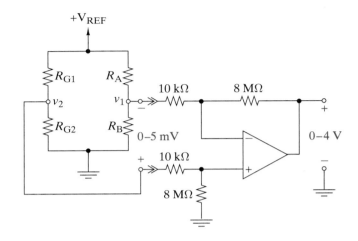

The input resistance of the subtractor circuit must be large to avoid loading the bridge circuit. The Thévenin resistance looking back into the bridge circuit is

$$R_T = R_{G1}\|R_{G2} + R_A\|R_B$$
$$= R_G\|R_G + R_G\|R_G$$
$$= R_G = 120 \ \Omega$$

This value is small compared with the 10-kΩ input resistance of the subtractor's inverting input.

Comment: The transducer in this example is the resistor R_{G2}. In the unstressed state the voltage across this resistor is $v_2 = 12.5$ V. In the stressed state the voltage is $v_2 = 12.5$ V plus a 5-mV signal. In other words, the transducer's 5-mV signal component is accompanied by a much larger bias component. We cannot afford to amplify the 12.5-V bias component by $K = 800$ before subtracting it out. The bias is eliminated at the input by using a bridge circuit in which $v_1 = 12.5$ V, and then processing the differential signal $v_2 - v_1$. The situation illustrated in this example is quite common. Consequently, the first stage in most instrumentation systems is a differential amplifier that removes the transducer bias. ∎

D DESIGN EXERCISE: 4–16

Design an OP AMP circuit to translate the temperature range from –20°C to +120°C onto a 0- to 1-V output signal using the transducer characteristics shown in Figure 4–62(a). The transducer has a Thévenin resistance of 880 Ω.

Answer: One possible design is shown in Figure 4–62(b).

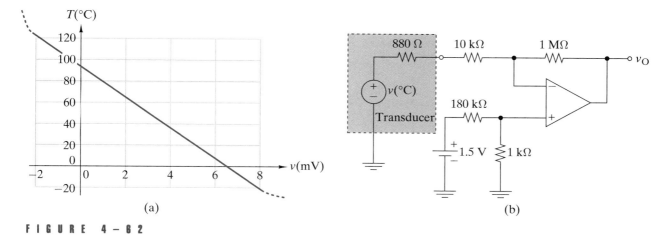

FIGURE 4–62

DISCUSSION: *It is important for the input resistor of the inverting amplifier to be at least 10 times as large as the Thévenin resistance of the transducer. Our design uses an input resistance of 10 kΩ compared with a transducer Thévenin resistance of 880 Ω.*

RESISTOR TOLERANCES

Parameter **tolerance** is the total amount by which the value of a parameter is allowed to vary. Specifying a tolerance of, say, 10% means that the parameter is allowed to vary by ±10% from its nominal value. In linear OP AMP circuits the gain tolerance is determined by the resistors in the feedback path. A logical question is what tolerance must be placed on these resistors to ensure that the gain remains within a specified tolerance.

We use **worst-case analysis** to introduce the component tolerance assignment problem. Under worst-case analysis, it is assumed that every component value lies at one of its extreme values: either $X_{MAX} = X(1 + \Delta X)$ or $X_{MIN} = X(1 - \Delta X)$, where X is the nominal value and ΔX is the parameter tolerance. Moreover, it is assumed that these extreme values combine in such a way as to produce the worst possible outcome, hence the term *worst case*.

To use a concrete example, consider the gain of the noninverting amplifier configuration in Figure 4–40.

$$K = \frac{R_1 + R_2}{R_2} = 1 + \frac{R_1}{R_2} \qquad (4\text{–}39)$$

Clearly the gain K will be a maximum when R_1 is at a maximum and R_2 is at a minimum. In worst-case analysis notation the maximum gain is written as

$$K(1 + \Delta K) = 1 + \frac{R_1(1 + \Delta R)}{R_2(1 - \Delta R)}$$

where ΔK is the gain tolerance resulting from a resistor tolerance ΔR.

Looking back at Eq. (4–39), we see that $R_1/R_2 = K - 1$, so the preceding equation can be written as

$$K(1 + \Delta K) = 1 + (K - 1)\frac{1 + \Delta R}{1 - \Delta R}$$

Solving this equation for ΔK yields

$$\Delta K = \frac{K - 1}{K}\frac{2 \Delta R}{1 - \Delta R} \quad \text{for } K_{\text{MAX}} \tag{4–40}$$

This equation yields the worst-case gain tolerance when the component tolerances combine to produce the maximum gain.

By similar reasoning, in Eq. (4–39) the minimum gain occurs when R_1 is at a minimum and R_2 is at a maximum. In worst-case notation the minimum gain is

$$K(1 - \Delta K) = 1 + \frac{R_1(1 - \Delta R)}{R_2(1 + \Delta R)}$$

Again, recognizing that $R_1/R_2 = K - 1$ and solving for ΔK yields

$$\Delta K = \frac{K - 1}{K}\frac{2\Delta R}{1 + \Delta R} \quad \text{for } K_{\text{MIN}} \tag{4–41}$$

This equation yields the worst-case gain tolerance when the component tolerances combine to produce the minimum gain.

Comparing Eqs. (4–40) and (4–41), we see that the "worst of the worst" occurs when the gain is at a maximum. Since $(K - 1)/K < 1$, we can place an upper bound on the worst-case gain tolerance of

$$\Delta K \leq \frac{2\Delta R}{1 - \Delta R} \quad \text{worst case} \tag{4–42}$$

We used the noninverting amplifier to obtain this result. It turns out that the same bound applies to the inverting amplifier as well.

For $\Delta R \ll 1$, Eq. (4–42) reduces to $\Delta K < 2\Delta R$, which suggests the following rule of thumb:

> *The worst-case gain tolerance for a single-stage OP AMP amplifier circuit is roughly twice the resistor tolerance.*

If we use 1% resistors, then the worst-case gain tolerance will be about 2%. Conversely, if we want a worst case gain tolerance of 3%, the rule of thumb suggests a resistor tolerance of 1.5%. Unfortunately, commercially available standard resistors have tolerances of 0.5%, 1%, 2%, 5%, 10%, and 20%. Hence, for practical reasons we are forced to use 1% resistors to achieve 3% gain tolerance.

The following example illustrates the impact of multistage designs on worst-case analysis.

D E S I G N E X A M P L E 4 – 2 4

Design an OP AMP circuit with an overall gain of $K_{\text{TOTAL}} = +10^4 \pm 5\%$. The circuit must be designed to accommodate general-purpose OP AMPs with $A = 10^5$.

SOLUTION:

The overall gain cannot be achieved in a single-stage design since the circuit gain would not be small compared with the OP AMP's gain. One simple approach is a two-stage design using identical noninverting amplifiers, each realizing $K = 100$. The cascade connection of two such amplifiers yields an overall gain of $K^2 = K_{TOTAL} = 10^4$. A gain $K = 100$ can be obtained in the noninverting OP AMP configuration in Figure 4–40 using $R_1 = 99$ kΩ and $R_2 = 1$ kΩ.

What tolerance should be placed on these resistors to achieve an overall gain tolerance of 5%? Our rule of thumb suggests 2% resistor tolerance. However, Eq. (4–40) points out that 2% resistor tolerance yields a worst-case gain tolerance of

$$\Delta K = \frac{99}{100} \frac{2 \times 0.02}{1 - 0.02} = 0.040$$

For a resistor tolerance of 2%, the worst-case gain per stage is 100 (1 + 0.04). Hence, a cascade connection of two such amplifiers yields a worst-case overall gain of

$$100 \, (1 + 0.04) \times 100 \, (1 + 0.04) \approx 10^4 \, (1 + 0.08)$$

In summary, 2% resistors produce a worst-case tolerance of 8% on the overall gain. Thus, to achieve a 5% gain tolerance in a two-stage design, we must use 1% resistors.　　　　　　　　　　　　　　　■

In the preceding analysis and example, it is a simple matter to determine the component values that lead to worst-case performance. In more complicated circuits it is generally not practical to determine analytically the combination of extreme values that produce the worst case. For this reason, circuit analysis programs like SPICE include a worst-case analysis option. Under this option user-specified components are sequentially assigned their extreme values and the worst-case combination identified. Finding the worst of the worst takes two passes, one to find the worst maximum and the other to find the worst minimum. Component tolerances derived from worst-case analysis are very conservative since they depend on an unlikely combination of extreme value events.

4–7　THE COMPARATOR

Up to this point we have treated circuits in which negative feedback keeps the OP AMP in the linear mode with $v_P \approx v_N$. When v_P and v_N differ by more than a few millivolts, the OP AMP is driven into one of its two saturation modes:

1. +Saturation with $v_O = +V_{CC}$ when $(v_P - v_N) > 0$
2. −Saturation with $v_O = -V_{CC}$ when $(v_P - v_N) < 0$.

With no feedback the OP AMP operates in one of these two saturation modes whenever the inputs are not equal. Under saturation conditions the output voltage indicates whether $v_P > v_N$ or $v_P < v_N$. A device that discriminates between two unequal voltages is called a **comparator**.

FIGURE　4–63 *An OP AMP comparator.*

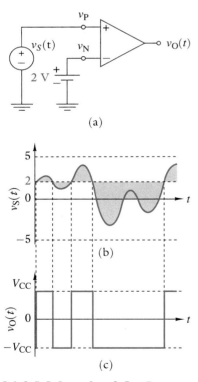

FIGURE 4–64 *Input and output signals of an OP AMP comparator.*

FIGURE 4–65 *Comparator with a 2-V offset: (a) Circuit diagram. (b) Input voltage. (c) Output voltage.*

Figure 4–63 shows an example of an OP AMP operating as a comparator. First note that there is no feedback and $v_N = 0$, since the inverting input is grounded. The voltage source connected to the noninverting input means that when $v_P = v_S > 0$, the OP AMP is driven into +saturation with $v_O = +V_{CC}$. Conversely, when $v_P = v_S < 0$, the OP AMP is in –saturation with $v_O = -V_{CC}$. Figure 4–64 shows an example of a plot of $v_S(t)$ versus time plus the resulting comparator output. The circuit in Figure 4–63 is called a **zero crossing detector** because the comparator output changes its saturation state whenever the input v_S crosses through zero. Notice that the comparator output is not proportional to the input, which shows that the comparator is a nonlinear device.

A modified version of the zero-crossing detector is shown in Figure 4–65. In this circuit a constant 2-V source is applied at the inverting input and the input signal v_S from Figure 4–64 is applied to the noninverting input. The input signal is now compared with 2 V rather than with zero. When $v_P = v_S > 2$ V, the OP AMP is in +saturation with $v_O = +V_{CC}$, and when $v_P = v_S < 2$ V, the OP AMP is in –saturation with $v_O = -V_{CC}$. A plot of the resulting output voltage is shown in Figure 4–65(c). The value of the fixed source determines the input signal level at which the comparator switches from one saturation state to the other. For example, connecting a 10-V fixed source to the inverting input causes the comparator output to remain at $-V_{CC}$, since v_S never exceeds 10 V.

Although ordinary OP AMPs can be used as comparators, there are integrated circuit devices specifically designed to operate in saturation. These comparator devices are designed to switch rapidly from one saturation state to the other and to have output saturation levels that are compatible with digital circuits.

Using digital circuit terminology, the input-output characteristics of an **ideal comparator** can be written as

$$\text{If } v_P > v_N, \text{ then } v_O = V_{OH}, \text{ else}$$
$$\text{If } v_P < v_N, \text{ then } v_O = V_{OL} \qquad (4\text{--}43)$$

where V_{OH} and V_{OL} are the high and low saturation levels of the element. For example, a comparator with $V_{OH} = +5$ V and $V_{OL} = 0$ V is compatible with commonly used TTL (transistor–transistor logic) digital circuits. Note that the equal sign is omitted in the conditional part of the statement because we cannot be certain what the output will be if $v_P = v_N$.

EXAMPLE 4–25

Figure 4–66 shows an OP AMP whose output is connected to two 15-V lamps. Analyze the circuit and find the input signal condition required to turn each of the lamps on.

SOLUTION:

The circuit has no feedback so the OP AMP operates as a comparator. Figure 4–66 shows the power supply inputs to the OP AMP. The positive supply terminal is connected to a +15-V source and the negative supply terminal is grounded. Therefore, the two output saturation levels are $V_{OH} = +15$ V and $V_{OL} = 0$ V.

FIGURE 4-66

Because the 5-V source is connected to the inverting input, the comparator changes its state whenever the input voltage v_S passes through 5 V. When $v_S < 5$ V, the comparator is in negative saturation with $v_O = V_{OL} = 0$ V. In this case, the voltages across lamps 1 and 2 are 15 V and 0 V, respectively. Conversely, when $v_S > 5$ V, the comparator is in positive saturation with $v_O = V_{OH} = +15$ V, and hence the voltages across lamps 1 and 2 are 0 V and 15 V, respectively. These observations can be summarized as follows:

INPUT RANGE	COMPARATOR STATE	OUTPUT VOLTAGE	LAMP 1	LAMP 2
$v_S < 5$ V	−Saturation	$v_O = 0$ V	On	Off
$v_S > 5$ V	+Saturation	$v_O = 15$ V	Off	On

APPLICATION NOTE: EXAMPLE 4-26

The circuit in Figure 4–67 is an analog-to-digital converter (ADC) that changes the analog voltage v_S into a 3-bit digital output v_{O1}, v_{O2}, and v_{O3}. The inverting inputs to the comparators are connected to a voltage divider. Using voltage division, we see that $v_{N1} = 1$ V, $v_{N2} = 3$ V, and $v_{N3} = 5$ V. The analog input v_S is applied simultaneously to the noninverting inputs of all three comparators. When $v_S < 1$ V, all three comparators are in −saturation with $v_{O1} = v_{O2} = v_{O3} = V_{OL} = 0$. When the analog input is in the range $3 > v_S > 1$, comparator 1 switches to +saturation and the three outputs are $v_{O1} = V_{OH} = 5$ V and $v_{O2} = v_{O3} = V_{OL} = 0$ V. When $5 > v_S > 3$, comparators 1 and 2 are in +saturation so the three outputs are $v_{O1} = v_{O2} = V_{OH} = 5$ V and $v_{O3} = V_{OL} = 0$ V. Finally, when $v_S > 5$ V, all three comparators are in +saturation with $v_{O1} = v_{O2} = v_{O3} = V_{OH} = 5$ V. These observations can be summarized as follows:

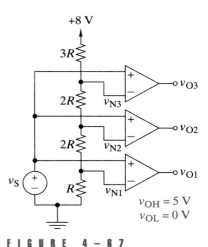

FIGURE 4-67

INPUT RANGE	COMPARATOR 1 V_{O1}	COMPARATOR 2 V_{O2}	COMPARATOR 3 V_{O3}
$1 > v_S$	0	0	0
$3 > v_S > 1$	5	0	0
$5 > v_S > 3$	5	5	0
$v_S > 5$	5	5	5

Note that each analog voltage range is converted into a unique 3-bit digital word. This 3-bit code can be converted to a 2-bit binary code using standard combinatorial logic design methods. The circuit is called a flash converter because the comparators operate in parallel and the analog-to-digital conversion takes place almost instantaneously.

SUMMARY

- The output of a dependent source is controlled by a signal in a different part of the circuit. Linear dependent sources are circuit elements used to model active devices and are represented in this text by a diamond-shaped source symbol. Each type of controlled source is characterized by a single-gain parameter μ, β, r, or g.

- The Thévenin resistance of a circuit containing dependent sources can be found using the open-circuit voltage and the short-circuit current, or by directly solving for the interface i–v characteristic. The active Thévenin resistance may be significantly different from the passive lookback resistance.

- The large-signal model divides the transistor operating characteristics into cutoff, active, and saturation modes. Each mode has a unique set of i–v characteristics. The operating mode can be found by assuming the transistor is in the active mode and checking the signs of i_B and v_{CE}. The operating mode is active if $i_B > 0$ and $v_{CE} > 0$, cutoff if $i_B < 0$, and saturation if $v_{CE} < 0$.

- The OP AMP is an active device with at least five terminals: the inverting input, the noninverting input, the output, and two power supply terminals. The power supply terminals are not usually shown in circuit diagrams. The integrated circuit OP AMP is a differential amplifier with a very high voltage gain.

- The OP AMP can operate in a linear mode when there is a feedback path from the output to the inverting input. To remain in the linear mode, the output voltage is limited to the range $-V_{CC} \leq v_O \leq +V_{CC}$, where $\pm V_{CC}$ are the supply voltages.

- The i–v characteristics of the ideal model of an OP AMP are $i_P = i_N = 0$ and $v_P = v_N$. The ideal OP AMP has an infinite voltage gain, an infinite input resistance, and zero output resistance. The ideal model is a good approximation to real devices as long as the circuit gain is much smaller than the OP AMP gain.

- Four basic signal-processing functions performed by OP AMP circuits are the inverting amplifier, noninverting amplifier, inverting summer, and differential amplifier. These arithmetic operations can also be represented in block diagram form.

- OP AMP circuits can be connected in cascade to obtain more complicated signal-processing functions. The analysis and design of the individual stages in the cascade can be treated separately, provided the input resistance of the following stage is kept sufficiently high.

- OP AMP circuits are easily treated using node analysis. A node voltage is identified at each OP AMP output, but a node equation is not written at these nodes. Node equations are then written at the remaining nodes, and the ideal OP AMP input voltage constraint ($v_N = v_P$) is used to reduce the number of unknowns.

- The comparator is a nonlinear signal-processing device obtained by operating an OP AMP device without feedback. The comparator has two analog inputs and a two-state digital output.

PROBLEMS

ERO 4–1 LINEAR ACTIVE CIRCUITS (SECTS. 4–1, 4–2)

Given a circuit containing linear resistors, dependent sources, and independent sources,

(a) Find selected output signal variables, input-output relationships, or equivalent circuits.
(b) Select circuit parameters so that a specified signal is delivered to a load.

4–1 For the circuit in Figure P4–1,
  (a) Find the voltage gain v_O/v_S.
  (b) Find the current gain i_O/i_x.
  (c) For $v_S = 0.1$ V, find the power supplied by the input source v_S and the power delivered to the 1-kΩ load resistor.

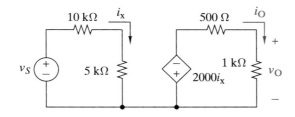

FIGURE P 4 – 1

4–2 For the circuit in Figure P4–2,
  (a) Find the voltage gain v_O/v_1.
  (b) Find the current gain i_O/i_S.
  (c) For $i_S = 2$ mA, find the power supplied by the input source i_S and the power delivered to the 2-kΩ load resistor.

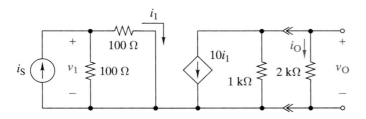

FIGURE P 4 – 2

4–3 The circuit in Figure P4–3 is a dependent-source model of a two-stage amplifier. For $v_S = 1$ mV, find the output voltage v_3 and the current gain i_3/i_1.

FIGURE P4–3

4–4 The circuit in Figure P4–4 is an ideal voltage amplifier with negative feedback provided via the resistor R_F.
  (a) For $v_S = 10$ mV and $R_F = 50$ kΩ, find the output voltage v_2 and the current gain i_2/i_1.
  (b) Find the input resistance $R_{IN} = v_1/i_1$.

FIGURE P4–4

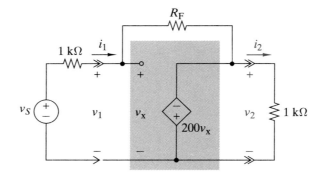

4–5 The circuit in Figure P4–5 is an ideal current amplifier with negative feedback via the resistor R_E.
  (a) For $i_S = 25$ µA and $R_E = 250$ Ω, find the output current i_2 and the voltage gain v_2/v_1.
  (b) Find the input resistance $R_{IN} = v_1/i_1$.

FIGURE P4–5

4–6 For the circuit in Figure P4–6,
  (a) Find the voltage gain v_O/v_S.
  (b) Find the current gain i_O/i_S.

(c) For $v_S = 20$ mV, find the power supplied by the input source v_S, the power delivered to the 500-Ω output resistor, and the power gain p_O/p_S in decibels.

4-7   Find the voltage gain v_O/v_S of the circuit in Figure P4-7.

4-8   The circuit in Figure P4-8 is a model of a feedback amplifier using two identical transistors. Formulate either node-voltage or mesh-current equations for this circuit. Use these equations to solve for the input-output relationship $v_O = Kv_S$ using $r_\pi = 1$ kΩ, $R_E = 200$ Ω, $R_C = 10$ kΩ, $R_L = 5$ kΩ, $R_F = 5$ kΩ, and $\beta = 100$.

4-9   Find the voltage v_x in Figure P4-9.

4-10   **D** Select the value of R_F in Figure P4-4 so that the voltage gain $K = v_2/v_S = -10$.

4-11   **D** The circuit in Figure P4-8 is a model of a feedback amplifier using two identical transistors. Using $r_\pi = 1$ kΩ, $R_C = 10$ Ω,

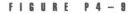

$R_L = 5$ kΩ, $R_F = 5$ kΩ, and $\beta = 50$, select the value of R_E so that the voltage gain $K = v_O/v_S = 50$.

4–12   Find the Thévenin equivalent circuit seen by the resistor R_L in Figure P4–12.

FIGURE   P 4 - 1 2

ERO 4–2 TRANSISTOR CIRCUITS (SECT. 4–3)

Given a circuit containing linear resistive elements and one transistor, use the large signal model to

(a) Find the transistor operating mode and selected circuit responses or transfer characteristics.
(b) Select circuit parameters to obtain a specified transistor operating mode.

4–13   The parameters of the circuit in Figure P4–13 are $R_B = 250$ kΩ, $R_C = 2$ kΩ, $\beta = 100$, $V_\gamma = 0.7$ V, and $V_{CC} = 10$ V:
   (a) Find i_B, i_C, and v_{CE} for $v_S = 8$ V.
   (b) Repeat part (a) with $\beta = 200$.

4–14   The parameters of the circuit in Figure P4–13 are $R_B = 100$ kΩ, $R_C = 5$ kΩ, $\beta = 50$, $V_\gamma = 0.6$ V, and $V_{CC} = 20$ V:
   (a) Find the range of values of v_S for which the transistor operates in the active mode.
   (b) Repeat part (a) with $\beta = 100$.

4–15   The parameters of the circuit in Figure P4–13 are $R_B = 250$ kΩ, $R_C = 4.7$ kΩ, $\beta = 150$, $V_\gamma = 0.7$ V, and $V_{CC} = 15$ V:
   (a) Find the range of values of v_S for which the transistor operates in the active mode.
   (b) Repeat part (a) with $\beta = 250$.

4–16   **D** The parameters of the circuit in Figure P4–13 are $R_C = 4.7$ kΩ, $\beta = 100$, $V_\gamma = 0.7$ V, and $V_{CC} = 15$ V:
   (a) With $v_S = 2.5$ V, select a value of R_B so that the transistor is in the active mode and $v_{CE} = V_{CC}/2$.
   (b) Repeat part (a) with $\beta = 200$.

4–17   **D** The parameters of the circuit in Figure P4–13 are $R_C = 3.5$ kΩ, $\beta = 50$, $V_\gamma = 0.7$ V, and $V_{CC} = 15$ V:
   (a) Select a value of R_B so that the transistor is in the active mode when 1 V $\le v_S \le 9$ V and is in the saturation mode for $v_S \ge 12$ V.
   (b) For the value of R_B found in (a), plot the transfer characteristic v_{CE} versus v_S as the input voltage sweeps across the range $0 \le v_S \le 15$ V.

4–18   **E** A circuit of the form of Figure P4–13 must saturate when $v_S \ge 5$ V. The transistor previously used is no longer available. The replacement transistor must operate with the existing circuit parameters $R_B = 100$ kΩ, $R_C = 2$ kΩ, and $V_{CC} = 5$ V. The data sheets for a candidate

FIGURE   P 4 - 1 3

replacement states that $V_\gamma = 0.7$ V and $50 \leq \beta \leq 100$. Does this transistor meet the requirements?

4–19 The element values for the circuit in Figure P4–13 are $R_B = 800$ kΩ, $R_C = 10$ kΩ, and $V_{CC} = 12$ V. The following data are measured in the laboratory:

v_S (V)	1	2	5	10
v_{CE} (V)	11.6	10.6	7.6	2.6

Sketch the circuit transfer characteristic and estimate the values of the transistor parameters V_γ and β.

4–20 The input source in Figure P4–20 is a series connection of a dc source V_{BB} and a signal source v_S. The circuit parameters are $R_B = 500$ kΩ, $R_C = 5$ kΩ, β = 100, $V_\gamma = 0.7$ V, and $V_{CC} = 15$ V.
(a) With $v_S = 0$, select the value of V_{BB} so that the transistor is in the active mode with $v_{CE} = V_{CC}/2$.
(b) Using the value of V_{BB} found in (a), find the range of values of the signal voltage v_S for which the transistor remains in the active mode.
(c) Plot the transfer characteristic v_{CE} versus v_S as the signal voltage sweeps across the range from −10 V to +10 V.

4–21 The output of the transistor circuit of Figure P4–21 is defined across the resistor R_E connected in series with the emitter. For $R_B = 100$ kΩ, $R_E = 1$ kΩ, $V_{CC} = 15$ V, β = 50, and $V_\gamma = 0.7$ V, plot the transfer characteristic v_O versus v_S as the input sweeps across the range $0 \leq v_S \leq 50$ V.

4–22 The parameters of the transistor in Figure P4–22 are $V_\gamma = 0.7$ V and β = 100. Select the value of R_B so that the circuit operates in a linear mode with $v_O = -5v_S$ when the input is in the range $0 \leq v_S \leq 3$ V. Sketch transfer characteristics of your design for $0 \leq v_S \leq 5$ V to verify that the circuit operates in a linear mode for the specified input range.

FIGURE P4–20

FIGURE P4–21

FIGURE P4–22

ERO 4–3 OP AMP CIRCUIT ANALYSIS (SECTS. 4–4, 4–5)

Given a circuit consisting of linear resistors, OP AMPs, and independent sources, find selected output signals or input-output relationships in equation or block diagram form.

FIGURE P4 – 23

4–23 For the circuit in Figure P4–23,
  (a) Find the relationship between the output v_O and input v_S.
  (b) For $v_S = 1.5$ V, find the power delivered to the 5-kΩ load resistor and the power supplied by the input voltage source.
  (c) For $V_{CC} = \pm12$ V, sketch the circuit transfer characteristic v_O versus v_S for an input voltage on the range -10 V $< v_S < 10$ V.

4–24 For the circuit in Figure P4–24,
  (a) Derive a relationship between the output v_O and the inputs v_{S1} and v_{S2}.
  (b) Draw a block diagram of the input-output relationship found in (a).
  (c) For $V_{CC} = \pm15$ V and $v_{S1} = 10$ V, find the allowable range of v_{S2} for linear operation.
  (d) If v_{S1} is known to fall in the range from 0 to 10 V, select a value for v_{S2} that ensures that v_O remains in the range from -10 V to $+10$ V.

FIGURE P4 – 24

4–25 The input-output relationship for a 3-input inverting summer is

$$v_O = - [v_1 + 3v_2 + 6v_3]$$

The resistance of the feedback resistor is 54 kΩ and the supply voltages are $V_{CC} = \pm15$ V.
  (a) Find the values of the input resistors R_1, R_2, and R_3.
  (b) For $v_2 = 0.5$ V and $v_3 = -1$ V, find the allowable range of v_1 for linear operation.

4–26 For the circuit in Figure P4–26,
  (a) Find the input-output relationship between v_O and v_S in terms of R_1, R_2, and V_{BB}.
  (b) Draw a block diagram of the relationship found in part (a).
  (c) For $V_{CC} = \pm15$ V, $V_{BB} = 5$ V, and $R_1 = R_2$, sketch the circuit transfer relationship v_O versus v_S for v_S over the range from -15 V to $+15$ V.

FIGURE P4 – 26

4–27 For the circuit in Figure P4–27,
  (a) Find the maximum voltage, current, and power available from the 5-V source.
  (b) Find the voltage, current, and power delivered to the 5-kΩ load resistor.
  (c) Explain why the voltage, current, and power found in part (b) exceed the maximum available values found in part (a).

Source

FIGURE P4-27

4-28 **E** It is claimed that the OP AMP circuit in Figure P4-28 operates as a voltage follower with $v_O = v_S$ when the switch S is open and operates as an inverter with $v_O = -v_S$ when the switch S is closed. Prove or disprove this claim.

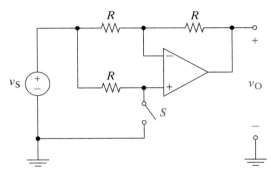

FIGURE P4-28

4-29 The circuit in Figure P4-29 has $V_{CC} = \pm15$ V.
  (a) Show that when the OP AMP is in the linear mode, the circuit to the left of the interface operates as an ideal current source with $i_L = v_S/R$ regardless of the value of load resistance.
  (b) For $v_S = 2$ V, find the maximum load voltage that the current source can deliver without saturating the OP AMP.
  (c) For $i_L = 0.5$ mA and $R = 10$ kΩ, find the maximum load resistance R_L that the current source can drive without saturating the OP AMP.

4-30 Show that the input-output relationship for the circuit in Figure P4-30 is $v_O = 2v_S$. For power supply voltages $V_{CC} = \pm15$ V, plot the transfer characteristics of the circuit for the input on the range -20 V $< v_S < +20$ V. Explain why neither OP AMP saturates when the output voltage falls outside the power supply voltage range.

4-31 For the OP AMP circuit in Figure P4-31, find the output v_2 in terms of the input v_1.

FIGURE P4-31

FIGURE P4-29

FIGURE P4-30

4–32 For the OP AMP circuit in Figure P4–32, find the output v_2 in terms of the input v_1.

4–33 For the OP AMP circuit in Figure P4–33, find the output v_O in terms of the input v_S.

4–34 The circuit in Figure P4–34 is a simplified diagram of a standard instrumentation amplifier that is commercially available in a single integrated circuit package. The objective is to show that the circuit is a differential amplifier whose input-output relationship is of the form $v_O = K(v_{S2} - v_{S1})$ and to find the gain K in terms of the circuit resistances.

(a) Write node-voltage equations at nodes B, C, E, and F.

(b) Solve the equations at nodes E and F to show that the circuit to the right of nodes A and D is a subtractor with $v_O = v_D - v_A$.

(c) Solve the equations at nodes B and C for v_A and v_D in terms of v_{S1} and v_{S2}.

(d) Substitute the results from (c) into the result found in (b) to derive the overall input-output relationship of the entire circuit.

4–35 For the OP AMP circuit in Figure P4–35, find the output v_O in terms of the inputs v_1 and v_2.

FIGURE P4-35

ERO 4–4 OP AMP CIRCUIT DESIGN (SECT. 4–6)

Given an input-output relationship, use resistors and OP AMPs to design one or more circuits that implement the relationship within stated constraints.

4–36 ◖D◗ Show how to interconnect a single OP AMP and the R-$2R$ resistor array shown in Figure P4–36 to obtain voltage gains of +1/4, +4, −1/3, and −3.

4–37 ◖D◗ Design circuits using resistors and OP AMPs to implement each of the following input-output relationships:

(a) $v_O = 3v_1 - 2v_2$

(b) $v_O = 2v_1 + v_2$

4–38 ◖D◗ The purpose of this problem is to design a signal conditioning circuit for the temperature transducer whose characteristics are shown in Figure P4–38. The conditioning circuit must convert the transducer output for temperatures between −300°C and −100°C onto an analog-to-digital converter input range of 0 to 5 V. The voltage gain for all stages must be less than 1000.

4–39 ◖D◗ The resistance of a pressure transducer varies from 5 kΩ to 20 kΩ when the pressure varies over its specified operating range. Design a signal conditioning circuit to convert the sensor resistance variation to a voltage signal on the range 0 to 5 V.

4–40 ◖D◗ Design an OP AMP circuit that implements the block diagram in Figure P4–40 using only standard resistance values for ±5% tolerance (see Appendix A, Table A–1).

4–41 ◖D◗ Design an OP AMP circuit with an inverting voltage gain $K = -20$ and an input resistance of at least 500 Ω using only 1-kΩ, 10-kΩ, 100-kΩ resistors.

4–42 ◖D◗ A heart monitoring system combines the outputs of three transducers into a single output using gains of −22, 5, and 18. Each transducer has a Thévenin resistance of 600 Ω and a Thévenin voltage in the range ±120 mV. Design an OP AMP summer to meet these requirements.

FIGURE P4-36

T (°C)

FIGURE P4-38

FIGURE P4-40

4–43 **D** A requirement exists for an amplifier with a gain of −12,000 and an input resistance of at least 300 kΩ. Design an OP AMP circuit that meets the requirements using general-purpose OP AMPs with voltage gains of $A = 2 \times 10^5$, input resistance of $R_I = 4 \times 10^8$ Ω, and output resistances $R_O = 20$ Ω.

4–44 **D** A noninverting amplifier with a nominal voltage gain of 200 is required. The actual gain must be between 197 and 204 when resistor tolerances are considered. Specify the allowable resistor tolerance using worst case analysis.

4–45 A noninverting amplifier with a voltage gain of 25 is to be designed using 5% resistors. Calculate the limits on the worst case gain variation.

ERO 4–5 THE COMPARATOR (SECT. 4–7)

(a) Given a circuit with one or more comparators, find the circuit input-output relationship.
(b) Given an input-output relationship, use resistors, voltage sources, and comparators to design circuits that realize the desired relationship.

4–46 For the circuit in Figure P4–46,
   (a) Estimate the values of V_{OH} and V_{OL}.
   (b) Determine the input voltage ranges for which $v_O = V_{OH}$ and $v_O = V_{OL}$.
   (c) Sketch the circuit transfer characteristics for v_S over the range −15 V to +15 V.

4–47 Repeat Problem 4–46 for the circuit of Figure P4–47.

FIGURE P4–46

FIGURE P4–47

4–48 The OP AMP circuit in Figure P4–48 is called a window detector. In this circuit the OP AMPs saturation levels are $\pm V_{CC}$. For $V_{CC} =$

FIGURE P4–48

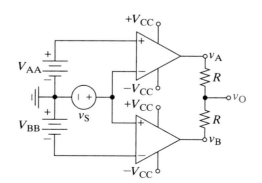

15 V, $V_{AA} = 10$ V, and $V_{BB} = 5$ V, show that the output is $v_O = 15$ V when $-V_{BB} < v_S < V_{AA}$ and $v_O = 0$ otherwise.

4–49 ⬤ The circuit in Figure P4–49 is an automatic tester that lights a 15-V green lamp when a resistor inserted at the interface is in the range 56 kΩ ±5% and a 15-V red lamp otherwise. Select the values of the resistors in the Wheatstone bridge and design an interface circuit to light the appropriate lamp.

FIGURE P4–49

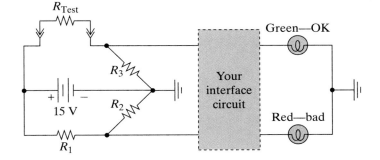

4–50 ⬤ An instrumentation system for a temperature transducer produces the characteristics shown in Figure P4–50. Design a circuit that lights a green lamp when the temperature is less than 120°C, a yellow caution light when the temperature is greater than 120°C but less than 150°C, and a red lamp when the temperature is greater than 150°C.

FIGURE P4–50

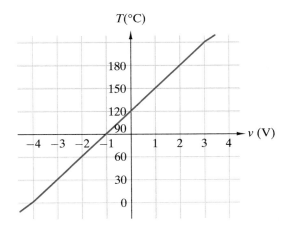

4–51 Sketch the output voltage v_O of the circuit in Figure P4–51 for $t > 0$ when the input is $v_S = 10t$ V and $V_{CC} = 15$ V.

4–52 Repeat problem 4–51 when the connections to inverting and noninverting inputs are reversed

FIGURE P4–51

INTEGRATING PROBLEMS

4–53 **A D E** BRIDGED-T INVERTING AMPLIFIER

Using the basic inverting OP AMP configuration to obtain a large voltage gain requires a small input resistor, a large feedback resistor, or both. Small input resistors load the input source, and large ($R > 1$ MΩ) feedback resistors have more noise and exhibit greater life-cycle variations. The circuit in Figure P4–53 circumvents these problems by using a bridged-T circuit in the feedback path. Note that R_3 occurs twice in this diagram.

(a) **A** Show that the gain of the circuit in Figure P4–53 can be written as $K = -R_{\text{FDBK}}/R_1$, where R_{FDBK} is the effective feedback resistance.

$$R_{\text{FDBK}} = R_3 \left(2 + \frac{R_3}{R_2} \right)$$

(b) **D** Design a basic inverting amplifier to achieve $K = -400$ and $R_{\text{IN}} \geq 20$ kΩ.

(c) **D** Design a bridged-T inverting amplifier to achieve $K = -400$ and $R_{\text{IN}} \geq 20$ kΩ.

(d) **E** Compare the two designs in terms of element count, element spread (ratio of largest over smallest resistance), and total resistance (sum of all resistances).

4–54 **A** HYBRID CIRCUIT ANALYSIS

The two-port interface circuit in Figure P4–54 is a small-signal model of a bipolar junction transistor using what are called hybrid parameters. The analysis of this circuit illustrates that it can be useful to use a mixture (or hybrid) of mesh-current and node-voltage equations.

(a) Using the symbolic notation in the figure, write a mesh-current equation for the input circuit and a node-voltage equation for the output circuit.

(b) Using $R_S = 2$ kΩ, $R_1 = 5$ kΩ, $R_2 = 50$ kΩ, $R_L = 150$ kΩ, $\mu = 10^{-3}$, and $\beta = 50$, solve the equations from (a) for the input current i_1 and the output voltage v_2 in terms of the input v_S.

(c) Using the results from (b), solve for the input resistance and the voltage gain of the circuit.

(d) Calculate the power gain defined as the ratio of the power delivered to R_L divided by the power supplied by v_S.

FIGURE P4-54

4–55 **A D** WHEATSTONE BRIDGE AMPLIFIER

The circuit in Figure P4–55 shows a Wheatstone bridge consisting of three equal resistors and a fourth resistor that is a transducer whose resistance is $R + \Delta R$, where $\Delta R << R$. The bridge is excited by a constant reference voltage source V_{REF} and is connected to the inputs of an OP AMP. For $\Delta R << R$ the OP AMP output voltage can be expressed in the form

$$v_O = K \left[\frac{\Delta R}{R} \right] V_{REF}$$

(a) **A** Verify that this expression is correct and express K in terms of circuit parameters.

(b) **D** For $V_{REF} = 15$ V, $R = 100 \ \Omega$ and $\Delta R/R$ in the range $\pm 0.04\%$, select value R_F so that the output voltage v_O falls in the range ± 3 V.

FIGURE P4-55

4–56 **A** OP AMP POWER BOOST

Inserting an OP AMP voltage follower at an interface isolates the source and load, but may result in a load current that exceeds the OP AMP rating. In such cases we can use the voltage follower with the transistor power boost shown in Figure P4–56. In this circuit the OP AMP output current is reduced because the transistor supplies the load current. Note that the transistor base-emitter port is included inside the OP AMP feedback path. For this analysis use the behavioral model of the transistor with a forward current gain β and turnout voltage V_γ. Use the ideal OP AMP model with output saturation levels of $\pm V_{CC}$.

When the transistor is in the active mode and the OP AMP is in the linear mode,

(a) Show that the load current is given by $i_L = v_S/R_L$.

(b) Show that the OP AMP output current is $i_L/(\beta + 1)$

(c) Show that the OP AMP output voltage is $v_S + V_\gamma$.

(d) Show that the transistor is in the active mode, and that the OP AMP is in the linear mode when the input is in the range $0 < v_S < V_{CC} - V_\gamma$.

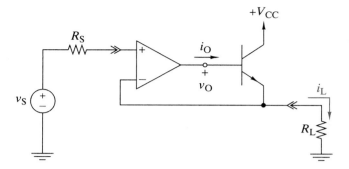

4–57 **A D E** SUBTRACTOR CIRCUITS

Both circuits in Figure P4–57 have input-output relationships of the form $v_O = K_2v_2 + K_1v_1$.

(a) **A** For circuits C1 and C2 determine the constants K_1 and K_2 in terms of circuit parameters.

C1

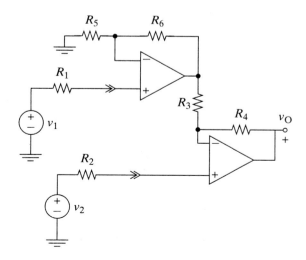

C2

(b) **D** In circuit C1 with $R_1 = R_2 = 1$ kΩ and $R_3 = R_5 = 10$ kΩ, select the values R_4 and R_6 that produce $v_O = 5(v_2 - v_1)$.

(c) **D** Repeat part (b) for circuit C2.

(d) **E** Compare these two designs on the basis of the number of devices required and the load each presents to the input signal sources.

4–58 **A D E** PHOTORESISTOR INSTRUMENTATION

Both circuits in Figure P4–58 contain a photoresistor R_X whose resistance varies inversely with the intensity of the incident light. In complete darkness its resistance is 10 kΩ. In bright sunlight its resistance is 2 kΩ. At any given light level the circuit is linear, so its input-output relationship is of the form $v_O = Kv_1$.

(a) **A** For circuits C1 and C2 determine the constant K in terms of circuit resistances.

(b) **D** For circuit C1 with $v_1 = +15$ V, select the values of R and R_F so that $v_O = -10$ V in bright sunlight and $+10$ V in complete darkness.

(c) **D** Repeat part (b) for circuit C2.

(d) **E** Compare these two designs on the basis of the number of devices required and the total power dissipated in each design.

C1

C2

FIGURE P4–58

4–59 **A E** DESIGN EVALUATION

As chief engineer of a small electronics company, you find yourself with a dilemma. Your two engineering interns have worked independently and have produced different solutions for the design problem you gave them. Their proposed solutions are shown in Figure P4–59.

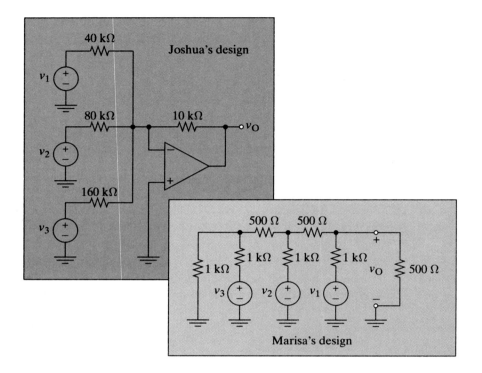

You asked them to design a circuit with three inputs v_1, v_2, v_3, and an output v_O. The output is to be proportional to $v_1/16 + v_2/8 + v_3/4$. The output can be either positive or negative, but must be able to drive a 500-Ω load. A ± 15 V dc power supply is available within the system without added cost. You need to manufacture 2 million of these circuits, so part costs are a concern, but performance is the top priority. Electrical characteristics and cost data for available parts are as follows.

PART TYPE	PART CHARACTERISTICS	RELATIVE COST
Standard resistors	1.0, 1.5, 2.2, 3.3, 4.7, 6.8 Ω and multiples of 10 thereof	$4000 per 100,000
Custom resistors	Values to three significant figures from 100 Ω to 200 kΩ	$4500 per 100,000
OP AMPs	Minimum voltage gain 100,000	$11,500 per 100,000
	Output voltage range ± 15 V	
	Maximum output current 10 mA	

(a) Verify whether both circuits comply with the performance requirements. If not, explain how one or both can be modified to comply.

(b) Which of the two circuits would you select for production and why?

4–60 ⦿ COMPUTER-AIDED CIRCUIT DESIGN

Use computer-aided circuit analysis to find the value of R_F in Figure P4–60 that causes the input resistance seen by i_S to be 50 Ω. Find the current gain i_O/i_S for this value of R_F. Use β = 100, r_π = 1.1 kΩ, R_C = 10 kΩ, R_E = 100 Ω, and R_L = 100 Ω.

FIGURE P4-60

4–61 ⦿ ⓔ DIGITAL TO ANALOG CONVERSION

A 5-bit parallel binary-coded digital signal (v_1, v_2, v_3, v_4, v_5) is to be converted into an analog signal. The digital signal represents a logic "1" as 0 V and a logic "0" as +5 V. When all 5 bits are one (11111) the analog output voltage must be v_O = 23.25 V. When all 5 bits are zero (00000) the anlog output must be v_O = 0 V. The digital signals are written with the most significant bit first. Use OP AMPs with V_{CC} = ±25 V and resistors with ±1% tolerance.

(a) Determine the input-output relationship required to implement the D-to-A conversion.

(b) Design an OP AMP circuit that implements the required relationship.

(c) Check your design by applying the digital inputs corresponding to 00000, 01111, and 11111 and comparing the analog output with the result predicted by your input-output relationship.

(d) Calculate the difference between analog outputs corresponding to 01111 and 10000, and comment on the adequacy of ±1% tolerance on the resistors.

4–62 ⓐ ⓓ ⓔ ANALOG VOLTMETER DESIGN

Figure P4–62(a) shows a voltmeter circuit in which a D'Arsonval meter with resistance R_M is connected in the feedback path of a noninverting amplifier circuit. The resistor R_S is to be selected so that meter current reaches full-scale deflection current I_{FS} when the voltage to be measured $v_x = V_{max}$ is applied at the noninverting input.

(a) Derive an expression for the required value of R_S in terms of V_{max} and the meter full-scale deflection current I_{FS}.

(b) Show that the OP AMP does not saturate if $V_{max} < V_{CC} - I_{FS}R_M$, where V_{CC} is the power supply voltage.

(c) Select a value of R_S so that V_{max} = 10 V produces a full-scale deflection of a meter with I_{FS} = 0.08 mA and R_M = 100 Ω.

(d) It is claimed that the voltmeter circuit in Figure P4–62(b) meets the requirements in (c). Verify this claim.

(e) Discuss the advantages and disadvantages of each of the voltmeters.

FIGURE P4-62

CHAPTER 5

SIGNAL WAVEFORMS

Under the sea, under the sea
mark how the telegraph motions to me.
Under the sea, under the sea
signals are coming along.

James Clerk Maxwell, 1873,
Scottish Physicist and
Occasional Humorous Poet

5–1 Introduction
5–2 The Step Waveform
5–3 The Exponential Waveform
5–4 The Sinusoidal Waveform
5–5 Composite Waveforms
5–6 Waveform Partial Descriptors
Summary
Problems
Integrating Problems

James Clerk Maxwell (1831–1879) is considered the unifying founder of the mathematical theory of electromagnetics. This genial Scotsman often communicated his thoughts to friends and colleagues via whimsical poetry. In the preceding short excerpt, Maxwell reminds us that the purpose of a communication system (the submarine cable telegraph in this case) is to transmit signals and that those signals must be changing, or *in motion,* as he put it.

This chapter marks the beginning of a new phase of our study of circuits. Up to this point we have dealt with resistive circuits, in which voltages and currents are constant (for example, +15 V or –3 mA). From this point forward we will be dealing with dynamic circuits, in which voltages and currents vary as functions of time. To analyze dynamic circuits, we need models for the time-varying signals and models for devices that describe the effects of time-varying signals in circuits. Signal models are introduced in this chapter. The dynamic circuit elements called capacitance and inductance are introduced in the next chapter. Chapters 7, 8, 9, and 10 then show how we combine signal models and device models to analyze dynamic circuits.

This chapter introduces the basic signal models used in the remainder of the book. We devote a chapter to this so you can master these models prior to launching into the complexities of dynamic circuits. We first introduce three key waveforms: the step, exponential, and sinusoid functions. By combining these three models we can build composite waveforms for all signals encountered in this book. Descriptors used to classify and describe waveforms are introduced because they highlight important signal attributes.

5–1 INTRODUCTION

We normally think of a signal as an electrical current $i(t)$ or voltage $v(t)$. The time variation of the signal is called a waveform. More formally,

> A **waveform** *is an equation or graph that defines the signal as a function of time.*

Up to this point our study has been limited to the type of waveform shown in Figure 5–1. Waveforms that are constant for all time are called **dc signals**. The abbreviation *dc* stands for direct current, but it applies to either voltage or current. Mathematical expressions for a dc voltage $v(t)$ or current $i(t)$ take the form

$$\left. \begin{array}{l} v(t) = V_0 \\ i(t) = I_0 \end{array} \right\} \quad \text{for} \;\; -\infty < t < \infty \qquad (5\text{–}1)$$

This equation is only a model. No physical signal can remain constant forever. It is a useful model, however, because it approximates the signals produced by physical devices such as batteries.

There are two matters of notation and convention that must be discussed before continuing. First, quantities that are constant (non-time-

F I G U R E   5 – 1   *A constant or dc waveform.*

varying) are usually represented by uppercase letters (V_A, I, T_O) or lowercase letters in the early part of the alphabet (a, b_7, f_0). Time-varying electrical quantities are represented by the lowercase letters i, v, p, q, and w. The time variation is expressly indicated when we write these quantities as $v_1(t)$, $i_A(t)$, or $w_C(t)$. Time variation is implicit when they are written as v_1, i_A, or w_C.

Second, in a circuit diagram signal variables are normally accompanied by the reference marks (+, −) for voltage and (→) for current. It is important to remember that these reference marks *do not* indicate the polarity of a voltage or the direction of current. The marks provide a baseline for determining the sign of the numerical value of the actual waveform. When the actual voltage polarity or current direction coincides with the reference directions, the signal has a positive value. When the opposite occurs, the value is negative. Figure 5–2 shows examples of voltage waveforms, including some that assume both positive and negative values. The bipolar waveforms indicate that the actual voltage polarity is changing as a function of time.

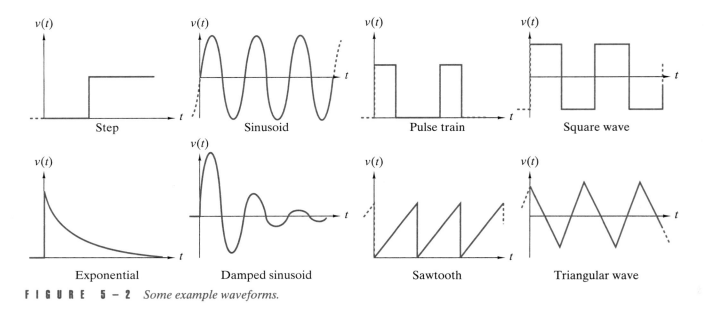

FIGURE 5 – 2 *Some example waveforms.*

The waveforms in Figure 5–2 are examples of signals used in electrical engineering. Since there are many such signals, it may seem that the study of signals involves the uninviting task of compiling a lengthy catalog of waveforms. However, it turns out that a long list is not needed. In fact, we can derive most of the waveforms of interest using just three basic signal models: the step, exponential, and sinusoidal functions. The small number of basic signals illustrates why models are so useful to engineers. In reality, waveforms are very complex, but their time variation can be approximated adequately using only a few basic building blocks.

Finally, in this chapter we will generally use voltage $v(t)$ to represent a signal waveform. Remember, however, that a signal can be either a voltage $v(t)$ or current $i(t)$.

5–2 THE STEP WAVEFORM

The first basic signal in our catalog is the step waveform. The general step function is based on the **unit step function** defined as

$$u(t) = \begin{cases} 0 & \text{for } t < 0 \\ 1 & \text{for } t > 0 \end{cases} \tag{5–2}$$

The step function waveform is equal to zero when its argument t is negative, and is equal to unity when its argument is positive. Mathematically, the function $u(t)$ has a jump discontinuity at $t = 0$.

Strictly speaking, it is impossible to generate a true step function since signal variables like current and voltage cannot jump from one value to another in zero time. Practically speaking, we can generate very good approximations to the step function. What is required is that the transition time be short compared with other response times in the circuit. Actually, the generation of approximate step functions is an everyday occurrence since people frequently turn things like TVs, stereos, and lights on and off.

On the surface, it may appear that the step function is not a very exciting waveform or, at best, only a source of temporary excitement. However, the step waveform is a versatile signal used to construct a wide range of useful waveforms. Multiplying $u(t)$ by a constant V_A produces the waveform

$$V_A u(t) = \begin{cases} 0 & \text{for } t < 0 \\ V_A & \text{for } t \geqslant 0 \end{cases} \tag{5–3}$$

Replacing t by $(t - T_S)$ produces a waveform $V_A u(t - T_S)$, which takes on the values

$$V_A u(t - T_S) = \begin{cases} 0 & \text{for } t < T_S \\ V_A & \text{for } t \geqslant T_S \end{cases} \tag{5–4}$$

The **amplitude** V_A scales the size of the step discontinuity, and the **time-shift** parameter T_S advances or delays the time at which the step occurs, as shown in Figure 5–3.

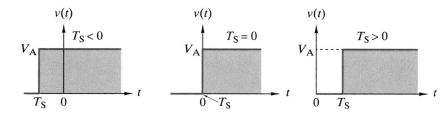

FIGURE 5−3 *Effect of changing time shifting on step function waveform.*

Amplitude and time-shift parameters are required to define the general step function. The amplitude V_A carries the units of volts. The amplitude of step function in electric current is I_A and carries the units of amperes. The constant T_S carries the units of time, usually seconds. The parameters V_A (or I_A) and T_S can be positive, negative, or zero. By combining several step functions, we can represent a number of important waveforms. One possibility is illustrated in the following example:

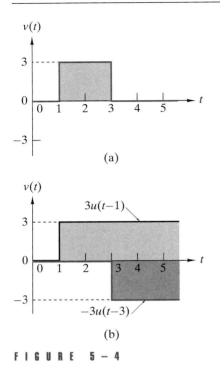

(a)

(b)

F I G U R E   5 - 4

EXAMPLE 5–1

Express the waveform in Figure 5–4(a) in terms of step functions.

SOLUTION:

The amplitude of the pulse jumps to a value of 3 V at $t = 1$ s; therefore, $3u(t - 1)$ is part of the equation for the waveform. The pulse returns to zero at $t = 3$ s, so an equal and opposite step must occur at $t = 3$ s. Putting these observations together, we express the rectangular pulse as

$$v(t) = 3u(t - 1) - 3u(t - 3)$$

Figure 5–4(b) shows how the two step functions combine to produce the given rectangular pulse. ■

THE IMPULSE FUNCTION

The generalization of Example 5–1 is the waveform

$$v(t) = V_A[u(t - T_1) - u(t - T_2)]$$

This waveform is a rectangular pulse of amplitude V_A that turns on at $t = T_1$ and off at $t = T_2$. The pulse train and square wave signals in Figure 5–2 can be generated by a series of these pulses. Pulses that turn on at some time T_1 and off at some later time T_2 are sometimes called **gating functions** because they are used in conjunction with electronic switches to enable or inhibit the passage of another signal.

A unit-area pulse centered on $t = 0$ is written in terms of step functions as

$$v(t) = \frac{1}{T}\left[u\left(t + \frac{T}{2}\right) - u\left(t - \frac{T}{2}\right)\right] \tag{5–5}$$

The pulse in Eq. (5–5) is zero everywhere except in the range $-T/2 \le t \le T/2$, where its value is $1/T$. The area under the pulse is 1 because its scale factor is inversely proportional to its duration. As shown in Figure 5–5(a), the pulse becomes narrower and higher as T decreases but maintains its unit area. In the limit as $T \to 0$ the scale factor approaches infinity but the area remains 1. The function obtained in the limit is called a **unit impulse**, symbolized as $\delta(t)$. The graphical representation of $\delta(t)$ is shown in Figure 5–5(b). The impulse is an idealized model of a large-amplitude, short-duration pulse.

A formal definition of the unit impulse is

$$\delta(t) = 0 \text{ for } t \ne 0 \quad \text{and} \quad \int_{-\infty}^{t} \delta(x)dx = u(t) \tag{5–6}$$

The first condition says the impulse is zero everywhere except at $t = 0$. The second condition suggests that the unit impulse is the derivative of a unit step function:

$$\delta(t) = \frac{du(t)}{dt} \tag{5–7}$$

The conclusion in Eq. (5–7) cannot be justified using elementary mathematics since the function $u(t)$ has a discontinuity at $t = 0$ and its derivative at that point does not exist in the usual sense. However, the concept can be justified using limiting conditions on continuous functions, as discussed in texts on signals and systems.[1] Accordingly, we defer the question of mathematical rigor to later courses and think of the unit impulse as the derivative of a unit step function. Note that this means that the unit impulse $\delta(t)$ has units of reciprocal time, or s^{-1}.

An impulse of strength K is denoted $v(t) = K\delta(t)$. Consequently, the scale factor K has the units of V-s and is the area under the impulse $K\delta(t)$. In the graphical representation of the impulse the value of K is written in parentheses beside the arrow, as shown in Figure 5–5(b).

EXAMPLE 5–2

Calculate and sketch the derivative of the pulse in Figure 5–6(a).

SOLUTION:
In Example 5–1 the pulse waveform was written as

$$v(t) = 3u(t - 1) - 3u(t - 3) \text{ V}$$

Using the derivative property of the step function, we write

$$\frac{dv(t)}{dt} = 3\delta(t - 1) - 3\delta(t - 3)$$

The derivative waveform consists of a positive-going impulse at $t = 1$ s and a negative-going impulse at $t = 3$ s. Figure 5–6(b) shows how the impulse train is represented graphically. The waveform $v(t)$ has the units of volts (V), so its derivative $dv(t)/dt$ has the units of V/s. ∎

THE RAMP FUNCTION

The **unit ramp** is defined as the integral of a step function:

$$r(t) = \int_{-\infty}^{t} u(x)dx = tu(t) \qquad (5\text{–}8)$$

The unit ramp waveform $r(t)$ in Figure 5–7(a) is zero for $t < 0$ and is equal to t for $t > 0$. Notice that the slope of $r(t)$ is 1 and has the units of time, or s. A ramp of strength K is denoted $v(t) = Kr(t)$, where the scale factor K has the units of V/s and is the slope of the ramp. The general ramp waveform shown in Figure 5–7(b) written as $v(t) = Kr(t - T_S)$ is zero for $t < T_S$ and equal to $K(t - T_S)$ for $t \geq T_S$. By adding a sequence of ramps, we can create the triangular and sawtooth waveforms shown in Figure 5–2.

SINGULARITY FUNCTIONS

The unit impulse, unit step, and unit ramp form a triad of related signals that are referred to as **singularity functions**. They are related by integration as

1 For example, see Alan V. Oppenheim and Allan S. Willsky, *Signals and Systems Analysis*, (Englewood Cliffs, N.J.: Prentice Hall, 1983), pp. 22–23.

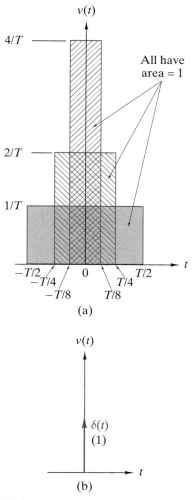

FIGURE 5–5 *Rectangular pulse waveforms and the impulse.*

FIGURE 5–6

FIGURE 5–7 *(a) Unit ramp waveform. (b) General ramp waveform.*

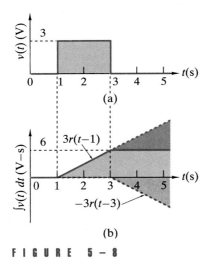

FIGURE 5–8

$$u(t) = \int_{-\infty}^{t} \delta(x)dx$$

$$r(t) = \int_{-\infty}^{t} u(x)dx \qquad (5\text{–}9)$$

or by differentiation as

$$\delta(t) = \frac{du(t)}{dt}$$

$$u(t) = \frac{dr(t)}{dt} \qquad (5\text{–}10)$$

These signals are used to generate other waveforms and as test inputs to linear systems to characterize their responses. When applying the singularity functions in circuit analysis, it is important to remember that $u(t)$ is a dimensionless function. But Eqs. (5–9) and (5–10) point out that $\delta(t)$ carries the units of s^{-1} and $r(t)$ carries units of seconds.

EXAMPLE 5–3

Derive an expression for the waveform for the integral of the pulse in Figure 5–8(a).

SOLUTION:
In Example 5–1 the pulse waveform was written as

$$v(t) = 3u(t - 1) - 3u(t - 3) \text{ V}$$

Using the integration property of the step function, we write

$$\int_{-\infty}^{t} v(x)dx = 3r(t - 1) - 3r(t - 3)$$

The integral is zero for $t < 1$ s. For $1 < t < 3$ the waveform is $3(t - 1)$. For $t > 3$ it is $3(t - 1) - 3(t - 3) = 6$. These two ramps produce the pulse integral shown in Figure 5–8(b). The waveform $v(t)$ has the units of volts (V), so the units of its integral are V-s. ■

EXAMPLE 5–4

Figure 5–9(a) shows an ideal electronic switch whose input is a ramp $2r(t)$, where the scale factor $K = 2$ carries the units of V/s. Find the switch output $v_O(t)$ when the gate function in Example 5–1 is applied to the control terminal (G) of the switch.

SOLUTION:
In Example 5–1 the gate function was written as

$$v_G(t) = 3u(t - 1) - 3u(t - 3) \text{ V}$$

The gate function turns the switch on at $t = 1$ s and off at $t = 3$ s. The output voltage of the switch is

$$v_O(t) = \begin{cases} 0 & t < 1 \\ 2t & 1 < t < 3 \\ 0 & 3 < t \end{cases}$$

Only the portion of the input waveform within the gate interval appears at the output. Figures 5–9(b), 5–9(c), and 5–9(d) show how the gate function $v_G(t)$ controls the passage of the input signal through the electronic switch. ■

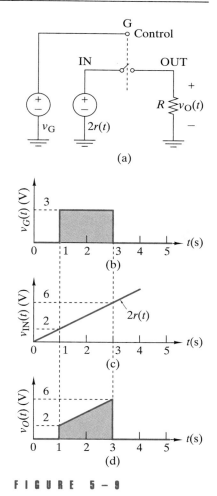

(a)

(b)

(c)

(d)

FIGURE 5 – 9

Exercise 5–1

Express the following signals in terms of singularity functions:

(a) $v_1(t) = \begin{cases} 0 & t < 2 \\ 4 & 2 < t < 4 \\ -4 & 4 < t \end{cases}$ (b) $v_2(t) = \begin{cases} 0 & t < 2 \\ 4 & 2 < t < 4 \\ -2t + 12 & 4 < t \end{cases}$

(c) $v_3(t) = \displaystyle\int_{-\infty}^{t} v_1(x)\,dx$ (d) $v_4(t) = \dfrac{dv_2(t)}{dt}$

Answers:

(a) $v_1(t) = 4\,u(t-2) - 8\,u(t-4)$ (b) $v_2(t) = 4\,u(t-2) - 2\,r(t-4) + 8\,u(t-4)$
(c) $v_3(t) = 4\,r(t-2) - 8\,r(t-4)$ (d) $v_4(t) = 4\,\delta(t-2) - 2\,u(t-4)$

Exercise 5–2

(a) Write an expression for a rectangular pulse with an amplitude of 15 V that begins at $t = -5$ s and ends at $t = 10$ s.
(b) Write an expression for the derivative of the pulse defined in (a).
(c) Write an expression for the integral of the pulse in (a).

Answers:

(a) $15[u(t+5) - u(t-10)]$
(b) $15[\delta(t+5) - \delta(t-10)]$
(c) $15(t+5)u(t+5) - 15(t-10)u(t-10) = 15[r(t+5) - r(t-10)]$

5–3 THE EXPONENTIAL WAVEFORM

The **exponential waveform** is a step function whose amplitude factor gradually decays to zero. The equation for this waveform is

$$v(t) = [V_A e^{-t/T_C}]\,u(t) \qquad (5\text{–}11)$$

A graph of $v(t)$ versus t/T_C is shown in Figure 5–10. The exponential starts out like a step function. It is zero for $t < 0$ and jumps to a maximum amplitude of V_A at $t = 0$. Thereafter it monotonically decays toward zero as time marches on. The two parameters that define the waveform are the **amplitude** V_A (in volts) and the **time constant** T_C (in seconds). The amplitude of a current exponential would be written I_A and carry the units of amperes.

The time constant is of special interest, since it determines the rate at which the waveform decays to zero. An exponential decays to about 37%

FIGURE 5 – 1 0 *The exponential wave-form.*

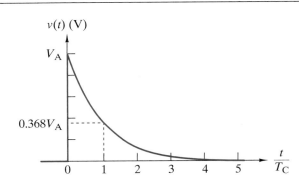

of its initial amplitude $v(0) = V_A$ in one time constant, because at $t = T_C$, $v(T_C) = V_A e^{-1}$, or approximately $0.368 \times V_A$. At $t = 5T_C$, the value of the waveform is $V_A e^{-5}$, or approximately $0.00674\ V_A$. An exponential signal decays to less than 1% of its initial amplitude in a time span of five time constants. In theory, an exponential endures forever, but practically speaking after about $5T_C$ the waveform amplitude becomes negligibly small. We define the **duration** of a waveform to be the interval of time outside of which the waveform is everywhere less than a stated value. Using this concept, we say the duration of an exponential waveform is $5T_C$.

EXAMPLE 5–5

Plot the waveform $v(t) = [-17e^{-100t}]u(t)$ V.

SOLUTION:

From the form of $v(t)$, we recognize that $V_A = -17$ V and $T_C = 1/100$ s or 10 ms. The minimum value of $v(t)$ is $v(0) = -17$ V, and the maximum value is approximately 0 V as t approaches $5T_C = 50$ ms. These observations define appropriate scales for plotting the waveform. Spreadsheet programs are especially useful for the repetitive calculations and graphical functions involved in waveform plotting. Figure 5–11 shows how this example can be handled using Excel. We begin by filling in the "A" column with time values ranging from $t = 0$ to $t = 50$ ms. The data in the "B" column are obtained by calculating the value of $-17e^{-1001t}$ for each of the values of t (ms) in the "A" column. Then using the graphing tool, we create a plot showing that the waveform starts out at $v(t) = -17$ V at $t = 0$ and then increases toward $v(t) = 0$ as $t \to 50$ ms. ■

PROPERTIES OF EXPONENTIAL WAVEFORMS

The **decrement property** describes the decay rate of an exponential signal. For $t > 0$ the exponential waveform is given by

$$v(t) = V_A e^{-t/T_C} \tag{5–12}$$

The step function can be omitted since it is unity for $t > 0$. At time $t + \Delta t$ the amplitude is

$$v(t + \Delta t) = V_A e^{(t + \Delta t)/T_C} = V_A e^{-t/T_C} e^{-\Delta t/T_C} \tag{5–13}$$

FIGURE 5–11

The ratio of these two amplitudes is

$$\frac{v(t + \Delta t)}{v(t)} = \frac{V_A e^{-t/T_C} e^{-\Delta t/T_C}}{V_A e^{-t/T_C}} = e^{-\Delta t/T_C} \tag{5–14}$$

The decrement ratio is independent of amplitude and time. In any fixed time period Δt, the fractional decrease depends only on the time constant. The decrement property states that the same percentage decay occurs in equal time intervals.

The slope of the exponential waveform (for $t > 0$) is found by differentiating Eq. (5–12) with respect to time:

$$\frac{dv(t)}{dt} = -\frac{V_A}{T_C} e^{-t/T_C} = -\frac{v(t)}{T_C} \tag{5–15}$$

The **slope property** states that the time rate of change of the exponential waveform is inversely proportional to the time constant. Small time constants lead to large slopes or rapid decays, while large time constants produce shallow slopes and long decay times.

Equation (5–15) can be rearranged as

$$\frac{dv(t)}{dt} + \frac{v(t)}{T_C} = 0 \tag{5–16}$$

When $v(t)$ is an exponential of the form in Eq. (5–12), then $dv/dt + v/T_C = 0$. That is, the exponential waveform is a solution of the first-order linear differential equation in Eq. (5–16). We will make use of this fact in Chapter 7.

The time-shifted exponential waveform is obtained by replacing t in Eq. (5–11) by $t - T_S$. The general exponential waveform is written as

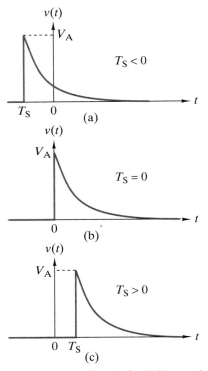

$$v(t) = [V_A e^{-(t-T_S)/T_C}] u(t - T_S) \qquad (5\text{--}17)$$

where T_S is the time-shift parameter for the waveform. Figure 5–12 shows exponential waveforms with the same amplitude and time constant but different values of T_S. Time shifting translates the waveform to the left or right depending on whether T_S is negative or positive. *Caution:* The factor $t - T_S$ must appear in both the argument of the step function and the exponential, as shown in Eq. (5–17).

FIGURE   5 – 1 2 *Effect of time shifting on the exponential waveform.*

EXAMPLE 5–6

An oscilloscope is a laboratory instrument that displays the instantaneous value of waveform versus time. Figure 5–13 shows an oscilloscope display of a portion of an exponential waveform. In the figure the vertical (amplitude) axis is calibrated at 2 V per division, and the horizontal (time) axis is calibrated at 1 ms per division. Find the time constant of the exponential.

SOLUTION:

For $t > 0$ the general expression for an exponential in Eq. (5–17) becomes

$$v(t) = V_A e^{-(t-T_S)/T_C}$$

We have only a portion of the waveform, so we do not know the location of the $t = 0$ time origin; hence, we cannot find the amplitude V_A or the time shift T_S from the display. But, according to the decrement property, we should be able to determine the time constant since the decrement ratio is independent of amplitude and time. Specifically, Eq. (5–14) points out that

$$\frac{v(t + \Delta t)}{v(t)} = e^{-\Delta t/T_C}$$

Solving for the time constant T_C yields

$$T_C = \frac{\Delta t}{\ln\left[\dfrac{v(t)}{v(t + \Delta t)}\right]}$$

FIGURE   5 – 1 3

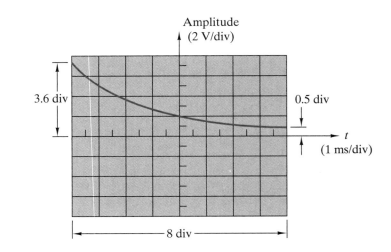

Taking the starting point at the left edge of the oscilloscope display yields

$$v(t) = (3.6 \text{ div})(2 \text{ V/div}) = 7.2 \text{ V}$$

Next, defining Δt to be the full width of the display produces

$$\Delta t = (8 \text{ div})(1 \text{ ms/div}) = 8 \text{ ms}$$

and

$$v(t + \Delta t) = (0.5 \text{ div})(2 \text{ V/div}) = 1 \text{ V}$$

As a result, the time constant of the waveform is found to be

$$T_C = \frac{\Delta t}{\ln\left[\dfrac{v(t)}{v(t + \Delta t)}\right]} = \frac{8 \times 10^{-3}}{\ln(7.2/1)} = 4.05 \text{ ms} \qquad \blacksquare$$

Exercise 5–3

(a) An exponential waveform has $v(0) = 1.2$ V and $v(3) = 0.5$ V. What are V_A and T_C for this waveform?

(b) An exponential waveform has $v(0) = 5$ V and $v(2) = 1.25$ V. What are values of $v(t)$ at $t = 1$ and $t = 4$?

(c) An exponential waveform has $v(0) = 5$ and an initial ($t = 0$) slope of -25 V/s. What are V_A and T_C for this waveform?

(d) An exponential waveform decays to 10% of its initial value in 3 ms. What is T_C for this waveform?

(e) A waveform has $v(2) = 4$ V, $v(6) = 1$ V, and $v(10) = 0.5$ V. Is it an exponential waveform?

Answers:

(a) $V_A = 1.2$ V, $T_C = 3.43$ s

(b) $v(1) = 2.5$ V, $v(4) = 0.3125$ V

(c) $V_A = 5$ V, $T_C = 200$ ms

(d) $T_C = 1.303$ ms

(e) No, it violates the decrement property

Exercise 5–4

Find the amplitude and time constant of the following exponential signals:

(a) $v_1(t) = [-15e^{-1000t}]u(t)$ V

(b) $v_2(t) = [+12e^{-t/10}]u(t)$ mV

(c) $i_3(t) = [15e^{-500t}]u(-t)$ mA

(d) $i_4(t) = [4e^{-200(t-100)}]u(t-100)]$ A

Answers:

(a) $V_A = -15$ V, $T_C = 1$ ms   (b) $V_A = 12$ mV, $T_C = 10$ s

(c) $I_A = 15$ mA, $T_C = 2$ ms   (d) $I_A = 4$ A, $T_C = 5$ ms

5–4 THE SINUSOIDAL WAVEFORM

The cosine and sine functions are important in all branches of science and engineering. The corresponding time-varying waveform in Figure 5–14 plays an especially prominent role in electrical engineering.

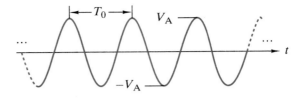

In contrast with the step and exponential waveforms studied earlier, the sinusoid, like the dc waveform in Figure 5–1, extends indefinitely in time in both the positive and negative directions. The sinusoid has neither a beginning nor an end. Of course, real signals have finite durations. They were turned on at some finite time in the past and will be turned off at some time in the future. While it may seem unrealistic to have a signal model that lasts forever, it turns out that the eternal sinewave is a very good approximation in many practical applications.

The sinusoid in Figure 5–14 is an endless repetition of identical oscillations between positive and negative peaks. The **amplitude** V_A (in volts) defines the maximum and minimum values of the oscillations. The **period** T_0 (usually seconds) is the time required to complete one cycle of the oscillation. The sinusoid can be expressed mathematically using either the sine or the cosine function. The choice between the two depends on where we choose to define $t = 0$. If we choose $t = 0$ at a point where the sinusoid is zero, then it can be written as

$$v(t) = V_A \sin(2\pi t/T_0) \tag{5–18a}$$

On the other hand, if we choose $t = 0$ at a point where the sinusoid is at a positive peak, we can write an equation for it in terms of a cosine function:

$$v(t) = V_A \cos(2\pi t/T_0) \tag{5–18b}$$

Although either choice will work, it is common practice to choose $t = 0$ at a positive peak; hence Eq. (5–18b) applies. Thus, we will continue to call the waveform a sinusoid even though we use a cosine function to describe it.

As in the case of the step and exponential functions, the general sinusoid is obtained by replacing t by $(t - T_S)$. Inserting this change in Eq. (5–18) yields a general expression for the sinusoid as

$$v(t) = V_A \cos[2\pi(t - T_S)/T_0] \tag{5–19}$$

where the constant T_S is the time-shift parameter. Figure 5–15 shows that the sinusoid shifts to the right when $T_S > 0$ and to the left when $T_S < 0$. In effect, time shifting causes the positive peak nearest the origin to occur at $t = T_S$.

The time-shifting parameter can also be represented by an angle:

$$v(t) = V_A \cos[2\pi t/T_0 + \phi] \tag{5–20}$$

The parameter ϕ is called the **phase angle**. The term *phase angle* is based on the circular interpretation of the cosine function. We think of the period as being divided into 2π radians. In this sense the phase angle is the angle between $t = 0$ and the nearest positive peak. Comparing Eqs. (5–19) and (5–20), we find the relation between T_S and ϕ to be

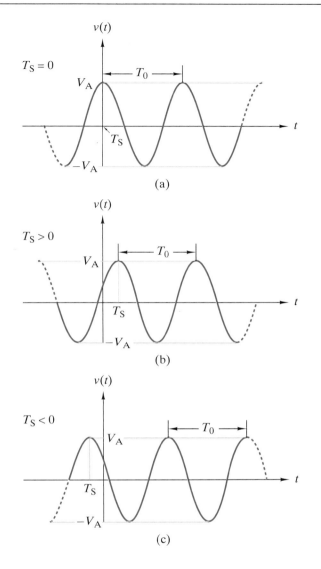

FIGURE 5 – 1 5 *Effect of time shifting on the sinusoidal waveform.*

$$\phi = -2\pi \frac{T_S}{T_0} \tag{5–21}$$

Changing the phase angle moves the waveform to the left or right, revealing different phases of the oscillating waveform (hence the name *phase angle*).

The phase angle should be expressed in radians, but is often reported in degrees. Care must be used when numerically evaluating the argument of the cosine $(2\pi t/T_0 + \phi)$ to ensure that both terms have the same units. The term $2\pi t/T_0$ has the units of radians, so it is necessary to convert ϕ to radians when it is given in degrees.

An alternative form of the general sinusoid is obtained by expanding Eq. (5–20) using the identity $\cos(x + y) = \cos(x)\cos(y) - \sin(x)\sin(y)$,

$$v(t) = [V_A \cos \phi] \cos (2\pi t/T_0) + [-V_A \sin \phi] \sin (2\pi t/T_0)$$

The quantities inside the brackets in this equation are constants; therefore, we can write the general sinusoid in the form

$$v(t) = a \cos(2\pi t/T_0) + b \sin(2\pi t/T_0) \tag{5-22}$$

The two amplitudelike parameters a and b have the same units as the waveform (volts in this case) and are called Fourier coefficients. By definition, the Fourier coefficients are related to the amplitude and phase parameters by the equations

$$a = V_A \cos \phi$$
$$b = -V_A \sin \phi \tag{5-23}$$

The inverse relationships are obtained by squaring and adding the expressions in Eq. (5–23):

$$V_A = \sqrt{a^2 + b^2} \tag{5-24}$$

and by dividing the second expression in Eq. (5–23) by the first:

$$\phi = \tan^{-1} \frac{-b}{a} \tag{5-25}$$

Caution: The inverse tangent function has a ±180° ambiguity that can be resolved by considering the signs of the Fourier coefficients a and b.

It is customary to describe the time variation of the sinusoid in terms of a frequency parameter. **Cyclic frequency** f_0 is defined as the number of periods per unit time. By definition, the period T_0 is the number of seconds per cycle; consequently, the number of cycles per second is

$$f_0 = \frac{1}{T_0} \tag{5-26}$$

where f_0 is the cyclic frequency or simply the frequency. The unit of frequency (cycles per second) is the **hertz** (Hz). The **angular frequency** ω_0 in radians per second is related to the cyclic frequency by the relationship

$$\omega_0 = 2\pi f_0 = \frac{2\pi}{T_0} \tag{5-27}$$

because there are 2π radians per cycle.

There are two ways to express the concept of sinusoidal frequency: cyclic frequency (f_0, hertz) and angular frequency (ω_0, radians per second). When working with signals, we tend to use the former. For example, radio stations transmit carrier signals at frequencies specified as 690 kHz (AM band) or 101 MHz (FM band). Radian frequency is more convenient when describing the characteristics of circuits driven by sinusoidal inputs.

In summary, there are several equivalent ways to describe the general sinusoid:

$$v(t) = V_A \cos\left[\frac{2\pi(t - T_S)}{T_0} \right] = V_A \cos\left(\frac{2\pi t}{T_0} + \phi \right) = a \cos\left(\frac{2\pi t}{T_0} \right) + b \sin\left(\frac{2\pi t}{T_0} \right)$$

$$= V_A \cos[2\pi f_0(t - T_S)] = V_A \cos(2\pi f_0 t + \phi) = a \cos(2\pi f_0 t) + b \sin(2\pi f_0 t)$$

$$= V_A \cos[\omega_0(t - T_S)] = V_A \cos(\omega_0 t + \phi) = a \cos(\omega_0 t) + b \sin(\omega_0 t)$$

To use any one of these expressions, we need three types of parameters:

1. *Amplitude:* either V_A or the Fourier coefficients a and b
2. *Time shift:* either T_S or the phase angle ϕ
3. *Time/frequency:* either T_0, f_0, or ω_0.

In different parts of this book we use different forms to represent a sinusoid. Therefore, it is important for you to understand thoroughly the relationships among the various parameters in Eqs. (5–21) through (5–27).

EXAMPLE 5−7

Figure 5–16 shows an oscilloscope display of a sinusoid. The vertical axis (amplitude) is calibrated at 5 V per division, and the horizontal axis (time) is calibrated at 0.1 ms per division. Derive an expression for the sinusoid displayed in Figure 5–16.

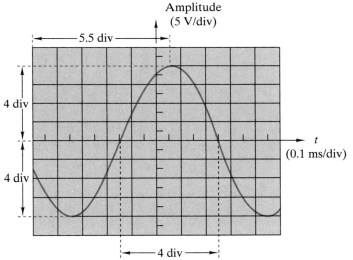

FIGURE   5 − 1 6

SOLUTION:

The maximum amplitude of the waveform is seen to be four vertical divisions; therefore,

$$V_A = (4 \text{ div})(5 \text{ V/div}) = 20 \text{ V}$$

There are four horizontal divisions between successive zero crossings, which means there are a total of eight divisions in one cycle. The period of the waveform is

$$T_0 = (8 \text{ div})(0.1 \text{ ms/div}) = 0.8 \text{ ms}$$

The two frequency parameters are $f_0 = 1/T_0 = 1.25$ kHz and $\omega_0 = 2\pi f_0 = 7854$ rad/s. The parameters V_A, T_0, f_0, and ω_0 do not depend on the location of the $t = 0$ axis.

To determine the time shift T_S, we need to define a time origin. The $t = 0$ axis is arbitrarily taken at the left edge of the display in Figure 5–16. The positive peak shown in the display is 5.5 divisions to the right of $t = 0$, which is more than half a cycle (four divisions). The positive peak closest to $t = 0$ is not shown in Figure 5–16 because it must lie beyond the left edge of the display. However, the positive peak shown in the display is located at $t = T_S + T_0$ since it is one cycle after $t = T_S$. We can write

$$T_S + T_0 = (5.5 \text{ div})(0.1 \text{ ms/div}) = 0.55 \text{ ms}$$

which yields $T_S = 0.55 - T_0 = -0.25$ ms. As expected, T_S is negative because the nearest positive peak is to the left of $t = 0$.

Given T_S, we can calculate the remaining parameters of the sinusoid as follows:

$$\phi = -\frac{2\pi T_S}{T_0} = 1.96 \text{ rad or } 112.5°$$

$$a = V_A \cos \phi = -7.65 \text{ V}$$

$$b = -V_A \sin \phi = -18.5 \text{ V}$$

Finally, the three alternative expressions for the displayed sinusoid are

$$v(t) = 20 \cos[(7854 \, (t + 0.25 \times 10^{-3})]$$

$$= 20 \cos(7854 \, t + 112.5°)$$

$$= -7.65 \cos 7854t - 18.5 \sin 7854t \qquad ∎$$

<hr>

Exercise 5–5

Derive an expression for the sinusoid displayed in Figure 5–16 when $t = 0$ is placed in the middle of the display.

Answer: $v(t) = 20 \cos(7854t - 22.5°)$

<hr>

PROPERTIES OF SINUSOIDS

In general, a waveform is said to be **periodic** if

$$v(t + T_0) = v(t)$$

for all values of t. The constant T_0 is called the period of the waveform if it is the smallest nonzero interval for which $v(t + T_0) = v(t)$. Since this equality must be valid for all values of t, it follows that periodic signals must have eternal waveforms that extend indefinitely in time in both directions. Signals that are not periodic are called **aperiodic**.

The sinusoid is a periodic signal since

$$v(t + T_0) = V_A \cos[2\pi \, (t + T_0)/T_0 + \phi]$$

$$= V_A \cos[2\pi \, (t)/T_0 + \phi + 2\pi]$$

But $\cos(x + 2\pi) = \cos(x)$. Consequently,

$$v(t + T_0) = V_A \cos(2\pi t/T_0 + \phi) = v(t)$$

for all t.

The **additive property** of sinusoids states that summing two or more sinusoids with the same frequency yields a sinusoid with different amplitude and phase parameters but the same frequency. To illustrate, consider two sinusoids

$$v_1(t) = a_1 \cos(2\pi f_0 t) + b_1 \sin(2\pi f_0 t)$$
$$v_2(t) = a_2 \cos(2\pi f_0 t) + b_2 \sin(2\pi f_0 t)$$

The waveform $v_3(t) = v_1(t) + v_2(t)$ can be written as

$$v_3(t) = (a_1 + a_2) \cos(2\pi f_0 t) + (b_1 + b_2) \sin(2\pi f_0 t)$$

because cosine and sine are linearly independent functions. We obtain the Fourier coefficients of the sum of two sinusoids by adding their Fourier coefficients, provided the two have the same frequency. *Caution:* The summation must take place with the sinusoids in Fourier coefficient form. Sums of sinusoids *cannot* be found by adding amplitudes and phase angles.

The **derivative** and **integral** properties state that when we differentiate or integrate a sinusoid, the result is another sinusoid with the same frequency:

$$\frac{d(V_A \cos \omega t)}{dt} = -\omega V_A \sin \omega t = \omega V_A \cos(\omega t + \pi/2)$$

$$\int V_A \cos(\omega t) \, dt = \frac{V_A}{\omega} \sin \omega t = \frac{V_A}{\omega} \cos(\omega t - \pi/2)$$

These operations change the amplitude and phase angle but do not change the frequency. The fact that differentiation and integration preserve the underlying waveform is a key property of the sinusoid. No other periodic waveform has this shape-preserving property.

EXAMPLE 5–8

(a) Find the periods cyclic and radian frequencies of the sinusoids
$v_1(t) = 17 \cos(2000t - 30°)$
$v_2(t) = 12 \cos(2000t + 30°)$.
(b) Find the waveform of $v_3(t) = v_1(t) + v_2(t)$.

SOLUTION:
(a) The two sinusoids have the same frequency $\omega_0 = 2000$ rad/s since a term $2000t$ appears in the arguments of $v_1(t)$ and $v_2(t)$. Therefore, $f_0 = \omega_0/2\pi = 318.3$ Hz and $T_0 = 1/f_0 = 3.14$ ms.
(b) We use the additive property, since the two sinusoids have the same frequency. Beyond this checkpoint, the frequency plays no further role in the calculation. The two sinusoids must be converted to the Fourier coefficient form using Eq. (5–23).

$$a_1 = 17 \cos(-30°) = +14.7 \text{ V}$$
$$b_1 = -17 \sin(-30°) = +8.50 \text{ V}$$
$$a_2 = 12 \cos(30°) = +10.4 \text{ V}$$
$$b_2 = -12 \sin(30°) = -6.00 \text{ V}$$

The Fourier coefficients of the signal $v_3 = v_1 + v_2$ are found as

$$a_3 = a_1 + a_2 = 25.1 \text{ V}$$
$$b_3 = b_1 + b_2 = 2.50 \text{ V}$$

The amplitude and phase angle of $v_3(t)$ are found using Eqs. (5–24) and (5–25):

$$V_A = \sqrt{a_3^2 + b_3^2} = 25.2 \text{ V}$$
$$\phi = \tan^{-1}(2.5/25.1) = 5.69°$$

Two equivalent representations of $v_3(t)$ are

$$v_3(t) = 25.1 \cos(2000t) + 2.5 \sin(2000t) \text{ V}$$

and

$$v_3(t) = 25.2 \cos(2000t + 5.69°) \text{ V} \qquad \blacksquare$$

EXAMPLE 5–9

The balanced three-phase voltages used in electrical power systems can be written as

$$v_A(t) = V_m \cos(2\pi f_0 t)$$
$$v_B(t) = V_m \cos(2\pi f_0 t + 120°)$$
$$v_C(t) = V_m \cos(2\pi f_0 t + 240°)$$

Show that the sum of these voltages is zero.

SOLUTION:
The three voltages are given in amplitude/phase angle form. They can be converted to the Fourier coefficient form using Eq. (5–23):

$$v_A(t) = \; + V_m \cos(2\pi f_0 t)$$
$$v_B(t) = \; -\frac{V_m}{2} \cos(2\pi f_0 t) + \frac{\sqrt{3}V_m}{2} \sin(2\pi f_0 t)$$
$$v_C(t) = \; -\frac{V_m}{2} \cos(2\pi f_0 t) - \frac{\sqrt{3}V_m}{2} \sin(2\pi f_0 t)$$

The sum of the cosine and sine terms is zero; consequently, $v_A(t) + v_B(t) + v_C(t) = 0$. The zero sum occurs because the three sinusoids have equal amplitudes and are disposed at 120° intervals. If the amplitudes are not equal or the phase angles do not differ by 120°, then the voltages are said to be unbalanced. $\qquad \blacksquare$

Exercise 5–6

Write an equation for the waveform obtained by integrating and differentiating the following signals:

$$\text{(a) } v_1(t) = 30 \cos(10t - 60°)$$

$$\text{(b) } v_2(t) = 3 \cos(4000\pi t) - 4 \sin(4000\pi t)$$

Answers:

(a) $\dfrac{dv_1}{dt} = 300 \cos(10t + 30°)$

$\displaystyle\int v_1(t)\, dt = 3 \cos(10t - 150°)$

(b) $\dfrac{dv_2}{dt} = 2\pi \times 10^4 \cos(4000\pi t + 143.1°)$

$\displaystyle\int v_2 \, dt = \dfrac{1}{800\pi} \cos(4000\pi t - 36.87°)$

Exercise 5–7

A sinusoid has a period of 5 μs. At $t = 0$ the amplitude is 12 V. The waveform reaches its first positive peak after $t = 0$ at $t = 4$ μs. Find its amplitude, frequency, and phase angle.

Answers: $V_A = 38.8$ V, $f_0 = 200$ kHz, $\phi = +72°$

5–5 COMPOSITE WAVEFORMS

In the previous sections we introduced the step, exponential, and sinusoidal waveforms. These waveforms are basic signals because they can be combined to synthesize all other signals used in this book. Signals generated by combining the three basic waveforms are called **composite signals**. This section provides examples of composite waveforms.

EXAMPLE 5–10

Characterize the composite waveform generated by

$$v(t) = V_A u(t) - V_A u(-t)$$

SOLUTION:
The first term in this waveform is simply a step function of amplitude V_A that occurs at $t = 0$. The second term involves the function $u(-t)$, whose waveform requires some discussion. Strictly speaking, the general step function $u(x)$ is unity when $x > 0$ and zero when $x < 0$. That is, $u(x)$ is unity when its argument is positive and zero when it is negative. Under this rule the function $u(-t)$ is unity when $-t > 0$ and zero when $-t < 0$, that is,

$$u(-t) = \begin{cases} 1 & \text{for } t < 0 \\ 0 & \text{for } t > 0 \end{cases}$$

which is the reverse of the step function $u(t)$. Figure 5–17 shows how the two components combine to produce a composite waveform that extends indefinitely in both directions and has a jump discontinuity of $2V_A$ at $t = 0$. This composite waveform is called a **signum** function.

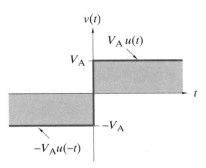

FIGURE 5 – 1 7 *The signum waveform.*

EXAMPLE 5–11

Characterize the composite waveform generated by subtracting an exponential from a step function with the same amplitude.

SOLUTION:

The equation for this composite waveform is

$$v(t) = V_A u(t) - [V_A e^{-t/T_C}]u(t)$$
$$= V_A[1 - e^{-t/T_C}]u(t)$$

For $t < 0$ the waveform is zero because of the step function. At $t = 0$ the waveform is still zero since the step and exponential cancel:

$$v(0) = V_A[1 - e^0](1) = 0$$

For $t \gg T_C$ the waveform approaches a constant value V_A because the exponential term decays to zero. For practical purposes $v(t)$ is within less than 1% of its final value V_A when $t = 5T_C$. At $t = T_C$, $v(T_C) = V_A(1 - e^{-1}) = 0.632V_A$. The waveform rises to about 63% of its final value in one time constant. All of the observations are summarized in the plot shown in Figure 5–18. This waveform is called an **exponential rise**. It is also sometimes referred to as a "charging exponential," since it represents the behavior of signals that occur during the buildup of voltage in resistor-capacitor circuits studied in Chapter 7. ■

F I G U R E   5 – 1 8 *The exponential rise waveform.*

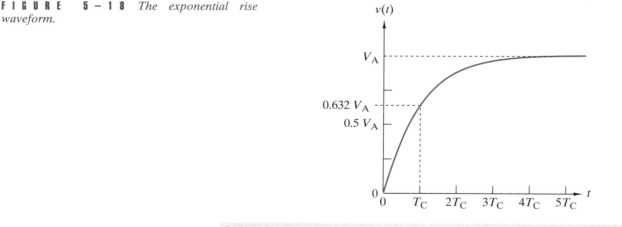

EXAMPLE 5–12

Characterize the composite waveform obtained by multiplying the ramp $r(t)/T_C$ times an exponential.

SOLUTION:

The equation for this composite waveform is

$$v(t) = \frac{r(t)}{T_C}[V_A e^{-t/T_C}]u(t)$$

$$= V_A[(t/T_C)\,e^{-t/T_C}]u(t)$$

For $t < 0$ the waveform is zero because of the step function. At $t = 0$ the waveform is zero because $r(0) = 0$. For $t > 0$ there is a competition between two effects—the ramp increases linearly with time while the exponential decays to zero. Since the composite waveform is the product of these terms, it is important to determine which effect dominates. In the limit, as $t \rightarrow \infty$, the product of the ramp and exponential takes on the indeterminate form of infinity times zero. A single application of *l'Hôpital's* rule, then, shows that the exponential dominates, forcing the $v(t)$ to zero as t becomes large. That is, the exponential decay overpowers the linearly increasing ramp, as shown by the graph in Figure 5–19. The waveform obtained by multiplying a ramp by a decaying exponential is called a **damped ramp**. ■

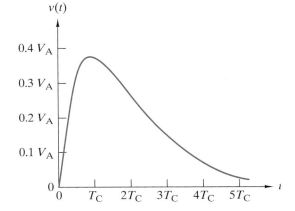

FIGURE 5 – 1 9   *The damped ramp waveform.*

EXAMPLE 5–13

Characterize the composite waveform obtained by multiplying sin $\omega_0 t$ by an exponential.

SOLUTION:
In this case the composite waveform is expressed as

$$v(t) = \sin \omega_0 t [V_A e^{-t/T_C}] u(t)$$

$$= V_A [e^{-t/T_C} \sin \omega_0 t] u(t)$$

Figure 5–20 shows a graph of this waveform for $T_0 = 2T_C$. For $t < 0$ the step function forces the waveform to be zero. At $t = 0$, and periodically thereafter, the waveform passes through zero because sin $(n\pi) = 0$. The waveform is not periodic, however, because the decaying exponential gradually reduces the amplitude of the oscillation. For all practical purposes the oscillations become negligibly small for $t > 5T_C$. The waveform obtained by multiplying a sinusoid by a decaying exponential is called a **damped sine**. ■

EXAMPLE 5–14

Characterize the composite waveform obtained as the difference of two exponentials with the same amplitude.

F I G U R E   5 – 2 0   *The damped sine waveform.*

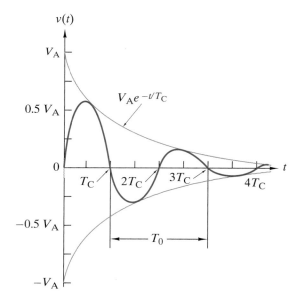

SOLUTION:
The equation for this composite waveform is

$$v(t) = [V_A\, e^{-t/T_1}]\, u(t) - [V_A\, e^{-t/T_2}]\, u(t)$$
$$= V_A\, (e^{-t/T_1} - e^{-t/T_2})\, u(t)$$

For $T_1 > T_2$ the resulting waveform is illustrated in Figure 5–21 (plotted for $T_1 = 2T_2$). For $t < 0$ the waveform is zero. At $t = 0$ the waveform is still zero, since

$$v(0) = V_A\, (e^{-0} - e^{-0})$$
$$= V_A\, (1 - 1) = 0$$

F I G U R E   5 – 2 1   *The double exponential waveform.*

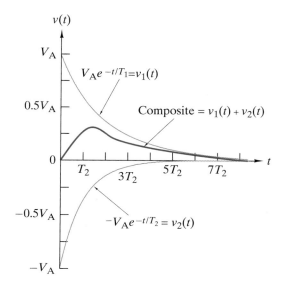

For $t >> T_1$ the waveform returns to zero because both exponentials decay to zero. For $5T_1 > t > 5T_2$ the second exponential is negligible and the waveform essentially reduces to the first exponential. Conversely, for $t << T_1$ the first exponential is essentially constant, so the second exponential determines the early time variation of the waveform. The waveform is called a **double exponential**, since both exponential components make important contributions to the waveform. ∎

EXAMPLE 5–15

Characterize the composite waveform defined by

$$v(t) = 5 - \frac{10}{\pi} \sin(2\pi 500\, t) - \frac{10}{2\pi} \sin(2\pi 1000\, t) - \frac{10}{3\pi} \sin(2\pi 1500\, t)$$

SOLUTION:

The waveform is the sum of a constant (dc) term and three sinusoids at different frequencies. The first sinusoidal component is called the **fundamental** because it has the lowest frequency. As a result the frequency $f_0 = 500$ Hz is called the **fundamental frequency**. The other sinusoidal terms are said to be harmonics because their frequencies are integer multiples of f_0. Specifically, the second sinusoidal term is called the **second harmonic** ($2f_0 = 1000$ Hz) while the third term is the **third harmonic** ($3f_0 = 1500$ Hz). Figure 5–22 shows a plot of this waveform. Note that the waveform is periodic with a period equal to that of the fundamental component, namely, $T_0 = 1/f_0 = 2$ ms. The decomposition of a periodic waveform into a sum of harmonic sinusoids is called a **Fourier series**, a topic we will study in detail in Chapter 13. In fact, the waveform in this example is the first four terms in the Fourier series for a 10-V sawtooth wave of the type shown in Figure 5–1. ∎

FIGURE   5–22

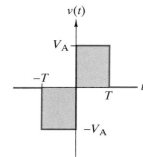

$v(t)$

V_A

$-T$

T

t

$-V_A$

FIGURE 5-23

Exercise 5-8

Write an expression for the pulse in Figure 5-23 using step functions.

Answer:

$$v(t) = V_A[u(t) - u(t - T) - u(-t) + u(-t - T)]$$
$$\text{or}$$
$$v(t) = V_A[-u(t + T) + 2u(t) - u(t - T)]$$

Exercise 5-9

Find the maximum amplitude and the approximate duration of the following composite waveforms:

(a) $v_1(t) = [25 \sin 1000t][u(t) - u(t - 10)]$ V
(b) $v_2(t) = [50 \cos 1000t][e^{-200t}]u(t)$ V
(c) $i_3(t) = [3000te^{-1000t}]u(t)$ mA
(d) $i_4(t) = [10e^{5000t}]u(-t) + [10e^{-5000t}]u(t)$ A

Answers:

(a) 25 V, 10 s
(b) 50 V, 25 ms
(c) 1.10 mA, 5 ms
(d) 10 A, 2 ms

5-6   WAVEFORM PARTIAL DESCRIPTORS

An equation or graph defines a waveform for all time. The value of a waveform $v(t)$ or $i(t)$ at time t is called the **instantaneous value** of the waveform. We often use parameters called **partial descriptors** that characterize important features of a waveform but do not give a complete description. These partial descriptors fall into two categories: (1) those that describe temporal features and (2) those that describe amplitude features.

TEMPORAL DESCRIPTORS

Temporal descriptors identify waveform attributes relative to the time axis. For example, waveforms that repeat themselves at fixed time intervals are said to be **periodic**. Stated formally,

> *A signal v(t) is periodic if v(t + T$_0$) = v(t) for all t, where the period T$_0$ is the smallest value that meets this condition. Signals that are not periodic are called aperiodic.*

The fact that a waveform is periodic provides important information about the signal, but does not specify all of its characteristics. Thus, the fact that a signal is periodic is itself a partial description, as is the value of the period. The eternal sinewave is the premier example of a periodic signal. The square wave and triangular wave in Figure 5-2 are also periodic. Examples of aperiodic waveforms are the step function, exponential, and damped sine.

Waveforms that are identically zero prior to some specified time are said to be **causal**. Stated formally,

A signal v(t) is causal if there exists a value of **T** *such that* v(t)
≡ *0 for all* t < T; *otherwise it is noncausal.*

It is usually assumed that a causal signal is zero for $t < 0$, since we can always use time shifting to make the starting point of a waveform at $t = 0$.
Examples of causal waveforms are the step function, exponential, and
damped sine. The eternal sinewave is, of course, noncausal.

   Causal waveforms play a central role in circuit analysis. When the input
driving force $x(t)$ is causal, the circuit response $y(t)$ must also be causal.
That is, a physically realizable circuit cannot anticipate and respond to an
input before it is applied. Causality is an important temporal feature, but
only a partial description of the waveform.

AMPLITUDE DESCRIPTORS

Amplitude descriptors are positive scalars that describe signal strength.
Generally, a waveform varies between two extreme values denoted as
V_{MAX} and V_{MIN}. The **peak-to-peak value** (V_{pp}) describes the total excursion
of $v(t)$ and is defined as

$$V_{pp} = V_{MAX} - V_{MIN} \qquad (5\text{–}28)$$

Under this definition V_{pp} is always positive even if V_{MAX} and V_{MIN} are both
negative. The **peak value** (V_p) is the maximum of the absolute value of the
waveform. That is,

$$V_p = \text{Max}\{|V_{MAX}|, \ |V_{MIN}|\} \qquad (5\text{–}29)$$

The peak value is a positive number that indicates the maximum absolute
excursion of the waveform from zero. Figure 5–24 shows examples of these
two amplitude descriptors.

FIGURE 5 – 24 *Peak value (V$_p$) and peak-to-peak value (V$_{pp}$).*

The peak and peak-to-peak values describe waveform variation using the extreme values. The average value smooths things out to reveal the underlying waveform baseline. Average value is the area under the waveform over some period of time T, divided by that time period. Mathematically, we define **average value** (V_{avg}) over the time interval T as

$$V_{avg} = \frac{1}{T} \int_t^{t+T} v(x)dx \qquad (5\text{--}30)$$

For periodic signals the period T_0 is used as the averaging interval T.

For some periodic waveforms the integral in Eq. (5–30) can be estimated graphically. The net area under the waveform is the area above the time axis minus the area below the time axis. For example, the two waveforms in Figure 5–24 obviously have nonzero average values. The top waveform has a negative average value because the negative area below the time axis more than cancels the area above the axis. Similarly, the bottom waveform clearly has a positive average value.

The average value indicates whether the waveform contains a constant, non-time-varying component. The average value is also called the **dc component** because dc signals are constant for all t. On the other hand, the **ac components** have zero average value and are periodic. For example, the waveform in Example 5–15

$$v(t) = 5 - \frac{10}{\pi} \sin (2 \pi 500\, t) - \frac{10}{2 \pi} \sin (2 \pi 1000\, t) - \frac{10}{3 \pi} \sin (2 \pi 1500\, t)$$

has a 5-V average value due to its dc component. The three sinusoids are ac components because they are periodic and have zero average value. Sinusoids have zero average value because over any given cycle the positive area above the time axis is exactly canceled by the negative area below.

EXAMPLE 5–16

Find the peak, peak-to-peak, and average values of the periodic input and output waveforms in Figure 5–25.

SOLUTION:
The input waveform is a sinusoid whose amplitude descriptors are

$$V_{pp} = 2V_A \quad V_p = V_A \quad V_{avg} = 0$$

The output waveform is obtained by clipping off the negative half-cycle of the input sinusoid. The amplitude descriptors of the output waveform are

$$V_{pp} = V_p = V_A$$

The output has a nonzero average value, since there is a net positive area under the waveform. The upper limit in Eq. (5–30) can be taken as $T_0/2$, since the waveform is zero from $T_0/2$ to T_0.

$$V_{avg} = \frac{1}{T_0} \int_0^{T_0/2} V_A \sin(2\pi t/T_0)dt = \left. -\frac{V_A}{2\pi} \cos(2\pi\, t/T_0) \right|_0^{T_0/2}$$

$$= \frac{V_A}{\pi}$$

The signal processor produces an output with a dc value from an input with no dc component. Rectifying circuits described in electronics courses produce waveforms like the output in Figure 5–25. ■

ROOT-MEAN-SQUARE VALUE

The **root-mean-square value** (V_{rms}) is a measure of the average power carried by the signal. The instantaneous power delivered to a resistor R by a voltage $v(t)$ is

$$p(t) = \frac{1}{R}[v(t)]^2 \qquad (5\text{–}31)$$

The average power delivered to the resistor in time span T is defined as

$$P_{avg} = \frac{1}{T}\int_{t}^{t+T} p(t)dt \qquad (5\text{–}32)$$

Combining Eqs. (5–31) and (5–32) yields

$$P_{avg} = \frac{1}{R}\left[\frac{1}{T}\int_{t}^{t+T}[v(t)]^2\, dt\right] \qquad (5\text{–}33)$$

The quantity inside the large brackets in Eq. (5–33) is the average value of the square of the waveform. The units of the bracketed term are volts squared. The square root of this term defines the amplitude partial descriptor V_{rms}:

$$V_{rms} = \sqrt{\frac{1}{T}\int_{t}^{t+T}[v(t)]^2\, dt} \qquad (5\text{–}34)$$

The amplitude descriptor V_{rms} is called the root-mean-square (rms) value because it is obtained by taking the square root of the average (mean) of the square of the waveform. For periodic signals the averaging interval is one cycle since such a waveform repeats itself every T_0 seconds.

We can express the average power delivered to a resistor in terms of V_{rms} as

$$P_{avg} = \frac{1}{R}V_{rms}^2 \qquad (5\text{–}35)$$

The equation for average power in terms of V_{rms} has the same form as the power delivered by a dc signal. For this reason the rms value was originally called the **effective value**, although this term is no longer common. If the waveform amplitude is doubled, its rms value is doubled, and the average power is quadrupled. Commercial electrical power systems use transmission voltages in the range of several hundred kilovolts (rms).

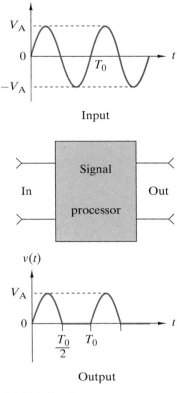

Input

Output

FIGURE 5-25

EXAMPLE 5–17

Find the average and rms values of the sinusoid and sawtooth in Figure 5–26.

FIGURE 5 – 26

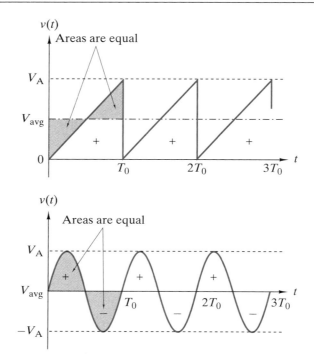

SOLUTION:

As noted previously, the sinusoid has an average value of zero. The sawtooth clearly has a positive average value. By geometry, the net area under one cycle of the sawtooth waveform is $V_A T_0/2$, so its average value is $(1/T_0)(V_A T_0/2) = V_A/2$. To obtain the rms value of the sinusoid we apply Eq. (5–34) as

$$V_{rms} = \sqrt{\frac{(V_A)^2}{T_0} \int_0^{T_0} sin^2(2\pi t/T_0)dt}$$

$$= \sqrt{\frac{(V_A)^2}{T_0} \left[\frac{t}{2} - \frac{sin(4\pi t/T_0)}{8\pi/T_0}\right]_0^{T_0}} = \frac{V_A}{\sqrt{2}}$$

For the sawtooth the rms value is found as:

$$V_{rms} = \sqrt{\frac{1}{T_0} \int_0^{T_0} (V_A t/T_0)^2 \, dt} = \sqrt{\frac{(V_A)^2}{T_0^3} \left[\frac{t^3}{3}\right]_0^{T_0}} = \frac{V_A}{\sqrt{3}} \qquad \blacksquare$$

APPLICATION NOTE: **EXAMPLE 5–18**

In the United States residential ac power is supplied at 60 Hz and voltages of 115 and 230 V. These voltages are the rms values of the sinusoidal waveforms supplied by the power company. An oscilloscope displaying the waveform at a 115 V (rms) outlet indicates that the peak value of the sinusoid is larger than 115 V. Example 5–16 showed that the rms value of a sinusoid equals its peak divided by $\sqrt{2}$. If the rms value is 115 V, then the peak value is $115\sqrt{2}$ or about 163 V. Most ac voltmeters are calibrated to indicate the rms value of a sinusoid. If we measure a sinusoid with a peak value of 200 V using an ac voltmeter, the meter will indicate 200/√2, or about 141 V (rms). In general, ac volt-meters are designed to measure pure sinusoids and will not correctly measure the rms value of nonsinusoidal waveforms such as a square wave.

The concept of rms value applies to sinusoidal currents as well. You are invited to return to Eq. (5–31) and write the instantaneous power delivered to the resistor in terms of the current through it. Continuing the derivation in terms of current shows that

$$I_{rms} = \sqrt{\frac{1}{T} \int_t^{t+T} [i(x)]^2 \, dx}$$

and

$$P_{avg} = RI_{rms}^2$$

This power relationship has the same form as the dc case studied in Chapter 2. Most ac ammeters are calibrated to display the rms value of the current. Similarly, the ampere ratings of devices such as fuses, mo-tors, and circuit breakers refer to rms current.

Since rms values relate to the average power, it should come as no surprise to discover that ac power ratings of devices refer to average power. A 75-W lightbulb will draw an average power of 75 W at its rated (rms) voltage. The instantaneous power delivered to the bulb has a constant average value and an oscillatory component. As a result, its light output fluctuates slightly, although the variation is not noticeable to most people. We will return to the concept of average power again in Chapter 8 when we study maximum power transfer conditions. The consequences of the oscillatory component of the instantaneous power will also be studied in Chapter 16.

EXAMPLE 5–19

A pulse train consists of a periodic sequence of exponential waveforms. The first cycle of the pulse train can be written as

$$v(t) = [u(t) - u(t - T_0)] V_A e^{-t/T_C}$$

Use MATLAB to calculate the V_{rms} of the pulse train for $V_A = 10$ V, $T_C = 2$ ms, and $T_0 = 5T_C$.

SOLUTION:
The MATLAB environment includes a number of common math functions such as $\sin(x)$, $\cos(x)$, and $abs(x)$. In the present example we need to evaluate the expression

$$V_{rms} = \sqrt{\frac{1}{T_0} \int_0^{T_0} [v(t)]^2 \, dt}$$

MATLAB has built-in functions for the square root operation [sqrt(x)] and the integration [quad('v²',0,T₀)]. However, we must supply the function that calculates the square of the voltage. User defined functions are created in what are called M-files. The M-file function required here can be written as

```
function y=Vsq(t)

VA=10;TC=0.002;

y=VA^2*exp(-2*t/TC);
```

The reserve word `function` in the first line tells us this is a function whose name is `Vsq(t)`. The next line assigns values to the parameters V_A and T_C that define the exponential voltage $v(t)$. The third line calculates the square of $v(t)$, where we have squared the exponential function by doubling its exponent. This M-file is written in the **editor/debugger** screen accessible through the **File** command in the MATLAB command line. Once created, the file is saved as `Vsq.M`. Such files are called M-files because they all have an "M" extention. The user-defined function `Vsq.M` is now available for use in the interactive mode in the MATLAB command window, where the function is called as follows:

```
VA=10;TC=0.002;T0=5*TC;

Vrms=sqrt(1/T0*quad('Vsq',0,T0))

Vrms=
```

3.1622

The MATLAB command `quad('Vsq',0,T0)` integrates the user-defined function `Vsq(t)` from $t = 0$ to $t = T_0$. Notice that in the command window we must again assign values to the parameters `VA` and `TC`. The values assigned in Vsq.M are local identifiers that are not available outside the function `Vsq(t)`. ∎

Exercise 5–10

Find the peak, peak-to-peak, average, and rms values of the periodic waveform in Figure 5–27.

Answers:

$$V_p = 2V_A, \quad V_{pp} = 3V_A, \quad V_{avg} = \frac{V_A}{4}, \quad V_{rms} = \frac{\sqrt{5}}{2}V_A$$

Exercise 5–11

Classify each of the following signals as periodic or aperiodic and causal or non-causal. Then calculate the average and rms values of the periodic waveforms, and the peak and peak-to-peak values of the other waveforms.

FIGURE 5–27

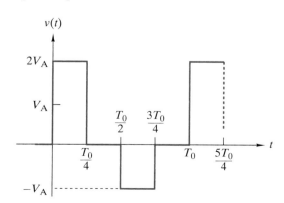

(a) $v_1(t) = 99 \cos 3000t - 132 \sin 3000t$ V
(b) $v_2(t) = 34 [\sin 800\pi t][u(t) - u(t - 0.03)]$ V
(c) $i_3(t) = 120[u(t + 5) - u(t - 5)]$ mA
(d) $t_4(t) = 50$ A

Answers:
(a) Periodic, noncausal, $V_{avg} = 0$ and $V_{rms} = 117$ V
(b) Aperiodic, causal, $V_p = 34$ V and $V_{pp} = 68$ V
(c) Aperiodic, causal, $V_p = V_{pp} = 120$ mA
(d) Aperiodic, noncausal, $V_p = 50$ A and $V_{pp} = 0$

Exercise 5–12

Construct waveforms that have the following characteristics:

(a) Aperiodic and causal with $V_p = 8$ V and $V_{pp} = 15$ V
(b) Periodic and noncausal with $V_{avg} = 10$ V and $V_{pp} = 50$ V
(c) Periodic and noncausal with $V_{avg} = V_p/2$
(d) Aperiodic and causal with $V_p = V_{pp} = 10$ V

Answers: There are many possible correct answers since the given parameters are only partial descriptions of the required waveforms. Some examples that meet the requirements are as follows:

(a) $v(t) = 8 u(t) - 15 u(t - 1) + 7 u(t - 2)$
(b) $v(t) = 10 + 25 \sin 1000t$
(c) A sawtooth wave
(d) $v(t) = 10[e^{-100t}]u(t)$

SUMMARY

- A waveform is an equation or graph that describes a voltage or current as a function of time. Most signals of interest in electrical engineering can be derived using three basic waveforms: the step, exponential, and sinusoid.

- The step function is defined by its amplitude and time-shift parameters. The impulse, step, and ramp are called singularity functions and are often used as test inputs for circuit analysis purposes.

- The exponential waveform is defined by its amplitude, time constant, and time-shift parameter. For practical purposes, the duration of the exponential waveform is five time constants.

- A sinusoid can be defined in terms of three types of parameters: *amplitude* (either V_A or the Fourier coefficients a and b), *time shift* (either T_S or the phase angle ϕ), and *time/frequency* (either T_0 or f_0 or ω_0).

- Many composite waveforms can be derived using the three basic waveforms. Some examples are the impulse, ramp, damped ramp, damped sinusoid, exponential rise, and double exponential.

- Partial descriptors are used to classify or describe important signal attributes. Two important temporal attributes are periodicity and causality. Periodic waveforms repeat themselves every T_0 seconds. Causal signals are zero for $t < 0$. Some important amplitude descriptors are peak value V_p, peak-to-peak value V_{pp}, average value V_{avg}, and root-mean-square value V_{rms}.

- A spectrum is an equation or graph that defines the amplitudes and phase angles of sinusoidal components contained in a signal. The signal bandwidth (B) is a partial descriptor that defines the range of frequencies outside of which the component amplitudes are less than a specified value. Periodic signals can be resolved into a dc component and a sum of ac components at harmonic frequencies.

PROBLEMS

ERO 5–1 BASIC WAVEFORMS (SECTS. 5–2, 5–3, 5–4)

Given an equation, graph, or word description of a step, exponential, or sinusoid waveform,

(a) Construct an alternative description of the waveform.
(b) Find new waveforms by summing, integrating, or differentiating the given waveform.

5–1 Graph the following step function waveforms:
  (a) $v_1(t) = -10u(t)$
  (b) $v_2(t) = 5u(t - 2)$
  (c) $v_3(t) = -5u(t + 1)$
  (d) $v_4(t) = 10u(t - 2)$

5–2 Graph the waveforms obtained by adding the waveforms in Problem 5–1.
  (a) $v_A(t) = v_1(t) + v_2(t)$
  (b) $v_B(t) = v_1(t) + v_3(t)$
  (c) $v_C(t) = v_1(t) + v_4(t)$
  (d) $v_D(t) = v_2(t) + v_3(t)$

5–3 Graph the waveforms obtained by integrating and differentiating each of the waveforms in Problem 5–1.

5–4 A waveform $v(t)$ is zero for t in the range 5 ms $\geq t \geq$ 3 ms and +5 V elsewhere. Write an equation for the waveform using step functions.

5–5 A waveform $v(t)$ is zero for $t = 0$, rises linearly to 5 V at $t = 2$ ms, and abruptly drops to zero thereafter. Write an equation for the waveform using singularity functions.

5–6 Determine the amplitude and time constant of each of the following exponential waveforms. Graph each of the waveforms.
  (a) $v_1(t) = [10\,e^{-2t}]u(t)$ (b) $v_2(t) = [10\,e^{-t/2}]u(t)$
  (c) $v_3(t) = [-10\,e^{-20t}]u(t)$ (d) $v_4(t) = [-10\,e^{-t/20}]u(t)$

5–7 Graph the exponential waveform obtained by differentiating the exponential waveforms in Problem 5–6.

5–8 An exponential waveform starts at $t = 0$ and decays to 5 V at $t = 4$ ms and 3 V at $t = 6$ ms. Find the amplitude and time constant of the waveform.

5–9   The amplitude of an exponential waveform is 5 V at $t = 5$ ms. At $t = 7$ ms its amplitude is 3.5 V. What is the waveform amplitude at $t = 2$ ms?

5–10   Show that the exponential waveform $v(t) = V_A e^{-\alpha t} u(t)$ is a solution of the following first-order differential equation:

$$\frac{dv(t)}{dt} + \alpha v(t) = 0$$

5–11   Determine the period, frequency, amplitude, time shift, and phase angle of the following sinusoids. Graph the waveform.
   (a) $v_1(t) = 10 \cos(2000\pi t) + 10 \sin(2000\pi t)$
   (b) $v_2(t) = -30 \cos(2000\pi t) - 20 \sin(2000\pi t)$
   (c) $v_3(t) = 10 \cos(2\pi t/10) - 10 \sin(2\pi t/10)$
   (d) $v_4(t) = -20 \cos(800\pi t) + 30 \sin(800\pi t)$

5–12   Determine the frequency, period, amplitude, time shift, and phase angle of the sum of the first two sinusoids in Problem 5–11.

5–13   Write an equation for a sinusoid with an amplitude of 15 V, a period of 75 ms, and its first positive peak at $t = 50$ ms. Graph the waveform.

5–14   A sinusoid has a frequency of 5 MHz, a value of 10 V at $t = 0$, and reaches its first positive peak at $t = 25$ ns. Determine its amplitude, phase angle, and Fourier coefficients.

5–15   Determine the frequency, period, and Fourier coefficients of the following sinusoids:
   (a) $v_1(t) = 10 \cos(400\pi t + 270°)$
   (b) $v_2(t) = 15 \cos(400\pi t - 45°)$
   (c) $v_3(t) = 45 \cos(2\pi t/40 - 30°)$
   (d) $v_4(t) = 45 \sin(200\pi t + 60°)$

5–16   Determine the frequency, period, phase angle, and amplitude of the sum of the first two sinusoids in Problem 5–15.

5–17   Show that the sinusoid $v(t) = a \cos \omega t + b \sin \omega t$ is a solution of the following second-order differential equation:

$$\frac{d^2 v(t)}{dt^2} + \omega^2 v(t) = 0$$

ERO 5–2 COMPOSITE WAVEFORMS (SECT. 5–5)

Given an equation, graph, or word description of a composite waveform,

(a) Construct an alternative description of the waveform.
(b) Find new waveforms by summing, integrating, differentiating, or clipping the given waveform.

5–18   Graph the following waveforms:
   (a) $v_1(t) = 10 \left[1 - 2e^{-200t} \sin(1000\pi t) \right] u(t)$
   (b) $v_2(t) = \left[20 - 10e^{-1000t} \right] u(t)$
   (c) $v_3(t) = 10 \left[2 - \sin(1000\pi t) \right] u(t)$
   (d) $v_4(t) = 10 \left[4 - 2e^{-1000t} \right] u(t)$

5–19 Graph the following ramp waveforms. Then graph and write an equation for the derivative of each waveform.
(a) $v_1(t) = tu(t) - (t-2)u(t-2)$
(b) $v_2(t) = tu(t) - 2(t-1)u(t-1) + (t-2)u(t-2)$

5–20 A waveform is known to be of the form $v(t) = V_A - V_B e^{-\alpha t}$. At $t = 0$ the value of the waveform is 5 V, at $t = 5$ μs its value is 7.5 V, and it approaches 10 V for large values of t. Find the values of V_A, V_B, and α, and then graph the waveform.

5–21 A waveform is known to be of the form $v(t) = V_A e^{-\alpha t} \sin \beta t$. The waveform periodically passes through zero every 2.5 ms. At $t = 1$ ms its value is 3.5 V and at $t = 2$ ms it is 0.8 V. Find the values of V_A, α, and β, and then graph the waveform.

5–22 The periodic waveform in Figure P5–22 is applied at the input of an inverting amplifier with a voltage gain of $K = -3$. Sketch the amplifier output waveform.

5–23 The periodic waveform in Figure P5–23 is applied at the input of a noninverting OP AMP amplifier. The circuit has a voltage gain of $K = 10$, and the OP AMP saturates at $V_{CC} = \pm15$ V. Write an equation for the first cycle of the amplifier output waveform and then graph the output waveform.

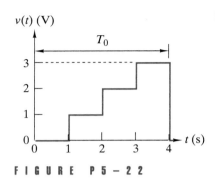

FIGURE P 5 – 2 2

FIGURE P 5 – 2 3

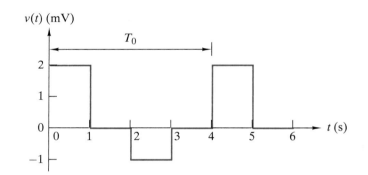

5–24 A waveform is known to be of the form $v(t) = V_A - V_B \sin \beta t$. At $t = 0$ the value of the waveform is 10 V, and it periodically reaches a minimum value of –2 V every 25 μs. Find the value of V_A, V_B, and β, and then graph the waveform.

5–25 Write an equation for a waveform in Figure P5–25. Graph the waveform obtained by differentiating the waveform.

FIGURE P 5 – 2 5

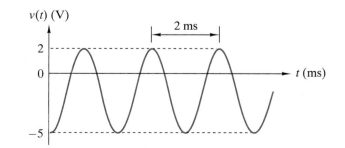

5–26 The SPICE circuit simulation language defines a composite exponential waveform as

$$v(t) = \begin{vmatrix} V_0 + (V_1 - V_0)(1 - e^{-t/T_{C1}}) & 0 \le t \le T_D \\ V_0 + (V_1 - V_0)(1 - e^{-T_D/T_{C1}})e^{-(t-T_D)/T_{C2}} & T_D < t \end{vmatrix}$$

What is the value of the waveform at $t = 0$? At $t = T_D$? What is the value of the waveform as t approaches infinity? What is the maximum amplitude of the waveform? Sketch the waveform on the range from $t = 0$ to $t = T_D + 2T_{C2}$. What is the practical duration of the waveform?

5–27 Write an equation for the first cycle of the waveform in Figure P5–27 and then graph the waveform obtained by differentiating the signal.

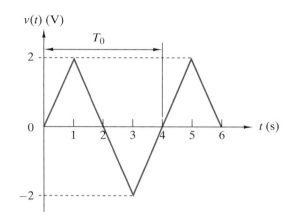

FIGURE P5-27

5–28 The exponential rise waveform $v(t) = V_A(1 - e^{-\alpha t})u(t)$ is 5 V at $t = 3$ ms and 10 V at $t = \infty$. Find the values of V_A and α.

5–29 Figure P5–29 shows a plot of the waveform $v(t) = V_A[e^{-\alpha t} \sin \beta t]u(t)$. Find the values of V_A, α, and β.

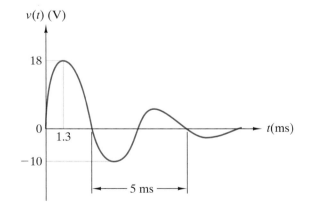

FIGURE P5-29

5–30 The step function $u(\sin \pi t)$ defines a periodic composite waveform. Sketch the waveform and determine its period. The basic waveform $u(t)$ is zero for $t < 0$. Is the composite waveform $u(\sin \pi t)$ zero for

$t < 0$? If not, why not? How would you modify the waveform to change its period and peak amplitude?

ERO 5–3 WAVEFORM PARTIAL DESCRIPTORS (SECT. 5–6)

Given a complete description of a basic or composite waveform,

(a) Classify the waveform as periodic or aperiodic and causal or noncausal.
(b) Find the applicable partial waveform descriptors.

5–31 Classify the following waveforms as periodic or aperiodic and causal or noncausal. Determine V_p and V_{pp} for all of the waveforms. Determine V_{rms} and V_{avg} for those waveforms that are periodic. Determine the duration of those waveforms that are causal.

    (a) $v_1(t) = [15 \sin(10000\pi t)]u(t)$
    (b) $v_2(t) = 40[e^{-200t} \sin(2000\pi t)]u(t)$
    (c) $v_3(t) = 10 \cos(200\pi t) + 15 \sin(200\pi t)$
    (d) $v_4(t) = [5 - 10e^{-40t}]u(t)$

5–32 Determine V_p, V_{pp}, V_{rms}, and V_{avg} for each of the following periodic waveforms.

    (a) $v_1(t) = 20 \cos(400\pi t) - 30 \sin(400\pi t)$
    (b) $v_2(t) = -30 \sin(2000\pi t) + 20 \cos(2000\pi t)$
    (c) $v_3(t) = 10 \cos(2\pi t/50) - 20 \sin(2\pi t/50)$
    (d) $v_4(t) = 10 \cos(200\pi t) - 10 \sin(200\pi t)$

5–33 Classify the waveforms as periodic or aperiodic and causal or noncausal. Determine V_p and V_{pp} for all of the waveforms. Determine V_{rms} and V_{avg} for those waveforms that are periodic.

    (a) $v_1(t) = 20 [1 + e^{-1000t}]u(t)$
    (b) $v_2(t) = [-10 + 20e^{-500t}]u(t)$
    (c) $v_3(t) = 20 [1 + \cos(100\pi t)]$
    (d) $v_4(t) = 20 [1 - \sin(100\pi t)]u(t)$

5–34 Determine V_p, V_{pp}, V_{rms}, and V_{avg} for the waveform in Figure P5–22.

5–35 Determine V_p, V_{pp}, V_{rms}, and V_{avg} for the waveform in Figure P5–23.

5–36 A signal of the form $v(t) = V_A + V_B \cos(2000t)$ has $V_p = 15$ V and $V_{avg} = 5$ V. Find the values of V_A and V_B.

5–37 The waveform $v(t) = V_0 - 10 \cos(2000\pi t)$ is applied at the input of an OP AMP voltage follower with $V_{CC} = \pm 12$ V. What range of values of the dc component V_0 ensures that the OP AMP does not saturate?

5–38 The first cycle ($t > 0$) of a periodic waveform with $T_0 = 70$ ms can be expressed as

$$v(t) = u(t) - 3u(t - 0.01) + 5u(t - 0.06)   \text{V}$$

Sketch the waveform and find V_{max}, V_{min}, V_p, V_{pp}, and V_{avg}.

5–39 A periodic waveform can be expressed as

$$v(t) = 100 - 200 \cos 2000 \pi t - 75 \sin 40000 \pi t + 35 \cos 80000 \pi t \text{mV}$$

Use a math solver or spreadsheet to plot $v(t)$ and determine V_{max}, V_{min}, V_p, V_{pp}, and V_{avg}. What is the amplitude of the fundamental component? The dc component? The fourth harmonic component?

5–40 The first cycle ($t > 0$) of a periodic pulse train with $T_0 = 10$ ms can be expressed as

$$v(t) = V_A[u(t) - u(t - T)]$$

A meter measures the average value of the waveform to be 5 V and the rms value to be 15 V. Find the values of V_A and T.

INTEGRATING PROBLEMS

5–41 **A** ENERGY SIGNALS

The normalized energy carried by a signal $v(t)$ is defined as

$$W = \int_{-\infty}^{\infty} v^2(x)dx$$

If $W < \infty$ the signal is said to be an energy signal. Which of the following are energy signals?
(a) $v_1(t) = 10 \cos(2000\pi t) + 10 \sin(2000\pi t)$
(b) $v_2(t) = [10\ e^{-2t}]u(t)$
(c) $v_3(t) = 20\ [1 - e^{-1000t}]u(t)$
(d) $v_4(t) = 20\ [e^{-20t} \sin(100\pi t)]u(t)$

🔖 5–42 **A** ANALOG-TO-DIGITAL CONVERSION

Figure P5–42 shows a circuit diagram of an analog-to-digital converter based on a voltage divider and OP AMPs operating without feedback. Each of the OP AMPs operates as a comparator whose output is V_{OH} = 10 V when $v_P > v_N$ and V_{OL} = 0 V when $v_P < v_N$. The input signal $v_S(t)$ is applied to all of the noninverting inputs simultaneously. The voltages applied to the inverting inputs come from the four-resistor voltage divider shown. For an input voltage of $v_S(t) = 25e^{-2000t}$ V, express the output voltages $v_A(t)$, $v_B(t)$, $v_C(t)$, and $v_D(t)$ in terms of step functions.

5–43 **A** DIGITAL-TO-ANALOG CONVERSION

Figure P5–43 shows a 3-bit digital-to-analog converter of the type discussed in Sect. 4–6. The circuit consists of an R-$2R$ ladder connected to the inverting input to the OP AMP. The ladder circuit is driven by three digital input signals $v_1(t)$, $v_2(t)$, and $v_3(t)$.
(a) Show that the input-output relationship for the OP AMP circuit is

$$v_O(t) = -\frac{R_F}{8R}[4v_1 + 2v_2 + v_3]$$

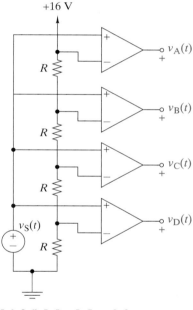

FIGURE P 5 – 4 2

FIGURE P 5 – 4 3

(b) The digital input signals can be expressed in terms of step functions as follows:

$$v_1(t) = 5u(t) - 5u(t - 2) \text{ V}$$

$$v_2(t) = 5u(t) - 5u(t - 4) + 5u(t - 6) - 5u(t - 10) \text{ V}$$

$$v_3(t) = 5u(t - 1) - 5u(t - 3) + 5u(t - 8) \text{ V}$$

Plot each of the digital waveforms separately and stack the plots on top of each other with $v_1(t)$ at the bottom and $v_3(t)$ at the top. A stacked plot of digital signals is called a timing diagram.

(c) For $R_F = R$, use the input-output relationship in (a) and the input timing diagram in (b) to obtain a plot of the output voltage in the range $0 < t < 10$ s.

5–44 **A** ANALOG-TO-DIGITAL-TO-ANALOG CONVERSION

Figure P5–44 shows a block diagram of a 3-bit analog-to-digital converter (ADC) whose digital outputs D_0, D_1, and D_2 are inputs to a 3-bit digital-to-analog converter (DAC). The ADC implements the algorithm

if $0 \leq v_S < 1$ then $D_0 = 0$ and $D_1 = 0$ and $D_2 = 0$ else
if $1 \leq v_S < 2$ then $D_0 = 1$ and $D_1 = 0$ and $D_2 = 0$ else
if $2 \leq v_S < 3$ then $D_0 = 1$ and $D_1 = 1$ and $D_2 = 0$ else
if $3 \leq v_S$ then $D_0 = 1$ and $D_1 = 1$ and $D_2 = 1$.

The input-output relationship of the DAC is

$$v_O = D_0 + D_1 + D_2$$

(a) For $v_S(t) = 4\sin(\pi t)$ sketch the analog input waveform for $0 \leq t \leq 1$.
(b) Express D_0, D_1, and D_2 in terms of unit step functions and sketch their waveforms for $0 \leq t \leq 1$.
(c) Sketch the analog output waveform $v_O(t)$ for $0 \leq t \leq 1$.
(d) Comment on the degree to which the analog output duplicates the analog input.

FIGURE P 5 – 4 4

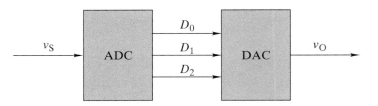

5–45 **A** FOURIER SERIES

Reasonably well behaved periodic waveforms can be expressed as an infinite sum of sinusoidal components called a Fourier series. For example, the sum

$$v(t) = \sum_{n=1}^{n=\infty} \frac{8V_A}{[(2n - 1)\pi]^2} \cos[2\pi(2n - 1)t/T_0]$$

is the Fourier series of one of the periodic waveforms in Figure 5–2. Use a spreadsheet or math solver to plot the sum of the first 10 terms of this series for $V_A = 5$ V and $T_0 = 200$ μs. Use a time interval $0 \leq t \leq 2T_0$. Can you recognize the waveform in Figure 5–2?

5–46 ⓐ VOLTMETER CALIBRATION

An ordinary dc voltmeter measures the average value of the unknown signal. The basic dc meter can be adapted to measure the rms value of a sinusoid by passing the unknown signal through a rectifier circuit whose output is the absolute value of the sinusoid. The rectifier output is applied to a dc meter that measures the average absolute value of the unknown sinusoid. The meter scale is calibrated to indicate the rms value of the unknown sinusoid. Thus, a calibration factor is needed to convert the average absolute value into the rms value. What is the required calibration factor for a sinusoid? Would the same calibration factor apply to a square wave?

5–47 TWO-SIDED EXPONENTIAL

The waveform defined as $v(t) = V_A[u(t)e^{-\alpha t} + u(-t)e^{+\alpha t}]$ is called a two-sided exponential.

(a) Sketch the waveform and its time derivative.
(b) Is the signal causal or noncausal? Periodic or aperiodic?
(c) What are the peak and peak-to-peak values of the waveform?
(d) Theoretically the waveform extends from $t = -\infty$ to $t = +\infty$. What is the practical duration of the waveform?

5–48 ⓐ STEP RESPONSE PARTIAL DESCRIPTORS

The zero-state step response of a first-order circuit is an exponential rise waveform written as

$$y(t) = FV(1 - e^{-t/T_C}) \quad t > 0$$

where FV is the final value of the response and T_C is the circuit time constant. Partial descriptors of this waveform include **rise time**, **delay time**, and **settling time**. Rise time is defined to be the time required for the response to rise from 10% to 90% of its final value. Delay time is defined as the time required for the waveform to reach 50% of its final value. Settling time is defined as the time required for the response to enter and remain within 2% of its final value. Derive the relationships between these three partial descriptors and the circuit time constant.

CHAPTER 6

CAPACITANCE
AND INDUCTANCE

From the foregoing facts, it appears that a current of electricity is produced, for an instant, in a helix of copper wire surrounding a piece of soft iron whenever magnetism is induced in the iron; also that an instantaneous current in one or the other direction accompanies every change in the magnetic intensity of the iron.

Joseph Henry, 1831,
American Physicist

6–1 The Capacitor

6–2 The Inductor

6–3 Dynamic OP AMP Circuits

6–4 Equivalent Capacitance and Inductance

Summary

Problems

Integrating Problems

Joseph Henry (1797–1878) and the British physicist Michael Faraday (1791–1867) independently discovered magnetic induction almost simultaneously. The foregoing quotation is Henry's summary of the experiments leading to his discovery of magnetic induction. Although Henry and Faraday used similar apparatus and observed almost the same results, Henry was the first to recognize the importance of the discovery. The unit of circuit inductance (henry) honors Henry, while the mathematical generalization of magnetic induction is called Faraday's law. Michael Faraday had wide-ranging interests and performed many fundamental experiments in chemistry and electricity. His electrical experiments often used capacitors, or Leyden jars as they were called in those days. Faraday was a meticulous experimenter, and his careful characterization of these devices may be the reason that the unit of capacitance (the farad) honors Faraday.

The dynamic circuit responses involve memory effects that cannot be explained by the circuit elements we included in resistance circuits. The two new circuit elements required to explain memory effects are the capacitor and the inductor. The purpose of this chapter is to introduce the *i–v* characteristics of these dynamic elements and to explore circuit applications that illustrate dynamic behavior.

The first two sections of this chapter develop the device constraints for the linear capacitor and inductor. The third section is a circuit application in which resistors, capacitors, and operational amplifiers (OP AMPs) perform waveform integration and differentiation. The fourth section shows that capacitors or inductors connected in series or parallel can be replaced by a single equivalent capacitance or inductance.

6-1 THE CAPACITOR

A capacitor is a dynamic element involving the time variation of an electric field produced by a voltage. Figure 6–1(a) shows the parallel plate capacitor, which is the simplest physical form of a capacitive device. Figure 6–1 also shows two alternative circuit symbols and sketches of actual devices. Some of the physical features of commercially available devices are given in Appendix A.

Electrostatics shows that a uniform electric field $\mathcal{E}(t)$ exists between the metal plates in Figure 6–1(a) when a voltage exists across the capacitor.[1] The electric field produces charge separation with equal and opposite charges appearing on the capacitor plates. When the separation *d* is small compared with the dimension of the plates, the electric field between the plates is

$$\mathcal{E}(t) = \frac{q(t)}{\varepsilon A} \qquad (6\text{–}1)$$

where ε is the permittivity of the dielectric, *A* is the area of the plates, and $q(t)$ is the magnitude of the electric charge on each plate. The relationship between the electric field and the voltage across the capacitor $v_C(t)$ is given by

(a)

(b)

FIGURE 6 – 1 *The capacitor: (a) Parallel plate device. (b) Circuit symbol. (c) Example devices.*

---

1 An electric field is a vector quantity. In Figure 6–1(a) the field is confined to the space between the two plates and is perpendicular to the plates.

$$\mathcal{E}(t) = \frac{v_C(t)}{d} \tag{6-2}$$

Substituting Eq. (6–2) into Eq. (6–1) and solving for the charge $q(t)$ yields

$$q(t) = \left[\frac{\varepsilon A}{d}\right] v_C(t) \tag{6-3}$$

The proportionality constant inside the brackets in this equation is the **capacitance** C of the capacitor. That is, by definition,

$$C = \frac{\varepsilon A}{d} \tag{6-4}$$

The unit of capacitance is the **farad** (F), a term that honors the British physicist Michael Faraday. Values of capacitance range from a few pF (10^{-12} F) in semiconductor devices to tens of mF (10^{-3} F) in industrial capacitor banks. Using Eq. (6–4), the defining relationship for the capacitor becomes

$$q(t) = C v_C(t) \tag{6-5}$$

Figure 6–2 graphically displays the element constraint in Eq. (6–5). The graph points out that the capacitor is a linear element since the defining relationship between voltage and charge is a straight line through the origin.

I–V RELATIONSHIP

To express the element constraint in terms of voltage and current, we differentiate Eq. (6–5) with respect to time t:

$$\frac{dq(t)}{dt} = \frac{d[C v_C(t)]}{dt}$$

Since C is constant and $i_C(t)$ is the time derivative of $q(t)$, we obtain a capacitor i–v relationship in the form

$$i_C(t) = C \frac{dv_C(t)}{dt} \tag{6-6}$$

The relationship assumes that the reference marks for the current and voltage follow the passive sign convention shown in Figure 6–3.

The time derivative in Eq. (6–6) means the current is zero when the voltage across the capacitor is constant, and vice versa. In other words, the capacitor acts like an open circuit ($i_C = 0$) when dc excitations are applied. The capacitor is a dynamic element because the current is zero unless the voltage is changing. However, a discontinuous change in voltage would require an infinite current, which is physically impossible. Therefore, the capacitor voltage must be a continuous function of time.

Equation (6–6) relates the capacitor current to the rate of change of the capacitor voltage. To express the voltage in terms of the current, we multiply both sides of Eq. (6–6) by dt, solve for the differential dv_C, and integrate:

$$\int dv_C = \frac{1}{C} \int i_C(t)\, dt$$

q(t)

F I G U R E   6 – 2 *Graph of the defining relationship of a linear capacitor.*

F I G U R E   6 – 3 *Capacitor current and voltage.*

Selecting the integration limits requires some discussion. We assume that at some time t_0 the voltage across the capacitor $v_C(t_0)$ is known and we want to determine the voltage at some later time $t > t_0$. Therefore, the integration limits are

$$\int_{v_C(t_0)}^{v_C(t)} dv_C = \int_{t_0}^{t} i_C(x)dx$$

where x is a dummy integration variable. Integrating the left side of this equation yields

$$v_C(t) = v_C(t_0) + \frac{1}{C}\int_{t_0}^{t} i_C(x)dx \qquad (6\text{–}7)$$

In practice, the time t_0 is established by a physical event such as closing a switch or the start of a particular clock pulse. Nothing is lost in the integration in Eq. (6–7) if we arbitrarily define t_0 to be zero. Using $t_0 = 0$ in Eq. (6–7) yields

$$v_C(t) = v_C(0) + \frac{1}{C}\int_{0}^{t} i_C(x)dx \qquad (6\text{–}8)$$

Equation (6–8) is the integral form of the capacitor i–v constraint. Both the integral form and the derivative form in Eq. (6–6) assume that the reference marks for current and voltage follow the passive sign convention in Figure 6–3.

POWER AND ENERGY

With the passive sign convention the capacitor power is

$$p_C(t) = i_C(t)v_C(t) \qquad (6\text{–}9)$$

Using Eq. (6–6) to eliminate $i_C(t)$ from Eq. (6–9) yields the capacitor power in the form

$$p_C(t) = Cv_C(t)\frac{dv_C(t)}{dt} = \frac{d}{dt}[^1/_2Cv_C^2(t)] \qquad (6\text{–}10)$$

This equation shows that the power can be either positive or negative because the capacitor voltage and its time rate of change can have opposite signs. With the passive sign convention, a positive sign means the element absorbs power, while a negative sign means the element delivers power. The ability to deliver power implies that the capacitor can store energy.

To determine the stored energy, we note that the expression for power in Eq. (6–10) is a perfect derivative. Since power is the time rate of change of energy, the quantity inside the brackets must be the energy stored in the capacitor. Mathematically, we can infer from Eq. (6–10) that the energy at time t is

$$w_C(t) = {}^1/_2Cv_C^2(t) + \text{constant}$$

The constant in this equation is the value of stored energy at some instant t when $v_C(t) = 0$. At such an instant the electric field is zero; hence the stored energy is also zero. As a result, the constant is zero and we write the capacitor energy as

$$w_C(t) = \frac{1}{2}Cv_C^2(t) \qquad (6-11)$$

The stored energy is never negative, since it is proportional to the square of the voltage. The capacitor absorbs power from the circuit when storing energy and returns previously stored energy when delivering power to the circuit.

The relationship in Eq. (6–11) also implies that voltage is a continuous function of time, since an abrupt change in the voltage implies a discontinuous change in energy. Since power is the time derivative of energy, a discontinuous change in energy implies infinite power, which is physically impossible. The capacitor voltage is called a **state variable** because it determines the energy state of the element.

To summarize, the capacitor is a dynamic circuit element with the following properties:

1. *The current through the capacitor is zero unless the voltage is changing. The capacitor acts like an open circuit to dc excitations.*

2. *The voltage across the capacitor is a continuous function of time. A discontinuous change in capacitor voltage would require infinite current and power, which is physically impossible.*

3. *The capacitor absorbs power from the circuit when storing energy and returns previously stored energy when delivering power. The net energy transfer is nonnegative, indicating that the capacitor is a passive element.*

The following examples illustrate these properties.

EXAMPLE 6–1

The voltage in Figure 6–4(a) appears across a ½-μF capacitor. Find the current through the capacitor.

SOLUTION:

The capacitor current is proportional to the time rate of change of the voltage. For $0 < t < 2$ ms the slope of the voltage waveform has a constant value

$$\frac{dv_C}{dt} = \frac{10}{2 \times 10^{-3}} = 5000 \text{ V/s}$$

The capacitor current during this interval is

$$i_C(t) = C\frac{dv_C}{dt} = (0.5 \times 10^{-6}) \times (5 \times 10^3) = 2.5 \text{ mA}$$

For $2 < t < 3$ ms the rate of change of the voltage is -5000 V/s. Since the rate of change of voltage is negative, the current changes direction and takes on the value $i_C(t) = -2.5$ mA. For $t > 3$ ms, the voltage is constant, so its slope is zero; hence the current is zero. The resulting current waveform

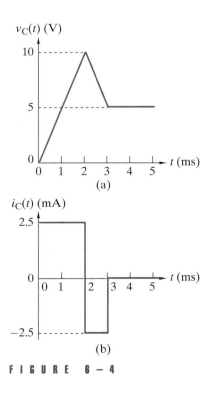

F I G U R E 6 – 4

is shown in Figure 6–4(b). Note that the voltage across the capacitor (the state variable) is continuous, but the capacitor current can be, and in this case is, discontinuous. ∎

EXAMPLE 6–2

The $i_C(t)$ in Figure 6–5(a) is given by

$$i_C(t) = I_0(e^{-t/T_C})u(t)$$

Find the voltage across the capacitor if $v_C(0) = 0$ V.

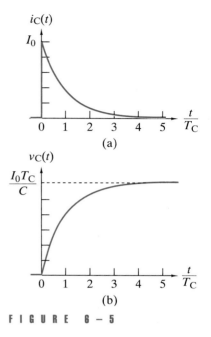

(a)

(b)

FIGURE 6–5

SOLUTION:
Using the capacitor i–v relationship in integral form,

$$v_C(t) = v_C(0) + \frac{1}{C}\int_0^t i_C(x)dx$$

$$= 0 + \frac{1}{C}\int_0^t I_0 e^{-x/T_C}\,dx = \frac{I_0 T_C}{C}(-e^{-x/T_C})\Big|_0^t$$

$$= \frac{I_0 T_C}{C}(1 - e^{-t/T_C})$$

The graphs in Figure 6–5(b) show that the voltage is continuous while the current is discontinuous. ∎

Exercise 6–1

(a) The voltage across a 10-μF capacitor is $25[\sin 2000t]u(t)$. Derive an expression for the current through the capacitor.

(b) At $t = 0$ the voltage across a 100-pF capacitor is –5 V. The current through the capacitor is $10[u(t) - u(t - 10^{-4})]$ μA. What is the voltage across the capacitor for $t > 0$?

Answers:

(a) $i_C(t) = 0.5\,[\cos 2000t]u(t)$ A.

(b) $v_C(t) = -5 + 10^5 t$ V for $0 < t < 0.1$ ms and $v_C(t) = 5$ V for $t > 0.1$ ms

Exercise 6–2

For $t \geq 0$ the voltage across a 200-pF capacitor is $5e^{-4000t}$ V.

(a) What is the charge on the capacitor at $t = 0$ and $t = +\infty$?

(b) Derive an expression for the current through the capacitor for $t \geq 0$.

(c) For $t > 0$ is the device absorbing or delivering power?

Answers:

(a) 1 nC and 0 C

(b) $i_C(t) = -4e^{-4000t}$ μA

(c) Delivering

EXAMPLE 6–3

Figure 6–6(a) shows the voltage across 0.5-µF capacitor. Find the capacitor's energy and power.

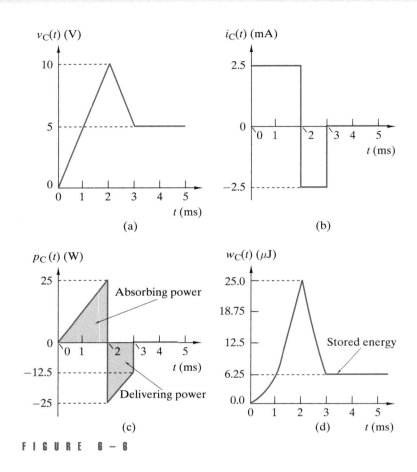

FIGURE 6–6

SOLUTION:

The current through the capacitor was found in Example 6–1. The power waveform is the point-by-point product of the voltage and current waveforms. The energy is found by either integrating the power waveform or by calculating $\frac{1}{2}C[v_C(t)]^2$ point by point. The current, power, and energy are shown in Figures 6–6(b), 6–6(c), and 6–6(d). Note that the capacitor energy increases when it is absorbing power $[p_C(t) > 0]$ and decreases when delivering power $[p_C(t) < 0]$. ■

EXAMPLE 6–4

The current through a capacitor is given by

$$i_C(t) = I_0[e^{-t/T_C}]u(t)$$

Find the capacitor's energy and power.

SOLUTION:
The current and voltage were found in Example 6–2 and are shown in Figures 6–7(a) and 6–7(b). The power waveform is found as the product of current and voltage:

$$p_C(t) = i_C(t)v_C(t)$$

$$= [I_0 e^{-t/T_C}]\left[\frac{I_0 T_C}{C}(1 - e^{-t/T_C})\right]$$

$$= \frac{I_0^2 T_C}{C}(e^{-t/T_C} - e^{-2t/T_C})$$

The waveform of the power is shown in Figure 6–7(c). The energy is

$$w_C(t) = \frac{1}{2}Cv_C^2(t) = \frac{(I_0 T_C)^2}{2C}(1 - e^{-t/T_C})^2$$

The time history of the energy is shown in Figure 6–7(d). In this example both power and energy are always positive. ∎

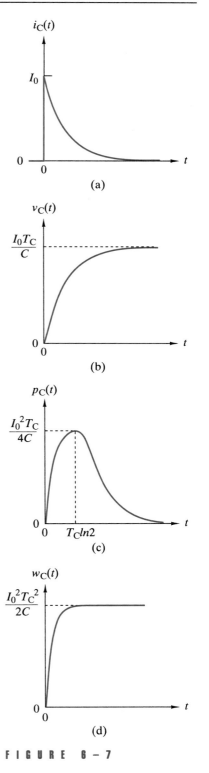

Figure 6–7 labels: $i_C(t)$, I_0, (a); $v_C(t)$, $\frac{I_0 T_C}{C}$, (b); $p_C(t)$, $\frac{I_0^2 T_C}{4C}$, $T_C ln2$, (c); $w_C(t)$, $\frac{I_0^2 T_C^2}{2C}$, (d)

FIGURE 6–7

Exercise 6–3

Find the power and energy for the capacitors in Exercise 6–1.

Answers:

(a) $p_C(t) = 6.25[\sin 4000t]u(t)$ W
     $w_C(t) = 3.125 \sin^2 2000t$ mJ
(b) $p_C(t) = -0.05 + 10^3 t$ mW for $0 < t < 0.1$ ms
     $p_C(t) = 0$ for $t > 0.1$ ms
     $w_C(t) = 1.25 - 5 \times 10^4 t + 5 \times 10^8 t^2$ nJ for $0 < t < 0.1$ ms
     $w_C(t) = 1.25$ nJ for $t > 0.1$ ms

Exercise 6–4

Find the power and energy for the capacitor in Exercise 6–2.

Answers:

$$p_C(t) = -20e^{-8000t} \ \mu W$$

$$w_C(t) = 2.5e^{-8000t} \ nJ$$

APPLICATION NOTE: **EXAMPLE 6–5**

A track-hold circuit is usually found at the input to an analog-to-digital (A/D) converter. The purpose of the circuit is to follow continuously a time-varying input voltage and then to hold a constant value during the time needed to perform the A/D conversion. This example shows that a capacitor can perform the track-hold function.

Figure 6–8(a) shows the simplest form of a capacitor track-hold circuit. When the switch is closed, the circuit is in the track mode and the voltage across the capacitor follows or tracks the input source voltage.

FIGURE 6 – 8

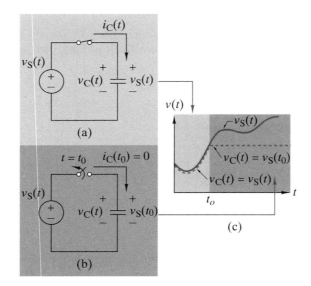

(a)

(b)

(c)

At some time, $t = t_0$ the switch opens and the circuit changes to the hold mode shown in Figure 6–8(b). For $t > t_0$, the source voltage continues to change, producing the waveform $v_S(t)$ shown in Figure 6–8(c). However, Figure 6–8(c) shows in black that the capacitor voltage $v_C(t)$ is constant for $t > t_0$. Opening the switch at $t = t_0$ forces the capacitor current to zero, and zero current means the voltage is constant. The capacitor holds or remembers the voltage existing across it at the instant the switch was opened—namely, $v_S(t_0)$.

Theoretically, the capacitor holds $v_S(t_0)$ forever. In reality, some leakage occurs through the dielectric, so the capacitor gradually loses stored energy and the voltage decreases. The stored value in a track-hold circuit must be occasionally refreshed for long-term storage. The capacitor can hold or remember past events because it can store energy.

6–2   THE INDUCTOR

The inductor is a dynamic circuit element involving the time variation of the magnetic field produced by a current. Magnetostatics shows that a magnetic flux ϕ surrounds a wire carrying an electric current. When the wire is wound into a coil, the lines of flux concentrate along the axis of the coil, as shown in Figure 6–9. In a linear magnetic medium, the flux is proportional to both the current and the number of turns in the coil. Therefore, the total flux is

$$\phi(t) = k_1 N i_L(t) \tag{6–12}$$

where k_1 is a constant of proportionality.

The magnetic flux intercepts or links the turns of the coil. The flux linkage in a coil is represented by the symbol λ, with units of webers (Wb), named after the German scientist Wilhelm Weber (1804–1891). The flux linkage is proportional to the number of turns in the coil and to the total magnetic flux, so $\lambda(t)$ is

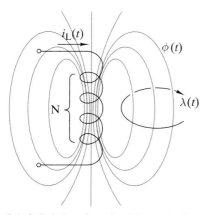

FIGURE 6 – 9   *Magnetic flux surrounding a current carrying coil.*

$$\lambda(t) = N\phi(t) \qquad (6\text{–}13)$$

Substituting Eq. (6–12) into Eq. (6–13) gives

$$\lambda(t) = [k_1 N^2] i_L(t) \qquad (6\text{–}14)$$

The proportionality constant inside the brackets in this equation is the **inductance** L of the coil. That is, by definition

$$L = k_1 N^2 \qquad (6\text{–}15)$$

The unit of inductance is the henry (H) (plural henrys), a name that honors American scientist Joseph Henry. Figure 6–10 shows the circuit symbol for an inductor and some examples of actual devices.

Using Eq. (6–15), the defining relationship for the inductor becomes

$$\lambda(t) = L i_L(t) \qquad (6\text{–}16)$$

Figure 6–11 graphically displays the inductor's element constraint in Eq. (6–16). The graph points out that the inductor is a linear element since the defining relationship is a straight line through the origin.

I–V RELATIONSHIP

Equation (6–16) is the inductor element constraint in terms of current and flux linkage. To obtain the element characteristic in terms of voltage and current, we differentiate Eq. (6–16) with respect to time:

$$\frac{d[\lambda(t)]}{dt} = \frac{d[L i_L(t)]}{dt} \qquad (6\text{–}17)$$

The inductance L is a constant. According to Faraday's law, the voltage across the inductor is equal to the time rate of change of flux linkage. Therefore, we obtain an inductor *i–v* relationship in the form

$$v_L(t) = L\frac{di_L(t)}{dt} \qquad (6\text{–}18)$$

The time derivative in Eq. (6–18) means that the voltage across the inductor is zero unless the current is time varying. Under dc excitation the current is constant and $v_L = 0$, so the inductor acts like a short circuit. The inductor is a dynamic element because only a changing current produces a nonzero voltage. However, a discontinuous change in current would produce an infinite voltage, which is physically impossible. Therefore, the current $i_L(t)$ must be a continuous function of time t.

Equation (6–18) relates the inductor voltage to the rate of change of the inductor current. To express the inductor current in terms of the voltage, we multiply both sides of Eq. (6–18) by dt, solve for the differential di_L, and integrate:

$$\int di_L = \frac{1}{L}\int v_L(t)dt \qquad (6\text{–}19)$$

To set the limits of integration, we assume that the inductor current $i_L(t_0)$ is known at some time t_0. Under this assumption the integration limits are

(a)

(b)

FIGURE 6–10 *The inductor: (a) Circuit symbol. (b) Example devices.*

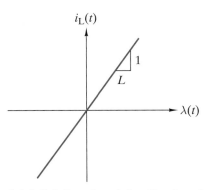

FIGURE 6–11 *Graph of the defining relationship of a linear inductor.*

$$\int_{i_L(t_0)}^{i_L(t)} di_L = \frac{1}{L} \int_{t_0}^{t} v_L(x)dx \qquad (6\text{--}20)$$

where x is a dummy integration variable. The left side of Eq. (6–20) integrates to produce

$$i_L(t) = i_L(t_0) + \frac{1}{L} \int_{t_0}^{t} v_L(x)dx \qquad (6\text{--}21)$$

The reference time t_0 is established by some physical event, such as closing or opening a switch. Without losing any generality, we can assume $t_0 = 0$ and write Eq. (6–21) in the form

$$i_L(t) = i_L(0) + \frac{1}{L} \int_{0}^{t} v_L(x)dx \qquad (6\text{--}22)$$

Equation (6–22) is the integral form of the inductor i–v characteristic. Both the integral form and the derivative form in Eq. (6–18) assume that the reference marks for the inductor voltage and current follow the passive sign convention shown in Figure 6–10.

POWER AND ENERGY

With the passive sign convention the inductor power is

$$p_L(t) = i_L(t)v_L(t) \qquad (6\text{--}23)$$

Using Eq. (6–18) to eliminate $v_L(t)$ from this equation puts the inductor power in the form

$$p_L(t) = [i_L(t)]\left[L\frac{di_L(t)}{dt}\right] = \frac{d}{dt}[^1/_2 Li_L^2(t)] \qquad (6\text{--}24)$$

This expression shows that power can be positive or negative because the inductor current and its time derivative can have opposite signs. With the passive sign convention a positive sign means the element absorbs power, while a negative sign means the element delivers power. The ability to deliver power indicates that the inductor can store energy.

To find the stored energy, we note that the power relation in Eq. (6–24) is a perfect derivative. Since power is the time rate of change of energy, the quantity inside the brackets must represent the energy stored in the magnetic field of the inductor. From Eq. (6–24), we infer that the energy at time t is

$$w_L(t) = \frac{1}{2}Li_L^2(t) + \text{constant}$$

As is the case with capacitor energy, the constant in this expression is zero since it is the energy stored at an instant t at which $i_L(t) = 0$. As a result, the energy stored in the inductor is

$$w_L(t) = \frac{1}{2}Li_L^2(t) \qquad (6\text{--}25)$$

The energy stored in an inductor is never negative because it is proportional to the square of the current. The inductor stores energy when absorbing power and returns previously stored energy when delivering power, so that the net energy transfer is never negative.

Equation (6–25) implies that inductor current is a continuous function of time because an abrupt change in current causes a discontinuity in the energy. Since power is the time derivative of energy, an energy discontinuity implies infinite power, which is physically impossible. Current is called the **state variable** of the inductor because it determines the energy state of the element.

In summary, the inductor is a dynamic circuit element with the following properties:

1. *The voltage across the inductor is zero unless the current through the inductor is changing. The inductor acts like a short circuit for dc excitations.*

2. *The current through the inductor is a continuous function of time. A discontinuous change in inductor current would require infinite voltage and power, which is physically impossible.*

3. *The inductor absorbs power from the circuit when storing energy and delivers power to the circuit when returning previously stored energy. The net energy is nonnegative, indicating that the inductor is a passive element.*

EXAMPLE 6-6

The current through a 2-mH inductor is $i_L(t) = 4 \sin 1000t + 1 \sin 3000t$ A. Find the resulting inductor voltage.

SOLUTION:
The voltage is found from the derivative form of the *i–v* relationship:

$$v_L(t) = L\frac{di_L(t)}{dt} = 0.002\,[4 \times 1000 \cos 1000\,t + 1 \times 3000 \cos 3000\,t]$$

$$= 8 \cos 1000\,t + 6 \cos 3000\,t \ \text{V}$$

The current and voltage waveforms are shown in Figure 6–12. Note that the current and voltage each contain two sinusoids at different frequencies. However, the relative amplitudes of the two sinusoids are different. In $i_L(t)$ the ratio of the amplitude of the ac component at $\omega = 3$ krad/s to the ac component at $\omega = 1$ krad/s is 1-to-3, whereas in $v_L(t)$ this ratio is 3-to-4. The fact that the ac responses of energy storage elements depend on frequency allows us to create frequency selective signal processors called filters. ∎

EXAMPLE 6-7

Figure 6–13 shows the current through and voltage across an unknown energy storage element.
(a) What is the element and what is its numerical value?
(b) If the energy stored in the element at $t = 0$ is zero, how much energy is stored in the element at $t = 1$ s?

(a) Current

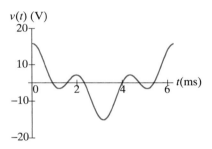

(b) Voltage

FIGURE 6-12 *(a) Current. (b) Voltage.*

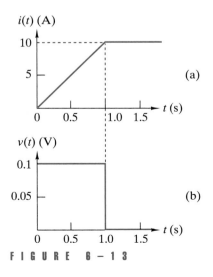

FIGURE 6-13

SOLUTION:

(a) By inspection, the voltage across the device is proportional to the derivative of the current, so the element is a linear inductor. During the interval $0 < t < 1$ s, the slope of the current waveform is 10 A/s. During the same interval the voltage is a constant 100 mV. Therefore, the inductance is

$$L = \frac{v}{di/dt} = \frac{0.1 \text{ V}}{10 \text{ A/s}} = 10 \text{ mH}$$

(b) The energy stored at $t = 1$ s is

$$w_{\text{L}}(1) = \frac{1}{2}Li_{\text{L}}^2(1) = 0.5(0.01)(10)^2 = 0.5 \text{ J} \qquad ■$$

EXAMPLE 6–8

The current through a 2.5-mH inductor is a damped sine $i(t) = 10e^{-500t}$ sin $2000t$. Plot the waveforms of the element current, voltage, power, and energy.

SOLUTION:

Figure 6–14 shows a Mathcad file that produces the required graphs. The first two lines in the file define the inductance and the current. The inductor voltage is found by differentiation, the power as the product of voltage and current, and the energy by integrating the power. In the plots shown in Figure 6–14, note that the current, voltage, and power alternate signs, whereas the total energy is always positive.

In Mathcad the time scale statement shown in the figure uses the following format:

$$t = T_{\text{Start}}, \ T_{\text{Start}} + \Delta t \dots T_{\text{Stop}}$$

Numerical values for T_{Start}, Δt, and T_{Stop} are chosen by considering the damped sine waveform. The period of the sinusoid is

$$T_0 = \frac{2 \times \pi}{2000} = 0.00314 \text{ s}$$

The plots start at $T_{\text{Start}} = 0$. To include at least one cycle requires an end time of $T_{\text{End}} > T_0 = 0.00314$ s; hence we select $T_{\text{Stop}} = 0.004$ s. To include at least 20 points per cycle requires a time increment ($t < T_O/20 = 0.000158$; hence we select $\Delta t = 0.0001$ s. ■

Exercise 6–5

For $t > 0$, the voltage across a 4-mH inductor is $v_{\text{L}}(t) = 20e^{-2000t}$ V. The initial current is $i_{\text{L}}(0) = 0$.

(a) What is the current through the inductor for $t > 0$?
(b) What is the power and (c) what is the energy for $t > 0$?

FIGURE 6–14

```
Mathcad PLUS - [EX6-8.MCD]
File   Edit   Text   Math   Graphics   Symbolic   Window   Books   Help

Example 6-8

L := 2.5·10⁻³              <---Given Inductance

i(t) := 10·e⁻⁵⁰⁰·ᵗ·sin(2000·t)    <---Given inductor current

v(t) := L· d/dt i(t)       <---Inductor voltage

p(t) := v(t)·i(t)          <---Inductor power

          ⌠ᵗ
w(t) :=   ⎮   p(t) dt      <---Stored energy
          ⌡₀

t := 0, .0001 .. .004      <---Time scale

auto                Page 1
```

```
Mathcad PLUS - [EX6-8.MCD]
File   Edit   Text   Math   Graphics   Symbolic   Window   Books   Help
```

Answers:

(a) $i_L(t) = 2.5(1 - e^{-2000t})$ A
(b) $p_L(t) = 50(e^{-2000t} - e^{-4000t})$ W
(c) $w_L(t) = 12.5(1 - 2e^{-2000t} + e^{-4000t})$ mJ

Exercise 6–6

For $t < 0$, the current through a 100-mH inductor is zero. For $t \geq 0$, the current is
$i_L(t) = 20e^{-2000t} - 20e^{-4000t}$ mA.

(a) Derive an expression for the voltage across the inductor for $t > 0$.
(b) Find the time $t > 0$ at which the inductor voltage passes through zero.
(c) Derive an expression for the inductor power for $t > 0$.
(d) Find the time interval over which the inductor absorbs power and the interval over which it delivers power.

Answers:

(a) $v_L(t) = -4e^{-2000t} + 8e^{-4000t}$ V

(b) $t = 0.347$ ms

(c) $p_L(t) = -80e^{-4000t} + 240e^{-6000t} - 160e^{-8000t}$ mW

(d) Absorbing for $0 < t < 0.347$ ms, delivering for $t > 0.347$ ms

---

MORE ABOUT DUALITY

The capacitor and inductor characteristics are quite similar. Interchanging C and L, and i and v converts the capacitor equations into the inductor equations, and vice versa. This interchangeability illustrates the principle of duality. The dual concepts seen so far are as follows:

KVL	$\leftrightarrow$	KCL
Loop	$\leftrightarrow$	Node
Resistance	$\leftrightarrow$	Conductance
Voltage source	$\leftrightarrow$	Current source
Thévenin	$\leftrightarrow$	Norton
Short circuit	$\leftrightarrow$	Open circuit
Series	$\leftrightarrow$	Parallel
Capacitance	$\leftrightarrow$	Inductance
Flux linkage	$\leftrightarrow$	Charge

The term in one column is the dual of the term in the other column. The **principle of duality** states that

> *If every electrical term in a correct statement about circuit behavior is replaced by its dual, then the result is another correct statement.*

This principle may help beginners gain confidence in their understanding of circuit analysis. When the concept in one column is understood, the dual concept in the other column becomes easier to remember and apply.

6–3 DYNAMIC OP AMP CIRCUITS

The dynamic characteristics of capacitors and inductors produce signal processing functions that cannot be obtained using resistors. The OP AMP circuit in Figure 6–15 is similar to the inverting amplifier circuit except for the capacitor in the feedback path. To determine the signal-processing function of the circuit, we need to find its input-output relationship.

We begin by writing a KCL equation at node A.

$$i_R(t) + i_C(t) = i_N(t)$$

The resistor and capacitor device equations are written using their i–v relationships and the fundamental property of node voltages:

$$i_C(t) = C\frac{d[v_O(t) - v_A(t)]}{dt}$$

$$i_R(t) = \frac{1}{R}[v_S(t) - v_A(t)]$$

FIGURE 6–15 *The inverting OP AMP integrator.*

The ideal OP AMP device equations are $i_N(t) = 0$ and $v_A(t) = 0$. Substituting all of the element constraints into the KCL connection constraint produces

$$\frac{v_S(t)}{R} + C\frac{dv_O(t)}{dt} = 0$$

To solve for the output $v_O(t)$, we multiply this equation by dt, solve for the differential dv_O, and integrate:

$$\int dv_0 = -\frac{1}{RC}\int v_S(t)dt$$

Assuming the output voltage is known at time $t_0 = 0$, the integration limits are

$$\int_{v_0(0)}^{v_0(t)} dv_O = -\frac{1}{RC}\int_0^t v_S(x)dx$$

which yields

$$v_O(t) = v_O(0) - \frac{1}{RC}\int_0^t v_S(x)dx$$

The initial condition $v_O(0)$ is actually the voltage on the capacitor at $t = 0$, since by KVL, we have $v_C(t) = v_O(t) - v_A(t)$. But $v_A = 0$ for the OP AMP, so in general $v_O(t) = v_C(t)$. When the voltage on the capacitor is zero at $t = 0$, the circuit input-output relationship reduces to

$$v_O(t) = -\frac{1}{RC}\int_0^t v_S(x)dx \qquad (6-26)$$

The output voltage is proportional to the integral of the input voltage when the initial capacitor voltage is zero. The circuit in Figure 6–15 is an **inverting integrator** since the proportionality constant is negative. The constant $1/RC$ has the units of reciprocal seconds (s^{-1}) so that both sides of Eq. (6–26) have the units of volts.

Interchanging the resistor and capacitor in Figure 6–15 produces the OP AMP differentiator in Figure 6–16. To find the input-output relationship of this circuit, we start by writing the element and connection equations. The KCL connection constraint at node A is

$$i_R(t) + i_C(t) = i_N(t)$$

The device equations for the input capacitor and feedback resistor are

$$i_C(t) = C\frac{d[v_S(t) - v_A(t)]}{dt}$$

$$i_R(t) = \frac{1}{R}[v_O(t) - v_A(t)]$$

The device equations for the OP AMP are $i_N(t) = 0$ and $v_A(t) = 0$. Substituting all of these element constraints into the KCL connection constraint produces

$$\frac{v_O(t)}{R} + C\frac{dv_S(t)}{dt} = 0$$

FIGURE 6-16 *The inverting OP AMP differentiator.*

Circuit	Block diagram	Gains

F I G U R E 6 − 1 7 *Summary of basic OP AMP signal processing circuits.*

Solving this equation for $v_O(t)$ produces the circuit input-output relationship:

$$v_O(t) = -RC\frac{dv_S(t)}{dt} \tag{6–27}$$

The output voltage is proportional to the derivative of the input voltage. The circuit in Figure 6–16 is an **inverting differentiator** since the proportionality constant ($-RC$) is negative. The units of the constant RC are seconds so that both sides of Eq. (6–27) have the units of volts.

There are OP AMP inductor circuits that produce the inverting integrator and differentiator functions; however, they are of little practical interest because of the physical size and resistive losses in real inductor devices.

Figure 6–17 shows OP AMP circuits and block diagrams for the inverting integrator and differentiator, together with signal-processing functions studied in Chapter 4. The term *operational amplifier* results from the various mathematical operations implemented by these circuits. The following examples illustrate using the collection of circuits in Figure 6–17 in the analysis and design of signal-processing functions.

EXAMPLE 6–9

The input to the circuit in Figure 6–18 is $v_S(t) = 10u(t)$. Derive an expression for the output voltage. The OP AMP saturates when $v_O(t) = \pm15$ V.

20 kΩ   20 nF

$v_S(t)$ 5 V $v_O(t)$

FIGURE 6–18

SOLUTION:
The circuit is the inverting integrator with an initial voltage of 5 V across the capacitor. For the reference marks shown in Figure 6–18, this means that $v_O(0) = +5$ V. Assuming the OP AMP is operating in the linear mode, the output voltage is

$$v_O(t) = v_O(0) - \frac{1}{RC}\int_0^t v_S\, dt$$

$$= 5 - 2500 \int_0^t 10\, dt$$

$$= 5 - 25{,}000t \quad t > 0$$

The output contains a negative going ramp because the circuit is an inverting integrator. The ramp output response is valid only as long as the OP AMP remains in its linear range. Negative saturation will occur when $5 - 25{,}000t = -15$, or at $t = 0.8$ ms. For $t > 0.8$ ms, the OP AMP is in the negative saturation mode with $v_O = -15$ V.

This example illustrates that dynamic circuits with bounded inputs may have unbounded responses. The circuit input here is a 10-V step function that has a bounded amplitude. The circuit output is a ramp whose output would be unbounded except that the OP AMP saturates.   ∎

Exercise 6–7

The input to the circuit in Figure 6–18 is $v_S(t) = 10\,[e^{-5000t}]u(t)$ V.

(a) For $v_C(0) = 0$, derive an expression for the output voltage, assuming the OP AMP is in its linear range.

(b) Does the OP AMP saturate with the given input?

Answers:

(a) $v_O(t) = 5(e^{-5000t} - 1)u(t)$

(b) Does not saturate

EXAMPLE 6–10

The input to the circuit in Figure 6–19(a) is a trapezoidal waveform shown in (b). Find the output waveform. The OP AMP saturates when $v_O(t) = \pm 15$ V.

FIGURE 6 – 19

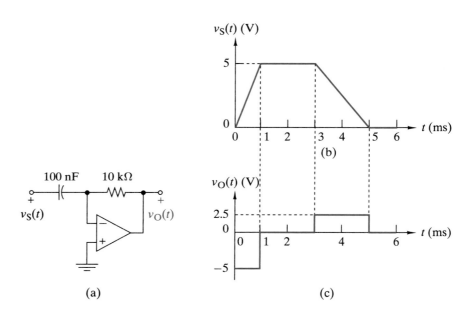

(a)

(b)

(c)

SOLUTION:

The circuit is the inverting differentiator with the following input-output relationship:

$$v_O(t) = -RC\frac{dv_S(t)}{dt} = -\frac{1}{1000}\frac{dv_S(t)}{dt}$$

The output voltage is constant over each of the following three time intervals:

1. For $0 < t < 1$ ms, the input slope is 5000 V/s and the output is $v_O = -5$ V.
2. For $1 < t < 3$ ms, the input slope is zero, so the output is zero as well.
3. For $3 < t < 5$ ms, the input slope is -2500 V/s and the output is $+2.5$ V.

The resulting output waveform is shown in Figure 6–19(c).

The output voltage remains within ±15-V limits, so the OP AMP operates in the linear mode. ∎

The input to the circuit in Figure 6–19 is $v_S(t) = V_A \cos 2000t$. The OP AMP saturates when $v_O = \pm15$ V.

(a) Derive an expression for the output, assuming that the OP AMP is in the linear mode.
(b) What is the maximum value of V_A for linear operation?

Answers:
(a) $v_O(t) = 2V_A \sin 2000t$
(b) $|V_A| \le 7.5$ V

EXAMPLE 6–11

Determine the input-output relationship of the circuit in Figure 6–20(a).

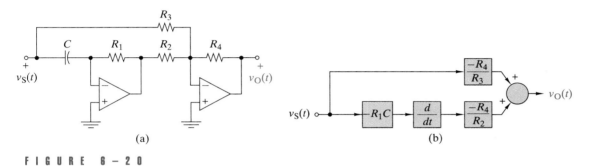

(a)   (b)

FIGURE 6-20

SOLUTION:
The circuit contains an inverting differentiator and an inverting summer. To find the input-output relationship, it is helpful to develop a block diagram for the circuit. Figure 6–20(b) shows a block diagram using the functional blocks in Figure 6–17. The product of the gains along the lower path yields its contribution to the output as

$$(-R_1 C)\left[\frac{d}{dt}\left(-\frac{R_4}{R_2}\right)v_S(t)\right]$$

The upper path contributes $(-R_4/R_3)v_S(t)$ to the output. The total output is the sum of the contributions from each path:

$$v_O(t) = \left(\frac{R_1 C R_4}{R_2}\right)\frac{dv_S(t)}{dt} - \left(\frac{R_4}{R_3}\right)v_S(t)$$

This equation assumes that both OP AMPs remain within their linear range.   ■

Find the input-output relationship of the circuit in Figure 6–21.

Answer:

$$v_O(t) = v_O(0) + \frac{1}{RC}\int_0^t (v_{S1} - v_{S2})dt$$

FIGURE 6 – 21

DESIGN EXAMPLE 6–12

Use the functional blocks in Figure 6–17 to design an OP AMP circuit to implement the following input-output relationship:

$$v_O(t) = 10v_S(t) + \int_0^t v_S \, dt$$

FIGURE 6 – 22

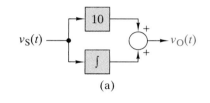

(a)

Inverting amplifier   Inverting summer

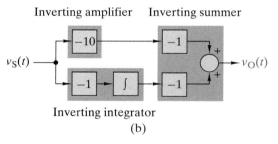

Inverting integrator

(b)

SOLUTION:

There is no unique solution to this design problem. We begin by drawing the block diagram in Figure 6–22(a), which shows that we need a gain block, an integrator, and a summer. However, the integrator and summer in Figure 6–17 are inverting circuits. Figure 6–22(b) shows how to overcome this problem by including an even number of signal inversions in each path. The inverting building blocks are realizable using OP AMP circuits in Figure 6–17, and the overall transfer characteristic is noninverting as required. One (of many) possible circuit realization of these processors is shown in Figure 6–23. The element parameter constraint on this circuit is $RC = 1$. Selecting the OP AMPs and the values of R and C depends on many additional factors, such as accuracy, internal resistance of the input source, and output load. ∎

FIGURE 6 – 23

6–4   EQUIVALENT CAPACITANCE AND INDUCTANCE

In Chapter 2 we found that resistors connected in series or parallel can be replaced by equivalent resistances. The same principle applies to connections of capacitors and inductors—for example, to the parallel connection of capacitors in Figure 6–24(a). Applying KCL at node A in Figure 6–24(a) yields

$$i(t) = i_1(t) + i_2(t) + \ldots + i_N(t)$$

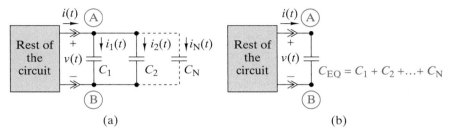

(a)   (b)

F I G U R E   6 – 2 4 *Capacitors connected in parallel. (a) Given circuit. (b) Equivalent circuit.*

Since the elements are connected in parallel, KVL requires

$$v_1(t) = v_2(t) = \ldots = v_N(t) = v(t)$$

Because the capacitors all have the same voltage, their *i–v* relationships are all of the form $i_k(t) = C_k dv(t)/dt$. Substituting the *i–v* relationships into the KCL equation yields

$$i(t) = C_1\frac{dv(t)}{dt} + C_2\frac{dv(t)}{dt} + \ldots + C_N\frac{dv(t)}{dt}$$

Factoring the derivative out of each term produces

$$i(t) = (C_1 + C_2 + \ldots + C_N)\frac{dv(t)}{dt}$$

This equation states that the responses $v(t)$ and $i(t)$ in Figure 6–24(a) do not change when the N parallel capacitors are replaced by an equivalent capacitance:

$$C_{EQ} = C_1 + C_2 + \ldots + C_N \quad \text{(parallel connection)} \tag{6–28}$$

The equivalent capacitance simplification is shown in Figure 6–24(b). The initial voltage, if any, on the equivalent capacitance is $v(0)$, the common voltage across all of the original N capacitors at $t = 0$.

Next consider the series connection of N capacitors in Figure 6–25(a). Applying KVL around loop 1 in Figure 6–25(a) yields the equation

$$v(t) = v_1(t) + v_2(t) + \ldots + v_N(t)$$

Since the elements are connected in series, KCL requires

$$i_1(t) = i_2(t) = \ldots = i_N(t) = i(t)$$

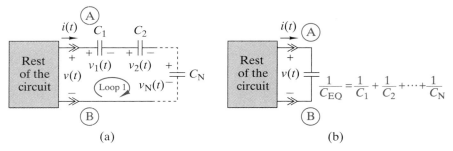

FIGURE 6 – 2 5 *Capacitors connected in series. (a) Given circuit. (b) Equivalent circuit.*

Since the same current exists in all capacitors, their *i–v* relationships are all of the form

$$v_k(t) = v_k(0) + \frac{1}{C_k}\int_0^t i(x)dx$$

Substituting these *i–v* relationships into the loop 1 KVL equation yields

$$v(t) = v_1(0) + \frac{1}{C_1}\int_0^t i(x)dx + v_2(0) + \frac{1}{C_2}\int_0^t i(x)dx$$

$$+ \cdots + v_N(0) + \frac{1}{C_N}\int_0^t i(x)dx$$

We can factor the integral out of each term to obtain

$$v(t) = [v_1(0) + v_2(0) + \ldots + v_N(0)] + \left(\frac{1}{C_1} + \frac{1}{C_2} + \ldots + \frac{1}{C_N}\right)\int_0^t i(x)dx$$

This equation indicates that the responses $v(t)$ and $i(t)$ in Figure 6–25(a) do not change when the N series capacitors are replaced by an equivalent capacitance:

$$\frac{1}{C_{EQ}} = \frac{1}{C_1} + \frac{1}{C_2} + \ldots + \frac{1}{C_N} \qquad \text{(series connection)} \qquad (6\text{--}29)$$

The equivalent capacitance is shown in Figure 6–25(b). The initial voltage on the equivalent capacitance is the sum of the initial voltages on each of the original N capacitors.

The equivalent capacitance of a parallel connection is the sum of the individual capacitances. The reciprocal of the equivalent capacitance of a series connection is the sum of the reciprocals of the individual capacitances. Since the capacitor and inductor are dual elements, the corresponding results for inductors are found by interchanging the series and parallel equivalence rules for the capacitor. That is, in a series connection the equivalent inductance is the sum of the individual inductances:

$$L_{EQ} = L_1 + L_2 + \ldots + L_N \qquad \text{(series connection)} \qquad (6\text{--}30)$$

For the parallel connection, the reciprocals add to produce the reciprocal of the equivalent inductance:

$$\frac{1}{L_{EQ}} = \frac{1}{L_1} + \frac{1}{L_2} + \ldots + \frac{1}{L_N} \quad \text{(parallel connection)} \qquad (6\text{–}31)$$

Derivation of Eqs. (6–30) and (6–31) uses the approach given previously for the capacitor except that the roles of voltage and current are interchanged. Completion of the derivation is left as a problem for the reader.

EXAMPLE 6–13

Find the equivalent capacitance and inductance of the circuits in Figure 6–26.

SOLUTION:

(a) For the circuit in Figure 6–26(a), the two 0.5-μF capacitors in parallel combine to yield an equivalent $0.5 + 0.5 = 1\text{-}\mu F$ capacitance. This 1-μF equivalent capacitance is in series with a 1-μF capacitor, yielding an overall equivalent of $C_{EQ} = 1/(1/1 + 1/1) = 0.5\ \mu F$.

(b) For the circuit of Figure 6–26(b), the 10-mH and the 30-mH inductors are in series and add to produce an equivalent inductance of 40 mH. This 40-mH equivalent inductance is in parallel with the 80-mH inductor. The equivalent inductance of the parallel combination is $L_{EQ} = 1/(1/40 + 1/80) = 26.67$ mH.

(c) The circuit of Figure 6–26(c) contains both inductors and capacitors. In later chapters, we will learn how to combine all of these into a single equivalent element. For now, we combine the inductors and the capacitors separately. The 5-pF capacitor in parallel with the 0.1-μF capacitor yields an equivalent capacitance of 0.100005 μF. For all practical purposes, the 5-pF capacitor can be ignored, leaving two 0.1-μF capacitors in series with equivalent capacitance of 0.05 μF. Combining this equivalent capacitance in parallel with the remaining 0.05-μF capacitor yields an overall equivalent capacitance of 0.1 μF. The parallel 700-μH and 300-μH inductors yield an equivalent inductance of $1/(1/700 + 1/300) = 210$ μH. This equivalent inductance is effectively in series with the 1-mH inductor at the bottom, yielding $1000 + 210 = 1210$ μH as the overall equivalent inductance.

Figure 6–27 shows the simplified equivalent circuits for each of the circuits of Figure 6–26. ∎

(a)

(b)

(c)

FIGURE 6–26

Exercise 6–10

The current through a series connection of two 1-μF capacitors is a rectangular pulse with an amplitude of 2 mA and a duration of 10 ms. At $t = 0$ the voltage across the first capacitor is +10 V and across the second is zero.

(a) What is the voltage across the series combination at $t = 10$ ms?
(b) What is the maximum instantaneous power delivered to the series combination?
(c) What is the energy stored on the first capacitor at $t = 0$ and $t = 10$ ms?

Answers:

(a) 50 V
(b) 100 mW at $t = 10$ ms
(c) 50 μJ and 450 μJ

(a)

(b)

(c)

FIGURE 6-27

(a)

(b)

FIGURE 6-28

FIGURE 6-29

DC EQUIVALENT CIRCUITS

Sometimes we need to find the dc response of circuits containing capacitors and inductors. In the first two sections of this chapter, we found that under dc conditions a capacitor acts like an open circuit and an inductor acts like a short circuit. In other words, under dc conditions, an equivalent circuit for a capacitor is an open circuit and an equivalent circuit of an inductor is a short circuit.

To determine dc responses, we replace capacitors by open circuits and inductors by short circuits and analyze the resulting resistance circuit using any of the methods in Chapters 2 through 4. The circuit analysis involves only resistance circuits and yields capacitor voltages and inductor currents along with any other variables of interest. Computer programs like SPICE use this type of dc analysis to find the initial operating point of a circuit to be analyzed. The dc capacitor voltages and inductor currents become initial conditions for a transient response that begins at $t = 0$ when something in the circuit changes, such as the position of a switch.

EXAMPLE 6-14

Determine the voltage across the capacitors and current through the inductors in Figure 6-28(a).

SOLUTION:

The circuit is driven by a 5-V dc source. Figure 6-28(b) shows the equivalent circuit under dc conditions. The current in the resulting series circuit is $5/(50 + 50) = 50$ mA. This dc current exists in both inductors, so $i_{L1} = i_{L2} = 50$ mA. By voltage division the voltage across the 50-Ω output resistor is $v = 5 \times 50/(50 + 50) = 2.5$ V; therefore, $v_{C1}(0) = 2.5$ V. The voltage across C_2 is zero because of the short circuits produced by the two inductors. ∎

Exercise 6-11

Find the OP AMP output voltage in Figure 6-29.

Answer:

$$v_O = \frac{R_2 + R_1}{R_1} V_{dc}$$

SUMMARY

• The linear capacitor and inductor are dynamic circuit elements that can store energy. The instantaneous element power is positive when they are storing energy and negative when they are delivering previously stored energy. The net energy transfer is never negative because inductors and capacitors are passive elements.

- The current through a capacitor is zero unless the voltage is changing. A capacitor acts like an open circuit to dc excitations.

- The voltage across an inductor is zero unless the current is changing. An inductor acts like a short circuit to dc excitations.

- Capacitor voltage and inductor current are called state variables because they define the energy state of a circuit. Circuit state variables are continuous functions of time as long as the circuit driving forces are finite.

- OP AMP capacitor circuits perform signal integration or differentiation. These operations, together with the summer and gain functions, provide the building blocks for designing dynamic input-output characteristics.

- Capacitors or inductors in series or parallel can be replaced with an equivalent element found by adding the individual capacitances or inductances or their reciprocals. The dc response of a dynamic circuit can be found by replacing all capacitors with open circuits and all inductors with short circuits.

PROBLEMS

ERO 6–1 CAPACITOR AND INDUCTOR RESPONSES (SECTS. 6–1, 6–2)

Given the current through (voltage across) a capacitor or an inductor,

(a) Determine the voltage across (current through) the element.
(b) Determine the power absorbed and energy stored by the element.

6–1   The voltage across a 4-μF capacitor is $v_C(t) = 3\ e^{-2000t}$ V. Derive expressions for, and sketch, $i_C(t)$, $p_C(t)$, and $w_C(t)$. Is the capacitor absorbing power, delivering power, or both?

6–2   The voltage across a 100-nF capacitor is $v_C(t) = 30\cos(5000t)u(t)$ V. Derive expressions for, and sketch, $i_C(t)$, $p_C(t)$, and $w_C(t)$. Is the capacitor absorbing power, delivering power, or both?

6–3   The current through a 125-nF capacitor is $i_C(t) = 50e^{-5000t}u(t)$ mA. Derive expressions for, and sketch, $v_C(t)$, $p_C(t)$, and $w_C(t)$. Assume $v_C(0) = 0$. Is the capacitor absorbing power, delivering power, or both?

6–4   The current through a 15-μF capacitor is $i_C(t) = 50[u(t) - u(t - 4 \times 10^{-3})]$ mA. Derive expressions for, and sketch, $v_C(t)$, $p_C(t)$, and $w_C(t)$. Is the capacitor absorbing power, delivering power, or both?

6–5   The initial voltage across a 250-nF capacitor is $v_C(0) = 25$ V. For $t > 0$, the current through the capacitor is $i_C(t) = 200\cos 10^5 t$ mA. Derive expressions for, and sketch, $v_C(t)$, $p_C(t)$, and $w_C(t)$. Is the capacitor absorbing power, delivering power, or both?

6–6   The voltage across a 0.5-μF capacitor is shown in Figure P6–6. Derive expressions for, and sketch, $i_C(t)$, $p_C(t)$, and $w_C(t)$. Is the capacitor absorbing power, delivering power, or both?

6–7   The current through a 1-μF capacitor is shown in Figure P6–7. Given that $v_C(0) = 0$, find the value of $v_C(t)$ at $t = 5$, 10, and 20 μs.

FIGURE   P6–6

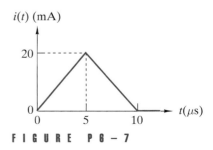

FIGURE   P6–7

6–8 The current through a 2-mH inductor is $i_L(t) = [1 - e^{-1000t}]u(t)$ A. Derive expressions for, and sketch, $v_L(t)$, $p_L(t)$, and $w_L(t)$. Is the capacitor absorbing power, delivering power, or both?

6–9 For $t > 0$, the current through a 500-mH inductor is $i_L(t) = 10 \, e^{-2000t} \sin 1000t$ mA. Derive expressions for, and sketch, $v_L(t)$, $p_L(t)$, and $w_L(t)$. Is the inductor absorbing power, delivering power, or both?

6–10 A voltage pulse $v_L(t) = 5 \cos(1000t) - 2 \sin(3000t)$ V appears across a 50-mH inductor. Derive expressions for and sketch, $i_L(t)$. Assume $i_L(0) = 0$. Discuss the effect of frequency on the relative amplitudes of the sinusoidal components in $v_L(t)$ and $i_L(t)$.

6–11 The voltage across a 50-mH inductor is $v_L(t) = e^{-1000t}u(t)$ V. Derive expressions for, and sketch, $i_L(t)$, $p_L(t)$, and $w_L(t)$. Assume $i_L(0) = 0$. Is the inductor absorbing power, delivering power, or both?

6–12 The current through a 25-mH inductor is shown in Figure P6–7. Derive expressions for, and sketch, $v_L(t)$, $p_L(t)$, and $w_L(t)$.

6–13 The voltage across a 100-μH inductor is shown in Figure P6–6. Given that $i_L(0) = 0$, find the value of $i_L(t)$ at $t = 5$, 10, and 20 μs. What is the value of $i_L(t)$ at $t > 40$ μs? Is the inductor absorbing power, delivering power, or both?

F I G U R E   P 6 – 1 4

6–14 The capacitor in Figure P6–14 carries an initial voltage $v_C(0) = 25$ V. At $t = 0$, the switch is closed, and thereafter the voltage across the capacitor is $v_C(t) = 25 \, e^{-5000t}$ V. For $t > 0$, derive expressions for $i_C(t)$ and $p_C(t)$. Is the capacitor absorbing power or delivering power?

6–15 The capacitor in Figure P6–14 carries an initial voltage $v_C(0) = 0$ V. At $t = 0$, the switch is closed, and thereafter the voltage across the capacitor is $v_C(t) = 15(1 - e^{-2000t})$ V. For $t > 0$, derive expressions for $i_C(t)$ and $p_C(t)$. Is the capacitor absorbing power from or delivering power to the rest of the circuit?

F I G U R E   P 6 – 1 6

6–16 The inductor in Figure P6–16 carries an initial current of $i_L(0) = 50$ mA. At $t = 0$, the switch opens, and thereafter the current into the rest of the circuit is $i(t) = -50 \, e^{-1000t}$ mA. For $t > 0$, derive expressions for $v_L(t)$ and $p_L(t)$. Is the inductor absorbing power from or delivering power to the rest of the circuit?

6–17 The inductor in Figure P6–16 carries an initial current of $i_L(0) = 20$ mA. At $t = 0$, the switch opens, and thereafter the current into the rest of the circuit is $i(t) = 40 - 20 \, e^{-500t}$ mA. For $t > 0$, derive expressions for $v_L(t)$ and $p_L(t)$. Is the inductor absorbing power from or delivering power to the rest of the circuit?

6–18 A 2.2-μF capacitor is connected in series with a 100-Ω resistor. The voltage across the capacitor is $v_C(t) = 10 \cos(1000t)$ V. What is the voltage across the resistor?

6–19 A 300-mH inductor is connected in parallel with a 10-kΩ resistor. The current through the inductor is $i_L(t) = 10 \, e^{-1000t}$ mA. What is the current through the resistor?

6–20 Using the passive sign convention, the voltage across and current through a linear energy storage element are $v(t) = 100 \, (1 - e^{-1000t})u(t)$ V and $i(t) = 0.1[e^{-1000t}]u(t)$ A. Is the element an inductor or a capacitor? What is the value of its inductance or capacitance?

ERO 6–2 DYNAMIC OP AMP CIRCUITS (SECT. 6–3)

(a) Given a circuit consisting of resistors, capacitors, inductors, and OP AMPs, determine its input-output relationship and use the relationship to find the output for specified inputs.
(b) Design an OP AMP circuit to implement a given input-output relationship or a block diagram.

FIGURE P6–21

6–21 Show that the OP AMP capacitor circuit in Figure P6–21 is a charge amplifier whose input-output relationship is

$$v_O(t) = -\frac{q_S(t)}{C}$$

where $q_S(t)$ is the charge carried by input current $i_S(t)$.

6–22 Show that the OP AMP capacitor circuit in Figure P6–22 is a noninverting integrator whose input-output relationship is

$$v_O(t) = \frac{1}{RC}\int_0^t v_S(x)dx$$

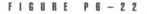

FIGURE P6–22

Hint: By voltage division, the voltage at the inverting input is $v_N(t) = v_O(t)/2$.

6–23 Find the input-output relationship of the OP AMP circuit in Figure P6–23.

6–24 In Figure P6–24, the voltage across the capacitor at $t = 0$ is such that $v_O(0) = +10$ V. The input signal is $v_S(t) = 5u(t)$ V. Derive an equation for the output voltage when the OP AMP is in its linear range. If the OP AMP saturates at ±15 V, find the time $t > 0$ when the OP AMP saturates.

FIGURE P6–23

6–25 In Figure P6–24, the voltage across the capacitor at $t = 0$ is 0 V. The input signal is $v_S(t) = 5[u(t) - u(t - 0.005)]$. The OP AMP saturates at ±15 V. Does the input signal saturate the OP AMP?

6–26 At $t = 0$, the voltage across the capacitor in Figure P6–24 is zero. The OP AMP saturates at ±15 V. For $v_S(t) = V_A[\cos 2000t]u(t)$ V, derive an expression for the output voltage when the OP AMP operates in its linear range. What is the maximum value of V_A for linear operation?

FIGURE P6–24

6–27 The input to the circuit in Figure P6–27 is $v_S(t) = V_A[\sin 10^5 t]u(t)$ V. Derive an expression for the output voltage when the OP AMP operates in its linear range. The OP AMP saturates at ±12 V. What is the maximum value of V_A for linear operation?

6–28 The input to the circuit in Figure P6–27 is $v_S(t) = 5[\sin 10^6 t]u(t)$ V. Derive an expression for the output voltage when the OP AMP operates in its linear range. The OP AMP saturates at ±15 V. What is the maximum value of the feedback resistor for linear operation?

6–29 The input to the circuit in Figure P6–27 is $v_S(t) = 5[e^{-\alpha t}]u(t)$ V. Derive an expression for the output voltage when the OP AMP operates in its linear range. If the OP AMP saturates at ±15 V, what is the largest value of α for linear operation?

FIGURE P6–27

FIGURE   P 6 – 3 2

C1

C2

FIGURE   P 6 – 3 6

FIGURE   P 6 – 4 0

6–30 Construct a block diagram for the following input-output relationship:

$$v_O(t) = -2v_S(t) + \frac{1}{50}\frac{dv_S(t)}{dt}$$

6–31 Construct a block diagram for the following input-output relationship:

$$v_O(t) = 2\,v_S(t) - 40\int_0^t v_S(x)dx$$

6–32 What is the input-output relationship of the block diagram in Figure P6–32?

6–33 **D** Design an OP AMP circuit to implement the input-output relationship in Problem 6–30.

6–34 **D** Design an OP AMP circuit to implement the input-output relationship in Problem 6–31.

6–35 **D** Design OP AMP *RC* circuits to implement the input-output relationship of the block diagram in Figure P6–32.

ERO 6–3 EQUIVALENT INDUCTANCE AND CAPACITANCE (SECT. 6–4)

(a) Derive equivalence properties of inductors and capacitors or use equivalence properties to simplify *LC* circuits.
(b) Solve for currents and voltages in *RLC* circuits with dc input signals.

6–36 Find a single equivalent element for each circuit in Figure P6–36.

6–37 Verify Eqs. (6–30) and (6–31).

6–38 A 3-mH inductor is connected in series with a 100-μF inductor and the combination connected in parallel with a 12-mH inductor. Find the equivalent inductance of the connection.

6–39 Two 6.8-μF capacitors are connected in series and the combination is connected in parallel with a series connection of a 3.3-μF capacitor and a 4.7-μF capacitor. Find the equivalent capacitance of the connection.

6–40 Find the equivalent capacitance between terminals A and B in Figure P6–40.

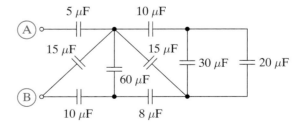

6–41 Standard capacitors of 2.2 nF and 3.3 nF are connected in series. The worst case tolerance on each capacitor is ±20%. What is the worst case tolerance on the capacitance of the series combination?

6–42 **D** A capacitor bank is required that can be charged to 5 kV and store at least 250 J of energy. Design a series/parallel combination that meets the voltage and energy requirements using 20-μF capacitors each rated at 1.5 kV max.

6–43 Reduce the circuit in Figure P6–43 to as few equivalent elements as possible.

6–44 The circuits in Figure P6–44 are driven by dc sources. Find the voltage across capacitors and the current through inductors.

6–45 The OP AMP circuits in Figure P6–45 are driven by dc sources. Find the output voltage v_O.

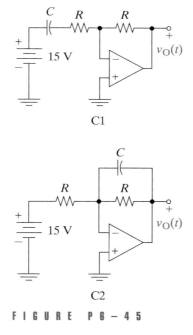

INTEGRATING PROBLEMS

6–46 ▲ CAPACITIVE DISCHARGE PULSER

A capacitor bank for a large pulse generator consists of 11 capacitor strings connected in parallel. Each string consists of 16 1.5-mF capacitors connected in series. The purpose of this problem is to calculate important characteristics of the pulser.

(a) What is the total equivalent capacitance of the bank?

(b) If each capacitor in a series string is charged to 300 V, what is the total energy stored in the bank?

(c) In the discharge mode, the voltage across the capacitor bank is $v(t) = 4.8[e^{-500t}]u(t)$ kV. What is the peak power delivered by the capacitor bank?

(d) For practical purposes the capacitor bank is completely discharged after about five time constants. What is the average power delivered during that interval?

6–47 **A** LC CIRCUIT RESPONSE

In the circuit shown in Figure P6–47, the initial capacitor voltage is $v_C(0) = 30$ V. At $t = 0$ the switch is closed, and thereafter the current into the rest of the circuit is

$$i(t) = 0.3\, e^{-1000t} \sin (2000\,t) \quad \text{A}$$

The purpose of this problem is to find the voltage $v(t)$ and the equivalent resistance looking into the rest of the circuit.

(a) Use the inductor's i–v characteristic to find $v_L(t)$ for $t \geq 0$. Find the value of $v_L(t)$ at $t = 0$.

(b) Use the capacitor's i–v characteristic to find $v_C(t)$ for $t \geq 0$. Find the value of $v_C(t)$ at $t = 0$. Does this value agree with the initial condition given in the problem statement? If not, you need to review your work to find the error.

(c) Use KVL and the results from (a) and (b) to find the voltage $v(t)$ delivered to the rest of the circuit. What is the value of $v(t)$ at $t = 0$?

(d) The $v(t)$ found in (c) should be proportional to the $i(t)$ given in the problem statement. If so, what is the equivalent resistance looking into the rest of the circuit?

FIGURE   P6 – 47

6–48 **A** CAPACITOR MULTIPLIER

It is claimed that the i–v characteristic at the input interface in Figure P6–48 is

$$i(t) = C_{EQ} \frac{dv(t)}{dt}$$

where $C_{EQ} = (1 + R_2/R_1)C$. Since $C_{EQ} > C$ the circuit is called a capacitor multiplier.

FIGURE   P6 – 48

(a) Prove or disprove this claim.

(b) If the initial capacitor voltage $v_C(0) = V_0$, what is the initial value of the input voltage $v(0)$?

(c) If the initial energy stored in the capacitor is $w_C(0) = W_0$, what is the initial energy stored in equivalent capacitor C_{EQ}?

6–49 **A D** INDUCTOR SIMULATION CIRCUIT

It is difficult to produce useful inductance using integrated circuit fabrication methods. Circuits that simulate an inductor can be produced using resistors, OP AMPs, and a capacitor. It is claimed that the i–v characteristic at the input interface in Figure P6–49 is

$$i(t) = \frac{1}{L_{EQ}} \int_0^t v(x)dx$$

where $L_{EQ} = R_1 R_2 C$.

(a) Prove or disprove this claim.

(b) If the initial capacitor voltage $v_C(0) = V_0$, what is the initial value of the input current $i(0)$?

(c) If the initial energy stored in the capacitor is $w_C(0) = W_0$, what is the initial energy stored in equivalent inductor L_{EQ}?

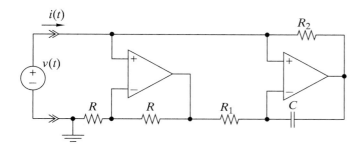

FIGURE P6-49

6–50 **A D E** RC OP AMP CIRCUIT DESIGN

An upgrade to one of your company's robotics products requires a proportional plus integral compensator that implements the input-output relationship

$$v_O(t) = v_S(t) + 50 \int_0^t v_S(x)dx$$

The input voltage $v_S(t)$ comes from an OP AMP, and the output voltage $v_O(t)$ drives a 10-kΩ resistive load. As the junior engineer in the company, you have been given the responsibility of developing a preliminary design.

(a) **D** Design a circuit that implements the relationship using the standard OP AMP building blocks in Figure 6–17. Minimize the parts count in your design.

(b) **A** The RonAl Corporation (founded by two well-known authors) has given you an unsolicited proposal claiming that their standard DIFF AMP-10 product can realize the required relationship. Their proposal is shown in Figure P6–50. Verify their claim.

FIGURE P6–50

RonAl Corporation

Note: External connections shown in color

(c) **E** Compare your design in part (a) with the RonAl proposal in part (b) in terms of the total parts costs. Parts costs in your company are as follows: resistors, $400 per 10,000; capacitors, $1500 per 10,000; and OP AMPs, $2500 per 10,000. RonAl's proposal offers to provide the company's standard DIFF AMP product for $5000 per 10,000. Note that for the RonAl product, you must supply the capacitor shown in Figure P6–50.

6–51 **A** RL CIRCUIT ANALYSIS

A 1-kΩ resistor is connected in parallel with a 500-mH inductor. For $t \geq 0$ the current through the inductor is $i(t) = 3e^{-2000t}$ mA.
(a) Find the voltage $v_R(t)$ across the resistor.
(b) Find the voltage $v_L(t)$ across the inductor.
(c) Show that $v_R(t) + v_R(t) = 0$ for all $t > 0$.
(d) Find the energy stored in the inductor at $t = 0$.
(e) Calculate the energy dissipated in the resistor as

$$w_R = \int_0^\infty Ri^2 dt$$

(f) The energy dissipated in the resistor should equal the initial energy stored in the inductor. Can you explain why?

6–52 **A** WAVEFORM GENERATOR

The inverting amplifier and integrator in Figure P6–52 have the same input. Their outputs drive the inputs of a comparator. The comparator has an output $+V_{CC}$ when $v_P > v_N$ and an output $-V_{CC}$ when $v_P < v_N$. Assume that there is no initial voltage on the capacitor at $t = 0$.
(a) Show that when an input $v_1 = +V_{CC}$ is applied at $t = 0$ the comparator output is $+V_{CC}$ until $t = RC$ and is $-V_{CC}$ thereafter.

(b) Show that when an input $v_1 = -V_{CC}$ is applied at $t = 0$ the comparator output is $-V_{CC}$ until $t = RC$ and is $+V_{CC}$ thereafter.

(c) Now force the condition $v_1 = v_O$ by making a feedback connection from the comparator output to the input. Assume that the initial comparator output is $+V_{CC}$. Show that the comparator output is a square wave with a period of $4RC$ and the integrator output is a triangular wave with the same period.

CHAPTER 7

FIRST- AND SECOND-ORDER CIRCUITS

When a mathematician engaged in investigating physical actions and results has arrived at his own conclusions, may they not be expressed in common language as fully, clearly and definitely as in mathematical formula? If so, would it not be a great boon to such as we to express them so—translating them out of their hieroglyphics that we also might work upon them by experiment.

Michael Faraday, 1857,
British Physicist

7–1 *RC* and *RL* Circuits

7–2 First-order Circuit Step Response

7–3 Initial and Final Conditions

7–4 First-order Circuit Sinusoidal Response

7–5 The Series *RLC* Circuit

7–6 The Parallel *RLC* Circuit

7–7 Second-order Circuit Step Response

7–8 Some Other Second-order Circuits

Summary

Problems

Integrating Problems

Michael Faraday (1791–1867) was appointed a Fellow in the Royal Society at age 32 and was a lecturer at the Royal Institution in London for more than 50 years. During this time he published over 150 papers on chemistry and electricity. The most important of these papers was the series *Experimental Researches in Electricity*, which included a description of his discovery of magnetic induction. A gifted experimentalist, Faraday had no formal education in mathematics and apparently felt that mathematics obscured the physical truths he discovered through experimentation. The foregoing quotation is taken from a letter written by Faraday to James Clerk Maxwell in 1857. Faraday's comments are perhaps ironic because Maxwell is best known today for his formulation of the mathematical theory of electromagnetics, which he published in 1873.

Since Faraday's time, tremendous strides have been made in correlating mathematical theory and experiment. Electrical circuit analysis is one of the areas in which there is a close relationship between mathematical theory and laboratory measurements. The nearly one-to-one correspondence between the predictions of circuit analysis and hardware results enormously simplifies the design and development of electrical and electronic systems.

The *i–v* characteristics of capacitors and inductors were treated in Chapter 6. This chapter combines these device characteristics with the connection constraints to derive the differential equations governing the response of dynamic circuits. Mathematical methods for analytically solving these differential equations are also treated. The first four sections treat the formulation and solution of the first-order differential equations governing circuits with one capacitor or inductor. The next four sections treat circuits with two dynamic elements that lead to second-order differential equations.

7–1   *RC* AND *RL* CIRCUITS

The flow diagram in Figure 7–1 shows the two major steps in the analysis of a dynamic circuit. In the first step we use device and connection equations to formulate a differential equation describing the circuit. In the second step we solve the differential equation to find the circuit response. In this chapter we examine basic methods of formulating circuit differential equations and the time-honored, classical methods of solving for responses. Solving for the responses of simple dynamic circuits gives us insight into the physical behavior of the basic modules of the complex networks in subsequent chapters. This insight will help us correlate circuit behavior with the results obtained by other methods of dynamic circuit analysis. The treatment of other methods includes phasor circuit analysis (Chapter 8) and the Laplace transform methods (beginning in Chapter 9).

FORMULATING *RC* AND *RL* CIRCUIT EQUATIONS

RC and *RL* circuits contain linear resistors and a single capacitor or a single inductor. Figure 7–2 shows how we can divide *RC* and *RL* circuits into two parts: (1) the dynamic element and (2) the rest of the circuit, containing only linear resistors and sources. To formulate the equation governing either of these circuits, we replace the resistors and sources by their Thévenin and Norton equivalents shown in Figure 7–2.

F I G U R E   7 – 1 *Flow diagram for dynamic circuit analysis.*

FIGURE 7 - 2 *First-order circuits: (a)* RC
circuit. (b) RL *circuit.*

F I G U R E 7 – 2 *First-order circuits: (a)* RC
circuit. (b) RL *circuit.*

Dealing first with the *RC* circuit in Figure 7–2(a), we note that the
Thévenin equivalent source is governed by the constraint

$$R_T i(t) + v(t) = v_T(t) \tag{7-1}$$

The capacitor *i–v* constraint is

$$i(t) = C\frac{dv(t)}{dt} \tag{7-2}$$

Substituting the *i–v* constraint into the source constraint yields

$$R_T C\frac{dv(t)}{dt} + v(t) = v_T(t) \tag{7-3}$$

Combining the source and element constraints produces the equation governing the *RC* series circuit. The unknown in Eq. (7–3) is the capacitor
voltage $v(t)$ that determines the amount of energy stored in the *RC* circuit
and is referred to as the **state variable**.

Mathematically, Eq. (7–3) is a first-order linear differential equation
with constant coefficients. The equation is first order because the first
derivative of the dependent variable is the highest order derivative in the
equation. The product $R_T C$ is a constant coefficient because it depends on
fixed circuit parameters. The signal $v_T(t)$ is the Thévenin equivalent of the
independent sources driving the circuit. The voltage $v_T(t)$ is the input, and
the capacitor voltage $v(t)$ is the circuit response.

The Norton equivalent source in the *RL* circuit in Figure 7–2(b) is governed by the constraint

$$G_N v(t) + i(t) = i_N(t) \tag{7-4}$$

The element constraint for the inductor can be written

$$v(t) = L\frac{di(t)}{dt} \tag{7-5}$$

Combining the element and source constraints produces the differential equation for the *RL* circuit:

$$G_N L \frac{di(t)}{dt} + i(t) = i_N(t) \qquad (7\text{–}6)$$

The response of the *RL* circuit is also governed by a first-order linear differential equation with constant coefficients. The dependent variable in Eq. (7–6) is the inductor current. The circuit parameters enter as the constant product $G_N L$, and the driving forces are represented by a Norton equivalent current $i_N(t)$. The unknown in Eq. (7–6) is the inductor current $i(t)$. This current determines the amount of energy stored in the *RL* circuit and is referred to as the **state variable**.

The state variables in first-order circuits are the capacitor voltage in the *RC* circuit and the inductor current in the *RL* circuit. As we will see, these state variables contain sufficient information about the past to determine future circuit responses.

We observe that Eqs. (7–3) and (7–6) have the same form. In fact, interchanging the quantities

$$R_T \leftrightarrow G_N \qquad C \leftrightarrow L \qquad v \leftrightarrow i \qquad v_T \leftrightarrow i_N$$

converts one equation into the other. This interchange is another example of the principle of duality. Because of duality we do not need to study the *RC* and *RL* circuits as independent problems. Everything we learn by solving the *RC* circuit can be applied to the *RL* circuit as well.

We refer to the *RC* and *RL* circuits as **first-order circuits** because they are described by a first-order differential equation. The first-order differential equations in Eqs. (7–3) and (7–6) describe general *RC* and *RL* circuits shown in Figure 7–2. Any circuit containing a single capacitor or inductor and linear resistors and sources is a first-order circuit.

ZERO-INPUT RESPONSE OF FIRST-ORDER CIRCUITS

The response of a first-order circuit is found by solving the circuit differential equation. For the *RC* circuit the response $v(t)$ must satisfy the differential equation in Eq. (7–3) and the initial condition $v(0)$. By examining Eq. (7–3) we see that the response depends on three factors:

1. The inputs driving the circuit $v_T(t)$
2. The values of the circuit parameters R_T and C
3. The value of $v(t)$ at $t = 0$ (i.e., the initial condition).

The first two factors apply to any linear circuit, including resistance circuits. The third factor relates to the initial energy stored in the circuit. The initial energy can cause the circuit to have a nonzero response even when the input $v_T(t) = 0$ for $t \geq 0$. The existence of a response with no input is something new in our study of linear circuits.

To explore this discovery we find the **zero-input response**. Setting all independent sources in Figure 7–2 to zero makes $v_T = 0$ in Eq. (7–3):

$$R_T C \frac{dv}{dt} + v = 0 \qquad (7\text{–}7)$$

Mathematically, Eq. (7–7) is a **homogeneous equation** because the right side is zero. The classical approach to solving a linear homogeneous differential equation is to try a solution in the form of an exponential

$$v(t) = Ke^{st} \tag{7-8}$$

where K and s are constants to be determined.

The form of the homogenous equation suggests an exponential solution for the following reasons. Equation (7–7) requires that $v(t)$ plus R_TC times its derivative must add to zero for all time $t \geq 0$. This can only occur if $v(t)$ and its derivative have the same form. In Chapter 5 we saw that an exponential signal and its derivative are both of the form $e^{-t/Tc}$. Therefore, the exponential is a logical starting place.

If Eq. (7–8) is indeed a solution, then it must satisfy the differential equation in Eq. (7–7). Substituting the trial solution into Eq. (7–7) yields

$$R_TCKse^{st} + Ke^{st} = 0$$

or

$$Ke^{st}(R_TCs + 1) = 0$$

The exponential function e^{st} cannot be zero for all t. The condition $K = 0$ is a trivial solution because it implies that $v(t)$ is zero for all time t. The only nontrivial way to satisfy the equation involves the condition

$$R_TCs + 1 = 0 \tag{7-9}$$

Equation (7–9) is the circuit **characteristic equation** because its root determines the attributes of $v(t)$. The characteristic equation has a single root at $s = -1/R_TC$ so the zero-input response of the RC circuit has the form

$$v(t) = Ke^{-t/R_TC} \qquad t \geq 0$$

The constant K can be evaluated using the value of $v(t)$ at $t = 0$. Using the notation $v(0) = V_0$ yields

$$v(0) = Ke^0 = K = V_0$$

The final form of the zero-input response is

$$v(t) = V_0e^{-t/R_TC} \qquad t \geq 0 \tag{7-10}$$

The zero-input response of the RC circuit is the familiar exponential waveform shown in Figure 7–3. At $t = 0$ the exponential response starts out at $v(0) = V_0$ and then decays to zero at $t \to \infty$. The time constant $T_C = R_TC$ depends only on fixed circuit parameters. From our study of the exponential signals in Chapter 5, we know that the $v(t)$ decays to about 37% of its initial amplitude in one time constant and to essentially zero after about five time constants. The zero-input response of the RC circuit is determined by two quantities: (1) the circuit time constant and (2) the value of the capacitor voltage at $t = 0$.

The zero-input response of the RL circuit in Figure 7–2(b) is found by setting the Norton current $i_N(t) = 0$ in Eq. (7–6).

$$G_NL\frac{di}{dt} + i = 0 \tag{7-11}$$

F I G U R E   7 – 3   *First-order RC circuit zero-input response.*

The unknown in this homogeneous differential equation is the inductor current $i(t)$. Equation (7–11) has the same form as the homogeneous equation for the *RC* circuit, which suggests a trial solution of the form

$$i(t) = Ke^{st}$$

where K and s are constants to be determined. Substituting the trial solution into Eq. (7–11) yields the *RL* circuit characteristic equation.

$$G_N Ls + 1 = 0 \qquad (7\text{–}12)$$

The root of this equation is $s = -1/G_N L$. Denoting the initial value of the inductor current by I_0, we evaluate the constant K:

$$i(0) = I_0 = Ke^0 = K$$

The final form of the zero-input response of the *RL* circuit is

$$i(t) = I_0 e^{-t/G_N L} \qquad t \geqslant 0 \qquad (7\text{–}13)$$

For the *RL* circuit the zero-input response of the state variable $i(t)$ is an exponential function with a time constant of $T_C = G_N L = L/R_T$. This response connects the initial state $i(0) = I_0$ with the final state $i(\infty) = 0$.

The zero-input responses in Eqs. (7–10) and (7–13) show the duality between first-order *RC* and *RL* circuits. These results point out that the zero-input response in a first-order circuit depends on two quantities: (1) the circuit time constant and (2) the value of the state variable at $t = 0$. Capacitor voltage and inductor current are called state variables because they determine the amount of energy stored in the circuit at any time t. The following examples show that the zero-input response of the state variable provides enough information to determine the zero-input response of every other voltage and current in the circuit.

EXAMPLE 7–1

The switch in Figure 7–4 is closed at $t = 0$, connecting a capacitor with an initial voltage of 30 V to the resistances shown. Find the responses $v_C(t)$, $i(t)$, $i_1(t)$, and $i_2(t)$ for $t \geq 0$.

FIGURE 7-4

SOLUTION:

This problem involves the zero-input response of an RC circuit since there is no independent source in the circuit. To find the required responses, we first determine the circuit time constant with the switch closed ($t \geq 0$). The equivalent resistance seen by the capacitor is

$$R_{EQ} = 10 + (20\|20) = 20 \text{ k}\Omega$$

For $t \geq 0$ the circuit time constant is

$$T_C = R_T C = 20 \times 10^3 \times 0.5 \times 10^{-6} = 10 \text{ ms}$$

The initial capacitor voltage is given by $V_0 = 30$ V. Using Eq. (7–10), the zero-input response of the capacitor voltage is

$$v_C(t) = 30e^{-100t} \text{ V} \quad t \geq 0$$

The capacitor voltage provides the information needed to solve for all other zero-input responses. The current $i(t)$ through the capacitor is

$$i(t) = C\frac{dv_C}{dt} = (0.5 \times 10^{-6})(30)(-100)\, e^{-100t}$$
$$= -1.5 \times 10^{-3}\, e^{-100t} \text{ A} \quad t \geq 0$$

The minus sign means the actual current direction is opposite of the reference direction shown in Figure 7–4. The minus sign makes physical sense because the initial voltage on the capacitor is positive, which forces current into the resistances to the right of the switch. The other current responses are found by current division.

$$i_1(t) = i_2(t) = \frac{20}{20 + 20}i(t) = -0.75 \times 10^{-3}e^{-100t} \text{ A} \quad t \geq 0$$

Notice the analysis pattern. We first determine the zero-input response of the capacitor voltage. The state variable response together with resistance circuit analysis techniques were then used to find other voltages and currents. The circuit time constant and the value of the state variable at $t = 0$ provide enough information to determine the zero-input response of every voltage or current in the circuit. ∎

Given circuit

Equivalent circuit

FIGURE 7-5

EXAMPLE 7-2

Find the response of the state variable of the RL circuit in Figure 7–5 using $L_1 = 10$ mH, $L_2 = 30$ mH, $R_1 = 2$ kΩ, $R_2 = 6$ kΩ, and $i_L(0) = 100$ mA.

SOLUTION:
The inductors are connected in series and can be replaced by an equivalent inductor

$$L_{EQ} = L_1 + L_2 = 10 + 30 = 40\ \text{mH}$$

Likewise, the resistors are connected in parallel and the conductance seen by L_{EQ} is

$$G_{EQ} = G_1 + G_2 = 10^{-3}/2 + 10^{-3}/6 = 2 \times 10^{-3}/3\ \text{S}$$

Figure 7–5 shows the resulting equivalent circuit. The interface signals $v(t)$ and $i(t)$ are the voltage across and current through $L_{EQ} = L_1 + L_2$. The time constant of the equivalent RL circuit is

$$T_C = G_{EQ}L_{EQ} = 8 \times 10^{-5}/3\ \text{s} = 1/37500\ \text{s}$$

The initial current through L_{EQ} is $i_L(0) = 0.1$ A. Using Eq. (7–13) with $I_0 = 0.1$ yields the zero-state response of the inductor current.

$$i(t) = 0.1e^{-37500t}\ \text{A} \qquad t \geqslant 0$$

Given the state variable response, we can find every other response in the original circuit. For example, by KCL and current division the currents through R_1 and R_2 are

$$i_{R_1}(t) = \frac{R_2}{R_1 + R_2}i(t) = 0.075\ e^{-37500t}\ \text{A} \qquad t \geqslant 0$$

$$i_{R_2}(t) = \frac{R_1}{R_1 + R_2}i(t) = 0.025\ e^{-37500t}\ \text{A} \qquad t \geqslant 0 \qquad \blacksquare$$

Example 7–2 illustrates an important point. The RL circuit in Figure 7–5 is a first-order circuit even though it contains two inductors. The two inductors are connected in series and can be replaced by a single equivalent inductor. In general, capacitors or inductors in series and parallel can be replaced by a single equivalent element. Thus, any circuit containing the *equivalent* of a single inductor or a single capacitor is a first-order circuit.

Sometimes it may be difficult to determine the Thévenin or Norton equivalent seen by the dynamic element in a first-order circuit. In such cases we use other circuit analysis techniques to derive the differential equation in terms of a more convenient signal variable. For example, the OP AMP RC circuit in Figure 7–6 is a first-order circuit because it contains a single capacitor.

From previous experience we know that the key to analyzing an inverting OP AMP circuit is to write a KCL equation at the inverting input. The sum of currents entering the inverting input is

$$\underbrace{G_1(v_S - v_N)}_{i_1(t)} + \underbrace{G_2(v_O - v_N)}_{i_2(t)} + \underbrace{C\frac{d(v_O - v_N)}{dt}}_{i_C(t)} - i_N(t) = 0$$

The element equations for the OP AMP are $i_N(t) = 0$ and $v_N(t) = v_P(t)$. However, the noninverting input is grounded; hence $v_N(t) = v_P(t) = 0$. Substituting the OP AMP element constraints into the KCL constraint yields

FIGURE   7 – 6   *First-order OP AMP RC circuit.*

$$G_1 v_S + G_2 v_O + C \frac{dv_O}{dt} = 0$$

which can be rearranged in standard form as

$$R_2 C \frac{dv_O}{dt} + v_O = -\frac{R_2}{R_1} v_S(t) \tag{7-14}$$

The unknown Eq. (7–14) is the OP AMP output voltage rather than the capacitor voltage. The form of the differential equation indicates that the circuit time constant is $T_C = R_2 C$.

EXAMPLE 7–3

Use PSpice to calculate the response $v_C(t)$ in the circuit in Figure 7–4 for $t \geq 0$.

SOLUTION:

This problem was solved analytically in Example 7–1. The problem is repeated here to introduce computer simulation of dynamic circuits. The simulation option used is called **transient analysis**, which predicts the variation of currents and voltages as a function of time. Figure 7–7 shows the circuit diagram as drawn in PSpice. This diagram does not include the switch in Figure 7–4 since for $t \geq 0$ the switch is closed. The net list in Figure 7–7 shows that the 0.5 µF connected between nodes 1 and 0 (ground) carries an initial condition of IC = 30 V. The Part Attribute dialog box in Figure 7–7 shows how the capacitor's initial condition attribute is given a numerical value. In circuit simulation programs the user must specify both

FIGURE 7 – 7 *Schematic, net list, and part dialog box.*

the element value attributes and the initial condition attributes for dynamic elements.

Once the circuit diagram is complete, we must set up a transient analysis run. Selecting **Analysis/Setup/Transient** from the schematics menu bar brings up the dialog box in Figure 7–8. The key entry in this box is the **Final Time**, which specifies the time duration of simulation since a PSpice transient analysis starts at $t = 0$ (always) and runs to $t = $ **Final Time**. In the present case we know that the circuit time constant is 10 ms (see Example 7–1), so we can safely specify **Final Time** $= 5T_C = 50$ ms.[1]

The **Print Step** parameter in Figure 7–8 is not important to us since we observe the calculated response using Probe. The **Step Ceiling** parameter is important since it controls the minimum number of points included in the Probe plots. A useful rule of thumb is to let **Step Ceiling** = **Final Time**/100. This rule generates at least 101 points (counting $t = 0$) in the Probe plot, which is often adequate. In the present case this means that **Step Ceiling** $= 0.050/100 = 0.5$ ms. We must also assign a numerical value to **Print Step** even though this parameter does not affect the calculated responses observed in Probe. The reason is that originally SPICE outputs were produced by a printer, so SPICE-based programs will not run unless $0 < $ **Print Step** $< $ **Final Time**.

The transient analysis dialog box in Figure 7–8 includes a box labeled **Use Init. Conditions**. When this box is checked, PSpice uses the initial conditions specified when the circuit is created in Schematics. When this box is not checked, PSpice automatically calculates the initial conditions for all dynamic elements using the dc analysis method discussed in Chapter 6. The initial conditions calculated by PSpice can be different from those specified by the user in Schematics. For example, the circuit in Figure 7–7 has no dc sources, so PSpice dc analysis would report that the initial capacitor voltage is zero, whereas we specified IC = 30 V in Schematics. It is the user's responsibility to tell PSpice how to obtain the initial conditions by checking or not checking the **Use Init. Conditions** box.

With the circuit defined and the transient analysis run set up, we can start PSpice by selecting **Analysis/Simulate** from the Schematics menu bar. Probe is automatically started once PSpice successfully completes the transient simulation. Using the Probe graphical interface produces the plots shown in Figure 7–9. The legend at the bottom of the figure indicates that there are two traces in the plot. The trace indicated by the □ markers is the node voltage $v(1)$, which is the PSpice solution for the capacitor voltage. The trace indicated by the ◇ markers is generated by the expression 30*exp(–100*time), which is the Probe implementation of the analytical solution found in Example 7–1. On this scale of observation the two traces appear to be congruent, indicating close agreement between the PSpice numerical solution and the analytical solution.

---

1 Specifying a value for **Final Time** presents a dilemma. We must have some knowledge about the response to specify an appropriate simulation time interval. If we have no knowledge, we may specify an interval that is too short or too long to observe the response. If we have complete knowledge, then there is no reason to run the simulation. Effectively using circuit simulation tools requires that we operate in a gray area between no knowledge and complete knowledge.

F I G U R E   **7 – 8**   *Analysis setup.*

C1

C2

C3

C4

FIGURE 7 – 10

FIGURE 7 – 11

FIGURE 7 – 9 *Response plots.*

Exercise 7–1

What are the time constants of the circuits shown in Figure 7–10?

Answers:

C1: RC C2: $2L/3R$ C3: $RC/4$ C4: $L/4R$

Exercise 7–2

The switch in Figure 7–11 closes at $t = 0$. For $t \geq 0$ the current through the resistor is $i_R(t) = e^{-100t}$ mA.

(a) What is the capacitor voltage at $t = 0$?
(b) Write an equation for $v(t)$ for $t \geq 0$.
(c) Write an equation for the power absorbed by the resistor for $t \geq 0$.
(d) How much energy does the resistor dissipate for $t \geq 0$?
(e) How much energy is stored in the capacitor at $t = 0$?

Answers:

(a) 10 V
(b) $v(t) = 10e^{-100t}$ V
(c) $p_R(t) = 10e^{-200t}$ mW
(d) 50 μJ
(e) 50 μJ

Exercise 7–3

For $t > 0$ the current through the 40-mH inductor in a first-order circuit is $20e^{-500t}$ mA.

(a) What is the circuit time constant?
(b) How much energy is stored in the inductor at $t = 0$, $t = T_C$, and $t = 5T_C$?
(c) Write an equation for the voltage across the inductor.
(d) What is the equivalent resistance seen by the inductor?

Answers:

(a) 2 ms
(b) 8, 1.08, 0.000363 μJ
(c) $-400e^{-500t}$ mV
(d) 20 Ω

7–2 FIRST-ORDER CIRCUIT STEP RESPONSE

Linear circuits are often characterized by applying step function and sinusoid inputs. This section introduces the step response of first-order circuits. Later in this chapter we treat the sinusoidal response of first-order
circuits and step response of second-order circuits. The step response
analysis introduces the concepts of forced, natural, and zero-state responses that appear extensively in later chapters.

Our development of first-order step response treats the RC circuit in
detail and then summarizes the corresponding results for its dual, the RL
circuit. When the input to the RC circuit in Figure 7–2 is a step function,
we can write the Thévenin source as $v_T(t) = V_A u(t)$. The circuit differential
equation in Eq. (7–3) becomes

$$R_T C \frac{dv}{dt} + v = V_A u(t) \qquad\qquad (7\text{–}15)$$

The step response is a function $v(t)$ that satisfies this differential equation
for $t \geq 0$ and meets the initial condition $v(0)$. Since $u(t) = 1$ for $t \geq 0$ we can
write Eq. (7–15) as

$$R_T C \frac{dv(t)}{dt} + v(t) = V_A \qquad \text{for } t \geq 0 \qquad\qquad (7\text{–}16)$$

Mathematics provides a number of approaches to solving this equation,
including separation of variables and integrating factors. However, because the circuit is linear we chose a method that uses superposition to divide solution $v(t)$ into two components:

$$v(t) = v_N(t) + v_F(t) \qquad\qquad (7\text{–}17)$$

The first component $v_N(t)$ is the **natural response** and is the general solution of Eq. (7–16) when the input is set to zero. The natural response has
its origin in the physical characteristic of the circuit and does not depend
on the form of the input. The component $v_F(t)$ is the **forced response** and is
a particular solution of Eq. (7–16) when the input is the step function. We
call this the forced response because it represents what the circuit is com-
pelled to do by the form of the input.

Finding the natural response requires the general solution of Eq. (7–16)
with the input set to zero:

$$R_T C \frac{dv_N(t)}{dt} + v_N(t) = 0 \qquad t \geq 0$$

But this is the homogeneous equation that produces the zero-input response in Eq. (7–8). Therefore, we know that the natural response takes the form

$$v_{\mathrm{N}}(t) = Ke^{-t/R_{\mathrm{T}}C} \qquad t \geq 0 \tag{7–18}$$

This is a general solution of the homogeneous equation because it contains an arbitrary constant K. At this point we cannot evaluate K from the initial condition, as we did for the zero-input response. The initial condition applies to the total response (natural plus forced), and we have yet to find the forced response.

Turning now to the forced response, we seek a particular solution of the equation

$$R_{\mathrm{T}}C\frac{dv_{\mathrm{F}}(t)}{dt} + v_{\mathrm{F}}(t) = V_{\mathrm{A}} \qquad t \geq 0 \tag{7–19}$$

The equation requires that a linear combination of $v_{\mathrm{F}}(t)$ and its derivative equal a constant V_{A} for $t \geq 0$. Setting $v_{\mathrm{F}}(t) = V_{\mathrm{A}}$ meets this condition since $dv_{\mathrm{F}}/dt = dV_{\mathrm{A}}/dt = 0$. Substituting $v_{\mathrm{F}} = V_{\mathrm{A}}$ into Eq. (7–19) reduces it to the identity $V_{\mathrm{A}} = V_{\mathrm{A}}$.

Now combining the forced and natural responses, we obtain

$$v(t) = v_{\mathrm{N}}(t) + v_{\mathrm{F}}(t)$$
$$= Ke^{-t/R_{\mathrm{T}}C} + V_{\mathrm{A}} \qquad t \geq 0$$

This equation is the general solution for the step response because it satisfies Eq. (7–16) and contains an arbitrary constant K. This constant can now be evaluated using the initial condition:

$$v(0) = V_0 = Ke^0 + V_{\mathrm{A}} = K + V_{\mathrm{A}}$$

The initial condition requires that $K = (V_0 - V_{\mathrm{A}})$. Substituting this conclusion into the general solution yields the step response of the RC circuit.

$$v(t) = (V_0 - V_{\mathrm{A}})e^{-t/R_{\mathrm{T}}C} + V_{\mathrm{A}} \qquad t \geq 0 \tag{7–20}$$

A typical plot of $v(t)$ is shown in Figure 7–12.

F I G U R E   7 – 1 2 *Step response of a first-order* RC *circuit.*

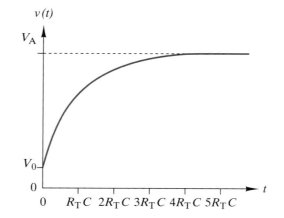

The RC circuit step response in Eq. (7–20) starts out at the initial condition V_0 and is driven to a final condition V_{A}, which is determined by the amplitude of the step function input. That is, the initial and final values of the response are

$$\lim_{t \to 0+} v(t) = (V_0 - V_A)e^{-0} + V_A = V_0$$

$$\lim_{t \to \infty} v(t) = (V_0 - V_A)e^{-\infty} + V_A = V_A$$

The path between the two end points is an exponential waveform whose time constant is the circuit time constant. We know from our study of exponential signals that the step response will reach its final value after about five time constants. In other words, after about five time constants the natural response decays to zero and we are left with a constant forced response caused by the step function input.

The RL circuit in Figure 7–2 is the dual of the RC circuit, so the development of its step responses follows the same pattern discussed previously. Briefly sketching the main steps, the Norton equivalent input is a step function $I_A u(t)$, and for $t \geq 0$ the RL circuit differential equation Eq. (7–6) becomes

$$G_N L \frac{di(t)}{dt} + i(t) = I_A \qquad t \geq 0 \qquad (7\text{--}21)$$

The solution of this equation is found by superimposing the natural and forced components. The natural response is the solution of the homogeneous equation [right side of Eq. (7–21) set to zero] and takes the same form as the zero-input response found in the previous section.

$$i_N(t) = K e^{-t/G_N L} \qquad t \geq 0$$

where K is a constant to be evaluated from the initial condition once the complete response is known. The forced response is a particular solution of the equation

$$G_N L \frac{di_F(t)}{dt} + i_F(t) = I_A \qquad t \geq 0$$

Setting $i_F = I_A$ satisfies this equation since $dI_A/dt = 0$.

Combining the forced and natural responses, we obtain the general solution of Eq. (7–21) in the form

$$i(t) = i_N(t) + i_F(t)$$

$$= K e^{-t/G_N L} + I_A \qquad t \geq 0$$

The constant K is now evaluated from the initial condition:

$$i(0) = I_0 = K e^{-0} + I_A = K + I_A$$

The initial condition requires that $K = I_0 - I_A$, so the step response of the RL circuit is

$$i(t) = (I_0 - I_A)e^{-t/G_N L} + I_A \qquad t \geq 0 \qquad (7\text{--}22)$$

The RL circuit step response has the same form as the RC circuit step response in Eq. (7–20). At $t = 0$ the starting value of the response is $i(0) = I_0$, as required by the initial condition. The final value is the forced response $i(\infty) = i_F = I_A$, since the natural response decays to zero as time increases.

A step function input to the *RC* or *RL* circuit drives the state variable from an initial value determined by what happened prior to $t = 0$ to a final value determined by amplitude of the step function applied at $t = 0$. The time needed to transition from the initial to the final value is about $5T_C$, where T_C is the circuit time constant. We conclude that the step response of a first-order circuit depends on three quantities:

1. The amplitude of the step input (V_A or I_A)
2. The circuit time constant ($R_T C$ or $G_N L$)
3. The value of the state variable at $t = 0$ (V_0 or I_0).

$V_A = 100$ V
at $t = 0$
$V_{01} = 5$ V
$V_{02} = 10$ V

$C_1 = 0.1\ \mu$F
$C_2 = 0.5\ \mu$F
$R_1 = 30$ kΩ
$R_2 = 10$ kΩ

FIGURE 7–13

EXAMPLE 7–4

Find the response of the *RC* circuit in Figure 7–13.

SOLUTION:
The circuit is first order, since the two capacitors in series can be replaced by a single equivalent capacitor

$$C_{EQ} = \frac{1}{\dfrac{1}{C_1} + \dfrac{1}{C_2}} = 0.0833\ \mu\text{F}$$

The initial voltage on C_{EQ} is the sum of the initial voltages on the original capacitors.

$$V_0 = V_{01} + V_{02} = 5 + 10 = 15\ \text{V}$$

To find the Thévenin equivalent seen by C_{EQ}, we first find the open-circuit voltage. Disconnecting the capacitors in Figure 7–13 and using voltage division at the interface yields

$$v_T = v_{OC} = \frac{R_2}{R_1 + R_2} V_A u(t) = \frac{10}{40} 100 u(t) = 25 u(t)\ \text{V}$$

Replacing the voltage source by a short circuit and looking to the left at the interface, we see R_1 in parallel with R_2. The Thévenin resistance of this combination is

$$R_T = \frac{1}{\dfrac{1}{R_1} + \dfrac{1}{R_2}} = 7.5\ \text{k}\Omega$$

The circuit time constant is

$$T_C = R_T C_{EQ} = (7.5 \times 10^3)(8.33 \times 10^{-8}) = \frac{1}{1600}\ \text{s}$$

For the Thévenin equivalent circuit, the initial capacitor voltage is $V_0 = 15$ V, the step input is $25u(t)$, and the time constant is 1/1600 s. Using the *RC* circuit step response in Eq. (7–20) yields

$$v(t) = (15 - 25)\, e^{-1600t} + 25$$
$$= 25 - 10\, e^{-1600t}\ \text{V} \qquad t \geq 0$$

The initial ($t = 0$) value of $v(t)$ is $25 - 10 = 15$ V, as required. The equivalent capacitor voltage is driven to a final value of 25 V by the step input in the Thévenin equivalent circuit. For practical purposes, $v(t)$ reaches 25 V after about $5T_C = 3.125$ ms. ■

EXAMPLE 7–5

Find the step response of the *RL* circuit in Figure 7–14(a). The initial condition is $i(0) = I_0$.

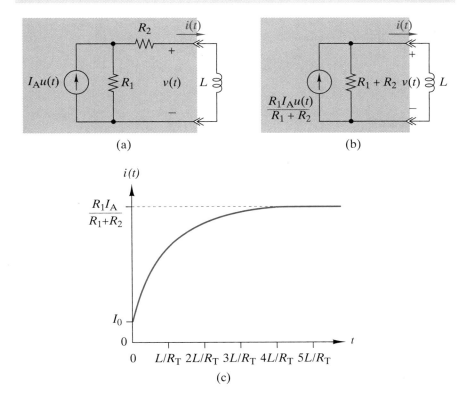

FIGURE 7 – 1 4

(a)

(b)

(c)

SOLUTION:

We first find the Norton equivalent to the left of the interface. By current division, the short-circuit current at the interface is

$$i_{SC}(t) = \frac{R_1}{R_1 + R_2} I_A u(t)$$

Looking to the left at the interface with the current source off (replaced by an open circuit), we see R_1 and R_2 in series producing a Thévenin resistance

$$R_T = \frac{1}{G_N} = R_1 + R_2$$

The time constant of the Norton equivalent circuit in Figure 7–14(b) is

$$T_C = G_N L = \frac{L}{(R_1 + R_2)}$$

The natural response of the Norton equivalent circuit is

$$i_N(t) = Ke^{-(R_1 + R_2)t/L} \qquad t \geq 0$$

The short-circuit current $i_{SC}(t)$ is the step function input in the Norton circuit. Therefore, the forced response is

$$i_F(t) = i_{SC}(t) = \frac{R_1}{(R_1 + R_2)}I_A u(t)$$

Superimposing the natural and forced responses yields

$$i(t) = Ke^{-(R_1 + R_2)t/L} + \frac{R_1 I_A}{R_1 + R_2} \qquad t \geq 0$$

The constant K can be evaluated from the initial condition:

$$i(0) = I_0 = K + \frac{R_1 I_A}{R_1 + R_2}$$

which requires that

$$K = I_0 - \frac{R_1 I_A}{R_1 + R_2}$$

So circuit step response is

$$i(t) = \left[I_0 - \frac{R_1 I_A}{R_1 + R_2} \right] e^{-(R_1 + R_2)t/L} + \frac{R_1 I_A}{R_1 + R_2} \qquad t \geq 0$$

An example of this response is shown in Figure 7–14(c). ∎

EXAMPLE 7–6

The state variable response of a first-order RC circuit for a step function input is

$$v_C(t) = 20e^{-200t} - 10 \text{ V} \qquad t \geq 0$$

(a) What is the circuit time constant?
(b) What is the initial voltage across the capacitor?
(c) What is the amplitude of the forced response?
(d) At what time is $v_C(t) = 0$?

SOLUTION:
(a) The natural response of a first-order circuit is of the form Ke^{-t/T_C}. Therefore, the time constant of the given responses is $T_C = 1/200 = 5$ ms.
(b) The initial ($t = 0$) voltage across the capacitor is

$$v_C(0) = 20e^{-0} - 10 = 20 - 10 = 10 \text{ V}$$

(c) The natural response decays to zero, so the forced response is the final value $v_C(t)$.

$$v_C(\infty) = 20e^{-\infty} - 10 = 0 - 10 = -10 \text{ V}$$

(d) The capacitor voltage must pass through zero at some intermediate time, since the initial value is positive and the final value negative. This time is found by setting the step response equal to zero:

$$20e^{-200t} - 10 = 0$$

which yields the condition $e^{200t} = 2$ or $t = \ln 2/200 = 3.47$ ms.   ∎

Exercise 7–4

Given the following first-order circuit step response

$$v_C(t) = 20 - 20e^{-1000t} \text{ V} \quad t \geq 0$$

(a) What is the amplitude of the step input?
(b) What is the circuit time constant?
(c) What is the initial value of the state variable?
(d) What is the circuit differential equation?

Answers:
(a) 20 V
(b) 1 ms
(c) 0 V
(d) $10^{-3} \, dv_C/dt + v_C = 20u(t)$

Exercise 7–5

Find the solution of the following first-order differential equations:

(a) $10^{-4}\dfrac{dv_C}{dt} + v_C = -5u(t)$ $\qquad v_C(0) = 5$ V

(b) $5 \times 10^{-2}\dfrac{di_L}{dt} + 2000 \, i_L = 10u(t)$ $\qquad i_L(0) = -5$ mA

Answers:
(a) $v_C(t) = -5 + 10e^{-10000t}$ V $\qquad t \geq 0$
(b) $i_L(t) = 5 - 10e^{-40000t}$ mA $\qquad t \geq 0$

ZERO-STATE RESPONSE

Additional properties of dynamic circuit responses are revealed by rearranging the RC and RL circuit step responses in Eqs. (7–20) and (7–22) in the following way:

$$RC \text{ circuit: } v(t) = \underbrace{V_0 e^{-t/R_TC}}_{\substack{\text{Zero-input} \\ \text{response}}} + \underbrace{V_A(1 - e^{-t/R_TC})}_{\substack{\text{Zero-state} \\ \text{response}}} \qquad t \geq 0$$

$$RL \text{ circuit: } i(t) = \overbrace{I_0 e^{-t/G_NL}} + \overbrace{I_A(1 - e^{-t/G_NL})} \qquad t \geq 0$$

We recognize the first term on the right in each equation as the zero-input response discussed in Sect. 7–1. By definition, the **zero-input response** occurs when the input is zero ($V_A = 0$ or $I_A = 0$). The second term on the right in each equation is called the **zero-state response** because this part occurs when the initial state of the circuit is zero ($V_0 = 0$ or $I_0 = 0$).

The zero-state response is proportional to the amplitude of the input step function (V_A or I_A). However, the total response (zero input plus zero state) is not directly proportional to the input amplitude. When the initial state is not zero, the circuit appears to violate the proportionality property of linear circuits. However, bear in mind that the proportionality property applies to linear circuits with only one input.

The RC and RL circuits can store energy and have memory. In effect, they have two inputs: (1) the input that occurred before $t = 0$, and (2) the step function applied at $t = 0$. The first input produces the initial energy state of the circuit at $t = 0$, and the second causes the zero-state response for $t \geq 0$. In general, for $t \geq 0$, the total response of a dynamic circuit is the sum of two responses: (1) the zero-input response caused by the initial conditions produced by inputs applied before $t = 0$, and (2) the zero-state response caused by inputs applied after $t = 0$.

EXAMPLE 7–7

Each clock pulse in a particular digital system can be modeled as a rectangular pulse of the form

$$v_S(t) = V_A[u(t) - u(t - T)]$$

where $V_A = 5$ V and $T = 10$ ns. The clock pulse drives the input of a digital device that can be modeled by the RC circuit in Figure 7–15(a) with $RC = 10$ ns. Find the zero-state response of the voltage $v(t)$.

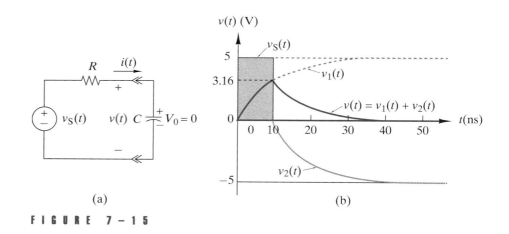

(a)   (b)

FIGURE 7 – 15

SOLUTION:
The rectangular pulse input is shown by the shading in Figure 7–15(b). Since the initial condition is zero (zero state), the total response is the superposition of responses caused by two inputs:

1. A positive 5-V step function applied at $t = 0$.

2. A negative 5-V step function applied at $t = 10$ ns.

The first input causes a zero-state response of

$$v_1(t) = V_A(1 - e^{-t/RC})u(t)$$
$$= 5(1 - e^{-10^8 t})u(t)$$

The second input causes a zero-state response of

$$v_2(t) = -V_A(1 - e^{-(t-T)/RC})u(t - T)$$
$$= -5[1 - e^{-10^8(t - 10^{-8})}]u(t - 10^{-8})$$

Notice that $v_2(t) = -v_1(t - T)$, that is, $v_2(t)$ is obtained by inverting and delaying $v_1(t)$ by $T = 10$ ns. The total response is the superposition of these two responses.

$$v(t) = v_1(t) + v_2(t)$$

Figure 7–15(b) shows how the two responses combine to produce the overall pulse response of the circuit. The response $v_1(t)$ begins at zero and eventually reaches a final value of +5 V. At $t = T = 10$ ns the first response reaches $v_1(T) = 5(1 - e^{-1}) = 3.16$ V. The second response $v_2(t)$ begins at $t = T = 10$ ns, and thereafter is equal and opposite to $v_1(t)$ except that it is delayed by $T = 10$ ns. The final value of the second response is –5 V, so the two responses exactly cancel in the long run. As a result, the total response initially rises until $t = T = 10$ ns, but then returns to zero as the second response comes into play. The example illustrates that a sharply defined rectangular pulse gets distorted when processed by a circuit whose time constant is comparable to the pulse duration. ■

Exercise 7–6

The switch in Figure 7–16 closes at $t = 0$.

(a) Find the zero-state response of the capacitor voltage.
(b) Find the zero-state response of the current through the 10-V voltage source.

Answers:
(a) $v_C = 2.5(1 - e^{-200t})$ V
(b) $i_S = -0.4545(1 - e^{-200t})$ mA

FIGURE 7–16

D DESIGN EXERCISE: 7–7

Given the following first-order circuit zero-state step responses:
(1) $v_C(t) = 10(1 - e^{-2000t})u(t)$ V
(2) $i_L(t) = 5(1 - e^{-5000t})u(t)$ mA
  (a) Find the characteristic equation for each circuit.
  (b) Find the circuit time constants.
  (c) Design first-order circuits that realize each of these responses.

Answers:
(1) (a) $(s + 2000) = 0$
    (b) 0.5 ms
    (c) There are no unique answers to the (c) part of this exercise because the time constants only specify the products RC and GL. Figure 7–17 shows some possible answers.

FIGURE 7–17

(2) (a) $(s + 5000) = 0$
     (b) 0.2 ms
     (c) See (c) above.

DISCUSSION: *The response in (1) is an RC circuit with* $V_A = 10$ V *and a time* $RC = 5 \times 10^{-4}$ s. *Selecting* $R = 10$ kΩ *yields* $C = 50$ nF, *producing the circuit in Figure 7–17(a). The response in (2) is an RL circuit with* $I_A = 5$ mA *and a time constant of* $GL = 2 \times 10^{-4}$ s. *Selecting* $R = 1$ kΩ *yields* $L = 200$ mH. *A source conversion changes the 5-mA current source in parallel with* R *into a* $(5 \times 10^{-3})(10^{3}) = 5$-V *voltage source in series with* R, *as shown in Figure 7–17(b).*

7–3   INITIAL AND FINAL CONDITIONS

Reviewing the first-order step responses of the last section shows that for $t \geq 0$ the state variable responses can be written in the form

$$RC \text{ circuit: } v_C(t) = [v_C(0) - v_C(\infty)]e^{-t/T_C} + v_C(\infty) \qquad t \geq 0$$

$$RL \text{ circuit: } i_L(t) = [i_L(0) - i_L(\infty)]e^{-t/T_C} + i_L(\infty) \qquad t \geq 0$$

(7–23)

In both circuits the step response is of the general form

$$
\text{The state variable response} = \left[\text{The initial value of the state variable} - \text{The final value of the state variable} \right] \times e^{-\frac{t}{T_C}} + \text{The final value of the state variable}
$$

To determine the step response of a first-order circuit, we need three quantities: the initial value of the state variable, the final value of the state variable, and the time constant. Since we know how to get the time constant directly from the circuit, it would be useful to have a direct way to determine the initial and final values by inspecting the circuit itself.

The final value can be calculated directly from the circuit by observing that for $t > 5T_C$ the step responses approach a constant value or dc value. Under dc conditions a capacitor acts like an open circuit and an inductor acts like a short circuit. As a result, the final value of the state variable is found by applying dc analysis methods to the circuit configuration for $t > 0$, with capacitors replaced by open circuits and inductors replaced by short circuits.

We can also use dc analysis to determine the initial value in many practical situations. A common situation is a circuit containing dc sources and a switch that is in one position for a period of time much greater than the circuit time constant, and then is moved to a new position at $t = 0$. For example, if the switch is closed for a long period of time, then the dc sources drive the state variable to a final value. If the switch is now opened at $t = 0$, then the dc sources drive the state variable to a new final condition appropriate to the new circuit configuration for $t > 0$.

Note: The initial condition at $t = 0$ is the dc value of the state variable for the circuit configuration that existed before the switch changed positions at $t = 0$. The switching action cannot cause an instantaneous change in the initial condition because capacitor voltage and inductor current are continuous functions of time. In other words, opening a switch at $t = 0$ marks the boundary between two eras. The final condition of the state variable for the $t < 0$ era is the initial condition for the $t > 0$ era that follows.

The usual way to state a switched circuit problem is to say that a switch has been closed (open) for a long time and then is opened (closed) at $t = 0$. In this context, a long time means at least five time constants. Time constants rarely exceeds a few hundred milliseconds in electrical circuits, so a long time passes rather quickly.

The state variable response in switched dynamic circuits is found using the following steps:

STEP 1: Find the initial value by applying dc analysis to the circuit configuration for $t < 0$.

STEP 2: Find the final value by applying dc analysis to the circuit configuration for $t > 0$.

STEP 3: Find the time constant T_C of the circuit in the configuration for $t > 0$.

STEP 4: Write the step response directly using Eq. (7–23) without formulating and solving the circuit differential equation.

For example, the switch in Figure 7–18(a) has been closed for a long time and is opened at $t = 0$. We want to find the capacitor voltage $v(t)$ for $t \geq 0$.

STEP 1: The initial condition is found by dc analysis of the circuit configuration in Figure 7–18(b), where the switch is closed. Using voltage division, the initial capacitor voltage in found to be

$$v(0) = \frac{R_2 V_A}{R_1 + R_2}$$

STEP 2: The final condition is found by dc analysis of the circuit configuration in Figure 7–18(c), where the switch is open. When the switch is open the circuit has no dc excitation, so the final value of the capacitor voltage is zero.

STEP 3: The circuit in Figure 7–18(c) also gives us the time constant. Looking back at the interface, we see an equivalent resistance of R_2, since R_1 is connected in series with an open switch. For $t \geq 0$ the time constant is R_2C. Using Eq. (7–23), the capacitor voltage for $t \geq 0$ is

$$v(t) = [v(0) - v(\infty)]e^{-t/T_C} + v_C(\infty)$$

$$= \frac{R_2 V_A}{R_1 + R_2}e^{-t/R_2C} \qquad t \geq 0$$

The result is a zero-input response, since there is no excitation for $t \geq 0$. But now we see how the initial condition for the zero-input response could be produced physically by opening a switch that has been closed for a long time.

To continue the analysis, we find the capacitor current using its element constraint:

$$i(t) = C\frac{dv}{dt} = -\frac{V_A}{R_1 + R_2}e^{-t/R_2C} \qquad t \geq 0$$

(a)

(b)

(c)

FIGURE 7-18 *Solving a switched dynamic circuit using the initial and final conditions.*

FIGURE 7–19 *Two responses in the RC circuit of Figure 7–18.*

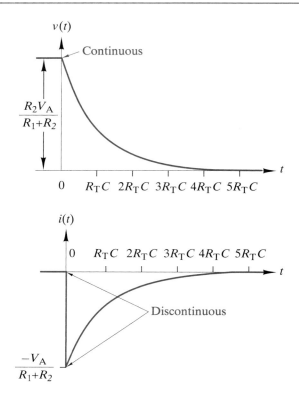

This is the capacitor current for $t \geq 0$. For $t < 0$ the circuit in Figure 7–18(b) points out that the capacitor current is zero since the capacitor acts like an open circuit.

The capacitor voltage and current responses are plotted in Figure 7–19. The capacitor voltage is continuous at $t = 0$, but the capacitor current has a jump discontinuity at $t = 0$. In other words, state variables are continuous, but nonstate variables can have discontinuities at $t = 0$. Since the state variable is continuous, we first find the circuit state variable and then solve for other circuit variables using the element and connection constraints.

EXAMPLE 7–8

The switch in Figure 7–20(a) has been open for a long time and is closed at $t = 0$. Find the inductor current for $t > 0$.

SOLUTION:

We first find the initial condition using the circuit in Figure 7–20(b). By series equivalence the initial current is

$$ i(0) = \frac{V_A}{R_1 + R_2} $$

The final condition and the time constant are determined from the circuit in Figure 7–17(c). Closing the switch shorts out R_2, and the final condition and time constant for $t > 0$ are

FIGURE 7–20

$$i(\infty) = \frac{V_A}{R_1} \quad \text{and} \quad T_C = G_N L = \frac{L}{R_1}$$

Using Eq. (7–23), the inductor current for $t \geq 0$ is

$$i(t) = [i(0) - i(\infty)]e^{-t/T_C} + i(\infty)$$

$$= \left[\frac{V_A}{R_1 + R_2} - \frac{V_A}{R_1}\right]e^{-R_1 t/L} + \frac{V_A}{R_1} \qquad t \geq 0 \qquad \blacksquare$$

(a)

EXAMPLE 7-9

The switch in Figure 7–21(a) has been closed for a long time and is opened at $t = 0$. Find the voltage $v_O(t)$.

SOLUTION:

The problem asks for voltage $v_O(t)$, which is not the circuit state variable. Our approach is first to find the state variable response and then use this response to solve for the required nonstate variable.

For $t < 0$ the circuit in Figure 7–21(b) applies. By voltage division, the initial capacitor voltage is

$$v(0) = \frac{R_1 V_A}{R_1 + R_2}$$

The final value and time constant are found from the circuit in Figure 7–18(c).

$$v(\infty) = V_A \quad \text{and} \quad T_C = R_T C = R_2 C$$

Using Eq. (7–23), the capacitor voltage for $t \geq 0$ is

$$v(t) = [v(0) - v(\infty)]e^{-t/T_C} + v(\infty)$$

$$= \left[\frac{R_1 V_A}{R_1 + R_2} - V_A\right]e^{-t/R_2 C} + V_A$$

$$= V_A + \frac{R_2 V_A}{R_1 + R_2}e^{-t/R_2 C} \qquad t \geq 0$$

(b)

(c)

FIGURE 7 - 21

Given the state variable, we can find the voltage $v_O(t)$ by writing a KVL equation around the perimeter of the circuit in Figure 7–21(a):

$$-V_A + v(t) + v_O(t) = 0$$

or

$$v_O(t) = V_A - v(t) = -\frac{R_2 V_A}{R_1 + R_2}e^{-t/R_2 C} \qquad t \geq 0$$

The output voltage response looks like a zero-input response even though the circuit input is not zero for $t \geq 0$. However, $v_O(t)$ is not the state variable but the voltage across the resistor R_2. The voltage across R_2 is proportional to the capacitor current, which eventually decays to zero in the final circuit configuration in Figure 7–21(c) because the capacitor acts like an open circuit. $\blacksquare$

FIGURE 7-22

EXAMPLE 7-10

The switch in Figure 7–22 has been in position A for quite a while and is moved to position B at $t = 0$. Find the OP AMP output voltage $v_O(t)$.

SOLUTION:

Since $v_O(t)$ does not appear to be a state variable, we need to show that Eq. (7–23) applies to this problem. Applying the fundamental principle of node voltages to the capacitor voltage yields

$$v_C(t) = v_O(t) - v_N(t)$$

For the ideal OP AMP model, $v_N = v_P = 0$ since the noninverting input is grounded. In other words, the OP AMP output voltage equals the capacitor voltage, which is the state variable. Hence, we can use Eq. (7–23) to find $v_O(t)$. In Sect. 7–1 we showed that this OP AMP circuit is a first-order circuit with a time constant of R_2C. When the capacitor is replaced by an open circuit, the configuration is an inverting amplifier with a gain of $-R_2/R_1$. For $t < 0$ the switch is in position A and the initial output (and capacitor) voltage is

$$v_O(0) = -\frac{R_2}{R_1}V_A$$

For $t \geq 0$ the switch is in position B and the final output (and capacitor) voltage is

$$v_O(\infty) = -\frac{R_2}{R_1}(-V_A) = -v_O(0)$$

Applying Eq. (7–23), the step response is

$$v_O(t) = -2\frac{R_2}{R_1}V_A e^{-t/R_2C} + \frac{R_2}{R_1}V_A$$

$$= \frac{R_2}{R_1}V_A(1 - 2e^{-t/R_2C}) \qquad t \geq 0 \qquad \blacksquare$$

◐ DESIGN EXAMPLE 7-11

Design an RC circuit whose response is $v_C(t) = -5 + 15e^{-2000t}$ V for $t \geq 0$. The circuit can include one switch that is thrown at $t = 0$.

SOLUTION:

The required initial and final conditions for the specified response are

$$v_C(0) = 15e^{-0} - 5 = +10 \text{ V}$$

$$v_C(\infty) = 15e^{-\infty} - 5 = -5 \text{ V}$$

The time constant of the specified response is $T_C = 1/2000 = 0.5$ ms. The time constant of a simple RC circuit is $T_C = RC = 0.5$ ms. Choosing $R = 10$ kΩ yields $C = 50$ nF. Figure 7–23(a) shows a simple RC circuit that produces the required $v_C(0)$. Similarly, the RC circuit in Figure 7–23(b) produces the required $v_C(\infty)$. One switch is allowed, so we combine these two circuits, as indicated in Figure 7–23(c). Many other designs are possible. ■

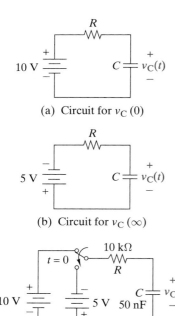

(a) Circuit for $v_C(0)$

(b) Circuit for $v_C(\infty)$

(c) Candidate design

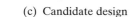

FIGURE 7-23

In each circuit shown in Figure 7–24 the switch has been in position A for a long time and is moved to position B at $t = 0$. Find the circuit state variable for each circuit for $t \geq 0$.

Answers:

$$\text{(a) } v_C(t) = V_A e^{-t/(R_1 + R_2)C}$$

$$\text{(b) } i_L(t) = \frac{V_A}{R_2} e^{-(R_1 + R_2)t/L}$$

(a)

(b)

FIGURE 7 - 2 4

In each circuit shown in Figure 7–24 the switch has been in position B for a long time and is moved to position A at $t = 0$. Find the circuit state variable for each circuit for $t \geq 0$.

Answers:

$$\text{(a) } v_C(t) = V_A(1 - e^{-t/R_2 C})$$

$$\text{(b) } i_L(t) = \frac{V_A}{R_2}(1 - e^{-R_2 t/L})$$

In the circuit in Figure 7–25 the switch has been in position A for a long time and is moved to position B at $t = 0$. For $t \geq 0$ find

(a) The output voltage $v_O(t)$
(b) The current $i_S(t)$ through the switch.

FIGURE 7 - 2 5

Answers:
(a) $v_O(t) = -4e^{-200t}$ V
(b) $i_S(t) = -0.2 + 0.24e^{-200t}$ A

7–4 FIRST-ORDER CIRCUIT SINUSOIDAL RESPONSE

The response of linear circuits to sinusoidal inputs is one of the central themes of electrical engineering. In this introduction to the concept we treat the sinusoidal response of first-order circuits using differential equations. In later chapters we see that sinusoidal response can be found using other techniques, such as phasors and Laplace transforms. But for the moment, we concentrate on the classical method of finding the forced response from the circuit differential equation.

If the input to the RC circuit in Figure 7–2 is a casual sinusoid, then the circuit differential equation in Eq. (7–3) is written as

$$R_T C \frac{dv(t)}{dt} + v(t) = V_A[\cos \omega t]u(t) \qquad \text{(7–24)}$$

The input on the right side of Eq. (7–24) is *not* an eternal sinewave but a casual sinusoid that starts at $t = 0$, through some action such as closing a switch. We seek a solution function $v(t)$ that satisfies Eq. (7–24) for $t \geq 0$ and that meets the prescribed initial condition $v(0) = V_0$.

As with the step response, we find the solution in two parts: natural response and forced response. The natural response is of the form

$$v_N(t) = K e^{-t/R_T C} \qquad t \geq 0$$

The natural response of a first-order circuit always has this form because it is a general solution of the homogeneous equation with input set to zero. The form of the natural response depends on the physical characteristics of the circuit and is independent of the input.

The forced response depends on both the circuit and the nature of the forcing function. The forced response is a particular solution of the equation

$$R_T C \frac{dv_F(t)}{dt} + v_F(t) = V_A \cos \omega t \qquad t \geq 0$$

This equation requires that $v_F(t)$ plus $R_T C$ times its first derivative add to produce a cosine function for $t \geq 0$. The only way this can happen is for $v_F(t)$ and its derivative to be sinusoids of the same frequency. This requirement brings to mind the derivative property of the sinusoid. So we try a solution in the form of a general sinusoid. As noted in Chapter 5, a general sinusoid can be written in amplitude and phase angle form as

$$v_F(t) = V_F \cos(\omega t + \phi) \qquad \text{(7–25a)}$$

or in terms of Fourier coefficients as

$$v_F(t) = a \cos \omega t + b \sin \omega t \qquad \text{(7–25b)}$$

While either form will work, it is somewhat easier to work with the Fourier coefficient format.

The approach we are using is called the method of undetermined coefficients, where the unknown coefficients are the Fourier coefficients a and b in Eq. (7–25b). To find these unknowns we insert the proposed forced response in Eq. (7–25b) into the differential equation to obtain

$$R_T C \frac{d}{dt}(a \cos \omega t + b \sin \omega t) + (a \cos \omega t + b \sin \omega t) = V_A \cos \omega t \qquad t \geq 0$$

Performing the differentiation gives

$$R_T C(-\omega a \sin \omega t + \omega b \cos \omega t) + (a \cos \omega t + b \sin \omega t) = V_A \cos \omega t$$

We next gather all sine and cosine terms on one side of the equation.

$$[R_T C\omega b + a - V_A]\cos \omega t + [-R_T C\omega a + b]\sin \omega t = 0$$

The left side of this equation is zero for all $t \geq 0$ only when the coefficients of the cosine and sine terms are identically zero. This requirement yields two linear equations in the unknown coefficients a and b:

$$a + (R_T C\omega)b = V_A$$

$$-(R_T C\omega)a + b = 0$$

The solutions of these linear equations are

$$a = \frac{V_A}{1 + (\omega R_T C)^2} \qquad b = \frac{\omega R_T C V_A}{1 + (\omega R_T C)^2}$$

These equations express the unknowns a and b in terms of known circuit parameters ($R_T C$) and known input signal parameters (ω and V_A).

We combine the forced and natural responses as

$$v(t) = Ke^{-t/R_T C} + \frac{V_A}{1 + (\omega R_T C)^2}(\cos \omega t + \omega R_T C \sin \omega t) \qquad t \geq 0$$

$$(7\text{–}26)$$

The initial condition requires

$$v(0) = V_0 = K + \frac{V_A}{1 + (\omega R_T C)^2}$$

which means K is

$$K = V_0 - \frac{V_A}{1 + (\omega R_T C)^2}$$

We substitute this value of K into Eq. (7–26) to obtain the function $v(t)$ that satisfies the differential equation and the initial conditions.

$$v(t) = \underbrace{\left[V_0 - \frac{V_A}{1 + (\omega R_T C)^2}\right]e^{-t/R_T C}}_{\text{Natural response}} +$$

$$\underbrace{\frac{V_A}{1 + (\omega R_T C)^2}(\cos \omega t + \omega R_T C \sin \omega t)}_{\text{Forced response}} \qquad t \geq 0$$

This expression seems somewhat less formidable when we convert the forced response to an amplitude and phase angle format

$$v(t) = \underbrace{\left[V_0 - \frac{V_A}{1 + (\omega R_T C)^2}\right]e^{-t/R_T C}}_{\text{Natural Response}} + \underbrace{\frac{V_A}{\sqrt{1 + (\omega R_T C)^2}}\cos(\omega t + \theta)}_{\text{Forced Response}} \qquad t \geq 0$$

$$(7\text{–}27)$$

where

$$\theta = \tan^{-1}(-b/a) = \tan^{-1}(-\omega R_\text{T} C)$$

Equation (7–27) is the complete response of the *RC* circuit for an initial condition V_0 and a sinusoidal input $[V_\text{A} \cos \omega t]u(t)$. Several aspects of the response deserve comment:

1. After roughly five time constants the natural response decays to zero but the sinusoidal forced response persists.

2. The forced response is a sinusoid with the same frequency (ω) as the input but with a different amplitude and phase angle.

3. The forced response is proportional to V_A. This means that the amplitude of the forced component has the proportionality property because the circuit is linear.

In the terminology of electrical engineering, the forced component is called the **sinusoidal steady-state response**. The words *steady state* may be misleading since together they seem to imply a constant or "steady" value, whereas the forced response is a sustained oscillation. To electrical engineers *steady state* means the conditions reached after the natural response has died out. The sinusoidal steady-state response is also called the **ac steady-state response**. Often the words *steady state* are dropped and it is called simply the **ac response**. Hereafter, ac response, sinusoidal steady-state response, and the forced response for a sinusoidal input will be used interchangeably.

Finally, the forced response due to a step function input is called the **zero-frequency** or **dc steady-state response**. The zero-frequency terminology means that we think of a step function as a cosine $V_\text{A}[\cos \omega t]u(t)$ with $\omega = 0$. The reader can easily show that inserting $\omega = 0$ reduces Eq. (7–27) to the *RC* circuit step response in Eq. (7–20).

F I G U R E 7 – 2 6

EXAMPLE 7–12

The switch in Figure 7–26 has been open for a long time and is closed at $t = 0$. Find the voltage $v(t)$ for $t \geq 0$ when $v_\text{S}(t) = [20 \sin 1000t]u(t)$ V.

SOLUTION:

We first derive the circuit differential equation. By voltage division, the Thévenin voltage seen by the capacitor is

$$v_\text{T}(t) = \frac{4}{4 + 4}v_\text{S}(t) = 10 \sin 1000t \quad \text{V}$$

The Thévenin resistance (switch closed and source off) looking back into the interface is two 4-kΩ resistors in parallel, so $R_\text{T} = 2$ kΩ. The circuit time constant is

$$T_\text{C} = R_\text{T} C = (2 \times 10^3)(1 \times 10^{-6}) = 2 \times 10^{-3} = 1/500 \text{ s}$$

Given the Thévenin equivalent seen by the capacitor and the circuit time constant, the circuit differential is

$$2 \times 10^{-3}\frac{dv(t)}{dt} + v(t) = 10 \sin 1000t \qquad t \geq 0$$

Note that the right side of the circuit differential equation is the Thévenin voltage $v_T(t)$ and not the original source input $v_S(t)$. The natural response is of the form

$$v_N(t) = Ke^{-500t} \qquad t \geq 0$$

The forced response with undetermined Fourier coefficients is

$$v_F(t) = a \cos 1000t + b \sin 1000t$$

Substituting the forced response into the differential equation produces

$$2 \times 10^{-3}(-1000\, a \sin 1000t + 1000b \cos 1000t) +$$

$$a \cos 1000t + b \sin 1000t = 10 \sin 1000t$$

Collecting all sine and cosine terms on one side of this equation yields

$$(a + 2b) \cos 1000t + (-2a + b - 10)\sin 1000t = 0$$

The left side of this equation is zero for all $t \geq 0$ only when the coefficient of the sine and cosine terms vanish:

$$a + 2b = 0$$

$$-2a + b = 10$$

The solutions of these two linear equations are $a = -4$ and $b = 2$. We combine the forced and natural responses

$$v(t) = Ke^{-500t} - 4 \cos 1000t + 2 \sin 1000t \qquad t \geq 0$$

The constant K is found from the initial conditions

$$v(0) = V_0 = K - 4$$

The initial condition is $V_0 = 0$ because with the switch open the capacitor had no input for a long time prior to $t = 0$. The initial condition $v(0) = 0$ requires $K = 4$, so we can now write the complete response in the form

$$v(t) = 4e^{-500t} - 4 \cos 1000t + 2 \sin 1000t \text{ V} \qquad t \geq 0$$

or, in an amplitude, phase angle format as

$$v(t) = 4e^{-500t} + 4.47 \cos(1000t + 153°) \text{ V} \qquad t \geq 0$$

Figure 7–27 shows an Excel worksheet that generates plots of the natural response, forced response, and total response. Column A is the time at 0.25-ms intervals. Columns B and C calculate the natural response ($4e^{-500t}$) and the forced response ($-4 \cos 1000t + 2 \sin 1000t$) at each of the times given in column A. The total response in column D is the sum of the entries in columns B and C. The plots show that the total response merges into the sinusoidal forced response since the natural response decays to zero after about $5T_C = 10$ ms. That is, after about 10 ms or so the circuit settles down to an ac steady-state condition. ∎

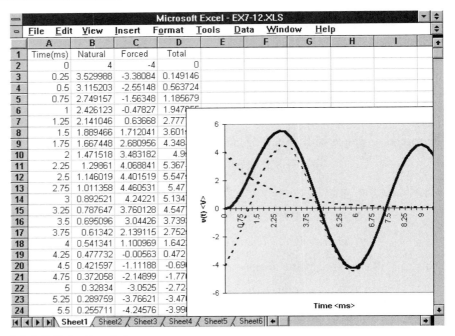

	A	B	C	D	E	F	G	H	I
1	Time(ms)	Natural	Forced	Total					
2	0	4	-4	0					
3	0.25	3.529988	-3.38084	0.149146					
4	0.5	3.115203	-2.55148	0.563724					
5	0.75	2.749157	-1.56348	1.185679					
6	1	2.426123	-0.47827	1.947					
7	1.25	2.141046	0.63668	2.777					
8	1.5	1.889466	1.712041	3.601					
9	1.75	1.667448	2.680956	4.348					
10	2	1.471518	3.483182	4.9					
11	2.25	1.29861	4.068841	5.367					
12	2.5	1.146019	4.401519	5.547					
13	2.75	1.011358	4.460531	5.47					
14	3	0.892521	4.24221	5.134					
15	3.25	0.787647	3.760128	4.547					
16	3.5	0.695096	3.04426	3.739					
17	3.75	0.61342	2.139115	2.752					
18	4	0.541341	1.100969	1.642					
19	4.25	0.477732	-0.00563	0.472					
20	4.5	0.421597	-1.11188	-0.69					
21	4.75	0.372058	-2.14899	-1.77					
22	5	0.32834	-3.0525	-2.72					
23	5.25	0.289759	-3.76621	-3.47					
24	5.5	0.255711	-4.24576	-3.99					

Sheet1 / Sheet2 / Sheet3 / Sheet4 / Sheet5 / Sheet6

FIGURE 7–27

FIGURE 7–28

EXAMPLE 7–13

Find the sinusoidal steady-state response of the output voltage $v_O(t)$ in Figure 7–28 when the input current is $i_S(t) = [I_A \cos \omega t]u(t)$.

SOLUTION:

In keeping with our general analysis approach, we first find the steady-state response of the state variable $i(t)$ and use it to determine the required output voltage. The differential equation of the circuit in terms of the inductor current is

$$GL\frac{di}{dt} + i = I_A \cos \omega t \qquad t \geq 0$$

To find the steady-state response, we need to find the unknown Fourier coefficients a and b in the forced component:

$$i_F(t) = a \cos \omega t + b \sin \omega t \qquad t \geq 0$$

Substituting this expression into the differential equation yields

$$GL(-a\omega \sin \omega t + b\omega \cos \omega t) +$$

$$a \cos \omega t + b \sin \omega t = I_A \cos \omega t$$

Collecting sine and cosine terms produces

$$(a + GLb\omega - I_A)\cos \omega t + (-GLa\omega + b)\sin \omega t = 0$$

The left side of this equation is zero for all $t \geq 0$ only if

$$a + (GL\omega)b = I_A$$
$$-(GL\omega)a + b = 0$$

The solutions of these linear equations are

$$a = \frac{I_A}{1 + (GL\omega)^2} \qquad b = \frac{\omega GLI_A}{1 + (GL\omega)^2}$$

Therefore, the forced component of the inductor current is

$$i_F(t) = \frac{I_A}{1 + (\omega GL)^2}(\cos \omega t + \omega GL \sin \omega t) \qquad t \geq 0$$

The prescribed output is the voltage across the inductor. The steady-state output voltage is found using the inductor element equation:

$$v_O = L\frac{di_F}{dt} = \left[\frac{I_A L}{1 + (\omega GL)^2}\right]\frac{d}{dt}[\cos \omega t + \omega GL \sin \omega t]$$

$$= \left[\frac{I_A L}{1 + (\omega GL)^2}\right][-\omega \sin \omega t + \omega^2 GL \cos \omega t]$$

$$= \frac{I_A \omega L}{\sqrt{1 + (\omega GL)^2}} \cos (\omega t + \theta) \qquad t \geq 0$$

where $\theta = \tan^{-1}(1/\omega GL)$. The output voltage is a sinusoid with the same frequency as the input signal, but with a different amplitude and phase angle. In fact, in the sinusoidal steady state every voltage and current in a linear circuit is sinusoidal with the same frequency.

Notice that the amplitude of the steady-state output takes the form

$$\frac{I_A \omega L}{\sqrt{1 + (\omega GL)^2}}$$

Thus, the amplitude changes with the frequency of the sinusoidal current input. At $\omega = 0$ the input $i_S(t) = I_A \cos(0) = I_A$ is a constant dc waveform and the steady-state output is $v_O = 0$. This makes sense because at dc the inductor acts like a short circuit that forces the steady-state output in Figure 7–28 to be zero. At very high frequencies ($\omega GL \gg 1$) the steady-state output approaches $v_O = I_A R$. This also makes sense because at very high frequency the inductor acts like an open circuit that forces all of the input current to pass through the resistor in Figure 7–28. In between these two extremes the input current divides between the two paths in a manner that depends on the frequency. We will study frequency dependent responses in detail in later chapters. ∎

Exercise 7–11

Find the forced component solution of the differential equation

$$10^{-3}\frac{dv}{dt} + v = 10 \cos \omega t$$

for the following frequencies:

(a) $\omega = 500$ rad/s
(b) $\omega = 1000$ rad/s
(c) $\omega = 2000$ rad/s

Answers:
(a) $v_F(t) = 8 \cos 500t + 4 \sin 500t$ $t \geq 0$
(b) $v_F(t) = 5 \cos 1000t + 5 \sin 1000t$ $t \geq 0$
(c) $v_F(t) = 2 \cos 2000t + 4 \sin 2000t$ $t \geq 0$

DISCUSSION: *Converting these answers to an amplitude and phage angle as*
(a) $v_F(t) = 8.94 \cos (500t - 26.6°)$ $t \geq 0$
(b) $v_F(t) = 7.07 \cos (1000t - 45°)$ $t \geq 0$
(c) $v_F(t) = 4.47 \cos (2000t - 63.4°)$ $t \geq 0$

we see that increasing the frequency of the input sinusoid decreases the amplitude and phase angle of the sinusoidal steady-state output of the circuit.

FIGURE 7-29

Exercise 7-12

The circuit in Figure 7–29 is operating in the sinusoidal steady state with

$$v_O(t) = 10 \cos(100t - 45°) \text{ V}$$

Find the source voltage $v_S(t)$.

Answer:
$v_S(t) = 10\sqrt{2} \cos 100t \text{ V}$

7-5 THE SERIES *RLC* CIRCUIT

Second-order circuits contain two energy storage elements that cannot be replaced by a single equivalent element. They are called **second-order circuits** because the circuit differential equation involves the second derivative of the dependent variable. Although there is an endless number of such circuits, in this chapter we will concentrate on two classical forms: (1) the series *RLC* circuit and (2) the parallel *RLC* circuit. These two circuits illustrate almost all of the basic concepts of second-order circuits and serve as vehicles for studying the solution of second-order differential equations. In subsequent chapters we use Laplace transform techniques to analyze any second-order circuit.

FORMULATING SERIES *RLC* CIRCUIT EQUATIONS

We begin with the circuit in Figure 7–30(a), where the inductor and capacitor are connected in series. The source-resistor circuit can be reduced to the Thévenin equivalent shown in Figure 7–25(b). The result is a circuit in which a voltage source, resistor, inductor, and capacitor are connected in series (hence the name **series *RLC* circuit**).

The first task is to develop the equations that describe the series *RLC* circuit. The Thévenin equivalent to the left of the interface in Figure 7–30(b) produces the KVL constraint

$$v + R_T i = v_T \tag{7–28}$$

Applying KVL around the loop on the right side of the interface yields

$$v = v_L + v_C \tag{7–29}$$

Finally, the *i–v* characteristics of the inductor and capacitor are

(a)

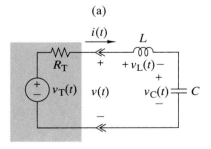

(b)

FIGURE 7-30 *The series RLC circuit.*

$$v_L = L\frac{di}{dt} \tag{7-30}$$

$$i = C\frac{dv_C}{dt} \tag{7-31}$$

Equations (7–28) through (7–31) are four independent equations in four unknowns (i, v, v_L, v_C). Collectively, this set of equations provides a complete description of the dynamics of the series *RLC* circuit. To find the circuit response using classical methods, we must derive a circuit equation containing only one of these unknowns.

We use circuit state variables as solution variables because they are continuous functions of time. In the series *RLC* circuit in Figure 7–25(b), there are two state variables: (1) the capacitor voltage $v_C(t)$ and (2) the inductor current $i(t)$. We first show how to describe the circuit using the capacitor voltage as the solution variable:

To derive a single equation in $v_C(t)$, we substitute Eqs. (7–29) and (7–31) into Eq. (7–28).

$$v_L + v_C + R_T C\frac{dv_C}{dt} = v_T \tag{7-32}$$

These substitutions eliminate the unknowns except v_C and v_L. To eliminate the inductor voltage, we substitute Eq. (7–31) into Eq. (7–30) to obtain

$$v_L = LC\frac{d^2 v_C}{dt^2}$$

Substituting this result into Eq. (7–32) produces

$$LC\frac{d^2 v_C}{dt^2} + R_T C\frac{dv_C}{dt} + v_C = v_T \tag{7-33}$$

$$v_L \quad + \quad v_R \quad + v_C = v_T$$

In effect, this is a KVL equation around the loop in Figure 7–30(b), where the inductor and resistor voltages have been expressed in terms of the capacitor voltage.

Equation (7–33) is a second-order linear differential equation with constant coefficients. It is a second-order equation because the highest order derivative is the second derivative of the dependent variable $v_C(t)$. The coefficients are constant because the circuit parameters L, C, and R_T do not change. The Thévenin voltage $v_T(t)$ is a known driving force. The initial conditions

$$v_C(0) = V_0 \text{ and } \frac{dv_C}{dt}(0) = \frac{1}{C}i(0) = \frac{I_0}{C} \tag{7-34}$$

are determined by the values of the capacitor voltage and inductor current at $t = 0$, V_0 and I_0.

In summary, the second-order differential equation in Eq. (7–33) characterizes the response of the series *RLC* circuit in terms of the capacitor voltage $v_C(t)$. Once the solution $v_C(t)$ is found, we can solve for every other voltage or current, including the inductor current, using the element and connection constraints in Eqs. (7–28) to (7–31).

Alternatively, we can characterize the series RLC circuit using the inductor current. We first write the capacitor i–v characteristics in integral form:

$$v_C(t) = \frac{1}{C}\int_0^t i(x)dx + v_C(0) \tag{7-35}$$

Equations (7–35), (7–30), and (7–29) are inserted into the interface constraint of Eq. (7–28) to obtain a single equation in the inductor current $i(t)$:

$$L\frac{di}{dt} + \frac{1}{C}\int_0^t i(x)dx + v_C(0) + R_T i = v_T \tag{7-36}$$

$$\quad v_L + \qquad\qquad v_C \qquad\qquad + v_R \; = v_T$$

In effect, this a KVL equation around the loop in Figure 7–25(b), where the capacitor and resistor voltages have been expressed in terms of the inductor current.

Equation (7–36) is a second-order linear integro-differential equation with constant coefficients. It is second order because it involves the first derivative and the first integral of the dependent variable $i(t)$. The coefficients are constant because the circuit parameters L, C, and R_T do not change. The Thévenin equivalent voltage $v_T(t)$ is a known driving force, and the initial conditions are $v_C(0) = V_0$ and $i(0) = I_0$.

Equations (7–33) and (7–36) involve the same basic ingredients: (1) an unknown state variable, (2) three circuit parameters (R_T, L, C), (3) a known input $v_T(t)$, and (4) two initial conditions (V_0 and I_0). The only difference is that one expresses the sum of voltages around the loop in terms of the capacitor voltage, while the other uses the inductor current. Either equation characterizes the dynamics of the series RLC circuit because once a state variable is found, every other voltage or current can be found using the element and connection constraints.

ZERO-INPUT RESPONSE OF THE SERIES RLC CIRCUIT

The circuit dynamic response for $t \geq 0$ can be divided into two components: (1) the zero-input response caused by the initial conditions and (2) the zero-state response caused by driving forces applied after $t = 0$. Because the circuit is linear, we can solve for these responses separately and superimpose them to get the total response. We first deal with the zero-input response.

With $v_T = 0$ (zero-input) Eq. (7–33) becomes

$$LC\frac{d^2 v_C}{dt^2} + R_T C\frac{dv_C}{dt} + v_C = 0 \tag{7-37}$$

This result is a second-order homogeneous differential equation in the capacitor voltage. Alternatively, we set $v_T = 0$ in Eq. (7–36) and differentiate once to obtain the following homogeneous differential equation in the inductor current:

$$LC\frac{d^2 i}{dt^2} + R_T C\frac{di}{dt} + i = 0 \tag{7-38}$$

We observe that Eqs. (7–37) and (7–38) have exactly the same form except that the dependent variables are different. The zero-input response of the capacitor voltage and inductor current have the same general form. We do not need to study both to understand the dynamics of the series *RLC* circuit. In other words, in the series *RLC* circuit we can use either state variable to describe the zero-input response.

In the following discussion we will concentrate on the capacitor voltage response. Equation (7–37) requires the capacitor voltage, plus *RC* times its first derivative, plus *LC* times its second derivative to add to zero for all $t \geq 0$. The only way this can happen is for $v_C(t)$, its first derivative, and its second derivative to have the same waveform. No matter how many times we differentiate an exponential of the form e^{st}, we are left with a signal with the same waveform. This observation, plus our experience with first-order circuits, suggests that we try a solution of the form

$$v_C(t) = Ke^{st}$$

where the parameters K and s are to be evaluated. When the trial solution is inserted in Eq. (7–37), we obtain the condition

$$Ke^{st}(LCs^2 + R_TCs + 1) = 0$$

The function e^{st} cannot be zero for all $t \geq 0$. The condition $K = 0$ is not allowed because it is a trivial solution declaring that $v_C(t)$ is zero for all t. The only useful way to meet the condition is to require

$$LCs^2 + R_TCs + 1 = 0 \qquad\qquad \text{(7–39)}$$

Equation (7–39) is the **characteristic equation** of the series *RLC* circuit. The characteristic equation is a quadratic because the circuit contains two energy storage elements. Inserting Ke^{st} into the homogeneous equation of the inductor current in Eq. (7–38) produces the same characteristic equation. Thus, Eq. (7–39) relates the zero-input response to circuit parameters for both state variables (hence the name *characteristic equation*).

In general, a quadratic characteristic equation has two roots:

$$s_1, s_2 = \frac{-R_TC \pm \sqrt{(R_TC)^2 - 4LC}}{2LC} \qquad\qquad \text{(7–40)}$$

From the form of the expression under the radical in Eq. (7–40), we see that there are three distinct possibilities:

Case A: If $(R_TC)^2 - 4LC > 0$, there are two real, unequal roots ($s_1 = -\alpha_1 \neq s_2 = -\alpha_2$).

Case B: If $(R_TC)^2 - 4LC = 0$, there are two real, equal roots ($s_1 = s_2 = -\alpha$).

Case C: If $(R_TC)^2 - 4LC < 0$, there are two complex conjugate roots ($s_1 = -\alpha - j\beta$ and $s_2 = -\alpha + j\beta$).

Where the symbol j represents the imaginary number $\sqrt{-1}$.[2] Before dealing with the form of the zero-input response for each case, we consider an example.

---

2 Mathematicians use the letter i to represent $\sqrt{-1}$. Electrical engineers use j, since the letter i represents electric current.

EXAMPLE 7-14

A series RLC circuit has $C = 0.25\ \mu\text{F}$ and $L = 1$ H. Find the roots of the characteristic equation for $R_T = 8.5$ kΩ, 4 kΩ, and 1 kΩ.

SOLUTION:

For $R_T = 8.5$ kΩ, the characteristic equation is

$$0.25 \times 10^{-6}s^2 + 2.125 \times 10^{-3}s + 1 = 0$$

whose roots are

$$s_1, s_2 = -4250 \pm \sqrt{(3750)^2} = -500, \quad -8000$$

These roots illustrate case A. The quantity under the radical is positive, and there are two real, unequal roots at $s_1 = -500$ and $s_2 = -8000$.

For $R_T = 4$ kΩ, the characteristic equation is

$$0.25 \times 10^{-6}s^2 + 10^{-3}s + 1 = 0$$

whose roots are

$$s_1,\ s_2 = -2000 \pm \sqrt{4 \times 10^6 - 4 \times 10^6} = -2000$$

This is an example of case B. The quantity under the radical is zero, and there are two real, equal roots at $s_1 = s_2 = -2000$.

For $R_T = 1$ kΩ the characteristic equation is

$$0.25 \times 10^{-6}s^2 + 0.25 \times 10^{-3}\ s + 1 = 0$$

whose roots are

$$s_1,\ s_2 = -500 \pm 500\sqrt{-15}$$

The quantity under the radical is negative, illustrating case C.

$$s_1,\ s_2 = -500 \pm j500\sqrt{15}$$

In case C the two roots are complex conjugates.   ■

Exercise 7-13

For a series RLC circuit:

(a) Find the roots of the characteristic equation when $R_T = 2$ kΩ, $L = 100$ mH, and $C = 0.4\ \mu\text{F}$.
(b) For $L = 100$ mH, select the values of R_T and C so the roots of the characteristic equation are $s_1, s_2 = -1000 \pm j2000$.
(c) Select the values of R_T, L, and C so $s_1 = s_2 = -10^4$.

Answers:
(a) $s_1 = -1340$, $s_2 = -18{,}660$
(b) $R_T = 200$ Ω, $C = 2\ \mu\text{F}$
(c) There is no unique answer to part (c) since the requirement

$$(10^{-4}s + 1)^2 = LCs^2 + R_TCs + 1$$
$$= 10^{-8}s^2 + 10^{-4}s + 1$$

gives two equations $R_TC = 10^{-4}$ and $LC = 10^{-8}$ in three unknowns. One solution is to select $C = 1\ \mu\text{F}$, which yields $L = 10$ mH and $R_T = 200$ Ω.

We have not introduced complex numbers simply to make things complex. Complex numbers arise quite naturally in practical physical situations involving nothing more than factoring a quadratic equation. The ability to deal with complex numbers is essential to our study. For those who need a review of such matters, there is a concise discussion in Appendix C.

FORM OF THE ZERO-INPUT RESPONSE

Since the characteristic equation has two roots, there are two solutions to the homogeneous differential equation:

$$v_{C1}(t) = K_1 e^{s_1 t}$$
$$v_{C2}(t) = K_2 e^{s_2 t}$$

That is,

$$LC \frac{d^2}{dt^2}(K_1 e^{s_1 t}) + R_T C \frac{d}{dt}(K_1 e^{s_1 t}) + K_1 e^{s_1 t} = 0$$

and

$$LC \frac{d^2}{dt^2}(K_2 e^{s_2 t}) + R_T C \frac{d}{dt}(K_2 e^{s_2 t}) + K_2 e^{s_2 t} = 0$$

The sum of these two solutions is also a solution since

$$LC \frac{d^2}{dt^2}(K_1 e^{s_1 t} + K_2 e^{s_2 t}) + R_T C \frac{d}{dt}(K_1 e^{s_1 t} + K_2 e^{s_2 t}) + K_1 e^{s_1 t} + K_2 e^{s_2 t} = 0$$

Therefore, the general solution for the zero-input response is of the form

$$v_C(t) = K_1 e^{s_1 t} + K_2 e^{s_2 t} \tag{7–41}$$

The constants K_1 and K_2 can be found using the initial conditions given in Eq. (7–34). At $t = 0$ the condition on the capacitor voltage yields

$$v_C(0) = V_0 = K_1 + K_2 \tag{7–42}$$

To use the initial condition on the inductor current, we differentiate Eq. (7–41).

$$\frac{dv_C}{dt} = K_1 s_1 e^{s_1 t} + K_2 s_2 e^{s_2 t}$$

Using Eq. (7–34) to relate the initial value of the derivative of the capacitor voltage to the initial inductor current $i(0)$ yields

$$\frac{dv_C(0)}{dt} = \frac{I_0}{C} = K_1 s_1 + K_2 s_2 \tag{7–43}$$

Equations (7–42) and (7–43) provide two equations in the two unknown constants K_1 and K_2:

$$K_1 + K_2 = V_0$$
$$s_1 K_1 + s_2 K_2 = I_0 / C$$

The solutions of these equations are

$$K_1 = \frac{s_2 V_0 - I_0/C}{s_2 - s_1} \text{ and } K_2 = \frac{-s_1 V_0 + I_0/C}{s_2 - s_1}$$

Inserting these solutions back into Eq. (7–41) yields

$$v_C(t) = \frac{s_2 V_0 - I_0/C}{s_2 - s_1} e^{s_1 t} + \frac{-s_1 V_0 + I_0/C}{s_2 - s_1} e^{s_2 t} \qquad t \geq 0 \qquad (7\text{–}44)$$

Equation (7–44) is the general zero-input response of the series RLC circuit. The response depends on two initial conditions V_0 and I_0, and the circuit parameters R_T, L, and C since s_1 and s_2 are the roots of the characteristic equation $LCs^2 + R_T Cs + 1 = 0$. The response takes on different forms depending on whether the roots s_1 and s_2 fall under case A, B, or C.

For case A the two roots are real and distinct. Using the notation $s_1 = -\alpha_1$ and $s_2 = -\alpha_2$, the form zero-input response for $t \geq 0$ is

$$v_C(t) = \left[\frac{\alpha_2 V_0 + I_0/C}{\alpha_2 - \alpha_1} \right] e^{-\alpha_1 t} + \left[\frac{-\alpha_1 V_0 - I_0/C}{\alpha_2 - \alpha_1} \right] e^{-\alpha_2 t} \qquad (7\text{–}45)$$

For case A the response is the sum of two exponential functions similar to the double exponential signal treated in Example 5–14. The function has two time constants $1/\alpha_1$ and $1/\alpha_2$. The time constants can be greatly different, or nearly equal, but they cannot be equal because we would have case B.

With case B the roots are real and equal. Using the notation $s_1 = s_2 = -\alpha$, the general form in Eq. (7–44) becomes

$$v_C(t) = \frac{(\alpha V_0 + I_0/C)e^{-\alpha t} + (-\alpha V_0 - I_0/C)e^{-\alpha t}}{\alpha - \alpha}$$

We immediately see a problem here because the denominator vanishes. However, a closer examination reveals that the numerator vanishes as well, so the solution reduces to the indeterminate form 0/0. To investigate the indeterminacy, we let $s_1 = -\alpha$ and $s_2 = -\alpha + x$, and we explore the situation as x approaches zero. Inserting s_1 and s_2 in this notation in Eq. (7–44) produces

$$v_C(t) = V_0 e^{-\alpha t} + \left[\frac{-\alpha V_0 - I_0/C}{x} \right] e^{-\alpha t} + \left[\frac{\alpha V_0 + I_0/C}{x} \right] e^{-\alpha t} e^{xt}$$

which can be arranged in the form

$$v_C(t) = e^{-\alpha t} \left[V_0 - (\alpha V_0 + I_0/C) \frac{1 - e^{xt}}{x} \right]$$

We see that the indeterminacy comes from the term $(1 - e^{xt})/x$, which reduces to 0/0 as x approaches zero. Application of l'Hôpital's rule reveals

$$\lim_{x \to 0} \frac{1 - e^{xt}}{x} = \lim_{x \to 0} \frac{-te^{xt}}{1} = -t$$

This result removes the indeterminacy, and as x approaches zero the zero-input response reduces to

$$v_C(t) = V_0 e^{-\alpha t} + (\alpha V_0 + I_0/C) te^{-\alpha t} \qquad t \geq 0 \qquad (7\text{–}46)$$

For case B the response includes an exponential and the damped ramp studied in Example 5–12. The damped ramp is required, rather than two exponentials, because in case B the two equal roots produce the same exponential function.

Case C produces complex conjugate roots of the form

$$s_1 = -\alpha - j\beta \quad \text{and} \quad s_2 = -\alpha + j\beta$$

Inserting these roots into Eq. (7–44) yields

$$v_C(t) = \left[\frac{(-\alpha + j\beta) V_0 - I_0/C}{j2\beta}\right] e^{-\alpha t} e^{-j\beta t} + \left[\frac{(\alpha + j\beta) V_0 + I_0/C}{j2\beta}\right] e^{-\alpha t} e^{j\beta t}$$

which can be arranged in the form

$$v_C(t) = V_0 e^{-\alpha t}\left[\frac{e^{j\beta t} + e^{-j\beta t}}{2}\right] + \frac{\alpha V_0 + I_0/C}{\beta} e^{-\alpha t}\left[\frac{e^{j\beta t} - e^{-j\beta t}}{j2}\right] \qquad (7\text{–}47)$$

The expressions within the brackets have been arranged in a special way for the following reasons. Euler's relationships for an imaginary exponential are written as

$$e^{j\theta} = \cos \theta + j \sin \theta$$

and

$$e^{-j\theta} = \cos \theta - j \sin \theta$$

When we add and subtract these equations, we obtain

$$\cos \theta = \frac{e^{j\theta} + e^{-j\theta}}{2} \quad \text{and} \quad \sin \theta = \frac{e^{j\theta} - e^{-j\theta}}{2j}$$

Comparing these expressions for $\sin \theta$ and $\cos \theta$ with the complex terms in Eq. (7–47) reveals that we can write $v_C(t)$ in the form

$$v_C(t) = V_0 e^{-\alpha t} \cos \beta t + \frac{\alpha V_0 + I_0/C}{\beta} e^{-\alpha t} \sin \beta t \qquad t \geqslant 0$$

For case C the response contains the damped sinusoid studied in Example 5–13. The real part of the roots (α) provides the exponent coefficient in the exponential function, while the imaginary part (β) defines the frequency of the sinusoidal oscillation.

In summary, the roots of the characteristic equation affect the form of the zero-input response in the following ways. In case A the two roots are real and unequal ($s_1 = -\alpha_1 \neq s_2 = -\alpha_2$) and the zero-input response is the sum of two exponentials of the form

$$v_C(t) = K_1 e^{-\alpha_1 t} + K_2 e^{-\alpha_2 t} \qquad (7\text{–}48a)$$

In case B the two roots are real and equal ($s_1 = s_2 = -\alpha$) and the zero-input response is the sum of an exponential and a damped ramp.

$$v_C(t) = K_1 e^{-\alpha t} + K_2 t e^{-\alpha t} \qquad (7\text{–}48b)$$

In case C the two roots are complex conjugates ($s_1 = -\alpha - j\beta$, $s_2 = -\alpha + j\beta$) and the zero-input response is the sum of a damped cosine and a damped sine.

$$v_C(t) = K_1 e^{-\alpha t} \cos \beta t + K_2 e^{-\alpha t} \sin \beta t \qquad (7\text{--}48c)$$

In determining the zero-input response we use the parameters s, α, and β. At various points in the development these parameters appear in expressions such as e^{st}, $e^{-\alpha t}$, and $e^{j\beta t}$. Since the exponent of e must be dimensionless, the parameters s, α, and β all have the dimensions of the reciprocal of time, or equivalently, frequency. Collectively, we say that s, α, and β define the **natural frequencies** of the circuit. When it is necessary to distinguish between these three parameters we say that s is the **complex frequency**, α is the **neper frequency**, and β is the **radian frequency**. The importance of this notation will become clear as we proceed through subsequent chapters of this book. To be consistent with expressions such as $s = -\alpha + j\beta$, we specify numerical values of s, α, and β in units of radians per second (rad/s).[3]

The constants K_1 and K_2 in Eqs. (7–48a), (7–48b), and (7–48c) are determined by the initial conditions on two state variables, as illustrated in the following example.

FIGURE 7-31

EXAMPLE 7-15

The circuit of Figure 7–31 has $C = 0.25\ \mu F$ and $L = 1\ H$. The switch has been open for a long time and is closed at $t = 0$. Find the capacitor voltage for $t \geq 0$ for (a) $R = 8.5\ k\Omega$, (b) $R = 4\ k\Omega$, and (c) $R = 1\ k\Omega$. The initial conditions are $I_0 = 0$ and $V_0 = 15\ V$.

SOLUTION:

The roots of the characteristic equation for these three values of resistance are found in Example 7–14. We are now in a position to use those results to find the corresponding zero-input responses.

(a) In Example 7–14 the value $R = 8.5\ k\Omega$ yields case A with roots $s_1 = -500$ and $s_2 = -8000$. The corresponding zero-input solution takes the form in Eq. (7–48a).

$$v_C(t) = K_1 e^{-500t} + K_2 e^{-8000t}$$

The initial conditions yield two equations in the constants K_1 and K_2:

$$v_C(0) = V_0 = 15 = K_1 + K_2$$

$$\frac{dv_C(0)}{dt} = \frac{I_0}{C} = 0 = -500K_1 - 8000K_2$$

Solving these equations yields $K_1 = 16$ and $K_2 = -1$, so that the zero-input response is

$$v_C(t) = 16e^{-500t} - e^{-8000t}\ V \qquad t \geq 0$$

---

3 The term *neper frequency* honors the sixteenth-century mathematician John Napier, who invented the base e or natural logarithms. The term *complex frequency* was apparently first used about 1900 by the British engineer Oliver Heaviside.

(b) In Example 7–14 the value $R = 4$ kΩ yields case B with roots $s_1 = s_2 = -2000$. The zero-input response takes the form in Eq. (7–48b):

$$v_C(t) = K_1 e^{-2000t} + K_2 t e^{-2000t}$$

The initial conditions yield two equations in the constants K_1 and K_2:

$$v_C(0) = V_0 = 15 = K_1$$

$$\frac{dv_C(0)}{dt} = \frac{I_0}{C} = 0 = -2000 K_1 + K_2$$

Solving these equations yields $K_1 = 15$ and $K_2 = 2000 \times 15$, so the zero-input response is

$$v_C(t) = 15 e^{-2000t} + 15(2000t) e^{-2000t} \text{ V} \qquad t \geq 0$$

(c) In Example 7–14 the value $R = 1$ kΩ yields case C with roots s_1, $s_2 = -500 \pm j500\sqrt{15}$. The zero-input response takes the form in Eq. (7–48c):

$$v_C(t) = K_1 e^{-500t} \cos(500\sqrt{15})t + K_2 e^{-500t} \sin(500\sqrt{15})t$$

The initial conditions yield two equations in the constants K_1 and K_2:

$$v_C(0) = V_0 = 15 = K_1$$

$$\frac{dv_C(0)}{dt} = \frac{I_0}{C} = 0 = -500 K_1 + 500\sqrt{15} K_2$$

which yield $K_1 = 15$ and $K_2 = \sqrt{15}$, so the zero-input response is

$$v_C(t) = 15 e^{-500t} \cos(500\sqrt{15})t + \sqrt{15} \, e^{-500t} \sin(500\sqrt{15})t \text{ V} \qquad t \geq 0$$

Figure 7–32 shows plots of these responses. All three responses start out at 15 V (the initial condition) and all eventually decay to zero. The temporal decay of the responses is caused by energy loss in the circuit and is called **damping**. The case A response does not change sign and is called the **overdamped** response. The case C response undershoots and then oscillates about the final value. This response is said to be **underdamped** because there is not enough damping to prevent these oscillations. The case B response is said to be **critically damped** since it is a special case at the boundary between overdamping and underdamping.

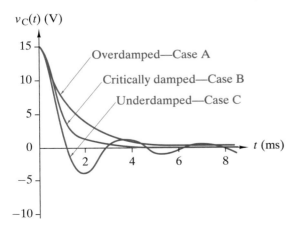

FIGURE 7–32

For a series RLC circuit, $R = 5$ kΩ, $C = 0.1$ μF, $L = 100$ mH, $V_0 = 0$ V, and $I_0 = 25$ mA.

(a) What is the capacitor voltage for $t \geq 0$?
(b) What is the inductor current for $t \geq 0$?
(c) What is the capacitor voltage if R is changed to 100 Ω?

Answers:
(a) $v_C(t) = 5.45(e^{-2087t} - e^{-47913t})$ V
(b) $i_L(t) = -1.14e^{-2087t} + 26.14e^{-47913t}$ mA
(c) $v_C(t) = 25.03e^{-500t} \sin 9987t$ V

In a series RLC circuit, the zero-input responses are:

$$v_C(t) = 10e^{-1000t} \sin 1000t \text{ V}$$

$$i_L(t) = 10e^{-1000t}(\cos 1000t - \sin 1000t) \text{ mA}$$

(a) What is the circuit characteristic equation?
(b) What are the initial values of the state variables?
(c) What are the values of R, L, and C?

Answers:
(a) $s^2 + 2000s + 2 \times 10^6 = 0$
(b) $V_0 = 0$ V, $I_0 = 10$ mA
(c) $R = 1$ kΩ, $L = 0.5$ H, $C = 1$ μF

7–6 THE PARALLEL RLC CIRCUIT

The inductor and capacitor in Figure 7–33(a) are connected in parallel. The source-resistor circuit can be reduced to the Norton equivalent shown in Figure 7–33(b). The result is a **parallel RLC circuit** consisting of a current source, resistor, inductor, and capacitor. Our first task is to develop a differential equation for this circuit. We expect to find a second-order differential equation because there are two energy storage elements.

The Norton equivalent to the left of the interface introduces the constraint

$$i + G_N v = i_N \tag{7–49}$$

Writing a KCL equation at the interface yields

$$i = i_L + i_C \tag{7–50}$$

The i–v characteristics of the inductor and capacitor are

$$i_C = C\frac{dv}{dt} \tag{7–51}$$

$$v = L\frac{di_L}{dt} \tag{7–52}$$

(a)

(b)

FIGURE 7–33   *The parallel RLC circuit.*

Equations (7–49) through (7–52) provide four independent equations in four unknowns (i, v, i_L, i_C). Collectively these equations describe the dynamics of the parallel *RLC* circuit. To solve for the circuit response using classical methods, we must derive a circuit equation containing only one of these four variables.

We prefer using state variables because they are continuous. To obtain a single equation in the inductor current, we substitute Eqs. (7–50) and (7–52) into Eq. (7–49):

$$i_L + i_C + G_N L \frac{di_L}{dt} = i_N \tag{7-53}$$

The capacitor current can be eliminated from this result by substituting Eq. (7–52) into Eq. (7–51) to obtain

$$i_C = LC \frac{d^2 i_L}{dt^2} \tag{7-54}$$

Inserting this equation into Eq. (7–53) produces

$$LC \frac{d^2 i_L}{dt^2} + G_N L \frac{di_L}{dt} + i_L = i_N \tag{7-55}$$

$$i_C \quad + \quad i_R \quad + i_L = i_N$$

This result is a KCL equation in which the resistor and capacitor currents are expressed in terms of the inductor current.

Equation (7–55) is a second-order linear differential equation of the same form as the series *RLC* circuit equation in Eq. (7–33). In fact, if we interchange the following quantities:

$$v_C \leftrightarrow i_L \qquad L \leftrightarrow C \qquad R_T \leftrightarrow G_N \qquad v_T \leftrightarrow i_N$$

we change one equation into the other. The two circuits are duals, which means that the results developed for the series case apply to the parallel circuit with the preceding duality interchanges.

However, it is still helpful to outline the major features of the analysis of the parallel *RLC* circuit. The initial conditions in the parallel circuit are the initial inductor current I_0 and capacitor voltage V_0. The initial inductor current provides the condition $i_L(0) = I_0$ for the differential equation in Eq. (7–55). By using Eq. (7–52), the initial capacitor voltage specifies the initial rate of change of the inductor current as

$$\frac{di_L(0)}{dt} = \frac{1}{L} v_C(0) = \frac{1}{L} V_0$$

These initial conditions are the dual of those obtained for the series *RLC* circuit in Eq. (7–34).

To solve for the zero-input response, we set $i_N = 0$ in Eq. (7–55) and obtain a homogeneous equation in the inductor current:

$$LC \frac{d^2 i_L}{dt^2} + G_N L \frac{di_L}{dt} + i_L = 0$$

A trial solution of the form $i_L = K e^{st}$ leads to the characteristic equation

$$LC s^2 + G_N L s + 1 = 0 \tag{7-56}$$

The characteristic equation is quadratic because there are two energy storage elements in the parallel RLC circuit. The characteristic equation has two roots:

$$s_1, s_2 = \frac{-G_N L \pm \sqrt{(G_N L)^2 - 4LC}}{2LC}$$

and, as in the series case, there are three distinct cases:

Case A: If $(G_N L)^2 - 4LC > 0$, there are two unequal real roots $s_1 = -\alpha_1$ and $s_2 = -\alpha_2$ and the zero-input response is the overdamped form

$$i_L(t) = K_1 e^{-\alpha_1 t} + K_2 e^{-\alpha_2 t} \qquad t \geqslant 0 \qquad (7\text{--}57)$$

Case B: If $(G_N L)^2 - 4LC = 0$, there are two real equal roots $s_1 = s_2 = -\alpha$ and the zero-input response is the critically damped form

$$i_L(t) = K_1 e^{-\alpha t} + K_2 t e^{-\alpha t} \qquad t \geqslant 0 \qquad (7\text{--}58)$$

Case C: If $(G_N L)^2 - 4LC < 0$, there are two complex, conjugate roots $s_1, s_2 = -\alpha \pm j\beta$ and the zero-input response is the underdamped form

$$i_L(t) = K_1 e^{-\alpha t} \cos \beta t + K_2 e^{-\alpha t} \sin \beta t \qquad t \geqslant 0 \qquad (7\text{--}59)$$

The analysis results for the series RLC circuit apply to the parallel RLC case with the appropriate duality replacements. In particular, the form of overdamped, critically damped, and underdamped response applies to both circuits. The forms of the responses in Eqs. (7–57), (7–58), and (7–59) have been written with two arbitrary constants K_1 and K_2. The following example shows how to evaluate these constants using the initial conditions for the two state variables.

E X A M P L E 7 – 1 6

In a parallel RLC circuit $R_T = 1/G_N = 500 \ \Omega$, $C = 1 \ \mu F$, $L = 0.2 \ H$. The initial conditions are $I_0 = 50 \ mA$ and $V_0 = 0$. Find the zero-input response of inductor current, resistor current, and capacitor voltage.

S O L U T I O N :
From Eq. (7–56) the circuit characteristic equation is

$$LCs^2 + G_N Ls + 1 = 2 \times 10^{-7}s^2 + 4 \times 10^{-4}s + 1 = 0$$

The roots of the characteristic equation are

$$s_1, s_2 = \frac{-4 \times 10^{-4} \pm \sqrt{16 \times 10^{-8} - 8 \times 10^{-7}}}{4 \times 10^{-7}} = -1000 \pm j2000$$

Since roots are complex conjugates, we have the underdamped case. The zero-input response of the inductor current takes the form of Eq. (7–59).

$$i_L(t) = K_1 e^{-1000t} \cos 2000t + K_2 e^{-1000t} \sin 2000t \qquad t \geqslant 0$$

The constants K_1 and K_2 are evaluated from the initial conditions. At $t = 0$ the inductor current reduces to

$$i_L(0) = I_0 = K_1 e^0 \cos 0 + K_2 e^0 \sin 0 = K_1$$

We conclude that $K_1 = I_0 = 50$ mA. To find K_2 we use the initial capacitor voltage. In a parallel *RLC* circuit the capacitor and inductor voltages are equal.

$$L\frac{di_L}{dt} = v_C(t)$$

In this example the initial capacitor voltage is zero, so the initial rate of change of inductor current is zero at $t = 0$. Differentiating the zero-input response produces

$$\frac{di_L}{dt} = -2000K_1 e^{-1000t}\sin 2000t - 1000K_1 e^{-1000t}\cos 2000t$$
$$- 1000K_2 e^{-1000t}\sin 2000t + 2000K_2 e^{-1000t}\cos 2000t$$

Evaluating this expression at $t = 0$ yields

$$\frac{di_L}{dt}(0) = -2000K_1 e^0 \sin 0 - 1000K_1 e^0 \cos 0$$
$$- 1000K_2 e^0 \sin 0 + 2000K_2 e^0 \cos 0$$
$$= -1000K_1 + 2000K_2 = 0$$

The derivative initial condition gives condition $K_2 = K_1/2 = 25$ mA. Given the values of K_1 and K_2, the zero-input response of the inductor current is

$$i_L(t) = 50e^{-1000t}\cos 2000t + 25e^{-1000t}\sin 2000t \text{ mA} \qquad t \geq 0$$

The zero-input response of the inductor current allows us to solve for every voltage and current in the parallel *RLC* circuit. For example, using the *i–v* characteristic of the inductor, we obtain the inductor voltage:

$$v_L(t) = L\frac{di_L}{dt} = -25e^{-1000t}\sin 2000t \text{ V} \qquad t \geq 0$$

Since the elements are connected in parallel, we obtain the capacitor voltage and resistor current as

$$v_C(t) = v_L(t) = L\frac{di_L}{dt} = -25e^{-1000t}\sin 2000t \text{ V} \qquad t \geq 0$$

$$i_R(t) = \frac{v_L(t)}{R} = -50e^{-1000t}\sin 2000t \text{ mA} \qquad t \geq 0 \qquad \blacksquare$$

EXAMPLE 7 – 17

The switch in Figure 7–34 has been open for a long time and is closed at $t = 0$.

(a) Find the initial conditions at $t = 0$.
(b) Find the inductor current for $t \geq 0$.
(c) Find the capacitor voltage and current through the switch for $t \geq 0$.

FIGURE 7–34

SOLUTION:

(a) For $t < 0$ the circuit is in the dc steady state, so the inductor acts like a short circuit and the capacitor like an open circuit. Since the inductor shorts out the capacitor, the initial conditions just prior to closing the switch at $t = 0$ are

$$v_C(0) = 0 \qquad i_L(0) = \frac{9}{250 + 50} = 30 \text{ mA}$$

(b) For $t \geq 0$ the circuit is a zero-input parallel RLC circuit with initial conditions found in (a). The circuit characteristic equation is

$$LCs^2 + GLs + 1 = 4 \times 10^{-6}s^2 + 2 \times 10^{-2}s + 1 = 0$$

The roots of this equation are

$$s_1 = -50.51 \quad \text{and} \quad s_2 = -4950$$

The circuit is overdamped (case A), since the roots are real and unequal. The general form of the inductor current zero-input response is

$$i_L(t) = K_1 e^{-50.51t} + K_2 e^{-4950t} \qquad t \geq 0$$

The constants K_1 and K_2 are found using the initial conditions. At $t = 0$ the zero-input response is

$$i_L(0) = K_1 e^0 + K_2 e^0 = K_1 + K_2 = 30 \times 10^{-3}$$

The initial capacitor voltage establishes an initial condition on the derivative of the inductor current since

$$L\frac{di_L}{dt}(0) = v_C(0) = 0$$

The derivative of the inductor response at $t = 0$ is

$$\frac{di_L}{dt}(0) = (-50.51K_1 e^{-50.51t} - 4950K_2 e^{-4950t})\Big|_{t=0}$$

$$= -50.51K_1 - 4950K_2 = 0$$

The initial conditions on inductor current and capacitor voltage produce two equations in the unknown constants K_1 and K_2:

$$K_1 + K_2 = 30 \times 10^{-3}$$

$$-50.51K_1 - 4950K_2 = 0$$

Solving these equations yields $K_1 = 30.3$ mA and $K_2 = -0.309$ mA. The zero-input response of the inductor current is

$$i_L(t) = 30.3e^{-50.51t} - 0.309e^{-4950t} \text{ mA} \qquad t \geq 0$$

(c) Given the inductor current in (b), the capacitor voltage is

$$v_C(t) = L\frac{di_L}{dt} = -1.53e^{-50.51t} + 1.53e^{-4950t} \text{ V} \qquad t \geq 0$$

For $t \geq 0$ the current $i_{SW}(t)$ is the current through the 50-Ω resistor plus the current through the 250-Ω resistor.

$$i_{SW}(t) = i_{250} + i_{50} = \frac{9}{250} + \frac{v_C(t)}{50}$$
$$= 36 - 30.6e^{-50.51t} + 30.6e^{-4950t} \text{ mA} \qquad t \geq 0 \qquad \blacksquare$$

Exercise 7–16

The zero-input responses of a parallel RLC circuit are observed to be

$$i_L(t) = 10te^{-2000t} \text{ A}$$
$$v_C(t) = 10e^{-2000t} - 20000te^{-2000t} \text{ V} \qquad t \geq 0$$

(a) What is the circuit characteristic equation?
(b) What are the initial values of the state variables?
(c) What are the values of R, L, and C?
(d) Write an expression for the current through the resistor.

Answers:
(a) $s^2 + 4000s + 4 \times 10^6 = 0$
(b) $i_L(0) = 0$, $v_C(0) = 10$ V
(c) $L = 1$ H, $C = 0.25$ μF, $R = 1$ kΩ
(d) $i_R(t) = 10e^{-2000t} - 20000te^{-2000t}$ mA $\qquad t \geq 0$

7–7 SECOND-ORDER CIRCUIT STEP RESPONSE

The step response provides important insights into the response of dynamic circuits in general. So it is natural that we investigate the step response of second-order circuits. In Chapter 11 we will develop general techniques for determining the step response of any linear circuit. However, in this introduction we use classical methods of solving differential equations to find the step response of second-order circuits.

The general second-order linear differential equation with a step function input has the form

$$a_2\frac{d^2y(t)}{dt^2} + a_1\frac{dy(t)}{dt} + a_0y(t) = Au(t) \qquad (7\text{–}60)$$

where $y(t)$ is a voltage or current response, $Au(t)$ is the step function input, and a_2, a_1, and a_0 are constant coefficients. The step response is the general solution of this differential equation for $t \geq 0$. The step response can be found by partitioning $y(t)$ into forced and natural components:

$$y(t) = y_N(t) + y_F(t) \qquad (7\text{–}61)$$

The natural response $y_N(t)$ is the general solution of the homogeneous equation (input set to zero), while the forced response $y_F(t)$ is a particular solution of the equation

$$a_2 \frac{d^2 y_F}{dt^2} + a_1 \frac{dy_F}{dt} + a_0 y_F = A \qquad t \geq 0$$

Since A is a constant, it follows that dA/dt and d^2A/dt^2 are both zero, so it is readily apparent that $y_F = A/a_0$ is a particular solution of this differential equation. So much for the forced response.

Turning now to the natural response, we seek a general solution of the homogeneous equation. The natural response has the same form as the zero-state response studied in the previous section. In a second-order circuit the zero-state and natural responses take one of the three possible forms: overdamped, critically damped, or underdamped. To describe the three possible forms, we introduce two new parameters: ω_0 (omega zero) and ζ (zeta). These parameters are defined in terms of the coefficients of the general second-order equation in Eq. (7–60):

$$\omega_0^2 = \frac{a_0}{a_2} \quad \text{and} \quad 2\zeta\omega_0 = \frac{a_1}{a_2} \qquad \text{(7–62)}$$

The parameter ω_0 is called the **undamped natural frequency** and ζ is called the **damping ratio**. Using these two parameters, the general homogeneous equation is written in the form

$$\frac{d^2 y_N(t)}{dt^2} + 2\zeta\omega_0 \frac{dy_N(t)}{dt} + \omega_0^2 y_N(t) = 0 \qquad \text{(7–63)}$$

The left side of Eq. (7–63) is called the **standard form** of the second-order linear differential equation. When a second-order equation is arranged in this format, we can determine its damping ratio and undamped natural frequency by equating its coefficients with those in the standard form. For example, in standard form the homogeneous equation for the series RLC circuit in Eq. (7–37) is

$$\frac{d^2 v_C}{dt^2} + \frac{R_T}{L} \frac{dv_C}{dt} + \frac{1}{LC} v_C = 0$$

Equating like terms yields

$$\omega_0^2 = \frac{1}{LC} \quad \text{and} \quad 2\zeta\omega_0 = \frac{R_T}{L}$$

for the series RLC circuit. In an analysis situation the circuit element values determine the value of the parameters ω_0 and ζ. In a design situation we select L and C to obtain a specified ω_0 and then select R_T to obtain a specified ζ.

To determine the form of the natural response using ω_0 and ζ, we insert a trial solution $y_N(t) = Ke^{st}$ into the standard form in Eq. (7–63). The trial function Ke^{st} is a solution provided that

$$Ke^{st}[s^2 + 2\zeta\omega_0 s + \omega_0^2] = 0$$

Since $K = 0$ is the trivial solution and $e^{st} \neq 0$ for all $t \geq 0$, the only useful way for the right side of this equation to be zero for all t is for the quadratic expression within the brackets to vanish. The quadratic expression is the characteristic equation for the general second-order differential equation:

$$s^2 + 2\zeta\omega_0 s + \omega_0^2 = 0$$

The roots of the characteristic equation are

$$s_1, s_2 = \omega_0(-\zeta \pm \sqrt{\zeta^2 - 1})$$

We begin to see the advantage of using the parameters ω_0 and ζ. The constant ω_0 is a scale factor that designates the size of the roots. The expression under the radical defines the form of the roots and depends only on the damping ratio ζ. As a result, we can express the three possible forms of the natural response in terms of the damping ratio.

Case A: For $\zeta > 1$ the discriminant is positive, there are two unequal, real roots

$$s_1, s_2 = -\alpha_1, -\alpha_2 = \omega_0(-\zeta \pm \sqrt{\zeta^2 - 1}) \qquad \text{(7–64a)}$$

and the natural response has the overdamped form

$$y_N(t) = K_1 e^{-\alpha_1 t} + K_2 e^{-\alpha_2 t} \qquad t \geqslant 0 \qquad \text{(7–64b)}$$

Case B: For $\zeta = 1$ the discriminant vanishes, there are two real, equal roots

$$s_1 = s_2 = -\alpha = -\zeta\omega_0 \qquad \text{(7–65a)}$$

and the natural response has the critically damped form

$$y_N(t) = K_1 e^{-\alpha t} + K_2 t e^{-\alpha t} \qquad t \geqslant 0 \qquad \text{(7–65b)}$$

Case C: For $\zeta < 1$ the discriminant is negative, leading to two complex, conjugate roots $s_1, s_2 = -\alpha \pm j\beta$, where

$$\alpha = \zeta\omega_0 \quad \text{and} \quad \beta = \omega_0\sqrt{1 - \zeta^2} \qquad \text{(7–66a)}$$

and the natural response has the underdamped form

$$y_N(t) = K_1 e^{-\alpha t} \cos \beta t + K_2 e^{-\alpha t} \sin \beta t \qquad t \geqslant 0 \qquad \text{(7–66b)}$$

Equations (7–64a), (7–65a), and (7–66a) provide relationships between the natural frequency parameters α and β and the new parameters ζ and ω_0. The reasons for using two equivalent sets of parameters to describe the natural frequencies of a second-order circuit will become clear as we continue our study of dynamic circuits. Since the units of complex frequency s are radians per second, the standard form of the characteristic equation $s^2 + 2\zeta\omega_0 s + \omega_0^2$ shows that ω_0 is specified in radians per second and ζ is dimensionless.

Combining the forced and natural responses yields the step response of the general second-order differential equation in the form

$$y(t) = y_N(t) + A/a_0 \qquad t \geqslant 0 \qquad \text{(7–67)}$$

The factor A/a_0 is the forced response. The natural response $y_N(t)$ takes one of the forms in Eqs. (7–64b), (7–65b), or (7–66b), depending on the value of the damping ratio. The constants K_1 and K_2 in natural response can be evaluated from the initial conditions.

In summary, the step response of a second-order circuit is determined by

1. The amplitude of the step function input $Au(t)$
2. The damping ratio ζ and natural frequency ω_0
3. The initial conditions $y(0)$ and dy/dt (0).

In this regard the damping ratio and natural frequency play the same role for second-order circuits that the time constant plays for first-order circuits. That is, these circuit parameters determine the basic form of the natural response, just as the time constant defines the form of the natural response in a first-order circuit. It is not surprising that a second-order circuit takes two parameters, since it contains two energy storage elements.

EXAMPLE 7–18

The series RLC circuit in Figure 7–35 is driven by a step function and is in the zero state at $t = 0$. Find the capacitor voltage for $t \geq 0$.

$V_A = 10$ V $C = 0.5\ \mu$F
$R = 1$ kΩ $L = 2$ H

FIGURE 7–35

SOLUTION:
This is a series RLC circuit, so the differential equation for the capacitor voltage is

$$10^{-6}\frac{d^2v_C}{dt^2} + 0.5 \times 10^{-3}\frac{dv_C}{dt} + v_C = 10 \qquad t \geq 0$$

By inspection, the forced response is $v_{CF} = 10$ V. In standard format the homogeneous equation is

$$\frac{d^2v_{CN}}{dt^2} + 500\frac{dv_{CN}}{dt} + 10^6\,v_{CN} = 0 \qquad t \geq 0$$

Comparing this format to the standard form in Eq. (7–63) yields

$$\omega_0^2 = 10^6 \quad \text{and} \quad 2\zeta\omega_0 = 500$$

so $\omega_0 = 1000$ and $\zeta = 0.25$. Since $\zeta < 1$, the natural response is underdamped (case C). Using Eqs. (7–66a) and (7–66b), we have

$$\alpha = \zeta\omega_0 = 250$$
$$\beta = \omega_0\sqrt{1 - \zeta^2} = 968$$
$$v_{CN}(t) = K_1 e^{-250t}\cos 968t + K_2 e^{-250t}\sin 968t$$

The general solution of the circuit differential equation is the sum of the forced and natural responses:

$$v_C(t) = 10 + K_1 e^{-250t}\cos 968t + K_2 e^{-250t}\sin 968t \qquad t \geq 0$$

The constants K_1 and K_2 are determined by the initial conditions. The circuit is in the zero state at $t = 0$, so the initial conditions are $v_C(0) = 0$ and $i_L(0) = 0$. Applying the initial condition constraints to the general solution yields two equations in the constants K_1 and K_2:

$$v_C(0) = 10 + K_1 = 0$$

$$\frac{dv_C}{dt}(0) = -250\,K_1 + 968\,K_2 = 0$$

These equations yield $K_1 = -10$ and $K_2 = -2.58$. The step response of the capacitor voltage step response is

$$v_C(t) = 10 - 10e^{-250t} \cos 968t - 2.58e^{-250t} \sin 968t \ \text{V} \qquad t \geq 0$$

A plot of $v_C(t)$ versus time is shown in Figure 7–36. The response and its first derivative at $t = 0$ satisfy the initial conditions. The natural response decays to zero, so the forced response determines the final value of $v_C(\infty)$ = 10 V. Beginning at $t = 0$ the response climbs rapidly but overshoots the mark several times before eventually settling down to the final value. The damped sinusoidal behavior results from the fact that $\zeta < 1$, producing an underdamped natural response. ■

$v_C(t)$

FIGURE   7 – 36

EXAMPLE 7–19

The step response of the series RLC circuit in Figure 7–35 is found analytically in Example 7–18. In this example we explore the effect of varying the resistance on the step response of this series RLC circuit. Use Electronics Workbench to calculate the zero-state step response of the capacitor voltage as R sweeps across the range from 1 kΩ to 8 kΩ with $L = 2$ H, $C = 0.5$ μF, and $V_A = 10$ V.

SOLUTION:

The Electronics Workbench circuit diagram in Figure 7–37 includes numerical values for the inductor, capacitor, voltage source, and the starting value of the resistance. The 10-V dc source simulates the step function input for $t \geq 0$. To get the zero-state response the initial inductor current and capacitor voltage must be set to zero. Note that the node voltage at node 3 is the voltage across the capacitor.

To obtain step responses as the resistance varies, we select the **Parameter Sweep** option from the **Analysis** menu. This brings up the parameter

FIGURE   7 – 37

FIGURE 7-38

(a)

(b)

sweep dialog box shown at the top of Figure 7–38. Here we select resistance **R1** as the component to be varied, specify an analysis **Start value** of 1 $k\Omega$ and an **End value** of 8 $k\Omega$. Specifying an octave **Sweep type** means that the parameter sweep values are obtained by doubling the start value until the end value is reached. That is, separate analyses will be run for R1 = 1, 2, 4, and 8 $k\Omega$. Finally, the **Output node** to be monitored is set to node 3, which is the voltage across the capacitor in the series RLC circuit in Figure 7–37.

With the parameter sweep conditions set, we next press the **Set transient analysis** button, which brings up the transient analysis dialog box shown at the bottom of Figure 7–38. In this box the **Initial conditions** are set to zero (to get the zero-state response), the **Start time** is set to zero, the **End time** set to 0.02 s = 20 ms, and the **Minimum number of time points** set to 100. Pressing the **Accept** button returns us to the parameter sweep dialog box, where pressing the **Simulate** button launches a sequence of transient analysis runs as R1 sweeps across the specified range.

FIGURE 7 - 39

Figure 7–39 shows the family of step responses for R1 = 1, 2, 4, and 8 kΩ. As should be expected, increasing the resistance increases the damping so that the step response goes from the underdamped case at R1 = 1 kΩ to being overdamped at R1 = 8 kΩ.

We can qualitatively check these results by calculating ω_0 and ζ.

$$\omega_0 = \frac{1}{\sqrt{LC}} = \frac{1}{\sqrt{2 \times 0.5 \times 10^{-6}}} = 1000 \text{ rad/s}$$

and

$$\zeta = \frac{R_1}{2\omega_0 L} = \frac{R_1}{4000}$$

Hence, $R_1 = 1$ kΩ yields $\zeta = 0.25$, $R_1 = 2$ kΩ yields $\zeta = 0.5$, $R_1 = 4$ kΩ yields $\zeta = 1$, and $R_1 = 8$ kΩ yields $\zeta = 2$. These calculations confirm that sweeping the resistance from 1 kΩ to 8 kΩ takes the step response from an underdamped case to an overdamped case, passing through the critically damped case ($\zeta = 1$) at $R_1 = 4$ kΩ. ■

Exercise 7–17

Find the zero-state solution of the following differential equations:

(a) $10^{-4}\dfrac{d^2v}{dt^2} + 2 \times 10^{-2}\dfrac{dv}{dt} + v = 100u(t)$

(b) $\dfrac{d^2i}{dt^2} + 2500\dfrac{di}{dt} + 4 \times 10^6 i = 10^5 u(t)$

Answers:

(a) $v(t) = 100 - 100e^{-100t} - 10^4 te^{-100t}$ V $t \geqslant 0$

(b) $i(t) = \dfrac{1}{40} - \dfrac{1}{40}e^{-1250t} \cos 1561t - \dfrac{1}{50}e^{-1250t} \sin 1561t$ A $t \geqslant 0$

Exercise 7–18

The step response of a series RLC circuit is observed to be

$$v_C(t) = 15 - 15e^{-1000t} \cos 1000t \text{ V} \qquad t \geq 0$$

$$i_L(t) = 45e^{-1000t} \cos 1000t + 45e^{-1000t} \sin 1000t \text{ mA} \qquad t \geq 0$$

(a) What is the circuit characteristic equation?
(b) What are the initial values of the state variables?
(c) What is the amplitude of the step input?
(d) What are the values of R, L, and C?
(e) What is the voltage across the resistor?

Answers:
(a) $s^2 + 2000s + 2 \times 10^6 = 0$
(b) $v_C(0) = 0$, $i_L(0) = 45$ mA
(c) $V_A = 15$ V
(d) $R = 333 \ \Omega$, $L = 167$ mH, $C = 3 \ \mu$F
(e) $v_R(t) = 15e^{-1000t} \cos 1000t + 15e^{-1000t} \sin 1000t$ V

7–8 SOME OTHER SECOND-ORDER CIRCUITS

Up to this point, the series and parallel RLC circuits are the only second-order circuits we have considered. Using the classical differential equation methods, we found that the form of the natural (zero-input) response can be overdamped, critically damped, or underdamped depending on the roots of the characteristic equation. We now show that these concepts apply to other second-order circuits besides the series and parallel RLC.

The OP AMP circuit in Figure 7–40 contains three equal resistors and two unequal capacitors. The circuit is second order because the two capacitors are not in series or parallel and cannot be replaced by a single equivalent capacitor. We use the basic concept of device and connection equations to formulate the circuit differential equation.

FIGURE 7–40 *A second-order OP AMP RC circuit.*

From our experience with OP AMP resistance circuits, we know that two node equations describe an OP AMP circuit of this type. We formulate these equations by writing KCL connection constraints at nodes A and B:

$$\text{Node A: } i_1 + i_2 + i_3 - i_4 = 0$$

$$\text{Node B: } \quad -i_2 + i_5 - i_N = 0$$

The resistor and capacitor element constraints in terms of the node voltages are

$$i_1 = G(v_S - v_A) \quad i_4 = C_1 \frac{dv_A}{dt}$$

$$i_2 = G(v_B - v_A) \quad i_5 = C_2 \frac{d(v_O - v_B)}{dt}$$

$$i_3 = G(v_O - v_A)$$

The ideal OP AMP model requires $i_N = 0$ and $v_B = 0$ because the noninverting input is connected to ground. Substituting all of these element equations into the KCL connection equations at nodes A and B results in

$$-3Gv_A - C_1\frac{dv_A}{dt} + Gv_O + Gv_S = 0$$

$$Gv_A + C_2\frac{dv_O}{dt} = 0$$

Solving the second of these equations for v_A yields

$$v_A = -RC_2\frac{dv_O}{dt}$$

Inserting this result into the first equation produces the circuit differential equation:

$$R^2C_1C_2\frac{d^2v_O}{dt^2} + 3RC_2\frac{dv_O}{dt} + v_O = -v_S$$

This expression is a second-order linear differential equation in which the OP AMP output voltage is the dependent variable.

In standard format the homogeneous equation for this circuit is

$$\frac{d^2v_O}{dt^2} + \frac{3}{RC_1}\frac{dv_O}{dt} + \frac{1}{R^2C_1C_2}v_O = 0$$

Comparing this expression with the standard form yields

$$\omega_0^2 = \frac{1}{R^2C_1C_2} \quad \text{and} \quad 2\zeta\omega_0 = \frac{3}{RC_1}$$

or

$$\omega_0 = \frac{1}{\sqrt{RC_1RC_2}} \quad \text{and} \quad \zeta = \frac{3}{2}\sqrt{\frac{C_2}{C_1}}$$

The natural frequency ω_0 of the circuit is determined by RC products and the damping ratio ζ is determined by C_2 and C_1. In particular, we see that the natural response has three forms.

Case A: If $9C_2 > 4C_1$ then $\zeta > 1$ and the natural response is overdamped.

Case B: If $9C_2 = 4C_1$ then $\zeta = 1$ and the natural response is critically damped.

Case C: If $9C_2 < 4C_1$ then $\zeta < 1$ and the natural response is underdamped.

The *RC* OP AMP circuit in Figure 7–40 produces the same response forms as the series and parallel *RLC* circuits without using an inductor. The next example shows that active *RC* circuits can produce responses that cannot be produced by passive *RLC* circuits.

FIGURE 7–41

EXAMPLE 7–20

Find the damping ratio and natural frequency of the second-order circuit in Figure 7–41.

SOLUTION:

The circuit differential equation is obtained from the connection and element constraints. Writing KCL equations at nodes A and B, we obtain the following connection constraints:

$$\text{Node A: } i_1 - i_2 + i_3 = 0$$

$$\text{Node B: } \quad i_2 - i_4 = 0$$

The element constraints are

$$i_1 = G(v_S - v_A) \quad i_3 = C\frac{d(v_O - v_A)}{dt}$$

$$i_2 = G(v_A - v_B) \quad i_4 = C\frac{dv_B}{dt}$$

$$v_O = \mu v_B$$

Substituting the element equations into the connection equations yields

$$\text{Node A: } Gv_S - 2Gv_A + Gv_B + \mu C\frac{dv_B}{dt} - C\frac{dv_A}{dt} = 0$$

$$\text{Node B: } \qquad\qquad Gv_A - Gv_B - C\frac{dv_B}{dt} = 0$$

We have two coupled first-order differential equations in node voltages v_A and v_B. To derive a single second-order equation, we need to eliminate one of these node voltages. Solving the node B equation for v_A produces

$$v_A = v_B + RC\frac{dv_B}{dt}$$

Substituting this result into the node A equation and using the fact that $v_O = \mu v_B$ yields the circuit differential equation.

$$\frac{d^2v_O}{dt^2} + \frac{3-\mu}{RC}\frac{dv_O}{dt} + \frac{1}{(RC)^2}v_O = \frac{\mu}{(RC)^2}v_S$$

Comparing this equation with the standard form of the second-order differential equation in Eq. (7–63) identifies the damping ratio and undamped natural frequency of the circuit.

$$\omega_0 = \frac{1}{RC} \qquad \zeta = \frac{3-\mu}{2}$$

The natural frequency is determined by the RC product and the damping ratio by the dependent source gain μ. There are the usual three responses:

Case A: If $\mu < 1$ then $\zeta > 1$ and the circuit is overdamped.

Case B: If $\mu = 1$ then $\zeta = 1$ and the circuit is critically damped.

Case C: If $3 > \mu > 1$ then $0 < \zeta < 1$ and the circuit is underdamped.

In addition, this active circuit has two response forms that are not possible in passive RLC circuits. If $\mu = 3$ then $\zeta = 0$ and there is no damping. The natural response has the form

$$v_{ON}(t) = K_1 \cos \omega_0 t + K_2 \sin \omega_0 t \qquad t \geq 0$$

No damping corresponds to a sustained oscillation where the energy to overcome the losses in the resistors comes from the dependent source. Note that when $\zeta = 0$ the frequency of the undamped sinusoid is ω_0 (hence the name undamped natural frequency).

Even more dramatic is the case $\mu > 3$ and $\zeta < 0$. When $\zeta < 0$ there is negative damping, which means that the exponentials in the natural response

$$v_{ON}(t) = K_1 e^{-\zeta \omega_0 t} \cos \beta t + K_2 e^{-\zeta \omega_0 t} \sin \beta t \qquad t \geq 0$$

grow without bound. A circuit whose natural response increases without bound is said to be **unstable**.   ∎

EXAMPLE 7−21

(a) Select element values in the active circuit in Figure 7−41 so the circuit is unstable with $\zeta = -0.1$ and $\omega_0 = 1000$ rad/s.
(b) Use Electronics Workbench to plot the zero-state output voltage for a unit step function input.

SOLUTION:

(a) In Example 7−20 we found that $\omega_0 = 1/RC$ and $\zeta = (3 - \mu)/2$. Selecting $R = 1$ kΩ, we can solve for the capacitance and gain as

$$C = \frac{1}{R\omega_0} = \frac{1}{1000 \times 1000} = 1\mu F$$

$$\mu = 3 - 2\zeta = 3.2$$

(b) Figure 7−42 shows the circuit diagram as constructed using Electronics Workbench schematic editor. The 1-V dc source simulates the step function input for $t \geq 0$. Note that the circuit output voltage is at node 4 in this diagram. Selecting **transient** from the **Analysis** menu brings up the transient analysis dialog box shown in Figure 7−42. In this box the **Initial conditions** are set to zero to obtain the zero-state response. We need to select an appropriate simulation interval to be able to observe this response. However, in this case the step response does not settle down on a final value because the circuit is unstable. The following analysis suggests a way to estimate an appropriate simulation interval when the response does not have a finite duration.

FIGURE 7–42

In this circuit the period of the undamped natural frequency is $2\pi/\omega_0$, or about $T_0 = 6.2$ ms. To observe about four cycles of the unstable response, we need a simulation time of $4T_0$ or about 25 ms. Therefore, in the Transient Analysis dialog box, we set **End Time (TSTOP)** = 0.025 (25 ms) and **Plotting increment (TSTEP)** = **End Time**/100 = 0.00025 (0.25 ms). Finally, to obtain the output voltage we add node 4 to the **Nodes for analysis** box. With the analysis setup completed, we press the **Simulate** button in the dialog box to initiate a transient analysis, the results of which are shown in the output response plot in Figure 7–42. Note that the step response is a negatively damped oscillation that, theoretically, increases in amplitude without bound. Practically speaking, the active element in the circuit will saturate at some finite level, leading to a sustained oscillation whose form cannot be predicted by the linear models we are using here. ■

SUMMARY

- Circuits containing linear resistors and the equivalent of one capacitor or one inductor are described by first-order differential equations in which the unknown is the circuit state variable.

- The zero-input response in a first-order circuit is an exponential whose time constant depends on circuit parameters. The amplitude of the exponential is equal to the initial value of the state variable.

- The natural response is the general solution of the homogeneous differential equation obtained by setting the input to zero. The forced response is a particular solution of the differential equation for the given input. For linear circuits the total response is the sum of the forced and natural responses.

- For linear circuits the total response is the sum of the zero-input and zero-state responses. The zero-input response is caused by the initial energy stored in capacitors or inductors. The zero-state response results from the input driving forces.

- The initial and final values of the step response of a first- and second-order circuit can be found by replacing capacitors by open circuits and inductors by short circuits and then using resistance circuit analysis methods.

- For a sinusoidal input the forced response is called the sinusoidal steady-state response, or the ac response. The ac response is a sinusoid with the same frequency as the input but with a different amplitude and phase angle. The ac response can be found from the circuit differential equation using the method of undetermined coefficients.

- Circuits containing linear resistors and the equivalent of two energy storage elements are described by second-order differential equations in which the dependent variable is one of the state variables. The initial conditions are the values of the two state variables at $t = 0$.

- The zero-input response of a second-order circuit takes different forms depending on the roots of the characteristic equation. Unequal real roots produce the overdamped response, equal real roots produce the critically damped response, and complex conjugate roots produce underdamped responses.

- The circuit damping ratio ζ and undamped natural frequency ω_0 determine the form of the zero-input and natural responses of any second-order circuit. The response is overdamped if $\zeta > 1$, critically damped if $\zeta = 1$, and underdamped if $\zeta < 1$. Active circuits can produce undamped ($\zeta = 0$) and unstable ($\zeta < 0$) responses.

- Computer-aided circuit analysis programs can generate numerical solutions for circuit transient responses. Some knowledge of analytical methods and an estimate of the general form of the expected response are necessary to use these analysis tools.

PROBLEMS

ERO 7-1 FIRST-ORDER CIRCUIT ANALYSIS (SECTS. 7-1, 7-2, 7-3, 7-4)

Given a first-order RC or RL circuit,

(a) Find the circuit differential equation, the circuit characteristic equation, the circuit time constant, and the initial conditions (if not given).
(b) Find the zero-input response.
(c) Find the forced and natural responses for step function and sinusoidal inputs.

7–1 For each of the following differential equations, find the function that satisfies the equation and the initial condition.

$$\text{(a)}\ \frac{dv(t)}{dt} + 1500v(t) = 0, \quad v(0) = -15\ \text{V}$$

$$\text{(b)}\ 10^{-4}\frac{di(t)}{dt} + 10^{-1}i(t) = 10^{-2}u(t), \quad i(0) = -20\ \text{mA}$$

7–2 Find the function that satisfies the following differential equation and the initial condition:

$$20\frac{dv(t)}{dt} + 500v(t) = 100\,(\cos 10t)u(t), \quad v(0) = 0\ \text{V}$$

7–3 Determine the time constant of each of the circuits in Figure P7–3.
7–4 Determine the time constant of each of the circuits in Figure P7–4.

FIGURE P7–3

FIGURE P7–4

FIGURE P7–5

7–5 The switch in Figure P7–5 has been closed for a long time and is opened at $t = 0$.
(a) Determine the value of the state variable at $t = 0$ and the value of the state variable as $t \to \infty$.
(b) Determine the circuit time constant.
(c) Write an expression for the state variable for $t \geq 0$ and sketch its waveform.
(d) Repeat (a), (b), and (c) when the capacitor is replaced by an inductor L.

7–6 Repeat Problem 7–5 when the switch has been open for a long time and then is closed at $t = 0$.

7–7 The switch in each circuit in Figure P7–7 has been in position A for a long time and is moved to position B at $t = 0$. For each circuit write an expression for the state variable for $t \geq 0$ and sketch its waveform.

7–8 Repeat Problem 7–7 when the switch in each circuit has been in position B for a long time and is moved to position A at $t = 0$.

FIGURE P7–7

7–9   The circuit in Figure P7–9 is in the zero state when the step function $V_Au(t)$ is applied. Find the output voltage response $v_O(t)$ for $t \geq 0$. Identify the forced and natural components and sketch their wave-forms.

7–10   The circuit in Figure P7–10 is in the zero state when the step function $V_Au(t)$ is applied. Find the output voltage response $v_O(t)$ for $t \geq 0$. Identify the forced and natural components and sketch their wave-forms.

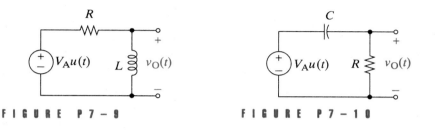

FIGURE P7-9   **FIGURE P7-10**

7–11   The switch in Figure P7–11 has been in position A for a long time and is moved to position B at $t = 0$. Write an expression for $v_C(t)$ for $t \geq 0$, identify the forced and natural components, and plot its wave-form.

FIGURE P7-11

7–12   Repeat Problem 7–11 when the switch has been in position B for a long time and is moved to position A at $t = 0$.

7–13   The input in Figure P7–13 is $v_S(t) = 15$ V. The switch has been open for a long time and is closed at $t = 0$. Solve for $v_C(t)$ for $t \geq 0$, identify the forced and natural components, and plot its waveform.

FIGURE P7-13

7–14   The input in Figure P7–13 is $v_S(t) = 15$ V. The switch has been open for a long time and is closed at $t = 0$. The switch is subsequently re-opened at $t = 200$ μs. Solve for $v_C(t)$ for $t \geq 0$ and plot its waveform.

7–15   Repeat Problem 7–13 for an input $v_S(t) = 20 \cos 10^4 t$.

7–16 The input in Figure P7–16 is $v_S(t) = 12$ V. The switch has been closed for a long time and is opened at $t = 0$. Solve for $i_O(t)$ for $t \geq 0$ and plot its waveform.

FIGURE P7–16

7–17 The input in Figure P7–16 is $v_S(t) = 10(\cos 10^6 t)$. The switch has been open for a long time and is closed at $t = 0$. Solve for $i_O(t)$ for $t \geq 0$ and plot its waveform.

7–18 The switch in Figure P7–18 has been in position A for a long time and is moved to position B at $t = 0$. Find $i_L(t)$ for $t \geq 0$ and plot its waveform.

7–19 The switch in Figure P7–19 has been in position A for a long time and is moved to position B at $t = 0$. Find $v_C(t)$ for $t \geq 0$ and plot its waveform.

FIGURE P7–18

FIGURE P7–19

7–20 Switches 1 and 2 in Figure P7–20 have both been in position A for a long time. Switch 1 is moved to position B at $t = 0$, and Switch 2 is moved to position B at $t = 20$ ms. Find the voltage across the 0.1-μF capacitor for $t > 0$ and plot its waveform.

FIGURE P7–20

ERO 7–2 FIRST-ORDER CIRCUIT RESPONSES (SECTS. 7–1, 7–2, 7–3)

Given the response of a first-order RC or RL circuit,

(a) Find the circuit characteristic equation, the circuit time constant, and the initial conditions.
(b) Design a first-order circuit that produces the given response.

7–21 The following function is the state variable response of a first-order *RC* circuit:

$$v_C(t) = 5 - 10e^{-1000t} \text{ V} \quad t \geq 0$$

(a) Determine the circuit characteristic equation, the circuit time constant, and the initial condition of the state variable.
(b) Determine the Thévenin equivalent input voltage.

7–22 The following function is the state variable response of a first-order *RL* circuit:

$$i_L(t) = 15 + 5e^{-2000t} \text{ mA} \quad t \geq 0$$

(a) Determine the circuit characteristic equation, the circuit time constant, and the initial condition of the state variable.
(b) Determine the Norton equivalent input current.

7–23 The following function is the state variable response of a first-order *RC* circuit:

$$v_C(t) = 10 \cos 1000t + 10 \sin 1000t - 5e^{-1000t} \text{ V} \quad t \geq 0$$

(a) Determine the circuit characteristic equation, the circuit time constant, and the initial condition.
(b) Determine the amplitude and phase angle of the Thévenin equivalent input voltage.

7–24 The following function is the state variable response of a first-order *RL* circuit:

$$i_L(t) = -12 + 6e^{-100t} \text{ mA} \quad t \geq 0$$

(a) Determine the circuit characteristic equation, the circuit time constant, and the initial condition.
(b) Determine the Norton equivalent input current.

7–25 The following function is the zero-input response of a first-order *RC* circuit. Note that the given response is *not* a state variable.

$$i_C(t) = -20e^{-1000t} \text{ mA} \quad t \geq 0$$

The initial value of the state variable is $v_C(0) = 12$ V.
(a) Determine the circuit characteristic equation and the circuit time constant.
(b) Determine the value of the circuit parameters R and C.

7–26 The following function is the zero-input response of a first-order *RL* circuit. Note that the given response is *not* a state variable.

$$v_L(t) = -10e^{-5000t} \text{ V} \quad t \geq 0$$

The initial value of the state variable is $i_L(0) = 5$ mA.
(a) Determine the circuit characteristic equation and the circuit time constant.
(b) Determine the value of the circuit parameters R and L.

7–27 For t > 0 the voltage across and current through the energy storage element in a first-order circuit are

$$v(t) = 10e^{-1000t} \text{ V} \text{ and } i(t) = 5 - 10e^{-1000t} \text{ mA}$$

(a) Is this an *RC* or an *RL* circuit?
(b) What is the initial value of the state variable?

(c) What is the Thévenin equivalent resistance seen by the energy storage element?

7–28 A 15-V step function input is applied to an RC circuit in the zero state. For $t > 0$ the output response is observed to be

$$v_O(t) = 15e^{-5000t} \text{ V}$$

What is the voltage across the capacitor?

7–29 ⓓ Design a first-order RC circuit that produces the following zero-state response.

$$v_C(t) = 10 - 10e^{-100t} \text{ V} \quad t \geqslant 0$$

7–30 ⓓ Design a first-order RL circuit that produces the response given in Problem 7–26.

ERO 7–3 SECOND-ORDER CIRCUIT ANALYSIS (SECTS. 7–5, 7–6, 7–7, 7–8)

Given a second-order circuit,

(a) Find the circuit differential equation, the circuit characteristic equation, circuit damping ratio and undamped natural frequency, and the initial conditions (if not given).

(b) Find the zero-input response.

(c) Find the forced and natural responses for a step function input.

7–31 Find the response $v(t)$ that satisfies the following differential equation and meets the initial conditions.

$$\frac{d^2v}{dt^2} + 8\frac{dv}{dt} + 16v = 0, \quad v(0) = 0 \text{ V}, \quad \frac{dv}{dt}(0) = 15 \text{ V/s}$$

7–32 Find the function that satisfies the following differential equation and the initial conditions.

$$\frac{d^2v}{dt^2} + 10\frac{dv}{dt} + 125v = 250u(t), \quad v(0) = 5 \text{ V}, \quad \frac{dv}{dt}(0) = 25 \text{ V/s}$$

7–33 Determine the numerical value of the circuit damping ratio ζ and the undamped natural frequency ω_0 for each circuit in Figure P7–33. Specify whether the circuits are overdamped, underdamped, or critically damped.

FIGURE P7-33

FIGURE P7-34

⌨ 7–34 The switch in Figure P7–34 has been open for a long time and is closed at $t = 0$. The circuit parameters are $L = 0.1$ H, $C = 25$ nF, $R_1 = 5$ kΩ, $R_2 = 5$ kΩ, and $V_A = 10$ V.

(a) Find the initial values of v_C and i_L at $t = 0$ and the final values of v_C and i_L as $t \to \infty$.

(b) Find the differential equation for $v_C(t)$ and the circuit characteristic equation.

(c) Find $v_C(t)$ and $i_L(t)$ for $t \geq 0$, identify the forced and natural components of the responses, and plot $v_C(t)$ and $i_L(t)$. Is the circuit overdamped or underdamped?

⌨ 7–35 The switch in Figure P7–34 has been closed for a long time and is opened at $t = 0$. The circuit parameters are $L = 32$ mH, $C = 50$ nF, $R_1 = 1$ kΩ, $R_2 = 1$ kΩ, and $V_A = 10$ V.

(a) Find the initial values of v_C and i_L at $t = 0$ and the final values of v_C and i_L as $t \to \infty$.

(b) Find the differential equation for $v_C(t)$ and the circuit characteristic equation.

(c) Find $v_C(t)$ and $i_L(t)$ for $t \geq 0$, identify the forced and natural components of the responses, and plot $v_C(t)$ and $i_L(t)$. Is the circuit overdamped or underdamped?

7–36 The switch in Figure P7–36 has been open for a long time and is closed at $t = 0$. The circuit parameters are $L = 0.5$ H, $C = 25$ nF, $R_1 = 5$ kΩ, $R_2 = 12$ kΩ, and $V_A = 10$ V.

(a) Find the initial values of v_C and i_L at $t = 0$ and the final values of v_C and i_L as $t \to \infty$.

(b) Find the differential equation for $v_C(t)$ and the circuit characteristic equation.

(c) Find $v_C(t)$ and $i_L(t)$ for $t \geq 0$, identify the forced and natural components of the responses, and $v_C(t)$ and $i_L(t)$. Is the circuit overdamped or underdamped?

FIGURE P7-36

7–37 The switch in Figure P7–36 has been closed for a long time and is opened at $t = 0$. The circuit parameters are $L = 0.5$ H, $C = 0.4$ μF, $R_1 = 1$ kΩ, $R_2 = 2$ kΩ, and $V_A = 15$ V.

(a) Find the initial values of v_C and i_L at $t = 0$ and the final values of v_C and i_L as $t \to \infty$.

(b) Find the differential equation for $v_C(t)$ and the circuit characteristic equation.

(c) Find $v_C(t)$ and $i_L(t)$ for $t \geq 0$, identify the forced and natural components of the responses, and plot $v_C(t)$ and $i_L(t)$. Is the circuit overdamped or underdamped?

7–38 The switch in Figure P7–38 has been open for a long time and is closed at $t = 0$. The circuit parameters are $L = 0.1$ H, $C = 0.1$ μF, $R_1 = 3$ kΩ, $R_2 = 3$ kΩ, and $V_A = 10$ V.

(a) Find the initial values of v_C and i_L at $t = 0$ and the final values of v_C and i_L as $t \to \infty$.

(b) Find the differential equation for $i_L(t)$ and the circuit characteristic equation.

(c) Find $v_C(t)$ and $i_L(t)$ for $t \geq 0$, identify the forced and natural components of the responses, and plot $v_C(t)$ and $i_L(t)$. Is the circuit overdamped or underdamped?

FIGURE P7-38

7–39 The switch in Figure P7–38 has been closed for a long time and is opened at $t = 0$. The circuit parameters are $L = 1$ H, $C = 10$ nF, $R_1 = 1$ kΩ, $R_2 = 3$ kΩ, and $V_A = 10$ V.

(a) Find the initial values of v_C and i_L at $t = 0$ and the final values of v_C and i_L as $t \to \infty$.

(b) Find the differential equation for $i_L(t)$ and the circuit characteristic equation.

(c) Find $v_C(t)$ and $i_L(t)$ for $t \geq 0$, identify the forced and natural components of the responses, and plot $v_C(t)$ and $i_L(t)$. Is the circuit overdamped or underdamped?

7–40 The switch in Figure P7–38 has been open for a long time and is closed at $t = 0$. The circuit parameters are $R_1 = 1$ kΩ, $R_2 = 5$ kΩ, $L = 100$ mH, $C = 50$ nF, and $V_A = 12$ V. Find $v_C(t)$ and $i_L(t)$ for $t > 0$, identify the forced and natural components of the responses, and plot $v_C(t)$ and $i_L(t)$. Is the circuit overdamped or underdamped?

FIGURE P.7-42

FIGURE P7-44

FIGURE P7-48

7–41 The switch in Figure P7–38 has been closed for a long time and is opened at $t = 0$. The circuit parameters are $R_1 = 5$ kΩ, $R_2 = 5$ kΩ, $L = 800$ mH, $C = 10$ nF, and $V_A = 12$ V. Find $v_C(t)$ and $i_L(t)$ for $t > 0$, identify the forced and natural components of the responses, and plot $v_C(t)$ and $i_L(t)$. Is the circuit overdamped or underdamped?

7–42 The switch in the circuit shown in Figure P7–42 has been in position A for a long time. At $t = 0$ it is moved to position B. The circuit parameters are $R_1 = 1$ kΩ, $R_2 = 4$ kΩ, $L = 0.625$ mH, $C = 6.25$ nF, and $V_A = 15$ V. Find $v_C(t)$ and $i_L(t)$ for $t > 0$, identify the forced and natural components of the responses, and plot $v_C(t)$ and $i_L(t)$. Is the circuit overdamped or underdamped?

7–43 The switch in the circuit shown in Figure P7–42 has been in position B for a long time. At $t = 0$ it is moved to position A. The circuit parameters are $R_1 = 1$ kΩ, $R_2 = 4$ kΩ, $L = 200$ mH, $C = 2$ nF, and $V_A = 15$ V. Find $v_C(t)$ and $i_L(t)$ for $t > 0$, identify the forced and natural components of the responses, and plot $v_C(t)$ and $i_L(t)$. Is the circuit overdamped or underdamped?

7–44 The switch in Figure P7–44 has been in position A for a long time and is moved to position B at $t = 0$. The circuit parameters are $R_1 = 1$ kΩ, $R_2 = 4$ kΩ, $L = 200$ mH, $C = 50$ nF, and $V_A = 5$ V. Find $v_C(t)$ and $i_L(t)$ for $t > 0$, identify the forced and natural components of the responses, and plot $v_C(t)$ and $i_L(t)$. Is the circuit overdamped or underdamped?

7–45 The switch in Figure P7–44 has been in position B for a long time and is moved to position A at $t = 0$. The circuit parameters are $R_1 = 500$ Ω, $R_2 = 2$ kΩ, $L = 10$ mH, $C = 10$ nF, and $V_A = 5$ V. Find $v_C(t)$ and $i_L(t)$ for $t > 0$, identify the forced and natural components of the responses, and plot $v_C(t)$ and $i_L(t)$. Is the circuit overdamped or underdamped?

7–46 The circuit in Figure P7–46 is in the zero state when the step function input is applied. The element values are $L = 50$ mH, $C = 50$ nF, $R = 4$ kΩ, and $V_A = 100$ V. Find the output $v_O(t)$.

7–47 The circuit in Figure P7–47 is in the zero state when the step function input is applied. The element values are $R = 4$ kΩ, $L = 160$ mH, $C = 25$ nF, and $V_A = 100$ V. Find the output $v_O(t)$.

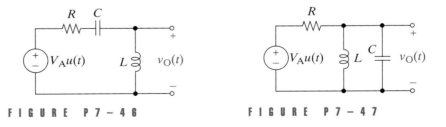

FIGURE P7-46   **FIGURE P7-47**

7–48 The circuit in Figure P7–48 is in the zero state when the step function input is applied. The element values are $R = 4$ kΩ, $L = 400$ mH, $C = 12.5$ nF, and $V_A = 100$ V. Find the output $v_O(t)$.

7–49 Derive expressions for the damping ratio and natural frequency of the *RLC* circuit in Figure P7–48 in terms of circuit parameters.

7–50 It is claimed that the circuit in Figure P7–50 is always critically damped regardless of the values of *R* and *C*. Prove or disprove this claim.

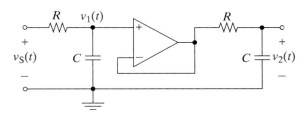

ERO 7–4 SECOND-ORDER CIRCUIT RESPONSES (SECTS. 7–5, 7–6, 7–7, 7–8)

Given the response of a second-order *RLC* circuit,

(a) Find the circuit differential equation, the circuit characteristic equation, circuit damping ratio and undamped natural frequency, and the initial conditions (if not given).

(b) Design a second-order circuit that produces the given response.

7–51 The following functions are the step responses of a series *RLC* circuit:

$$v_C(t) = 10 - 5e^{-3000t} - 5e^{-6000t} \text{ V} \quad t \ge 0$$

$$i_L(t) = 2e^{-3000t} + 4e^{-6000t} \text{ mA} \quad t \ge 0$$

(a) Determine the circuit damping ratio and undamped natural frequency.

(b) Find state variable initial and final conditions and the Thévenin equivalent input.

(c) Determine the value of *C*.

7–52 The following functions are the step responses of a series *RLC* circuit:

$$v_C(t) = 10 - 5e^{-100t} \sin 100t \text{ V} \quad t \ge 0$$

$$i_L(t) = 0.5e^{-100t}(\sin 100t - \cos 100t) \text{ mA} \quad t \ge 0$$

(a) Determine the circuit damping ratio and undamped natural frequency.

(b) Find state variable initial and final conditions and the Thévenin equivalent input.

(c) Determine the value of *C*.

7–53 The following functions are the step responses of a parallel *RLC* circuit:

$$v_C(t) = e^{-100t}(5 \cos 500t + 25 \sin 500t) \text{ V} \quad t \ge 0$$

$$i_L(t) = 20 - 25e^{-100t}\cos 500t \text{ mA} \quad t \ge 0$$

(a) Determine the circuit damping ratio and undamped natural frequency.

(b) Find state variable initial and final conditions and the Norton equivalent input.

(c) Determine the value of L.

7–54 The following functions are the zero-input responses of an *RLC* circuit:

$$v_C(t) = 2e^{-2000t} \cos 1000t - 4e^{-2000t} \sin 1000t \text{ V} \quad t \geq 0$$

$$i_L(t) = -80e^{-2000t} \cos 1000t + 60e^{-2000t} \sin 1000t \text{ mA} \quad t \geq 0$$

(a) Determine the circuit damping ratio and the undamped natural frequency.

(b) Is this a series *RLC* or a parallel *RLC* circuit?

7–55 The following functions are zero-input responses of a series *RLC* circuit. The initial value of the state variables are $v_C(0) = 0$ V and $i_L(0) = 2$ mA. Note that these responses are *not* state variables.

$$i_C(t) = 2e^{-2000t} - 2e^{-4000t} \text{ mA}$$

$$v_L(t) = -4e^{-2000t} + 8e^{-4000t} \text{ V}$$

(a) Determine the circuit characteristic equation, damping ratio, and undamped natural frequency.

(b) Use the inductor *i–v* characteristic to find the value of L.

(c) Use the value of L from (b) and the characteristic equation from (a) to find the values of R_T and C.

7–56 The response of a series *RLC* circuit with $\omega_0 = 2000$ rad/s and $\zeta = 1.25$ is

$$v_C(t) = 10e^{-500t} + 5e^{-1000t} - 10e^{-4000t} \text{ V} \quad t \geq 0$$

Find the Thévenin equivalent input.

7–57 The response of a parallel *RLC* circuit with $v_C(0) = -10$ V is

$$i_L(t) = 10e^{-10000t} \cos(20000t) \text{ mA} \quad t \geq 0$$

Find the values of L and C.

7–58 **D** Design a series *RLC* circuit whose zero-state output is

$$v_C(t) = V_A - 2V_A e^{-200t} + V_A e^{-800t} \text{ V} \quad t > 0$$

when the Thévenin equivalent input voltage is $v_T(t) = V_A u(t)$.

7–59 **D** Design a parallel *RLC* circuit whose zero-state output is

$$i_L(t) = I_A - I_A e^{-500t} - I_A(100t)e^{-500t} \text{ V} \quad t > 0$$

when the Norton equivalent input current is $i_N(t) = I_A u(t)$.

7–60 **D** The circuit in Figure P7–60 is a simplified diagram of a pulser that simulates lightning transients. When the switch closes the short-circuit current delivered by the pulser must be of the form $i_{SC}(t) = I_A e^{-\alpha t} \cos \beta t$, with $\alpha = 200$ krad/s, $\beta = 10$ Mrad/s, and $I_A = 2$ kA. Select the values of L, C, and the initial charge on the capacitor.

Pulser

FIGURE P7–60

INTEGRATING PROBLEMS

7–61 **D** *RC* CIRCUIT DESIGN

Design an *RC* circuit whose step response fits neatly in the unshaded region in Figure P7–61. The source you must use is a 1.5-V battery.

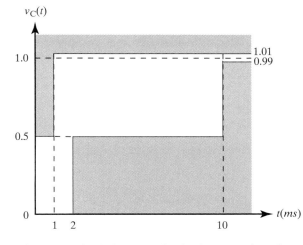

Your design must include a method of generating the required step function and can use resistors in the range from 1 kΩ to 100 kΩ. Test your design by finding its step response and calculating its value at $t = 1$ ms, 2 ms, and 10 ms.

7–62 🅐 🅓 RC CIRCUIT STEP RESPONSE

Figure P7–62 shows a general first-order RC circuit. In effect, the linear resistive circuit has two inputs, $v_S(t)$ the circuit input and $v_C(t)$ the voltage across the capacitor C. The superposition principle tells us that the circuit output can be written as $v_O(t) = K_1 v_S(t) + K_2 v_C(t)$. If the input and the capacitor voltages are

$$v_S(t) = V_A u(t) \quad \text{and} \quad v_C(t) = V_A(1 - e^{-t/RC})u(t)$$

what are the values of the proportionality constants K_1 and K_2 when the circuit output is

(a) $v_O(t) = V_A(1 - e^{-t/RC})u(t)$
(b) $v_O(t) = V_A e^{-t/RC}u(t)$
(c) $v_O(t) = V_A(1 - 0.5\, e^{-t/RC})u(t)$
(d) Devise a linear resistive circuit that produces each of these responses.

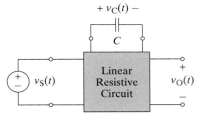

7–63 🅐 EXPERIMENTAL SECOND-ORDER RESPONSE

Figure P7–63 shows an oscilloscope display of the voltage across the resistor in a series RLC circuit.
(a) Estimate the values of α and β of the damped sine signal.
(b) Use the values of α and β from (a) to write the characteristic equation of the circuit.
(c) The resistor is known to be 2.2 kΩ. Use the characteristic equation from (b) to determine the values of L and C.
(e) If the display shows the zero-input response of the circuit, what are the values of the initial conditions $v_C(0)$ and $i_L(0)$?

7–64 🅐 SAMPLE HOLD CIRCUIT

Figure P7–64 shows a sample-hold circuit at the input of an analog-to-digital converter. In the sample mode the converter commands the

FIGURE P7 – 63

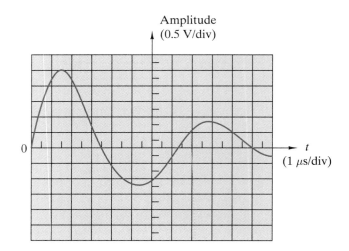

analog switch to close (ON) and the capacitor charges up to the value of the input signal. In the hold mode the converter commands the analog switch to open (OFF), and the capacitor holds and feeds the input signal value to the converter via the OP AMP voltage follower. When conversion is completed the analog switch is turned ON again and the sample-hold cycle repeats.

(a) The series resistances of the analog switch are $R_{ON} = 50\ \Omega$ and $R_{OFF} = 100\ M\Omega$. When $C = 20$ pF, what is the time constant in the sample mode and the time constant in the hold mode?

(b) The number of sample-hold cycles per second must be at least twice the highest frequency in the analog input signal. What is the minimum number of sample-hold cycles per second for an input $v_S(t) = 5 + 5 \sin 2\pi 1000t$?

(c) Sampling at 10 times the minimum number of sample-hold cycles per second, what is the duration of the sample mode if the hold mode lasts nine times as long as the sample mode?

(d) For the input in (b), will the capacitor voltage reach a steady-state condition during the sample mode?

(e) What fraction of the capacitor voltage will be lost during the hold mode?

FIGURE P7 – 64

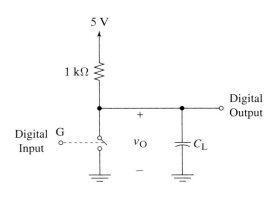

7–65 Ⓐ DIGITAL INVERTER RESPONSE TIMES

Figure P7–65 is a simplified circuit for a digital inverter driving a capacitive load $C_L = 100$ pF. In the steady-state the inverter output is low when the input is high and vice versa. Because of the capacitive load the transition from low-to-high or high-to-low does not occur instantaneously. The purpose of this problem is to determine the time required for the inverter to change its output state. The switch resistances are $R_{ON} = 100$ Ω (closed) and $R_{OFF} = 10^{11}$ Ω (open). The output is defined to be in the low state when $v_O < 0.5$ V and in the high state when $v_O > 3.5$ V.

(a) If the switch has been closed for a long time and is opened at $t = 0$, how long does it take for the output to reach the high state?

(b) If the switch has been open for a long time and is closed at $t = 0$, how long does it take for the output to reach the low state?

(c) Are the times found in (a) and (b) the same? If not, why not?

7–66 Ⓐ Ⓓ Ⓔ SECOND-ORDER CIRCUIT RESPONSES

When the input $x(t) = 5u(t)$ is applied to a linear second-order circuit the response of a state variable is

$$y(t) = 10e^{-100t} - 5e^{-200t} + 5 \quad t > 0$$

(a) What are the initial conditions $y(0)$ and $y'(0)$?

(b) For the same initial conditions, what is the response when the input is $x(t) = 20u(t)$?

(c) Design a simple RLC circuit that would produce all of these responses.

(d) It is claimed that the zero-state response of the circuit for an $x(t) = 10e^{-50t} u(t)$ is

$$y(t) = 10e^{-50t} - 15e^{-100t} + 5e^{-200t} \quad t > 0$$

Prove or disprove the claim.

CHAPTER 8

SINUSOIDAL STEADY-STATE RESPONSE

The vector diagram of sine waves gives the best insight into the mutual relationships of alternating currents and emf's.

Charles P. Steinmetz, 1893,
American Engineer

8-1 Sinusoids and Phasors

8-2 Phasor Circuit Analysis

8-3 Basic Circuit Analysis with Phasors

8-4 Circuit Theorems with Phasors

8-5 General Circuit Analysis with Phasors

8-6 Energy and Power

Summary

Problems

Integrating Problems

In this chapter we introduce the vector representation of sinusoids as a tool for finding the response of a linear circuit. The vector model was first discussed in detail by Charles Steinmetz (1865–1923) at the International Electric Congress of 1893. Although there is some evidence of earlier use by Oliver Heaviside, Steinmetz is generally credited with popularizing the vector approach by demonstrating its many applications to alternating current devices and systems. By the turn of the century, the concept was well established in engineering practice and education. In the 1950s the Steinmetz vector representation of sinusoids came to be called *phasors* to avoid possible confusion with the space vectors used to describe electromagnetic fields.

The response of a linear circuit consists of a natural component and a forced component. In a stable circuit the natural response eventually vanishes, leaving a sustained response caused by the input driving force. When the input is a sinusoid, the forced response is a sinusoid, with the same frequency as the input but with a different amplitude and phase angle. In electrical engineering the forced sinusoidal component remaining after the natural component disappears is called the **sinusoidal steady-state response**. The sinusoidal steady-state response is also called the **ac response** since the driving force is an alternating current signal.

The phasor defined in the first section of this chapter is the key concept in sinusoidal steady-state analysis. The second section examines the form taken by the device and connection constraints when sinusoidal voltages and currents are represented by phasors. It turns out that the device and connection constraints have the same form as resistive circuits. As a result, we can analyze circuits in the sinusoidal steady state using the analysis tools developed in Chapters 2, 3, and 4. The basic circuit analysis techniques, such as series equivalence and voltage division, are treated in the third section. The phasor domain versions of circuit theorems, such as superposition and Thévenin's theorem, are treated in the fourth section. The fifth section treats general phasor analysis methods using node voltages and mesh currents. The last section treats power transfer, an important application of phasor circuit analysis.

8-1   SINUSOIDS AND PHASORS

The phasor concept is the foundation for the analysis of linear circuits in the sinusoidal steady state. Simply put, a **phasor** is a complex number representing the amplitude and phase angle of a sinusoidal voltage or current. The connection between sinewaves and complex numbers is provided by Euler's relationship:

$$e^{j\theta} = \cos \theta + j \sin \theta \tag{8-1}$$

Equation (8–1) relates the sine and cosine functions to the complex exponential $e^{j\theta}$. To develop the phasor concept, it is necessary to adopt the point of view that the cosine and sine functions can be written in the form

$$\cos \theta = \text{Re}\{e^{j\theta}\} \tag{8-2}$$

and

$$\sin \theta = \text{Im}\{e^{j\theta}\} \tag{8-3}$$

where Re stands for the "real part of" and Im for the "imaginary part of." Development of the phasor concept can begin with either Eq. (8–2) or (8–3). The choice between the two involves deciding whether to describe the eternal sinewave using a sine or cosine function. In Chapter 5 we chose the cosine, so we will reference phasors to the cosine function.

When Eq. (8–2) is applied to the general sinusoid defined in Chapter 5, we obtain

$$
\begin{aligned}
v(t) &= V_A \cos(\omega t + \phi) \\
&= V_A \text{Re}\{e^{j(\omega t + \phi)}\} = V_A \text{Re}\{e^{j\omega t}e^{j\phi}\} \\
&= \text{Re}\{(V_A e^{j\phi})e^{j\omega t}\}
\end{aligned}
\tag{8–4}
$$

In the last line of Eq. (8–4), moving the amplitude V_A inside the real part operation does not change the final result because it is a real constant.

By definition, the quantity $V_A e^{j\phi}$ in the last line of Eq. (8–4) is the **phasor representation** of the sinusoid $v(t)$. The phasor $\mathbf{V}$ is written as

$$
\mathbf{V} = V_A e^{j\phi} = V_A \cos\phi + jV_A \sin\phi
\tag{8–5}
$$

Note that $\mathbf{V}$ is a complex number determined by the amplitude and phase angle of the sinusoid. Figure 8–1 shows a graphical representation commonly called a phasor diagram.

The phasor is a complex number that can be written in either polar or rectangular form. An alternative way to write the polar form is to replace the exponential $e^{j\phi}$ by the shorthand notation $\angle\phi$. In subsequent discussions, we will often express phasors as $\mathbf{V} = V_A\angle\phi$, which is equivalent to the polar form in Eq. (8–5).

Two features of the phasor concept need emphasis:

1. Phasors are written in boldface type like $\mathbf{V}$ or $\mathbf{I}_1$ to distinguish them from signal waveforms such as $v(t)$ and $i_1(t)$.

2. A phasor is determined by amplitude and phase angle and does not contain any information about the frequency of the sinusoid.

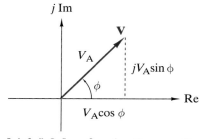

F I G U R E 8 – 1 *Phasor diagram.*

The first feature points out that signals can be described in different ways. Although the phasor $\mathbf{V}$ and waveform $v(t)$ are related concepts, they have different physical interpretations and our notation must clearly distinguish between them. The absence of frequency information in the phasors results from the fact that in the sinusoidal steady state, all currents and voltages are sinusoids with the same frequency. Carrying frequency information in the phasor would be redundant, since it is the same for all phasors in any given steady-state circuit problem.

In summary, given a sinusoidal signal $v(t) = V_A \cos(\omega t + \phi)$, the corresponding phasor representation is $\mathbf{V} = V_A e^{j\phi}$. Conversely, given the phasor $\mathbf{V} = V_A e^{j\phi}$, the corresponding sinusoid is found by multiplying the phasor by $e^{j\omega t}$ and reversing the steps in Eq. (8–4) as follows:

$$
\begin{aligned}
v(t) &= \text{Re}\{\mathbf{V}e^{j\omega t}\} = \text{Re}\{(V_A e^{j\phi})e^{j\omega t}\} \\
&= V_A \text{Re}\{e^{j(\omega t + \phi)}\} = V_A \text{Re}\{\cos(\omega t + \phi) + j\sin(\omega t + \phi)\} \\
&= V_A \cos(\omega t + \phi)
\end{aligned}
\tag{8–6}
$$

The frequency ω in the complex exponential $\mathbf{V}e^{j\omega t}$ in Eq. (8–6) must be expressed or implied in a problem statement, since by definition it is not contained in the phasor. Figure 8–2 shows a geometric interpretation of the complex exponential $\mathbf{V}e^{j\omega t}$ as a vector in the complex plane of length V_A, which rotates counterclockwise with a constant angular velocity of ω. The real part operation projects the rotating vector onto the horizontal (real) axis and thereby generates $v(t) = V_A \cos(\omega t + \phi)$. The complex exponential is sometimes called a **rotating phasor**, and the phasor $\mathbf{V}$ is viewed as a snapshot of the situation at $t = 0$.

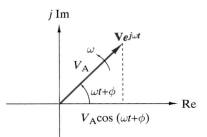

F I G U R E   8 – 2   *Complex exponential $\mathbf{V}e^{j\omega t}$.*

PROPERTIES OF PHASORS

Two important properties of phasors play key roles in circuit analysis. First, the **additive property** states that the phasor representing a sum of sinusoids of the same frequency is obtained by adding the phasor representations of the component sinusoids. To establish this property we write the expression

$$v(t) = v_1(t) + v_2(t) + \ldots + v_N(t)$$
$$= \text{Re}\{\mathbf{V}_1 e^{j\omega t}\} + \text{Re}\{\mathbf{V}_2 e^{j\omega t}\} + \ldots + \text{Re}\{\mathbf{V}_N e^{j\omega t}\} \qquad (8\text{–}7)$$

where $v_1(t)$, $v_2(t)$, . . . and $v_N(t)$ are sinusoids of the same frequency whose phasor representations are $\mathbf{V}_1$, $\mathbf{V}_2$, . . . and $\mathbf{V}_N$. The real part operation is additive, so the sum of real parts equals the real part of the sum. Consequently, Eq. (8–7) can be written in the form

$$v(t) = \text{Re}\{\mathbf{V}_1 e^{j\omega t} + \mathbf{V}_2 e^{j\omega t} + \ldots + \mathbf{V}_N e^{j\omega t}\}$$
$$= \text{Re}\{(\mathbf{V}_1 + \mathbf{V}_2 + \ldots + \mathbf{V}_N)e^{j\omega t}\} \qquad (8\text{–}8)$$

Comparing the last line in Eq. (8–8) with the definition of a phasor, we conclude that the phasor $\mathbf{V}$ representing $v(t)$ is

$$\mathbf{V} = \mathbf{V}_1 + \mathbf{V}_2 + \ldots + \mathbf{V}_N \qquad (8\text{–}9)$$

The result in Eq. (8–9) applies only if the component sinusoids all have the same frequency so that $e^{j\omega t}$ can be factored out as shown in the last line in Eq. (8–8).

In Chapter 5 we found that the time derivative of a sinusoid is another sinusoid with the same frequency. Since they have the same frequency the signal and its derivative can be represented by phasors. The **derivative property** of phasors allows us easily to relate the phasor representing a sinusoid to the phasor representing its derivative.

Equation (8–6) relates a sinusoid function and its phasor representation as

$$v(t) = \text{Re}\{\mathbf{V}e^{j\omega t}\}$$

Differentiating this equation with respect time t yields

$$\frac{dv(t)}{dt} = \frac{d}{dt}\text{Re}\{\mathbf{V}e^{j\omega t}\} = \text{Re}\left\{\mathbf{V}\frac{d}{dt}e^{j\omega t}\right\}$$
$$= \text{Re}\{(j\omega\mathbf{V})e^{j\omega t}\} \qquad (8\text{–}10)$$

From the definition of a phasor, we see that the quantity $(j\omega\mathbf{V})$ on the right side of this equation is the phasor representation of the time derivative of the sinusoidal waveform. This phasor can be written in the form

$$j\omega\mathbf{V} = (\omega e^{j90°})(V_A e^{j\theta})$$

$$= \omega V_A e^{j(\theta + 90°)}$$

(8–11)

which points out that differentiating a sinusoid changes its amplitude by a multiplicative factor ω and shifts the phase angle by 90°.

In summary, the additive property states that adding phasors is equivalent to adding sinusoids of the same frequency. The derivative property states that multiplying a phasor by $j\omega$ is equivalent to differentiating the corresponding sinusoid. The following examples show applications of these two properties of phasors.

EXAMPLE 8–1

(a) Construct the phasors for the following signals:

$$v_1(t) = 10\cos(1000t - 45°) \text{ V}$$

$$v_2(t) = 5\cos(1000t + 30°) \text{ V}$$

(b) Use the additive property of phasors and the phasors found in (a) to find $v(t) = v_1(t) + v_2(t)$.

SOLUTION:

(a) The phasor representations of $v_1(t)$ and $v_2(t)$ are

$$\mathbf{V}_1 = 10e^{-j45°} = 10\cos(-45°) + j10\sin(-45°)$$

$$= 7.07 - j7.07$$

$$\mathbf{V}_2 = 5\,e^{+j30°} = 5\cos(30°) + j5\sin(30°)$$

$$= 4.33 + j2.5$$

(b) The two sinusoids have the same frequency so the additive property of phasors can be used to obtain their sum:

$$\mathbf{V} = \mathbf{V}_1 + \mathbf{V}_2 = 11.4 - j4.57 = 12.3e^{-j21.8°}$$

The waveform corresponding to this phasor sum is

$$v(t) = \text{Re}\{(12.3e^{-j21.8°})e^{j1000t}\}$$

$$= 12.3\cos(1000t - 21.8°) \text{ V}$$

The phasor diagram in Figure 8–3 shows that summing sinusoids can be viewed geometrically in terms of phasors. ∎

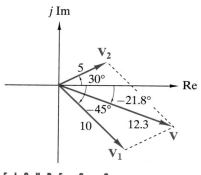

FIGURE 8 – 3

EXAMPLE 8–2

(a) Construct the phasors representing the following signals:

$$i_A(t) = 5 \cos(377t + 50°) \text{ A}$$

$$i_B(t) = 5 \cos(377t + 170°) \text{ A}$$

$$i_C(t) = 5 \cos(377t - 70°) \text{ A}$$

(b) Use the additive property of phasors and the phasors found in (a) to find the sum of these waveforms.

SOLUTION:

(a) The phasor representation of the three sinusoidal currents are

$$\mathbf{I}_A = 5e^{j50°} = 5 \cos(50°) + j5 \sin(50°) = 3.21 + j3.83 \text{ A}$$

$$\mathbf{I}_B = 5e^{j170°} = 5 \cos(170°) + j5 \sin(170°) = -4.92 + j0.87 \text{ A}$$

$$\mathbf{I}_C = 5e^{-j70} = 5 \cos(-70°) + j5 \sin(-70°) = 1.71 - j4.70 \text{ A}$$

(b) The currents have the same frequency, so the additive property of phasors applies. The phasor representing the sum of these currents is

$$\mathbf{I}_A + \mathbf{I}_B + \mathbf{I}_C = (3.21 - 4.92 + 1.71) + j(3.83 + 0.87 - 4.70)$$

$$= 0 + j0 \text{ A}$$

It is not obvious by examining the waveforms that these three currents add to zero. However, the phasor diagram in Figure 8–4 makes this fact clear, since the sum of any two phasors is equal and opposite to the third. Phasors of this type occur in balanced three-phase power systems, which we study in Chapter 16. The balanced condition occurs when three equal-amplitude phasors are displaced in phase by exactly 120°. ■

EXAMPLE 8–3

Use the derivative property of phasors to find the time derivative of $v(t) = 15 \cos(200t - 30°)$ V.

SOLUTION:

The phasor for the sinusoid is $\mathbf{V} = 15\angle-30°$. According to the derivative property, the phasor representing the dv/dt is found by multiplying $\mathbf{V}$ by $j\omega$.

$$(j200) \times (15\angle-30°) = (200e^{j90°}) \times (15e^{-j30°}) = 3000e^{j60°}$$

The sinusoid corresponding to the phasor $j\omega\mathbf{V}$ is

$$\frac{dv(t)}{dt} = \text{Re}\{(3000e^{j60°})e^{j200t}\} = 3000 \text{ Re}\{e^{j(200t + 60°)}\}$$

$$= 3000 \cos(200t + 60°) \text{ V/s}$$

Finding the derivative of a sinusoid is easily carried out in phasor form, since it only involves manipulating complex numbers. ■

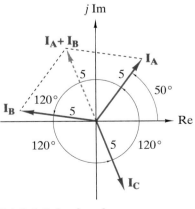

FIGURE 8–4

EXAMPLE 8–4

Use phasors to find the forced response of a first-order RC circuit with an input $v_T = V_A \cos \omega t$.

SOLUTION:

In Chapter 7 we found that the forced response $v_F(t)$ must satisfy the differential equation

$$R_T C \frac{dv_F}{dt} + v_F = V_A \cos \omega t$$

In Chapter 7 the forced response is found by substituting a trial solution $v_F = a \cos \omega t + b \sin \omega t$ into this differential equation and then solving for the unknown Fourier coefficients a and b. The forced response can also be written in phasor form as $v_F = \mathrm{Re}\,\{\mathbf{V}_F e^{j\omega t}\}$, where $\mathbf{V}_F$ is a phasor whose amplitude and phase angle are unknown. Substituting the phasor trial solution into the differential equation yields

$$(R_T C) \times \mathrm{Re}\{j\omega \mathbf{V}_F e^{j\omega t}\} + \mathrm{Re}\{\mathbf{V}_F e^{j\omega t}\} = \mathrm{Re}\{V_A e^{j\omega t}\}$$

where we use the derivative property to get the phasor for dv_F/dt. The unknown in this equation is the phasor $\mathbf{V}_F$. To solve for $\mathbf{V}_F$, we move the real constant $R_T C$ inside the real part operation. Likewise, the real part operation is additive, so the sum of real parts equals the real part of the sum. Applying these observations reduces the left side of this equation to

$$\mathrm{Re}\{(j\omega R_T C + 1)\mathbf{V}_F e^{j\omega t}\} = \mathrm{Re}\{V_A e^{j\omega t}\}$$

The only way this equality can hold is for the quantities multiplying $e^{j\omega t}$ on each side of this equation to be equal. Hence,

$$(j\omega R_T C + 1)\mathbf{V}_F = V_A$$

Solving for the unknown phasor yields

$$\mathbf{V}_F = \frac{V_A}{j\omega R_T C + 1}$$

$$= \frac{V_A e^{j\theta}}{\sqrt{1 + (\omega R_T C)^2}}$$

where $\theta = -\tan^{-1}(\omega R_T C)$. Given the amplitude and phase of the phasor forced response, we can now obtain the forced response as

$$v_F(t) = \mathrm{Re}\{\mathbf{V}_F e^{j\omega t}\} = \mathrm{Re}\left\{ \frac{V_A e^{j(\omega t + \theta)}}{\sqrt{1 + (\omega R_T C)^2}} \right\}$$

$$= \frac{V_A}{\sqrt{1 + (\omega R_T C)^2}} \cos(\omega t + \theta)$$

This solution agrees with the result given in Eq. (7–27) of Sect. 7–4. To see the advantages of using phasors, compare the classical methodology leading to Eq. (7–27) to the phasor approach used here. ∎

Exercise 8–1

Convert the following sinusoids to phasors in polar and rectangular form:

(a) $v(t) = 20 \cos(150t - 60°)$ V
(b) $v(t) = 10 \cos(1000t + 180°)$ V
(c) $i(t) = -4 \cos 3t + 3 \cos(3t - 90°)$ A

  Answers:

(a) $\mathbf{V} = 20 \angle -60° = 10 - j17.3$ V
(b) $\mathbf{V} = 10 \angle 180° = -10 + j0$ V
(c) $\mathbf{I} = 5 \angle -143° = -4 - j3$ A

Exercise 8–2

Convert the following phasors to sinusoids:

(a) $\mathbf{V} = 169 \angle -45°$ V at $f = 60$ Hz
(b) $\mathbf{V} = 10 \angle 90° + 66 - j10$ V at $\omega = 10$ krad/s
(c) $\mathbf{I} = 15 + j5 + 10 \angle 180°$ mA at $\omega = 1000$ rad/s

  Answers:

(a) $v(t) = 169 \cos(377t - 45°)$ V
(b) $v(t) = 66 \cos 10^4 t$ V
(c) $i(t) = 7.07 \cos(1000t + 45°)$ mA

Exercise 8–3

Express the phasors for $V_A \cos \omega t$, $V_A \sin \omega t$, $-V_A \cos \omega t$, and $-V_A \sin \omega t$ in rectangular form. *Hint:* Use the identity $\sin \theta = \cos(\theta - 90°)$.

  Answer:

$$V_A + j0,\; 0 - jV_A,\; -V_A + j0,\; 0 + jV_A.$$

Exercise 8–4

Express the phasors for the time derivatives of $V_A \cos \omega t$, $V_A \sin \omega t$, $-V_A \cos \omega t$, and $-V_A \sin \omega t$ in rectangular form. *Hint:* Use the results from Exercise 8–3.

  Answers:

$$0 + j\omega V_A,\; \omega V_A + j0,\; 0 - j\omega V_A,\; -\omega V_A + j0.$$

8–2   PHASOR CIRCUIT ANALYSIS

Phasor circuit analysis is a method of finding sinusoidal steady-state re-sponses directly from the circuit without using differential equations. How do we perform phasor circuit analysis? At several points in our study we have seen that circuit analysis is based on two kinds of constraints: (1) con-nection constraints (Kirchhoff's laws), (2) device constraints (element equations). To analyze phasor circuits, we must see how these constraints are expressed in phasor form.

CONNECTION CONSTRAINTS IN PHASOR FORM

The sinusoidal steady-state condition is reached after the circuit's natural response decays to zero. In the steady state all of the voltages and currents are sinusoids with the same frequency as the driving force. Under these conditions, the application of KVL around a loop could take the form

$$V_1 \cos(\omega t + \phi_1) + V_2 \cos(\omega t + \phi_2) \dots + V_N \cos(\omega t + \phi_N) = 0$$

These sinusoids have the same frequency but have different amplitudes and phase angles. The additive property of phasors discussed in the preceding section shows that there is a one-to-one correspondence between waveform sums and phasor sums. Therefore, if the sum of the waveforms is zero, then the corresponding phasors must also sum to zero.

$$\mathbf{V}_1 + \mathbf{V}_2 \dots + \mathbf{V}_N = 0$$

Clearly the same result applies to phasor currents and KCL. In other words, we can state Kirchhoff's laws in phasor form as follows:

> **KVL: *The algebraic sum of phasor voltages around a loop is zero.***

> **KCL: *The algebraic sum of phasor currents at a node is zero.***

DEVICE CONSTRAINTS IN PHASOR FORM

Turning now to the device constraints, we note that the i–v characteristics of the three passive elements are

$$\text{Resistor: } v_R(t) = R i_R(t)$$

$$\text{Inductor: } v_L(t) = L \frac{d i_L(t)}{dt} \qquad (8\text{–}12)$$

$$\text{Capacitor: } i_C(t) = C \frac{d v_C(t)}{dt}$$

In the sinusoidal steady state, all of these currents and voltages are sinusoids. Given that the signals are sinusoid, how do these i–v relationships constrain the corresponding phasors?

In the sinusoidal steady state, the voltage and current of the resistor can be written in terms of phasors as $v_R(t) = \text{Re } \{\mathbf{V}_R e^{j\omega t}\}$ and $i_R(t) = \text{Re } \{\mathbf{I}_R e^{j\omega t}\}$. Consequently, the resistor i–v relationship in Eq. (8–12) can be expressed in terms of phasors as follows:

$$\text{Re}\{\mathbf{V}_R e^{j\omega t}\} = R \times \text{Re}\{\mathbf{I}_R e^{j\omega t}\}$$

Since R is a real constant, moving it inside the real part operation on the right side of this equation does not change things:

$$\text{Re}\{\mathbf{V}_R e^{j\omega t}\} = \text{Re}\{R \mathbf{I}_R e^{j\omega t}\}$$

This relationship holds only if the phasor voltage and current for a resistor are related as

$$\mathbf{V}_R = R \mathbf{I}_R \qquad (8\text{–}13)$$

To explore this relationship, we assume that the current through a resistor is $i_R(t) = I_A \cos(\omega t + \phi)$. Then the phasor current is $\mathbf{I}_R = I_A e^{j\phi}$, and according to Eq. (8–13), the phasor voltage across the resistor is

$$\mathbf{V}_R = RI_A e^{j\phi}$$

This result shows that the voltage has the same phase angle (ϕ) as the current. Phasors with the same phase angle are said to be **in phase**; otherwise they are said to be **out of phase**. Figure 8–5 shows the phasor diagram of the resistor current and voltage. Two scale factors are needed to construct a phasor diagram showing both voltage and current, since the two phasors do not have the same dimensions.

F I G U R E   8 – 5 *Phasor   i–v characteristics of the resistor.*

In the sinusoidal steady state, the voltage and phasor current for the inductor can be written in terms of phasors as $v_L(t) = \mathrm{Re}\,\{\mathbf{V}_L e^{j\omega t}\}$ and $i_L(t) = \mathrm{Re}\,\{\mathbf{I}_L e^{j\omega t}\}$. Using the derivative property of phasors, the inductor i–v relationship in Eq. (8–12) can be expressed as follows:

$$\mathrm{Re}\{\mathbf{V}_L e^{j\omega t}\} = L \times \mathrm{Re}\{j\omega \mathbf{I}_L e^{j\omega t}\}$$
$$= \mathrm{Re}\{j\omega \mathbf{I}_L e^{j\omega t}\}$$

Since L is a real constant, moving it inside the real part operation does not change things. Written this way, we see that the phasor voltage and current for an inductor are related as

$$\mathbf{V}_L = j\omega L \mathbf{I}_L \qquad\qquad (8\text{–}14)$$

When the current is $i_L(t) = I_A \cos(\omega t + \phi)$ the corresponding phasor is $\mathbf{I}_L = I_A e^{j\phi}$ and the i–v constraint in Eq. (8–14) yields

$$\mathbf{V}_L = j\omega L \mathbf{I}_L = (\omega L e^{j90°})\,(I_A e^{j\phi})$$
$$= \omega L I_A e^{j(\phi + 90°)}$$

The resulting phasor diagram in Figure 8–6 shows that the inductor voltage and current are 90° out of phase. The voltage phasor is advanced by 90° counterclockwise, which is in the direction of rotation of the complex exponential $e^{j\omega t}$. When the voltage phasor is advanced counterclockwise (that is, ahead of the rotating current phasor), we say that the voltage phasor *leads* the current phasor by 90° or, equivalently, the current *lags* the voltage by 90°.

F I G U R E   8 – 6 *Phasor   i–v characteristics of the inductor.*

Finally, the capacitor voltage and current in the sinusoidal steady state can be written in terms of phasors as $v_C(t) = \mathrm{Re}\,\{\mathbf{V}_C e^{j\omega t}\}$ and $i_C(t) = \mathrm{Re}\,\{\mathbf{I}_C e^{j\omega t}\}$. Using the derivative property of phasors, the i–v relationship of the capacitor in Eq. (8–12) becomes

$$\mathrm{Re}\{\mathbf{I}_C e^{j\omega t}\} = C \times \mathrm{Re}\{j\omega \mathbf{V}_C e^{j\omega t}\}$$
$$= \mathrm{Re}\{j\omega C \mathbf{V}_C e^{j\omega t}\}$$

Moving the real constant C inside the real part operation does not change the final result, so we conclude that the phasor voltage and current for a capacitor are related as

$$\mathbf{I}_C = j\omega C \mathbf{V}_C$$

Solving for $\mathbf{V}_C$ yields

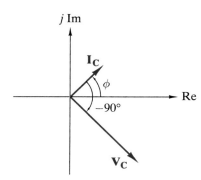

F I G U R E 8 – 7 *Phasor i–v characteristics of the capacitor.*

$$\mathbf{V}_C = \frac{1}{j\omega C}\mathbf{I}_C \qquad (8\text{–}15)$$

When $i_C(t) = I_A \cos(\omega t + \phi)$, then according to Eq. (8–15), the phasor voltage across the capacitor is

$$\mathbf{V}_C = \frac{1}{j\omega C}\mathbf{I}_C = \left(\frac{1}{\omega C}e^{-j90°}\right)(I_A e^{j\phi})$$

$$= \frac{I_A}{\omega C}e^{j(\phi - 90°)}$$

The resulting phasor diagram in Figure 8–7 shows that voltage and current are 90° out of phase. In this case, the voltage phasor is retarded by 90° clockwise, which is in a direction opposite to rotation of the complex exponential $e^{j\omega t}$. When the voltage is retarded clockwise (that is, behind the rotating current phasor), we say that the voltage phasor *lags* the current phasor by 90° or, equivalently, the current *leads* the voltage by 90°.

THE IMPEDANCE CONCEPT

The phasor **I–V** constraints in Eqs. (8–13), (8–14), and (8–15) are all of the form

$$\mathbf{V} = Z\mathbf{I} \qquad (8\text{–}16)$$

where Z is called the impedance of the element. Equation (8–16) is analogous to Ohm's law in resistive circuits. **Impedance** is the proportionality constant relating phasor voltage and phasor current in linear, two-terminal elements. The impedances of the three passive elements are

$$\text{Resistor: } Z_R = R$$

$$\text{Inductor: } Z_L = j\omega L \qquad (8\text{–}17)$$

$$\text{Capacitor: } Z_C = \frac{1}{j\omega C} = -\frac{j}{\omega C}$$

Since impedance relates phasor voltage to phasor current, it is a complex quantity whose units are ohms. Although impedance can be a complex number, it is not a phasor. Phasors represent sinusoidal signals, while impedances characterize circuit elements in the sinusoidal steady state. Finally, it is important to remember that the generalized two-terminal device constraint in Eq. (8–16) assumes that the passive sign convention is used to assign the reference marks to the voltage and current.

EXAMPLE 8–5

The circuit in Figure 8–8 is operating in the sinusoidal steady state with $v_S(t) = 50 \cos(4000t - 20°)$ V and $i_1(t) = 0.5 \cos(4000t)$ A. Find the impedance of the elements in the rectangular box.

SOLUTION:

The purpose of this example is to show that we can apply basic circuit analysis tools to phasors. The phasors representing the given signals are $\mathbf{V}_S$

FIGURE 8-8

$R_S = 50 \ \Omega$
$R_L = 100 \ \Omega$

$= 50\angle{-20°}$ V and $\mathbf{I}_1 = 0.5\angle0°$ A. Using Ohm's law in phasor form, we find that $\mathbf{V}_1 = R_S \mathbf{I}_1 = 25\angle0°$ V. Applying KVL around loop A yields $\mathbf{V}_S - \mathbf{V}_1 - \mathbf{V}_2 = 0$. Solving for the unknown voltage yields

$$\mathbf{V}_2 = \mathbf{V}_S - \mathbf{V}_1 = 50\angle{-20°} - 25\angle0° = 27.8\angle{-37.9°} \text{ V}$$

This voltage appears across the load resistor R_L; hence $\mathbf{I}_3 = \mathbf{V}_2/R_L = 0.278\angle{-37.9°}$ A. Applying KCL at node A yields $\mathbf{I}_2 = \mathbf{I}_1 - \mathbf{I}_3 = 0.5\angle0° - 0.278\angle{-37.9°} = 0.328\angle31.4°$ A. Given the phasor voltage across and current through the unknown element, we find the impedance to be

$$Z = \frac{\mathbf{V}_2}{\mathbf{I}_2} = \frac{27.8\angle{-37.9°}}{0.328\angle31.4} = 84.8\angle{-69.3°}$$

$$= 30.1 - j79.2 \ \Omega$$

This complex number is the impedance of the unknown elements inside the rectangular box in Figure 8–8. ■

Exercise 8-5

The circuit in Figure 8–9 is operating in the sinusoidal steady state with $v_A(t) = 10 \cos \omega t$ V, $v_B(t) = 10 \sin \omega t$ V, $i_1 = \sqrt{2} \cos(\omega t + 135°)$ A, and $i_4 = \cos \omega t$ A. Use KCL and KVL to find the phasor voltage across and current through each element. Use the passive sign convention.

Answers:

$\mathbf{V}_1 = 10\angle0°$ V, $\mathbf{V}_2 = 10 + j\,10$ V, $\mathbf{V}_3 = \mathbf{V}_4 = 10\angle{-90°}$ V, $\mathbf{I}_1 = -1 + j$ A, $\mathbf{I}_2 = 1 - j$ A,

$\mathbf{I}_3 = 1\angle{-90°}$ A, $\mathbf{I}_4 = 1 + j0$ A

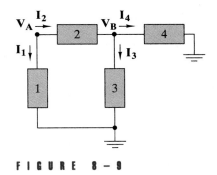

FIGURE 8-9

Exercise 8-6

The current through a 12-mH inductor is $i_L(t) = 20 \cos(10^6 t)$ mA. Determine

(a) The impedance of the inductor
(b) The phasor voltage across the inductor
(c) The waveform of the voltage across the inductor.

Answers:
(a) $Z_L = j12 \ k\Omega$
(b) $\mathbf{V}_L = 240\angle90°$ V
(c) $v_L = 240 \cos(10^6 t + 90°)$ V

Exercise 8–7

The current through a 20-pF capacitor is $i_C(t) = 0.3 \cos(10^6 t)$ mA.

(a) Find the impedance of the capacitor.
(b) Find the phasor voltage across the capacitor.
(c) Find the waveform of the voltage across the capacitor.

Answers:
(a) $Z_C = -j50$ kΩ
(b) $\mathbf{V}_C = 15\angle{-90°}$ V
(c) $v_C = 15 \cos(10^6 t - 90°)$ V

8–3 B A S I C   C I R C U I T   A N A L Y S I S   W I T H   P H A S O R S

Functions of time like $v(t) = V_A \cos(\omega t + \phi_V)$ and $i(t) = I_A \cos(\omega t + \phi_I)$ are time-domain representations of sinusoidal signals. Producing the corresponding phasors can be thought of as a transformation that carries $v(t)$ and $i(t)$ into a complex-number domain where signals are represented as phasors **V** and **I**. We call this complex-number domain the **phasor domain**. When we analyze circuits in this phasor domain, we obtain sinusoidal steady-state responses in terms of phasors like **V** and **I**. Performing the inverse phasor transformation as $v(t) = \text{Re}\{\mathbf{V}e^{j\omega t}\}$ and $i(t) = \text{Re}\{\mathbf{I}e^{j\omega t}\}$ carries the responses back into the time domain. To perform ac circuit analysis in this way, we obviously need to develop methods of analyzing circuits in the phasor domain.

In the preceding section, we showed that KVL and KCL apply in the phasor domain and that the phasor element constraints all have the form **V** = Z**I**. These element and connection constraints have the same format as the underlying constraints for resistance circuit analysis as developed in Chapters 2, 3, and 4. Therefore, familiar algebraic circuit analysis tools, such as series and parallel equivalence, voltage and current division, proportionality and superposition, and Thévenin and Norton equivalent circuits, are applicable in the phasor domain. In other words, we do not need new analysis techniques to handle circuits in the phasor domain. The only difference is that circuit responses are phasors (complex numbers) rather than dc signals (real numbers).

We can think of phasor domain circuit analysis in terms of the flow diagram in Figure 8–10. The analysis begins in the time domain with a linear circuit operating in the sinusoidal steady state and involves three major steps:

STEP 1: The circuit is transformed into the phasor domain by representing the input and response sinusoids as phasors and the passive circuit elements by their impedances.

STEP 2: Standard algebraic circuit analysis techniques are applied to solve the phasor domain circuit for the desired unknown phasor responses.

STEP 3: The phasor responses are inverse transformed back into time-domain sinusoids to obtain the response waveforms.

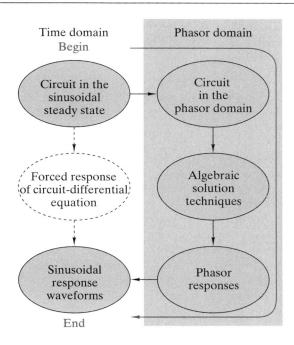

FIGURE 8 – 10 *Flow diagram for phasor circuit analysis.*

The third step assumes that the required end product is a time-domain waveform. However, a phasor is just another representation of a sinusoid. With some experience, we learn to think of the response as a phasor without converting it back into a time-domain waveform.

Figure 8–10 points out that there is another route to time-domain response using the classical differential equation method from Chapter 7. However, the phasor-domain method works directly with the circuit model and is far simpler. More important, phasor-domain analysis provides insights into ac circuit analysis that are essential to understanding much of the terminology and viewpoint of electrical engineering.

SERIES EQUIVALENCE AND VOLTAGE DIVISION

We begin the study of phasor-domain analysis with two basic analysis tools—series equivalence and voltage division. In Figure 8–11 the two-terminal elements are connected in series, so by KCL, the same phasor current **I** exists in impedances $Z_1, Z_2, \ldots Z_N$. Using KVL and the element constraints, the voltage across the series connection can be written as

$$\mathbf{V} = \mathbf{V}_1 + \mathbf{V}_2 + \ldots + \mathbf{V}_N$$
$$= Z_1\mathbf{I} + Z_2\mathbf{I} + \ldots + Z_N\mathbf{I} \tag{8–18}$$
$$= (Z_1 + Z_2 + \ldots + Z_N)\mathbf{I}$$

FIGURE 8 – 11 *A series connection of impedances.*

The last line in this equation points out that the phasor responses $\mathbf{V}$ and $\mathbf{I}$ do not change when the series connected elements are replaced by an equivalent impedance:

$$Z_{EQ} = Z_1 + Z_2 + \ldots + Z_N \qquad (8-19)$$

In general, the equivalent impedance Z_{EQ} is a complex quantity of the form

$$Z_{EQ} = R + jX$$

where R is the real part and X is the imaginary part. The real part of Z is called **resistance** and the imaginary part (X, not jX) is called **reactance**. Both resistance and reactance are expressed in ohms (Ω), and both can be functions of frequency (ω). For passive circuits, resistance is always positive, while reactance X can be either positive or negative. A positive X is called an **inductive** reactance because the reactance of an inductor is ωL, which is always positive. A negative X is called a **capacitive** reactance because the reactance of a capacitor is $-1/\omega C$, which is always negative.

Combining Eqs. (8–18) and (8–19), we can write the phasor voltage across the kth element in the series connection as

$$\mathbf{V}_k = Z_k \mathbf{I}_k = \frac{Z_k}{Z_{EQ}} \mathbf{V} \qquad (8-20)$$

Equation (8–20) is the phasor version of the voltage division principle. The phasor voltage across any element in a series connection is equal to the ratio of its impedance to the equivalent impedance of the connection times the total phasor voltage across the connection.

FIGURE 8 – 12

EXAMPLE 8–6

The circuit in Figure 8–12 is operating in the sinusoidal steady state with $v_S(t) = 35 \cos 1000t$ V.

(a) Transform the circuit into the phasor domain.
(b) Solve for the phasor current $\mathbf{I}$.
(c) Solve for the phasor voltage across each element.
(d) Construct the waveforms corresponding to the phasors found in (b) and (c).

SOLUTION:

(a) The phasor representing the input source voltage is $\mathbf{V}_S = 35\angle 0°$. The impedances of the three passive elements are

$$Z_R = R = 50 \ \Omega$$

$$Z_L = j\omega L = j1000 \times 25 \times 10^{-3} = j25 \ \Omega$$

$$Z_C = \frac{1}{j\omega C} = \frac{1}{j1000 \times 10^{-5}} = -j100 \ \Omega$$

Using these results, we obtain the phasor-domain circuit in Figure 8–13.

(b) The equivalent impedance of the series connection is

$$Z_{EQ} = 50 + j25 - j100 = 50 - j75 = 90.1\angle -56.3° \ \Omega$$

FIGURE 8 – 13

The current in the series circuit is

$$\mathbf{I} = \frac{\mathbf{V}_S}{Z_{EQ}} = \frac{35\angle 0°}{90.1\angle -56.3°} = 0.388\angle 56.3° \text{ A}$$

(c) The current $\mathbf{I}$ exists in all three series elements, so the voltage across each passive element is

$$\mathbf{V}_R = Z_R\mathbf{I} = 50 \times 0.388\angle 56.3° = 19.4\angle 56.3° \text{ V}$$

$$\mathbf{V}_L = Z_L\mathbf{I} = j25 \times 0.388\angle 56.3° = 9.70\angle 146.3° \text{ V}$$

$$\mathbf{V}_C = Z_C\mathbf{I} = -j100 \times 0.388\angle 56.3° = 38.8\angle -33.7° \text{ V}$$

(d) The sinusoidal steady-state waveforms corresponding to the phasors in (b) and (c) are

$$i(t) = \text{Re}\{0.388e^{j56.3°}e^{j1000t}\} = 0.388 \cos(1000t + 56.3°) \text{ A}$$

$$v_R(t) = \text{Re}\{19.4e^{j56.3°}e^{j1000t}\} = 19.4 \cos(1000t + 56.3°) \text{ V}$$

$$v_L(t) = \text{Re}\{9.70e^{j146.3°}e^{j1000t}\} = 9.70 \cos(1000t + 146.3°) \text{ V}$$

$$v_C(t) = \text{Re}\{38.8e^{-j33.7°}e^{j1000t}\} = 38.8 \cos(1000t - 33.7°) \text{ V} \qquad ■$$

◄D DESIGN EXAMPLE 8–7

Design the voltage divider in Figure 8–14 so that an input $v_S = 15 \cos 2000t$ V produces a steady-state output $v_O(t) = 2 \sin 2000t$ V.

FIGURE 8 – 1 4

SOLUTION:
Using voltage division, we can relate the input and output phasor as follows:

$$\mathbf{V}_O = \frac{Z_2}{Z_1 + Z_2}\mathbf{V}_S$$

The phasor representation of the input voltage is $\mathbf{V}_S = 15\angle 0 = 15 + j0$. Using the identity $\cos(x - 90°) = \sin x$, we write the required output phasor as $\mathbf{V}_O = 2\angle -90° = 0 - j2$. The design problem is to select the impedances Z_1 and Z_2 so that

$$0 - j2 = \frac{Z_2}{Z_1 + Z_2}(15 + j0)$$

Solving this design constraint for Z_1 yields

$$Z_1 = \frac{15 + j2}{-j2}Z_2$$

Evidently, we can choose Z_2 and then solve for Z_1. In making this choice, we must keep some physical realizability conditions in mind. In general, an impedance has the form $Z = R + jX$. The reactance X can be either positive (an inductor) or negative (a capacitor), but the resistance R must be positive. With these constraints in mind, we select $Z_2 = -j1000$ (a capacitor) and solve for $Z_1 = 7500 + j1000$ (a resistor in series with an inductor). Figure 8–15 shows the resulting phasor circuit. To find the values of L and

FIGURE 8 – 1 5

C, we note that the input is $v_S = 15 \cos 2000t$ hence, the frequency is $\omega = 2000$. The inductive reactance $\omega L = 1000$ requires $L = 0.5$ H, while the capacitive reactance requires $-(\omega C)^{-1} = -1000$ or $C = 0.5$ μF. Other possible designs are obtained by selecting different values of Z_2. To be physically realizable, the selected value of Z_2 must produce $R \geq 0$ for Z_1 and Z_2. ∎

FIGURE 8 – 16 *Impedance bridge.*

APPLICATION NOTE:

EXAMPLE 8 – 8

The purpose of the impedance bridge in Figure 8–16 is to measure the unknown impedance Z_X by adjusting known impedances Z_1, Z_2, and Z_3 until the detector voltage $\mathbf{V}_{DET}$ is zero. The circuit consists of a sinusoidal source $\mathbf{V}_S$ driving two voltage dividers connected in parallel. Using the voltage division principle, we find that the detector voltage is

$$\mathbf{V}_{DET} = \mathbf{V}_A - \mathbf{V}_B = \frac{Z_2}{Z_1 + Z_2}\mathbf{V}_S - \frac{Z_X}{Z_3 + Z_X}\mathbf{V}_S$$

$$= \left[\frac{Z_2 Z_3 - Z_1 Z_X}{(Z_1 + Z_2)(Z_3 + Z_X)} \right]\mathbf{V}_S$$

This equation shows that the detector voltage will be zero when $Z_2 Z_3 = Z_1 Z_X$. When the branch impedances are adjusted so that the detector voltage is zero, the unknown impedance can be written in terms of the known impedances as follows:

$$Z_X = R_X + jX_X = \frac{Z_2 Z_3}{Z_1}$$

This equation is called the bridge balance condition. Since the equality involves complex quantities, at least two of the known impedances must be adjustable to balance both the resistance R_X and the reactance X_X of the unknown impedance. In practice, bridges are designed assuming that the sign of the unknown reactance is known. Bridges that measure only positive reactance are called inductance bridges, while those that measure only negative reactance are called capacitance bridges.

The Maxwell inductance bridge in Figure 8–17 is used to measure the resistance R_X and inductance L_X of an inductive device by alternately adjusting resistances R_1 and R_2 to balance the bridge circuit. The impedances of the legs of this bridge are

$$Z_1 = \frac{1}{j\omega C_1 + \dfrac{1}{R_1}}$$

$$Z_2 = R_2 \quad Z_3 = R_3$$

FIGURE 8 – 17 *Maxwell bridge.*

For the Maxwell bridge, the balance condition $Z_X = Z_2 Z_3 / Z_1$ yields

$$R_X + j\omega L_X = \frac{R_2 R_3}{R_1} + j\omega C_1 R_2 R_3$$

Equating the real and imaginary parts on each side of this equation yields the parameters of the unknown impedance in terms of the known impedances:

$$R_X = \frac{R_2 R_3}{R_1} \quad \text{and} \quad L_X = R_2 R_3 C_1$$

Note that adjusting R_1 only affects R_X. The Maxwell bridge measures inductance by balancing the positive reactance of an unknown inductive device with a calibrated fraction of negative reactance of the known capacitor C_1. If the reactance of the unknown device is actually capacitive (negative), then the Maxwell bridge cannot be balanced.

PARALLEL EQUIVALENCE AND CURRENT DIVISION

In Figure 8–18 the two-terminal elements are connected in parallel, so the same phasor voltage **V** appears across the impedances $Z_1, Z_2, \ldots Z_N$. Using the phasor element constraints, the current through each impedance is $\mathbf{I}_k = \mathbf{V}/Z_k$. Next, using KCL, the total current entering the parallel connection is

$$\mathbf{I} = \mathbf{I}_1 + \mathbf{I}_2 + \ldots + \mathbf{I}_N$$

$$= \frac{\mathbf{V}}{Z_1} + \frac{\mathbf{V}}{Z_2} + \ldots + \frac{\mathbf{V}}{Z_N} \tag{8–21}$$

$$= \left(\frac{1}{Z_1} + \frac{1}{Z_2} + \ldots + \frac{1}{Z_N} \right) \mathbf{V}$$

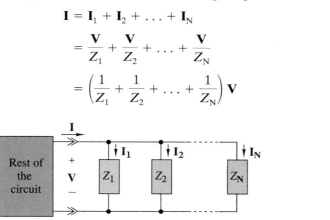

FIGURE 8 – 1 8 *Parallel connection of impedances.*

The same phasor responses **V** and **I** exist when the parallel connected elements are replaced by an equivalent impedance.

$$\frac{1}{Z_{EQ}} = \frac{\mathbf{I}}{\mathbf{V}} = \frac{1}{Z_1} + \frac{1}{Z_2} + \ldots + \frac{1}{Z_N} \tag{8–22}$$

These results can also be written in terms of admittance Y, which is defined as the reciprocal of impedance:

$$Y = \frac{1}{Z} = G + jB$$

The real part of Y is called **conductance** and the imaginary part B is called **susceptance** both of which are expressed in units of siemens (S).

Using admittances to rewrite Eq. (8–21) yields

$$\mathbf{I} = \mathbf{I}_1 + \mathbf{I}_2 + \ldots + \mathbf{I}_N$$

$$= Y_1 \mathbf{V} + Y_2 \mathbf{V} + \ldots + Y_N \mathbf{V} \tag{8–23}$$

$$= (Y_1 + Y_2 + \ldots + Y_N) \mathbf{V}$$

Hence, the equivalent admittance of the parallel connection is

$$Y_{EQ} = \frac{I}{V} = Y_1 + Y_2 + \ldots + Y_N \tag{8–24}$$

Combining Eqs. (8–23) and (8–24), we find that the phasor current through the kth element in the parallel connection is

$$I_k = Y_k V_k = \frac{Y_k}{Y_{EQ}} I \tag{8–25}$$

Equation (8–25) is the phasor version of the current division principle. The phasor current through any element in a parallel connection is equal to the ratio of its admittance to the equivalent admittance of the connection times the total phasor current entering the connection.

FIGURE 8–19

EXAMPLE 8–9

The circuit in Figure 8–19 is operating in the sinusoidal steady state with $i_S(t) = 50 \cos 2000t$ mA.

(a) Transform the circuit into the phasor domain.
(b) Solve for the phasor voltage **V**.
(c) Solve for the phasor current through each element.
(d) Construct the waveforms corresponding to the phasors found in (b) and (c).

SOLUTION:

(a) The phasor representing the input source current is $I_S = 0.05\angle 0°$ A. The impedances of the three passive elements are

$$Z_R = R = 500 \ \Omega$$

$$Z_L = j\omega L = j2000 \times 0.5 = j1000 \ \Omega$$

$$Z_C = \frac{1}{j\omega C} = \frac{1}{j2000 \times 10^{-6}} = -j500 \ \Omega$$

Using these results, we obtain the phasor-domain circuit in Figure 8–20.

FIGURE 8–20

(b) The admittances of the two parallel branches are

$$Y_1 = \frac{1}{-j500} = j2 \times 10^{-3} \text{ S}$$

$$Y_2 = \frac{1}{500 + j1000} = 4 \times 10^{-4} - j8 \times 10^{-4} \text{ S}$$

The equivalent admittance of the parallel connection is

$$Y_{EQ} = Y_1 + Y_2 = 4 \times 10^{-4} + j12 \times 10^{-4}$$
$$= 12.6 \times 10^{-4} \angle 71.6° \text{ S}$$

and the voltage across the parallel circuit is

$$\mathbf{V} = \frac{\mathbf{I}_S}{Y_{EQ}} = \frac{0.05 \angle 0°}{12.6 \times 10^{-4} \angle 71.6°}$$
$$= 39.7 \angle -71.6° \text{ V}$$

(c) The current through each parallel branch is

$$\mathbf{I}_1 = Y_1 \mathbf{V} = j2 \times 10^{-3} \times 39.7 \angle -71.6° = 79.4 \angle 18.4° \text{ mA}$$
$$\mathbf{I}_2 = Y_2 \mathbf{V} = (4 \times 10^{-4} - j8 \times 10^{-4}) \times 39.7 \angle -71.6°$$
$$= 35.5 \angle -135° \text{ mA}$$

(d) The sinusoidal steady-state waveforms corresponding to the phasors in (b) and (c) are

$$v(t) = \text{Re}\{39.7e^{-j71.6°}e^{j2000t}\} = 39.7 \cos(2000t - 71.6°) \text{ V}$$
$$i_1(t) = \text{Re}\{79.4e^{j18.4°}e^{j2000t}\} = 79.4 \cos(2000t + 18.4°) \text{ mA}$$
$$i_2(t) = \text{Re}\{35.5e^{-j135°}e^{j2000t}\} = 35.5 \cos(2000t - 135°) \text{ mA} \quad \blacksquare$$

EXAMPLE 8-10

Find the steady-state currents $i(t)$, $i_C(t)$, and $i_R(t)$ in the circuit of Figure 8-21 for $v_S = 100 \cos 2000t$ V, $L = 250$ mH, $C = 0.5$ μF, and $R = 3$ kΩ.

FIGURE 8-21

SOLUTION:
The phasor representation of the input voltage is $100 \angle 0°$. The impedances of the passive elements are

$$Z_L = j500 \ \Omega \quad Z_C = -j1000 \ \Omega \quad Z_R = 3000 \ \Omega$$

Figure 8-22(a) shows the phasor-domain circuit.

To solve for the required phasor responses, we reduce the circuit using a combination of series and parallel equivalence. Using parallel equivalence, we find that the capacitor and resistor can be replaced by an equivalent impedance

$$Z_{EQ1} = \frac{1}{Y_{EQ1}} = \frac{1}{\dfrac{1}{-j1000} + \dfrac{1}{3000}}$$
$$= 300 - j900 \ \Omega$$

The resulting circuit reduction is shown in Figure 8-22(b). The equivalent impedance Z_{EQ1} is connected in series with the impedance $Z_L = j500$. This series combination can be replaced by an equivalent impedance

$$Z_{EQ2} = j500 + Z_{EQ1} = 300 - j400 \ \Omega$$

This step reduces the circuit to the equivalent input impedance shown in Figure 8-22(c). The phasor input current in Figure 8-22(c) is

FIGURE   8 – 2 2

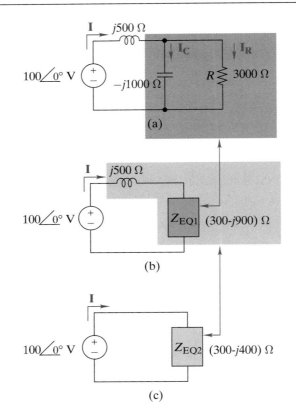

$$\mathbf{I} = \frac{100\angle 0°}{Z_{EQ2}} = \frac{100\angle 0°}{300 - j400} = 0.12 + j0.16 = 0.2\angle 53.1° \text{ A}$$

Given the phasor current $\mathbf{I}$, we use current division to find $\mathbf{I}_C$:

$$\mathbf{I}_C = \frac{Y_C}{Y_C + Y_R}\mathbf{I} = \frac{\dfrac{1}{-j1000}}{\dfrac{1}{-j1000} + \dfrac{1}{3000}} 0.2\angle 53.1°$$

$$= 0.06 + j0.18 = 0.19\angle 71.6° \text{ A}$$

By KCL, $\mathbf{I} = \mathbf{I}_C + \mathbf{I}_R$, so the remaining unknown current is

$$\mathbf{I}_R = \mathbf{I} - \mathbf{I}_C = 0.06 - j0.02 = 0.0632\angle -18.4° \text{ A}$$

The waveforms corresponding to the phasor currents are

$$i(t) = \text{Re}\{\mathbf{I}e^{j2000t}\} = 0.2 \cos(2000t + 53.1°) \text{ A}$$

$$i_C(t) = \text{Re}\{\mathbf{I}_C e^{j2000t}\} = 0.19 \cos(2000t + 71.6°) \text{ A}$$

$$i_R(t) = \text{Re}\{\mathbf{I}_R e^{j2000t}\} = 0.0632 \cos(2000t - 18.4°) \text{ A} \qquad \blacksquare$$

APPLICATION NOTE:

EXAMPLE 8–11

In general, the equivalent impedance seen at any pair of terminals can be written in rectangular form as

$$Z_{EQ} = R_{EQ} + jX_{EQ} \qquad (8\text{–}26)$$

In a passive circuit the equivalent resistance R_{EQ} must always be non-negative, that is, $R_{EQ} \geq 0$. However, the equivalent reactance X_{EQ} can be either positive (inductive) or negative (capacitive). When inductance and capacitance are both present, their reactances may exactly cancel at certain frequencies. When $X_{EQ} = 0$, the impedance is purely resistive and the circuit is said to be in **resonance**. The frequency at which this occurs is called a **resonant frequency**, denoted by ω_0.

For example, suppose we want to find the resonant frequency of the circuit in Figure 8–23. We first find the equivalent impedance of the parallel resistor and capacitor:

$$Z_{RC} = \frac{1}{Y_R + Y_C} = \frac{1}{\dfrac{1}{R} + j\omega C} = \frac{R}{1 + j\omega RC}$$

FIGURE 8 – 2 3

This expression can be put into rectangular form by multiplying and dividing by the conjugate of the denominator:

$$Z_{RC} = \frac{R}{1 + j\omega RC} \frac{1 - j\omega RC}{1 - j\omega RC} = \frac{R}{1 + (\omega RC)^2} - j\frac{\omega R^2 C}{1 + (\omega RC)^2}$$

The impedance Z_{RC} is connected in series with the inductor. Therefore, the overall equivalent impedance Z_{EQ} is

$$Z_{EQ} = Z_L + Z_{RC}$$

$$= \frac{R}{1 + (\omega RC)^2} + j\left[\omega L - \frac{\omega R^2 C}{1 + (\omega RC)^2}\right]$$

$$= \underbrace{ R_{EQ} }_{} \quad + \quad \underbrace{ jX_{EQ} }_{}$$

Note that the equivalent resistance R_{EQ} is positive for all ω. However, the equivalent reactance X_{EQ} can be positive or negative. The resonant frequency is found by setting the reactance to zero

$$X_{EQ}(\omega_0) = \omega_0 L - \frac{\omega_0 R^2 C}{1 + (\omega_0 RC)^2} = 0$$

and solving for the resonant frequency:

$$\omega_0 = \sqrt{\frac{1}{LC} - \frac{1}{(RC)^2}} \qquad (8\text{–}27)$$

Note the reactance X_{EQ} is inductive (positive) when $\omega > \omega_0$ and capacitive (negative) with $\omega < \omega_0$.

10 mH

100 Ω

C

Z

FIGURE   8 - 2 4

EXAMPLE 8–12

The circuit in Figure 8–24 is operating in the sinusoidal steady state with $\omega = 5$ krad/s.

(a) Find the value of capacitance C that causes the input impedance Z to be purely resistive.
(b) Find the real part of the input impedance for this value of C.

SOLUTION:
(a) At the specified frequency the impedance is

$$Z = j5000 \times 0.01 + \cfrac{1}{\cfrac{1}{100} + j5000\,C}$$

which can be written as

$$Z = j50 + \frac{100}{1 + j5 \times 10^5\,C} \times \frac{1 - j5 \times 10^5\,C}{1 - j5 \times 10^5\,C}$$

$$= j50 + \frac{100}{1 + (5 \times 10^5\,C)^2} - j\,\frac{5 \times 10^7\,C}{1 + (5 \times 10^5\,C)^2}$$

A purely resistive input impedance means that the imaginary part of Z is zero, which requires that

$$50 - \frac{5 \times 10^7\,C}{1 + (5 \times 10^5\,C)^2} = 0 \quad \text{or} \quad 25 \times 10^{10}\,C^2 - 10^6\,C + 1 = 0$$

This quadratic has a double root at $C = 2 \times 10^{-6}$ F.
(b) For this value of capacitance the real part of Z is

$$Re\{Z\} = R_{EQ} = \frac{100}{1 + (5 \times 10^5\,C)^2} = 50\ \Omega$$

Exercise 8–8

(a) Find the equivalent impedance Z in Figure 8–25 at $\omega = 1$ krad/s.
(b) Repeat (a) with $\omega = 4$ krad/s.

FIGURE   8 - 2 5

Z   200 mH

1 µF   1 kΩ

Answers:
(a) $Z = 500 - j300\ \Omega$
(b) $Z = 58.8 + j565\ \Omega$

Exercise 8–9

A voltage source $v_S(t) = 15 \cos 2000t$ V is applied at the input in Figure 8–25.

(a) Find the steady-state current through the inductor.
(b) Find the steady-state voltage across the 1-kΩ resistor.

Answers:
(a) $75 \cos 2000t$ mA
(b) $33.5 \cos (2000t – 63.4°)$ V

Exercise 8–10

The circuit in Figure 8–26 is operating at $\omega = 10$ krad/s.

(a) Find the equivalent impedance Z.
(b) What element should be connected in series with Z to make the total reactance zero?

FIGURE 8 – 2 6

Answers:
(a) $Z = 256 + j195 = 322\angle37.3°$ Ω
(b) A capacitor with $C = 513$ nF

Exercise 8–11

In Figure 8–27 $v_S(t) = 12.5 \cos 1000t$ V and $i_S(t) = 0.2 \cos (1000t – 36.9°)$ A. What is the impedance seen by the voltage source and what element is in the box?

Answer:
$Z = 50 + j37.5$ Ω and the element is a 37.5-mH inductor.

FIGURE 8 – 2 7

8–4 CIRCUIT THEOREMS WITH PHASORS

In this section we treat basic properties of phasor circuits that parallel the resistance circuit theorems developed in Chapter 3. Circuit linearity is the foundation for all of these properties. The proportionality and superposition properties are two fundamental consequences of linearity.

PROPORTIONALITY

The **proportionality** property states that phasor output responses are proportional to the input phasor. Mathematically, proportionality means that

$$\mathbf{Y} = K\mathbf{X} \tag{8–28}$$

where **X** is the input phasor, **Y** is the output phasor, and K is the proportionality constant. In phasor circuit analysis, the proportionality constant is generally a complex number.

The unit output method discussed in Chapter 3 is based on the proportionality property and is applicable to phasors. To apply the unit output method in the phasor domain, we assume that the output is a unit phasor **Y** $= 1\angle 0°$. By successive application of KCL, KVL, and the element impedances, we solve for the input phasor required to produce the unit output. Because the circuit is linear, the proportionality constant relating input and output is

$$K = \frac{\text{Output}}{\text{Input}} = \frac{1\angle 0°}{\text{Input phasor for unit output}}$$

Once we have the constant K, we can find the output for any input or the input required to produce any specified output.

The next example illustrates the unit output method for phasor circuits.

EXAMPLE 8–13

Use the unit output method to find the input impedance, current **I**$_1$, output voltage **V**$_C$, and current **I**$_3$ of the circuit in Figure 8–28 for **V**$_S$ = $10\angle 0°$ V.

SOLUTION:
The following steps implement the unit output method for the circuit in Figure 8–28:

FIGURE 8–28

1. Assume a unit output voltage **V**$_C$ = $1 + j0$ V.

2. By Ohm's law, **I**$_3$ = **V**$_C$/50 = $0.02 + j0$ A.

3. By KVL, **V**$_B$ = **V**$_C$ + $(j100)$**I**$_3$ = $1 + j2$ V.

4. By Ohm's law, **I**$_2$ = **V**$_B$/$(-j50)$ = $-0.04 + j0.02$ A.

5. By KCL, **I**$_1$ = **I**$_2$ + **I**$_3$ = $-0.02 + j0.02$ A.

6. By KVL, **V**$_S$ = $(50 + j100)$**I**$_1$ + **V**$_B$ = $-2 + j1$ V.

Given **V**$_S$ and **I**$_1$, the input impedance is

$$Z_{\text{IN}} = \frac{\mathbf{V}_S}{\mathbf{I}_1} = \frac{-2 + j1}{-0.02 + j0.02} = 75 + j25 \ \Omega$$

The proportionality factor between the input $\mathbf{V}_S$ and output voltage $\mathbf{V}_C$ is

$$K = \frac{1}{\mathbf{V}_S} = \frac{1}{-2 + j} = -0.4 - j0.2$$

Given K and Z_{IN}, we can now calculate the required responses for an input $\mathbf{V}_S = 10\angle 0°$:

$$\mathbf{V}_C = K\mathbf{V}_S = -4 - j2 = 4.47\angle -153° \text{ V}$$

$$\mathbf{I}_1 = \frac{\mathbf{V}_S}{Z_{IN}} = 0.12 - j0.04 = 0.126\angle -18.4° \text{ A}$$

$$\mathbf{I}_3 = \frac{\mathbf{V}_C}{50} = -0.08 - j0.04 = 0.0894\angle -153° \text{ A} \qquad \blacksquare$$

SUPERPOSITION

The superposition principle applies to phasor responses only if all of the independent sources driving the circuit have the *same frequency*. That is, when the input sources have the same frequency, we can find the phasor response due to each source acting alone and obtain the total response by adding the individual phasors. If the sources have different frequencies, then superposition can still be used but its application is different. With different frequency sources, each source must be treated in a separate steady-state analysis because the element impedances change with frequency. The phasor response for each source must be changed into waveforms and then superposition applied in the time domain. In other words, the superposition principle always applies in the time domain. It also applies in the phasor domain when all independent sources have the same frequency. The following examples illustrate both cases.

EXAMPLE 8–14

Use superposition to find the steady-state voltage $v_R(t)$ in Figure 8–29 for $R = 20 \ \Omega$, $L_1 = 2 \text{ mH}$, $L_2 = 6 \text{ mH}$, $C = 20 \ \mu\text{F}$, $v_{S1} = 100 \cos 5000t$ V, and $v_{S2} = 120 \cos(5000t + 30°)$ V.

FIGURE   8 – 2 9

SOLUTION:

In this example, the two sources operate at the same frequency. Figure 8–30(a) shows the phasor domain circuit with source no. 2 turned off and replaced by a short circuit. The three elements in parallel in Figure 8–30(a) produce an equivalent impedance of

$$Z_{EQ} = \frac{1}{\dfrac{1}{20} + \dfrac{1}{-j10} + \dfrac{1}{j30}} = 7.20 - j9.60 \ \Omega$$

By voltage division, the phasor response $\mathbf{V}_{R1}$ is

$$\mathbf{V}_{R1} = \frac{Z_{EQ1}}{j10 + Z_{EQ1}} 100\angle 0°$$

$$= 92.3 - j138 = 166\angle -56.3° \text{ V}$$

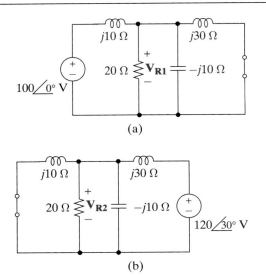

(a)

(b)

Figure 8–30(b) shows the phasor-domain circuit with source no. 1 turned off and source no. 2 on. The three elements in parallel in Figure 8–30(b) produce an equivalent impedance of

$$Z_{EQ2} = \cfrac{1}{\cfrac{1}{20} + \cfrac{1}{-j10} + \cfrac{1}{j10}} = 20 - j0 \ \Omega$$

By voltage division, the response $\mathbf{V}_{R2}$ is

$$\mathbf{V}_{R2} = \frac{Z_{EQ2}}{j30 + Z_{EQ2}} 120\angle 30°$$

$$= 59.7 - j29.5 = 66.6\angle{-26.3°} \text{ V}$$

Since the sources have the same frequency, the total response can be found by adding the individual phasor responses $\mathbf{V}_{R1}$ and $\mathbf{V}_{R2}$:

$$\mathbf{V}_R = \mathbf{V}_{R1} + \mathbf{V}_{R2} = 152 - j167 = 226\angle{-47.8°} \text{ V}$$

The time-domain function corresponding to the phasor sum is

$$v_R(t) = \text{Re}\{\mathbf{V}_R e^{j5000t}\} = 226 \cos(5000t - 47.8°) \text{ V}$$

The overall response can also be obtained by adding the time-domain functions corresponding to the individual phasor responses $\mathbf{V}_{R1}$ and $\mathbf{V}_{R2}$:

$$v_R(t) = \text{Re}\{\mathbf{V}_{R1} e^{j5000t}\} + \text{Re}\{\mathbf{V}_{R2} e^{j5000t}\}$$

$$= 166 \cos(5000t - 56.3°) + 66.6 \cos(5000t - 26.3°) \text{ V}$$

You are encouraged to show that the two expressions for $v_R(t)$ are equivalent using the additive property of sinusoids. ∎

EXAMPLE 8–15

Use superposition to find the steady-state current $i(t)$ in Figure 8–31 for $R = 10 \text{ k}\Omega$, $L = 200 \text{ mH}$, $v_{S1} = 24 \cos 20000t$ V, and $v_{S2} = 8 \cos(60000t + 30°)$ V.

SOLUTION:

In this example the two sources operate at different frequencies. With source no. 2 off, the input phasor is $\mathbf{V}_{S1} = 24\angle0°$ V at a frequency of $\omega = 20$ krad/s. At this frequency the equivalent impedance of the inductor and resistor is

$$Z_{EQ1} = R + j\omega L = 10 + j4 \text{ k}\Omega$$

The phasor current due to source no. 1 is

$$\mathbf{I}_1 = \frac{\mathbf{V}_{S1}}{Z_{EQ1}} = \frac{24\angle0°}{10000 + j4000} = 2.23\angle-21.8° \text{ mA}$$

With source no. 1 off and source no. 2 on, the input phasor $\mathbf{V}_{S2} = 8\angle30°$ V at a frequency of $\omega = 60$ krad/s. At this frequency the equivalent impedance of the inductor and resistor is

$$Z_{EQ2} = R + j\omega L = 10 + j12 \text{ k}\Omega$$

The phasor current due to source no. 2 is

$$\mathbf{I}_2 = \frac{\mathbf{V}_{S2}}{Z_{EQ2}} = \frac{8\angle30°}{10000 + j12000} = 0.512\angle-20.2° \text{ mA}$$

The two input sources operate at different frequencies, so the phasors responses $\mathbf{I}_1$ and $\mathbf{I}_2$ cannot be added to obtain the overall response. In this case the overall response is obtained by adding the corresponding time-domain functions.

$$i(t) = \text{Re}\{\mathbf{I}_1 e^{j20000t}\} + \text{Re}\{\mathbf{I}_2 e^{j60000t}\}$$

$$= 2.23 \cos(20000t - 21.8°) + 0.512 \cos(60000t - 20.2°) \text{ mA} \quad \blacksquare$$

EXAMPLE 8–16

The voltage source in Figure 8–32 produces a 60-Hz sinusoid with a peak amplitude of 200 V plus a 180-Hz third harmonic with a peak amplitude of 10 V. The purpose of the LC circuit is to reduce the relative size of the third harmonic component delivered to the 100-Ω load resistor. Use PSpice to calculate the amplitude of the 60-Hz and 180-Hz components in the output v_L.

FIGURE 8 – 3 2

SOLUTION:

Figure 8–33 shows the circuit diagram as constructed in PSpice. The value AC = 1 means that $\mathbf{V}_S$ is an AC voltage source with an amplitude of 1 V. The printer symbol attached to node 4 causes MicroSim PSpice to write the magnitude and phase of the node voltage V(4) in the output file. The problem involves ac analysis with input sinusoids at 60 Hz and 180 Hz. **Selecting Analysis/Setup/AC Sweep** from the Schematics menu bar brings up the AC sweep menu shown in Figure 8–33. We select a two-point linear sweep that starts at 60 Hz and ends at 180 Hz. Returning to the main menu and selecting **Analysis/Simulate** causes PSpice to make two AC analysis runs, first at $f = 60$ Hz and then again at $f = 180$ Hz. The input phasor for each run is $\mathbf{V}_S = 1\angle0°$. The relevant portion of the output file is shown at

FIGURE 8 – 33

the bottom in Figure 8–33. The symbol VM(4) is the magnitude of the output voltage phasor in volts and VP(4) is the phase angle in degrees. These values are for the normalized input $\mathbf{V}_S = 1\angle 0°$. To get the actual output, we must scale the normalized results by the actual input amplitudes at each frequency.

FREQUENCY (HZ)	INPUT (V)	OUTPUT (V)
60	200	$200 \times 0.9994 = 199.9$
180	10	$10 \times 0.005534 = 0.05534$

The amplitude of the 60-Hz component is virtually unchanged, while the amplitude of the 180-Hz component is greatly reduced. This frequency selectivity occurs because the impedances of the inductor and capacitor change with frequency. A frequency selective signal processor is called a filter, a topic that will be treated extensively in later chapters. ■

FIGURE 8–34

Exercise 8–12

The two sources in Figure 8–34 have the same frequency. Use superposition to find the phasor current $\mathbf{I}_X$.

Answer:

$$\mathbf{I}_X = 0.206\angle{-158}° \text{ A}$$

THÉVENIN AND NORTON EQUIVALENT CIRCUITS

In the phasor domain, a two-terminal circuit containing linear elements and sources can be replaced by the Thévenin or Norton equivalent circuits shown in Figure 8–35. The general concept of Thévenin's and Norton's theorems and their restrictions are the same as in the resistive circuit studied in Chapter 3. The important difference here is that the signals $\mathbf{V}_T$, $\mathbf{I}_N$, $\mathbf{V}$, and $\mathbf{I}$ are phasors, and $Z_T = 1/Y_N$ and Z_L are complex numbers representing the source and load impedances.

Finding the Thévenin or Norton equivalent of a phasor circuit involves the same process as for resistance circuits, except that now we must manipulate complex numbers. The Thévenin and Norton circuits are equivalent to each other, so their circuit parameters are related as follows:

(a)

$$\mathbf{V}_{OC} = \mathbf{V}_T = \mathbf{I}_N Z_T$$

$$\mathbf{I}_{SC} = \frac{\mathbf{V}_T}{Z_T} = \mathbf{I}_N \qquad (8-29)$$

$$Z_T = \frac{1}{Y_N} = \frac{\mathbf{V}_{OC}}{\mathbf{I}_{SC}}$$

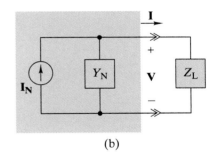

(b)

FIGURE 8–35 *Thévenin and Norton equivalent circuits in the phasor analysis.*

Algebraically, the results in Eq. (8–29) are identical to the corresponding equations for resistance circuits. The important difference is that these equations involve phasors and impedances rather than waveforms and resistances. These equations point out that we can determine a Thévenin or Norton equivalent by finding any two of the following quantities: (1) the open-circuit voltage $\mathbf{V}_{OC}$, (2) the short-circuit current $\mathbf{I}_{SC}$, and (when there are no dependent sources) (3) the impedance Z_T looking back into the source circuit with all independent sources turned off.

The relationships in Eq. (8–29) define source transformations that allow us to convert a voltage source in series with an impedance into a current source in parallel with the same impedance, or vice versa. Phasor-domain source transformations simplify circuits and are useful in formulating general node-voltage or mesh-current equations, discussed in the next section.

The next two examples illustrate applications of source transformation and Thévenin equivalent circuits.

EXAMPLE 8–17

Both sources in Figure 8–36(a) operate at a frequency of $\omega = 5000$ rad/s. Find the steady-state voltage $v_R(t)$ using source transformations.

SOLUTION:

Example 8–14 solves this problem using superposition. In this example we use source transformations. We observe that the voltage sources in Figure 8–36(a) are connected in series with an impedance and can be converted into the following equivalent current sources:

$$\mathbf{I}_{EQ1} = \frac{100\angle 0°}{j10} = 0 - j10 \text{ A}$$

$$\mathbf{I}_{EQ2} = \frac{120\angle 30°}{j30} = 2 - j3.46 \text{ A}$$

FIGURE 8 – 36

(a)

(b)

Figure 8–36(b) shows the circuit after these two source transformations. The two current sources are connected in parallel and can be replaced by a single equivalent current source:

$$\mathbf{I}_{EQ} = \mathbf{I}_{EQ1} + \mathbf{I}_{EQ2} = 2 - j13.46 = 13.6\angle -81.5° \text{ A}$$

The four passive elements are connected in parallel and can be replaced by an equivalent impedance:

$$Z_{EQ} = \frac{1}{\dfrac{1}{20} + \dfrac{1}{-j10} + \dfrac{1}{j10} + \dfrac{1}{j30}} = 16.6\angle 33.7° \text{ } \Omega$$

The voltage across this equivalent impedance equals $\mathbf{V}_R$, since one of the parallel elements is the resistor R. Therefore, the unknown phasor voltage is

$$\mathbf{V_R} = \mathbf{I_{EQ}Z_{EQ}} = (13.6\angle -81.5°) \times (16.6\angle 33.7°) = 226\angle -47.8° \text{ V}$$

The value of $\mathbf{V_R}$ is the same as found in Example 8–14 using superposition. The corresponding time-domain function is

$$v_R(t) = \text{Re}\{\mathbf{V_R}e^{j5000t}\} = 226 \cos(5000t - 47.8°) \text{ V} \qquad \blacksquare$$

EXAMPLE 8–18

Use Thévenin's theorem to find the current $\mathbf{I_X}$ in the bridge circuit shown in Figure 8–37.

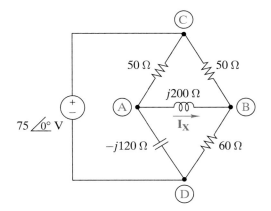

FIGURE 8 - 37

SOLUTION:

Disconnecting the impedance $j200$ from the circuit in Figure 8–37 produces the circuit shown in Figure 8–38(a). The voltage between nodes A and B is the Thévenin voltage since removing the impedance $j200$ leaves an open circuit. The voltages at nodes A and B can each be found by voltage division. Since the open-circuit voltage is the difference between these node voltages, we have

$$\mathbf{V_T} = \mathbf{V_A} - \mathbf{V_B}$$

$$= \frac{-j120}{50 - j120}75\angle 0° - \frac{60}{60 + 50}75\angle 0°$$

$$= 23.0 - j26.6 \text{ V}$$

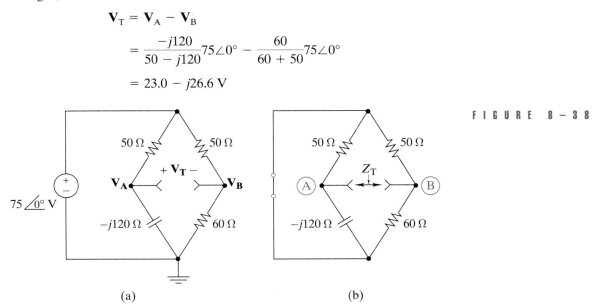

FIGURE 8 - 38

(a)   (b)

Turning off the voltage source in Figure 8–38(a) and replacing it by a short circuit produces the situation shown in Figure 8–38(b). The lookback impedance seen at the interface is a series connection of two pairs of elements connected in parallel. The equivalent impedance of the series/parallel combination is

$$Z_T = \frac{1}{\dfrac{1}{50} + \dfrac{1}{-j120}} + \frac{1}{\dfrac{1}{50} + \dfrac{1}{60}} = 69.9 - j17.8 \ \Omega$$

Given the Thévenin equivalent circuit, we treat the impedance $j200$ as a load connected at the interface and calculate the resulting load current $\mathbf{I}_X$ as

$$\mathbf{I}_X = \frac{\mathbf{V}_T}{Z_T + j200} = \frac{23.0 - j26.6}{69.9 + j182} = 0.180\angle{-118°} \ \text{A} \qquad \blacksquare$$

EXAMPLE 8–19

In the steady state, the open-circuit voltage at an interface is observed to be

$$v_{OC}(t) = 12 \cos 2000t \quad \text{V}$$

When a 50-mH inductor is connected across the interface, the interface voltage is observed to be

$$v(t) = 17 \cos(2000t + 45°) \quad \text{V}$$

Find the Thévenin equivalent circuit at the interface.

SOLUTION:
The phasors for $v_{OC}(t)$ and $v(t)$ are $\mathbf{V}_{OC} = 12\angle 0°$ and $\mathbf{V} = 17\angle 45°$. The phasor Thévenin voltage at the interface is $\mathbf{V}_T = \mathbf{V}_{OC} = 12\angle 0°$. The impedance of the inductor is $Z_L = j\omega L = j2000 \times 0.050 = j100 \ \Omega$. When the inductor load is connected across the interface, we use voltage division to express the interface voltage as

$$\mathbf{V} = \frac{Z_L}{Z_T + Z_L} \mathbf{V}_T$$

Inserting the known numerical values yields

$$17\angle 45° = \frac{j100}{Z_T + j100} 12\angle 0°$$

Solving for Z_T, we have

$$Z_T = j100 \times \frac{12\angle 0°}{17\angle 45°} - j100 = 49.9 - j50.1 \ \Omega$$

The Thévenin equivalent circuit at the interface is defined by $\mathbf{V}_T = 12\angle 0°$ and $Z_T = 49.9 - j50.1 \ \Omega$. $\qquad \blacksquare$

Exercise 8–13

(a) Find the Thévenin equivalent circuit seen by the inductor in Figure 8–34.
(b) Use the Thévenin equivalent to calculate the current $\mathbf{I}_X$.

Answers:
(a) $\mathbf{V}_T = -15.4 - j22.6$ V, $Z_T = 109.9 - j0.990$ Ω
(b) $\mathbf{I}_X = 0.206\angle-158°$ A

Exercise 8–14

By inspection, determine the Thévenin equivalent circuit seen by the capacitor in Figure 8–28 for $\mathbf{V}_S = 10\angle0°$ V.

Answer:
$\mathbf{V}_T = 5\angle0°$ V, $Z_T = 25 + j50$ Ω

Exercise 8–15

In the steady state the short-circuit current at an interface is observed to be

$$i_{SC}(t) = 0.75 \sin \omega t \text{ A}$$

When a 150-Ω resistor is connected across the interface, the interface current is observed to be

$$i(t) = 0.6 \cos(\omega t - 53.1°) \text{ A}$$

Find the Norton equivalent phasor circuit at the interface.

Answer:

$$\mathbf{I}_N = 0 - j0.75 \text{ A}, Z_N = 0 + j200 \text{ }\Omega$$

8–5 GENERAL CIRCUIT ANALYSIS WITH PHASORS

The previous sections discuss basic analysis methods based on equivalence, reduction, and circuit theorems. These methods are valuable because we work directly with element impedances and thereby gain insight into steady-state circuit behavior. We also need general methods, such as node and mesh analysis, to deal with more complicated circuits than the basic methods can easily handle. These general methods use node-voltage or mesh-current variables to reduce the number of equations that must be solved simultaneously.

Node-voltage equations involve selecting a reference node and assigning a node-to-datum voltage to each of the remaining nonreference nodes. Because of KVL, the voltage between any two nodes equals the difference of the two node voltages. This fundamental property of node voltages plus the element impedances allow us to write KCL constraints at each of the nonreference nodes.

For example, consider node A in Figure 8–39. The sum of currents leaving this node can be written as

$$\mathbf{I}_{S2} - \mathbf{I}_{S1} + \frac{\mathbf{V}_A}{Z_1} + \frac{\mathbf{V}_A - \mathbf{V}_B}{Z_2} + \frac{\mathbf{V}_A - \mathbf{V}_C}{Z_3} = 0$$

Rewriting this equation with unknowns grouped on the left and known inputs on the right yields

F I G U R E   8 – 3 9   *An example node.*

$$\left[\frac{1}{Z_1} + \frac{1}{Z_2} + \frac{1}{Z_3}\right]\mathbf{V}_A - \frac{1}{Z_2}\mathbf{V}_B - \frac{1}{Z_3}\mathbf{V}_C = \mathbf{I}_{S1} - \mathbf{I}_{S2}$$

Expressing this result in terms of admittances produces the following equation.

$$[Y_1 + Y_2 + Y_3]\mathbf{V}_A - [Y_2]\mathbf{V}_B - [Y_3]\mathbf{V}_C = \mathbf{I}_{S1} - \mathbf{I}_{S2}$$

This equation has a familiar pattern. The unknowns $\mathbf{V}_A$, $\mathbf{V}_B$, and $\mathbf{V}_C$ are the node-voltage phasors. The coefficient $[Y_1 + Y_2 + Y_3]$ of $\mathbf{V}_A$ is the sum of the admittances of all of the elements connected to node A. The coefficient $[Y_2]$ of $\mathbf{V}_B$ is admittance of the elements connected between nodes A and B, while $[Y_3]$ is the admittance of the elements connected between nodes A and C. Finally, $\mathbf{I}_{S1}$ and $\mathbf{I}_{S2}$ are the phasor current sources connected to node A, with $\mathbf{I}_{S1}$ directed into and $\mathbf{I}_{S2}$ directed away from the node. These observations suggest that we can write node-voltage equations for phasor circuits by inspection, just as we did with resistive circuits.

Circuits that can be drawn on a flat surface with no crossovers are called **planar** circuits. The mesh-current variables are the loop currents assigned to each mesh in a planar circuit. Because of KCL, the current through any two-terminal element is equal to the difference of the two adjacent meshes. This fundamental property of mesh currents together with the element impedances allow us to write KVL constraints around each of the meshes.

For example, the sum of voltages around mesh A in Figure 8–40 is

$$Z_1\mathbf{I}_A + Z_2[\mathbf{I}_A - \mathbf{I}_B] + Z_3[\mathbf{I}_A - \mathbf{I}_C] - \mathbf{V}_{S1} + \mathbf{V}_{S2} = 0$$

The mesh A equation is obtained by equating this sum to the sum of the source voltages produced in mesh A. Arranging this equation in standard form yields

$$[Z_1 + Z_2 + Z_3]\mathbf{I}_A - [Z_2]\mathbf{I}_B - [Z_3]\mathbf{I}_C = \mathbf{V}_{S1} - \mathbf{V}_{S2}$$

This equation also displays a familiar pattern. The unknowns $\mathbf{I}_A$, $\mathbf{I}_B$, and $\mathbf{I}_C$ are mesh-current phasors. The coefficient $[Z_1 + Z_2 + Z_3]$ of $\mathbf{I}_A$ is the sum of the impedances in mesh A. The coefficient $[Z_2]$ of $\mathbf{I}_B$ is the impedance in both mesh A and mesh B, while $[Z_3]$ is the impedance common to meshes A and C. Finally, $\mathbf{V}_{S1}$ and $\mathbf{V}_{S2}$ are the phasor voltage sources in mesh A. These observations allow us to write mesh-current equations for phasor circuits by inspection.

The preceding discussion assumes that the circuit contains only current sources in the case of node analysis and voltage sources in mesh analysis. If there is a mixture of sources, we may be able to use the source transformations discussed in Sect. 8–4 to convert from voltage to current sources, or vice versa. A source transformation is possible only when there is an impedance connected in series with a voltage source or an admittance in parallel with a current source. When a source transformation is not possible, we use the phasor version of the modified node- and mesh-analysis methods described in Chapter 3.

Formulating a set of equilibrium equations in phasor form is a straightforward process involving concepts that we have used before in Chapters 3 and 4. Once formulated, we use Cramer's rule or Gaussian reduction to solve these equations for phasor responses, although this requires manipu-

F I G U R E 8 – 4 0 *An example mesh.*

lating linear equations with complex coefficients. In principle, the solution process can be done by hand, but as a practical matter circuits with more than three nodes or meshes are best handled using computer tools. Modern hand-held scientific calculators and math analysis programs like Mathcad can deal with sets of linear equations with complex coefficients. Circuit analysis programs such as SPICE have ac analysis options that handle steady-state circuit analysis problems.

If computer tools are required for all but the simplest circuits, why bother with the hand solution at all? Why not always use SPICE or Mathcad? The answer is that hand analysis and computer-aided analysis are complementary rather than competitive. Computer-aided analysis excels at generating numerical responses when numerical values of circuit parameters are given. With hand analysis we can generate responses in symbolic form for all possible element values and input frequencies. Solving simple circuits in symbolic form gives insight as to how various circuit elements affect the steady-state response which helps us intelligently use the numerical profusion generated by computer-based "what if" analysis of large-scale circuits.

EXAMPLE 8−20

Use node analysis to find the node voltages $\mathbf{V_A}$ and $\mathbf{V_B}$ in Figure 8–41(a).

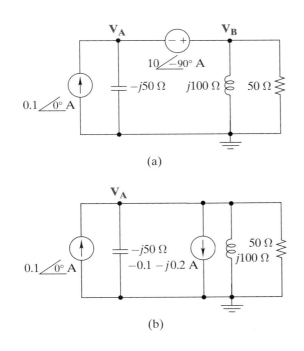

(a)

(b)

FIGURE 8 − 41

SOLUTION:

The voltage source in Figure 8–41(a) is connected in series with an impedance consisting of a resistor and inductor connected in parallel. The equivalent impedance of this parallel combination is

$$Z_{EQ} = \cfrac{1}{\cfrac{1}{50} + \cfrac{1}{j100}} = 40 + j20 \ \Omega$$

Applying a source transformation produces an equivalent current source of

$$\mathbf{I}_{EQ} = \frac{10\angle -90°}{40 + j20} = -0.1 - j0.2 \ A$$

Figure 8–41(b) shows the circuit produced by the source transformation. Note that the transformation eliminates node B. The node-voltage equation at the remaining nonreference node in Figure 8–41(b) is

$$\left(\frac{1}{-j50} + \frac{1}{j100} + \frac{1}{50} \right) \mathbf{V}_A = 0.1\angle 0° - (-0.1 - j0.2)$$

Solving for $\mathbf{V}_A$ yields

$$\mathbf{V}_A = \frac{0.2 + j0.2}{0.02 + j0.01} = 12 + j4 = 12.6\angle 18.4° \ V$$

Referring to Figure 8–41(a), we see that KVL requires $\mathbf{V}_B = \mathbf{V}_A + 10\angle -90°$. Therefore, $\mathbf{V}_B$ is found to be

$$\mathbf{V}_B = (12 + j4) + 10\angle -90° = 12 - j6 = 13.4\angle -26.6° \ V \qquad ■$$

EXAMPLE 8–21

Use node analysis to find the current $\mathbf{I}_X$ in Figure 8–42.

FIGURE 8–42

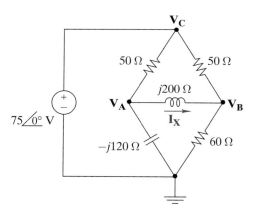

SOLUTION:
In this example we use node analysis on a problem solved in Example 8–18 using a Thévenin equivalent circuit. The voltage source cannot be replaced by source transformation because it is not connected in series with an impedance. By inspection, the node equations at nodes A and B are

$$\text{Node A:} \ \frac{\mathbf{V}_A}{-j120} + \frac{\mathbf{V}_A - \mathbf{V}_B}{j200} + \frac{\mathbf{V}_A - \mathbf{V}_C}{50} = 0$$

$$\text{Node B:} \ \frac{\mathbf{V}_B}{60} + \frac{\mathbf{V}_B - \mathbf{V}_A}{j200} + \frac{\mathbf{V}_B - \mathbf{V}_C}{50} = 0$$

A node equation at node C is not required because the voltage source forces the condition $\mathbf{V}_C = 75\angle0°$. Substituting this constraint into the equations of nodes A and B and arranging the equations in standard form yields two equations in two unknowns:

$$\text{Node A:} \left(\frac{1}{50} + \frac{1}{-j120} + \frac{1}{j200}\right)\mathbf{V}_A - \left(\frac{1}{j200}\right)\mathbf{V}_B = \left(\frac{75\angle0°}{50}\right)$$

$$\text{Node B:} -\left(\frac{1}{j200}\right)\mathbf{V}_A + \left(\frac{1}{50} + \frac{1}{60} + \frac{1}{j200}\right)\mathbf{V}_B = \left(\frac{75\angle0°}{50}\right)$$

Solving these equations for $\mathbf{V}_A$ and $\mathbf{V}_B$ yields

$$\mathbf{V}_A = 70.4 - j21.4 \text{ V}$$

$$\mathbf{V}_B = 38.6 - j4.33 \text{ V}$$

Using these values for $\mathbf{V}_A$ and $\mathbf{V}_B$, the unknown current is found to be

$$\mathbf{I}_X = \frac{\mathbf{V}_A - \mathbf{V}_B}{j200} = \frac{31.8 - j17.1}{j200} = 0.180\angle-118° \text{ A}$$

This value of $\mathbf{I}_X$ is the same as the answer obtained in Example 8–18. ■

EXAMPLE 8–22

Use node-voltage analysis to determine the phasor input-output relationship of the OP AMP circuit in Figure 8–43.

FIGURE 8–43

SOLUTION:
In the sinusoidal steady state the sum of currents leaving the inverting input node is

$$\frac{\mathbf{V}_N - \mathbf{V}_S}{Z_1} + \frac{\mathbf{V}_N - \mathbf{V}_O}{Z_2} + \mathbf{I}_N = 0$$

This is the only required node equation, since input source forces the condition $\mathbf{V}_A = \mathbf{V}_S$ and no node equation is ever required at an OP AMP output. In the time domain the i–v relationships of an ideal OP AMP are $v_P(t) = v_N(t)$ and $i_P(t) = i_N(t) = 0$. In the sinusoidal steady state these equations are written in phasor form as $\mathbf{V}_P = \mathbf{V}_N$ and $\mathbf{I}_P = \mathbf{I}_N = 0$. In the present case this means $\mathbf{V}_N = 0$ since the noninverting input is grounded. When the ideal OP AMP constraints are inserted in the node equation, we can solve for the OP AMP input-output relationship as

$$\mathbf{V}_O = -\frac{Z_2}{Z_1}\mathbf{V}_S$$

This result is the phasor domain version of the inverting amplifier configuration. In the phasor domain, the "gain" $K = -Z_2/Z_1$ is determined by a ratio of impedances rather than resistances. Thus, the gain affects both the amplitude and the phase angle of the steady-state output. ■

◁ DESIGN AND EVALUATION EXAMPLE 8 – 2 3

(a) Use the inverting amplifier circuit in Figure 8–43 to design an OP AMP circuit that produces a steady-state output $v_O(t) = 2 \sin(2000t)$ V when the input is $v_S(t) = 15 \cos(2000t)$ V. Use only resistors and capacitors in Z_1 and Z_2.

(b) Compare the OP AMP circuit design in this example with the passive RLC circuit design developed in Example 8–7.

SOLUTION:

(a) The phasor representation of the input voltage is $\mathbf{V}_S = 15\angle 0 = 15 + j0$. Using the identity $\cos(x - 90°) = \sin x$, we write the required output phasor as $\mathbf{V}_O = 2\angle -90° = 0 - j2$. The design problem is to select the impedances Z_1 and Z_2 so that

$$0 - j2 = -\frac{Z_2}{Z_1}(15 + j0)$$

FIGURE 8 – 44

Solving for Z_1 yields $Z_1 = -j7.5 \times Z_2$. Selecting a value for Z_2 determines Z_1. The design problem statement limits us to resistors and capacitors. If we choose $Z_2 = 10^4$ (a 10-kΩ resistor) then $Z_1 = -j7.5 \times 10^4$, which is a capacitive reactance. Since $X_C = -(\omega C)^{-1}$, the required capacitance is $C = (7.5 \times 10^4 \times 2000) = 6.67$ nF. Figure 8–44 shows the OP AMP circuit design, which we recognize as the differentiating circuit from Chapter 6.

(b) The passive voltage divider designed in Example 8–7 meets the same design requirements used in this example. The two designs are summarized as follows:

EXAMPLE	FIGURE	DESCRIPTION	R	L	C	OP AMP
8–7	8–14	RLC voltage divider	1	1	1	0
8–16	8–44	RC inverting amplifier	1	0	1	1

Selecting a final design involves evaluation using additional factors, such as the cost of the power supply required by the OP AMP circuit versus the fact that it can drive reasonable loads. ∎

EXAMPLE 8 – 2 4

The circuit in Figure 8–45 is an equivalent circuit of an ac induction motor. The current $\mathbf{I}_S$ is called the stator current, $\mathbf{I}_R$ the rotor current, and $\mathbf{I}_M$ the magnetizing current. Use the mesh-current method to solve for the branch currents $\mathbf{I}_S$, $\mathbf{I}_R$, and $\mathbf{I}_M$.

FIGURE 8–45

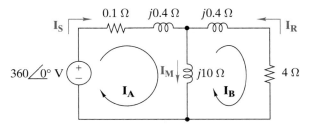

SOLUTION:
Applying KVL to the sum of voltages around each mesh in Figure 8–45 yields

Mesh A: $-360\angle 0° + (0.1 + j0.4)\mathbf{I}_A + j10(\mathbf{I}_A - \mathbf{I}_B) = 0$

Mesh B: $j10(\mathbf{I}_B - \mathbf{I}_A) + (4 + j0.4)\mathbf{I}_B \qquad = 0$

Arranging these equations in standard form yields

$$(0.1 + j10.4)\mathbf{I}_A - (j10)\mathbf{I}_B = 360\angle 0°$$

$$-(j10)\mathbf{I}_A + (4 + j10.4)\mathbf{I}_B = 0$$

Solving these equations for $\mathbf{I}_A$ and $\mathbf{I}_B$ produces

$$\mathbf{I}_A = 79.0 - j48.2 \text{ A}$$

$$\mathbf{I}_B = 81.7 - j14.9 \text{ A}$$

The required stator, rotor, and magnetizing currents are related to these mesh currents, as follows:

$$\mathbf{I}_S = \mathbf{I}_A = 92.5\angle -31.4° \text{ A}$$

$$\mathbf{I}_R = -\mathbf{I}_B = -81.8 + j14.7 = 83.0\angle 170° \text{ A}$$

$$\mathbf{I}_M = \mathbf{I}_A - \mathbf{I}_B = -2.68 - j33.3 = 33.4\angle -94.6° \text{ A} \qquad ■$$

EXAMPLE 8–25

Use the mesh-current method to solve for output voltage $\mathbf{V}_2$ and input impedance Z_{IN} of the circuit in Figure 8–46.

FIGURE 8–46

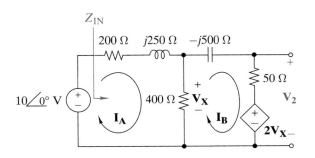

SOLUTION:

The circuit contains a voltage-controlled voltage source. We initially treat the dependent source as an independent source and use KCL to write the sum of voltages around each mesh:

$$\text{Mesh A:} \ -10\angle0° + (200 + j250)\mathbf{I}_A + 400(\mathbf{I}_A - \mathbf{I}_B) = 0$$

$$\text{Mesh B:} \ \ 400(\mathbf{I}_B - \mathbf{I}_A) + (50 - j500)\mathbf{I}_B + 2\mathbf{V}_X \ \ \ \ = 0$$

Arranging these equations in standard form produces

$$\text{Mesh A:} \ (600 + j250)\mathbf{I}_A - 400\mathbf{I}_B \ \ \ = 10\angle0°$$

$$\text{Mesh B:} \ -400\mathbf{I}_A + (450 - j500)\mathbf{I}_B = \ -2\mathbf{V}_X$$

Using Ohm's law, the control voltage $\mathbf{V}_X$ is

$$\mathbf{V}_X = 400(\mathbf{I}_A - \mathbf{I}_B)$$

Eliminating $\mathbf{V}_X$ from the mesh equations yields

$$\text{Mesh A:} \ (600 + j250)\mathbf{I}_A - 400\mathbf{I}_B \ \ \ = 10\angle0°$$

$$\text{Mesh B:} \ 400\mathbf{I}_A + (-350 - j500)\mathbf{I}_B = 0$$

Solving for the two mesh currents produces

$$\mathbf{I}_A = 10.8 - j11.1 \ \text{mA}$$

$$\mathbf{I}_B = \ -1.93 - j9.95 \ \text{mA}$$

Using these values of the mesh currents, the output voltage and input impedance are

$$\mathbf{V}_2 = 2\mathbf{V}_X + 50\mathbf{I}_B = 800(\mathbf{I}_A - \mathbf{I}_B) + 50\mathbf{I}_B$$

$$= 800\mathbf{I}_A - 750\mathbf{I}_B = 10.1 - j1.42$$

$$= 10.2\angle-8.00° \ \text{V}$$

$$Z_{IN} = \frac{10\angle0°}{\mathbf{I}_A} = \frac{10\angle0°}{0.0108 - j0.0111} = 450 + j463 \ \Omega \qquad \blacksquare$$

EXAMPLE 8–26

In the circuit in Figure 8–47 the input voltage is $v_S(t) = 10 \cos 10^5 t$ V. Use mesh equations and MATLAB to find the input impedance at the input interface and the proportionality constant relating the input voltage phasor to the phasor voltage across the 50-Ω load resistor.

SOLUTION:

In matrix form, the mesh-current equations for this circuit are

$$\begin{bmatrix} R_S + Z_{C1} + Z_{L2} & -Z_{L2} & -Z_{C1} \\ -Z_{L2} & Z_{L2} + Z_{C2} + R_L & -Z_{C2} \\ -Z_{C1} & -Z_{C2} & Z_{C1} + Z_{C2} + Z_{L1} \end{bmatrix} \begin{bmatrix} \mathbf{I}_A \\ \mathbf{I}_B \\ \mathbf{I}_C \end{bmatrix} = \begin{bmatrix} \mathbf{V}_S \\ 0 \\ 0 \end{bmatrix}$$

FIGURE   8–47

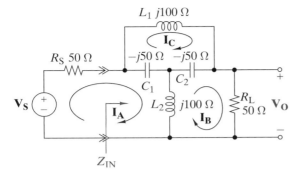

This matrix equation is of the form $\mathbf{ZI} = \mathbf{V}$, where $\mathbf{Z}$ is a 3×3 matrix of element impedance, $\mathbf{I}$ is the 3×1 mesh current vector, and $\mathbf{V}$ is the 3×1 input voltage vector. The entries on the main diagonal of the $\mathbf{Z}$ matrix are the total impedances in each mesh. The $\mathbf{Z}$ matrix is symmetrical because the off diagonal entries are the negatives of the impedances common to two meshes. For example, the entry in the 2nd row and 3rd column $(-Z_{C2})$ is equal to the entry in the 3rd row and 2nd column.

With the circuit equations formulated, we can use MATLAB to solve for the required circuit descriptors. To solve for the mesh current vector, we need the matrix product $\mathbf{Z}^{-1}\mathbf{V}$. We have used MATLAB for this purpose in previous examples. The only thing new here is that the matrices contain complex numbers.

One of the advantages of MATLAB is the ease with which it handles complex numbers. Complex inputs are identified using either the mathematical convention ($i = \sqrt{-1}$) or the electrical engineering convention ($j = \sqrt{-1}$). That is, MATLAB accepts complex number inputs written as z=x+yi or z=x+yj. Accordingly, in the MATLAB command window, we enter the element parameters as

```
VS=10; RS=50; RL=50;
ZC1=-50j; ZC2=-50j;
ZL1=100j; ZL2=100j;
```

the $\mathbf{Z}$ and $\mathbf{V}$ matrices as

```
Z=[RS+ZC1+ZL2      -ZL2         -ZC1
        -ZL2     ZL2+ZC2+RL     -ZC2
        -ZC1        -ZC2      ZC1+ZC2+ZL1];
V=[VS ; 0 ; 0];
```

and solve for the mesh currents as

```
I=inv(Z)*V
I=
          0.0100 - 0.0300i
         -0.0100 + 0.0300i
               0 - 0.1000i
```

The form of the output points out that MATLAB accepts inputs in either convention, but it reports complex number outputs using the mathematical convention. In MATLAB matrix notation, the mesh currents are $\mathbf{I}_A = \mathrm{I}(1)$, $\mathbf{I}_B = \mathrm{I}(2)$, and $\mathbf{I}_C = \mathrm{I}(3)$. We can now calculate the required parameters as

```
ZIN=(VS-I(1)*RS)/I(1)
ZIN =
      5.0000e+001 + 3.0000e+002i
K=I(2)*RL/VS
K=
     -0.05000 + 0.15000i
```

That is, in electrical engineering notation, $Z_{IN} = 50 + j300\ \Omega$ and $K = -0.05 + j0.15$. ∎

Exercise 8–16

Use the mesh-current or node-voltage method to find the branch currents $\mathbf{I}_1$, $\mathbf{I}_2$, and $\mathbf{I}_3$ in Figure 8–48.

FIGURE 8 – 4 8

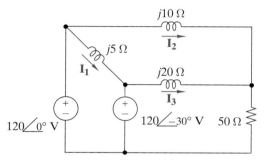

Answer:

$$\mathbf{I}_1 = 12.4\angle-15°\ A,\ \mathbf{I}_2 = 3.61\angle-16.1°\ A,\ \mathbf{I}_3 = 1.31\angle166°\ A$$

Exercise 8–17

Use the mesh-current or node-voltage method to find the output voltage $\mathbf{V}_2$ and input impedance Z_{IN} in Figure 8–49.

FIGURE 8 – 4 9

Answer:

$$\mathbf{V}_2 = 1.77\angle-135°\ V,\ Z_{IN} = 100 - j100\ \Omega$$

Use the mesh-current or node-voltage method to find the current $\mathbf{I}_X$ in Figure 8–50.

Answer:

$\mathbf{I}_X = 1.44\angle171°$ mA

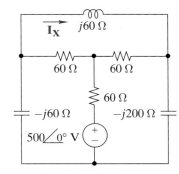

FIGURE 8 – 50

8–7 ENERGY AND POWER

In the sinusoidal steady state, ac power is transferred from sources to various loads. To study the transfer process, we must calculate the power delivered in the sinusoidal steady state to any specified load. It turns out that there is an upper bound on the available load power; hence, we need to understand how to adjust the load to extract the maximum power from the rest of the circuit. In this section the load is assumed to be made up of passive resistance, inductance, and capacitance. To reach our objectives, we must first study the power and energy delivered to these passive elements in the sinusoidal steady state.

In the sinusoidal steady state the current through a resistor can be expressed as $i_R(t) = I_A \cos(\omega t)$. The instantaneous power delivered to the resistor is

$$p_R(t) = Ri_R^2(t) = RI_A^2 \cos^2(\omega t) \qquad (8–30)$$

$$= \frac{RI_A^2}{2}[1 + \cos(2\omega t)]$$

where the identity $\cos^2(x) = \frac{1}{2}[1 + \cos(2x)]$ is used to obtain the last line in Eq. (8–30). The energy delivered for $t \geq 0$ is found to be

$$w_R(t) = \int_0^t p_R(x)dx = \frac{RI_A^2}{2}\int_0^t dx + \frac{RI_A^2}{2}\int_0^t \cos 2\omega x \, dx$$

$$= \frac{RI_A^2}{2}t + \frac{RI_A^2}{4\omega} \sin 2\omega t$$

Figure 8–51 shows the time variation of $p_R(t)$ and $w_R(t)$. Note that the power is a periodic function with twice the frequency of the current, that both $p_R(t)$ and $w_R(t)$ are always positive, and that $w_R(t)$ increases without bound. These observations remind us that a resistor is a passive element that dissipates energy.

In the sinusoidal steady state an inductor operates with a current $i_L(t) = I_A \cos(\omega t)$. The corresponding energy stored in the element is

$$w_L(t) = \frac{1}{2} Li_L^2(t) = \frac{1}{2} LI_A^2 \cos^2 \omega t$$

$$= \frac{1}{4} LI_A^2(1 + \cos 2\omega t)$$

where the identity $\cos^2(x) = \frac{1}{2}[1 + \cos(2x)]$ is again used to produce the last line. The instantaneous power delivered to the inductor is

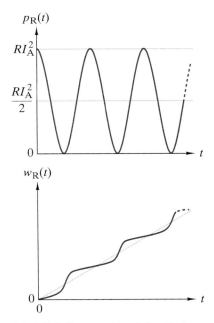

FIGURE   8 – 51 *Resistor power and energy in the sinusoidal steady state.*

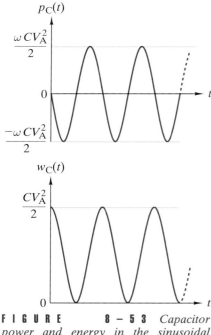

FIGURE 8–52 *Inductor power and energy in the sinusoidal steady state.*

FIGURE 8–53 *Capacitor power and energy in the sinusoidal steady state.*

$$p_L(t) = \frac{dw_L}{dt} = -\frac{\omega L I_A^2}{2}\sin(2\omega t) \qquad (8\text{–}31)$$

Figure 8–52 shows the time variation of $p_L(t)$ and $w_L(t)$. Observe that both $p_L(t)$ and $w_L(t)$ are periodic functions at twice the frequency of the ac current, that $p_L(t)$ is alternately positive and negative, and that $w_L(t)$ is never negative. Since $w_L(t) \geq 0$, the inductor does not deliver net energy to the rest of the circuit. Unlike the resistor's energy in Figure 8–51, the energy in the inductor is bounded by $\frac{1}{2} L I_A^2 \geq w_L(t)$, which means that the inductor does not dissipate energy. Finally, since $p_L(t)$ alternates signs, we see that the inductor stores energy during a positive half cycle and then returns the energy undiminished during the next negative half cycle. Thus, in the sinusoidal steady state there is a lossless interchange of energy between an inductor and the rest of the circuit.

In the sinusoidal steady state the voltage across a capacitor is $v_C(t) = V_A \cos(\omega t)$. The energy stored in the element is

$$w_C(t) = \frac{1}{2}Cv_C^2(t) = \frac{1}{2}CV_A^2 \cos^2 \omega t$$

$$= \frac{1}{4}CV_A^2(1 + \cos 2\omega t)$$

The instantaneous power delivered to the capacitor is

$$p_C(t) = \frac{dw_C}{dt} = -\frac{\omega C V_A^2}{2}\sin(2\omega t) \qquad (8\text{–}32)$$

Figure 8–53 shows the time variation of $p_C(t)$ and $w_C(t)$. Observe that these relationships are the duals of those found for the inductor. Thus, in the sinusoidal steady state the element power is sinusoidal and there is a lossless interchange of energy between the capacitor and the rest of the circuit.

AVERAGE POWER

We are now in a position to calculate the average power delivered to various loads. The instantaneous power delivered to any of the three passive elements is a periodic function that can be described by an average value. The **average power** in the sinusoidal steady state is defined as

$$P = \frac{1}{T_0} \int_0^{T_0} p(t)dt$$

The power variation of the inductor in Eq. (8–31) and capacitor in Eq. (8–32) have the same sinusoidal form. The average value of any sinusoid is zero since the areas under alternate cycles cancel. Hence, the average power delivered to an inductor or capacitor is zero:

$$\text{Inductor: } P_L = 0$$

$$\text{Capacitor: } P_C = 0$$

The resistor power in Eq. (8–30) has both a sinusoidal ac component and a constant dc component $\frac{1}{2} R I_A^2$. The average value of the ac component is zero, but the dc component yields

$$\text{Resistor: } P_{\text{R}} = \frac{1}{2}RI_{\text{A}}^2$$

To calculate the average power delivered to an arbitrary load $Z_{\text{L}} = R_{\text{L}} + jX_{\text{L}}$, we use phasor circuit analysis to find the phasor current $\mathbf{I}_{\text{L}}$ through Z_{L}. The average power delivered to the load is dissipated in R_{L}, since the reactance X_{L} represents the net inductance or capacitance of the load. Hence the average power to the load is

$$P = \frac{1}{2}R_{\text{L}}|\mathbf{I}_{\text{L}}|^2 \qquad (8\text{–}33)$$

Caution: When a circuit contains two or more sources, superposition applies only to the total load current and not to the total load power. You cannot find the total power to the load by summing the power delivered by each source acting alone.

The following example illustrates a power transfer calculation.

EXAMPLE 8–27

Find the average power delivered to the load to the right of the interface in Figure 8–54.

FIGURE 8–54

SOLUTION:
The equivalent impedance to the right of the interface is

$$Z_{\text{L}} = j250 + \cfrac{1}{\cfrac{1}{-j75} + \cfrac{1}{100}} = 36 + j202 \ \ \Omega$$

The current delivered to the load is

$$\mathbf{I}_{\text{L}} = \frac{150\angle 0°}{50 + Z_{\text{L}}} = 0.683\angle -66.9° \ \text{A}$$

Hence the average power delivered across the interface is

$$P = \frac{1}{2}R_{\text{L}}|\mathbf{I}_{\text{L}}|^2 = \frac{36}{2}|0.683|^2 = 8.40 \ \ \text{W}$$

Note: All of this power goes into the 100-Ω resistor since the inductor and capacitor do not absorb average power. ∎

Exercise 8–19

Find the average power delivered to the 25-Ω load resistor in Figure 8–55.

Answer: 234 W

MAXIMUM POWER

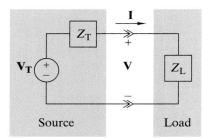

F I G U R E   8 – 5 6   *A source-load interface in the sinusoidal steady state.*

To address the maximum power transfer problem, we model the source/load interface as shown in Figure 8–56. The source circuit is represented by a Thévenin equivalent circuit with source voltage $\mathbf{V}_T$ and source impedance $Z_T = R_T + jX_T$. The load circuit is represented by an equivalent impedance $Z_L = R_L + jX_L$. In the maximum power transfer problem the source parameters $\mathbf{V}_T$, R_T, and X_T are given, and the objective is to adjust the load impedance R_L and X_L so that average power to the load is a maximum.

The average power to the load is expressed in terms of the phasor current and load resistance:

$$P = \frac{1}{2} R_L |\mathbf{I}|^2$$

Then, using series equivalence, we express the magnitude of the interface current as

$$|\mathbf{I}| = \left| \frac{\mathbf{V}_T}{Z_T + Z_L} \right| = \frac{|\mathbf{V}_T|}{|(R_T + R_L) + j(X_T + X_L)|}$$

$$= \frac{|\mathbf{V}_T|}{\sqrt{(R_T + R_L)^2 + (X_T + X_L)^2}}$$

Combining the last two equations yields the average power delivered across the interface:

$$P = \frac{1}{2} \frac{R_L |\mathbf{V}_T|^2}{(R_T + R_L)^2 + (X_T + X_L)^2} \tag{8–34}$$

The quantities $|\mathbf{V}_T|$, R_T, and X_T in Eq. (8–34) are fixed. Our problem is to select R_L and X_L to maximize P.

Clearly, for every value of R_L the denominator in Eq. (8–34) is minimized and P maximized when $X_L = -X_T$. This choice of X_L is possible because a reactance can be positive or negative. When the source Thévenin equivalent has an inductive reactance ($X_T > 0$), we modify the load to have a capacitive reactance of the same magnitude, and vice versa. This step reduces the net reactance of the series connection in Figure 8–56 to zero, creating a condition in which the net impedance seen by the Thévenin voltage source is purely resistive.

When the source and load reactances cancel out, the expression for average power in Eq. (8–34) reduces to

$$P = \frac{1}{2}\frac{R_L|\mathbf{V}_T|^2}{(R_T + R_L)^2}$$ (8–35)

This equation has the same form encountered in Chapter 3 in dealing with maximum power transfer in resistive circuits. From the derivation in Sect. 3–5, we know P is maximized when $R_L = R_T$. In summary, to obtain maximum power transfer in the sinusoidal steady state, we select the load resistance and reactance so that

$$R_L = R_T \quad \text{and} \quad X_L = -X_T$$ (8–36)

These conditions can be compactly expressed in the following way:

$$Z_L = Z_T^*$$ (8–37)

The condition for maximum power transfer is called a **conjugate match**, since the load impedance is the conjugate of the source impedance. When the conjugate-match conditions are inserted into Eq. (8–34), we find that the maximum average power available from the source circuit is

$$P_{\text{MAX}} = \frac{|\mathbf{V}_T|^2}{8R_T}$$ (8–38)

where $|\mathbf{V}_T|$ is the peak amplitude of the Thévenin equivalent voltage.

It is important to remember that conjugate matching applies when the source is fixed and the load is adjustable. These conditions arise frequently in power-limited communication systems. However, as we will see in Chapter 16, conjugate matching does not apply to electrical power systems because the power transfer constraints are different.

EXAMPLE 8–28

(a) Calculate the average power delivered to the load in the circuit shown in Figure 8–57 for $v_S(t) = 5 \cos 10^6 t$ V, $R = 200\ \Omega$, $R_L = 200\ \Omega$, and $C = 10$ nF.
(b) Calculate the maximum average power available at the interface and specify the load required to draw the maximum power.

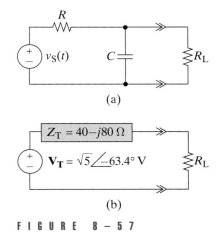

(a)

(b)

FIGURE 8–57

SOLUTION:

(a) To find the power delivered to the 200-Ω load resistor, we use a Thévenin equivalent circuit. By voltage division, the open-circuit voltage at the interface is

$$\mathbf{V}_T = \frac{Z_C}{Z_R + Z_C}\mathbf{V}_S = \frac{-j100}{200 - j100}5\angle0°$$

$$= 1 - j2 = \sqrt{5}\angle-63.4° \text{ V}$$

By inspection, the short-circuit current at the interface is

$$\mathbf{I}_N = \frac{5\angle0°}{200} = 0.025 + j0 \text{ A}$$

Given $\mathbf{V}_T$ and $\mathbf{I}_N$, we calculate the Thévenin source impedance.

$$Z_T = \frac{\mathbf{V}_T}{\mathbf{I}_N} = \frac{1 - j2}{0.025} = 40 - j80 \quad \Omega$$

Using the Thévenin equivalent shown in Figure 8–57(b), we find that the current through the 200-Ω resistor is

$$\mathbf{I} = \frac{\mathbf{V}_T}{Z_T + Z_L} = \frac{\sqrt{5}\angle -63.4°}{40 - j80 + 200} = 8.84\angle -45° \quad \text{mA}$$

and the average power delivered to the load resistor is

$$P = \frac{1}{2}R_L|\mathbf{I}|^2 = 100(8.84 \times 10^{-3})^2 = 7.81 \quad \text{mW}$$

(b) Using Eq. (8–38), the maximum average power available at the interface is

$$P_{\text{MAX}} = \frac{|\mathbf{V}_T|^2}{8R_T} = \frac{(\sqrt{5})^2}{(8)(40)} = 15.6 \quad \text{mW}$$

The 200-Ω load resistor in part (a) draws about half of the maximum available power. To extract maximum power, the load impedance must be

$$Z_L = Z_T^* = 40 + j80 \quad \Omega$$

This impedance can be obtained using a 40-Ω resistor in series with a reactance of +80 Ω. The required reactance is inductive (positive) and can be produced by an inductance of

$$L = \frac{|X_T|}{\omega} = \frac{80}{10^6} = 80 \;\mu\text{H} \qquad \blacksquare$$

Exercise 8–20

Calculate the maximum average power available at the interface in Figure 8–58.

Answer: 125 mW.

FIGURE 8-58

$0.1\angle 0°$ A
$j100 \; \Omega$
$50 \; \Omega$
$-j50 \; \Omega$

SUMMARY

- A phasor is a complex number representing a sinusoidal waveform. The magnitude and angle of the phasor correspond to the amplitude and phase angle of the sinusoid. The phasor does not provide frequency information.

- The additive property states that adding phasors is equivalent to adding sinusoids of the same frequency. The derivative property states that multiplying a phasor by $j\omega$ is equivalent to differentiating the corresponding sinusoid.

- In the sinusoidal steady state, phasor currents and voltages obey Kirchhoff's laws and the element i–v relationships are written in terms of impedances. Impedance can be defined as the ratio of phasor voltage over phasor current. The device and connection constraints for phasor circuit analysis have the same form as resistance circuits.

- Phasor circuit analysis techniques include series equivalence, parallel equivalence, Y-Δ transformations, circuit reduction, Thévenin's and Norton's theorems, unit output method, superposition, node-voltage analysis, and mesh-current analysis.

- In the sinusoidal steady state the equivalent impedance at a pair of terminals is $Z(j\omega) = R(\omega) + jX(\omega)$, where $R(\omega)$ is called resistance and $X(\omega)$ is called reactance. A frequency at which an equivalent impedance is purely real is called a resonant frequency. Admittance is the reciprocal of impedance.

- In the sinusoidal steady state the instantaneous power to a passive element is a periodic function at twice the frequency of the driving force. The average power delivered to an inductor or capacitor is zero. The average power delivered to a resistor is $\frac{1}{2} R|\mathbf{I}_R|^2$. The maximum average power is delivered by a fixed source to an adjustable load when the source and load impedances are conjugates.

PROBLEMS

ERO 8–1 SINUSOIDS AND PHASORS (SECT. 8–1)

Use the additive and derivative properties of phasors to convert sinusoidal waveforms into phasors and vice versa.

8–1 Transform the following sinusoids into phasor form. Use the additive property of phasors to find $v_1(t) + v_2(t)$. Construct a phasor diagram showing these voltages.

$$\text{(a)}\ v_1(t) = 100 \cos(\omega t + 45°)$$

$$\text{(b)}\ v_2(t) = 150 \cos \omega t + 200 \sin \omega t$$

8–2 Transform the following sinusoids into phasor form. Use the additive property of phasors to find $i_1(t) + i_2(t)$. Construct the phasor diagram showing these currents.

$$\text{(a)}\ i_1(t) = 6 \cos \omega t$$

$$\text{(b)}\ i_2(t) = 3 \sin (\omega t - 90°)$$

8–3 Convert the following phasors into sinusoidal waveforms at a frequency of 10 krad/s:

$$\text{(a)}\ \mathbf{V}_1 = 10e^{-j30°}\quad \text{(b)}\ \mathbf{V}_2 = 60e^{-j220°}$$

$$\text{(c)}\ \mathbf{I}_1 = 5e^{j90°}\qquad \text{(d)}\ \mathbf{I}_2 = 2e^{j270°}$$

8–4 Use the additive property on the phasors in Problem 8–3 to find $v_1(t) - v_2(t)$ and $i_1(t) - i_2(t)$. Construct phasor diagrams for the phasors and the sums.

8–5 Use the phasor derivative property to find the time derivative of the waveform $v(t)$ when $\mathbf{V} = 10 + j5$ V and $\omega = 500$ rad/s.

8–6 Convert the following phasors into sinusoids at a frequency of 200 rad/s:

$$\text{(a) } \mathbf{V}_1 = \frac{10 + j10}{3 - j2} \qquad \text{(b) } \mathbf{V}_2 = (8 - j3)5e^{-j60°}$$

$$\text{(c) } \mathbf{I}_1 = \frac{10}{3 + j1} \qquad \text{(d) } \mathbf{I}_2 = \frac{3 + j1}{1 - j3}$$

8–7   Use the additive and derivative properties of phasors to find the $v(t)$ that meets the condition

$$\frac{dv(t)}{dt} + 50v(t) = \sin 100t$$

8–8   Given the sinusoids

$$v_1(t) = 50 \cos(\omega t - 45°) \qquad \text{and} \qquad v_2(t) = 25 \sin \omega t$$

use the additive property of phasors to find $v_3(t)$ such that $v_1 + v_2 + v_3 = 0$.

8–9   The phasor $\mathbf{V}_1 = 6 + j4$ is rotated clockwise by 100°. Express the resulting phasor in rectangular form.

8–10   Given a phasor $\mathbf{V}_1 = -3 + j4$, use phasor methods to find a voltage $v_2(t)$ that leads $v_1(t)$ by 90° and has an amplitude of 10 V.

ERO 8–2 EQUIVALENT IMPEDANCE (SECTS. 8–2, 8–3)

Given a linear circuit, use series and parallel equivalence to find the equivalent impedance at a specified pair of terminals.

8–11   Express the equivalent impedance of the following circuits in rectangular and polar form:
  (a) A 50-Ω resistor in series with a 20-mH inductor at $\omega = 2000$ rad/s.
  (b) A 50-Ω resistor in parallel with a 20-μF capacitor at $\omega = 2000$ rad/s.
  (c) The circuit formed by connecting the circuits in (a) and (b) in parallel.
  (d) Repeat (c) at $\omega = 4000$ rad/s.

8–12   Find the equivalent impedance Z in Figure P8–12. Express the result in both polar and rectangular form.

8–13   Express the equivalent impedance of the following circuits in rectangular and polar form:
  (a) A 100-mH inductor in series with a 1-μF capacitor at $\omega = 2000$ rad/s.
  (b) A 30-mH inductor in parallel with 60-Ω resistor at $\omega = 2000$ rad/s.
  (c) The circuit formed by connecting the circuits in (a) and (b) in series.
  (d) Repeat (c) at $\omega = 4000$ rad/s.

8–14   Find the equivalent impedance Z in Figure P8–14. Express the result in both polar and rectangular form.

8–15   The voltage applied at the input to a linear circuit is $v(t) = 200 \cos(1000t - 45°)$ V. In the sinusoidal steady state the input current is observed to be $i(t) = 20 \cos 1000t$ mA.
  (a) Find the equivalent impedance at the input.
  (b) Find the steady-state current $i(t)$ for $v(t) = 150 \cos(1000t - 270°)$ V.

8–16   The circuit in Figure P8–16 is operating in the sinusoidal steady state with $\omega = 1$ krad/s. Find the equivalent impedance Z. Which element dominates the equivalent impedance? Repeat for $\omega = 100$ krad/s.

F I G U R E   P 8 – 1 2

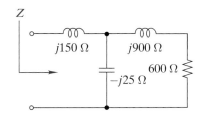

F I G U R E   P 8 – 1 4

F I G U R E   P 8 – 1 6

8–17 **D** An inductor L is connected in parallel with a 600-Ω resistor. The parallel combination is then connected in series with a capacitor C. Select the values of L and C so that the equivalent impedance of the combination is $100 + j0$ Ω at $\omega = 100$ krad/s.

8–18 A 200-Ω resistor is connected in parallel with an unknown reactive element. When the parallel combination is connected across a source $v_S(t) = 100 \cos(1000t)$ V the total current delivered is $0.707 \cos(1000t + 45°)$ A. Is the unknown element a capacitor or inductor and what is its value?

8–19 **D** A capacitor C is connected in series with a 50-Ω resistor. The series combination is then connected in parallel with an inductor L. Select the values of L and C so that the equivalent impedance of the combination is $600 + j0$ Ω at $\omega = 1$ Mrad/s.

8–20 A relay coil can be modeled as a resistor and inductor in series. When a certain coil is connected across a 110-V, 60-Hz source the amplitude of the relay current is observed to be 0.25 A. When the frequency is increased to 400 Hz the amplitude drops to 0.12 A. Find the resistance and inductance of the coil.

ERO 8–3 BASIC PHASOR CIRCUIT ANALYSIS (SECTS. 8–2, 8–3, 8–4)

Given a linear circuit operating in the sinusoidal steady state, find phasor responses using basic analysis methods such as series and parallel equivalence, voltage and current division, circuit reduction, Thévenin or Norton equivalent circuits, and proportionality or superposition.

8–21 A voltage $v(t) = 10 \cos 2500t$ V is applied across a series connection of a 100-Ω resistor and 40-mH inductor. Find the steady-state current $i(t)$ through the series connection. Draw a phasor diagram showing **V** and **I**.

8–22 The circuit in Figure P8–22 is operating in the sinusoidal steady state with $v_S(t) = V_A \sin \omega t$ V. Use phasor circuit reduction to derive a general expression for the steady-state response $i_L(t)$.

8–23 A current source delivering $i(t) = 300 \cos 377t$ mA is connected across a parallel combination of a 100-kΩ resistor and a 50-nF capacitor. Find the steady-state current $i_R(t)$ through the resistor and the steady-state current $i_C(t)$ through the capacitor. Draw a phasor diagram showing **I**, $\mathbf{I}_C$, and $\mathbf{I}_R$.

8–24 The circuit in Figure P8–24 is operating in the sinusoidal steady state with $i_S(t) = I_A \sin \omega t$ A. Use phasor circuit reduction to derive a general expression for the steady-state response $v_R(t)$.

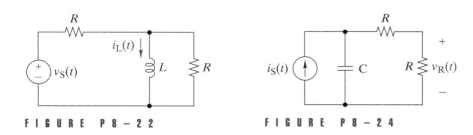

FIGURE P8–22   FIGURE P8–24

8–25 A voltage $v(t) = 50 \cos(1000t + 45°)$ V is applied across a parallel connection of a 5-kΩ resistor and a 200-nF capacitor. Find the steady-state current $i_C(t)$ through the capacitor and the steady-state current $i_R(t)$ through the resistor. Draw a phasor diagram showing **V**, **I**$_C$, and **I**$_R$.

8–26 The circuit in Figure P8–26 is operating in the sinusoidal steady state. Use circuit reduction to find the input impedance seen by the voltage source and the steady-state response $v_x(t)$.

8–27 The circuit in Figure P8–27 is operating in the sinusoidal steady state. Use circuit reduction to find the input impedance seen by the current source and steady-state response $v_x(t)$.

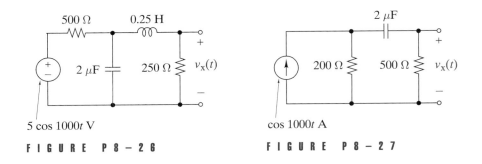

FIGURE P8–26   FIGURE P8–27

8–28 The circuit in Figure P8–28 is operating in the sinusoidal steady state. Use circuit reduction to find the input impedance seen by the voltage source and the steady-state phasor response **V**$_x$.

FIGURE P8–28

8–29 The circuit in Figure P8–29 is operating in the sinusoidal steady state. Use superposition to find the phasor response **I**$_x$.

FIGURE P8–29

8–30 The circuit in Figure P8–30 is operating in the sinusoidal steady state. Use superposition to find the phasor response $\mathbf{V}_x$.

FIGURE   P 8 – 3 0

8–31 The circuit in Figure P8–31 is operating in the sinusoidal steady state. Use superposition to find the response $v_x(t)$. *Note:* The sources do not have the same frequency.

8–32 The circuit in Figure P8–32 is operating in the sinusoidal steady state. Use the unit output method to find the input impedance seen by the voltage source and the steady-state response $v_x(t)$.

FIGURE   P 8 – 3 1

FIGURE   P 8 – 3 2

8–33 The circuit in Figure P8–33 is operating in the sinusoidal steady state. Use the unit output method to find the input impedance seen by the voltage source and the phasor response $\mathbf{V}_x$.

FIGURE   P 8 – 3 3

8–34 Find the phasor Thévenin equivalent of the source circuit to the left of the interface in Figure P8–34. Then use the equivalent circuit to find the steady-state voltage $v(t)$ and current $i(t)$ delivered to the load.

8–35 Find the phasor Thévenin equivalent of the source circuit to the left of the interface in Figure P8–35. Then use the equivalent circuit to find the phasor voltage $\mathbf{V}$ and current $\mathbf{I}$ delivered to the load.

FIGURE   P 8 – 3 4

FIGURE P8–35

FIGURE P8–36

8–36 The circuit in Figure P8–36 is operating in the sinusoidal steady state. When $Z_L = 0$ the phasor current at the interface is $\mathbf{I} = 3.6 - j4.8$ mA. When $Z_L = -j40$ kΩ the phasor interface current is $\mathbf{I} = 10 + j0$ mA. Find the Thévenin equivalent of the source circuit.

8–37 The circuit in Figure P8–36 is operating in the sinusoidal steady state with $\omega = 2$ Mrad/s. When $Z_L = 0$ the phasor short-circuit current is $\mathbf{I} = \mathbf{I}_{SC} = 0.4 - j0.8$ A. When a 1-nF capacitor is connected across the interface the current delivered is $\mathbf{I} = \mathbf{I}_{LOAD} = 2 + j0$ A. Note that the amplitude of the load current is greater than the amplitude of the short-circuit current. Find the Thévenin equivalent of the source circuit and then explain why $|\mathbf{I}_{LOAD}| > |\mathbf{I}_{SC}|$.

8–38 **D** Design a two-port circuit so that an input voltage $v_S(t) = 10 \cos(2000t)$ V produces a steady-state output voltage of $v_O(t) = 5 \cos(2000t - 45°)$ V.

8–39 **D** Design a two-port circuit so that an input voltage $v_S(t) = 100 \cos(10^4 t)$ V delivers a steady-state output current of $i_O(t) = 10 \cos(10^4 t - 35°)$ mA to a 50-Ω resistive load.

8–40 **D** Design a two-port circuit so that an input current $i_S(t) = 25 \cos(5000t)$ mA delivers a steady-state output voltage of $v_O(t) = 50 \cos(5000t - 45°)$ V and has an input impedance $|Z| > 5$ kΩ.

ERO 8–4 GENERAL CIRCUIT ANALYSIS (SECT. 8–5)

Given a linear circuit operating in the sinusoidal steady-state, find equivalent impedances and phasor responses using node-voltage or mesh-current analysis.

8–41 Use node-voltage analysis to find the sinusoidal steady-state response $v_x(t)$ in the circuit shown in Figure P8–41.

FIGURE P8–41

8–42 Use node-voltage analysis to find the steady-state phasor response $\mathbf{V}_O$ in the circuit shown in Figure P8–42.

FIGURE P 8 – 4 2

8–43 Use node-voltage analysis to find the input impedance Z_{IN} and gain $K = \mathbf{V}_O/\mathbf{V}_S$ of the circuit shown in Figure P8–43 with $\mu = 100$.

FIGURE P 8 – 4 3

8–44 Use mesh-current analysis to find the phasor branch currents $\mathbf{I}_1$, $\mathbf{I}_2$, and $\mathbf{I}_3$ in the circuit shown in Figure P8–44.

FIGURE P 8 – 4 4

8–45 Use mesh-current analysis to find the phasor response $\mathbf{V}_y$ and $\mathbf{I}_x$ in the circuit shown in Figure P8–45.

FIGURE P 8 – 4 5

8–46 Use mesh-current analysis to find the input impedance Z_{IN} and gain $K = \mathbf{V}_O/\mathbf{V}_S$ of the circuit shown in Figure P8–46.

FIGURE P8–46

8–47 Find the input impedance Z_{IN} and gain $K = \mathbf{V}_O/\mathbf{V}_S$ of the circuit shown in Figure P8–47.

8–48 Find the sinusoidal steady-state response $v_O(t)$ in the circuit shown in Figure P8–48. The element values are $\mathbf{V}_S = 80$ mV, $R_1 = 10$ kΩ, $R_P = 5$ kΩ, $R_F = 1$ MΩ, $R_C = 10$ kΩ, $R_L = 100$ kΩ, $C = 0.25$ nF, and $\beta = 50$. The frequency is $\omega = 500$ rad/s.

FIGURE P8–47

FIGURE P8–48

8–49 Find the phasor response $\mathbf{I}_{IN}$ and $\mathbf{V}_O$ in the circuit shown in Figure P8–49.

FIGURE P8–50

FIGURE P8–49

8–50 The OP AMP circuit in Figure P8–50 is operating in the sinusoidal steady state with $\omega = 100$ krad/s. Find the output phasor $\mathbf{V}_O$ when the input phasor $\mathbf{V}_S = 100 + j0$ V.

ERO 8–5 AVERAGE POWER AND MAXIMUM POWER TRANSFER (SECT. 8–6)

Given a linear circuit operating in the sinusoidal steady state,

(a) Find the average power delivered at a specified interface.
(b) Find the maximum average power available at a specified interface.
(c) Find the load impedance required to draw the maximum available power.

8–51 A load consisting of a 50-Ω resistor in parallel with a 100-mH inductor is connected across a voltage source $v_S(t) = 35 \cos 500t$ V. Find the phasor voltage, current, and average power delivered to the load.

8–52 A load consisting of a 50-Ω resistor in parallel with a 80-μF capacitor is connected across a voltage source $v_S(t) = 50 \cos 2500t$ V. Find the phasor voltage, current, and average power delivered to the load.

8–53 A load consisting of a resistor and capacitor connected in parallel draws an average power of 100 mW and a peak current of 44 mA when connected to voltage source $v_S(t) = 16 \cos 2500t$ V. Find the values of R and C.

8–54 A load consisting of a 600-Ω resistor in parallel with an inductor whose reactance is 800 Ω is connected to a current source whose peak amplitude is 26 mA. Find the average power delivered to the load.

8–55 (a) Find the average power delivered to the load in Figure P8–55.
(b) Find the maximum available average power at the interface shown in Figure P8–55.
(c) Specify the load required to extract the maximum average power.

FIGURE P8 – 55

8–56 A 100-V (rms) source delivers 500 W to an inductive load whose series resistance is 8 Ω. What is the load impedance?

8–57 (a) Find the maximum average power available at the interface in Figure P8–57.
(b) Specify the values of R and C that will extract the maximum power from the source circuit.

FIGURE P8 – 57

8–58 The steady-state open-circuit voltage at the output interface of a source circuit is $10\angle 0°$ V. When a 100-Ω resistor is connected across the interface, the steady-state output voltage is $4\angle{-}30°$ V. Find the maximum average power available at the interface and the resistance and reactance required to extract the maximum power.

8–59 The phasor Thévenin voltage of a certain source is $\mathbf{V}_T = 10 + j0$ V, and the Thévenin impedance is $Z_T = 100 + j50$ Ω. The problem is to extract as much average power as possible from this source using a single standard value resistor as the load. The values of resistance available are 1.0, 1.5, 2.2, 3.3, 4.7, 6.8, and multiples of 10 times these values.

(a) Find the maximum available average power from the source.

(b) Find the standard load resistor that extracts the most average power.

8–60 The 220-V (rms) source operating at $f = 400$ Hz delivers a current of 3 A (rms) and 600 W to a load consisting of a resistor and capacitor in parallel. Find the values of R and C.

INTEGRATING PROBLEMS

8–61 **A** AC VOLTAGE MEASUREMENT

An ac voltmeter measurement indicates the amplitude of a sinusoid and not its phase angle. The magnitude and phase can be inferred by making several measurements and using KVL. For example, Figure P8–61 shows a relay coil of unknown resistance and inductance. The following ac voltmeter readings are taken with the circuit operating in the sinusoidal steady state at $f = 1$ kHz: $|\mathbf{V}_S| = 10$ V, $|\mathbf{V}_1| = 4$ V, and $|\mathbf{V}_2| = 8$ V.

(a) Use these voltage magnitude measurements to solve for R and L.

(b) Determine the phasor voltage across each element and show that they satisfy KVL.

F I G U R E   P 8 – 6 1

8–62 **A** **D** OP AMP CIRCUITS

The characteristics of resistive OP AMP circuits can be extended to dynamic circuits operating in the sinusoidal steady state using the concept of impedance. For the circuit in Figure P8–62

(a) Derive an expression relating the output phasor $\mathbf{V}_O$ and the input phasor $\mathbf{V}_S$ in terms of the general impedances Z_1 and Z_2.

(b) If $Z_1 = 10$ kΩ and $Z_2 = 10 - j20$ kΩ, what is the maximum amplitude of the input sinusoidal for linear operation when $V_{CC} = \pm15$ V.

(c) If $Z_1 = 10$ kΩ, $Z_2 = 10 - j20$ kΩ, and $\mathbf{V}_S = 2 + j1$ V, what is the amplitude of the OP AMP output current?

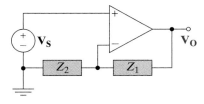

F I G U R E   P 8 · 6 2

8–63 **A** CIRCUIT FORCED RESPONSE

The purpose of this problem is to demonstrate that the steady-state response obtained using phasor circuit analysis is the forced component of the solution of the circuit differential equation.

(a) Transform the circuit in Figure P8–63 into the phasor domain and use voltage division to solve for phasor output **V**.

(b) Convert the phasor found in (a) into a sinusoid of the form $v(t) = a \cos \omega t + b \sin \omega t$.

(c) Formulate the circuit differential equation in the capacitor voltage $v(t)$.

(d) Show that the $v(t)$ found in part (b) satisfies the differential equation in (c).

FIGURE P 8 – 6 3

8–64 D AC CIRCUIT DESIGN

Select the element values in the circuit shown in Figure P8–64 so that an input $v_S(t) = 30 \cos 2000t$ V produces an output $v_O(t) = 5 \sin 2000t$ V.

FIGURE P 8 – 6 4

8–65 D AC CIRCUIT DESIGN

Select values for the reactances X_1 and X_2 in Figure P8–65 so that the input voltage source sees an equivalent impedance of $75 + j0$ Ω and the 600-Ω load resistor sees an equivalent source impedance of $600 + j0$ Ω. Select values of inductance and capacitance to produce these reactances when the frequency is $\omega = 10^6$ rad/s.

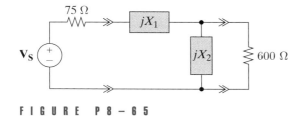

FIGURE P 8 – 6 5

8–66 A AC CIRCUIT ANALYSIS

Ten years after graduating with a BSEE you decide to go to graduate school for a master's degree. In desperate need of income you agree to sign on as a grader in the basic circuit analysis course. One of the problems asks the students to find $v(t)$ in Figure P8–66 when the circuit operates in the sinusoidal steady state. One of the students offers the following solution:

$$v(t) = (R + j\omega L) \times i(t)$$
$$= (20 + j20) \times 0.5 \cos 200t$$
$$= 10 \cos 200t + j10 \cos 200t$$
$$= 10\sqrt{2} \cos (200t + 45°)$$

Is the answer correct? If not, what grade would you give the student? If correct, what comments would you give the student about the method of solution?

FIGURE P 8 – 6 6

CHAPTER 9

LAPLACE
TRANSFORMS

My method starts with a complex integral; I fear this sounds rather formidable; but it is really quite simple . . . I am afraid that no physical people will ever try to make out my method: but I am hoping that it may give them confidence to try your methods.

Thomas John Bromwich, 1915,
British Mathematician

9–1 Signal Waveforms and Transforms

9–2 Basic Properties and Pairs

9–3 Pole-zero Diagrams

9–4 Inverse Laplace Transforms

9–5 Some Special Cases

9–6 Circuit Response Using Laplace Transforms

9–7 Initial Value and Final Value Properties

Summary

Problems

Integrating Problems

Laplace transforms have their roots in the pioneering work of the eccentric British engineer Oliver Heaviside (1850–1925). His operational calculus was essentially a collection of intuitive rules that allowed him to formulate and solve a number of the important technical problems of his day. Heaviside was a practical man with no interest in mathematical elegance. His intuitive approach drew bitter criticism from the mathematicians of his day. However, mathematicians like John Bromwich and others eventually recognized the importance of Heaviside's methods and began to supply the necessary mathematical foundations. The foregoing quotation is taken from a 1915 letter from Bromwich to Heaviside in which he described what we now call the Laplace transformation. The transformation is named for Laplace because a complete mathematical development of Heaviside's methods was eventually found in the 1780 writings of the French mathematician Pierre Simon Laplace.

The Laplace transformation provides a new and important method of representing circuits and signals. The transform approach offers a viewpoint and terminology that pervades electrical engineering, particularly in linear circuit analysis and design. The first two sections of this chapter present the basic properties of the Laplace transformation and the concept of converting signals from the time domain to the frequency domain. The third section introduces frequency domain signal description via the pole-zero diagram. The fourth and fifth sections treat the inverse procedure for transforming signals from the frequency domain back into the time domain. In the sixth section, transform methods are used to solve differential equations that describe the response of linear dynamic circuits. The last section treats additional properties of the Laplace transformation that provide important insights into the relationship between the time and frequency domains.

9–1   SIGNAL WAVEFORMS AND TRANSFORMS

A mathematical transformation employs rules to change the form of data without altering its meaning. An example of a transformation is the conversion of numerical data from decimal to binary form. In engineering circuit analysis, transformations are used to obtain alternative representations of circuits and signals. These alternate forms provide a different perspective that can be quite useful or even essential. Examples of the transformations used in circuit analysis are the Fourier transformation, the Z-transformation, and the Laplace transformation. These methods all involve specific transformation rules, make certain analysis techniques more manageable, and provide a useful viewpoint for circuit and system design.

This chapter deals with the Laplace transformation. The discussion of the Laplace transformation follows the path shown in Figure 9–1 by the solid arrow. The process begins with a linear circuit. We derive a differential equation describing the circuit response and then transform this equation into the frequency domain, where it becomes an algebraic equation. Algebraic techniques are then used to solve the transformed equation for the circuit response. The inverse Laplace transformation then changes the frequency domain response into the response waveform in the time domain. The dashed arrow in Figure 9–1 shows that there is another route to the time domain response using the classical techniques discussed in Chapter 7. The classical approach appears to be more direct, but the advantage of the Laplace transformation is that solving a differential equation becomes an algebraic process.

FIGURE 9–1 *Flow diagram dynamic circuit analysis with Laplace transforms.*

Symbolically, we represent the Laplace transformation as

$$\mathcal{L}\{f(t)\} = F(s) \tag{9–1}$$

This expression states that $F(s)$ is the Laplace transform of the waveform $f(t)$. The transformation operation involves two domains: (1) the time domain, in which the signal is characterized by its **waveform** $f(t)$, and (2) the complex frequency domain, in which the signal is represented by its **transform** $F(s)$.

The symbol s stands for the complex frequency variable, a notation we first introduced in Chapter 7 in connection with the zero-state response of linear circuits. The variable s has the dimensions of reciprocal time, or frequency, and is expressed in units of radians per second. In this chapter the complex frequency variable is written as $s = \sigma + j\omega$, where $\sigma = \text{Re}\{s\}$ is the real part and $\omega = \text{Im}\{s\}$ is the imaginary part. This variable is the independent variable in the s domain, just as t is the independent variable in the time domain. Although we cannot physically measure complex frequency in the same sense that we measure time, it is an extremely useful concept that pervades the analysis and design of linear systems.

A signal can be expressed as a waveform or a transform. Collectively, $f(t)$ and $F(s)$ are called a **transform pair**, where the pair involves two representations of the signal. To distinguish between the two forms, a lowercase letter denotes a waveform and an uppercase a transform. For electrical waveforms such as current $i(t)$ or voltage $v(t)$, the corresponding transforms are denoted $I(s)$ and $V(s)$. In this chapter we will use $f(t)$ and $F(s)$ to stand for signal waveforms and transforms in general.

The **Laplace transformation** is defined by the integral

$$F(s) = \int_{0-}^{\infty} f(t)e^{-st}dt \qquad (9\text{--}2)$$

Since the definition involves an improper integral (the upper limit is infinite), we must discuss the conditions under which the integral exists (converges). The integral exists if the waveform $f(t)$ is piecewise continuous and of exponential order. **Piecewise continuous** means that $f(t)$ has a finite number of steplike discontinuities in any finite interval. **Exponential order** means that constants K and b exist such that $|f(t)| < Ke^{bt}$ for all $t > 0$. As a practical matter, the signals encountered in engineering applications meet these conditions.

From the integral definition of the Laplace transformation, we see that when a voltage waveform $v(t)$ has units of volts (V) the corresponding voltage transform $V(s)$ has units of volt-seconds (V-s). Similarly, when a current waveform $i(t)$ has units of amperes (A), the corresponding current transform $I(s)$ has units of ampere-seconds (A-s). Thus, waveforms and transforms do not have the same units. Even so, we often refer to both $V(s)$ and $v(t)$ as voltages and both $I(s)$ and $i(t)$ as currents despite the fact that they have different units. The reason is simply that it is awkward to keep adding the words *waveform* and *transform* to statements when the distinction is clear from the context.

Equation (9–2) uses a lower limit denoted $t = 0-$ to indicate a time just a whisker before $t = 0$. We use $t = 0-$ because in circuit analysis $t = 0$ is defined by a discrete event, such as closing a switch. Such an event may cause a discontinuity in $f(t)$ at $t = 0$. To capture this discontinuity, we set the lower limit at $t = 0-$, just prior to the event. Fortunately, in many situations there is no discontinuity so we will not distinguish between $t = 0-$ and $t = 0$ unless it is crucial.

Equally fortunate is the fact that the number of different waveforms encountered in linear circuits is relatively small. The list includes the three basic waveforms from Chapter 5 (the step, exponential, and sinusoid), as well as composite waveforms such as the impulse, ramp, damped ramp, and damped sinusoid. Since the number of waveforms of interest is relatively small, we do not often use the integral definition in Eq. (9–2) to find Laplace transforms. Once a transform pair has been found, it can be cataloged in a table for future reference and use. Tables 9–2 and 9–3 in this chapter are sufficient for our purposes.

EXAMPLE 9–1

Show that the Laplace transform of the unit step function $f(t) = u(t)$ is $F(s) = 1/s$.

SOLUTION:
Applying Eq. (9–2) yields

$$F(s) = \int_{0}^{\infty} u(t)e^{-st}dt$$

Since $u(t) = 1$ throughout the range of integration this integral becomes

$$F(s) = \int_{0-}^{\infty} e^{-st}dt = \left. -\frac{e^{-st}}{s}\right|_{0-}^{\infty} = \left. -\frac{e^{-(\sigma + j\omega)t}}{\sigma + j\omega}\right|_{0-}^{\infty}$$

The last expression on the right side vanishes at the upper limit since e^{-st} goes to zero as t approaches infinity provided that $\sigma > 0$. At the lower limit the expression reduces to $1/s$. The integral used to calculate $F(s)$ is only valid in the region for which $\sigma > 0$. However, once evaluated, the result $F(s) = 1/s$ can be extended to neighboring regions provided that we avoid the point at $s = 0$ where the function blows up.   ■

EXAMPLE 9–2

Show that the Laplace transform $f(t) = [e^{-\alpha t}]u(t)$ is $F(s) = 1/(s + \alpha)$.

SOLUTION:
Applying Eq. (9–2) yields

$$F(s) = \int_{0}^{\infty} e^{-\alpha t}e^{-st}\, dt = \int_{0}^{\infty} e^{-(s+\alpha)t}dt = \left. \frac{e^{-(s+\alpha)t}}{-(s + \alpha)}\right|_{0}^{\infty}$$

The last term on the right side vanishes at the upper limit since $e^{-(s+\alpha)t}$ vanishes as t approaches infinity provided that $\sigma > \alpha$. At the lower limit the last term reduces to $1/(s + \alpha)$. Again, the integral is only valid for a limited region, but the result $F(s) = 1/(s + \alpha)$ can be extended outside this region if we avoid the point at $s = -\alpha$.   ■

EXAMPLE 9–3

Show that the Laplace transform of the impulse function $f(t) = \delta(t)$ is $F(s) = 1$.

SOLUTION:
Applying Eq. (9–2) yields

$$F(s) = \int_{0-}^{\infty} \delta(t)e^{-st}\, dt = \int_{0-}^{0+} \delta(t)e^{-st}\, dt = \int_{0-}^{0+} \delta(t)dt = 1$$

The difference between $t = 0-$ and $t = 0$ is important since the impulse is zero everywhere except at $t = 0$. To capture the impulse in the integration, we take a lower limit at $t = 0-$ and an upper limit at $t = 0+$. Since $e^{-st} = 1$ and $\int \delta(t)dt = 1$ on this integration interval, we find that $F(s) = 1$.   ■

INVERSE TRANSFORMATION

So far we have used the direct transformation to convert waveforms into transforms. But Figure 9–1 points out the need to perform the inverse transformation to convert transforms into waveforms. Symbolically, we represent the inverse process as

$$\mathcal{L}^{-1}\{F(s)\} = f(t) \tag{9–3}$$

This equation states that $f(t)$ is the inverse Laplace transform of $F(s)$. The **inverse Laplace transformation** is defined by the complex inversion integral

$$f(t) = \frac{1}{2\pi j} \int_{\alpha - j\infty}^{\alpha + j\infty} F(s)e^{st}\, ds \qquad (9\text{–}4)$$

The Laplace transformation is an integral transformation since both the direct process in Eq. (9–2) and the inverse process in Eq. (9–4) involve integrations.

Happily, formal evaluation of the complex inversion integral is not necessary because of the uniqueness property of the Laplace transformation. A symbolical statement of the **uniqueness property** is

$$\text{IF } \mathscr{L}\{f(t)\} = F(s) \text{ THEN } \mathscr{L}^{-1}\{F(s)\}(\ =\ )u(t)f(t)$$

The mathematical justification for this statement is beyond the scope of our treatment.[1] However, the notation (=) means "equal almost everywhere." The only points where equality may not hold is at the discontinuities of $f(t)$.

If we just look at the definition of the direct transformation in Eq. (9–2), we could conclude that $F(s)$ is not affected by the values of $f(t)$ for $t < 0$. However, when we use Eq. (9–2) we are not just looking for the Laplace transform of $f(t)$, but a Laplace transform pair such that $\mathscr{L}\{f(t)\} = F(s)$ and $\mathscr{L}^{-1}\{F(s)\} = f(t)$. The inverse Laplace transformation in Eq. (9–4) always produces a causal waveform, one that is zero for $t < 0$. Hence a transform pair $[f(t) \leftrightarrow F(s)]$ is unique if and only if $f(t)$ is causal. For instance, in Example 9–1 we show that $\mathscr{L}\{u(t)\} = 1/s$; hence by the uniqueness property we know that $\mathscr{L}^{-1}\{1/s\} (=) u(t)$.

For this reason Laplace transform–related waveforms are written as $[f(t)]u(t)$ to make their causality visible. For example, in the next section we find the Laplace transform of the sinusoid waveform $\cos \beta t$. In the context of Laplace transforms this signal is not an eternal sinusoid but a causal waveform $f(t) = [\cos \beta t]u(t)$. It is important to remember that causality and Laplace transforms go hand in hand when interpreting the results of circuit analysis.

9–2   BASIC PROPERTIES AND PAIRS

The previous section gave the definition of the Laplace transformation and showed that the transforms of some basic signals can be found using the integral definition. In this section we develop the basic properties of the Laplace transformation and show how these properties can be used to obtain additional transform pairs.

The **linearity** property of the Laplace transformation states that

$$\mathscr{L}\{Af_1(t) + Bf_2(t)\} = AF_1(s) + BF_2(s) \qquad (9\text{–}5)$$

---

1 See Wilber R. LePage, *Complex Variables and Laplace Transform for Engineering*, Dover Publishing Co., New York, 1980, p. 318.

where A and B are constants. This property is easily established using the integral definition in Eq. (9–2):

$$\mathcal{L}\{Af_1(t) + Bf_2(t)\} = \int_0^\infty [Af_1(t) + Bf_2(t)]e^{-st}dt$$

$$= A\int_0^\infty f_1(t)e^{-st}dt + B\int_0^\infty f_2(t)e^{-st}dt$$

$$= AF_1(s) + BF_2(s)$$

The integral definition of the inverse transformation in Eq. (9–4) is also a linear operation, so it follows that

$$\mathcal{L}^{-1}\{AF_1(s) + BF_2(s)\} = Af_1(t) + Bf_2(t) \tag{9–6}$$

An important consequence of linearity is that for any constant K

$$\mathcal{L}\{Kf(t)\} = KF(s) \text{ and } \mathcal{L}^{-1}\{KF(s)\} = Kf(t) \tag{9–7}$$

The linearity property is an extremely important feature that we will use many times in this and subsequent chapters. The next two examples show how this property can be used to obtain the transform of the exponential rise waveform and a sinusoidal waveform.

EXAMPLE 9–4

Show that the Laplace transform of $f(t) = A(1 - e^{-\alpha t})u(t)$ is

$$F(s) = \frac{A\alpha}{s(s + \alpha)}$$

SOLUTION:
This waveform is the difference between a step function and an exponential. We can use the linearity property of Laplace transforms to write

$$\mathcal{L}\{A(1 - e^{-\alpha t})u(t)\} = A\mathcal{L}\{u(t)\} - A\mathcal{L}\{e^{-\alpha t}u(t)\}$$

The transforms of the step and exponential functions were found in Examples 9–1 and 9–2. Using linearity, we find that the transform of the exponential rise is

$$F(s) = \frac{A}{s} - \frac{A}{s + \alpha} = \frac{A\alpha}{s(s + \alpha)} \qquad \blacksquare$$

EXAMPLE 9–5

Show that the Laplace transform of the sinusoid $f(t) = A[\sin(\beta t)]u(t)$ is $F(s) = A\beta/(s^2 + \beta^2)$.

SOLUTION:
Using Euler's relationship, we can express the sinusoid as a sum of exponentials.

$$e^{+j\beta t} = \cos \beta t + j \sin \beta t$$

$$e^{-j\beta t} = \cos \beta t - j \sin \beta t$$

Subtracting the second equation from the first yields

$$f(t) = A \sin \beta t = \frac{A(e^{j\beta t} - e^{-j\beta t})}{2j} = \frac{A}{2j}e^{j\beta t} - \frac{A}{2j}e^{-j\beta t}$$

The transform pair $\mathcal{L}\{e^{-\alpha t}\} = 1/(s + \alpha)$ in Example 9–2 is valid even if the exponent α is complex. Using this fact and the linearity property, we obtain the transform of the sinusoid as

$$\mathcal{L}\{A \sin \beta t\} = \frac{A}{2j}\mathcal{L}\{e^{j\beta t}\} - \frac{A}{2}\mathcal{L}\{e^{-j\beta t}\}$$

$$= \frac{A}{2j}\left[\frac{1}{s - j\beta} - \frac{1}{s + j\beta}\right]$$

$$= \frac{A\beta}{s^2 + \beta^2} \qquad \blacksquare$$

INTEGRATION PROPERTY

In the time domain the i–v relationships for capacitors and inductors involve integration and differentiation. Since we will be working in the s domain, it is important to establish the s-domain equivalents of these mathematical operations. Applying the integral definition of the Laplace transformation to a time-domain integration yields

$$\mathcal{L}\left[\int_0^t f(\tau)d\tau\right] = \int_0^\infty \left[\int_0^t f(\tau)d\tau\right]e^{-st} dt \qquad (9\text{–}8)$$

The right side of this expression can be integrated by parts using

$$y = \int_0^t f(\tau)d\tau \quad \text{and} \quad dx = e^{-st} dt$$

These definitions result in

$$dy = f(t)dt \quad \text{and} \quad x = \frac{-e^{-st}}{s}$$

Using these factors reduces the right side of Eq. (9–8) to

$$\mathcal{L}\left[\int_0^t f(\tau)d\tau\right] = \left[\frac{-e^{-st}}{s}\int_0^t f(\tau)d\tau\right]_0^\infty + \frac{1}{s}\int_0^\infty f(t)e^{-st} dt \qquad (9\text{–}9)$$

The first term on the right in Eq. (9–9) vanishes at the lower limit because the integral over a zero-length interval is zero provided that $f(t)$ is finite at $t = 0$. It vanishes at the upper limit because e^{-st} approaches zero as t goes to infinity for $\sigma > 0$. By the definition of the Laplace transformation, the second term on the right is $F(s)/s$. We conclude that

$$\mathcal{L}\left[\int_0^t f(\tau)d\tau\right] = \frac{F(s)}{s} \qquad (9\text{–}10)$$

The **integration property** states that time-domain integration of a waveform $f(t)$ can be accomplished in the s domain by the algebraic process of dividing its transform $F(s)$ by s. The next example applies the integration property to obtain the transform of the ramp function.

EXAMPLE 9–6

Show that the Laplace transform of the ramp function $r(t) = tu(t)$ is $1/s^2$.

SOLUTION:

From our study of signals, we know that the ramp waveform can be obtained from $u(t)$ by integration.

$$r(t) = \int_0^t u(\tau) \, d\tau$$

In Example 9–1 we found $\mathcal{L}\{u(t)\} = 1/s$. Using these facts and the integration property of Laplace transforms, we obtain

$$\mathcal{L}\{r(t)\} = \mathcal{L}\left[\int_0^t u(\tau) \, d\tau\right] = \frac{1}{s}\mathcal{L}\{u(t)\} = \frac{1}{s^2} \qquad \blacksquare$$

DIFFERENTIATION PROPERTY

The time-domain differentiation operation transforms into the s domain as follows:

$$\mathcal{L}\left[\frac{df(t)}{dt}\right] = \int_0^\infty \left[\frac{df(t)}{dt}\right] e^{-st} \, dt \qquad (9\text{–}11)$$

The right side of this equation can be integrated by parts using

$$y = e^{-st} \text{ and } dx = \frac{df(t)}{dt} \, dt$$

These definitions result in

$$dy = -se^{-st} \, dt \quad \text{and} \quad x = f(t)$$

Inserting these factors reduces the right side of Eq. (9–11) to

$$\mathcal{L}\left[\frac{df(t)}{dt}\right] = f(t)e^{-st}\Big|_{0-}^{\infty} + s \int_{0-}^{\infty} f(t)e^{-st} \, dt \qquad (9\text{–}12)$$

For $\sigma > 0$ the first term on the right side of Eq. (9–12) is zero at the upper limit because e^{-st} approaches zero as t goes to infinity. At the lower limit it reduces to $-f(0-)$. By the definition of the Laplace transform, the second term on the right side is $sF(s)$. We conclude that

$$\mathcal{L}\left[\frac{df(t)}{dt}\right] = sF(s) - f(0-) \qquad (9\text{–}13)$$

The **differentiation property** states that time-domain differentiation of a waveform $f(t)$ is accomplished in the s domain by the algebraic process of multiplying the transform $F(s)$ by s and subtracting the constant $f(0-)$. Note that the constant $f(0-)$ is the value of $f(t)$ at $t = 0-$ just prior to $t = 0$.

The s-domain equivalent of a second derivative is obtained by repeated application of Eq. (9–13). We first define a waveform $g(t)$ as

$$g(t) = \frac{df(t)}{dt} \quad \text{hence} \quad \frac{d^2f(t)}{dt^2} = \frac{dg(t)}{dt}$$

Applying the differentiation rule to these two equations yields

$$G(s) = sF(s) - f(0-) \text{ and } \mathcal{L}\left[\frac{d^2f(t)}{dt^2}\right] = sG(s) - g(0-)$$

Substituting the first of these equations into the second results in

$$\mathcal{L}\left[\frac{d^2f(t)}{dt^2}\right] = s^2F(s) - sf(0-) - f'(0-)$$

where

$$f'(0-) = \frac{df}{dt}\bigg|_{t=0-}$$

Repeated application of this procedure produces the nth derivative:

$$\mathcal{L}\left[\frac{d^n f(t)}{dt^n}\right] = s^n F(s) - s^{n-1}f(0-) - s^{n-2}f'(0-) \ldots - f^{(n)}(0-)$$

$$(9-14)$$

where $f^{(n)}(0-)$ is the nth derivative of $f(t)$ evaluated at $t = 0-$.

A hallmark feature of the Laplace transformation is the fact that time integration and differentiation change into algebraic operations in the s domain. This observation gives us our first hint as to why it is often easier to work with circuits and signals in the s domain. The next example shows how the differentiation rule can be used to obtain additional transform pairs.

EXAMPLE 9–7

Show that the Laplace transform of $f(t) = [\cos \beta t]u(t)$ is $F(s) = s/(s^2 + \beta^2)$.

SOLUTION:
We can express $\cos \beta t$ in terms of the derivative of $\sin \beta t$ as

$$\cos \beta t = \frac{1}{\beta}\frac{d}{dt}\sin \beta t$$

In Example 9–5 we found $\mathcal{L}\{\sin \beta t\} = \beta/(s^2 + \beta^2)$. Using these facts and the differentiation rule, we can find the Laplace transform of $\cos \beta t$ as follows:

$$\mathcal{L}\{\cos \beta t\} = \frac{1}{\beta}\mathcal{L}\left\{\frac{d}{dt}\sin \beta t\right\} = \frac{1}{\beta}\left[s\left(\frac{\beta}{s^2 + \beta^2}\right) - \sin(0-)\right]$$

$$= \frac{s}{s^2 + \beta^2} \qquad\qquad\blacksquare$$

TRANSLATION PROPERTIES

The **s-domain translation property** of the Laplace transformation is

$$\text{IF } \mathcal{L}\{f(t)\} = F(s) \text{ THEN } \mathcal{L}\{e^{-\alpha t}f(t)\} = F(s + \alpha)$$

This theorem states that multiplying $f(t)$ by $e^{-\alpha t}$ is equivalent to replacing s by $s + \alpha$ (that is, translating the origin in the s plane by an amount α). In engineering applications the parameter α is always a real number, but it can be either positive or negative so the origin in the s domain can be translated to the left or right. Proof of the theorem follows almost immediately from the definition of the Laplace transformation.

$$\mathcal{L}\{e^{-\alpha t}f(t)\} = \int_0^\infty e^{-\alpha t}f(t)e^{-st}dt$$

$$= \int_0^\infty f(t)e^{-(s+\alpha)t}\,dt$$

$$= F(s + \alpha)$$

The s-domain translation property can be used to derive transforms of damped waveforms from undamped prototypes. For instance, the Laplace transform of the ramp, cosine, and sine functions are

$$\mathcal{L}\{tu(t)\} = \frac{1}{s^2}$$

$$\mathcal{L}\{[\cos \beta t]u(t)\} = \frac{s}{s^2 + \beta^2}$$

$$\mathcal{L}\{[\sin \beta t]u(t)\} = \frac{\beta}{s^2 + \beta^2}$$

To obtain the damped ramp, damped cosine, and damped sine functions, we multiply each undamped waveform by $e^{-\alpha t}$. Using the s-domain translation property, we replace s by $s + \alpha$ to obtain transforms of the corresponding damped waveforms.

$$\mathcal{L}\{te^{-\alpha t}u(t)\} = \frac{1}{(s + \alpha)^2}$$

$$\mathcal{L}\{[e^{-\alpha t}\cos \beta t]u(t)\} = \frac{s + \alpha}{(s + \alpha)^2 + \beta^2}$$

$$\mathcal{L}\{[e^{-\alpha t}\sin \beta t]u(t)\} = \frac{\beta}{(s + \alpha)^2 + \beta^2}$$

This completes the derivation of a basic set of transform pairs.

The **time-domain translation property** of the Laplace transformation is

IF $\mathcal{L}\{f(t)\} = F(s)$ THEN for $a > 0$ $\mathcal{L}\{f(t - a)u(t - a)\} = e^{-as}F(s)$

The theorem states that multiplying $F(s)$ by e^{-as} is equivalent to shifting $f(t)$ to the right in the time domain by an amount $a > 0$. In other words, it is equivalent to delaying $f(t)$ in time by an amount $a > 0$. Proof of this property follows from the definition of the Laplace transformation.

$$\mathcal{L}\{f(t - a)u(t - a)\} = \int_{0-}^\infty f(t - a)u(t - a)e^{-st}dt = \int_a^\infty f(t - a)e^{-st}dt$$

In this equation we have used the fact that $u(t - a)$ is zero for $t < a$ and is unity for $t \geq a$. We now change the integration variable from t to $\tau = t - a$.

With this change of variable the last integral in this equation takes the form

$$\mathcal{L}\{f(t-a)u(t-a)\} = \int_0^\infty f(\tau)e^{-s\tau}e^{-as}d\tau$$

$$= e^{-as}\int_0^\infty f(\tau)e^{-s\tau}\,d\tau$$

$$= e^{-as}F(s)$$

which confirms the statement of the time-domain translation property. A simple application of this property is finding the Laplace transform of the delayed step function.

$$\mathcal{L}\{u(t-T)\} = e^{-sT}\mathcal{L}\{u(t)\} = \frac{e^{-sT}}{s}$$

In this section we derived the basic transform properties listed in Table 9–1.

The Laplace transformation has other properties that are useful in signal-processing applications. We treat two of these properties in the last section of this chapter. However, the basic properties in Table 9–1 are used frequently in circuit analysis and are sufficient for nearly all of the applications in this book.

Similarly, Table 9–2 lists a basic set of Laplace transform pairs that is sufficient for most of the applications in this book. All of these pairs were derived in the preceding two sections.

T A B L E 9–1   BASIC LAPLACE TRANSFORMATION PROPERTIES

PROPERTIES	TIME DOMAIN	FREQUENCY DOMAIN
Independent variable	t	s
Signal representation	$f(t)$	$F(s)$
Uniqueness	$\mathcal{L}^{-1}\{F(s)\} \,(=)\, [f(t)]u(t)$	$\mathcal{L}\{f(t)\} = F(s)$
Linearity	$Af_1(t) + Bf_2(t)$	$AF_1(s) + BF_2(s)$
Integration	$\displaystyle\int_0^t f(\tau)\,d\tau$	$\dfrac{F(s)}{s}$
Differentiation	$\dfrac{df(t)}{dt}$	$sF(s) - f(0-)$
	$\dfrac{d^2f(t)}{dt^2}$	$s^2F(s) - sf(0-) - f'(0-)$
	$\dfrac{d^3f(t)}{dt^3}$	$s^3F(s) - s^2f(0-) - sf'(0-) - f''(0-)$
s-Domain translation	$e^{-\alpha t}f(t)$	$F(s + \alpha)$
t-Domain translation	$f(t-a)u(t-a)$	$e^{-as}F(s)$

T A B L E   **9–2**   **BASIC LAPLACE TRANSFORM PAIRS**

SIGNAL	WAVEFORM f(t)	TRANSFORM F(s)
Impulse	$\delta(t)$	1
Step function	$u(t)$	$\dfrac{1}{s}$
Ramp	$tu(t)$	$\dfrac{1}{s^2}$
Exponential	$[e^{-\alpha t}]u(t)$	$\dfrac{1}{s + \alpha}$
Damped ramp	$[te^{-\alpha t}]u(t)$	$\dfrac{1}{(s + \alpha)^2}$
Sine	$[\sin \beta t]u(t)$	$\dfrac{\beta}{s^2 + \beta^2}$
Cosine	$[\cos \beta t]u(t)$	$\dfrac{s}{s^2 + \beta^2}$
Damped sine	$[e^{-\alpha t}\sin \beta t]u(t)$	$\dfrac{\beta}{(s + \alpha)^2 + \beta^2}$
Damped cosine	$[e^{-\alpha t}\cos \beta t]u(t)$	$\dfrac{(s + \alpha)}{(s + \alpha)^2 + \beta^2}$

All of the waveforms in Table 9–2 are causal. As a result, the Laplace transform pairs are unique and we can use the table in either direction. That is, given an $f(t)$ in the waveform column we find its Laplace transform in the right column, or given an $F(s)$ in the right column we find its inverse transform in the waveform column.

The last example in this section shows how to use the properties and pairs in Tables 9–1 and 9–2 to obtain the transform of a waveform not listed in the tables.

EXAMPLE 9–8

Find the Laplace transform of the waveform

$$f(t) = 2u(t) - 5[e^{-2t}]u(t) + 3[\cos 2t]u(t) + 3[\sin 2t]u(t)$$

SOLUTION:

Using the linearity property, we write the transform of $f(t)$ in the form

$$\mathscr{L}\{f(t)\} = 2\mathscr{L}\{u(t)\} - 5\mathscr{L}\{e^{-2t}u(t)\} + 3\mathscr{L}\{[\cos 2t]u(t)\} + 3\mathscr{L}\{[\sin 2t]u(t)\}$$

The transforms of each term in this sum are listed in Table 9–2:

$$F(s) = \frac{2}{s} - \frac{5}{s + 2} + \frac{3s}{s^2 + 4} + \frac{6}{s^2 + 4}$$

Normally, a Laplace transform is written as a quotient of polynomials rather than as a sum of terms. Rationalizing the preceding sum yields

$$F(s) = \frac{16(s^2 + 1)}{s(s + 2)(s^2 + 4)}$$ ∎

Exercise 9–1

Find the Laplace transforms of the following waveforms:

(a) $f(t) = [e^{-2t}]u(t) + 4tu(t) - u(t)$

(b) $f(t) = [2 + 2 \sin 2t - 2 \cos 2t]u(t)$

Answers:

(a) $F(s) = \dfrac{2(s + 4)}{s^2(s + 2)}$

(b) $F(s) = \dfrac{4(s + 2)}{s(s^2 + 4)}$

Exercise 9–2

Find the Laplace transforms of the following waveforms:

(a) $f(t) = [e^{-4t}]u(t) + 5 \displaystyle\int_0^t \sin 4x \, dx$

(b) $f(t) = 5[e^{-40t}]u(t) + \dfrac{d[5te^{-40t}]u(t)}{dt}$

Answers:

(a) $F(s) = \dfrac{s^3 + 36s + 80}{s(s + 4)(s^2 + 16)}$

(b) $F(s) = \dfrac{10s + 200}{(s + 40)^2}$

Exercise 9–3

Find the Laplace transforms of the following waveforms:

(a) $f(t) = A[\cos(\beta t - \phi)]u(t)$

(b) $f(t) = A[e^{-\alpha t} \cos (\beta t - \phi)]u(t)$

Answers:

(a) $F(s) = A \cos \phi \left[\dfrac{s + \beta \tan \phi}{s^2 + \beta^2} \right]$

(b) $F(s) = A \cos \phi \left[\dfrac{s + \alpha + \beta \tan \phi}{(s + \alpha)^2 + \beta^2} \right]$

Exercise 9–4

Find the Laplace transforms of the following waveforms:

$$(a)\ f(t) = Au(t) - 2Au(t - T) + Au(t - 2T)$$

$$(b)\ f(t) = Ae^{-\alpha(t-T)}u(t - T)$$

Answers:

$$(a)\ F(s) = \frac{A\ (1 - e^{-Ts})^2}{s}$$

$$(b)\ F(s) = \frac{Ae^{-Ts}}{s + \alpha}$$

9–3   POLE-ZERO DIAGRAMS

The transforms for signals in Table 9–2 are ratios of polynomials in the complex frequency variable s. Likewise, the transform found in Example 9–8 takes the form of a ratio of two polynomials in s. These results illustrate that the signal transforms of greatest interest to us usually have the form

$$F(s) = \frac{b_m s^m + b_{m-1} s^{m-1} + \ldots + b_1 s + b_0}{a_n s^n + a_{n-1} s^{n-1} + \ldots + a_1 s + a_0} \tag{9–15}$$

If numerator and denominator polynomials are expressed in factored form, then $F(s)$ is written as

$$F(s) = K \frac{(s - z_1)(s - z_2) \ldots (s - z_m)}{(s - p_1)(s - p_2) \ldots (s - p_n)} \tag{9–16}$$

where the constant $K = b_m/a_n$ is called the **scale factor**.

The roots of the numerator and denominator polynomials, together with the scale factor K, uniquely define a transform $F(s)$. The denominator roots are called **poles** because for $s = p_i$ ($i = 1, 2, \ldots n$) the denominator vanishes and $F(s)$ becomes infinite. The roots of the numerator polynomial are called **zeros** because the transform $F(s)$ vanishes for $s = z_i$ ($i = 1, 2, \ldots m$). Collectively the poles and zeros are called **critical frequencies** because they are values of s at which $F(s)$ does dramatic things, like vanish or blow up.

In the s domain we can specify a signal transform by listing the location of its critical frequencies together with the scale factor K. That is, in the frequency domain we describe signals in terms of poles and zeros. The description takes the form of a **pole-zero diagram**, which shows the location of poles and zeros in the complex s plane. The pole locations in such plots are indicated by an $\times$ and the zeros by an $\bigcirc$. The independent variable in the frequency domain is the complex frequency variable s, so the poles or zeros can be complex as well. In the s plane we use a horizontal axis to plot the value of the real part of s and a vertical j-axis to plot the imaginary part. The j-axis is an important boundary in the frequency domain because it divides the s plane into two distinct half planes. The real part of s is negative in the left half plane and positive in the right half plane. As we will soon see, the sign of the real part of a pole has a profound effect on the form of the corresponding waveform.

For example, Table 9–2 shows that the transform of the exponential waveform $f(t) = e^{-\alpha t}u(t)$ is $F(s) = 1/(s + \alpha)$. The exponential signal has a sin-

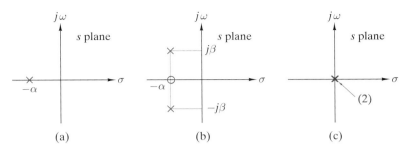

FIGURE 9–2 *Pole-zero diagrams in the s plane.*

gle pole at $s = -\alpha$ and no finite zeros. The pole-zero diagram in Figure 9–2(a) is the s-domain portrayal of the exponential signal. In this diagram the X identifies the pole located at $s = -\alpha + j0$, a point on the negative real axis in the left half plane.

The damped sinusoid $f(t) = [A \, e^{-\alpha t} \cos \beta t]u(t)$ is an example of a signal with complex poles. From Table 9–2 the corresponding transform is

$$F(s) = \frac{A(s + \alpha)}{(s + \alpha)^2 + \beta^2}$$

The transform $F(s)$ has a finite zero on the real axis at $s = -\alpha$. The roots of the denominator polynomial are $s = -\alpha \pm j\beta$. The resulting pole-zero diagram is shown in Figure 9–2(b). The poles of the damped cosine do not lie on either axis in the s plane because neither the real nor imaginary parts are zero.

Finally, the transform of a unit ramp $f(t) = tu(t)$ is $F(s) = 1/s^2$. This transform has no finite zeros and two poles at the origin ($s = 0 + j0$) in the s plane as shown in Figure 9–2(c). The poles in all of the diagrams of Figure 9–2 lie in the left half plane or on the j-axis boundary.

The diagrams in Figure 9–2 show the poles and zeros in the finite part of the s plane. Signal transforms may have poles or zeros at infinity as well. For example, the step function has a zero at infinity since $F(s) = 1/s$ approaches zero as $s \to \infty$. In general, a transform $F(s)$ given by Eq. 9–16 has a zero of order $n - m$ at infinity if $n > m$ and a pole of order $m - n$ at infinity if $n < m$. Thus, the number of zeros equals the number of poles if we include those at infinity.

The pole-zero diagram is the s-domain portrayal of the signal, just as a plot of the waveform versus time depicts the signal in the t domain. The utility of a pole-zero diagram as a description of circuits and signals will become clearer as we develop additional s-domain analysis and design concepts.

EXAMPLE 9–9

Find the poles and zeros of the waveform

$$f(t) = [e^{-2t} + \cos 2t - \sin 2t]u(t)$$

SOLUTION:

Using the linearity property and the basic pairs in Table 9–2, we write the transform in the form

$$F(s) = \frac{1}{s + 2} + \frac{s}{s^2 + 4} - \frac{2}{s^2 + 4}$$

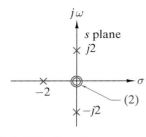

FIGURE 9 – 3

Rationalizing this expression yields $F(s)$.

$$F(s) = \frac{2s^2}{(s + 2)(s^2 + 4)} = \frac{2s^2}{(s + 2)(s + j2)(s - j2)}$$

This transform has three zeros and three poles. There are two zeros at $s = 0$ and one at $s = \infty$. There is a pole on the negative real axis at $s = -2 + j0$, and two poles on the imaginary axis at $s = \pm j2$. The resulting pole-zero diagram is shown in Figure 9–3. Reviewing the analysis, we can trace the poles to the components of $f(t)$. The pole on the real axis at $s = -2$ came from the exponential e^{-2t}, while the complex conjugate poles on the j-axis came from the sinusoid $\cos 2t - \sin 2t$. The zeros, however, are not traceable to specific components. Their locations depend on all three components. ■

EXAMPLE 9–10

Use a math analysis program to find the Laplace transform of the waveform

$$f(t) = [200te^{-25t} + 10e^{-50t} \sin (40t)]u(t)$$

and construct a pole-zero map of $F(s)$.

SOLUTION:
In this example we will illustrate the Laplace transform capabilities of MATLAB. Operating in the MATLAB command window, we first write the statement

```
syms t s
```

This defines the symbols t (for time) and s (for complex frequency) as symbolic variables rather than numerical quantities. We can then write the waveform $f(t)$ as

```
f=200*t*exp(-25*t)+10*exp(-50*t)*sin(40*t);
```

The Laplace transformation is performed by the statement

```
F=laplace (f, t, s);
```

This statement causes MATLAB to transform the previously defined waveform f from a domain where the independent variable is t to a domain in which the independent variable is s. The result of the transformation yields

```
F=
200/(s+25)^2+400/((s+50)^2+1600)
```

This result is clearly a term-by-term transformation of the waveform $f(t)$. The first term comes from the damped ramp and has a double pole at $s = -25$. The second term comes from the damped sine and has a pair of conjugate complex poles at $s = -50 \pm j40$. This set of poles can be predicted in advance from the form of the terms in the input waveform. To find the zeros of $F(s)$ we must express this result as a quotient of polynomials. To accomplish this we use a powerful MATLAB function called `simplify` that attempts to simplify an expression using many different mathematical tools. In the present case the result is

```
simplify(F)
ans=
200*(3*s^2+200*s+5350)/(s+25)^2/(s^2+100*s+4100)
```

Interpreting this result requires careful application of the MATLAB rules of precedence. Fortunately, MATLAB offers another way out. The function `pretty` attempts to place an expression in a form that more closely resembles type set mathematics. In the present case the result is

```
pretty(ans)

                   2
             3 s    + 200 s + 5350
     200  -----------------------------
                 2   2
            (s + 25) (s + 100 s + 4100)
```

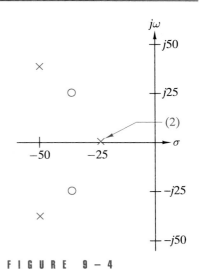

FIGURE 9-4

The zeros of $F(s)$ are roots of the numerator polynomial in this expression and are located at $s = -33.33 \pm j25.93$. Figure 9–4 shows the resulting pole-zero map. ∎

Exercise 9–5

Find the poles and zeros of transforms of each of the following waveforms:

(a) $f(t) = [-2e^{-t} - t + 2]u(t)$

(b) $f(t) = [4 - 3 \cos \beta t]u(t)$

(c) $f(t) = [e^{-\alpha t} \cos \beta t + (\alpha/\beta) e^{-\alpha t} \sin \beta t]u(t)$

Answers:
(a) Zeros: $s = 1, s = \infty$ (2); poles: $s = 0$ (2), $s = -1$
(b) Zeros: $s = \pm j2\beta, s = \infty$; poles: $s = 0, s = \pm j\beta$
(c) Zeros: $s = -2\alpha, s = \infty$; poles: $s = -\alpha \pm j\beta$

Exercise 9–6

A transform has poles at $s = -3 \pm j6$ and $s = -2$ and finite zeros at $s = 0$ and $s = -1$. Write $F(s)$ as a quotient of polynomials in s.

Answer:

$$F(s) = K\frac{s^2 + s}{s^3 + 8s^2 + 57s + 90}$$

9–4 INVERSE LAPLACE TRANSFORMS

The inverse transformation converts a transform $F(s)$ into the corresponding waveform $f(t)$. Applying the inverse transformation in Eq. (9–4) requires knowledge of a branch of mathematics called the complex variable. Fortunately, we do not need Eq. (9–4) because the uniqueness of the Laplace transform pairs in Table 9–2 allows us to go from a transform to a waveform. This may not seem like much help since it does not take a very complicated circuit or signal before we exceed the listing in Table 9–2, or even the more extensive tables that are available. However, there is a general method of expanding $F(s)$ into a sum of terms that are listed in Table 9–2.

For linear circuits the transforms of interest are ratios of polynomials in s. In mathematics such functions are called **rational functions**. To perform the inverse transformation, we must find the waveform corresponding to rational functions of the form

$$F(s) = K\frac{(s - z_1)(s - z_2) \ldots (s - z_m)}{(s - p_1)(s - p_2) \ldots (s - p_n)} \tag{9–17}$$

where the K is the scale factor, z_i ($i = 1, 2, \ldots m$) are the zeros, and p_i ($i = 1, 2, \ldots n$) are the poles of $F(s)$.

If there are more finite poles than finite zeros ($n > m$), then $F(s)$ is called a **proper rational function**. If the denominator in Eq. (9–17) has no repeated roots ($p_i \neq p_j$ for $i \neq j$), then $F(s)$ is said to have **simple poles**. In this section we treat the problem of finding the inverse transform of proper rational functions with simple poles. The problem of improper rational functions and multiple poles is covered in the next section.

If a proper rational function has only simple poles, then it can be decomposed into a partial fraction expansion of the form

$$F(s) = \frac{k_1}{s - p_1} + \frac{k_2}{s - p_2} + \ldots + \frac{k_n}{s - p_n} \tag{9–18}$$

In this case, $F(s)$ can be expressed as a linear combination of terms with one term for each of its n simple poles. The k's associated with each term are called **residues**.

Each term in the partial fraction decomposition has the form of the transform of an exponential signal. That is, we recognize that $\mathcal{L}^{-1}\{k/(s + \alpha)\} = [ke^{-\alpha t}]u(t)$. We can now write the corresponding waveform using the linearity property

$$f(t) = [k_1 e^{p_1 t} + k_2 e^{p_2 t} + \ldots + k_n e^{p_n t}]u(t) \tag{9–19}$$

In the time domain, the s-domain poles appear in the exponents of exponential waveforms and the residues at the poles become the amplitudes.

Given the poles of $F(s)$, finding the inverse transform $f(t)$ reduces to finding the residues. To illustrate the procedure, consider a case in which $F(s)$ has three simple poles and one finite zero.

$$F(s) = K\frac{(s - z_1)}{(s - p_1)(s - p_2)(s - p_3)} = \frac{k_1}{s - p_1} + \frac{k_2}{s - p_2} + \frac{k_3}{s - p_3}$$

We find the residue k_1 by first multiplying this equation through by the factor $(s - p_1)$:

$$(s - p_1)F(s) = K\frac{(s - z_1)}{(s - p_2)(s - p_3)} = k_1 + \frac{k_2(s - p_1)}{s - p_2} + \frac{k_3(s - p_1)}{s - p_3}$$

If we now set $s = p_1$, the last two terms on the right vanish, leaving

$$k_1 = (s - p_1)F(s)\big|_{s=p_1} = K\frac{(s - z_1)}{(s - p_2)(s - p_3)}\bigg|_{s=p_1}$$

Using the same approach for k_2 yields

$$k_2 = (s - p_2)F(s)\big|_{s=p_2} = K\frac{(s - z_1)}{(s - p_1)(s - p_3)}\bigg|_{s=p_2}$$

The technique generalizes so that the residue at any simple pole p_i is

$$k_i = (s - p_i)F(s)\big|_{s=p_i} \tag{9–20}$$

The process of determining the residue at any simple pole is sometimes called the **cover-up algorithm** because we temporarily remove (cover up) the factor $(s - p_i)$ in $F(s)$ and then evaluate the remainder at $s = p_i$.

EXAMPLE 9−11

Find the waveform corresponding to the transform

$$F(s) = 2\frac{(s + 3)}{s(s + 1)(s + 2)}$$

SOLUTION:

$F(s)$ is a proper rational function and has simple poles at $s = 0$, $s = -1$, $s = -2$. Its partial fraction expansion is

$$F(s) = \frac{k_1}{s} + \frac{k_2}{s + 1} + \frac{k_3}{s + 2}$$

The cover-up algorithm yields the residues as

$$k_1 = sF(s)\big|_{s=0} = \frac{2(s + 3)}{(s + 1)(s + 2)}\bigg|_{s=0} = 3$$

$$k_2 = (s + 1)F(s)\big|_{s=-1} = \frac{2(s + 3)}{s(s + 2)}\bigg|_{s=-1} = -4$$

$$k_3 = (s + 2)F(s)\big|_{s=-2} = \frac{2(s + 3)}{s(s + 1)}\bigg|_{s=-2} = 1$$

The inverse transform $f(t)$ is

$$f(t) = [3 - 4e^{-t} + e^{-2t}]u(t) \qquad ■$$

Exercise 9–7

Find the waveforms corresponding to the following transforms:

(a) $F_1(s) = \dfrac{4}{(s + 1)(s + 3)}$

(b) $F_2(s) = e^{-5s}\left[\dfrac{2s}{(s + 1)(s + 3)}\right]$

(c) $F_3(s) = \dfrac{4(s + 2)}{(s + 1)(s + 3)}$

Answers:

$$(a) \, f_1(t) = [2e^{-t} - 2e^{-3t}]u(t)$$

$$(b) \, f_2(t) = [-e^{-(t-5)} + 3e^{-3(t-5)}]u(t-5)$$

$$(c) \, f_3(t) = [2e^{-t} + 2e^{-3t}]u(t)$$

Comment: Note that

$$F_2(s) = e^{-5s}\left[\frac{2s}{(s+1)(s+3)}\right] = e^{-5s}\left[\frac{-1}{s+1} + \frac{3}{s+3}\right]$$

so that the delay factor e^{-5s} is not involved in the partial fraction expansion but simply flags the amount by which the resulting waveform $[-e^{-t} + 3e^{-3t}]u(t)$ is delayed to produce $f_2(t)$.

Exercise 9–8

Find the waveforms corresponding to the following transforms:

$$(a) \, F(s) = \frac{6(s+2)}{s(s+1)(s+4)}$$

$$(b) \, F(s) = \frac{4(s+1)}{s(s+1)(s+4)}$$

Answers:

$$(a) \, f(t) = [3 - 2e^{-t} - e^{-4t}]u(t)$$

$$(b) \, f(t) = [1 - e^{-4t}]u(t)$$

COMPLEX POLES

Special treatment is necessary when $F(s)$ has a complex pole. In physical situations the function $F(s)$ is a ratio of polynomials with real coefficients. If $F(s)$ has a complex pole $p = -\alpha + j\beta$, then it must also have a pole $p^* = -\alpha - j\beta$; otherwise the coefficients of the denominator polynomial would not be real. In other words, for physical signals the complex poles of $F(s)$ must occur in conjugate pairs. As a consequence, the partial fraction decomposition of $F(s)$ will contain two terms of the form

$$F(s) = \ldots + \frac{k}{s+\alpha-j\beta} + \frac{k^*}{s+\alpha+j\beta} + \ldots \qquad (9\text{–}21)$$

The residues k and k^* at the conjugate poles are themselves conjugates because $F(s)$ is a rational function with real coefficients. These residues can be calculated using the cover-up algorithm and, in general, they turn out to be complex numbers. If the complex residues are written in polar form as

$$k = |k|e^{j\theta} \text{ and } k^* = |k|e^{-j\theta}$$

then the waveform corresponding to the two terms in Eq. (9–21) is

$$f(t) = [\ldots + |k|e^{j\theta}e^{(-\alpha+j\beta)t} + |k|e^{-j\theta}e^{(-\alpha-j\beta)t} + \ldots]u(t)$$

This equation can be rearranged in the form

$$f(t) = \left[\ldots + 2|k|e^{-\alpha t} \left\{ \frac{e^{+j(\beta t + \theta)} + e^{-j(\beta t + \theta)}}{2} \right\} + \ldots \right] u(t) \qquad (9\text{–}22)$$

The expression inside the brackets is of the form

$$\cos x = \left\{ \frac{e^{+jx} + e^{-jx}}{2} \right\}$$

Consequently, we combine terms inside the braces as a cosine function with a phase angle:

$$f(t) = [\ldots + 2|k|e^{-\alpha t} \cos (\beta t + \theta) + \ldots] u(t) \qquad (9\text{–}23)$$

In summary, if $F(s)$ has a complex pole, then in physical applications there must be an accompanying conjugate complex pole. The inverse transformation combines the two poles to produce a damped cosine waveform. We only need to compute the residue at one of these poles because the residues at conjugate poles must be conjugates. Normally, we calculate the residue for the pole at $s = -\alpha + j\beta$ because its angle equals the phase angle of the damped cosine. Note that the imaginary part of this pole is positive, which means that the pole lies in the upper half of the s plane.

The inverse transform of a proper rational function with simple poles can be found by the partial fraction expansion method. The residues k at the simple poles can be found using the cover-up algorithm. The resulting waveform is a sum of terms of the form $[ke^{-\alpha t}]u(t)$ for real poles and $[2|k|e^{-\alpha t} \cos(\beta t + \theta)]u(t)$ for a pair of complex conjugate poles. The partial fraction expansion of the transform contains all of the data needed to construct the corresponding waveform.

EXAMPLE 9–12

Find the inverse transform of

$$F(s) = \frac{20(s + 3)}{(s + 1)(s^2 + 2s + 5)}$$

SOLUTION:
$F(s)$ has a simple pole at $s = -1$ and a pair of conjugate complex poles located at the roots of the quadratic factor

$$(s^2 + 2s + 5) = (s + 1 - j2)(s + 1 + j2)$$

The partial fraction expansion of $F(s)$ is

$$F(s) = \frac{k_1}{s + 1} + \frac{k_2}{s + 1 - j2} + \frac{k_2^*}{s + 1 + j2}$$

The residues at the poles are found from the cover-up algorithm.

$$k_1 = \left. \frac{20(s + 3)}{s^2 + 2s + 5} \right|_{s = -1} = 10$$

$$k_2 = \left. \frac{20(s + 3)}{(s + 1)(s + 1 + j2)} \right|_{s = -1 + j2} = -5 - j5 = 5\sqrt{2}e^{+j5\pi/4}$$

We now have all of the data needed to construct the inverse transform.

$$f(t) = [10e^{-t} + 10\sqrt{2}e^{-t}\cos(2t + 5\pi/4)]u(t)$$

In this example we used k_2 to obtain the amplitude and phase angle of the damped cosine term. The residue k_2^* is not needed, but to illustrate a point we note that its value is

$$k_2^* = (-5 - j5)^* = -5 + j5 = 5\sqrt{2}e^{-j5\pi/4}$$

If k_2^* is used instead, we get the same amplitude for the damped sine but the wrong phase angle. *Caution:* Remember that Eq. (9–23) uses the residue at the complex pole with a positive imaginary part. In this example this is the pole at $s = -1 + j2$, not the pole at $s = -1 - j2$.

Exercise 9–9

Find the inverse transforms of the following rational functions:

$$\text{(a) } F(s) = \frac{16}{(s + 2)(s^2 + 4)}$$

$$\text{(b) } F(s) = \frac{2(s + 2)}{s(s^2 + 4)}$$

Answers:

$$\text{(a) } f(t) = [2e^{-2t} + 2\sqrt{2}\cos(2t - 3\pi/4)]u(t)$$

$$\text{(b) } f(t) = [1 + \sqrt{2}\cos(2t - 3\pi/4)]u(t)$$

Exercise 9–10

Find the inverse transforms of the following rational functions:

$$\text{(a) } F(s) = \frac{8}{s(s^2 + 4s + 8)}$$

$$\text{(b) } F(s) = \frac{4s}{s^2 + 4s + 8}$$

Answers:

$$\text{(a) } f(t) = [1 + \sqrt{2}e^{-2t}\cos(2t + 3\pi/4)]u(t)$$

$$\text{(b) } f(t) = [4\sqrt{2}e^{-2t}\cos(2t + \pi/4)]u(t)$$

SUMS OF RESIDUES

The sum of the residues of a proper rational function are subject to certain conditions that are useful for checking the calculations in a partial fraction expansion. To derive these conditions, we multiply Eqs. (9–17) and (9–18) by s and take the limit as $s \to \infty$. These operations yield

$$\lim_{s \to \infty} sF(s) = \lim_{s \to \infty} \frac{Ks^{m+1}}{s^n} = \lim_{s \to \infty} \left(\frac{k_1 s}{s + p_1} + \dots + \frac{k_n s}{s + p_n} \right)$$

In the limit this equation reduces to

$$K \left[\lim_{s \to \infty} \frac{s^{m+1}}{s^n} \right] = k_1 + k_2 + \dots + k_n$$

Since $F(s)$ is a proper rational function with $n > m$, the limit process in this equation yields to following conditions:

$$k_1 + k_2 + \dots + k_n = \begin{cases} 0 \text{ if } n > m \\ K \text{ if } n = m \end{cases} \tag{9–24}$$

For a proper rational function with simple poles, the sum of residues is either zero or else equal to the transform scale factor K.

Exercise 9–11

Use the sum of residues to find the unknown residue in the following expansions:

(a) $\dfrac{21(s+5)}{(s+3)(s+10)} = \dfrac{6}{s+3} + \dfrac{k}{s+10}$

(b) $\dfrac{58s}{(s+2)(s^2+25)} = \dfrac{k}{s+2} + \dfrac{2+j5}{s+j5} + \dfrac{2-j5}{s-j5}$

Answers:
(a) $k = 15$
(b) $k = -4$

9–5 SOME SPECIAL CASES

Most of the transforms encountered in physical applications are proper rational functions with simple poles. The inverse transforms of such functions can be handled by the partial fraction expansion method developed in the previous section. This section covers the problem of finding the inverse transform when $F(s)$ is an improper rational function or has multiple poles. These matters are treated as special cases because they only occur for certain discrete values of circuit or signal parameters. However, some of these special cases are important, so we need to learn how to handle improper rational functions and multiple poles.

$F(s)$ is an **improper rational function** when the order of the numerator polynomial equals or exceeds the order of the denominator ($m \geq n$). For example, the transform

$$F(s) = \frac{s^3 + 6s^2 + 12s + 8}{s^2 + 4s + 3} \tag{9–25}$$

is improper because $m = 3$ and $n = 2$. This function can be reduced to a quotient and a remainder, which is a proper function using long division as follows:

$$\begin{array}{r}
s + 2 \\
s^2 + 4s + 3 \overline{\smash{\big)}\ s^3 + 6s^2 + 12s + 8} \\
\underline{s^3 + 4s^2 + 3s} \\
2s^2 + 9s + 8 \\
\underline{2s^2 + 8s + 6} \\
s + 2
\end{array}$$

which yields

$$F(s) = s + 2 + \frac{s + 2}{s^2 + 4s + 3}$$

$$= \text{Quotient} + \text{Remainder}$$

The remainder is a proper rational function, which can be expanded by partial fractions to produce

$$F(s) = s + 2 + \frac{1/2}{s + 1} + \frac{1/2}{s + 3}$$

All of the terms in this expansion are listed in Table 9–2 except the first term. The inverse transform of the first term is found using the transform of an impulse and the differentiation property. The Laplace transform of the derivative of an impulse is

$$\mathcal{L}\left[\frac{d\delta(t)}{dt}\right] = s\mathcal{L}[\delta(t)] - \delta(0-) = s$$

since $\mathcal{L}\{\delta(t)\} = 1$ and $\delta(0-) = 0$. By the uniqueness property of the Laplace transformation, we have $\mathcal{L}^{-1}\{s\} = d\delta(t)/dt$. The first derivative of an impulse is called a doublet. The inverse transform of the improper rational function in Eq. (9–25) is

$$f(t) = \frac{d\delta(t)}{dt} + 2\delta(t) + \left[\frac{1}{2}e^{-t} + \frac{1}{2}e^{-3t}\right]u(t)$$

The method illustrated by this example generalizes in the following way. When $m = n$, long division produces a quotient K plus a proper rational function remainder. The constant K corresponds to an impulse $K\delta(t)$, and the remainder can be expanded by partial fractions to find the corresponding waveform. If $m > n$, then long division yields a quotient with terms like $s, s^2, \ldots s^{m-n}$ before a proper remainder function is obtained. These higher powers of s correspond to derivatives of the impulse. These pathological waveforms are theoretically interesting, but they do not actually occur in real circuits.

Improper rational functions can arise during mathematical manipulation of signal transforms. When $F(s)$ is improper, it is essential to reduce it by long division prior to expansion; otherwise the resulting partial fraction expansion will be incomplete.

Exercise 9–12

Find the inverse transforms of the following functions:

$$\text{(a) } F(s) = \frac{s^2 + 4s + 5}{s^2 + 4s + 3}$$

$$\text{(b) } F(s) = \frac{s^2 - 4}{s^2 + 4}$$

Answers:

$$(a) f(t) = \delta(t) + [e^{-t} - e^{-3t}]u(t)$$

$$(b) f(t) = \delta(t) - [4\sin(2t)]u(t)$$

___

Exercise 9-13

Find the inverse transforms of the following functions:

$$(a)\ F(s) = \frac{2s^2 + 3s + 5}{s}$$

$$(b)\ F(s) = \frac{s^3 + 2s^2 + s + 3}{s + 2}$$

Answers:

$$(a)\ f(t) = 2\frac{d\delta(t)}{dt} + 3\delta(t) + 5u(t)$$

$$(b)\ f(t) = \frac{d^2\delta(t)}{dt^2} + \delta(t) + [e^{-2t}]u(t)$$

___

MULTIPLE POLES

Under certain special conditions, transforms can have multiple poles. For example, the transform

$$F(s) = \frac{K(s - z_1)}{(s - p_1)(s - p_2)^2} \tag{9-26}$$

has a simple pole at $s = p_1$ and a pole of order 2 at $s = p_2$. Finding the inverse transform of this function requires special treatment of the multiple pole. We first factor out one of the two multiple poles.

$$F(s) = \frac{1}{s - p_2}\left[\frac{K(s - z_1)}{(s - p_1)(s - p_2)}\right] \tag{9-27}$$

The quantity inside the brackets is a proper rational function with only simple poles and can be expanded by partial fractions using the method of the previous section.

$$F(s) = \frac{1}{s - p_2}\left[\frac{c_1}{s - p_1} + \frac{k_{22}}{s - p_2}\right]$$

We now multiply through by the pole factored out in the first step to obtain

$$F(s) = \frac{c_1}{(s - p_1)(s - p_2)} + \frac{k_{22}}{(s - p_2)^2}$$

The first term on the right is a proper rational function with only simple poles, so it too can be expanded by partial fractions as

$$F(s) = \frac{k_1}{s - p_1} + \frac{k_{21}}{s - p_2} + \frac{k_{22}}{(s - p_2)^2}$$

After two partial fraction expansions, we have an expression in which every term is available in Table 9-2. The first two terms are simple poles that lead to exponential waveforms. The third term is of the form $k/(s + \alpha)^2$, which is the transform of a damped ramp waveform $[kt \ e^{-\alpha t}]u(t)$. Therefore, the inverse transform of $F(s)$ in Eq. (9–26) is

$$f(t) = [k_1 e^{p_1 t} + k_{12} e^{p_2 t} + k_{22} t e^{p_2 t}]u(t) \qquad (9\text{--}28)$$

Caution: If $F(s)$ in Eq. (9–26) had another finite zero, then the term in the brackets in Eq. (9–27) would be an improper rational function. When this occurs, long division must be used to reduce the improper rational function before proceeding to the partial-fraction expansion in the next step.

As with simple poles, the *s*-domain location of multiple poles determines the exponents of the exponential waveforms. The residues at the poles are the amplitudes of the waveforms. The only difference here is that the double pole leads to two terms rather than a single waveform. The first term is an exponential of the form e^{pt}, and the second term is a damped ramp of the form te^{pt}.

EXAMPLE 9–13

Find the inverse transform of

$$F(s) = \frac{4(s + 3)}{s(s + 2)^2}$$

SOLUTION:
The given transform has a simple pole at $s = 0$ and a double pole at $s = -2$. Factoring out one of the multiple poles and expanding the remainder by partial fractions yields

$$F(s) = \frac{1}{s + 2}\left[\frac{4(s + 3)}{s(s + 2)}\right] = \frac{1}{s + 2}\left[\frac{6}{s} - \frac{2}{s + 2}\right]$$

Multiplying through by the removed factor and expanding again by partial fractions produces

$$F(s) = \frac{6}{s(s + 2)} - \frac{2}{(s + 2)^2} = \frac{3}{s} - \frac{3}{s + 2} - \frac{2}{(s + 2)^2}$$

The last expansion on the right yields the inverse transform as

$$f(t) = [3 - 3e^{-2t} - 2te^{-2t}]u(t) \qquad \blacksquare$$

In principle, the procedure illustrated in this example can be applied to higher-order poles, although the process rapidly becomes quite tedious. For example, an *n*th-order pole would require *n* partial-fraction expansions, which is not an idea with irresistible appeal. Mathematics offers other methods of determining multiple pole residues that reduce the computational burden somewhat. However, these advanced mathematical tools are probably not worth learning because computer tools like Mathcad and MATLAB readily handle multiple pole situations (see Example 9–14). Although we do encounter functions with high-order multiple poles in later chapters, we rarely need to find their inverse transforms.

Thus, for practical reasons our interest in multiple pole transforms is limited to two possibilities. First, a double pole on the negative real axis leads to the damped ramp:

$$\mathcal{L}^{-1}\left[\frac{k}{(s+\alpha)^2}\right] = [kte^{-\alpha t}]u(t) \qquad (9\text{--}29)$$

Second, a pair of double, complex poles leads to the damped cosine ramp:

$$\mathcal{L}^{-1}\left[\frac{k}{(s+\alpha-j\beta)^2} + \frac{k^*}{(s+\alpha+j\beta)^2}\right] = [2\,|k|\,te^{-\alpha t}\cos(\beta t + \angle k)]u(t)$$

$$(9\text{--}30)$$

These two cases illustrate a general principle: When a simple pole leads to a waveform $f(t)$, then a double pole at the same location leads to a waveform $tf(t)$. Multiplying a waveform by t tends to cause the waveform to increase without bound unless exponential damping is present. The following example illustrates a case in which the damping is not present.

EXAMPLE 9-14

Use a math analysis program to find the waveform corresponding to a transform with a zero at $s = -400$, a simple pole at $s = -1000$, a double pole at $s = j400$, a double pole at $s = -j400$, and a value at $s = 0$ of $F(0) = 2 \times 10^{-4}$. Plot the resulting waveform $f(t)$.

SOLUTION:

The symbolic analysis capability of Mathcad includes inverse Laplace transforms. Figure 9–5 is a worksheet demonstrating this capability. The specified critical frequencies define the $F(s)$ shown in the first line. The specified value at $s = 0$ allows us to evaluate the unknown scale factor K as shown in the second line. This leads directly to the transform in the third

FIGURE   9 - 5

line. After highlighting (clicking on) an "s" in $F(s)$ to identify the independent variable, we select **Symbolic, Transforms, Inverse Laplace Transform** from the menu bar. Mathcad responds by producing the $f(t)$ shown in the figure. This waveform contains five terms: The exponential e^{-1000t} came from the simple pole at $s = -1000$, and the four sinusoidal terms $\cos(400t)$, $\sin(400t)$, $t \times \cos(400t)$, and $t \times \sin(400t)$ all came from the double poles at $s = \pm j400$. As time increases, the terms $t \times \cos(400t)$ and $t \times \sin(400t)$ increase without bound, leading to the linearly increasing amplitude shown in the waveform plot. A key point to remember is that poles on the j-axis lead to sustained waveforms, like the step function and sinusoid, that do not decay. Consequently, double poles on the j-axis lead to $tf(t)$ waveforms whose amplitudes increase without bound, producing an unstable response.

Exercise 9–14

Find the inverse transforms of the following functions:

$$\text{(a) } F(s) = \frac{s}{(s + 1)(s + 2)^2}$$

$$\text{(b) } F(s) = \frac{16}{s^2(s + 4)}$$

$$\text{(c) } F(s) = \frac{800\, s(s + 1)}{(s + 2)(s + 10)^2}$$

Answers:

$$\text{(a) } f(t) = [-e^{-t} + e^{-2t} + 2te^{-2t}]u(t)$$

$$\text{(b) } f(t) = [4t - 1 + e^{-4t}]u(t)$$

$$\text{(c) } f(t) = [25e^{-2t} + 775e^{-10t} - 9000te^{-10t}]u(t)$$

9–6 CIRCUIT RESPONSE USING LAPLACE TRANSFORMS

The payoff for learning about the Laplace transformation comes when we use it to find the response of dynamic circuits. The pattern for circuit analysis is shown by the solid line in Figure 9–1. The basic analysis steps are as follows:

STEP 1: Develop the circuit differential equation in the time domain.

STEP 2: Transform this equation into the s domain and algebraically solve for the response transform.

STEP 3: Apply the inverse transformation to this transform to produce the response waveform.

FIGURE 9–6 *First-order RC circuit.*

The first-order RC circuit in Figure 9–6 will be used to illustrate these steps.

STEP 1

The KVL equation around the loop and the element i–v relationship or element equations are

$$\text{KVL: } -v_S(t) + v_R(t) + v_C(t) = 0$$

$$\text{Source: } v_S(t) = V_A u(t)$$

$$\text{Resistor: } v_R(t) = i(t)R$$

$$\text{Capacitor: } i(t) = C\frac{dv_C(t)}{dt}$$

Substituting the i–v relationships into the KVL equation and rearranging terms produces a first-order differential equation

$$RC\frac{dv_C(t)}{dt} + v_C(t) = V_A u(t) \tag{9–31}$$

with an initial condition $v_C(0-) = V_0$ V.

STEP 2

The analysis objective is to use Laplace transforms to find the waveform $v_C(t)$ that satisfies this differential equation and the initial condition. We first apply the Laplace transformation to both sides of Eq. (9–31):

$$\mathcal{L}\left[RC\frac{dv_C(t)}{dt} + v_C(t)\right] = \mathcal{L}[V_A u(t)]$$

Using the linearity property leads to

$$RC\mathcal{L}\left[\frac{dv_C(t)}{dt}\right] + \mathcal{L}[v_C(t)] = V_A\mathcal{L}[u(t)]$$

Using the differentiation property and the transform of a unit step function produces

$$RC[sV_C(s) - V_0] + V_C(s) = V_A\frac{1}{s} \tag{9–32}$$

This result is an algebraic equation in $V_C(s)$, which is the transform of the response we seek. We rearrange Eq. (9–32) in the form

$$(s + 1/RC)V_C(s) = \frac{V_A/RC}{s} + V_0$$

and algebraically solve for $V_C(s)$.

$$V_C(s) = \frac{V_A/RC}{s(s + 1/RC)} + \frac{V_0}{s + 1/RC} \quad \text{V-s} \tag{9–33}$$

The function $V_C(s)$ is the transform of the waveform $v_C(t)$ that satisfies the differential equation and the initial condition. The initial condition appears explicitly in this equation as a result of applying the differentiation rule to obtain Eq. (9–32).

STEP 3

To obtain the waveform $v_C(t)$, we find the inverse transform of the right side of Eq. (9–33). The first term on the right is a proper rational function with two simple poles on the real axis in the s plane. The pole at the origin was introduced by the step function input. The pole at $s = -1/RC$ came from the circuit. The partial fraction expansion of the first term in Eq. (9–33) is

$$\frac{V_A/RC}{s(s + 1/RC)} = \frac{k_1}{s} + \frac{k_2}{s + 1/RC}$$

The residues k_1 and k_2 are found using the cover-up algorithm.

$$k_1 = \left.\frac{V_A/RC}{s + 1/RC}\right|_{s=0} = V_A \quad \text{and} \quad k_2 = \left.\frac{V_A/RC}{s}\right|_{s=-1/RC} = -V_A$$

Using these residues, we expand Eq. (9–33) by partial fractions as

$$V_C(s) = \frac{V_A}{s} - \frac{V_A}{s + 1/RC} + \frac{V_0}{s + 1/RC} \tag{9–34}$$

Each term in this expansion is recognizable: The first is a step function and the next two are exponentials. Taking the inverse transform of Eq. (9–34) gives

$$\begin{aligned} v_C(t) &= [V_A - V_A e^{-t/RC} + V_0 e^{-t/RC}]u(t) \\ &= [V_A + (V_0 - V_A)e^{-t/RC}]u(t) \quad \text{V} \end{aligned} \tag{9–35}$$

The waveform $v_C(t)$ satisfies the differential equation in Eq. (9–31) and the initial condition $v_C(0-) = V_0$. The term $V_A u(t)$ is the forced response due to the step function input, and the term $[(V_0 - V_A)e^{t/RC}]u(t)$ is the natural response. The complete response depends on three parameters: the input amplitude V_A, the circuit time constant RC, and the initial condition V_0.

These results are identical to those found using the classical methods in Chapter 7. The outcome is the same, but the method is quite different. The Laplace transformation yields the complete response (forced and natural) by an algebraic process that inherently accounts for the initial conditions. The solid arrow in Figure 9–1 shows the overall procedure. Begin with Eq. (9–31) and relate each step leading to Eq. (9–35) to steps in Figure 9–1.

EXAMPLE 9–15

The switch in Figure 9–7 has been in position A for a long time. At $t = 0$ it is moved to position B. Find $i_L(t)$ for $t \geq 0$.

SOLUTION:

Step 1: The circuit differential equation is found by combining the KVL equation and element equations with the switch in position B.

$$\text{KVL: } v_R(t) + v_L(t) = 0$$

$$\text{Resistor: } v_R(t) = i_L(t)R$$

$$\text{Inductor: } v_L(t) = L\frac{di_L(t)}{dt}$$

FIGURE 9–7

Substituting the element equations into the KVL equation yields

$$L\frac{di_L(t)}{dt} + Ri_L(t) = 0$$

Prior to $t = 0$, the circuit was in a dc steady-state condition with the switch in position A. Under dc conditions the inductor acts like a short circuit, and the inductor current just prior to moving the switch is $i_L(0-) = I_0 = V_A/R$.

Step 2: Using the linearity and differentiation properties, we transform the circuit differential equation into the s domain as

$$L[sI_L(s) - I_0] + RI_L(s) = 0$$

Solving algebraically for $I_L(s)$ yields

$$I_L(s) = \frac{I_0}{s + R/L} \quad \text{A-s}$$

Step 3: The inverse transform of $I_L(s)$ is an exponential waveform:

$$i_L(t) = [I_0 e^{-Rt/L}]u(t) \quad \text{A}$$

where $I_0 = V_A/R$. Substituting $i_L(t)$ back into the differential equation yields

$$L\frac{di_L(t)}{dt} + Ri_L(t) = -RI_0 e^{-Rt/L} + RI_0 e^{-Rt/L} = 0$$

The waveform found using Laplace transforms does indeed satisfy the circuit differential equation and the initial condition. ∎

EXAMPLE 9−16

The switch in Figure 9–8 has been open for a long time. At $t = 0$ the switch is closed. Find $i(t)$ for $t \geq 0$.

$R = 400\ \Omega$ $C = 5\ \mu\text{F}$
$L = 1\ \text{H}$ $V_A = 10\ \text{V}$

F I G U R E   9 − 8

SOLUTION:

The governing equation for the second-order circuit in Figure 9–8 is found by combining the element equations and a KVL equation around the loop with the switch closed:

$$\text{KVL: } v_R(t) + v_L(t) + v_C(t) = 0$$

$$\text{Resistor: } v_R(t) = Ri(t)$$

$$\text{Inductor: } v_L(t) = L\frac{di(t)}{dt}$$

$$\text{Capacitor: } v_C(t) = \frac{1}{C}\int_0^t i(\tau)d\tau + v_C(0)$$

Substituting the element equations into the KVL equation yields

$$L\frac{di(t)}{dt} + Ri(t) + \frac{1}{C}\int_0^t i(\tau)d\tau + v_C(0) = 0$$

Using the linearity property, the differentiation property, and the integration property, we transform this second-order integro-differential equation into the s domain as

$$L[sI(s) - i_L(0)] + RI(s) + \frac{1}{C}\frac{I(s)}{s} + v_C(0)\frac{1}{s} = 0$$

Solving for $I(s)$ results in

$$I(s) = \frac{si_L(0) - v_C(0)/L}{s^2 + \frac{R}{L}s + \frac{1}{LC}} \quad \text{A-s}$$

Prior to $t = 0$, the circuit was in a dc steady-state condition with the switch open. In dc steady state, the inductor acts like a short circuit and the capacitor like an open circuit, so the initial conditions are $i_L(0-) = 0$ A and $v_C(0-) = V_A = 10$ V. Inserting the initial conditions and the numerical values of the circuit parameters into the equation for $I(s)$ gives

$$I(s) = -\frac{10}{s^2 + 400s + 2 \times 10^5} \quad \text{A-s}$$

The denominator quadratic can be factored as $(s + 200)^2 + 400^2$ and $I(s)$ written in the following form:

$$I(s) = -\frac{10}{400}\left[\frac{400}{(s + 200)^2 + (400)^2}\right] \quad \text{A-s}$$

Comparing the quantity inside the brackets with the entries in the $F(s)$ column of Table 9–2, we find that $I(s)$ is a damped sine with $\alpha = 200$ and $\beta = 400$. By linearity, the quantity outside the brackets is the amplitude of the damped sine. The inverse transform is

$$i(t) = [-0.025\,e^{-200t}\sin 400t]u(t) \quad \text{A}$$

Substituting this result back into the circuit integro-differential equation yields the following term-by-term tabulation:

$$L\frac{di(t)}{dt} = +5e^{-200t}\sin 400t - 10e^{-200t}\cos 400t$$

$$Ri(t) = -10e^{-200t}\sin 400t$$

$$\frac{1}{C}\int_0^t i(\tau)d\tau = +5e^{-200t}\sin 400t + 10e^{-200t}\cos 400t - 10$$

$$v_C(0) = +10$$

The sum of the right-hand sides of these equations is zero. This result shows that the waveform $i(t)$ found using Laplace transforms does indeed satisfy the circuit integro-differential equation and the initial conditions. ∎

Exercise 9–15

Find the transform $V(s)$ that satisfies the following differential equations and the initial conditions:

(a) $\dfrac{dv(t)}{dt} + 6v(t) = 4u(t)$ $v(0-) = -3$

(b) $4\dfrac{dv(t)}{dt} + 12v(t) = 16 \cos 3t,$ $v(0-) = 2$

Answers:

(a) $V(s) = \dfrac{4}{s(s+6)} - \dfrac{3}{s+6}$ V-s

(b) $V(s) = \dfrac{4s}{(s^2+9)(s+3)} + \dfrac{2}{s+3}$ V-s

Exercise 9–16

Find the $V(s)$ that satisfies the following equations:

(a) $\displaystyle\int_0^t v(\tau)d\tau + 10v(t) = 10u(t)$

(b) $\dfrac{d^2v(t)}{dt^2} + 4\dfrac{dv(t)}{dt} + 3v(t) = 5e^{-2t}$ $v'(0-) = 2$ $v(0-) = -2$

Answers:

(a) $V(s) = \dfrac{1}{s + 0.1}$ V-s

(b) $V(s) = \dfrac{5}{(s+1)(s+2)(s+3)} - \dfrac{2}{s+1}$ V-s

CIRCUIT RESPONSE WITH TIME-VARYING INPUTS

It is encouraging to find that the Laplace transformation yields results that agree with those obtained by classical methods. The transform method reduces solving circuit differential equations to an algebraic process that includes the initial conditions. However, before being overcome with euphoria we must remember that the Laplace transform method begins with the circuit differential equation and the initial conditions. It does not provide these quantities to us. The transform method simplifies the solution process, but it does not substitute for understanding how to formulate circuit equations.

The Laplace transform method is especially usefully when the circuit is driven by time-varying inputs. To illustrate, we return to the RC circuit in Figure 9–6 and replace the step function input by a general input signal denoted $v_S(t)$. The right side of the circuit differential equation in Eq. (9–31) changes to accommodate the new input by taking the form

$$RC\dfrac{dv_C(t)}{dt} + v_C(t) = v_S(t) \tag{9–36}$$

with an initial condition $v_C(0) = V_0$ V.

The only change here is that the driving force on the right side of the differential equation is a general time-varying waveform $v_S(t)$. The objective is to find the capacitor voltage $v_C(t)$ that satisfies the differential equation and the initial conditions. The classical methods of solving for the forced response depend on the form of $v_S(t)$. However, with the Laplace transform method we can proceed without actually specifying the form of the input signal.

We first transform Eq. (9–36) into the s domain:

$$RC[sV_C(s) - V_0] + V_C(s) = V_S(s)$$

The only assumption here is that the input waveform is Laplace transformable, a condition met by all causal signals of engineering interest. We now algebraically solve for the response $V_C(s)$:

$$V_C(s) = \frac{V_S(s)/RC}{s + 1/RC} + \frac{V_0}{s + 1/RC} \quad \text{V-s} \tag{9–37}$$

The function $V_C(s)$ is the transform of the response of the RC circuit in Figure 9–6 due to a general input signal $v_S(t)$. We have gotten this far without specifying the form of the input signal. In a sense, we have found the general solution in the s domain of the differential equation in Eq. (9–36) for any casual input signal.

All of the necessary ingredients are present in Eq. (9–37):

1. The transform $V_S(s)$ represents the applied input signal.

2. The pole at $s = -1/RC$ defines the circuit time constant.

3. The initial value $v_C(0-) = V_0$ summarizes all events prior to $t = 0$.

However, we must have a particular input in mind to solve for the waveform $v_C(t)$. The following examples illustrate the procedure for different input driving forces.

EXAMPLE 9–17

Find $v_C(t)$ in the RC circuit in Figure 9–6 when the input is the waveform $v_S(t) = [V_A e^{-\alpha t}]u(t)$.

SOLUTION:
The transform of the input is $V_S(s) = V_A/(s + \alpha)$. For the exponential input the response transform in Eq. (9–37) becomes

$$V_C(s) = \frac{V_A/RC}{(s + \alpha)(s + 1/RC)} + \frac{V_0}{s + 1/RC} \quad \text{V-s} \tag{9–38}$$

If $\alpha \neq 1/RC$ then the first term on the right is a proper rational function with two simple poles. The pole at $s = -\alpha$ came from the input and the pole at $s = -1/RC$ from the circuit. A partial fraction expansion of the first term has the form

$$\frac{V_A/RC}{(s + \alpha)(s + 1/RC)} = \frac{k_1}{s + \alpha} + \frac{k_2}{s + 1/RC}$$

The residues in this expansion are

$$k_1 = \left.\frac{V_A/RC}{s + 1/RC}\right|_{s = -\alpha} = \frac{V_A}{1 - \alpha RC}$$

$$k_2 = \left.\frac{V_A/RC}{s + \alpha}\right|_{s = -1/RC} = \frac{V_A}{\alpha RC - 1}$$

The expansion of the response transform $V_C(s)$ is

$$V_C(s) = \frac{V_A/(1 - \alpha RC)}{s + \alpha} + \frac{V_A/(\alpha RC - 1)}{s + 1/RC} + \frac{V_0}{s + 1/RC} \quad \text{V-s}$$

The inverse transform of $V_C(s)$ is

$$v_C(t) = \left[\frac{V_A}{1 - \alpha RC}e^{-\alpha t} + \frac{V_A}{\alpha RC - 1}e^{-t/RC} + V_0 e^{-t/RC}\right]u(t) \quad \text{V}$$

The first term is the forced response, and the last two terms are the natural response. The forced response is an exponential because the input introduced a pole at $s = -\alpha$. The natural response is also an exponential, but its time constant depends on the circuit's pole at $s = -1/RC$. In this case the forced and natural responses are both exponential signals with poles on the real axis. However, the forced response comes from the pole introduced by the input, while the natural response depends on the circuit's pole.

If $\alpha = 1/RC$, then the response just given is no longer valid (k_1 and k_2 become infinite). To find the response for this condition, we return to Eq. (9–38) and replace α by $1/RC$:

$$V_C(s) = \frac{V_A/RC}{(s + 1/RC)^2} + \frac{V_0}{s + 1/RC} \quad \text{V-s}$$

We now have a double pole at $s = -1/RC = -\alpha$. The double pole term is the transform of a damped ramp, so the inverse transform is

$$v_C(t) = \left[V_A\frac{t}{RC}e^{-t/RC} + V_0 e^{-t/RC}\right]u(t) \quad \text{V}$$

When $\alpha = 1/RC$ the s-domain poles of the input and the circuit coincide and the zero-state ($V_0 = 0$) response has the form $\alpha t e^{-\alpha t}$. We cannot separate this response into forced and natural components since the input and circuit poles coincide.   ■

EXAMPLE 9—18

Find $v_C(t)$ when the input to the RC circuit in Figure 9–6 is $v_S(t) = [V_A \cos \beta t]u(t)$.

SOLUTION:
The transform of the input is $V_S(s) = V_A s/(s^2 + \beta^2)$. For a cosine input the response transform in Eq. (9–37) becomes

$$V_C(s) = \frac{sV_A/RC}{(s^2 + \beta^2)(s + 1/RC)} + \frac{V_0}{s + 1/RC} \quad \text{V-s}$$

The sinusoidal input introduces a pair of poles located at $s = \pm j\beta$. The first term on the right is a proper rational function with three simple poles. The partial fraction expansion of the first term is

$$\frac{sV_A/RC}{(s - j\beta)(s + j\beta)(s + 1/RC)} = \frac{k_1}{s - j\beta} + \frac{k_1^*}{s + j\beta} + \frac{k_2}{s + 1/RC}$$

To find the response, we need to find the residues k_1 and k_2:

$$k_1 = \frac{sV_A/RC}{(s + j\beta)(s + 1/RC)}\bigg|_{s = j\beta} = \frac{V_A/2}{1 + j\beta RC} = |k_1|e^{j\theta}$$

where

$$|k_1| = \frac{V_A/2}{\sqrt{1 + (\beta RC)^2}} \quad \text{and} \quad \theta = -\tan^{-1}(\beta RC)$$

The residue k_2 at the circuit pole is

$$k_2 = \frac{sV_A/RC}{s^2 + \beta^2}\bigg|_{s = -1/RC} = -\frac{V_A}{1 + (\beta RC)^2}$$

We now perform the inverse transform to obtain the response waveform:

$$v_C(t) = [2|k_1|\cos(\beta t + \theta) + k_2 e^{-t/RC} + V_0 e^{-t/RC}]u(t)$$

$$= \left[\frac{V_A}{\sqrt{1 + (\beta RC)^2}} \cos(\beta t + \theta) - \frac{V_A}{1 + (\beta RC)^2} e^{-t/RC} + V_0 e^{-t/RC}\right]u(t) \quad \text{V}$$

The first term is the forced response, and the remaining two are the natural response. The forced response is sinusoidal because the input introduces poles at $s = \pm j\beta$. The natural response is an exponential with a time constant determined by the location of the circuit's pole at $s = -1/RC$. ∎

EXAMPLE 9–19

Find $v_C(t)$ in the RC circuit of Figure 9–6 when the input is a rectangular pulse

$$v_S(t) = V_A u(t) - V_A u(t - a) \quad \text{V}$$

SOLUTION:
The input is a rectangular pulse that begins at $t = 0$ and ends at $t = a$. Using the time-domain translation property, the transform of the input is

$$V_S(s) = \frac{V_A}{s} - \frac{V_A e^{-as}}{s} \quad \text{V-s}$$

Using this input the transform of the circuit response in Eq. (9–37) becomes

$$V_C(s) = \left[\frac{V_A/RC}{s(s + 1/RC)}\right] - e^{-as}\left[\frac{V_A/RC}{s(s + 1/RC)}\right] + \frac{V_0}{s + 1/RC} \quad \text{V-s}$$

The inverse transform of this reponse can be written as

$$v_C(t) = \mathcal{L}^{-1}\{F(s)\} + \mathcal{L}^{-1}\{e^{-as}F(s)\} + \mathcal{L}^{-1}\left\{\frac{V_0}{s + 1/RC}\right\}$$

where the rational function $F(s)$ is

$$F(s) = \frac{V_A/RC}{s(s + 1/RC)} = \frac{V_A}{s} - \frac{V_A}{s + 1/RC}$$

Notice that the factor e^{-as} does not change the partial fraction expansion of $e^{-as}F(s)$. It simply means that the corresponding waveform is delayed by a factor of $t = a$. Hence

$$\mathcal{L}^{-1}\{F(s)\} = f(t) = V_A[1 - e^{-t/RC}]u(t)$$

$$\mathcal{L}^{-1}\{e^{-as}F(s)\} = f(t - a)$$

$$\mathcal{L}^{-1}\left\{\frac{V_0}{s + 1/RC}\right\} = V_0 e^{-t/RC}u(t)$$

and the response waveform for a rectangular pulse input is

$$\begin{aligned} v_C(t) &= f(t) - f(t - a) + V_0 e^{-t/RC}u(t) \\ &= V_A[1 - e^{-t/RC}]u(t) - V_A[1 - e^{-(t-a)/RC}]u(t - a) + V_0 e^{-t/RC}\,u(t)\ \text{V} \end{aligned}$$

The first term in $v_C(t)$ is caused by the step function at $t = 0$, and the second by the step at $t = a$. The last term is the effect of the initial capacitor voltage at $t = 0$. Two plots of this response are shown in Figure 9–9 for $V_0 = 0$. For the short pulse, $a = RC$ and the negative step function at $t = a$ occurs before the circuit has fully responded to the positive step function at $t = 0$. For the long pulse, $a = 8RC$ and the step function at $t = a$ occurs after the circuit has essentially reached a steady-state condition due to the step at $t = 0$. The short pulse response in Figure 9–9 shows that a sharply defined rectangular pulse is distorted when the pulse duration is comparable to the circuit time constant. The long pulse response shows that there is much less distortion when the pulse duration is longer than the circuit time constant. ■

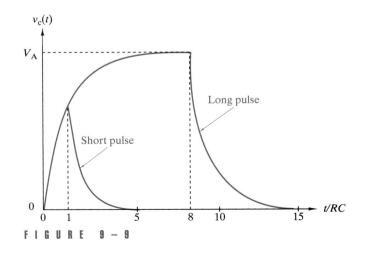

FIGURE 9–9

Exercise 9–17

Find $v_C(t)$ for the RC circuit in Figure 9–6 when the input is

(a) A ramp $v_S(t) = [V_A t/T]u(t)$
(b) A sinusoid $v_S(t) = [V_A \sin \beta t]u(t)$.

Answers:

(a) $v_C(t) = [V_A t/T - V_A RC/T + (V_0 + V_A RC/T)e^{-t/RC}]u(t)$ V

(b) $v_C(t) = \left[\dfrac{V_A}{\sqrt{1 + (\beta RC)^2}} \cos(\beta t + \theta) + \left(\dfrac{\beta RC V_A}{1 + (\beta RC)^2} + V_0 \right) e^{-t/RC} \right]u(t)$ V

where

$$\theta = -\pi/2 - \tan^{-1}(\beta RC)$$

9–7   I N I T I A L   V A L U E   A N D   F I N A L   V A L U E   P R O P E R T I E S

The **initial value** and **final value properties** can be stated as follows:

$$\text{Initial value: } \lim_{t \to 0+} f(t) = \lim_{s \to \infty} sF(s) \tag{9–39}$$

$$\text{Final value: } \lim_{t \to \infty} f(t) = \lim_{s \to 0} sF(s)$$

These properties display the relationship between the origin and infinity in the time and frequency domains. The value of $f(t)$ at $t = 0+$ in the time domain (initial value) is the same as the value of $sF(s)$ at infinity in the s plane. Conversely, the value of $f(t)$ as $t \to \infty$ (final value) is the same as the value of $sF(s)$ at the origin in the s plane.

Proof of both the initial value and final value properties starts with the differentiation property:

$$sF(s) - f(0-) = \int_{0-}^{\infty} \frac{df}{dt} e^{-st} \, dt \tag{9–40}$$

To establish the initial value property, we rewrite the integral on the right side of this equation and take the limit of both sides as $s \to \infty$.

$$\lim_{s \to \infty} [sF(s) - f(0-)] = \lim_{s \to \infty} \int_{0-}^{0+} \frac{df}{dt} e^{-st} dt + \lim_{s \to \infty} \int_{0+}^{\infty} \frac{df}{dt} e^{-st} \, dt \tag{9–41}$$

The first integral on the right side reduces to $f(0+) - f(0-)$ since e^{-st} is unity on the interval from $t = 0-$ to $t = 0+$. The second integral vanishes because e^{-st} goes to zero as $s \to \infty$. In addition, on the left side of Eq. (9–41) the $f(0-)$ is independent of s and can be taken outside the limiting process. Inserting all of these considerations reduces Eq. (9–41) to

$$\lim_{s \to \infty} sF(s) = \lim_{t \to 0+} f(t) \tag{9–42}$$

which completes the proof of the initial-value property.

Proof of the final value theorem begins by taking the limit of both sides of Eq. (9–40) as $s \to 0$:

$$\lim_{s \to 0} [sF(s) - f(0-)] = \lim_{s \to 0} \int_{0-}^{\infty} \frac{dt}{dt} e^{-st} \, dt \qquad (9\text{–}43)$$

The integral on the right side of this equation reduces to $f(\infty) - f(0-)$ because e^{-st} becomes unity as $s \to 0$. Again, the $f(0-)$ on the left side is independent of s and can be taken outside of the limiting process. Inserting all of these considerations reduces Eq. (9–43) to

$$\lim_{s \to 0} sF(s) = \lim_{t \to \infty} f(t) \qquad (9\text{–}44)$$

which completes the proof of the final value property.

A damped cosine waveform provides an illustration of the application of these properties. The transform of the damped cosine is

$$\mathcal{L}\{[Ae^{-\alpha t} \cos \beta t] u(t)\} = \frac{A(s + \alpha)}{(s + \alpha)^2 + \beta^2}$$

Applying the initial and final value limits, we obtain

$$\text{Initial value: } \lim_{t \to 0} f(t) = \lim_{t \to 0} Ae^{-\alpha t} \cos \beta t = A$$

$$\lim_{s \to \infty} sF(s) = \lim_{s \to \infty} \frac{sA(s + \alpha)}{(s + \alpha)^2 + \beta^2} = A$$

$$\text{Final value: } \lim_{t \to \infty} f(t) = \lim_{t \to \infty} Ae^{-\alpha t} \cos \beta t = 0$$

$$\lim_{s \to 0} sF(s) = \lim_{s \to 0} \frac{sA(s + \alpha)}{(s + \alpha)^2 + \beta^2} = 0$$

Note the agreement between the t-domain and s-domain limits in both cases.

There are restrictions on the initial and final value properties. The initial value property is valid when $F(s)$ is a proper rational function or, equivalently, when $f(t)$ does not have an impulse at $t = 0$. The final value property is valid when the poles of $sF(s)$ are in the left half plane or, equivalently, when $f(t)$ is a waveform that approaches a final value at $t \to \infty$. Note that the final value restriction allows $F(s)$ to have a simple pole at the origin since the limitation is on the poles of $sF(s)$.

Caution: The initial and final value properties will appear to work when the aforementioned restrictions are not met. In other words, these properties do not tell you they are giving nonsense answers when you violate their limitations. You must always check the restrictions on $F(s)$ before applying either of these properties.

For example, applying the final value property to a cosine waveform yields

$$\lim_{t \to \infty} \cos \beta t = \lim_{s \to 0} s \left[\frac{s}{s^2 + \beta^2} \right] = 0$$

The final value property appears to say that $\cos \beta t$ approaches zero as $t \to \infty$. This conclusion is incorrect since the waveform oscillates between

± 1. The problem is that the final value property does not apply to cosine waveform because $sF(s)$ has poles on the j-axis at $s = \pm j\beta$.

EXAMPLE 9–20

Find initial and final values of $v_C(t)$ in the RC circuit of Figure 9–6 when the input is a step function $v_S(t) = V_A u(t)$.

SOLUTION:
The input transform is V_A/s and the circuit response transform in Eq. (9–37) becomes

$$V_C(s) = \frac{V_A/RC}{s(s + 1/RC)} + \frac{V_0}{s + 1/RC} \quad \text{V}$$

The initial value property applies because $V_C(s)$ is a proper rational function. The initial value is

$$\lim_{t \to 0} v_C(t) = \lim_{s \to \infty} sV_C(s)$$

$$= \lim_{s \to \infty} \left[\frac{V_A/RC}{s + 1/RC}\right] + \lim_{s \to \infty} \left[\frac{sV_0}{s + 1/RC}\right]$$

$$= 0 + V_0 = v_C(0) \quad \text{V}$$

which is true by definition. The final value property applies since the pole of $sF(s)$ is located at $s = -1/RC$, which is in the left half plane. The final value is

$$\lim_{t \to \infty} v_C(t) = \lim_{s \to 0} sV_C(s)$$

$$= \lim_{s \to 0} \left[\frac{V_A/RC}{s + 1/RC}\right] + \lim_{s \to 0} \left[\frac{sV_0}{s + 1/RC}\right]$$

$$= V_A \quad \text{V}$$

A step function of amplitude V_A drives the circuit response from an initial value $v_C(0)$ to a final value V_A. ∎

Exercise 9–18

Find the initial and final values of the waveforms corresponding to the following transforms:

$$\text{(a) } F(s) = 100 \frac{s + 3}{s(s + 5)(s + 20)}$$

$$\text{(b) } F(s) = e^{-10s} \frac{s(s + 2)}{(s + 1)(s + 4)}$$

Answers:
(a) Initial value = 0, final value = 3
(b) Initial value = final value = 0

SUMMARY

- The Laplace transformation converts waveforms in the time domain to transforms in the s domain. The inverse transformation converts transforms into causal waveforms. A transform pair is unique if and only if $f(t)$ is casual.

- The Laplace transforms of basic signals like the step function, exponential, and sinusoid are easily derived from the integral definition. Other transform pairs can be derived using basic signal transforms and the uniqueness, linearity, time integration, time differentiation, and translation properties of the Laplace transformation.

- Proper rational functions with simple poles can be expanded by partial fraction to obtain inverse Laplace transforms. Simple real poles lead to exponential waveforms and simple complex poles to damped sinusoids. Partial-fraction expansions of improper rational functions and functions with multiple poles require special treatment.

- Using Laplace transforms to find the response of a linear circuit involves transforming the circuit differential equation into the s domain, algebraically solving for the response transform, and performing the inverse transformation to obtain the response waveform.

- The initial and final value properties determine the initial and final values of a waveform $f(t)$ from the value of $sF(s)$ at $s \to \infty$ and $s = 0$, respectively. The initial value property applies if $F(s)$ is a proper rational function. The final value property applies if all of the poles of $sF(s)$ are in the left half plane.

PROBLEMS

ERO 9–1 LAPLACE TRANSFORM (SECTS. 9–1, 9–2, 9–3)

Given a signal waveform, use the integral definition of the Laplace transformation or its basic properties and pairs to find the Laplace transform. Construct pole-zero diagrams.

9–1 Find the Laplace transforms of the following functions and plot their pole-zero diagrams for $\alpha > 0$:

$$\text{(a) } f(t) = A[2e^{-\alpha t} - 1]u(t)$$

$$\text{(b) } f(t) = A[(1 - 2\alpha t)e^{-\alpha t}]u(t)$$

9–2 Find the Laplace transform of $f(t) = A[\alpha t + e^{-\alpha t} - 1]u(t)$ and plot its pole-zero diagram for $\alpha > 0$.

9–3 Find the Laplace transform of $f(t) = A[2\sin(\beta t) - 1]u(t)$ for $\beta > 0$. Show that $F(s)$ has poles at $s = 0$ and $s = \pm j\beta$, and a double zero at $s = \beta$.

9–4 Find the Laplace transform of $f(t) = A[\cos(\beta t) - \sin(\beta t)]u(t)$. Show that $F(s)$ has poles $s = \pm j\beta$, and a zero at $s = \beta$.

9–5 Find the Laplace transform of $f(t) = A\delta(t) - A\beta e^{-\alpha t}\sin(\beta t)u(t)$ for $\alpha > 0$ and $\beta > 0$. Show that $F(s)$ has poles $s = -\alpha \pm j\beta$ and a double zero at $s = -\alpha$.

9–6   Given the Laplace transform of a step function, use the integration property of the Laplace transformation to find the transform of the ramp function $Ar(t) = Atu(t)$.

9–7   Find the Laplace transforms of the following functions and plot their pole-zero diagrams:

$$(a)\, f(t) = [5e^{-10t} - 10e^{-20t}]u(t)$$

$$(b)\, f(t) = 15\,[\cos(100t) - \cos(200t)]u(t)$$

9–8   Find the Laplace transforms of the following functions and plot their pole-zero diagrams:

$$(a)\, f(t) = \delta(t) + [5e^{-50t} - 15e^{-20t}]u(t)$$

$$(b)\, f(t) = [25 - 10\,\cos(500t)]u(t)$$

9–9   Find the Laplace transform of the following function and plot its pole-zero diagram:

$$f(t) = \delta(t) + [5e^{-50t} - 250te^{-50t}]u(t)$$

9–10   Find the Laplace transforms of the following function and plot its pole-zero diagrams:

$$f(t) = \delta(t) + [5e^{-50t} - 10 + 10\,\cos(200t)]u(t)$$

9–11   Find the Laplace transform of

$$f(t) = [-5e^{-10t} + 10 - 5\,\sin(5t) - 15\,\cos(5t)]u(t)$$

Locate the poles and zeros of $F(s)$.

9–12   The Laplace transform of $f(t) = 5[e^{-t}\sin 2t]u(t)$ is $F(s) = 10/(s^2 + 2s + 5)$.

   (a) Use the differentiation property of Laplace transforms to obtain the transform $G(s) = \mathcal{L}\{df(t)/dt\}$.

   (b) Check your result by first differentiating $f(t)$ and then transforming the resulting waveform $g(t)$.

   (c) Are there any poles or zeros in $G(s)$ that are not in $F(s)$? Explain.

9–13   **(a)** Write an expression for the waveform $f(t)$ in Figure P9–13 using only step functions and ramps.

   (b) Use the time-domain translation property to find the Laplace transform of the waveform found in part (a).

   (c) Verify your answer in (b) by applying the integral definition of the Laplace transformation to the $f(t)$ found in (a).

9–14   **(a)** Write an expression for the waveform $f(t)$ in Figure P9–14 using only step functions and ramps.

   (b) Use the time-domain translation property to find the Laplace transform of the waveform found in part (a).

   (c) Verify your answer in (b) by applying the integral definition of the Laplace transformation to the $f(t)$ found in (a).

9–15   Find the Laplace transform of

$$f(t) = [1600t + 8 + 75e^{-10t} + 17e^{-50t}]u(t)$$

Locate the poles and zeros of $F(s)$.

FIGURE P9–13

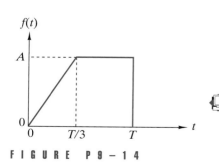

FIGURE P9–14

ERO 9–2 INVERSE TRANSFORMS (SECTS. 9–4, 9–5)

Given a rational signal transform, plot its pole-zero map, find the inverse transform, and sketch its waveform.

9–16 Find the inverse Laplace transforms of the following functions and sketch their waveforms for $\alpha > 0$:

$$\text{(a)}\ F_1(s) = \frac{s - \alpha}{s(s + \alpha)}$$

$$\text{(b)}\ F_2(s) = \frac{s^2 + \alpha^2}{s(s + \alpha)^2}$$

9–17 Find the inverse transforms of the following functions and sketch their waveforms for $\alpha > 0$:

$$\text{(a)}\ F_1(s) = \frac{\alpha^2}{s(s + \alpha)^2}$$

$$\text{(b)}\ F_2(s) = \frac{\alpha^2}{s^2(s + \alpha)}$$

9–18 Find the inverse transforms of the following functions and sketch their waveforms for $\alpha > 0$:

$$\text{(a)}\ F_1(s) = \frac{\beta(s + 2\alpha)}{(s + \alpha)^2 + \beta^2}$$

$$\text{(b)}\ F_2(s) = \frac{\beta(s - 2\alpha)}{(s + \alpha)^2 + \beta^2}$$

9–19 Find the inverse Laplace transforms of the following functions and sketch their waveforms for $\beta > 0$:

$$\text{(a)}\ F_1(s) = \frac{(s - \beta)}{s(s^2 + \beta^2)}$$

$$\text{(b)}\ F_2(s) = \frac{s(s - \beta)}{s^2 + \beta^2}$$

9–20 Find the inverse Laplace transforms of the following functions and sketch their waveforms for $\beta > 0$:

$$\text{(a)}\ F_1(s) = \frac{\beta}{s^2 - \beta^2}$$

$$\text{(b)}\ F_2(s) = \frac{s}{s^2 - \beta^2}$$

9–21 Plot the pole-zero diagrams of the following transforms, find the corresponding inverse transforms, and sketch their waveforms:

$$\text{(a)}\ F_1(s) = \frac{900}{s^2 + 65s + 900}$$

$$\text{(b)}\ F_2(s) = \frac{900}{s^2 + 60s + 900}$$

$$\text{(c)}\ F_3(s) = \frac{900}{s^2 + 36s + 900}$$

9–22 Find the inverse transforms for the following transforms:

$$(a)\ F(s) = \frac{16}{(s + 2)(s^2 + 12s + 36)}$$

$$(b)\ F(s) = \frac{2(s^2 + 2)}{s(s^2 + 4)}$$

9–23 Find the inverse transforms for the following transforms:

$$(a)\ F_1(s) = \frac{(s + 4)(s + 8)}{s(s + 2)(s + 6)}$$

$$(b)\ F_2(s) = \frac{(s^2 + 20)(s^2 + 60)}{s(s^2 + 45)(s^2 + 80)}$$

9–24 Find the inverse transforms for the following transforms:

$$(a)\ F(s) = \frac{6(s + 2)}{s(s^2 + 5s + 5)}$$

$$(b)\ F(s) = \frac{2s}{(s + 4)(s^2 + 4s + 8)}$$

9–25 Find the inverse transforms for the following transforms:

$$(a)\ F(s) = \frac{(s + 3)(s + 12)}{s^2(s^2 + 6s + 36)}$$

$$(b)\ F(s) = \frac{10(s + 13)(s + 18)}{s(s^2 + 20s + 100)}$$

9–26 Find the inverse transforms for the following transforms:

$$(a)\ F(s) = \frac{(s + 50)^2}{(s + 10)(s + 100)}$$

$$(b)\ F(s) = \frac{s + 1}{s^2 - 2s + 1}$$

9–27 A certain transform has a simple pole at $s = -30$ and a simple zero at $s = -\gamma$. The value of the transform at $s = 0$ is $F(0) = K$. Select values for K and γ so the inverse transform is

(a) $f(t) = \delta(t) - 10e^{-30t}$ (b) $f(t) = \delta(t)$ (c) $f(t) = \delta(t) + 10e^{-30t}$

9–28 A transform has the form

$$F(s) = K\frac{(s^2 + b_1s + b_0)}{(s^2 + 100)(s + 20)^2}$$

where the parameters K, b_1, and b_0 are real numbers. Select the values for K, b_1, and b_0 so that the inverse transform of $F(s)$ is

(a) $f(t) = [5te^{-20t}]u(t)$
(b) $f(t) = [10 \sin(10t)]u(t)$
(c) $f(t) = [e^{-20t} + \cos(10t)]u(t)$

9–29 Use Mathcad or MATLAB to find the inverse transform of the following function:

$$F(s) = \frac{s(s^2 + 3s + 4)}{(s + 2)(s^3 + 6s^2 + 16s + 16)}$$

9–30 Use Mathcad or MATLAB to find the inverse transform of the following function:

$$F(s) = \frac{40(s^3 + 2s^2 + s + 2)}{s(s^3 + 4s^2 + 4s + 16)}$$

ERO 9–3 CIRCUIT RESPONSE USING LAPLACE TRANSFORMS (SECT. 9–6)

Given a first- or second-order circuit,

(a) Determine the circuit differential equation and the initial conditions (if not given).
(b) Transform the differential equation into the s domain and solve for the response transform.
(c) Use the inverse transformation to find the response waveform.
(d) Identify the forced and natural components in the waveform and transform.

9–31 Use the Laplace transformation to find the $y(t)$ that satisfies the following first-order differential equations:

(a) $\dfrac{dy}{dt} + 10y = 0$ with $y(0-) = 5$

(b) $10^{-4}\dfrac{dy}{dt} + y = 10u(t)$ with $y(0-) = -10$

9–32 Use the Laplace transformation to find the $y(t)$ that satisfies the following first-order differential equations:

$$10^{-2}\frac{dy}{dt} + y = 5[\cos(100t)]u(t) \text{ with } y(0-) = 0$$

9–33 The switch in Figure P9–33 has been open for a long time and is closed at $t = 0$. For $R = 100\ \Omega$, $L = 100$ mH, and $V_A = 10$ V,
(a) Find the differential equation and the initial condition for the inductor current $i_L(t)$.
(b) Solve for $i_L(t)$ using the Laplace transformation.
(c) Identify the forced and natural components in the response waveform and transform.

9–34 The switch in Figure P9–33 has been closed for a long time and is opened at $t = 0$. For $R = 100\ \Omega$, $L = 100$ mH, and $V_A = 10$ V,
(a) Find the differential equation and the initial condition for the inductor current $i_L(t)$.
(b) Solve for $i_L(t)$ using the Laplace transformation.
(c) Identify the forced and natural components in the response waveform and transform.

FIGURE P9 – 3 3

9–35 The switch in Figure P9–35 has been open for a long time. At $t = 0$ the switch is closed.
(a) Find the differential equation for the circuit and the initial condition.
(b) Find $v_O(t)$ using the Laplace transformation for $v_S(t) = 10[e^{-1000t}]u(t)$.
(c) Identify the forced and natural components in the response waveform and transform.

FIGURE P9 – 3 5

9–36 Repeat Problem 9–35 for the input waveforms $v_S(t) = 10$ [$\sin 1000t$]$u(t)$.

9–37 Use the Laplace transformation to find the $y(t)$ that satisfies the following second-order differential equation:

$$\frac{d^2y}{dt^2} + 20\frac{dy}{dt} + 75y = 0$$

with $y(0-) = 10$ and $y'(0-) = 0$.

9–38 Use the Laplace transformation to find the $y(t)$ that satisfies the following second-order differential equation:

$$\frac{d^2y}{dt^2} + 11\frac{dy}{dt} + 10y = 100u(t)$$

with $y(0-) = 0$ and $y'(0-) = 0$.

9–39 The switch in Figure P9–39 has been open for a long time and is closed at $t = 0$. The circuit parameters are $R = 1$ kΩ, $L = 200$ mH, $C = 0.5$ μF, and $V_A = 10$ V.
 (a) Find the differential equation for the circuit and the initial conditions.
 (b) Use Laplace transforms to solve for the $i_L(t)$ for $t \geq 0$.

9–40 The switch in Figure P9–39 has been closed for a long time and is opened at $t = 0$. The circuit parameters are $R = 5$ kΩ, $L = 10$ mH, $C = 16$ pF, and $V_A = 10$ V.
 (a) Find the differential equation for the circuit and the initial conditions.
 (b) Use Laplace transforms to solve for the $i_L(t)$ for $t \geq 0$.

9–41 The switch in Figure P9–41 has been closed for a long time and is opened at $t = 0$. The circuit parameters are $L = 32$ mH, $C = 50$ nF, $R_1 = 1$ kΩ, $R_2 = 1$ kΩ, and $V_A = 10$ V.
 (a) Find the differential equation for the circuit and the initial conditions.
 (b) Use Laplace transforms to solve for the $i_L(t)$ for $t \geq 0$.

9–42 The switch in Figure P9–41 has been closed for a long time and is opened at $t = 0$, and then closed again at $t = 100$ μs. The circuit parameters are $L = 32$ mH, $C = 50$ nF, $R_1 = 1$ kΩ, $R_2 = 1$ kΩ, and $V_A = 10$ V.
 (a) Find the differential equation for the circuit and the initial conditions.
 (b) Use Laplace transforms to solve for the $i_L(t)$ for $t \geq 0$.

9–43 There is no initial energy storage in the circuit in Figure P9–43. Use Laplace transforms to find the current $i(t)$ when input is $v_S(t) = V_A u(t) - V_A u(t - T)$.

9–44 Repeat Problem 9–43 with $v_S(t) = V_A t/T[u(t) - u(t - T)]$.

FIGURE P9 – 39

FIGURE P9 – 41

FIGURE P9 – 43

ERO 9–4 INITIAL AND FINAL VALUE (SECT. 9–7)

Find the initial and final values of waveforms using the inital and final value properties of Laplace transforms.

9–45 Use the initial and final value properties to find the initial and final values of the waveform corresponding to the transforms in Problem 9–22. If either property is not applicable, explain why.

9–46 Use the initial and final value properties to find the initial and final values of the waveform corresponding to the transforms in Problem 9–23. If either property is not applicable, explain why.

9–47 Use the initial and final value properties to find the initial and final values of the waveform corresponding to the following transforms. If either property is not applicable, explain why.

$$\text{(a) } F_1(s) = \frac{(s^2 + 15s + 12)}{s(s^2 + 6s + 12)}$$

$$\text{(b) } F_2(s) = \frac{10(s^2 + 10s - 20)}{(s^2 - 20s + 100)}$$

9–48 Use the initial and final value properties to find the initial and final values of the waveform corresponding to the following transforms. If either property is not applicable, explain why.

$$\text{(a) } F_1(s) = \frac{(s^2 + 15s + 36)}{s^2(s^2 + 6s + 36)}$$

$$\text{(b) } F_2(s) = \frac{10(s^2 + 10s - 20)}{s(s^2 + 20s + 100)}$$

9–49 Use the initial and final value properties to find the initial and final values of the waveform corresponding to the transforms in Problem 9–24. If either property is not applicable, explain why.

9–50 Use the initial and final value properties to find the initial and final values of the waveform corresponding to the transforms in Problem 9–26. If either property is not applicable, explain why.

INTEGRATING PROBLEMS

9–51 A THE DOMINANT POLE APPROXIMATION

When a transform $F(s)$ has widely separated poles, then those closest to the j-axis tend to dominate the response because they have less damping. An approximation to the waveform can be obtained by ignoring the contributions of all except the dominant poles. We can ignore the nondominant poles by simply discarding their terms in the partial fraction expansion of $F(s)$. The purpose of this example is to examine a dominant pole approximation. Consider the transform

$$F(s) = 10^6 \frac{s + 1000}{(s + 4000)[(s + 25)^2 + 100^2]}$$

(a) Construct a partial fraction expansion of $F(s)$ and find $f(t)$.

(b) Construct a pole-zero diagram of $F(s)$ and identify the dominant poles.

(c) Construct a dominant pole approximation $g(t)$ by discarding the nondominant poles in the partial fraction expansion in (a).

(d) Plot $f(t)$ and $g(t)$ and comment on the accuracy of the approximation.

9–52   ▲ FIRST-ORDER CIRCUIT STEP RESPONSE

In Chapter 7 we found that the step response of a first-order circuit can be written as

$$f(t) = f(\infty) + [f(0) - f(\infty)]e^{-t/T_C} \quad t \geq 0$$

where $f(0)$ is the initial value, $f(\infty)$ is the final value, and T_C is the time constant. Show that the corresponding transform has the form

$$F(s) = K\left[\frac{s + \gamma}{s(s + \alpha)}\right]$$

and relate the time-domain parameters $f(0)$, $f(\infty)$, and T_C to the s-domain parameters K, γ, and α.

9–53   ▲ OP AMP *RC* CIRCUIT

In Chapter 6 we learned how to analyze and design the input-output relationships of dynamic OP AMP circuits in the time domain. In this problem we explore the same problem in the s domain.

(a) With zero initial conditions, transform the time-domain input-output relationship

$$v_O(t) = -2v_S(t) + \frac{1}{50}\frac{dv_S(t)}{dt}$$

into the s domain and draw a block diagram of the relationship between the output transform $V_O(s)$ and input transform $V_S(s)$. How does the time-domain differentiation involved show up in the s-domain block diagram? How are the poles of the output transform related to the poles of the input transform?

(b) Draw a block diagram of the s-domain input-output relationship

$$V_O(s) = 2\,V_S(s) - \frac{40}{s}V_S(s)$$

How are the poles of the output transform related to the poles of the input transform? Perform the inverse Laplace transform to obtain the t-domain input-output relationship. How do the time-domain signal processing functions show up in the s-domain block diagram?

9–54   ▲ LAPLACE TRANSFORMS AND PHASORS

In Chapter 8 we used phasors to solve for the sinusoidal steady-state response of a circuit. In this chapter we used Laplace transforms to solve for both the natural and the forced responses. The purpose of this problem is to explore the connection between transforms and phasors. In Chpater 8 we found that in the sinusoidal steady state an input phasor $\mathbf{V}_S = V_A\angle 0°$ produces an output phasor $\mathbf{V}_O = V_B\angle\phi$. In Chapter 9 we found that when a circuit is driven by a sinusoidal input $V_A \cos(\beta t)u(t)$ then the output transform $V_O(s)$ has poles at $s = \pm j\beta$. As a result, the partial fraction expansion of $V_O(s)$ must contain the terms

$$V_O(s) = \cdots\frac{k}{s - j\beta} + \frac{k^*}{s + j\beta}\cdots$$

Show that $2k = \mathbf{V}_O$. That is, show that the residue of the pole at $s = j\beta$ is equal to one-half the phasor response. *Hint:* review the definition of a phasor in Chapter 8 and then take a look at Eq. (9–23).

9–55 Ⓐ DAMPING RATIO AND POLE LOCATIONS

In Chapter 7 we found that the general second-order differential equation with a step function input can be written in terms of the damping ratio ζ and the undamped natural frequency ω_0 as

$$\frac{d^2y(t)}{dt^2} + 2\zeta\omega_0\frac{dy(t)}{dt} + \omega_0^2 y(t) = Au(t)$$

where $y(t)$ is a voltage or current response. In Chapter 7 we solved for the step response using classical time-domain methods. In this problem we explore the connections between classical and Laplace methods.

(a) With the initial conditions all zero, transform the differential equation into the s domain and solve for the response transform $Y(s)$.

(b) In Chapter 7 we found that the form of time-domain natural response $y_N(t)$ depends on the roots of the characteristic equation

$$s^2 + 2\zeta\omega_0 s + \omega_0^2 = 0$$

How are the roots of the characteristic equation related to the poles and zeros of the response transform $Y(s)$?

(c) In Chapter 7 we found that there are three possible forms of the natural response: (1) overdamped ($\zeta > 0$), (2) critically damped ($\zeta = 1$), and (3) underdamped ($\zeta < 1$). How are these three cases manifested in the poles of $Y(s)$?

(d) In Chapter 7 we found that the time-domain forced response $y_F(t) = A/\omega_0^2$. How is this result manifested in the partial fraction expansion of $Y(s)$?

9–56 Ⓐ COMBINING WAVEFORM AND TRANSFORM CONSTRAINTS

A Laplace transform pair $f(t)$ and $F(s)$ are alternative representations of a signal derived from the definitions of the transformation. However, it is possible to infer things about $F(s)$ from properties of $f(t)$ (and vice versa) without formally performing the transformation. The purpose of this problem is to explore this idea. Specifically, construct a transform pair $F(s) = \mathcal{L}\{f(t)\}$ that meets the following conditions.

(a) In the t domain: $f(10^{-3}) = 4.5$ and $f(t) \to 4$ as $t \to \infty$.

(b) In the s domain: $F(-\alpha) = \infty$ and $sF(s) \to 5$ as $s \to \infty$.

CHAPTER 10

s-DOMAIN CIRCUIT ANALYSIS

The resistance operator Z is a function of the electrical constants of the circuit components and of d/dt, *the operator of time-differentiation, which will in the following be denoted by* p *simply.*

Oliver Heaviside, 1887,
British Engineer

10–1 Transformed Circuits

10–2 Basic Circuit Analysis in the s Domain

10–3 Circuit Theorems in the s Domain

10–4 Node-voltage Analysis in the s Domain

10–5 Mesh-current Analysis in the s Domain

10–6 Summary of s-Domain Circuit Analysis

Summary

Problems

Integrating Problems

480

The Laplace transform techniques in this chapter have their roots in the works of Oliver Heaviside (1850–1925). The preceding quotation was taken from his book *Electrical Papers* originally published in 1887. His resistance operator *Z*, which he later called impedance, is a central theme for much of electrical engineering. Heaviside does not often receive the recognition he deserves because his intuitive approach to mathematics was not accepted by most Victorian scientists of his day. Mathematical justification for his methods was eventually supplied by John Bromwich and others. However, no important errors were found in Heaviside's results.

In this chapter we transform circuits directly from the time domain into the *s* domain without first developing a circuit differential equation. The process begins in the first section, where we find that the transformed circuit obeys Kirchhoff's laws and that the passive element *i–v* characteristics become linear algebraic equations in transform variables. The algebraic form of the element constraints leads to the concept of *s*-domain impedance as a generalization of resistance. In the *s* domain the underlying element and connection constraints are similar to those for resistance circuits. As a result, *s*-domain circuit analysis uses techniques like equivalence, superposition, and Thévenin's theorem, which parallel those developed for resistance circuits. Most of this chapter is devoted to extending resistance circuit analysis methods to cover dynamic circuits in the *s* domain. The basic analysis tools and circuit theorems are discussed in the second and third sections. The fourth and fifth sections develop general methods of circuit analysis using node voltages and mesh currents.

10–1   TRANSFORMED CIRCUITS

So far we have used the Laplace transformation to change waveforms into transforms and convert circuit differential equations into algebraic equations. These operations provide a useful introduction to the *s* domain. However, the real power of the Laplace transformation emerges when we transform the circuit itself and study its behavior directly in the *s* domain.

The solid arrow in Figure 10–1 indicates the analysis path we will be following in this chapter. The process begins with a linear circuit in the time domain. We transform the circuit into the *s* domain, write the circuit equations directly in that domain, and then solve these algebraic equations for response transform. The inverse Laplace transformation then produces the response waveform. However, the *s*-domain approach is not just another way to derive response waveforms. This approach allows us to work directly with the circuit model using analysis tools such as voltage division and equivalence. By working directly with the circuit model, we gain insights into the interaction between circuits and signals that cannot be obtained using the classical approach indicated by the dotted path in Figure 10–1.

How are we to transform a circuit? We have seen several times that circuit analysis is based on device and connection constraints. The connection constraints are derived from Kirchhoff's laws and the device constraints from the *i–v* relationships used to model the physical devices in the circuit. To transform circuits, we must see how these two types of constraints are altered by the Laplace transformation.

FIGURE 10-1 *Flow diagram for s-domain circuit analysis.*

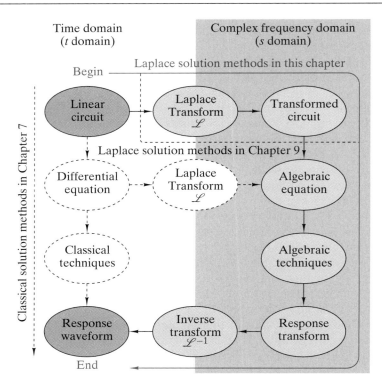

CONNECTION CONSTRAINTS IN THE *S* DOMAIN

A typical KCL connection constraint could be written as

$$i_1(t) + i_2(t) - i_3(t) + i_4(t) = 0$$

This connection constraint requires that the sum of the current waveforms at a node be zero for all times t. Using the linearity property, the Laplace transformation of this equation is

$$I_1(s) + I_2(s) - I_3(s) + I_4(s) = 0$$

In the s domain the KCL connection constraint requires that the sum of the current transforms be zero for all values of s. This idea generalizes to any number of currents at a node and any number of nodes. In addition, this idea obviously applies to Kirchhoff's voltage law as well. The form of the connection constraints do not change because they are linear equations and the Laplace transformation is a linear operation. In summary, KCL and KVL apply to waveforms in the t domain and to transforms in the s domain.

ELEMENT CONSTRAINTS IN THE *S* DOMAIN

Turning now to the element constraints, we first deal with the independent signal sources shown in Figure 10–2. The $i–v$ relationships for these elements are

FIGURE 1 0 – 2   s-*Domain models of independent sources.*

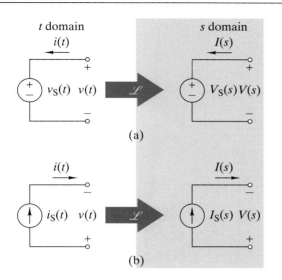

t domain

i(t)

$v_S(t)$ *v(t)*

(a)

s domain

I(s)

$V_S(s)$ *V(s)*

i(t)

$i_S(t)$ *v(t)*

(b)

I(s)

$I_S(s)$ *V(s)*

Voltage source: $v(t) = v_S(t)$

$$i(t) = \text{Depends on Circuit} \qquad (10\text{--}1)$$

Current source: $i(t) = i_S(t)$

$$v(t) = \text{Depends on Circuit}$$

Independent sources are two-terminal elements. In the *t* domain they constrain the waveform of one signal variable and adjust the unconstrained variable to meet the demands of the external circuit. We think of an independent source as a generator of a specified voltage or current waveform. The Laplace transformation of the expressions in Eq. (10–1) yields

Voltage source: $V(s) = V_S(s)$

$$I(s) = \text{Depends on Circuit} \qquad (10\text{--}2)$$

Current source: $I(s) = I_S(s)$

$$V(s) = \text{Depends on Circuit}$$

In the *s* domain independent sources function the same way as in the *t* domain, except that we think of them as generating voltage or current transforms rather than waveforms.

Next we consider the active elements in Figure 10–3. In the time domain the element constraints for linear dependent sources are linear algebraic equations. Because of the linearity property of the Laplace transformation, the forms of these constraints are unchanged when they are transformed into the *s* domain:

	t domain	*s* domain
Voltage-controlled voltage source	$v_2(t) = \mu v_1(t)$	$V_2(s) = \mu V_1(s)$
Current-controlled current source	$i_2(t) = \beta i_1(t)$	$I_2(s) = \beta I_1(s)$

FIGURE 10-3 s-*Domain models of dependent sources and OP AMPs.*

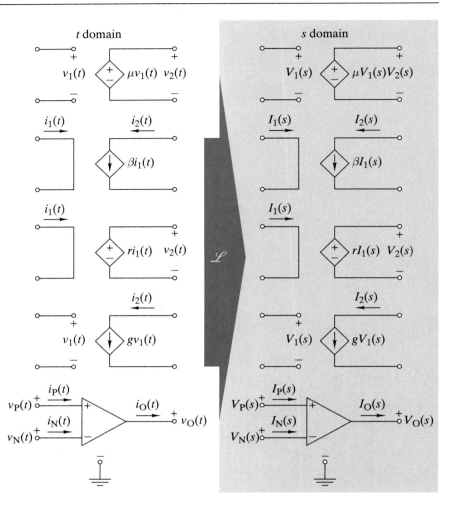

Current-controlled voltage source	$v_2(t) = ri_1(t)$	$V_2(s) = rI_1(s)$
Voltage-controlled current source	$i_2(t) = gv_1(t)$	$I_2(s) = gV_1(s)$

$$\text{(10–3)}$$

Similarly, the element constraints of the ideal OP AMP are linear algebraic equations that are unchanged in form by the Laplace transformation:

t domain	s domain	
$v_P(t) = v_N(t)$	$V_P(s) = V_N(s)$	
$i_N(t) = 0$	$I_N(s) = 0$	(10–4)
$i_P(t) = 0$	$I_P(s) = 0$	

Thus, for linear active devices the only difference is that in the s domain the ideal element constraints apply to transforms rather than waveforms.

Finally, we consider the two-terminal passive circuit elements shown in Figure 10–4. In the time domain their i–v relationships are

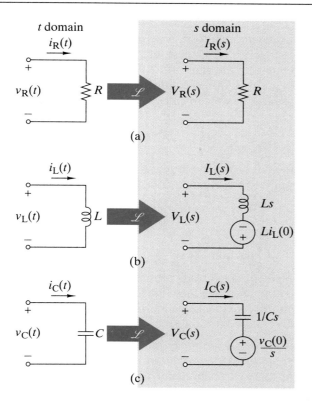

$$\text{Resistor: } v_R(t) = R\, i_R(t)$$

$$\text{Inductor: } v_L(t) = L\, \frac{di_L(t)}{dt} \qquad (10-5)$$

$$\text{Capacitor: } v_C(t) = \frac{1}{C} \int_0^t i_C(\tau)d\tau + v_C(0)$$

These element constraints are transformed into the s domain by taking the Laplace transform of both sides of each equation using the linearity, differentiation, and integration properties.

$$\text{Resistor: } V_R(s) = R\, I_R(s)$$

$$\text{Inductor: } V_L(s) = LsI_L(s) - Li_L(0) \qquad (10-6)$$

$$\text{Capacitor: } V_C(s) = \frac{1}{Cs} I_C(s) + \frac{v_C(0)}{s}$$

As expected, the element relationships are algebraic equations in the s domain. For the linear resistor the s domain version of Ohm's law says that the voltage transform $V_R(s)$ is proportional to the current transform $I_R(s)$. The element constraints for the inductor and capacitor also involve a proportionality between voltage and current, but include a term for the initial conditions as well.

The element constraints in Eq. (10–6) lead to the s-domain circuit models shown on the right side of Figure 10–4. The t-domain parameters L and

C are replaced by proportionality factors *Ls* and $1/Cs$ in the *s* domain. The initial conditions associated with the inductor and capacitor are modeled as voltage sources in series with these elements. The polarities of these sources are determined by the sign of the corresponding initial condition terms in Eq. (10–6). These initial condition voltage sources must be included when using these models to calculate the voltage transforms $V_L(s)$ or $V_C(s)$.

IMPEDANCE AND ADMITTANCE

The concept of impedance is a basic feature of *s*-domain circuit analysis. For zero initial conditions the element constraints in Eq. (10–6) reduce to

$$\text{Resistor: } V_R(s) = (R)\, I_R(s)$$

$$\text{Inductor: } V_L(s) = (Ls) I_L(s) \qquad (10\text{–}7)$$

$$\text{Capacitor: } V_C(s) = (1/Cs)\, I_C(s)$$

In each case the element constraints are all of the form $V(s) = Z(s)I(s)$, which means that in the *s* domain the voltage across the element is proportional to the current through it. The proportionality factor is called the element **impedance** $Z(s)$. Stated formally,

> **Impedance** *is the proportionality factor relating the transform of the voltage across a two-terminal element to the transform of the current through the element with all initial conditions set to zero.*

The impedances of the three passive elements are

$$\text{Resistor: } Z_R(s) = R$$

$$\text{Inductor: } Z_L(s) = (Ls) \qquad \text{with } i_L(0) = 0 \qquad (10\text{–}8)$$

$$\text{Capacitor: } Z_C(s) = (1/Cs) \qquad \text{with } v_C(0) = 0$$

It is important to remember that part of the definition of *s*-domain impedance is that the initial conditions are zero.

The *s*-domain impedance is a generalization of the *t*-domain concept of resistance. The impedance of a resistor is its resistance *R*. The impedance of the inductor and capacitor depend on the inductance *L* and capacitance *C* and the complex frequency variable *s*. Since a voltage transform has units of V-s and current transform has units of A-s, it follows that impedance has units of ohms since (V-s)/(A-s) = V/A = Ω.

Algebraically solving Eqs. (10–6) for the element currents in terms of the voltages produces alternative *s*-domain models.

$$\text{Resistor: } I_R(s) = \frac{1}{R}\, V_R(s)$$

$$\text{Inductor: } I_L(s) = \frac{1}{Ls}\, V_L(s) + \frac{i_L(0)}{s} \qquad (10\text{–}9)$$

$$\text{Capacitor: } I_C(s) = Cs\, V_C(s) - C\, v_C(0)$$

In this form, the *i*–*v* relations lead to the *s*-domain models shown in the Figure 10–5. The reference directions for the initial condition current sources

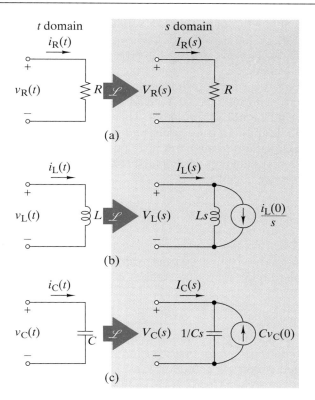

are determined by the sign of the corresponding terms in Eqs. (10–9). The initial condition sources are in parallel with the element impedance.

Admittance $Y(s)$ is the s-domain generalization of the t-domain concept of conductance and can be defined as the reciprocal of impedance.

$$Y(s) = \frac{1}{Z(s)} \tag{10–10}$$

Using this definition, the admittances of the three passive elements are

$$\text{Resistor: } Y_R(s) = \frac{1}{Z_R(s)} = \frac{1}{R} = G$$

$$\text{Inductor: } Y_L(s) = \frac{1}{Z_L(s)} = \frac{1}{Ls} \quad \text{with } i_L(0) = 0 \tag{10–11}$$

$$\text{Capacitor: } Y_C(s) = \frac{1}{Z_C(s)} = Cs \quad \text{with } v_C(0) = 0$$

Since $Y(s)$ is the reciprocal of impedance, its units are siemens since $\Omega^{-1} = \text{A/V} = \text{S}$.

In summary, to transform a circuit into the s domain we replace each element by an s-domain model. For independent sources, dependent sources, OP AMPs, and resistors, the only change is that these elements now constrain transforms rather than waveforms. For inductors and capacitors, we can use either the model with a series initial condition voltage source (Figure 10–4), or the model with a parallel initial condition current source (Figure 10–5). However, to avoid possible confusion we always write the inductor impedance Ls and capacitor impedance $1/Cs$ beside the transformed element regardless of which initial condition source is used.

To analyze the transformed circuit, we can use the tools developed for resistance circuits in Chapters 2 through 4. These tools are applicable because KVL and KCL apply to transforms, and the *s*-domain element constraints are linear equations similar to those for resistance circuits. These features make *s*-domain analysis of dynamic circuits an algebraic process that is akin to resistance circuit analysis.

EXAMPLE 10-1

The switch in Figure 10–6 has been in position 1 for a long time and is moved to position 2 at $t = 0$. For $t > 0$ transform the circuit into the *s* domain and use Laplace transforms to solve for the voltage $v_C(t)$.

SOLUTION:

We have solved this type of problem before using various methods. In this example, we find the response by first transforming the circuit itself. For $t > 0$ the transformed circuit takes the form in Figure 10–7, where we have used a parallel current source $Cv_C(0)$ to account for the initial condition.

Applying KCL to the current transforms at node A produces

$$-I_1(s) - I_2(s) + Cv_C(0) = 0 \qquad \text{A-s}$$

For $t < 0$ the switch in Figure 10–6 is in position 1 and the circuit is in a dc steady-state condition. As a result, we have $v_C(0) = V_A$. In the *s*-domain circuit in Figure 10–7, the two branch current transforms can be written in terms of the capacitor voltage and element impedances as

$$\text{Resistor:} \quad I_1(s) = \frac{V_C(s)}{R}$$

$$\text{Capacitor:} \quad I_2(s) = \frac{V_C(s)}{1/Cs} = Cs\, V_C(s)$$

Substituting these observations into the KCL equation and solving for $V_C(s)$ yields

FIGURE 10 - 6

FIGURE 10-7

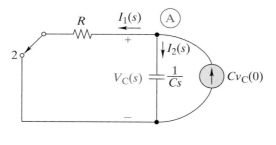

$$V_C(s) = \frac{CV_A}{Cs + \dfrac{1}{R}} = \frac{V_A}{s + \dfrac{1}{RC}} \quad \text{V-s}$$

Performing the inverse Laplace transformation leads to

$$v_C(t) = [V_A e^{-t/RC}]u(t) \quad \text{V}$$

The form of the response should be no great surprise. We could easily pre-dict this response using the classical differential equation methods studied in Chapter 7. What is important in this example is that we obtained the re-sponse using only basic circuit concepts applied in the s domain. ■

EXAMPLE 10-2

(a) Transform the circuit in Figure 10–8(a) into the s domain.
(b) Solve for the current transform $I(s)$.
(c) Perform the inverse transformation to the waveform $i(t)$.

SOLUTION:

(a) Figure 10–8(b) shows the transformed circuit using a series voltage source $Li_L(0)$ to represent the inductor initial condition. The imped-ances of the two passive elements are R and Ls. The independent source voltage $V_A u(t)$ transforms as V_A/s.

(b) By KVL, the sum of voltage transforms around the loop is

$$-\frac{V_A}{s} + V_R(s) + V_L(s) = 0$$

Using the impedance models, the s-domain element constraints are

$$\text{Resistor: } V_R(s) = RI(s)$$
$$\text{Inductor: } V_L(s) = LsI(s) - Li_L(0)$$

FIGURE 10-8

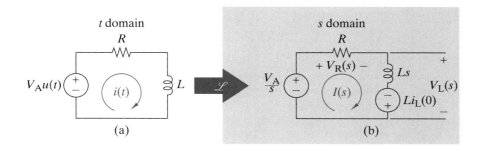

(a)   (b)

Substituting the element constraints into the KVL constraint and collecting terms yields

$$-\frac{V_A}{s} + (R + Ls)I(s) - Li_L(0) = 0$$

Solving for $I(s)$ produces

$$I(s) = \frac{V_A/L}{s(s + R/L)} + \frac{i_L(0)}{s + R/L} \quad \text{A-s}$$

The current $I(s)$ is the transform of the circuit response for a step function input. $I(s)$ is a rational function with simple poles at $s = 0$ and $s = -R/L$.

(c) To perform the inverse transformation, we expand $I(s)$ by partial fractions:

$$I(s) = \overbrace{\frac{V_A/R}{s}}^{\text{Forced}} \overbrace{- \frac{V_A/R}{s + R/L} + \frac{i_L(0)}{s + R/L}}^{\text{Natural}} \quad \text{A-s}$$

Taking the inverse transform of each term in this expansion gives

$$i(t) = \left[\overbrace{\frac{V_A}{R}}^{\text{Forced}} \overbrace{- \frac{V_A}{R}e^{-Rt/L} + i_L(0)e^{-Rt/L}}^{\text{Natural}}\right]u(t) \quad \text{A}$$

The forced response is caused by the step function input. The exponential terms in the natural response depend on the circuit time constant L/R. The step function and exponential components in $i(t)$ are directly related to the terms in the partial-fraction expansion of $I(s)$. The pole at the origin came from the step function input and leads to the forced response. The pole at $s = -R/L$ came from the circuit and leads to the natural response. Thus, in the s domain the forced response is that part of the total response that has the same poles as the input excitation. The natural response is that part of the total response whose poles came from the circuit. We say that the circuit contributes the natural poles because their locations depend on circuit parameters and not on the input. In other words, poles in the response do not occur by accident. They are present because the physical response depends on two things—(1) the input and (2) the circuit. ■

10–2   **B**ASIC **C**IRCUIT **A**NALYSIS IN THE *s* **D**OMAIN

In this section we develop the *s*-domain versions of series and parallel equivalence, and voltage and current division. These analysis techniques are the basic tools in *s*-domain circuit analysis, just as they are for resistance circuit analysis. These methods apply to circuits with elements connected in series or parallel. General analysis methods using node-voltage or mesh-current equations are covered later in Sects. 10–4 and 10–5.

SERIES EQUIVALENCE AND VOLTAGE DIVISION

The concept of a series connection applies in the *s* domain because Kirchhoff's laws do not change under the Laplace transformation. In Figure 10–9 the two-terminal elements are connected in series; hence by KCL the same current $I(s)$ exists in impedances $Z_1(s)$, $Z_2(s)$, ... $Z_N(s)$. Using KVL and the element constraints, the voltage across the series connection can be written as

$$V(s) = V_1(s) + V_2(s) + \cdots + V_N(s)$$

$$= Z_1(s)I(s) + Z_2(s)I(s) + \cdots + Z_N(s)I(s) \qquad (10\text{–}12)$$

$$= [Z_1(s) + Z_2(s) + \cdots + Z_N(s)]\, I(s)$$

The last line in this equation points out that the responses $V(s)$ and $I(s)$ do not change when the series-connected elements are replaced by an **equivalent impedance**:

$$Z_{EQ}(s) = Z_1(s) + Z_2(s) + \cdots + Z_N(s) \qquad (10\text{–}13)$$

In general, the equivalent impedance $Z_{EQ}(s)$ is a quotient of polynomials in the complex frequency variable of the form

$$Z_{EQ}(s) = \frac{b_m s^m + b_{m-1}s^{m-1} + \cdots + b_1 s + b_0}{a_n s^n + a_{n-1}s^{n-1} + \cdots + a_1 s + a_0} \qquad (10\text{–}14)$$

The roots of the numerator polynomial are the zeros of $Z_{EQ}(s)$, while the roots of the denominator are the poles.

Combining Eqs. (10–12) and (10–13), we can write the element voltages in the form

$$V_1(s) = \frac{Z_1(s)}{Z_{EQ}(s)}V(s) \quad V_2(s) = \frac{Z_2(s)}{Z_{EQ}(s)}V(s) \cdots V_N(s) = \frac{Z_N(s)}{Z_{EQ}(s)}V(s)$$

$$(10\text{–}15)$$

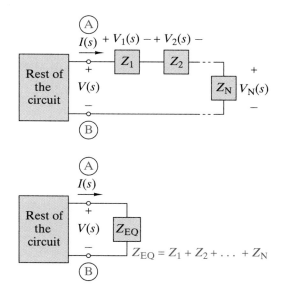

FIGURE 10−9 *Series equivalence in the* s *domain.*

These equations are the *s*-domain **voltage division principle**:

> *Every element voltage in a series connection is equal to its impedance divided by the equivalent impedance of the connection times the voltage across the series circuit.*

This statement parallels the corresponding rule for resistance circuits given in Chapter 4.

PARALLEL EQUIVALENCE AND CURRENT DIVISION

The parallel circuit in Figure 10–10 is the dual of the series circuit discussed previously. In this circuit the two-terminal elements are connected in parallel; hence by KVL the same voltage $V(s)$ appears across admittances $Y_1(s)$, $Y_2(s)$, ... $Y_N(s)$. Using KCL and the element constraints, the current into the parallel connection can be written as

$$I(s) = I_1(s) + I_2(s) + \cdots + I_N(s)$$
$$= Y_1(s)V(s) + Y_2(s)V(s) + \cdots + Y_N(s)V(s) \qquad (10\text{–}16)$$
$$= [Y_1(s) + Y_2(s) + \cdots + Y_N(s)]V(s)$$

The last line in this equation points out that the responses $V(s)$ and $I(s)$ do not change when the parallel connected elements are replaced by an **equivalent admittance**:

$$Y_{EQ}(s) = Y_1(s) + Y_2(s) + \cdots + Y_N(s) \qquad (10\text{–}17)$$

In general, the equivalent admittance $Y_{EQ}(s)$ is a quotient of polynomials in the complex frequency variable s. Since impedance and admittance are reciprocals, it turns out that if $Y_{EQ}(s) = p(s)/q(s)$, then the equivalent impedance at the same pair of terminals has the form $Z_{EQ}(s) = 1/Y_{EQ}(s) = q(s)/p(s)$. That is, at a given pair of terminals the poles of $Z_{EQ}(s)$ are zeros of $Y_{EQ}(s)$, and vice versa.

FIGURE 10–10 *Parallel equivalence in the* s *domain.*

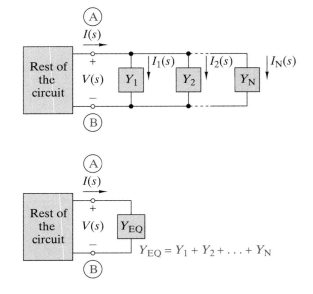

Combining Eqs. (10–16) and (10–17), we can write the element currents in the form

$$I_1(s) = \frac{Y_1(s)}{Y_{EQ}(s)}I(s) \quad I_2(s) = \frac{Y_2(s)}{Y_{EQ}(s)}I(s) \cdots I_N(s) = \frac{Y_N(s)}{Y_{EQ}(s)}I(s)$$

(10–18)

These equations are the *s*-domain **current division principle**:

> *Every element current in a parallel connection is equal to its admittance divided by the equivalent admittance of the connection times the current into the parallel circuit.*

This statement is the dual of the results for a series circuit and parallels the current division rule for resistance circuits.

We begin to see that *s*-domain circuit analysis involves basic concepts that parallel the analysis of resistance circuits in the *t* domain. Repeated application of series/parallel equivalence and voltage/current division leads to an analysis approach called circuit reduction, discussed in Chapter 2. The major difference here is that we use impedance and admittances rather than resistance and conductance, and the analysis yields voltage and current transforms rather than waveforms.

FIGURE 10-11

EXAMPLE 10-3

The inductor current and capacitor voltage in Figure 10–11 are zero at $t = 0$.

(a) Transform the circuit into the *s* domain and find the equivalent impedance between terminals A and B.
(b) Use voltage division to solve for the output voltage transform $V_2(s)$.

SOLUTION:
(a) Figure 10–12(a) shows the circuit in Figure 10–11 transformed into the *s* domain. As a first step we use parallel equivalence to find the equivalent impedance of the parallel resistor and capacitor.

$$Z_{EQ1}(s) = \frac{1}{Y_{EQ1}(s)} = \frac{1}{\dfrac{1}{R} + Cs} = \frac{R}{RCs + 1}$$

Figure 10–12(b) shows that the equivalent impedance $Z_{EQ1}(s)$ is connected in series with the inductor. This series combination can be replaced by an equivalent impedance

$$Z_{EQ}(s) = Ls + Z_{EQ1}(s) = Ls + \frac{R}{RCs + 1}$$

$$= \frac{RLCs^2 + Ls + R}{RCs + 1} \quad \Omega$$

as shown in Figure 10–12(c). The rational function $Z_{EQ}(s)$ is the impedance seen between terminals A and B in Figure 10–12(a).

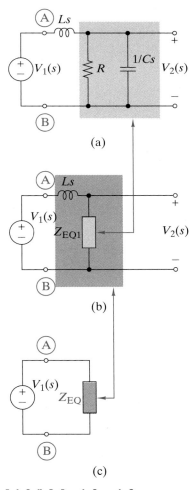

(a)

(b)

(c)

FIGURE 10-12

(a)

(b)

FIGURE 10 – 13

(b) Using voltage division in Figure 10–12(b), we find $V_2(s)$ as

$$V_2(s) = \left[\frac{Z_{EQ1}(s)}{Z_{EQ}(s)}\right]V_1(s) = \left[\frac{R}{RLCs^2 + Ls + R}\right]V_1(s)$$

Note that $Z_{EQ}(s)$ and $V_2(s)$ are rational functions of the complex frequency variable s. ∎

D EXAMPLE 10–4

In a circuit analysis problem we are required to find the poles and zeros of a circuit. In circuit design we are required to adjust circuit parameters to place the poles and zeros at specified s-plane locations. This example is a simple pole-placement design problem.

(a) Transform the circuit in Figure 10–13(a) into the s domain and find the equivalent impedance between terminals A and B.
(b) Select the values of R and C such that $Z_{EQ}(s)$ has a zero at $s = -5000$ rad/s.

SOLUTION:
(a) Figure 10–13(b) shows the circuit transformed to the s domain. The equivalent impedance $Z_{EQ1}(s)$ is

$$Z_{EQ1}(s) = \frac{1}{Y_R + Y_C} = \frac{1}{\frac{1}{R} + Cs}$$

$$= \frac{R}{RCs + 1}$$

Hence the equivalent impedance between terminals A and B is

$$Z_{EQ}(s) = R + Z_{EQ1}(s) = R + \frac{R}{RCs + 1}$$

$$= R\frac{RCs + 2}{RCs + 1} \quad \Omega$$

(b) For $Z_{EQ}(s)$ to have a zero at $s = -5000$ requires $2/RC = 5000$ or $RC = 4 \times 10^{-4}$. Selecting a standard value for the resistor $R = 10$ kΩ in turn requires $C = 40$ nF. ∎

Exercise 10–1

The inductor current and capacitor voltage in Figure 10–14 are zero at $t = 0$.

(a) Transform the circuit into the s domain and find the equivalent admittance between terminals A and B.
(b) Solve for the output current transform $I_2(s)$ in terms of the input current $I_1(s)$.

FIGURE 10 – 14

Answers:

$$\text{(a) } Y_{EQ}(s) = \frac{LCs^2 + RCs + 1}{Ls(RCs + 1)}$$

$$\text{(b) } I_2(s) = \left[\frac{LCs^2}{LCs^2 + RCs + 1}\right] I_1(s)$$

Exercise 10–2

The inductor current and capacitor voltage in Figure 10–15 are zero at $t = 0$.

(a) Transform the circuit into the s domain and find the equivalent impedance between terminals A and B.
(b) Solve for the output voltage transform $V_2(s)$ in terms of the input voltage $V_1(s)$.

Answers:

$$\text{(a) } Z_{EQ}(s) = \frac{(R_1 Cs + 1)(Ls + R_2)}{LCs^2 + (R_1 + R_2)Cs + 1}$$

$$\text{(b) } V_2(s) = \left[\frac{Ls}{Ls + R_2}\right] V_1(s)$$

FIGURE   10 – 15

10–3 CIRCUIT THEOREMS IN THE s DOMAIN

In this section we study the s-domain versions of proportionality, super-position, and Thévenin/Norton equivalent circuits. These theorems define fundamental properties that provide conceptual tools for the analysis and design of linear circuits. With some modifications, all of the theorems studied in Chapter 3 apply to linear dynamic circuits in the s domain.

PROPORTIONALITY

For linear resistance circuits the **proportionality theorem** states that any output y is proportional to the input x:

$$y = Kx \qquad (10\text{--}19)$$

The same concept applies to linear dynamic circuits in the s domain except that the proportionality factor is a rational function of s rather than a constant. For instance, in Example 10–3 we found the output voltage $V_2(s)$ to be

$$V_2(s) = \left[\frac{R}{RLCs^2 + Ls + R}\right] V_1(s) \qquad (10\text{--}20)$$

where $V_1(s)$ is the transform of the input voltage. The quantity inside the brackets is a rational function that serves as the proportionality factor between the input and output transforms.

In the s-domain rational functions that relate inputs and outputs are called **network functions**. We begin the formal study network functions in Chapter 11. In this chapter we will simply illustrate network functions by an example.

FIGURE 10 – 16

EXAMPLE 10–5

There is no initial energy stored in the circuit in Figure 10–16. Find the network functions relating $I_R(s)$ to $V_1(s)$ and $I_C(s)$ to $V_1(s)$.

SOLUTION:

The equivalent impedance seen by the voltage source is

$$Z_{EQ} = Ls + \frac{1}{\frac{1}{R} + Cs} = \frac{RLCs^2 + Ls + R}{RCs + 1}$$

Hence we can relate the $I_L(s)$ and $V_1(s)$ as

$$I_L(s) = \frac{V_1(s)}{Z_{EQ}(s)} = \left[\frac{RCs + 1}{RLCs^2 + Ls + R}\right] V_1(s)$$

Using s-domain current division, we can relate $I_R(s)$ and $I_C(s)$ to $I_L(s)$ as

$$I_R(s) = \frac{\frac{1}{R}}{\frac{1}{R} + Cs} I_L(s) = \left[\frac{1}{RCs + 1}\right] I_L(s)$$

$$I_C(s) = \frac{Cs}{\frac{1}{R} + Cs} I_L(s) = \left[\frac{RCs}{RCs + 1}\right] I_L(s)$$

Finally, using these relationships plus the relationship between $I_L(s)$ and $V_1(s)$ derived previously, we obtain the required network functions.

$$I_R(s) = \left[\frac{1}{RLCs^2 + Ls + R}\right] V_1(s)$$

$$I_C(s) = \left[\frac{RCs}{RLCs^2 + Ls + R}\right] V_1(s)$$ ∎

Exercise 10–3

In Figure 10–17 find the network function relating the output $V_2(s)$ to the input $I_1(s)$.

Answer:

$$V_2(s) = \left[\frac{R}{LCs^2 + RCs + 1}\right] I_1(s)$$

FIGURE 10 – 17

SUPERPOSITION

For linear resistance circuits the **superposition theorem** states that any output y of a linear circuit can be written as

$$y = K_1x_1 + K_2x_2 + K_3x_3 + \cdots \qquad (10\text{–}21)$$

where $x_1, x_2, x_3, \ldots$ are circuit inputs and $K_1, K_2. K_3, \ldots$ are weighting factors that depend on the circuit. The same concept applies to linear dynamic circuits in the s domain except that the weighting factors are rational functions of s rather than constants.

Superposition is usually thought of as a way to find the circuit response by adding the individual responses caused by each input acting alone. However, the principle applies to groups of sources as well. In particular, in the s domain there are two types of independent sources: (1) voltage and current sources representing the external driving forces for $t \geq 0$ and (2) initial condition voltage and current sources representing the energy stored at $t = 0$. As a result, the superposition principle states that the s-domain response can be found as the sum of two components: (1) the **zero-input response** caused by the initial condition sources with the external inputs turned off; or (2) the **zero-state response** caused by the external inputs with the initial condition sources turned off. Turning a source off means replacing voltage sources by short circuits $[V_S(s) = 0]$ and current sources by open circuits $[I_S(s) = 0]$.

The zero-input response is the response of a circuit to its initial conditions when the input excitations are set to zero. The zero-state response is the response of a circuit to its input excitations when all of the initial conditions are set to zero. The term *zero input* is self-explanatory. The term *zero state* is used because there is no energy stored in the circuit at $t = 0$.

The result is that voltage and current transform in a linear circuit can be found as the sum of two components of the form

$$V(s) = V_{zs}(s) + V_{zi}(s) \qquad I(s) = I_{zs}(s) + I_{zi}(s) \qquad (10-22)$$

where the subscript zs stands for zero state and zi for zero input. An important corollary is that the time-domain response can also be partitioned into zero-state and zero-input components because the inverse Laplace transformation is a linear operation.

We analyze the circuit treated in Example 10–2 to illustrate the superposition of zero-state and zero-input responses. The transformed circuit in Figure 10–18 has two independent voltage sources: (1) an input voltage source and (2) a voltage source representing the initial inductor current. The resistor and inductor are in series, so these two elements can be replaced by an impedance $Z_{EQ}(s) = Ls + R$.

First we turn off the initial condition source and replace its voltage source by a short circuit. Using the resulting zero-state circuit shown in Figure 10–18, we obtain the zero-state response:

$$I_{zs}(s) = \frac{V_A/s}{Z_{EQ}(s)} = \frac{V_A/L}{s(s + R/L)} \qquad (10-23)$$

The pole at $s = 0$ comes from the input source and the pole at $s = -R/L$ comes from the circuit. Next, we turn off the input source and use the zero-input circuit shown in Figure 10–18 to obtain the zero-input response:

$$I_{zi}(s) = \frac{Li_L(0)}{Z_{EQ}(s)} = \frac{i_L(0)}{s + R/L} \qquad (10-24)$$

The pole at $s = -R/L$ comes from the circuit. The zero-input response does not have a pole at $s = 0$ because the step function input is turned off.

Superposition states that the total response is the sum of the zero-state component in Eqs. (10–23) and the zero-input component in Eq. (10–24).

FIGURE 10–18 *Using superposition to find the zero-state and zero-input responses.*

$$I(s) = I_{zs}(s) + I_{zi}(s) = \frac{V_A/L}{s(s + R/L)} + \frac{i_L(0)}{s + R/L} \qquad (10\text{–}25)$$

The transform $I(s)$ in this equation is the same as found in Example 10–1. To derive the time-domain response, we expand $I(s)$ by partial fractions:

$$I(s) = \underbrace{\frac{V_A/R}{s} - \frac{V_A/R}{s + R/L}}_{\text{Zero State}} + \underbrace{\frac{i_L(0)}{s + R/L}}_{\text{Zero Input}} \qquad (10\text{–}26)$$

Performing the inverse transformation on each term yields

$$i(t) = \left[\overbrace{\frac{V_A}{R}}^{\text{Forced}} - \underbrace{\frac{V_A}{R} e^{-Rt/L}}_{} + \overbrace{i_L(0) e^{-Rt/L}}^{\text{Natural}} \right] u(t) \quad \text{A} \qquad (10\text{–}27)$$

$$\underbrace{\phantom{\frac{V_A}{R} - \frac{V_A}{R} e^{-Rt/L}}}_{\text{Zero State}} \underbrace{\phantom{i_L(0) e^{-Rt/L}}}_{\text{Zero Input}}$$

Using superposition to partition the waveform into zero-state and zero-input components produces the same result as Example 10–1. The zero-state component contains the forced response. The zero-state and zero-input components both contain an exponential term due to the natural pole at $s = -R/L$ because both the external driving force and the initial condition source excite the circuit's natural response.

The superposition theorem helps us understand the response of circuits with multiple inputs, including initial conditions. It is a conceptual tool that helps us organize our thinking about s-domain circuits in general. It is not necessarily the most efficient analysis tool for finding the response of a specific multiple-input circuit.

(a) *t* domain

(b) *s* domain

FIGURE 10–19

EXAMPLE 10–6

The switch in Figure 10–19(a) has been open for a long time and is closed at $t = 0$.

(a) Transform the circuit into the *s* domain.
(b) Find the zero-state and zero-input components of $V(s)$.
(c) Find $v(t)$ for $I_A = 1$ mA, $L = 2$ H, $R = 1.5$ kΩ, and $C = 1/6$ μF.

SOLUTION:

(a) To transform the circuit into the *s* domain, we must find the initial inductor current and capacitor voltage. For $t < 0$ the circuit is in a dc steady-state condition with the switch open. The inductor acts like a short circuit, and the capacitor acts like an open circuit. By inspection, the initial conditions at $t = 0-$ are $i_L(0) = 0$ and $v_C(0) = I_A R$. Figure 10–19(b) shows the *s*-domain circuit for these initial conditions. The current source version for the capacitor's initial condition is used here because the circuit elements are connected in parallel. The switch and constant current source combine to produce a step function $I_A u(t)$ whose transform is I_A/s.

(b) The resistor, capacitor, and inductor can be replaced by an equivalent impedance

$$Z_{EQ} = \frac{1}{Y_{EQ}} = \frac{1}{\dfrac{1}{Ls} + \dfrac{1}{R} + Cs}$$

$$= \frac{RLs}{RLCs^2 + Ls + R}$$

The zero-state response is found with the capacitor initial condition source replaced by an open circuit and the step function input source on:

$$V_{zs}(s) = Z_{EQ}(s)\frac{I_A}{s} = \left[\frac{RLs}{RLCs^2 + Ls + R}\right]\frac{I_A}{s} = \frac{I_A/C}{s^2 + \dfrac{s}{RC} + \dfrac{1}{LC}}$$

The pole in the input at $s = 0$ is canceled by the zero at the origin in $Z_{EQ}(s)$. As a result, the zero-state response does not have a forced pole at $s = 0$. The zero-input response is found by replacing the input source by an open circuit and turning the capacitor initial condition source on:

$$V_{zi}(s) = [Z_{EQ}(s)][CRI_A] = \frac{RI_A s}{s^2 + \dfrac{s}{RC} + \dfrac{1}{LC}}$$

(c) Inserting the given numerical values of the circuit parameters and expanding the zero-state and zero-input response transforms by partial fractions yields

$$V_{zs}(s) = \frac{6000}{(s + 1000)(s + 3000)} = \frac{3}{s + 1000} + \frac{-3}{s + 3000} \quad \text{V-s}$$

$$V_{zi}(s) = \frac{1.5\,s}{(s + 1000)(s + 3000)} = \frac{-0.75}{s + 1000} + \frac{2.25}{s + 3000} \quad \text{V-s}$$

The inverse transforms of these expansions are

$$v_{zs}(t) = [3e^{-1000t} - 3e^{-3000t}]\,u(t) \quad \text{V}$$

$$v_{zi}(t) = [-0.75e^{-1000t} + 2.25e^{-3000t}]\,u(t) \quad \text{V}$$

Note that the circuit responses contain only transient terms that decay to zero. There is no forced response because in the dc steady state the induc-

(a) *s*-domain circuit

(b) Voltage source OFF

(c) Current source OFF

FIGURE 10-20

tor acts like a short circuit, forcing $v(t)$ to zero for $t \to \infty$. From an *s* domain viewpoint there is no forced response because the forced pole at $s = 0$ is canceled by a zero in the network function. ∎

EXAMPLE 10-7

Use superposition to find the zero-state component of $I(s)$ in the *s*-domain circuit shown in Figure 10–20(a).

SOLUTION:
Turning the voltage source off produces the circuit in Figure 10–20(b). In this circuit the resistor and capacitor are connected in parallel, so current division yields $I_1(s)$ in the form

$$I_1(s) = \frac{Y_R}{Y_C + Y_R} \frac{I_A}{(s + \alpha)} = \frac{I_A}{(RCs + 1)(s + \alpha)}$$

Turning the voltage source on and the current source off produces the circuit in Figure 10–20(c). In this case the resistor and capacitor are connected in series, and series equivalence gives the current $I_2(s)$ as

$$I_2(s) = \frac{1}{Z_R + Z_C} \frac{V_A\beta}{s^2 + \beta^2} = \frac{CsV_A\beta}{(RCs + 1)(s^2 + \beta^2)}$$

Using superposition, the total zero-state response is

$$I_{zs}(s) = I_1(s) - I_2(s)$$

$$= \frac{I_A}{(RCs + 1)(s + \alpha)} - \frac{CsV_A\beta}{(RCs + 1)(s^2 + \beta^2)}$$

There is a minus in this equation because $I_1(s)$ and $I_2(s)$ were assigned opposite reference directions in Figures 10–20(b) and 10–20(c). The total zero-state response has four poles. The natural pole at $s = -1/RC$ came from the circuit. The forced pole at $s = -\alpha$ came from the current source, and the two forced poles at $s = \pm j\beta$ came from the voltage source.

In this example the time-domain response would have a transient component $Ke^{-t/RC}$ due to the natural pole, a forced component $Ke^{-\alpha t}$ due to the current source, and a forced component of the form $K_A \cos \beta t + K_B \sin \beta t$ due to the voltage source. We can infer these general conclusions regarding the time-domain response by simply examining the poles of the *s*-domain response. ∎

Exercise 10-4

The initial conditions for the circuit in Figure 10–21 are $v_C(0-) = 0$ and $i_L(0-) = I_0$. Transform the circuit into the *s* domain and find the zero-state and zero-input components of $V(s)$.

Answers:

$$V_{zs}(s) = \left[\frac{1}{LCs^2 + RCs + 1} \right] \frac{V_A}{s}$$

$$V_{zi}(s) = \frac{LI_0}{LCs^2 + RCs + 1}$$

FIGURE 10-21

FIGURE   10 – 22

Exercise 10–5

The initial conditions for the circuit in Figure 10–22 are $v_C(0-) = 0$ and $i_L(0-) = I_0$. Transform the circuit into the s domain and find the zero-state and zero-input components of $I(s)$.

Answers:

$$I_{zs}(s) = \left[\frac{LCs^2}{LCs^2 + RCs + 1}\right]\frac{I_A}{s}$$

$$I_{zi}(s) = \left[\frac{LCs^2}{LCs^2 + RCs + 1}\right]\frac{I_0}{s}$$

THÉVENIN AND NORTON EQUIVALENT CIRCUITS AND SOURCE TRANSFORMATIONS

When the source circuit in Figure 10–23(a) contains only independent sources and linear elements, it can be replaced by the Thévenin equivalent circuit in Figure 10–23(b) or the Norton equivalent circuit in Figure 10–23(c). In the s domain the methods of finding and using Thévenin or Norton equivalent circuits are similar to those for resistive circuits. The important differences are that the signals are transforms rather than waveforms and the circuit elements are impedances rather than resistances.

The open-circuit voltage and short-circuit current at the interface provided enough information to define the Thévenin and Norton equivalent circuits. With an open-circuit load connected in Figures 10–23(b) and 10–23(c), we see that the interface voltage is

$$V(s) = V_{OC}(s) = I_N(s)Z_N(s) = V_T(s) \qquad (10\text{–}28)$$

With a short-circuit load connected, the interface current is

$$I(s) = I_{SC}(s) = V_T(s)/Z_T(s) = I_N(s) \qquad (10\text{–}29)$$

Taken together, Eqs. (10–28) and (10–29) yield the conditions

$$V_T(s) = V_{OC}(s) \quad I_N(s) = I_{SC}(s) \qquad Z_T(s) = Z_N(s) = \frac{V_{OC}(s)}{I_{SC}(s)} \qquad (10\text{–}30)$$

Algebraically, the results in Eq. (10–30) are identical to the corresponding equations for resistance circuits. The important difference, as noted previously, is that these equations involve transforms and impedances rather than waveforms and resistances. In any case, the equations point out that the open-circuit voltage and short-circuit current are sufficient to define either equivalent circuit.

The Thévenin and Norton equivalent circuits are related by an s-domain source transformation. The s-domain **source transformation** relations in Eq. (10–30) allow us to transform a voltage source in series with an impedance into a current source in parallel with the same impedance, and vice versa. We use these source transformations when formulating node-voltage and mesh-current equations in the s domain.

FIGURE 10 – 23 *Thévenin and Norton equivalent circuits in the s domain.*

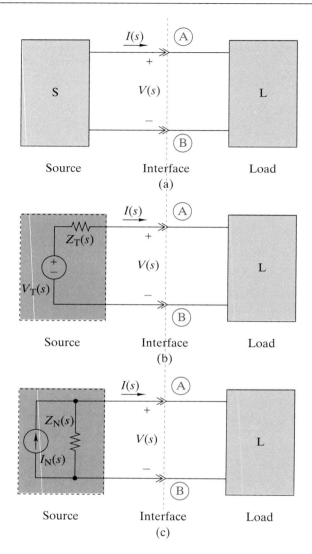

Consider the interface shown in Figure 10–24. The source circuit is given in the *s* domain with zero initial voltage on the capacitor. The open-circuit voltage at the interface is found by voltage division:

$$V_{\text{OC}}(s) = \frac{1/Cs}{R + 1/Cs} \frac{V_A}{s} = \frac{V_A}{s(RCs + 1)} \qquad (10\text{–}31)$$

The pole in $V_{\text{OC}}(s)$ at $s = 0$ comes from the step function voltage source and the pole at $s = -1/RC$ from the RC circuit. Connecting a short circuit at the interface effectively removes the capacitor from the circuit. The short-circuit current is

$$I_{\text{SC}}(s) = \frac{1}{R} \frac{V_A}{s} \qquad (10\text{–}32)$$

This current does not have a pole at $s = -1/RC$ because the capacitor is shorted out. Taking the ratio of Eqs. (10–31) and (10–32) yields the Thévenin impedance:

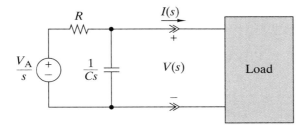

FIGURE 10-24 A source-load interface in the s domain.

$$Z_T(s) = \frac{V_{OC}(s)}{I_{SC}(s)} = \frac{R}{RCs + 1} \qquad (10\text{–}33)$$

When there are no dependent sources present, the Thévenin impedance can be found using the lookback method from Chapter 3. That is, the Thévenin impedance is the equivalent impedance seen looking back into the source circuit with the independent sources (external inputs and initial conditions) turned off. Replacing the voltage source in Figure 10–24 by a short circuit reduces the source circuit to a parallel combination of a resistor and capacitor. The equivalent lookback impedance of the combination is

$$Z_T(s) = \frac{1}{Y_R + Y_C} = \frac{1}{\dfrac{1}{R} + Cs} = \frac{R}{RCs + 1} \qquad (10\text{–}34)$$

which is the same result as Eq. (10–33).

This example illustrates that in the s domain, Thévenin or Norton theorems involve concepts similar to those discussed in Chapter 3. At a given interface these equivalent circuits are defined by any two of the following: (1) the open-circuit voltage, (2) the short-circuit current, or (3) the lookback impedance (when no dependent sources are involved). Thévenin and Norton equivalent circuits are useful tools that help us understand the operation of a circuit, especially when we must evaluate the circuit performance with different loads or parameter variations at an interface. They are not, in general, a way to reduce the computational burden of finding the response of a fixed circuit.

EXAMPLE 10-8

Find the interface voltage $v(t)$ in Figure 10–24 when the load is (a) a resistance R and (b) a capacitance C.

SOLUTION:
Figure 10–25 shows the s-domain circuit model of the source-load interface. The parameters of the Thévenin equivalent are given in Eqs. (10–31) and (10–33) as

$$V_T(s) = \frac{V_A}{s(RCs + 1)} \quad \text{and} \quad Z_T(s) = \frac{R}{RCs + 1}$$

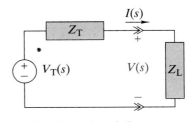

FIGURE 10-25

(a) For a resistance load, $Z_L(s) = R$ and the interface voltage is found by voltage division:

$$V(s) = \frac{Z_L(s)}{Z_T(s) + Z_L(s)} V_T(s)$$

$$= \left(\frac{R}{\dfrac{R}{RCs + 1} + R} \right) \frac{V_A}{s(RCs + 1)}$$

$$= \frac{V_A/RC}{s(s + 2/RC)} \quad \text{V-s}$$

The partial fraction expansion of $V(s)$ is

$$V(s) = \frac{V_A/2}{s} - \frac{V_A/2}{s + 2/RC}$$

Applying the inverse transform yields the interface voltage waveform as

$$v(t) = \frac{V_A}{2}[1 - e^{-2t/RC}]\, u(t) \quad \text{V}$$

(b) For a capacitance load, $Z_L(s) = 1/Cs$ and the interface voltage transform is

$$V(s) = \frac{Z_L(s)}{Z_T(s) + Z_L(s)} V_T(s) = \left(\frac{\dfrac{1}{Cs}}{\dfrac{R}{RCs + 1} + \dfrac{1}{Cs}} \right) \frac{V_A}{s(RCs + 1)}$$

$$= \frac{V_A/2RC}{s(s + 1/2\,RC)} \quad \text{V-s}$$

The partial fraction expansion of $V(s)$ is

$$V(s) = \frac{V_A}{s} - \frac{V_A}{s + 2/RC}$$

Applying the inverse transform to each term yields the interface voltage waveform for the capacitance load as

$$v(t) = V_A[1 - e^{-t/2RC}]\, u(t) \quad \text{V}$$

This example illustrates an important difference between resistive circuits and dynamic circuits. In a resistive circuit the load affects the amplitude but not the waveshape of the interface signals. In a dynamic circuit the load can change both amplitude and the waveshape. ∎

◗ DESIGN EXAMPLE 10–9

The shaded portion of the circuit in Figure 10–26(a) is a low-frequency model of a transistor. Select the value of C_E so that the circuit has a pole at $s = -300$ rad/s.

FIGURE 10-26

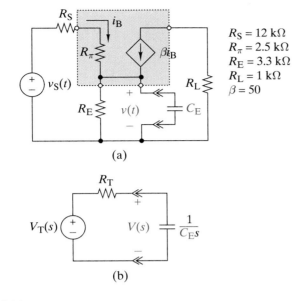

$R_S = 12 \text{ k}\Omega$
$R_\pi = 2.5 \text{ k}\Omega$
$R_E = 3.3 \text{ k}\Omega$
$R_L = 1 \text{ k}\Omega$
$\beta = 50$

(a)

(b)

SOLUTION:

Figure 10–26(b) shows the Thévenin equivalent seen by the capacitor C_E. In this circuit the voltage across the capacitor is

$$V_C(s) = \frac{1/C_E s}{R_T + 1/C_E s} V_T(s) = \frac{1}{R_T C_E s + 1} V_T(s)$$

The circuit has a simple pole at $s = -1/R_T C_E$, where R_T is the Thévenin equivalent resistance seen by the capacitor. To place the circuit pole at $s = -300$, we must select $C_E = 1/300 R_T$. Hence the problem reduces to finding the Thévenin equivalent resistance seen by the capacitor. With the capacitor in Figure 10–26(a) replaced by a short circuit, we find

$$i_{SC}(t) = \beta i_B(t) + i_B(t) = (\beta + 1)\left[\frac{v_S(t)}{R_S + R_\pi}\right]$$

$$= 3.52 v_S(t) \quad \text{mA}$$

Next, replacing the capacitor by an open circuit yields

$$v_{OC}(t) = i_B(t) R_E + \beta i_B(t) R_E$$

$$= (\beta + 1) R_E \left[\frac{v_S(t) - v_{OC}(t)}{R_S + R_\pi}\right]$$

Solving this expression for $v_{OC}(t)$, we obtain

$$v_{OC}(t) = \frac{(\beta + 1) R_E}{R_S + R_\pi + (\beta + 1) R_E} v_S(t)$$

$$= 0.921 v_S(t) \quad \text{V}$$

The Thévenin equivalent resistance seen by the capacitor is $R_T = v_{OC}/i_{SC} = 262 \ \Omega$, and the required value of capacitance is $C_E = 12.7 \ \mu\text{F}$. This example illustrates the general principle that we can select element values to position a circuit's poles at specified locations. ∎

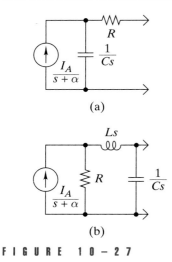

FIGURE 10–27

Find the Norton equivalent of the s-domain circuits in Figure 10–27.

Answers:

$$\text{(a) } I_N(s) = \frac{I_A}{(RCs + 1)(s + \alpha)} \qquad Z_N(s) = \frac{RCs + 1}{Cs}$$

$$\text{(b) } I_N(s) = \frac{RI_A}{(Ls + R)(s + \alpha)} \qquad Z_N(s) = \frac{Ls + R}{LCs^2 + RCs + 1}$$

10–4 NODE-VOLTAGE ANALYSIS IN THE S DOMAIN

The previous sections deal with basic analysis methods using equivalence, reduction, and circuit theorems. These methods are valuable because we work directly with the element impedances and thereby gain insight into s-domain circuit behavior. We also need general methods to deal with more complicated circuits that these basic methods cannot easily handle.

FORMULATING NODE-VOLTAGE EQUATIONS

Formulating node-voltage equations involves selecting a reference node and assigning a node-to-datum voltage to each of the remaining nonreference nodes. Because of KVL, the voltage across any two-terminal element is equal to the difference of two node voltages. This fundamental property of node voltages, together with element impedances, allows us to write KCL constraints at each of the nonreference nodes.

For example, consider the s-domain circuit in Figure 10–28. The sum of currents leaving node A can be written as

$$I_{S2}(s) - I_{S1}(s) + \frac{V_A(s)}{Z_1(s)} + \frac{V_A(s) - V_B(s)}{Z_2(s)} + \frac{V_A(s) - V_C(s)}{Z_3(s)} = 0$$

Rewriting this equation with unknown node voltages grouped on the left and inputs on the right yields

$$\left[\frac{1}{Z_1(s)} + \frac{1}{Z_2(s)} + \frac{1}{Z_3(s)}\right] V_A(s) - \frac{1}{Z_2(s)} V_B(s) - \frac{1}{Z_3(s)} V_C(s) = I_{S1}(s) - I_{S2}(s)$$

Expressing this result in terms of admittances produces the following equation:

$$[Y_1(s) + Y_2(s) + Y_3(s)] V_A(s) - [Y_2(s)] V_B(s)$$
$$- [Y_3(s)]V_C(s) = I_{S1}(s) - I_{S2}(s)$$

This equation has a familiar pattern. The unknowns are the node-voltage transforms $V_A(s)$, $V_B(s)$, and $V_C(s)$. The coefficient $[Y_1(s) + Y_2(s) + Y_3(s)]$ of $V_A(s)$ is the sum of the admittances of the elements connected to node A. The coefficient $[Y_2(s)]$ of $V_B(s)$ is the admittance of the elements connected between nodes A and B, while $[Y_3(s)]$ is the admittance of the ele-

FIGURE 10–28 *An example node.*

ments connected between nodes A and C. Finally, $I_{S1}(s) - I_{S2}(s)$ is the sum of the source currents directed into node A. These observations suggest that we can write node-voltage equations for *s*-domain circuits by inspection, just as we did with resistive circuits.

The formulation method just outlined assumes that there are no voltage sources in the circuit. When transforming the circuit we can always select the current source models to represent the initial conditions. However, the circuit may contain dependent or independent voltage sources. If so, they can be treated using the following methods:

> **Method 1:** If there is an impedance in series with the voltage source, use a source transformation to convert it into an equivalent current source.

> **Method 2:** Select the reference node so that one terminal of one or more of the voltage sources is connected to ground. The source voltage then determines the node voltage at the other source terminal, thereby eliminating an unknown.

> **Method 3:** Create a supernode surrounding any voltage source that cannot be handled by method 1 or 2.

Some circuits may require more than one of these methods.

Formulating a set of equilibrium equations in the *s* domain is a straightforward process involving concepts developed in Chapters 3 and 4 for resistance circuits. The following example illustrates the formulation process.

EXAMPLE 10–10

Formulate *s*-domain node-voltage equations for the circuit in Figure 10–29(a).

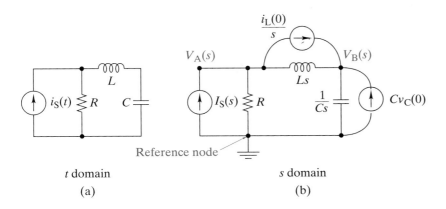

FIGURE 10–29

t domain

(a)

s domain

(b)

SOLUTION:

Figure 10–29(b) shows the circuit in the *s* domain. In transforming the circuit, we use current sources to represent the inductor and capacitor initial conditions. This choice facilitates writing node equations since the resulting *s*-domain circuit contains only current sources and fewer nodes. The sum of currents leaving nodes A and B can be written as

$$\text{Node A: } \frac{V_A(s)}{R} + \frac{V_A(s) - V_B(s)}{Ls} - I_S(s) + \frac{i_L(0)}{s} = 0$$

$$\text{Node B: } \frac{V_B(s)}{1/Cs} + \frac{V_B(s) - V_A(s)}{Ls} - \frac{i_L(0)}{s} - Cv_C(0) = 0$$

Rearranging these equations in the standard format with the unknowns on the left and the inputs on the right yields

$$\text{Node A: } \left(G + \frac{1}{Ls} \right) V_A(s) - \left(\frac{1}{Ls} \right) V_B(s) = I_S(s) - \frac{i_L(0)}{s}$$

$$\text{Node B: } -\left(\frac{1}{Ls} \right) V_A(s) + \left(\frac{1}{Ls} + Cs \right) V_B(s) = Cv_C(0) + \frac{i_L(0)}{s}$$

where $G = 1/R$ is the conductance of the resistor. Note that $(G + 1/Ls)$ is the sum of the admittance connected to node A, $(1/Ls + Cs)$ is the sum of the admittances connected to node B, and $1/Ls$ is the admittance connected between nodes A and B. The circuit is driven by an independent current source $I_S(s)$ and two initial condition sources $Cv_C(0)$ and $i_L(0)/s$. The terms on the right side of these equations are the sum of source currents directed into each node. With practice, we learn to write these equations by inspection. ∎

EXAMPLE 10–11

Formulate node-voltage equations for the circuit in Figure 10–30(a).

FIGURE 10 – 30

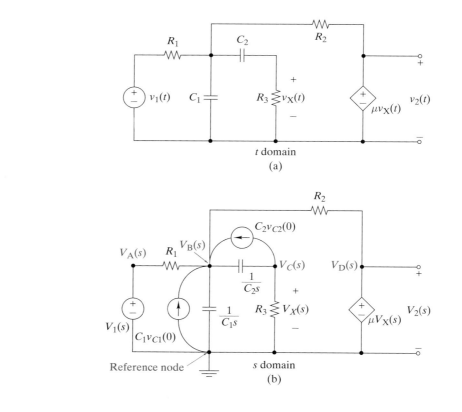

t domain
(a)

s domain
(b)

SOLUTION:

Figure 10–30(b) shows the circuit in the s domain using current sources for the two capacitor initial conditions. The indicated reference node was selected so that the two voltage sources are connected to ground. As a result, the voltage sources determine the voltages at nodes A and D. The only unknowns are the voltages at nodes B and C. The sums of currents leaving nodes B and C are

$$\text{Node B: } \frac{V_B(s) - V_A(s)}{R_1} + \frac{V_B(s) - V_D(s)}{R_2} + \frac{V_B(s)}{1/C_1 s} +$$

$$\frac{V_B(s) - V_C(s)}{1/C_2 s} - C_1 v_{C_1}(0) - C_2 v_{C_2}(0) = 0$$

$$\text{Node C: } \frac{V_C(s)}{R_3} + \frac{V_C(s) - V_B(s)}{1/C_2 s} + C_2 v_{C_2}(0) = 0$$

Writing these equations with node voltages on the left and the source terms on the right yields

$$\text{Node B: } -G_1 V_A(s) + [C_1 s + C_2 s + G_1 + G_2] V_B(s) -$$

$$C_2 s V_C(s) - G_2 V_D(s) = C_1 v_{C_1}(0) + C_2 v_{C_2}(0)$$

$$\text{Node C: } -C_2 s V_B(s) + [C_2 s + G_3] V_C(s) = -C_2 v_{C_2}(0)$$

where $G_1 = 1/R_1$, $G_2 = 1/R_2$, and $G_3 = 1/R_3$ are the conductances of the three resistors. In the node B equation the coefficient in the term $[C_1 s + C_2 s + G_1 + G_2]V_B(s)$ is the sum of the admittances connected to node B. The coefficients of the other terms in the same equation are the admittances connected between node B and the other nodes. The same pattern exists in the node B equation. Once we recognize this pattern, we can learn to write the foregoing node equations by inspection.

For the selected reference node, $V_A(s) = V_1(s)$ and $V_D(s) = \mu V_X(s) = \mu V_C(s)$. Substituting these relations into the foregoing node equations and arranging the result in standard form yields

$$\text{Node B: } (C_1 s + C_2 s + G_1 + G_2) V_B(s) - (C_2 s + \mu G_2) V_C(s) =$$

$$G_1 V_1(s) + C_1 v_{C_1}(0) + C_2 v_{C_2}(0)$$

$$\text{Node C: } -C_2 s V_B(s) + (C_2 s + G_3) V_C(s) = -C_2 v_{C_2}(0)$$

We now have two equations in the two unknown node voltages $V_B(s)$ and $V_C(s)$. The final node equations are not symmetrical because the dependent source is not a bilateral element like resistors, capacitors, and inductors. ◼

Exercise 10–7

Formulate node-voltage equations for the circuit in Figure 10–31. Assume that the initial conditions are zero.

FIGURE   10 – 31

Answer:

$$\left(G_1 + G_2 + \frac{1}{Ls}\right)V_B(s) - G_2V_C(s) = G_1V_S(s)$$

$$- G_2V_B(s) + (G_2 + C_S)V_C(s) = CsV_S(s)$$

Exercise 10–8

Formulate node-voltage equations for the circuit in Figure 10–31 when the initial conditions are not zero and a resistor R_3 is connected between nodes A and C.

Answer:

$$\left(G_1 + G_2 + \frac{1}{Ls}\right)V_B(s) - G_2V_C(s) = G_1V_S(s) - \frac{i_L(0)}{s}$$

$$- G_2V_B(s) + (G_2 + G_3 + Cs)V_C(s) = (G_3 + Cs)V_S(s) - Cv_C(0)$$

Exercise 10–9

Formulate node-voltage equations for the circuit in Figure 10–32. Assume that the initial conditions are zero.

Answer:

$$(G_1 + G_2 + C_2s)V_B(s) - (G_2 + \mu C_2s)V_C(s) = G_1V_S(s)$$

$$-G_2V_B(s) + (G_2 + C_1s)V_C(s) = 0$$

Exercise 10–10

Formulate node-voltage equations for the circuit in Figure 10–32 when a resistor R_3 is connected between nodes A and C. Assume that initial conditions are zero.

Answer:

$$(G_1 + G_2 + C_2s)V_B(s) - (G_2 + \mu C_2s)V_C(s) = G_1V_S(s)$$

$$-G_2V_B(s) + (G_2 + G_3 + C_1s)V_C(s) = G_3V_S(s)$$

FIGURE 10–32

SOLVING S-DOMAIN CIRCUIT EQUATIONS

Examples 10–10 and 10–11 show that node-voltage equations are linear algebraic equations in the unknown node voltages. In theory, solving these node equations is accomplished using techniques such as Cramer's rule or perhaps Gaussian reduction. In practice, we quickly lose interest in Gaussian reduction since coefficients in the equations are polynomials, making the algebra rather complicated. For hand analysis Cramer's rule is the better approach, especially when the element parameters are in symbolic form. For computer aided analysis we use the symbolic analysis capability of programs like Mathcad, MATLAB, and others to solve these linear equations.

We will illustrate the Cramer's rule solution process using an example from earlier in this section. In Example 10–10 we formulated the following node-voltage equations for the circuit in Figure 10–29:

$$\left(G + \frac{1}{Ls}\right) V_A(s) - \frac{1}{Ls} V_B(s) = I_S(s) - \frac{i_L(0)}{s}$$

$$-\frac{1}{Ls} V_A(s) + \left(Cs + \frac{1}{Ls}\right) V_B(s) = \frac{i_L(0)}{s} + Cv_C(0)$$

When using Cramer's rule, it is convenient first to find the determinant of these equations:

$$\Delta(s) = \begin{vmatrix} G + 1/Ls & -1/Ls \\ -1/Ls & Cs + 1/Ls \end{vmatrix}$$

$$= (G + 1/Ls)(Cs + 1/Ls) - (1/Ls)^2$$

$$= \frac{GLCs^2 + Cs + G}{Ls}$$

We call $\Delta(s)$ the **circuit determinant** because it only depends on the element parameters L, C, and $G = 1/R$. The determinant $\Delta(s)$ characterizes the circuit and does not depend on the input driving forces or initial conditions.

The node voltage $V_A(s)$ is found using Cramer's rule.

$$V_A(s) = \frac{\Delta_A(s)}{\Delta(s)} = \frac{\begin{vmatrix} I_S(s) - \dfrac{i_L(0)}{s} & -\dfrac{1}{Ls} \\ \dfrac{i_L(0)}{s} + Cv_C(0) & Cs + \dfrac{1}{Ls} \end{vmatrix}}{\Delta(s)} \qquad (10\text{–}35)$$

$$= \underbrace{\frac{(LCs^2 + 1) I_S(s)}{GLCs^2 + Cs + G}}_{\text{Zero State}} + \underbrace{\frac{-LCsi_L(0) + Cv_C(0)}{GLCs^2 + Cs + G}}_{\text{Zero Input}}$$

Solving for the other node voltage $V_B(s)$ yields

$$V_B(s) = \frac{\Delta_B(s)}{\Delta(s)} = \frac{\begin{vmatrix} G + \dfrac{1}{Ls} & I_S(s) - \dfrac{i_L(0)}{s} \\ -\dfrac{1}{Ls} & \dfrac{i_L(0)}{s} + Cv_C(0) \end{vmatrix}}{\Delta(s)} \qquad (10\text{–}36)$$

$$= \underbrace{\frac{I_S(s)}{GLCs^2 + Cs + G}}_{\text{Zero State}} + \underbrace{\frac{GLi_L(0) + (GLs + 1)Cv_C(0)}{GLCs^2 + Cs + G}}_{\text{Zero Input}}$$

Cramer's rule gives both the zero-input and zero-state components of the response transforms $V_A(s)$ and $V_B(s)$.

Cramer's rule yields the node voltages as a ratio of determinants of the form

$$V_X(s) = \frac{\Delta_X(s)}{\Delta(s)} \tag{10-37}$$

The response transform is a rational function of s whose poles are either zeros of the circuit determinant or poles of the determinant $\Delta_X(s)$. That is, $V_X(s)$ in Eq. (10–37) has poles when $\Delta(s) = 0$ or $\Delta_X(s) \to \infty$. The partial-fraction expansion of $V_X(s)$ will contain terms for each of these poles. We call the zeros of $\Delta(s)$ the **natural poles** because they depend only on the circuit and give rise to the natural response terms in the partial-fraction expansion. We call the poles of $\Delta_X(s)$ the **forced poles** because they depend on the form of the input signal and give rise to the forced response terms in the partial-fraction expansion.

EXAMPLE 10−12

Use computer-aided circuit analysis to find the s-domain, zero-state responses of the node voltages in Figure 10–33.

t domain
(a)

s domain
(b)

FIGURE 10 − 33

SOLUTION:
The node equations were formulated in Example 10–10 as

$$\text{Node A: } \left(G + \frac{1}{Ls}\right)V_A(s) - \left(\frac{1}{Ls}\right)V_B(s) = I_S(s) - \frac{i_L(0)}{s}$$

$$\text{Node B: } -\left(\frac{1}{Ls}\right)V_A(s) + \left(Cs + \frac{1}{Ls}\right)V_B(s) = \frac{i_L(0)}{s} + Cv_C(0)$$

In the zero state, these equations are written in matrix form as

$$\begin{bmatrix} G + \dfrac{1}{Ls} & -\dfrac{1}{Ls} \\ -\dfrac{1}{Ls} & Cs + \dfrac{1}{Ls} \end{bmatrix} \begin{bmatrix} V_A(s) \\ V_B(s) \end{bmatrix} = \begin{bmatrix} I_S(s) \\ 0 \end{bmatrix}$$

This matrix equation is of the form $\mathbf{YV} = \mathbf{I}$, where $\mathbf{Y}$ is a two by two matrix of admittances, $\mathbf{V}$ is a column vector containing the unknown node voltages, and **I** is a column vector containing the input driving force. The matrix equation is solved for the node voltage vector by forming the matrix product $\mathbf{Y}^{-1}\mathbf{I}$. We have used MATLAB in this way in several previous examples. What is new here is that the matrices contain symbolic variables rather than numerical values. To use MATLAB symbolic capability we must first define the symbolic variables as

```
syms G L C s IS
```

We next enter the **Y** and **I** matrices as

```
Y=[G+1/(L*s)   -1/(L*s);
   -1/(L*s)   Cs+1/(L*s)];
I = [IS;0];
```

and solve for the node voltage vector by the statement

```
V=inv(Y)*I

V=

[(C*s^2*L+1)/(G*L*s^2*C+G+C*s)*IS]
[          1/(G*L*s^2*C+G+C*s)*IS]
```

which means that the node voltages are

$$V_A(s) = \left(\frac{LCs^2 + 1}{GLCs^2 + Cs + G}\right)I_S(s)$$

$$V_B(s) = \left(\frac{1}{GLCs^2 + Cs + G}\right)I_S(s)$$

Note that the two node voltages have different zeros but the same natural poles, which are located at the roots of $GLCs^2 + Cs + G = 0$. We will explore this observation in the next example. ∎

EXAMPLE 10–13

Find the zero-state component for $v_A(t)$ and $v_B(t)$ in Figure 10–33 for $R = 1$ kΩ, $L = 0.5$ H, $C = 0.2$ μF, and $i_S(t) = 10u(t)$ mA.

SOLUTION:

In Example 10–12 we found the zero-state responses $V_A(s)$ and $V_B(s)$ in symbolic form. Inserting the specified numerical values into those expressions yields

$$V_A(s) = \left(\frac{10^{-7}s^2 + 1}{10^{-10}s^2 + 0.2 \times 10^{-6}s + 10^{-3}}\right)\frac{10^{-2}}{s}$$

$$= 10\,\frac{s^2 + 10^7}{s[(s + 1000)^2 + 3000^2]}\qquad \text{V-s}$$

$$V_B(s) = \left(\frac{1}{10^{-10}s^2 + 0.2 \times 10^{-6}s + 10^{-3}}\right)\frac{10^{-2}}{s}$$

$$= 10\,\frac{10^7}{s[(s + 1000)^2 + 3000^2]}\qquad \text{V-s}$$

Both response transforms have three poles: a forced pole at $s = 0$ and two natural poles at $s = -1000 \pm j3000$. The forced pole comes from the step function input, and the two natural poles are zeros of the circuit determinant. Expanding these rational functions as

$$V_A(s) = \frac{10}{s} - \frac{20}{3}\left(\frac{3000}{(s + 1000)^2 + 3000^2}\right)$$

$$V_B(s) = \frac{10}{s} - \frac{10}{3}\left(\frac{3000}{(s + 1000)^2 + 3000^2}\right) - 10\left(\frac{s + 1000}{(s + 1000)^2 + 3000^2}\right)$$

and taking the inverse transforms yields the required zero-state response waveforms:

$$v_A(t) = 10u(t) - 20e^{-1000t}\left[\frac{1}{3}\sin(3000\,t)\right]u(t) \qquad \text{V}$$

$$v_B(t) = 10u(t) - 10e^{-1000t}\left[\frac{1}{3}\sin(3000\,t) + \cos(3000\,t)\right]u(t) \qquad \text{V}$$

The step function in both responses is the forced response caused by the forced pole at $s = 0$. The damped sinusoids are natural responses determined by the natural poles.

Figure 10–34 shows part of an Excel spreadsheet that produces plots of $v_A(t)$ and $v_B(t)$. Spreadsheets are useful for generating graphs, especially when we wish to compare waveforms. The two plots show that the two response waveforms are different even though they have the same poles. In

FIGURE 10–34

other words, the basic form of a response is determined by the forced and natural poles, but the relative amplitudes are influenced by the zeros as well. ∎

APPLICATION NOTE: EXAMPLE 10-14

In s-domain circuit analysis and design, the location of complex poles is often specified in terms of the undamped natural frequency (ω_0) and damping ratio (ζ) parameters introduced in Chapter 7. Using these parameters, the standard form of a second-order factor is $s^2 + 2\zeta\omega_0 s + \omega_0^2$, which locates the poles at

$$s_{1,2} = \omega_0(-\zeta \pm \sqrt{\zeta^2 - 1})$$

The quantity under the radical depends only on the damping ratio ζ. When $\zeta > 1$ the quantity is positive and the two poles are real and distinct, and the second-order factor becomes the product of two first-order terms. If $\zeta = 1$ the quantity under the radical vanishes and there is a double pole as $s = -\omega_0$. If $\zeta < 1$ the quantity under the radical is negative and the two roots are complex conjugates.

The location of complex poles is also defined in terms of the two natural frequency parameters α and β. Using these parameters, the poles are at $s_{1,2} = -\alpha \pm j\beta$, and the standard form of a second-order factor is $(s + \alpha)^2 + \beta^2$.

In s-domain circuit design, we often need to convert from one set of parameters to the other. First, equating their standard forms

$$s^2 + 2\alpha s + \alpha^2 + \beta^2 = s^2 + 2\zeta\omega_0 s + \omega_0^2$$

and then equating the coefficients of like powers of s yields

$$\omega_0 = \sqrt{\alpha^2 + \beta^2} \quad \text{and} \quad \zeta = \frac{\alpha}{\sqrt{\alpha^2 + \beta^2}}$$

and conversely

$$\alpha = \zeta\omega_0 \quad \text{and} \quad \beta = \omega_0\sqrt{1 - \zeta^2}$$

Figure 10-35 shows how these parameters define the locations of complex poles in the s plane. The natural frequency parameters α and β define the rectangular coordinates of the poles. In a sense, the parameters ω_0 and ζ define the corresponding polar coordinates. The parameter ω_0 is the radial distance from the origin to the poles. The angle θ is determined by the damping ratio ζ alone, since $\theta = \cos^{-1}\zeta$.

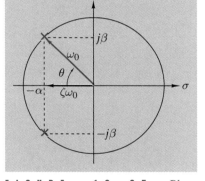

FIGURE 10-35 *s-Plane geometry relating α and β to ζ and ω_0.*

D DESIGN EXAMPLE 10-15

The circuit in Figure 10-36 is designed to produce a pair of complex poles. To simplify production, the final design must have equal element values $R_1 = R_2 = R_3 = R$ and $C_1 = C_2 = C$. Select the values of R, C, and the gain μ so that the circuit has natural poles defined by $\zeta = 0.5$ and $\omega_0 = 1000$ rad/s.

FIGURE 10 – 36

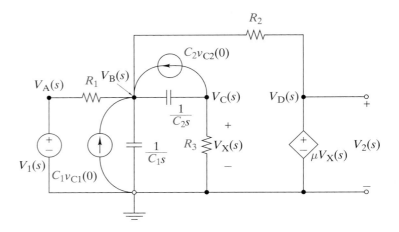

SOLUTION:

In Example 10–11 we formulated the following node-voltage equations for this circuit:

$$(C_1s + C_2s + G_1 + G_2)V_B(s) - (C_2s + \mu G_2)V_C(s)$$
$$= G_1V_1(s) + C_1v_{C_1}(0) + C_2v_{C_2}(0)$$
$$-C_2sV_B(s) + (C_2s + G_3)V_C(s) = -C_2v_{C_2}(0)$$

The natural poles are zeros of the circuit determinant:

$$\Delta(s) = \begin{vmatrix} C_1s + C_2s + G_1 + G_2 & -C_2s - \mu G_2 \\ -C_2s & C_2s + G_3 \end{vmatrix}$$
$$= C_1 C_2s^2 + (C_1G_3 + C_2G_1 + C_2G_2 + C_2G_3 - \mu G_2C_2)s$$
$$+ G_1G_3 + G_2G_3$$

For equal resistances $R_1 = R_2 = R_3 = R$ and equal capacitances $C_1 = C_2 = C$, the circuit determinant reduces to

$$\frac{\Delta(s)}{C^2} = s^2 + \frac{4 - \mu}{RC}s + \frac{2}{(RC)^2}$$

Comparing this second-order factor to the standard form $s^2 + 2\zeta \omega_0 s + \omega_0^2$ yields the following design constraints:

$$\omega_0 = \frac{\sqrt{2}}{RC} = 1000 \quad \text{and} \quad \zeta = \frac{4 - \mu}{\sqrt{2}} = 0.5$$

These constraints yield $RC = \sqrt{2} \times 10^{-3}$ and $\mu = 4 - \sqrt{2} = 2.586$. Selecting $R = 10\ k\Omega$ makes $C = 141.4$ nF. ∎

Exercise 10–11

(a) Solve for the determinant of the circuit in Figure 10–31 using the node equations in Exercise 10–7.
(b) Solve for the zero-state component of $V_B(s)$.

Answers:

(a) $\Delta(s) = \dfrac{(G_1 + G_2)\,LCs^2 + (G_1G_2L + C)s + G_2}{Ls}$

(b) $V_B(s) = \left[\dfrac{(G_1 + G_2)LCs^2 + G_1G_2Ls}{(G_1 + G_2)LCs^2 + (G_1G_2L + C)s + G_2}\right] V_S(s)$

Exercise 10–12

(a) Solve for the determinant of the circuit in Figure 10–32 using the node equations in Exercise 10–9.
(b) Solve for the zero-state component of $V_D(s)$.

Answers:

(a) $\Delta(s) = C_1C_2s^2 + (G_1C_1 + G_2C_1 + G_2C_2 - \mu G_2C_2)s + G_1G_2$

(b) $V_D(s) = \left[\dfrac{\mu G_1 G_2}{C_1C_2s^2 + (G_1C_1 + G_2C_1 + G_2C_2 - \mu G_2C_2)s + G_1G_2}\right] V_S(s)$

EXAMPLE 10–16

(a) Formulate *s*-domain node-voltage equations for the OP AMP circuit in Figure 10–37(a).
(b) Solve for the *s*-domain input-output relationship between the input $V_S(s)$ and the zero-state component of the output $V_O(s)$
(c) Solve for the zero-state output $v_O(t)$ when the input is a unit step function $v_S(t) = u(t)$.

(a)

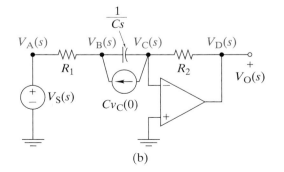

(b)

FIGURE   10–37

SOLUTION:

(a) Figure 10–37(b) shows the transformed circuit in the *s* domain. Node equations are not required at nodes A and D because the selected reference node makes $V_A(s) = V_S(s)$ and $V_D(s)$ is the OP AMP output. By inspection, the equations at nodes B and C are

$$\text{Node B:} \quad -G_1V_A(s) + (G_1 + Cs)V_B(s) - CsV_C(s) = Cv_C(0)$$

$$\text{Node C:} \quad -CsV_B(s) + (G_2 + Cs)V_C(s) - G_2V_D(s) = -Cv_C(0)$$

where $G_1 = 1/R_1$ and $G_2 = 1/R_2$. The preceding result yields two node equations in four node voltages. However, as noted previously, $V_A(s) = V_S(s)$ and $V_D(s) = V_O(s)$. The OP AMP voltage constraint requires $V_C(s) = 0$ since the noninverting input is grounded. Inserting the conditions $V_A(s) = V_S(s)$ and $V_C(s) = 0$ reduces node-voltage equations to the following form:

$$\text{Node B:} \quad (G_1 + Cs)V_B(s) = Cv_C(0) + G_1V_S(s)$$

$$\text{Node C:} \quad -CsV_B(s) - G_2V_O(s) = -Cv_C(0)$$

We now have two equations in two unknowns $V_B(s)$ and $V_O(s)$.

(b) For the zero-state response the initial condition $v_C(0) = 0$. Solving the node B equation for $V_B(s)$ in terms of $V_S(s)$,

$$V_B(s) = \left[\frac{G_1}{G_1 + Cs}\right]V_S(s)$$

Substituting this result into the node C equation and solving for $V_O(s)$ yields

$$V_O(s) = \left[\frac{G_1Cs/G_2}{G_1 + Cs}\right]V_S(s)$$

$$= -\left[\frac{R_2}{R_1}\frac{s}{s + 1/R_1C}\right]V_S(s) \quad \text{V-s}$$

This equation is the *s*-domain input-output relationship for the circuit. The circuit is linear, so the output transform is proportional to the input transform. The proportionality factor within the brackets is called a **network function**. It has a natural pole at $s = -1/R_1C$ and a zero at $s = 0$.

(c) A step function input $V_S(s) = 1/s$ produces a forced pole at $s = 0$. However, the zero in the network function cancels the forced pole. The net result is that the zero-state output contains only a natural pole and has the form

$$v_O(t) = \mathcal{L}^{-1}\left\{-\frac{R_2}{R_1}\frac{1}{s + 1/R_1C}\right\} = -\left[\frac{R_2}{R_1}e^{-t/R_1C}\right]u(t) \quad \text{V}$$

The general principle is that the forced response of a circuit can be zero even when the input driving force is not zero. In the *s* domain this occurs when the network function has a zero at the same location as the forced pole. ∎

10–5 MESH-CURRENT ANALYSIS IN THE *s* DOMAIN

We can use the mesh-current method only when the circuit can be drawn on a flat surface without crossovers. Such planar circuits have special loops called meshes that are defined as closed paths that do not enclose any elements. The mesh-current variables are the loop currents assigned to each mesh in a planar circuit. Because of KCL the current through any two-terminal element can be expressed as the difference of two adjacent mesh currents. This fundamental property of mesh currents, together with the element impedances, allows us to write KVL constraints around each of the meshes.

For example, in Figure 10–38 the sum of voltages around mesh A can be written as

$$Z_1(s)I_A(s) + Z_3[I_A(s) - I_C(s)] - V_{S1}(s)$$
$$+ Z_2[I_A(s) - I_B(s)] + V_{S2}(s) = 0$$

Rewriting this equation with unknown mesh currents grouped on the left and inputs on the right yields

$$(Z_1 + Z_2 + Z_3)I_A(s) - Z_2I_B(s) - Z_3I_C(s) = V_{S1}(s) - V_{S2}(s)$$

This equation displays the following pattern. The unknowns are the mesh-current transforms $I_A(s)$, $I_B(s)$, and $I_C(s)$. The coefficient $[Z_1(s) + Z_2(s) + Z_3(s)]$ of $I_A(s)$ is the sum of the impedances of the elements in mesh A. The coefficients $[Z_2(s)]$ of $I_B(s)$ and $[Z_3(s)]$ of $I_C(s)$ are the impedances common to mesh A and the other meshes. Finally, $V_{S1}(s) - V_{S2}(s)$ is the sum of the source voltages around mesh A. These observations suggest that we can write node-voltage equations for *s*-domain circuits by inspection, just as we did with resistive circuits.

The formulation approach just outlined assumes that there are no current sources in the circuit. When writing mesh-current equations, we select the voltage source model to represent the initial conditions. If the circuit contains dependent or independent current sources, they can be treated using the following methods:

Method 1: If there is an admittance in parallel with the current source, use a source transformation to convert it into an equivalent voltage source.

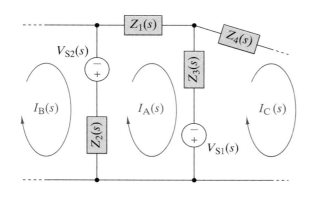

FIGURE 10 – 38 *An example mesh.*

Method 2: Draw the circuit diagram so that only one mesh current circulates through the current source. This mesh current is then determined by the source current.

Method 3: Create a supermesh for any current source that cannot be handled by methods 1 or 2.

Some circuits may require more than one of these methods.

The following examples illustrate the mesh-current method of *s*-domain circuit analysis.

EXAMPLE 10–17

(a) Formulate mesh-current equations for the circuit in Figure 10–39(a).
(b) Solve for the zero-input component of $I_A(s)$ and $I_B(s)$.
(c) Find the zero-input responses $i_A(t)$ and $i_B(t)$ for $R_1 = 200\ \Omega$, $R_2 = 300\ \Omega$, $L_1 = 50$ mH, and $L_2 = 100$ mH.

FIGURE 10–39

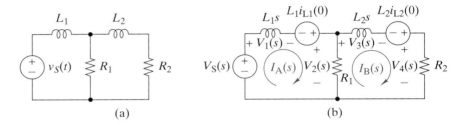

(a)   (b)

SOLUTION:

(a) Figure 10–39(b) shows the circuit transformed into the *s* domain. In transforming the circuit we used the voltage source model for the initial conditions. The net result is that the transformed circuit contains only voltage sources. The sum of voltages around meshes A and B can be written as

Mesh A: $\;-V_S(s) + L_1 s I_A(s) - L_1 i_{L_1}(0) + R_1[I_A(s) - I_B(s)] = 0$

Mesh B: $\;R_1[I_B(s) - I_A(s)] + L_2 s I_B(s) - L_2 i_{L_2}(0) + R_2 I_B(s) = 0$

Rearranging these equations in standard form yields

Mesh A: $\qquad (L_1 s + R_1) I_A(s) - R_1 I_B(s) = V_S(s) + L_1 i_{L_1}(0)$

Mesh B: $\;-R_1 I_A(s) + (L_2 s + R_1 + R_2) I_B(s) = L_2 i_{L_2}(0)$

These *s*-domain circuit equations are two linear algebraic equations in the two unknown mesh currents $I_A(s)$ and $I_B(s)$.

(b) To solve for the mesh equations, we first find the circuit determinant:

$$\Delta(s) = \begin{vmatrix} L_1 s + R_1 & -R_1 \\ -R_1 & L_2 s + R_1 + R_2 \end{vmatrix}$$

$$= L_1 L_2 s^2 + (R_1 L_2 + R_1 L_1 + R_2 L_1)s + R_1 R_2$$

To find the zero-input component of $I_A(s)$, we let $V_S(s) = 0$ and use Cramer's rule:

$$I_A(s) = \frac{\begin{vmatrix} L_1 i_{L_1}(0) & -R_1 \\ L_2 i_{L_2}(0) & L_2 s + R_1 + R_2 \end{vmatrix}}{\Delta(s)}$$

$$= \frac{(L_2 s + R_1 + R_2)L_1 i_{L_1}(0) + R_1 L_2 i_{L_2}(0)}{L_1 L_2 s^2 + (R_1 L_2 + R_1 L_1 + R_2 L_1)s + R_1 R_2}$$

Similarly, the zero-input component in $I_B(s)$ is

$$I_B(s) = \frac{\begin{vmatrix} L_1 s + R_1 & L_1 i_{L_1}(0) \\ -R_1 & L_2 i_{L_2}(0) \end{vmatrix}}{\Delta(s)}$$

$$= \frac{(L_1 s + R_1)L_2 i_{L_2}(0) + R_1 L_1 i_{L_1}(0)}{L_1 L_2 s^2 + (R_1 L_2 + R_1 L_1 + R_2 L_1)s + R_1 R_2}$$

(c) To find the time-domain response, we insert the numerical parameters into the preceding expressions to obtain

$$I_A(s) = \frac{(0.005s + 15)i_{L_1}(0) + 10 i_{L_2}(0)}{0.005 s^2 + 250 s + 20{,}000}$$

$$= \frac{(s + 3000)i_{L_1}(0) + 2000 i_{L_2}(0)}{(s + 1000)(s + 4000)}$$

$$I_B(s) = \frac{(0.005s + 100)i_{L_2}(0) + 5 i_{L_1}(0)}{0.005 s^2 + 250 s + 20{,}000}$$

$$= \frac{(s + 2000)i_{L_2}(0) + 1000 i_{L_1}(0)}{(s + 1000)(s + 4000)}$$

The circuit has natural poles at $s = -1000$ and -4000 rad/s. Expanding by partial fractions yields

$$I_A(s) = \frac{2}{3} \times \frac{i_{L_1}(0) + i_{L_2}(0)}{s + 1000} + \frac{1}{3} \times \frac{i_{L_1}(0) - 2 i_{L_2}(0)}{s + 4000} \quad \text{A-s}$$

$$I_B(s) = \frac{1}{3} \times \frac{i_{L_1}(0) + i_{L_2}(0)}{s + 1000} - \frac{1}{3} \times \frac{i_{L_1}(0) - 2 i_{L_2}(0)}{s + 4000} \quad \text{A-s}$$

The inverse transforms of these expansions are the required zero-state response waveforms:

$$i_A(t) = \left[\frac{2}{3}[i_{L_1}(0) + i_{L_2}(0)]e^{-1000t} + \frac{1}{3}[i_{L_1}(0) - 2 i_{L_2}(0)]e^{-4000t} \right] u(t) \quad \text{A}$$

$$i_B(t) = \left[\frac{1}{3}[i_{L_1}(0) + i_{L_2}(0)]e^{-1000t} - \frac{1}{3}[i_{L_1}(0) - 2 i_{L_2}(0)]e^{-4000t} \right] u(t) \quad \text{A}$$

Notice that if the initial conditions are $i_{L_1}(0) = -i_{L_2}(0)$, then both $I_A(s)$ and $I_B(s)$ have a zero at $s = -1000$. This zero effectively cancels the natural pole at $s = -1000$. As a result, this pole has zero residue in both partial fraction expansions, and the corresponding terms disappear from the time-domain responses. Likewise, if the initial conditions are $i_{L_1}(0) = 2 i_{L_2}(0)$, then both $I_A(s)$ and $I_B(s)$ have a zero at $s = -4000$, and the natural pole at $s = -4000$ disappears in the s-domain responses. The general principle is that all of the circuit's natural poles may not be present in a given response. When this happens the response transform has a zero at the same location as a natural pole, and we say that the natural pole is not observable in the specified response. ■

EXAMPLE 10–18

(a) Formulate mesh-current equations for the circuit in Figure 10–40(a).
(b) Solve for the zero-input component of $i_A(t)$ for $i_L(0) = 0$, $v_C(0) = 10$ V, $L = 250$ mH, $C = 1$ μF and $R = 1$ kΩ.

SOLUTION:

(a) Figure 10–40(a) is the s-domain circuit used in Example 10–10 to develop node equations. In this circuit each current source is connected in parallel with an impedance. Source transformations convert these current sources into the equivalent voltage sources shown in Figure 10–40(b). The circuit in Figure 10–40(b) is a series RLC circuit of the type treated in Chapter 7. By inspection, the KVL equation for the single mesh in this circuit is

$$\left[R + Ls + \frac{1}{Cs}\right] I_A(s) = RI_S(s) + Li_L(0) - \frac{v_C(0)}{s}$$

The circuit determinant is the factor $R + Ls + 1/Cs = (LCs^2 + RCs + 1)/Cs$. The zeros of the circuit determinant are roots of the quadratic equation $LCs^2 + RCs + 1 = 0$, which we recognize as the characteristic equation of a series RLC circuit. In Chapter 7 we called these roots natural frequencies. Thus, the natural poles of the circuit are its natural frequencies.

(b) Solving the mesh equation for the zero-input component yields

$$I_A(s) = \frac{LCsi_L(0) - Cv_C(0)}{LCs^2 + RCs + 1} \quad \text{A-s}$$

FIGURE 10 – 40

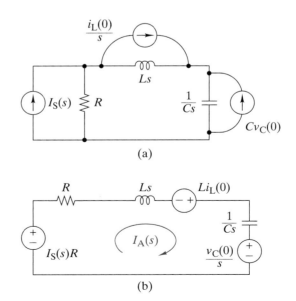

(a)

(b)

Inserting the given numerical values produces

$$I_A(s) = -\frac{10^{-5}}{0.25 \times 10^{-6}s^2 + 10^{-3}s + 1} = -\frac{40}{s^2 + 4 \times 10^3 s + 4 \times 10^6}$$

$$= -\frac{40}{(s + 2000)^2} \quad \text{A}$$

The zero-state response has two natural poles, both located at $s = -2000$. The inverse transform of $I_A(s)$ is a damped ramp waveform:

$$i_A(t) = -[40te^{-2000t}]\, u(t) \quad \text{A}$$

The damped ramp response indicates a critically damped second-order circuit. The minus sign means the direction of the actual current is opposite to the reference mark assigned to $I_A(s)$ in Figure 10–40. This sign makes sense physically since the capacitor initial condition source in Figure 10–40(b) tends to drive current in a direction opposite to the assigned reference mark. ■

▶ DESIGN EXAMPLE 10–19

Select the element values in Figure 10–41 so that the circuit has a real natural pole at $s = -20$ krad/s and a pair of complex poles with $\zeta = 0.5$ and $\omega_0 = 20$ krad/s.

FIGURE 10 – 41

SOLUTION:
To locate the circuit poles, we first write two mesh equations:

$$\text{Mesh A:} \quad \left(50 + \frac{1}{Cs} + L_1 s\right) I_A(s) - \frac{1}{Cs} I_B(s) = V_S(s)$$

$$\text{Mesh B:} \quad -\frac{1}{Cs} I_A(s) + \left(50 + \frac{1}{Cs} + L_2 s\right) I_B(s) = 0$$

The circuit determinant is

$$\Delta(s) = \left(50 + \frac{1}{Cs} + L_1 s\right)\left(50 + \frac{1}{Cs} + L_2 s\right) - \left(\frac{1}{Cs}\right)^2$$

$$= \frac{L_1 L_2 C s^3 + (50 L_1 C + 50\, L_2 C)s^2 + (L_1 + L_2 + 2500\, C)s + 100}{Cs}$$

The circuit characteristic equation in the numerator of $\Delta(s)$ can be placed in the following form:

$$q(s) = s^3 + \frac{(50L_1 + 50L_2)}{L_1 L_2} s^2 + \frac{(L_1 + L_2 + 2500C)}{L_1 L_2 C} s + \frac{100}{L_1 L_2 C}$$

To have the required pole positions, the circuit characteristic equation must have the form

$$q(s) = (s + 20{,}000)\,(s^2 + 20{,}000s + 20{,}000^2)$$
$$= s^3 + 4 \times 10^4 s^2 + 8 \times 10^8 s + 8 \times 10^{12}$$

Comparing the circuit's symbolic equation with the required polynomial leads to the following design constraints:

$$\frac{50L_1 + 50L_2}{L_1 L_2} = 4 \times 10^4 \qquad \frac{L_1 + L_2 + 2500C}{L_1 L_2 C} = 8 \times 10^8$$

$$\frac{100}{L_1 L_2 C} = 8 \times 10^{12}$$

The symbolic tools in MATLAB can solve equations like these. To use these tools the expressions need to be written in the form $f(x, y) = 0$. Accordingly, we rewrite the three design constraints as follows:

$$50L_1 + 50L_2 - 4 \times 10^4 L_1 L_2 = 0$$
$$L_1 + L_2 + 2500\,C - 8 \times 10^8 L_1 L_2 C = 0$$
$$100 - 8 \times 10^{12} L_1 L_2 C = 0$$

In the MATLAB command window we identify the symbolic variables and write the three design constraints as

```
syms  C  L1  L2
f1=50*L1+50*L2-4*10^4*L1*L2;
f2=L1+L2+2500*C-8*10^8*L1*L2*C;
f3=100-8*10^12*L1*L2*C;
```

and use the MATLAB `solve` function as

```
[C, L1, L2]=solve[f1, f2, f3]

C=

1/500000

L1=

1/400

L2=

1/400
```

Thus, the element values that produce the desired pole locations are

$$C = \frac{1}{500000} = 2\ \mu\text{F} \qquad \text{and} \qquad L_1 = L_2 = \frac{1}{400} = 2.5\ \text{mH}$$

In this case we were able to find a set of element values that meet the nonlinear design constraints. This may not always be possible. ∎

Exercise 10–13

(a) Formulate mesh-current equations for the circuit in Figure 10–42. Assume that the initial conditions are zero.

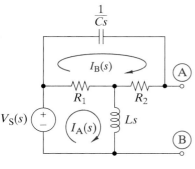

FIGURE 10 – 42

(b) Find the circuit determinant.
(c) Solve for the zero-state component of $I_B(s)$.

Answers:

(a) $\begin{cases} (R_1 + Ls)I_A(s) - R_1 I_B(s) = V_S(s) \\[2ex] -R_1 I_A(s) + (R_1 + R_2 + 1/Cs)I_B(s) = 0 \end{cases}$

(b) $\Delta(s) = \dfrac{(R_1 + R_2)LCs^2 + (R_1 R_2 C + L)s + R_1}{Cs}$

(c) $I_B(s) = \dfrac{R_1 Cs V_S(s)}{(R_1 + R_2)LCs^2 + (R_1 R_2 C + L)s + R_1}$

Exercise 10–14

Formulate mesh-current equations for the circuit in Figure 10–42 when a resistor R_3 is connected between nodes A and B. Assume that the initial conditions are zero.

Answer:

$$(R_1 + Ls)I_A(s) - R_1 I_B(s) - Ls I_C(s) = V_S(s)$$

$$-R_1 I_A(s) + \left(R_1 + R_2 + \frac{1}{Cs}\right)I_B(s) - R_2 I_C(s) = 0$$

$$-Ls I_A(s) - R_2 I_B(s) + (R_2 + R_3 + Ls)I_C(s) = 0$$

10–6 SUMMARY OF *s*-DOMAIN CIRCUIT ANALYSIS

At this point we review our progress and put *s*-domain circuit analysis into perspective. We have shown that linear circuits can be transformed from the time domain into the *s* domain. In this domain KCL and KVL apply to transforms and the passive element *i–v* characteristics become impedances with series or parallel initial condition sources. In relatively simple circuits we can use basic analysis methods, such as reduction, superposition, and voltage/current division. For more complicated circuits we use systematic procedures, such as the node-voltage or mesh-current methods, to solve for the circuit response.

In theory, we can perform *s*-domain analysis on circuits of any complexity. In practice, the algebraic burden of hand computations gets out of hand for circuits with more than three or four nodes or meshes. Of what practical use is an analysis method that becomes impractical at such a modest level of circuit complexity? Why not just appeal to computer-aided analysis tools in the first place?

Unquestionably, large-scale circuits are best handled by computer-aided analysis. Computer-aided analysis is probably the right approach even for small-scale circuits when numerical values for all circuit parameters are known and the desired end product is a plot or numerical listing of the response waveform. Simply put, *s*-domain circuit analysis is not a particularly efficient algorithm for generating numerical response data.

The purpose of *s*-domain circuit analysis is to gain insight into circuit behavior, not to grind out particular response waveforms. In this regard *s*-domain circuit analysis complements programs like PSpice. It offers a way of characterizing circuits in very general terms. It provides guidelines that allow us to use computer-aided analysis tools intelligently. Some of the useful general principles derived in this chapter are the following.

The response transform $Y(s)$[1] is a rational function whose partial fraction expansion leads directly to a response waveform of the form

$$y(t) = \sum_{j=1}^{\text{number of poles}} k_j\, e^{\,p_j\, t}$$

where k_j is the residue of the pole in $Y(s)$ located at $s = -p_j$. The location of the poles tells us a great deal about the form of the response. The pair of conjugate complex poles in Example 10–13 produced a damped sine waveform, the two distinct real poles in Example 10–17 produced exponential waveforms, and the double pole in Example 10–18 led to a damped ramp waveform. The general principle illustrated is as follows:

> *The poles of* **Y(s)** *are either real or complex conjugates. Simple real poles lead to exponentials, double real poles lead to a damped ramp, and complex conjugate poles lead to damped sinusoids.*

The poles in $Y(s)$ are introduced either by the circuit itself (natural poles) or by the input driving force (forced poles).

> *The natural poles are zeros of the circuit determinant and lead to the natural response. The forced poles are poles of the input* **X(s)** *and lead to the forced response.*

Figure 10–43 shows the waveforms corresponding to different pole locations in the *s* plane. Poles in the right half of the *s* plane produce waveforms whose amplitudes increase without bound, poles on the *j*-axis produce waveforms with constant amplitudes, and poles in the left half plane have amplitudes that decay to zero. A linear circuit is said to be stable if its natural response decays to zero as time marches on. As a result, we say that

> *A circuit is stable if all of its natural poles are located in the left half of the* **s** *plane.*

Some natural poles may be canceled by zeros in $Y(s)$, and the corresponding waveforms do not appear in the waveform $y(t)$.

> *Natural poles of a circuit that are not present in* **Y(s)** *are said to be not observable.*

---

1   In this context $Y(s)$ is not an admittance but the Laplace transform of the circuit output $y(t)$.

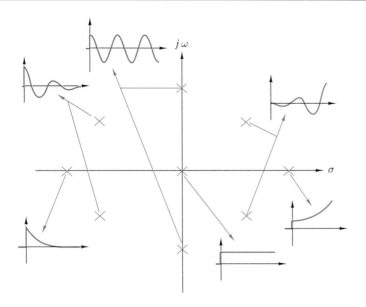

FIGURE 10–43 *Form of the natural response corresponding to different pole locations.*

Since natural poles play such a key role, we would like to find out how many to expect by simply examining the circuit. The following summary of this chapter's examples offers a hint:

EXAMPLE	NUMBER OF NATURAL POLES	NUMBER OF CAPACITORS	NUMBER OF INDUCTORS
10–5	1	0	1
10–9	1	1	0
10–13	2	1	1
10–14	2	2	0
10–16	2	0	2
10–19	3	1	2

These results suggest that the number of natural poles is equal to the number of energy storage elements in the circuit. This rule is a useful general guideline. There are exceptions, however, such as capacitors connected in parallel. As a result, the most we can say is that

The number of natural poles does not exceed the number of energy storage elements in a circuit.

In Design Examples 10–9, 10–14, and 10–19, we found that the parameters of a given circuit could be adjusted to produce specified locations for natural poles. This may not always be possible. That is, a given circuit may not be able to produce the specified pole locations no matter what element values we try. The general principle is as follows:

The general s-domain circuit design problem is to create a circuit arrangement that produces specified natural poles.

We address this type of design problem in the next chapter.

SUMMARY

- Kirchhoff's laws apply to voltage and current waveforms in the time domain and to the corresponding transforms in the *s* domain.

- The *s*-domain models for the passive elements include initial condition sources and the element impedance or admittance. Impedance is the proportionality factor in the expression $V(s) = Z(s)I(s)$ relating the voltage and current transforms. Admittance is the reciprocal of impedance.

- The impedances of the three passive elements are $Z_R(s) = R$, $Z_L(s) = Ls$, and $Z_C(s) = 1/Cs$.

- The *s*-domain circuit analysis techniques closely parallel the analysis methods developed for resistance circuits. Basic analysis techniques, such as circuit reduction, Thévenin's and Norton's theorems, the unit output method, or superposition, can be used in simple circuits. More complicated networks require a general approach, such as the node-voltage or mesh-current methods.

- Response transforms are rational functions whose poles are zeros of the circuit determinant or poles of the transform of the input driving forces. Poles introduced by the circuit determinant are called natural poles and lead to the natural response. Poles introduced by the input are called forced poles and lead to the forced response.

- In linear circuits, response transforms and waveforms can be separated into zero-state and zero-input components. The zero-state component is found by setting the initial capacitor voltages and inductor currents to zero. The zero-input component is found by setting all input driving forces to zero.

- The main purpose of *s*-domain circuit analysis is to gain insight into circuit performance without necessarily finding the time-domain response. The natural poles reveal the form, stability, and observability of the circuit's response. The number of natural poles is never greater than the number of energy storage elements in the circuit.

- The locations of the natural poles can be altered by adjusting circuit parameters. In some circuits there are limitations on the natural pole configuration that can be achieved by adjusting circuit parameters.

PROBLEMS

ERO 10–1 EQUIVALENT IMPEDANCE (SECTS. 10–1, 10–2)

Given a linear circuit, use series and parallel equivalence to find the equivalent impedance at a specified terminal pair. Select element values to obtain specified pole locations.

10–1 Find the equivalent impedance between terminals 1 and 2 in Figure P10–1. Express the impedance as a rational function and locate its poles and zeros. Select values of R and C so that the equivalent impedance has a pole at $s = -1000$ rad/s.

Z_{EQ}

FIGURE P10–1

10–2 A 2-kΩ resistor and a 0.1-μF capacitor are connected in series. The series combination is connected in parallel with a 0.5-H inductor. Find the equivalent impedance of the combination. Express the impedance as a rational function and locate its poles and zeros.

10–3 A 100-nF capacitor and a 0.5-H inductor are connected in series. The series combination is connected in parallel with a 5-kΩ resistor. Find the equivalent impedance of the combination. Express the impedance as a rational function and locate its poles and zeros.

10–4 Find the equivalent impedance between terminals 1 and 2 in Figure P10–4. Express the impedance as a rational function and locate its poles and zeros. Select values of R and L so that the impedance has a pole at $s = -5000$ rad/s and $Z_{EQ}(0) = 1$ kΩ.

10–5 A 1-kΩ resistor and a 500-mH inductor are connected in parallel. The parallel combination is connected in series with a 0.25-μF capacitor. Find the equivalent impedance of the combination. Express the impedance as a rational function and locate its poles and zeros.

10–6 Find the equivalent impedance between terminals 1 and 2 in the circuit of Figure P10–6. Express the impedance as a rational function and locate its poles and zeros. Evaluate $Z_{EQ}(s)$ at $s = 0$ and $s = \infty$. Explain your results in terms of the impedances of the inductor and capacitor.

10–7 Find the equivalent impedance between terminals 1 and 2 in the circuit of Figure P10–7. Express the impedance as a rational function and locate its poles and zeros.

FIGURE P10–4

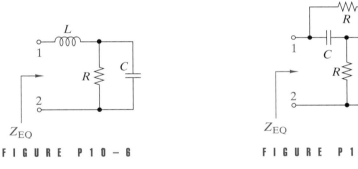

Z_{EQ}

FIGURE P10–6

Z_{EQ}

FIGURE P10–7

Z_{EQ}

FIGURE P10–9

10–8 Find the equivalent impedance between terminals 1 and 2 in the circuit of Figure P10–7 when a short circuit is connected between terminals 3 and 1. Express the impedance as a rational function and locate its poles and zeros.

10–9 Find the equivalent impedance between terminals 1 and 2 in the circuit of Figure P10–9. Select values of R and C so that $Z_{EQ}(s)$ has a pole at $s = -2000$ rad/s. Identify the location of the other poles and zeros of $Z_{EQ}(s)$ for your values of R and C.

10–10 Find the equivalent impedance between terminals 1 and 2 in the circuit of Figure P10–10. Select values of R and L so that $Z_{EQ}(s)$ has a pole at $s = -2000$ rad/s. Identify the location of the zeros of $Z_{EQ}(s)$ for your values of R and L.

Z_{EQ}

FIGURE P10–10

ERO 10–2 BASIC CIRCUIT ANALYSIS TECHNIQUES (SECTS. 10–2, 10–3)

Given a linear circuit,

(a) Determine the initial conditions (if not given) and transform the circuit into the *s* domain.
(b) Solve for zero-state and/or zero-input response transforms and waveforms using basic analysis methods such as circuit reduction, unit output, Thévenin/Norton equivalent circuits, or superposition.
(c) Locate the forced and natural poles or select circuit parameters to place the natural poles at specified locations.

10–11 The switch in Figure P10–11 has been open for a long time and is closed at $t = 0$. Transform the circuit into the *s* domain and solve for $I_L(s)$ and $i_L(t)$ in symbolic form.

10–12 The switch in Figure P10–11 has been closed for a long time and is opened at $t = 0$. Transform the circuit into the *s* domain and solve for $I_L(s)$ and $i_L(t)$ in symbolic form.

10–13 The switch in Figure P10–13 has been in position A for a long time and is moved to position B at $t = 0$. Transform the circuit into the *s* domain and solve for $V_C(s)$ and $v_C(t)$ in symbolic form.

10–14 The switch in Figure P10–13 has been in position B for a long time and is moved to position A at $t = 0$. Transform the circuit into the *s* domain and solve for $I_C(s)$ and $i_C(t)$ in symbolic form.

10–15 The switch in Figure P10–15 has been closed for a long time and is opened at $t = 0$. Transform the circuit into the *s* domain and solve for $I_L(s)$ and $i_L(t)$.

10–16 The switch in Figure P10–15 has been open for a long time and is closed at $t = 0$. Transform the circuit into the *s* domain and solve for $I_{SW}(s)$ and $i_{SW}(t)$.

10–17 The switch in Figure P10–17 has been in position A for a long time and is moved to position B at $t = 0$.
(a) Transform the circuit into the *s* domain and solve for $V_C(s)$ and $I_L(s)$ in symbolic form.
(b) Find $v_C(t)$ and $i_L(t)$ for $R_1 = R_2 = 500\ \Omega$, $L = 500$ mH, $C = 1\ \mu$F, and $V_A = 15$ V.

10–18 The switch in Figure P10–17 has been in position B for a long time and is moved to position A at $t = 0$.
(a) Transform the circuit into the *s* domain and solve for $V_C(s)$ and $I_L(s)$ in symbolic form.
(b) Find $v_C(t)$ and $i_L(t)$ for $R_1 = R_2 = 750\ \Omega$, $L = 500$ mH, $C = 1.6\ \mu$F, and $V_A = 15$ V.

10–19 The switch in Figure P10–19 has been in position A for a long time and is moved to position B at $t = 0$.

(a) Transform the circuit into the s domain and solve for $V_C(s)$ and $I_L(s)$ in symbolic form.

(b) Find $v_C(t)$ and $i_L(t)$ for $R_1 = 5$ kΩ, $R_2 = 2$ kΩ, $L = 1$ H, $C = 1$ μF, $V_A = 10$ V, and $V_B = 5$ V.

10–20 The switch in Figure P10–19 has been in position B for a long time and is moved to position A at $t = 0$.

(a) Transform the circuit into the s domain and solve for $V_C(s)$ and $I_L(s)$ in symbolic form.

(b) Find $v_C(t)$ and $i_L(t)$ for $R_1 = 5$ kΩ, $R_2 = 2$ kΩ, $L = 1$ H, $C = 10$ nF, $V_A = 10$ V, and $V_B = 5$ V.

10–21 There is no initial energy stored in the circuit in Figure P10–21.

(a) Transform the circuit into the s domain and use voltage division to solve for $V_C(s)$.

(b) Identify the forced and natural poles in $V_C(s)$ when $v_S(t) = 100$ $u(t)$ V, $L = 50$ mH, $C = 0.05$ μF, and $R = 5$ kΩ.

10–22 There is no initial energy stored in the circuit in Figure P10–21.

(a) Transform the circuit into the s domain and use voltage division to solve for $V_R(s)$.

(b) Identify the forced and natural poles in $V_R(s)$ when $v_S(t) = 10$ [cos (20000t)]$u(t)$ V, $L = 50$ mH, $C = 0.1$ μF, and $R = 500$ Ω.

10–23 There is no energy stored in the capacitor in Figure P10–23 at $t = 0$. Transform the circuit into the s domain and use current division to find $v_O(t)$ when the input is $i_S(t) = 0.1e^{-100t}u(t)$ A. Identify the forced and natural poles in $V_O(s)$.

10–24 Repeat Problem 10–23 when $i_S(t) = 0.1e^{-500t}u(t)$ A.

10–25 The initial capacitor voltage in Figure P10–25 is V_0.

(a) Transform the circuit into the s domain and find the Thévenin equivalent at the interface shown in the figure.

(b) Use the Thévenin equivalent to determine the zero-state and zero-input components of $I(s)$ when $i_S(t) = I_A u(t)$.

FIGURE P10–19

FIGURE P10–21

FIGURE P10–23

FIGURE P10–25   FIGURE P10–27

10–26 Repeat Problem 10–25 when the current source delivers $i_S(t) = I_A$ [cos βt]$u(t)$.

10–27 Adjust the value of C in Figure P10–27 so that the circuit has a natural pole at $s = -10$ Mrad/s when $R_S = 550$ Ω, $R_L = 2$ kΩ, and $g = 40$ mS.

10–28   There is no initial energy stored in the circuit in Figure P10–28. Transform the circuit into the *s* domain and solve for the zero-state component of $V(s)$. Identify the forced and natural poles in $V(s)$.

10–29   A series connection of a resistor and capacitor are connected across a voltage source with $v_S(t) = [10e^{-2000t}]u(t)$ V. The circuit response is observed to be $v_C(t) = [5e^{-2000t} - 3e^{-1000t}]u(t)$ V.
   (a) Find the zero-state and zero-input components of $v_C(t)$.
   (b) Identify the forced and natural poles in $V_C(s)$.
   (c) Find $v_C(t)$ when the input voltage is changed to $v_S(t) = [20e^{-2000t}]u(t)$ V.

10–30   There is no initial energy stored in the circuit in Figure P10–30. Use circuit reduction to find the output voltage $V_2(s)$ in terms of the input voltage $V_1(s)$.

ERO 10–3 GENERAL CIRCUIT ANALYSIS (SECTS. 10–4, 10–5, 10–6)

Given a linear circuit,

(a) Determine the initial conditions (if not given) and transform the circuit into the *s* domain.
(b) Solve for zero-state and/or zero-input response transforms and waveforms using node-voltage or mesh-current methods.
(c) Identify the forced and natural poles or select circuit parameters to place the natural poles at specified locations.

10–31   There is no initial energy stored in the circuit in Figure P10–31.
   (a) Transform the circuit into the *s* domain and formulate node-voltage equations.
   (b) Use these equations to find $V_2(s)$ and $I_2(s)$ in symbolic form.
   (c) Find $i_2(t)$ for $v_1(t) = 10u(t)$, $R_1 = 10$ kΩ, $R_2 = 2.2$ kΩ, $L = 100$ mH, and $C = 40$ nF.

10–32   The initial conditions in the circuit in Figure P10–31 are $v_C(0) = V_0$ and $i_L(0) = 0$.
   (a) Transform the circuit into the *s* domain and formulate node-voltage or mesh-current equations.
   (b) Use these equations to find the zero-input components of $V_2(s)$ and $I_2(s)$ in symbolic form.
   (c) Find $i_2(t)$ for $v_1(t) = 0$, $V_0 = 10$ V, $R_1 = 10$ kΩ, $R_2 = 2.2$ kΩ, $L = 100$ mH, and $C = 40$ nF.

10–33   There is no initial energy stored in the circuit in Figure P10–33.
   (a) Transform the circuit into the *s* domain and formulate node-voltage equations.
   (b) Use these equations to find $V_2(s)$ and $I_2(s)$ in symbolic form.
   (c) Find $v_2(t)$ for $v_1(t) = 10u(t)$, $R_1 = 10$ kΩ, $R_2 = 36$ kΩ, $C_1 = 550$ nF, and $C_2 = 400$ nF.

10–34   There is no initial energy stored in the circuit in Figure P10–33.
  (a) Transform the circuit into the s domain and formulate mesh-current equations.
  (b) Use these equations to find $V_2(s)$ and $I_2(s)$ in symbolic form.
  (c) Find $i_2(t)$ for $v_1(t) = 10u(t)$, $R_1 = 10\ \text{k}\Omega$, $R_2 = 5\ \text{k}\Omega$, $C_1 = 50\ \text{nF}$, and $C_2 = 40\ \text{nF}$.

10–35   In Figure P10–33 the initial conditions are $v_{C_1}(0) = V_0$ and $v_{C_2}(0) = 0$.
  (a) Transform the circuit into the s domain and formulate mesh-current or node-voltage equations.
  (b) Solve these equations for $I_2(s)$.

10–36   The circuit in Figure P10–36 is called a bridged-T.
  (a) Transform the circuit into the s domain and formulate either node-voltage or mesh-current equations.
  (b) Find the poles and zeros of the zero-state component of $V_2(s)$ for $R_1 = 10\ \text{k}\Omega$, $R_2 = 20\ \text{k}\Omega$, $C_1 = 100\ \text{nF}$, $C_2 = 50\ \text{nF}$, and $v_1(t) = V_A e^{-2000t}$.

10–37   The circuit in Figure P10–37 is an active RC filter with an OP AMP voltage follower output. Find the zero-state component of $v_2(t)$ for $R_1 = R_2 = 10\ \text{k}\Omega$, $C_1 = C_2 = 100\ \text{nF}$, and $v_1(t) = 10u(t)$ V.

10–38   The circuit in Figure P10–38 is an active RC filter with two feedback paths to the OP AMP's inverting input. Transform the circuit into the s domain and show that the circuit's natural poles are complex when $R_1 = R_2 = R$, $R_3 = 5R$, and $C_1 = C_2 = C$. Select values of R and C so that the natural poles are located at $s_{1,2} = -1000 \pm j3000$ rad/s.

10–39   ⓓ The circuit in Figure P10–39 represents an active RC highpass filter originally proposed by R. P. Sallen and E. K. Key in 1955.
  (a) Transform the circuit into the s domain and find the circuit determinant.
  (b) Select values of R, C, and μ so the circuit has natural poles at $s = \pm j5000$ rad/s.

FIGURE P10–36

FIGURE P10–37

FIGURE P10–38

FIGURE P10–39

10–40   ⓓ The circuit in Figure P10–40 represents an active RC bandpass filter proposed by J. J. Friend in 1970.
  (a) Transform the circuit into the s domain and find the circuit determinant.

FIGURE P10-40

(b) Locate the natural poles when $R_1 = 10$ kΩ, $R_2 = 20$ kΩ, $C_1 = C_2 = 10$ nF, and $\mu = 5$.

10–41 The circuit in Figure P10–41 is a low-frequency model of a transistor in the common emitter configuration.

(a) Transform the circuit into the s domain and solve for the zero-state output voltage $V_2(s)$ in terms of the input $V_1(s)$.

(b) Find the zero-state response $v_2(t)$ for $v_1(t) = 10u(t)$ mV, $R_1 = 10$ kΩ, $R_\pi = 1$ kΩ, $\beta = 100$, $R_O = 10$ kΩ, $C = 0.1$ μF, and $R_L = 100$ kΩ.

FIGURE P10-41

10–42 The circuit in Figure P10–41 is a low-frequency model of a transistor in the common emitter configuration. For $R_O = R_L = 1$ kΩ, select the value of C so the circuit has a natural pole at $s = -500$ rad/s.

10–43 The s-domain mesh equations for a circuit in the zero state are

$$(L_1 s + L_3 s + R_1)I_A(s) - L_3 s I_B(s) = V_S(s)$$
$$- L_3 s I_A(s) + (L_2 s + L_3 s + R_2)I_B(s) = 0$$

(a) Draw a circuit diagram of the circuit

(b) Find the natural poles of the circuit for $R_1 = 50$ Ω, $R_2 = 75$ Ω, $L_1 = 0.4$ mH, $L_2 = 1$ mH, and $L_3 = 0.15$ mH.

10–44 The switch in Figure P10–44 has been in position A for a long time and is moved to position B at $t = 0$. Solve for $V_C(s)$ and $v_C(t)$.

FIGURE P10-44

The following figure appears here:

A circuit diagram with a 10 V source, switch positions A and B with $t = 0$, resistors 1 kΩ, 2 kΩ, 1 kΩ, a $5e^{-2000t}$ source, an inductor 1 H, capacitor 50 nF with $v_C(t)$.

10–45 There is no energy stored in the circuit in Figure P10–45 at $t = 0$. Transform the circuit into the s domain and solve for $V_O(s)$ and $v_O(t)$.

10–46 The switch in Figure P10–46 has been open for a long time and is closed at $t = 0$. Transform the circuit into the s domain and solve for $I_O(s)$ and $i_O(t)$.

10–47 There is no initial energy stored in the circuit in Figure P10–47.
  (a) Transform the circuit into the s domain and solve for $V_O(s)$ in symbolic form.
  (b) Find $v_O(t)$ when $v_S(t) = 10^4 t u(t)$ V, $R_1 = R_2 = 200\ \Omega$, $C = 0.5\ \mu F$, and $L = 20$ mH.
10–48 There is no initial energy stored in the circuit in Figure P10–48.
  (a) Show that the circuit has natural poles at $s = -1000$ rad/s and $s = -2000$ rad/s when $R_1 = 10$ kΩ, $C_1 = 50$ nF, $R_2 = 10$ kΩ, and $C_2 = 100$ nF.
  (b) When the input is $v_S(t) = 10u(t)$ V, show that $V_1(s)$ has a forced pole at $s = 0$ but only one natural pole at $s = -2000$ rad/s.
  (c) For the input in (b), show that $V_2(s)$ has both of the circuit's natural poles but does not have a forced pole at $s = 0$.

10–49   The circuit in Figure P10–49 is a third-order lowpass filter. Show that the circuit has natural poles at $s = -4/RC$ and $s = (-2 \pm j2)/RC$ when $4L = R^2C$.

FIGURE P10–49

10–50   Select values of R_1, R_2, C_1, and C_2 such that the circuit in Figure P10–50 has natural poles located at $s = -100$ rad/s and $s = -500$ rad/s.

FIGURE P10–50

INTEGRATING PROBLEMS

10–51   **A** THÉVENIN'S THEOREM FROM TIME-DOMAIN DATA

A black box containing a linear circuit has an on-off switch and a pair of external terminals. When the switch is turned on, the open-circuit voltage between the external terminals is observed to be a sinewave with a period of 10π ms and a peak-to-peak amplitude of 20 V. When a short circuit is connected to the external terminals and the switch again turned on, the current is observed to be a steady 10 mA. What current would the box deliver to a 100-Ω resistive load?

10–52   **A** IMPEDANCE FROM TIME-DOMAIN DATA

A current of $u(t)$ amperes is injected at a two-terminal interface and the interface voltage is observed to be $[10 + 20 \sin 1000t]u(t)$ V. Find the waveform of the interface current when a unit-step function voltage is applied across the same pair of terminals.

10–53   **A** **D** RC CIRCUIT ANALYSIS AND DESIGN

The RC circuit in Figure P10–53 is driven by the exponential voltage shown.
(a)   **A**   Determine the zero-state response $v_O(t)$ for $C = 200$ nF. Identify the forced response and the natural response.
(b)   **A**   For $C = 200$ nF determine the initial condition $v_C(0)$ such that the natural response in $v_O(t)$ is identically zero.
(c)   **D**   Adjust the value of C until the zero-state response is of the form $Kte^{-\alpha t}$. What are the values of K and α?

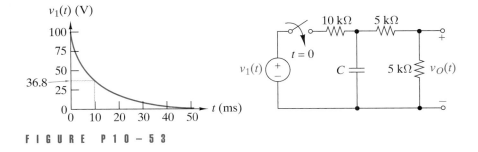

FIGURE P10-53

10–54 **A D** LIGHTNING PULSER DESIGN

The circuit in Figure P10–54 is a simplified circuit diagram of a pulser that is used to test equipment against lightning-induced transients. Select the values of V_0, R_1, R_2, C_1, and C_2 so that the pulser delivers an output pulse

$$v_O(t) = 500(e^{-2000t} - e^{-80000t}) \text{ V}$$

The pulser is completely discharged prior to delivering a pulse.

FIGURE P10-54

10–55 **E** PULSE CONVERSION CIRCUIT

The purpose of the test setup in Figure P10–55 is to deliver damped sine pulses to the test load. The excitation comes from a 1-Hz square wave generator. The pulse conversion circuit must deliver damped sine waveforms with $\zeta < 0.5$ and $\omega_0 > 10$ krad/s to 50-Ω and 600-Ω loads. The recommended values for the pulse conversion circuit are $L = 100$ mH and $C = 10$ nF. Verify that the test setup meets the specifications. If you find it does not meet the specifications, suggest a simple modification that will bring it within the specification limits.

FIGURE P10-55

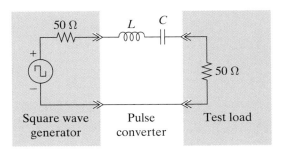

CHAPTER 11

NETWORK FUNCTIONS

The driving-point impedance of a network is the ratio of an impressed electromotive force at a point in a branch of the network to the resulting current at the same point.

Ronald M. Foster, 1924,

American Engineer

11–1 Definition of a Network Function

11–2 Network Functions of One- and Two-port Circuits

11–3 Network Functions and Impulse Response

11–4 Network Functions and Step Response

11–5 Network Functions and Sinusoidal Steady-state Response

11–6 Network Function Design

Summary

Problems

Integrating Problems

The concept of network functions emerged in the 1920s during the development of systematic methods of designing electric filters for long-distance telephone systems. The filter design effort eventually evolved into a formal realizability theory that came to be known as network synthesis. The objective of network synthesis is to design one or more networks that accomplish or realize a given network function. The foregoing quotation is taken from one of the first papers to approach network design from a synthesis viewpoint.[1] Ronald Foster, along with Sidney Darlington, Hendrik Bode, Wilhelm Cauer, and Otto Brune are generally considered the founders of the field of network synthesis. The classical synthesis methods they developed emphasized passive realizations using only resistors, capacitors, and inductors because the active elements of their era were costly and unreliable vacuum tubes. With the advent of semiconductor electronics, emphasis shifted to realizations using resistors, capacitors, and active devices such as transistors and OP AMPs.

In the last chapter we found that the transform of the zero-state output of a linear circuit is proportional to the transform of the input signal. In the s domain the proportionality factor is a rational function of s called a network function. The network function concept is used extensively in linear circuit analysis and design. This chapter begins a study of the properties of these functions that continues throughout the rest of this book.

The first two sections of this chapter show how to define and calculate network functions for different types of circuits. The next three sections discuss how to use network functions to obtain the impulse response, step response, and sinusoidal steady-state responses of a circuit. The final section treats some basic methods of synthesizing circuits that realize a specified network function.

11–1 DEFINITION OF A NETWORK FUNCTION

The proportionality property of linear circuits states that the output is proportional to the input. In Chapter 10 we noted that in the s domain the proportionality factor is a rational function of s called a network function. More formally, a network function is defined as the ratio of a zero-state response transform (output) to the excitation (input) transform.

$$\text{Network Function} = \frac{\text{Zero–state Response Transform}}{\text{Input Signal Transform}} \tag{11–1}$$

Note carefully that this definition specifies zero initial conditions and implies only one input.

To study the role of network functions in determining circuit responses, we write the s-domain input-output relationship as

$$Y(s) = T(s)X(s) \tag{11–2}$$

---

[1] R. M. Foster, "A Reactance Theorem," *Bell System Technical Journal*, No. 3, pp. 259–267, 1924.

FIGURE 11 – 1 *Block diagram for an s-domain input-output relationship.*

where $T(s)$ is a network function, $X(s)$ is the input signal transform, and $Y(s)$ is a zero-state response or output.[2] Figure 11–1 shows a block diagram representation of the s-domain input-output relationship in Eq. (11–2).

In an analysis problem, the circuit and input [$X(s)$ or $x(t)$] are specified. We determine $T(s)$ from the circuit, use Eq. (11–2) to find the response transform $Y(s)$, and use the inverse transformation to obtain the response waveform $y(t)$. In a design problem the circuit is unknown. The input and output are specified, or their ratio $T(s) = Y(s)/X(s)$ is given. The objective is to devise a circuit that realizes the specified input-output relationship. A linear circuit analysis problem has a unique solution, but a design problem may have one, many, or even no solutions.

Equation (11–2) points out that the poles of the response $Y(s)$ come from either the network function $T(s)$ or the input signal $X(s)$. When there are no repeated poles, the partial-fraction expansion of the right side of Eq. (11–2) takes the form

$$Y(s) = \underbrace{\sum_{j=1}^{N} \frac{k_j}{s - p_j}}_{\substack{\text{Natural} \\ \text{Poles}}} + \underbrace{\sum_{\ell=1}^{M} \frac{k_\ell}{s - p_\ell}}_{\substack{\text{Forced} \\ \text{Poles}}} \tag{11–3}$$

where p_j ($j = 1, 2, \ldots N$) are the poles of $T(s)$ and $s = p_\ell$ ($\ell = 1, 2, \ldots M$) are the poles of $X(s)$. The inverse transform of this expansion is

$$y(t) = \underbrace{\sum_{j=1}^{N} k_j e^{p_j t}}_{\substack{\text{Natural} \\ \text{Response}}} + \underbrace{\sum_{\ell=1}^{M} k_\ell e^{p_\ell t}}_{\substack{\text{Forced} \\ \text{Response}}} \tag{11–4}$$

The poles of *T(s)* lead to the natural response. In a stable circuit, the natural poles are all in the left half of the *s* plane, and all of the exponential terms in the natural response eventually decay to zero. The poles of $X(s)$ lead to the forced response. In a stable circuit, those elements in the forced response that do not decay to zero are called the **steady-state response**.

It is important to remember that the complex frequencies in the natural response are determined by the circuit and do not depend on input. Conversely, the complex frequencies in the forced response are determined by the input and do not depend on the circuit. However, the amplitude of its part of the response depends on the residues in the partial fraction expansion in Eq. (11–3). These residues are influenced by all of the poles and zeros, whether forced or natural. Thus, the amplitudes of the forced and natural responses depend on an interaction between the poles and zeros of $T(s)$ and $X(s)$.

The following example illustrates this discussion.

2　In this context $Y(s)$ is not an admittance, but the transform of the output waveform $y(t)$.

EXAMPLE 11–1

The transfer function of a circuit is

$$T(s) = \frac{V_2(s)}{V_1(s)} = \frac{2000\,(s + 2000)}{(s + 1000)\,(s + 4000)}$$

Find the zero-state response $v_2(t)$ when the input waveform is $v_1(t) = [20 + 15e^{-5000t}]u(t)$.

SOLUTION:

The transform of the input waveform is

$$V_1(s) = \frac{20}{s} + \frac{15}{s + 5000} = \frac{35\,s + 10^5}{s(s + 5000)}$$

Using the s-domain input-output relationship in Eq. (11–2), the transform of the response is

$$V_2(s) = \frac{10^4(s + 2000)\,(7s + 20000)}{(s + 1000)\,(s + 4000)s(s + 5000)}$$

Expanding by partial fractions,

$$V_2(s) = \underbrace{\frac{k_1}{s + 1000} + \frac{k_2}{s + 4000}}_{\text{Natural Poles}} + \underbrace{\frac{k_3}{s} + \frac{k_4}{s + 5000}}_{\text{Forced Poles}}$$

The two natural poles came from the circuit via the network function $T(s)$. The forced poles came from the step function and exponential inputs. Using the cover-up method to evaluate the residues yields

$$k_1 = \left.\frac{10^4(s + 2000)\,(7s + 20000)}{(s + 4000)s(s + 5000)}\right|_{s = -1000} = -\frac{65}{6}$$

$$k_2 = \left.\frac{10^4(s + 2000)\,(7s + 20000)}{(s + 1000)s(s + 5000)}\right|_{s = -4000} = \frac{40}{3}$$

$$k_3 = \left.\frac{10^4(s + 2000)\,(7s + 20000)}{(s + 1000)\,(s + 4000)\,(s + 5000)}\right|_{s = 0} = 20$$

$$k_4 = \left.\frac{10^4(s + 2000)\,(7s + 20000)}{(s + 1000)\,(s + 4000)s}\right|_{s = -5000} = -\frac{45}{2}$$

Collectively the residues depend on all of the poles and zeros. The inverse transform yields the zero-state response as

$$v_2(t) = \left[\underbrace{-\frac{65}{6}e^{-1000t} + \frac{40}{3}e^{-4000t}}_{\text{Natural Response}} + \underbrace{20 - \frac{45}{2}e^{-5000t}}_{\text{Forced Response}}\right]u(t)$$

The natural poles as $s = -1000$ and $s = -4000$ are in the left half of the s plane, so the natural response decays to zero. The forced pole as $s =$

−5000 leads to an exponential term that also decays to zero, leaving a steady-state response of $20u(t)$. ∎

TEST SIGNALS

While the transfer function is a useful concept, it is clear that we cannot find the circuit response until we are given an input signal. Here, we encounter a central paradox of circuit analysis. In practice, the input signal is a carrier of information and is therefore unpredictable. We could spend a lifetime studying a circuit for various inputs and still not treat all possible signals that might be encountered in practice. What we must do is calculate the responses due to certain standard test signals. Although these test signals may never occur as real input signals, their responses tell us enough to understand the signal processing capabilities of a circuit.

The two premier test signals used are the pulse and the sinusoid. The study of pulse response divides into two extreme cases, short and long. When the pulse is very short compared to the circuit response time, the sudden injection of energy causes a circuit response long after the input returns to zero. The short pulse is modeled by an impulse, and the resulting *impulse response* is treated in Sec. 11–3. At the other extreme, the long pulse has a duration that greatly exceeds the circuit response time. In this case, the circuit has ample time to be driven from the zero state to a new steady-state condition. The step function is used to model the long pulse input, and the resulting *step response* is studied in Section 11–4.

The impulse response is of great importance because it contains all of the information needed to calculate the response due to any other input. The step response is important because it describes how a circuit response transitions from one state to another. The signal transition requirements for circuits and systems are often stated in terms of the step response using partial waveform descriptors such as rise time, fall time, propagation delay, and overshoot.

The unique properties of the sinusoid make it a useful input for characterizing the signal-processing capabilities of linear circuits and systems. When a stable linear circuit is driven by a sinusoidal input, the steady-state output is a sinusoid with the same frequency, but with a different phase angle and amplitude. The frequency-dependent relationship between the sinusoidal input and the steady-state output is called *frequency response*, a signal-processing description that is often used to specify the performance of circuits and systems. The relationship between network functions and the sinusoidal steady-state response is studied in Sec. 11–5.

11–2 NETWORK FUNCTIONS OF ONE- AND TWO-PORT CIRCUITS

The two major types of network functions are driving-point impedance and transfer functions. A **driving-point impedance** relates the voltage and current at a pair of terminals called a port. The driving-point impedance $Z(s)$ of the one-port circuit in Figure 11–2 is defined as

$$Z(s) = \frac{V(s)}{I(s)} \tag{11–5}$$

FIGURE 11–2 *A one-port circuit.*

When the one port is driven by a current source the response is $V(s) = Z(s)I(s)$ and the natural frequencies in the response are the poles of impedance $Z(s)$. On the other hand, when the one port is driven by a voltage source the response is $I(s) = [Z(s)]^{-1}V(s)$ and the natural frequencies in the response are the poles of $1/Z(s)$; that is, the zeros of $Z(s)$. In other words, the driving-point impedance is a network function whether upside down or right side up.

The term *driving point* means that the circuit is driven at one port and the response is observed at the same port. The element impedances defined in Sect. 10–1 are elementary examples of driving-point impedances. The equivalent impedances found by combining elements in series and parallel are also driving-point impedances. Driving-point functions are the s-domain generalization of the concept of the input resistance. The terms *driving-point impedance*, *input impedance*, and *equivalent impedance* are synonymous.

The driving point impedance seen at a pair of terminals determines the loading effects that result when those terminals are connected to another circuit. When two circuits are connected together, these loading effects can profoundly alter the responses observed when the same two circuits operated in isolation. In an analysis situation it is important to be able to predict the response changes that occur when one circuit loads another. In design situations it is important to know when the circuits can be designed separately and then interconnected without encountering loading effects that alter their designed performance. The conditions under which loading can or cannot be ignored will be studied in this and subsequent chapters.

Transfer functions are usually of greater interest in signal-processing applications than driving-point impedances because they describe how a signal is modified by passing through a circuit. A **transfer function** relates an input and response (or output) at different ports in the circuit. Figure 11–3 shows the possible input-output configurations for a two-port circuit. Since the input and output signals can be either a current or a voltage, we can define four kinds of transfer function:

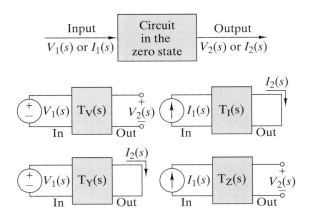

F I G U R E   1 1 – 3 *Two-port circuits and transfer functions.*

$$T_V(s) = \text{Voltage Transfer Function} = \frac{V_2(s)}{V_1(s)}$$

$$T_I(s) = \text{Current Transfer Function} = \frac{I_2(s)}{I_1(s)} \qquad (11\text{–}6)$$

$$T_Y(s) = \text{Transfer Admittance} = \frac{I_2(s)}{V_1(s)}$$

$$T_Z(s) = \text{Transfer Impedance} = \frac{V_2(s)}{I_1(s)}$$

The functions $T_V(s)$ and $T_I(s)$ are dimensionless since the input and output signals have the same units. The function $T_Z(s)$ has units of ohms and $T_Y(s)$ has units of siemens.

The functions in Eq. (11–6) are sometimes called forward transfer functions because they relate inputs applied at port 1 to outputs occurring at port 2. There are, of course, reverse transfer functions that relate inputs at port 2 to outputs at port 1. It is important to realize that a transfer function is only valid for a specified input port and output port. For example, the voltage transfer function $T_V(s) = V_2(s)/V_1(s)$ relates the input voltage applied at port 1 in Figure 11–3 to the voltage response observed at the output port. The reverse voltage transfer function for signal transmission from output to input is *not* $1/T_V(s)$. Unlike driving-point impedance, transfer functions are not network functions when they are turned upside down.

DETERMINING NETWORK FUNCTIONS

The rest of this section illustrates analysis techniques for deriving network functions. The application of network functions in circuit analysis and design begins in the next section and continues throughout the rest of this book. But first, we illustrate ways to find the network functions of a given circuit.

The divider circuits in Figure 11–4 occur so frequently that it is worth taking time to develop their transfer functions in general terms. Using s-domain voltage division in Figure 11–4(a), we can write

$$V_2(s) = \left[\frac{Z_2(s)}{Z_1(s) + Z_2(s)} \right] V_1(s)$$

Therefore, the voltage transfer function of a voltage divider circuit is

$$T_V(s) = \frac{V_2(s)}{V_1(s)} = \frac{Z_2(s)}{Z_1(s) + Z_2(s)} \qquad (11\text{–}7)$$

Similarly, using s-domain current division in Figure 11–4(b) yields the transfer function of a current divider circuit as

$$T_I(s) = \frac{I_2(s)}{I_1(s)} = \frac{Y_2(s)}{Y_1(s) + Y_2(s)} \qquad (11\text{–}8)$$

By series equivalence, the driving-point impedance at the input of the voltage divider is $Z_{EQ}(s) = Z_1(s) + Z_2(s)$. By parallel equivalence the driving-point impedance at the input of the current divider is $Z_{EQ}(s) = 1/(Y_1(s) + Y_2(s))$.

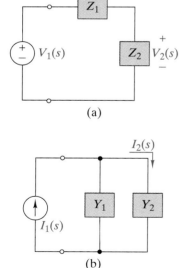

F I G U R E 1 1 – 4 *Basic divider circuits. (a) Voltage divider. (b) Current divider.*

Two other useful circuits are the inverting and noninverting OP AMP configurations shown in Figure 11–5. To determine the voltage transfer function of the inverting circuit in Figure 11–5(a), we write the sum of currents leaving node B:

$$\frac{V_B(s) - V_A(s)}{Z_1(s)} + \frac{V_B(s) - V_C(s)}{Z_2(s)} + I_N(s) = 0$$

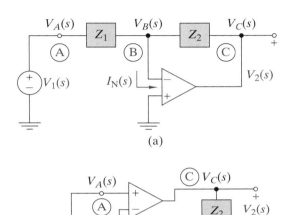

F I G U R E   1 1 – 5 *Basic OP AMP circuits. (a) Inverting amplifier. (b) Noninverting amplifier.*

(a)

(b)

But the ideal OP AMP constraints require that $I_N(s) = 0$ and $V_B(s) = 0$ since the noninverting input is grounded. By definition, the output voltage $V_2(s)$ equals node voltage $V_C(s)$ and the voltage source forces $V_A(s)$ to equal the input voltage $V_1(s)$. Inserting all of these considerations into the node equations and solving for the voltage transfer function yields

$$T_V(s) = \frac{V_2(s)}{V_1(s)} = -\frac{Z_2(s)}{Z_1(s)} \tag{11–9}$$

From the study of OP AMP circuits in Chapter 4, you should recognize Eq. (11–9) as the *s*-domain generalization of the inverting OP AMP circuit gain equation, $K = -R_2/R_1$.

The driving-point impedance at the input to the inverting circuit is

$$Z_{IN}(s) = \frac{V_S(s)}{[V_A(s) - V_B(s)]/Z_1(s)}$$

But $V_A(s) = V_S(s)$ and $V_B(s) = 0$; hence the input impedance is $Z_{IN}(s) = Z_1(s)$.

For the noninverting circuit in Figure 11–5(b), the sum of currents leaving node B is

$$\frac{V_B(s) - V_C(s)}{Z_2(s)} + \frac{V_B(s)}{Z_1(s)} + I_N(s) = 0$$

In the noninverting configuration the ideal OP AMP constraints require that $I_N(s) = 0$ and $V_B(s) = V_1(s)$. By definition, the output voltage $V_2(s)$ equals node voltage $V_C(s)$. Combining all of these considerations and solving for the voltage transfer function yields

$$T_V(s) = \frac{V_2(s)}{V_1(s)} = \frac{Z_1(s) + Z_2(s)}{Z_1(s)} \tag{11-10}$$

Equation (11–10) is the s-domain version of the noninverting amplifier gain equation, $K = (R_1 + R_2)/R_1$. The transfer function of the noninverting configuration is the reciprocal of the transfer function of the voltage divider in the feedback path. The ideal OP AMP draws no current at its input terminals, so theoretically the input impedance of the noninverting circuit is infinite.

The transfer functions of divider circuits and the basic OP AMP configurations are useful analysis and design tools in many practical situations. However, a general method is needed to handle circuits of greater complexity. One general approach is to formulate either node-voltage or mesh-current equations with all initial conditions set to zero. These equations are then solved for network functions using Cramer's rule for hand calculations or symbolic math analysis programs such as Mathcad. The algebra involved can be a bit tedious at times, even with Mathcad. But the tedium is reduced somewhat because we only need the zero-state response for a single input source.

The following examples illustrate methods of calculating network functions.

EXAMPLE 11–2

(a) Find the transfer functions of the circuits in Figure 11–6.
(b) Find the driving-point impedances seen by the input sources in these circuits.

SOLUTION:
(a) These are all divider circuits, so the required transfer functions can be obtained using Eq. (11–7) or (11–8).

For circuit C1: $Z_1 = R$, $Z_2 = 1/Cs$, and $T_V(s) = 1/(RCs + 1)$

For circuit C2: $Z_1 = Ls$, $Z_2 = R$, and $T_V(s) = 1/(GLs + 1)$

For circuit C3: $Y_1 = Cs$, $Y_2 = G$, and $T_I(s) = 1/(RCs + 1)$

These transfer functions are all of the form $1/(\tau s + 1)$, where τ is the circuit time constant.

(b) The driving-point impedances are found by series or parallel equivalence.

For circuit C1: $Z(s) = Z_1 + Z_2 = (RCs + 1)/Cs$

For circuit C2: $Z(s) = Z_1 + Z_2 = Ls + R$

For circuit C3: $Z(s) = 1/(Y_1 + Y_2) = 1/(Cs + G) = R/(RCs + 1)$

The three circuits have different driving-point impedances even though they have the same transfer functions.

C1

C2

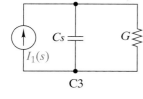

C3

FIGURE   11 – 6

The general principle illustrated here is that several different circuits can have the same transfer function. Put differently, a desired transfer function can be realized by several different circuits. This fact is important in design because circuits that produce the same transfer function offer alternatives that may differ in other features. In this example, they all have different input impedances. ∎

EXAMPLE 11–3

(a) Find the input impedance seen by the voltage source in Figure 11–7.
(b) Find the voltage transfer function $T_V(s) = V_2(s)/V_1(s)$ of the circuit.

SOLUTION:

(a) The circuit is a voltage divider. We first calculate the equivalent impedances of the two legs of the divider. The two elements in parallel combine to produce the series leg impedance $Z_1(s)$ as

$$Z_1(s) = \frac{1}{C_1 s + 1/R_1} = \frac{R_1}{R_1 C_1 s + 1}$$

The two elements in series combine to produce shunt leg impedance $Z_2(s)$:

$$Z_2(s) = R_2 + 1/C_2 s = \frac{R_2 C_2 s + 1}{C_2 s}$$

Using series equivalence, the driving-point impedance seen at the input is

$$Z_{EQ}(s) = Z_1(s) + Z_2(s)$$

$$= \frac{R_1 C_1 R_2 C_2 s^2 + (R_1 C_1 + R_2 C_2 + R_1 C_2)s + 1}{C_2 s (R_1 C_1 s + 1)}$$

(b) Using voltage division, the voltage transfer function is

$$T_V(s) = \frac{Z_2(s)}{Z_{EQ}(s)} = \frac{(R_1 C_1 s + 1)(R_2 C_2 s + 1)}{R_1 C_1 R_2 C_2 s^2 + (R_1 C_1 + R_2 C_2 + R_1 C_2)s + 1} \quad ∎$$

FIGURE   11–7

EXAMPLE 11–4

Find the driving-point impedance seen by the voltage source in Figure 11–8. Find the voltage transfer function $T_V(s) = V_2(s)/V_1(s)$ of the circuit.

FIGURE   11–8

SOLUTION:

The circuit is an inverting OP AMP configuration of the form in Figure 11–5(a). The input impedance of this circuit is

$$Z_1(s) = R_1 + \frac{1}{C_1 s} = \frac{R_1 C_1 s + 1}{C_1 s}$$

The impedance Z_2 in the feedback path is

$$Z_2(s) = \frac{1}{C_2 s + 1/R_2} = \frac{R_2}{R_2 C_2 s + 1}$$

and the voltage transfer function is

$$T_V(s) = -\frac{Z_2(s)}{Z_1(s)} = -\frac{R_2 C_1 s}{(R_1 C_1 s + 1)(R_2 C_2 s + 1)} \qquad \blacksquare$$

EXAMPLE 11–5

For the circuit in Figure 11–9 find the input impedance $Z(s) = V_1(s)/I_1(s)$, the transfer impedance $T_Z(s) = V_2(s)/I_1(s)$, and the voltage transfer function $T_V(s) = V_2(s)/V_1(s)$.

FIGURE 11 – 9

SOLUTION:

The circuit is not a simple voltage divider, so we use mesh-current equations to illustrate the general approach to finding network functions. By inspection, the mesh-current equations for this ladder circuit are

$$\left(R + \frac{1}{Cs}\right) I_A(s) - R I_B(s) = V_1(s)$$

$$-R I_A(s) + \left(2R + \frac{1}{Cs}\right) I_B(s) = 0$$

In terms of the mesh current, the input impedance is $Z(s) = V_1(s)/I_A(s)$. Using Cramer's rule to solve for $I_A(s)$ yields

$$I_A(s) = \frac{\Delta_A}{\Delta} = \frac{\begin{vmatrix} V_1(s) & -R \\ 0 & 2R + \dfrac{1}{Cs} \end{vmatrix}}{\begin{vmatrix} R + \dfrac{1}{Cs} & -R \\ -R & 2R + \dfrac{1}{Cs} \end{vmatrix}} = \frac{Cs(2RCs + 1)}{(RCs)^2 + 3RCs + 1} V_1(s)$$

The input impedance of the circuit is

$$Z(s) = \frac{V_1(s)}{I_A(s)} = \frac{(RCs)^2 + 3RCs + 1}{Cs(2RCs + 1)}$$

In terms of mesh current, the transfer impedance is $T_Z(s) = V_2(s)/I_A(s)$. The mesh-current equations do not yield the output voltage directly. But since $V_2(s) = I_B(s)Z_C(s)$, we can solve the second mesh equation for $I_B(s)$ in terms of as $I_A(s)$ as

$$I_B(s) = \frac{RCs}{2RCs + 1} I_A(s)$$

and obtain the specified transfer impedance as

$$T_Z(s) = \frac{I_B(s)[1/Cs]}{I_A(s)} = \frac{R}{2RCs + 1}$$

To obtain the specified voltage transfer function, we could use Cramer's rule to solve for $I_B(s)$ in terms of $V_1(s)$ and then use the fact that $V_2(s) = I_B(s)Z_C(s)$. But a moment's reflection reveals that

$$T_V(s) = \frac{V_2(s)}{V_1(s)} = \left[\frac{V_2(s)}{I_1(s)}\right]\left[\frac{I_1(s)}{V_1(s)}\right] = T_Z(s) \times \frac{1}{Z(s)}$$

Hence the specified voltage transfer function is

$$T_V(s) = \frac{R}{2RCs + 1} \times \frac{Cs(2RCs + 1)}{(RCs)^2 + 3RCs + 1}$$

$$= \frac{RCs}{(RCs)^2 + 3RCs + 1} \qquad \blacksquare$$

EXAMPLE 11–6

Find the voltage transfer function $T_V(s) = V_2(s)/V_1(s)$ of the circuit in Figure 11–10.

FIGURE 11–10

SOLUTION:
The voltage-controlled voltage source makes this an active *RC* circuit. We use node-voltage equations in this problem because the required output is a voltage. The circuit contains two voltage sources connected at a common node. Selecting this common node as the reference eliminates two unknowns since $V_A(s) = V_1(s)$ and $V_D(s) = \mu V_C(s) = V_2(s)$. The sums of currents leaving nodes B and C are

Node B: $\dfrac{V_B(s) - V_1(s)}{R_1} + \dfrac{V_B(s) - V_C(s)}{R_2} + \dfrac{V_B(s) - \mu V_C(s)}{1/C_1(s)} = 0$

Node C: $\dfrac{V_C(s) - V_B(s)}{R_2} + \dfrac{V_C(s)}{1/C_2 s} = 0$

Multiplying both equations by $R_1 R_2$ and rearranging terms produces

Node B: $(R_1 + R_2 + R_1 R_2 C_1 s) V_B(s) - (R_1 + \mu R_1 R_2 C_1 s) V_C(s) = R_2 V_1(s)$

Node C: $-V_B(s) + (1 + R_2 C_2 s) V_C(s) = 0$

Using the node C equation to eliminate $V_B(s)$ from the node B equation leaves

$$(R_1 + R_2 + R_1 R_2 C_1 s)(1 + R_2 C_2 s) V_C(s)$$
$$- (R_1 + \mu R_1 R_2 C_1 s) V_C(s) = R_2 V_1(s)$$

Since the output $V_2(s) = \mu V_C(s)$, the required transfer function is

$$T_V(s) = \frac{V_2(s)}{V_1(s)} = \frac{\mu}{R_1 R_2 C_1 C_2 s^2 + (R_1 C_1 + R_1 C_2 + R_2 C_2 - \mu R_1 C_1)s + 1}$$

This circuit is a member of the Sallen-Key family often used in filter design with $R_1 = R_2 = R$ and $C_1 = C_2 = C$, in which case the transfer function reduces to

$$T_V(s) = \frac{\mu}{(RCs)^2 + (3 - \mu)RCs + 1}$$

We will encounter this result again in later chapters. ∎

Exercise 11–1

Find the driving-point impedance seen by each of the voltage sources in Figure 11–11.

Answers:

Circuit C1: $Z(s) = \dfrac{RLCs^2 + Ls + R}{LCs^2 + 1}$

Circuit C2: $Z(s) = \dfrac{RCs + 1}{Cs}$

C1

C2

Exercise 11–2

Find the voltage transfer functions $T_V(s) = V_2(s)/V_1(s)$ of each of the circuits in Figure 11–11.

Answers:

Circuit C1: $T_V(s) = \dfrac{LCs^2 + 1}{LCs^2 + GLs + 1}$

Circuit C2: $T_V(s) = \dfrac{(RCs)^2}{(RCs + 1)^2}$

FIGURE 11-11

THE CASCADE CONNECTION AND THE CHAIN RULE

Signal-processing circuits often involve a **cascade connection** in which the output voltage of one circuit serves as the input to the next stage. In some cases, the overall voltage transfer function of the cascade can be related to the transfer functions of the individual stages by a **chain rule**

$$T_V(s) = T_{V1}(s)T_{V2}(s) \cdots T_{Vk}(s) \tag{11–11}$$

where T_{V1}, T_{V2}, . . . and T_{Vk} are the voltage transfer functions of the individual stages when operated separately. It is important to understand when the chain rule applies since it greatly simplifies the analysis and design of cascade circuits.

The chain rule in Eq. (11–11) applies if connecting the stages in cascade does not load (change) the output of any stage in the cascade. At any given stage, interface loading does not occur if (1) the Thévenin impedance of the source stage is zero or (2) the input impedance of the load stage is infinite. As a practical matter, however, the chain rule yields acceptable results when the Thévenin impedance of the source is simply much smaller than the input impedance of the load.

To illustrate the effect of loading, consider the two RC circuits or stages in Figure 11–12. When disconnected and operated separately, the transfer functions of each stage are easily found using voltage division as follows:

$$T_{V1}(s) = \frac{R}{R + 1/Cs} = \frac{RCs}{RCs + 1}$$

$$T_{V2}(s) = \frac{1/Cs}{R + 1/Cs} = \frac{1}{RCs + 1}$$

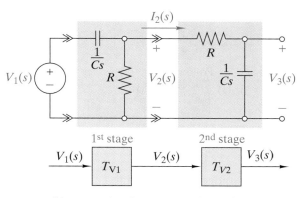

FIGURE 11–12 *Two-port circuits connected in cascade.*

When connected in cascade, the output of the first stage serves as the input to the second stage. If the chain rule applies, we would obtain the overall transfer function as

$$T_V(s) = \frac{V_3(s)}{V_1(s)} = \left(\frac{V_2(s)}{V_1(s)}\right)\left(\frac{V_3(s)}{V_2(s)}\right) = (T_{V1}(s))(T_{V2}(s)) \tag{11–12}$$

$$= \underbrace{\left(\frac{RCs}{RCs + 1}\right)}_{\substack{\text{First} \\ \text{Stage}}} \underbrace{\left(\frac{1}{RCs + 1}\right)}_{\substack{\text{Second} \\ \text{Stage}}} = \underbrace{\frac{RCs}{(RCs)^2 + 2RCs + 1}}_{\text{Overall}}$$

However, in Example 11–5, the overall transfer function of this circuit was found to be

$$T_V(s) = \frac{RCs}{(RCs)^2 + 3RCs + 1} \qquad (11–13)$$

which disagrees with the chain rule result in Eq. (11–12).

The reason for the discrepancy is that when they are connected in cascade, the second circuit "loads" the first circuit. That is, the voltage-divider rule requires that the interface current $I_2(s)$ in Figure 11–12 be zero. The no-load condition $I_2(s) = 0$ applies when the stages operate separately, but when connected in cascade, the interface current is not zero. The chain rule does not apply here because loading caused by the second stage changes the transfer function of the first stage.

However, Figure 11–13 shows how the loading problem goes away when an OP AMP voltage follower is inserted between the RC circuit stages. The follower does not draw any current from the first RC circuit [$I_2(s) = 0$] and applies $V_2(s)$ directly across the input of the second RC circuit. With this modification the chain rule in Eq. (11–11) applies because the voltage follower isolates the two circuits, thereby solving the loading problem.

F I G U R E   1 1 – 1 3 *Cascade   connection with voltage follower isolation.*

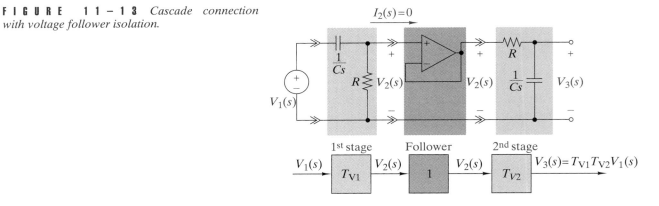

Thus, loading can be avoided by connecting an OP AMP voltage follower between stages, as in Figure 11–13. More important, loading does not occur if the output of the driving stage is the output of an OP AMP or controlled source. These elements act like ideal voltage sources whose outputs are unchanged by connecting the subsequent stage.

For example, the two circuits in Figure 11–14 contain the same two stages. The chain rule applies to the configuration in Figure 11–14(a) because the output of the first stage is an OP AMP. In Figure 11–14(b) the two stages are interchanged so that the output of the voltage divider drives the input to OP AMP stage. The chain rule does not apply to this configuration because the input impedance of OP AMP stage loads the output of the voltage divider stage.

FIGURE 11 – 14

(a)

(b)

EXAMPLE 11–7

Find the voltage transfer function of the circuit in Figure 11–14(a).

SOLUTION:

The chain rule applies to this circuit since the output of the first stage is an OP AMP. The transfer function of this inverting OP AMP stage is

$$T_{V1}(s) = -\frac{Z_2(s)}{Z_1(s)} = -\frac{R_2}{R_1 + 1/C_1s} = -\frac{R_2C_1s}{R_1C_1s + 1}$$

The second stage is a voltage divider whose transfer function is

$$T_{V2}(s) = \frac{Z_2(s)}{Z_2(s) + Z_1(s)} = \frac{R_3}{R_3 + 1/C_2s} = \frac{R_3C_2s}{R_3C_2s + 1}$$

and the chain rule yields the overall transfer function as

$$T_V(s) = T_{V1}(s) \times T_{V2}(s) = -\frac{R_2C_1R_3C_2s^2}{(R_1C_1s + 1)(R_3C_2s + 1)} \qquad \blacksquare$$

Exercise 11–3

Find the voltage transfer function of the circuit in Figure 11–14(b).

Answer:

$$T_V(s) = -\frac{R_2C_1R_3C_2s^2}{R_1C_1R_3C_2s^2 + (R_1C_1 + R_3C_2 + R_3C_1)s + 1} \neq T_{V2}(s) \times T_{V1}(s)$$

Note that the overall transfer function is not the product of the individual stage transfer functions because second stage loads the first stage.

Exercise 11–4

The transfer functions of the circuits in Figure 11–11 are

$$\text{Circuit C1: } T_V(s) = \frac{LCs^2 + 1}{LCs^2 + GLs + 1}$$

$$\text{Circuit C2: } T_V(s) = \frac{(RCs)^2}{(RCs + 1)^2}$$

Find the overall transfer function of a cascade connection in which circuit C2 is the first stage and circuit C1 is the second stage.

Answer:

$$T_V(s) = \frac{(RCs)^2(LCs^2 + 1)}{(RCs + 1)^2 (LCs^2 + GLs + 1)}$$

11–3 NETWORK FUNCTIONS AND IMPULSE RESPONSE

The **impulse response** is the zero-state response of a circuit when the driving force is a unit impulse applied at $t = 0$. When the input signal is $x(t) = \delta(t)$ then $X(s) = \mathcal{L}\{\delta(t)\} = 1$ and the input-output relationship in Eq. (11–2) reduces to

$$Y(s) = T(s) \times 1 = T(s)$$

The impulse response transform equals the network function, and we could treat $T(s)$ as if it is a signal transform. However, to avoid possible confusion between a network function (description of a circuit) and a transform (description of a signal), we denote the impulse response transform as $H(s)$ and use $h(t)$ to denote the corresponding waveform.[3] That is,

Impulse Response

Transform	Waveform	
$H(s) = T(s) \times 1$	$h(t) = \mathcal{L}^{-1}\{H(s)\}$	(11–14)

When there are no repeated poles, the partial fraction expansion of $H(s)$ is

$$H(s) = \underbrace{\frac{k_1}{s - p_1} + \frac{k_2}{s - p_2} + \cdots + \frac{k_N}{s - p_N}}_{\text{Natural Poles}}$$

where $p_1, p_2, \ldots p_N$ are the natural poles in the denominator of the transfer function $T(s)$. All of the poles of $H(s)$ are natural poles since the impulse excitation does not introduce any forced poles. The inverse transform gives the impulse response waveform as

$$h(t) = \underbrace{\left[k_1 e^{p_1 t} + k_2 e^{p_2 t} + \cdots + k_N e^{p_N t}\right]}_{\text{Natural Response}} u(t)$$

---

3 Not all books make this distinction. Books on signals and circuits often use $H(s)$ to represent both a transfer function and the impulse response transform.

When the circuit is stable, all of the natural poles are in the left half plane and the impulse response waveform $h(t)$ decays to zero as $t \to \infty$. A linear circuit whose impulse response ultimately returns to zero is said to be **asymptotically stable**. Asymptotic stability means that the impulse response has a finite time duration. That is, for every $\varepsilon > 0$ there exists a finite time duration T_D such that $|h(t)| < \varepsilon$ for all $t > T_D$.

It is important to note that the impulse response $h(t)$ contains all the information needed to determine the circuit response to any other input. That is, since $\mathcal{L}\{h(t)\} = T(s)$, we can calculate the output $y(t)$ for any Laplace transformable input $x(t)$ as

$$y(t) = \mathcal{L}^{-1}\{H(s)X(s)\}$$

This expression, known as the *convolution theorem*, states that the impulse response can be used to relate the input and output of a linear circuit. Thus, the impulse response $h(t)$ or $H(s)$ can be considered as a mathematical model of a linear circuit. Obviously it is important to be able to find the impulse response and to know how to use the impulse response to find the output for other inputs. The following examples illustrate both of these issues.

EXAMPLE 11–8

Find the response $v_2(t)$ in Figure 11–15 when the input is $v_1(t) = \delta(t)$. Use the element values $R_1 = 10$ kΩ, $R_2 = 12.5$ kΩ, $C_1 = 1$ μF, and $C_2 = 2$ μF.

FIGURE 11–15

SOLUTION:

In Example 11–3, the transfer function of this circuit was found to be

$$T_V(s) = \frac{V_2(s)}{V_1(s)} = \frac{(R_1C_1s + 1)(R_2C_2s + 1)}{R_1C_1R_2C_2s^2 + (R_1C_1 + R_2C_2 + R_1C_2)s + 1}$$

For the given element values, the impulse response transform is

$$H(s) = \frac{(s + 100)\,(s + 40)}{s^2 + 220s + 4000} = \frac{(s + 100)\,(s + 40)}{(s + 20)\,(s + 200)}$$

This $H(s)$ is not a proper rational function, so we use one step of a long division plus a partial-fraction expansion to obtain

$$H(s) = 1 + \frac{80/9}{s + 20} - \frac{800/9}{s + 200}$$

and the impulse response is

$$h(t) = \delta(t) + \frac{80}{9}[e^{-20t} - 10e^{-200t}]u(t)$$

In this case, the impulse response contains an impulse because the network function is not a proper rational function.   ∎

EXAMPLE 11–9

The impulse response of a linear circuit is $h(t) = 200e^{-100t}u(t)$. Find the output when the input is a unit ramp $r(t) = tu(t)$.

SOLUTION:
The circuit impulse response is

$$H(s) = \mathcal{L}\{h(t)\} = \mathcal{L}\{200e^{-100t}u(t)\} = \frac{200}{s + 100}$$

The Laplace transform of the unit ramp input is $1/s^2$, hence using the convolution theorem the response due to a ramp input is

$$
\begin{aligned}
y(t) &= \mathcal{L}^{-1}\left\{H(s)\frac{1}{s^2}\right\} = \mathcal{L}^{-1}\left\{\frac{200}{(s + 100)s^2}\right\} \\
&= \mathcal{L}^{-1}\left\{\frac{1/50}{s + 100} + \frac{2}{s^2} - \frac{1/50}{s}\right\} \\
&= \frac{1}{50}\,(e^{-100t} + 100t - 1)u(t)
\end{aligned}
$$

This example illustrates that the impulse response $h(t)$ contains all the information needed to calculate the response due to another input. ■

Exercise 11–5

Find the impulse response of the circuit in Figure 11–16.

Answer: $h(t) = 0.1\delta(t) + [90e^{-100t}]u(t)$

$v_1(t)$ 9 kΩ   1 kΩ   1 μF $v_2(t)$

FIGURE 11–16

11–4 NETWORK FUNCTIONS AND STEP RESPONSE

The **step response** is the zero-state response of the circuit output when the driving force is a unit step function applied at $t = 0$. When the input is $x(t) = u(t)$, then $X(s) = \mathcal{L}\{u(t)\} = 1/s$ and the s-domain input-output relationship in Eq. (11–2) yields $Y(s) = T(s)/s$. The step response transform and waveform will be denoted by $G(s)$ and $g(t)$, respectively. That is,

Step Response

| Transform | Waveform |

$$G(s) = \frac{T(s)}{s} \qquad g(t) = \mathcal{L}^{-1}\{G(s)\} \qquad (11\text{–}15)$$

The poles of $G(s)$ are the natural poles contributed by the network function $T(s)$ and a forced pole at $s = 0$ introduced by the step function input. The partial fraction expansion of $G(s)$ takes the form

$$G(s) = \underbrace{\frac{k_0}{s}}_{\substack{\text{Forced}\\\text{Pole}}} + \underbrace{\frac{k_1}{s - p_1} + \frac{k_2}{s - p_2} + \cdots + \frac{k_N}{s - p_N}}_{\substack{\text{Natural}\\\text{Poles}}}$$

where $p_1, p_2, \ldots p_N$ are the natural poles in $T(s)$. The inverse transformation gives the step response waveform as

$$g(t) = \underbrace{k_0 u(t)}_{\substack{\text{Forced} \\ \text{Response}}} + \underbrace{[k_1 e^{p_1 t} + k_2 e^{p_2 t} + \cdots + k_N e^{p_N t}] u(t)}_{\substack{\text{Natural} \\ \text{Response}}}$$

When the circuit is stable, the natural response decays to zero, leaving a forced component called the **dc steady-state response**. The amplitude of the steady-state response is the residue in the partial-fraction expansion of the forced pole at $s = 0$. By the cover-up method, this residue is

$$k_0 = sG(s)\big|_{s=0} = T(0)$$

For a unit step input the amplitude of the dc steady-state response equals the value of the transfer function at $s = 0$. By linearity, the general principle is that an input $Au(t)$ produces a dc steady-state output whose amplitude is $AT(0)$.

We next show the relationship between the impulse and step responses. First, combining Eqs. (11–14) and (11–15) gives

$$G(s) = \frac{H(s)}{s}$$

The step response transform is the impulse response transform divided by s. The integration property of the Laplace transform tells us that division by s in the s domain corresponds to integration in the time domain. Therefore, in the time domain, we can relate the impulse and step response waveforms by integration:

$$g(t) = \int_0^t h(\tau)d\tau \qquad (11\text{–}16)$$

Using the fundamental theorem of calculus, the impulse response waveform is expressed in terms of the step response waveform

$$h(t)(=)\frac{dg(t)}{dt} \qquad (11\text{–}17)$$

where the symbol $(=)$ means equal almost everywhere, a condition that excludes those points at which $g(t)$ has a discontinuity. In the time domain, the step response waveform is the integral of the impulse response waveform. Conversely, the impulse response waveform is (almost everywhere) the derivative of the step response waveform.

The key idea is that there are relationships between the network function $T(s)$ and the responses $H(s)$, $h(t)$, $G(s)$, and $g(t)$. If any one of these quantities is known, we can obtain any of the other four using relatively simple mathematical operations.

EXAMPLE 11–10

The element values for the circuit in Figure 11–17 are $R_1 = 10$ kΩ, $R_2 = 100$ kΩ, $C_1 = C_2 = 100$ nF, and $v_1(t) = u(t)$ V. Find the response $v_2(t)$.

SOLUTION:

For a unit step input $V_1(s) = 1/s$. In Example 11–4 the transfer function of this circuit is shown to be

$$T_V(s) = \frac{V_2(s)}{V_1(s)} = -\frac{R_2 C_1 s}{(R_1 C_1 s + 1)(R_2 C_2 s + 1)}$$

Substituting the numerical element values into this expression yields the step response transform as

$$V_2(s) = \frac{T_V(s)}{s} = -\frac{1000}{(s + 100)(s + 1000)}$$

$$= \frac{-10/9}{s + 100} + \frac{10/9}{s + 1000}$$

The inverse transform yields

$$v_2(t) = \frac{10}{9}(-e^{-100t} + e^{-1000t})u(t)$$

In this case the forced pole at $s = 0$ is cancelled by a zero of $T_V(s)$ and the final value of the step response is zero. Using the final value theorem, we have

$$g(\infty) = \underset{t \to \infty}{\text{Limit }} g(t) = \underset{s \to 0}{\text{Limit }} sG(s) = T(0)$$

the final steady-state value of the step response is equal to the value of the transfer function evaluated at $s = 0$. In the present case $T_V(0) = 0$, so the dc steady-state output of the circuit is zero. Recall from Chapter 6 that in the dc steady-state capacitors can be replaced by open circuits. Replacing C_1 in Figure 11–17 by an open circuit disconnects the input source from the OP AMP so no dc signals can be transferred through the circuit. A series capacitor that prevents the passage of dc signals is commonly called a *blocking capacitor*. ∎

EXAMPLE 11–11

The impulse response of a linear circuit is

$$h(t) = 5000 \, (e^{-1000t} \sin 2000 \, t)u(t)$$

Find the step response.

SOLUTION:
Using Eq. (11–16) we have

$$g(t) = \int_0^t h(\tau)d\tau = 5000\int_0^t e^{-1000\tau} \sin 2000\,\tau\,d\tau$$

$$= (-e^{-1000\tau} \sin 2000\,\tau - 2e^{-1000\tau} \cos 2000\,\tau)\Big|_0^t$$

$$= (2 - e^{-1000t} \sin 2000\,t - 2e^{-1000t} \cos 2000\,t)u(t)$$

We could have found $g(t)$ by first finding the Laplace transform of the impulse response $H(s)$ and then calculating the step response as $g(t) = \mathcal{L}^{-1}\{H(s)/s\}$. However, we choose to use Eq. (11–16), which means that the step response calculation took place entirely in the time domain. This is a simple example of a general time-domain method based on the *convolution integral,* a procedure that you will study in later courses in signal processing. ∎

APPLICATION NOTE: EXAMPLE 11–12

Three time-domain parameters often used to describe the step response are rise time, delay time, and overshoots. **Rise time (T_R)** is the time interval required for the step response to rise from 10% to 90% of its steady-state value $g(\infty)$. **Delay time (T_D)** is the time interval required for the step response to reach 50% of its steady-state value. **Overshoot** is the difference between the peak value of the step response and its steady-state value. Overshoot is usually expressed as a percentage of the steady-state value, namely

$$\text{Overshoot} = \frac{g_{\text{max}} - g(\infty)}{g(\infty)} \times 100$$

Figure 11–18 illustrates these descriptors for a typical step response.

F I G U R E   1 1 – 1 8 *Step response showing rise time (T_R), delay time (T_D), and overshoot.*

Step response descriptors are used to specify the performance of both analog and digital systems. Rise time governs how rapidly the system responds to an abrupt change in the input. Delay time controls the time between the application of an abrupt change and the appearance of a significant change in the output. Overshoot indicates the amount of damping present in the system. Lightly damped oscillations produce large overshoots and may cause erroneous state changes in digital systems.

Rise time, delay time, and overshoot can be determined experimentally or calculated using modern computer tools. For example, the MATLAB function step calculates and plots the step response of a transfer function of the form $T(s) = n(s)/d(s)$, where $n(s)$ is the numerator polynomial and $d(s)$ is the denominator polynomial. In the MATLAB command window these polynomials are defined by the row matrices num and den that contain the coefficients of the polynomials $n(s)$ and $d(s)$ in descending powers of s.

For example, suppose we need to evaluate the step response descriptors of a circuit whose transfer function is

$$T(s) = \frac{400(s + 100)}{(s + 100)^2 + 100^2}$$

The two coefficient matrices for this transfer function are entered in the MATLAB command window as

```
num=[400,400*100];
den=[1,2*100,100^2+100^2];
```

The first element in either matrix is the coefficient of the highest power of s, which is followed by the other coefficients in descending powers of s. The MATLAB statement

```
step(num,den)
```

produces the step response plot shown in Figure 11–19.

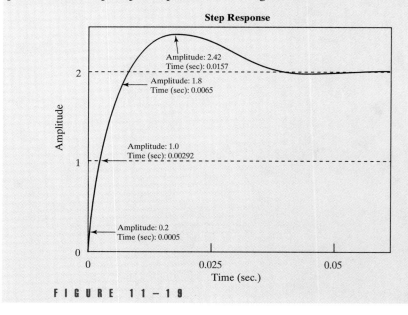

F I G U R E   1 1 – 1 9

First note that the steady-state response is $g(\infty) = 2$. Using the cursor in the MATLAB graphics window, we can find the points at which certain amplitudes are reached. The key points shown in the figure are:

(1) The 10% rise point (amplitude = 0.2 and time = 0.0005).
(2) The 50% rise point (amplitude = 1 and time = 0.00292).
(3) The 90% rise point (amplitude = 1.8 and time = 0.0065).
(4) The maximum point (amplitude = 2.42 and time = 0.0157).

Hence the step response descriptors for the circuit are:

Rise time: $T_R = 0.0065 - 0.005 = 6$ ms

Delay time: $T_D = 0.00292 = 2.92$ ms

Overshoot $= (2.42 - 2)/2 = 21\%$

Exercise 11–6

The impulse response of a circuit is $h(t) = 0.1\delta(t) + 90e^{-100t}u(t)$. Find the step response.

Answer: $g(t) = [1 - 0.9e^{-1000t}]u(t)$

Exercise 11–7

The step response of a linear circuit is $g(t) = 5[e^{-1000t}\sin(2000t)]u(t)$. Find the circuit transfer function $T(s)$.

Answer:

$$T(s) = \frac{10^4 s}{s^2 + 2000s + 5 \times 10^6}$$

11–5 NETWORK FUNCTIONS AND SINUSOIDAL STEADY-STATE RESPONSE

When a stable, linear circuit is driven by a sinusoidal input, the output contains a steady-state component that is a sinusoid of the same frequency as the input. This section deals with using the circuit transfer function to find the amplitude and phase angle of the sinusoidal steady-state response. To begin, we write a general sinusoidal input in the form

$$x(t) = X_A \cos(\omega t + \phi) \tag{11–18}$$

which can be expanded as

$$x(t) = X_A(\cos \omega t \cos \phi - \sin \omega t \sin \phi)$$

The waveforms $\cos \omega t$ and $\sin \omega t$ are basic signals whose transforms are given in Table 9–2 as $\mathcal{L}\{\cos \omega t\} = s/(s^2 + \omega^2)$ and $\mathcal{L}\{\sin \omega t\} = \omega/(s^2 + \omega^2)$. Therefore, the input transform is

$$X(s) = X_A\left[\frac{s}{s^2 + \omega^2}\cos\phi - \frac{\omega}{s^2 + \omega^2}\sin\phi\right]$$

$$= X_A\left[\frac{s\cos\phi - \omega\sin\phi}{s^2 + \omega^2}\right] \tag{11-19}$$

Equation (11–19) is the Laplace transform of the general sinusoidal waveform in Eq. (11–18).

Using Eq. (11–2), we obtain the response transform for a general sinusoidal input:

$$Y(s) = X_A\left[\frac{s\cos\phi - \omega\sin\phi}{(s - j\omega)(s + j\omega)}\right]T(s) \tag{11-20}$$

The response transform contains forced poles at $s = \pm j\omega$ because the input is a sinusoid. Expanding Eq. (11–20) by partial fractions,

$$Y(s) = \underbrace{\frac{k}{s - j\omega} + \frac{k^*}{s + j\omega}}_{\text{Forced Poles}} + \underbrace{\frac{k_1}{s - p_1} + \frac{k_2}{s - p_2} + \cdots + \frac{k_N}{s - p_N}}_{\text{Natural Poles}}$$

where $p_1, p_2, \ldots p_N$ are the natural poles contributed by the transfer function $T(s)$. To obtain the response waveform, we perform the inverse transformation:

$$y(t) = \underbrace{ke^{j\omega t} + k^*e^{-j\omega t}}_{\text{Forced Response}} + \underbrace{k_1e^{p_1 t} + k_2e^{p_2 t} + \cdots + k_Ne^{p_N t}}_{\text{Natural Response}}$$

When the circuit is stable, the natural response decays to zero, leaving a sinusoidal steady-state response due to the forced poles as $s = \pm j\omega$. The steady-state response is

$$y_{SS}(t) = ke^{j\omega t} + k^*e^{-j\omega t}$$

where the subscript SS identifies a steady-state condition.

To determine the amplitude and phase of the steady-state response, we must find the residue k. Using the cover-up method from Chapter 9, we find k to be

$$k = (s - j\omega)X_A\left[\frac{s\cos\phi - \omega\sin\phi}{(s - j\omega)(s + j\omega)}\right]T(s)\Big|_{s = j\omega}$$

$$= X_A\left[\frac{j\omega\cos\phi - \omega\sin\phi}{2j\omega}\right]T(j\omega)$$

$$= X_A\left[\frac{\cos\phi + j\sin\phi}{2}\right]T(j\omega) = \frac{1}{2}X_Ae^{j\phi}T(j\omega)$$

The complex quantity $T(j\omega)$ can be written in magnitude and angle form as $|T(j\omega)|e^{j\theta}$. Using these results, the residue becomes

$$k = \left[\frac{1}{2}X_Ae^{j\phi}\right]\left|T(j\omega)\right|e^{j\theta}$$

$$= \frac{1}{2}X_A\left|T(j\omega)\right|e^{j(\phi + \theta)}$$

The inverse transform yields the steady-state response in the form

$$y_{SS}(t) = 2|k| \cos(\omega t + \phi + \angle k)$$

$$= \underbrace{X_A |T(j\omega)|}_{\text{Amplitude}} \cos(\omega t + \underbrace{\phi + \theta}_{\text{Phase}}) \qquad (11\text{–}21)$$

In the development leading to Eq. (11–21), we treat frequency as a general variable where the symbol ω represents all possible input frequencies. In some cases the input frequency has a specific value, which we denote as ω_A. In this case the input is written as

$$x(t) = X_A \cos(\omega_A t + \phi)$$

To obtain the steady-state output, we evaluate the transfer function at the specific frequency (ω_A) of the input sinusoid, namely

$$T(j\omega_A) = |T(j\omega_A)| \angle T(j\omega_A)$$

and the steady-state output is expressed as

$$y_{SS}(t) = X_A |T(j\omega_A)| \cos[\omega_A t + \phi + \angle T(j\omega_A)]$$

This result emphasizes three things about the steady-state response:

(1) Output Frequency = Input Frequency = ω_A

(2) Output Amplitude = Input Amplitude $\times |T(j\omega_A)|$ (11–22)

(3) Output Phase = Input Phase + $\angle T(j\omega_A)$

The next two examples illustrate sinusoidal steady-state response calculations. In the first example, frequency is treated as a general variable and we examine the steady-state response as frequency varies over a wide range. In the second example, we evaluate the steady-state response at two specific frequencies.

EXAMPLE 11–13

Find the steady-state output in Figure 11–20 for a general input $v_1(t) = V_A \cos(\omega t + \phi)$.

FIGURE 11–20

SOLUTION:
In Example 11–2, the circuit transfer function is shown to be

$$T(s) = \frac{R}{Ls + R}$$

The magnitude and angle of $T(j\omega)$ are

$$|T(j\omega)| = \frac{R}{\sqrt{R^2 + (\omega L)^2}}$$

$$\theta(\omega) = -\tan^{-1}\left(\frac{\omega L}{R}\right)$$

Using Eq. (11–21), the sinusoidal steady-state output is

$$v_{2SS}(t) = \frac{V_A R}{\sqrt{R^2 + (\omega L)^2}} \cos[\omega t + \phi - \tan^{-1}(\omega L/R)]$$

Note that both the amplitude and phase angle of the steady-state response depend on the frequency of the input sinusoid. In particular, at $\omega = 0$ the amplitude of the steady-state output reduces to V_A, which is the same as the amplitude of the input sinusoid. This makes sense because at dc the inductor in Figure 11–20 acts like a short circuit that directly connects the output port to the input port. At very high frequency the amplitude of the steady-state output approaches zero. This also makes sense because at very high frequency the inductor acts like an open circuit that disconnects the output port from the input port. In between these two extremes the output amplitude decreases as the frequency increases. ■

EXAMPLE 11–14

The impulse response of a linear circuit is $h(t) = 5000[2e^{-1000t} \cos 2000t - e^{-1000t}\sin 2000t]u(t)$.
(a) Find the sinusoidal steady-state response when $x(t) = 5 \cos 1000t$.
(b) Repeat part (a) when $x(t) = 5 \cos 3000t$.

SOLUTION:

The transfer function corresponding to $h(t)$ is

$$T(s) = H(s) = \mathcal{L}\{h(t)\}$$

$$= 5000\left[\frac{2(s + 1000)}{(s + 1000)^2 + (2000)^2} - \frac{2000}{(s + 1000)^2 + (2000)^2}\right]$$

$$= \frac{10^4 s}{s^2 + 2000s + 5 \times 10^6}$$

(a) At $\omega_A = 1000$ rad/s, the value of $T(j\omega_A)$ is

$$T(j1000) = \frac{10^4(j1000)}{(j1000)^2 + 2000 (j1000) + 5 \times 10^6}$$

$$= \frac{j10^7}{(5 \times 10^6 - 10^6) + j2 \times 10^6} = \frac{j10}{4 + j2}$$

$$= \frac{10\ e^{j90°}}{\sqrt{20}e^{j26.6°}} = 2.24\ e^{j63.4°}$$

and the steady-state response for $x(t) = 5 \cos 1000t$ is

$$y_{SS}(t) = 5 \times 2.24 \cos(1000t + 0° + 63.4°)$$

$$= 11.2 \cos(1000t + 63.4°)$$

(b) At $\omega_A = 3000$ rad/s, the value of $T(j\omega_A)$ is

$$T(j3000) = \frac{10^4(j3000)}{(j3000)^2 + 2000\,(j3000) + 5 \times 10^6}$$

$$= \frac{j3 \times 10^7}{5 \times 10^6 - 9 \times 10^6 + j6 \times 10^6}$$

$$= \frac{j30}{-4 + j6} = \frac{30\,e^{j90°}}{\sqrt{52}e^{j123.7°}} = 4.16\,e^{-j33.7°}$$

and the steady-state response for $x(t) = 5 \cos 3000t$ is

$$y_{SS}(t) = 5 \times 4.16 \cos(3000t + 0° - 33.7°)$$

$$= 20.8 \cos(3000t - 33.7°)$$

Again note that the amplitude and phase angle of the steady-state response depend on the input frequency. ■

Exercise 11–8

The transfer function of a linear circuit is $T(s) = 5(s + 100)/(s + 500)$. Find the steady-state output for
(a) $x(t) = 3 \cos 100t$
(b) $x(t) = 2 \sin 500t$.

Answers:
(a) $y_{SS}(t) = 4.16 \cos(100t + 33.7°)$
(b) $y_{SS}(t) = 7.21 \cos(500t - 56.3°)$

Exercise 11–9

The impulse response of a linear circuit is $h(t) = \delta(t) - 100[e^{-100t}]u(t)$. Find the steady-state output for
(a) $x(t) = 25 \cos 100t$
(b) $x(t) = 50 \sin 100t$.

Answers:
(a) $y_{SS}(t) = 17.7 \cos(100t + 45°)$
(b) $y_{SS}(t) = 35.4 \cos(100t - 45°)$

NETWORK FUNCTIONS AND PHASOR CIRCUIT ANALYSIS

We have two methods of calculating the steady-state response $y_{SS}(t)$ when the input is a sinusoid of the form

$$x(t) = X_A \cos(\omega_A t + \phi)$$

The first method requires that we convert the input sinusoid into a phasor, calculate the impedance of the circuit elements at the frequency of the input, and then use the phasor circuit analysis methods developed in Chapter 8 to find the phasor response. The phasor response must then be converted into a waveform to obtain $y_{SS}(t)$. The method developed in this section requires that we find the network function $T(s)$ relating input and output, evaluate $T(j\omega_A)$ (where ω_A is the frequency of the input sinusoid), and then use Eq. (11–21) to find $y_{SS}(t)$. Under what conditions is one method preferable to the other?

Phasor circuit analysis works best when the circuit is driven at a single frequency and we need to find several voltages and currents, or perhaps average and maximum available power. The network function method works best when the circuit is driven at several frequencies and we only need to find a single response called the output. The network function is also the only method available when all we know about the circuit is its impulse or step response waveform. Thus, the preferred method depends on how the circuit is driven (single or multiple frequency) and what we need to find out (single or multiple response).

11–6 NETWORK FUNCTION DESIGN

Finding and using a network function of a given circuit is an *s*-domain **analysis** problem. An *s*-domain **synthesis** problem involves finding a circuit that realizes a given network function. For linear circuits an analysis problem always has a unique solution. In contrast, a synthesis problem may have many solutions because different circuits can have the same network function. A transfer function design problem involves synthesizing several circuits that realize a given function and evaluating the alternative designs, using criteria such as input impedance, cost, and power consumption.

In this section we develop methods of synthesizing first-order and second-order transfer functions using the circuits shown in Figure 11–21. In an analysis problem we start with a circuit diagram and transform the circuit into the *s* domain and the transfer function using the appropriate relationship in Figure 11–21. The analysis process proceeds from a circuit diagram to a network function and always has a unique solution.

The synthesis process proceeds in the opposite direction. We start with a transfer function, select one of the configurations in Figure 11–21, assign values to the impedances $Z_1(s)$ and $Z_2(s)$, and then draw the circuit diagram as the final step. In some cases it may not be possible to realize the given transfer function using the selected circuit configuration. In other cases, restrictions on the transfer function may be needed to ensure that $Z_1(s)$ and $Z_2(s)$ are physically realizable. In any event, the synthesis process proceeds from a network function to a circuit diagram and may have many solutions or possibly no solutions.

FIRST-ORDER VOLTAGE-DIVIDER CIRCUIT DESIGN

We begin our study of transfer function design by developing a voltage-divider realization of a first-order transfer function of the form $K/(s + \alpha)$. The impedances $Z_1(s)$ and $Z_2(s)$ are related to the given transfer function using the voltage-divider relationship in Figure 11–21:

$$T_V(s) = \frac{K}{s + \alpha} = \frac{Z_2(s)}{Z_1(s) + Z_2(s)} \tag{11–23}$$

To obtain a circuit realization, we must assign part of the given $T_V(s)$ to $Z_2(s)$ and the remainder to $Z_1(s)$. There are many possible realizations of $Z_1(s)$ and $Z_2(s)$ because there is no unique way to make this assignment.

Circuit	Transfer Function
Voltage divider	$T_v(s) = \dfrac{Z_2(s)}{Z_1(s)+Z_2(s)}$
Inverting OP AMP	$T_v(s) = -\dfrac{Z_2(s)}{Z_1(s)}$
Noninverting OP AMP	$T_v(s) = \dfrac{Z_1(s)+Z_2(s)}{Z_1(s)}$

F I G U R E   1 1 – 2 1 *Circuit design building blocks.*

(a) *RL* design

(b) *RC* design

F I G U R E   1 1 – 2 2 *Circuit realizations of* T(s) = K/(s + α) *for* K ≤ α.

For example, simply equating the numerators and denominators in Eq. (11–23) yields

$$Z_2(s) = K \quad \text{and} \quad Z_1(s) = s + \alpha - Z_2(s) = s + \alpha - K \qquad (11\text{–}24)$$

Inspecting this result, we see that $Z_2(s)$ is realizable as a resistance ($R_2 = K$ Ω) and $Z_1(s)$ as an inductance ($L_1 = 1$ H) in series with a resistance [$R_1 = (\alpha - K)$ Ω]. The resulting circuit diagram is shown in Figure 11–22(a). For $K = \alpha$ the resistance R_1 can be replaced by a short circuit because its resistance is zero. A gain restriction $K \le \alpha$ is necessary because a negative R_1 is not physically realizable as a single component.

An alternative synthesis approach involves factoring s out of the denominator of the given transfer function. In this case Eq. (11–23) is rewritten in the form

$$T_V(s) = \frac{K/s}{1 + \alpha/s} = \frac{Z_2(s)}{Z_1(s) + Z_2(s)} \qquad (11\text{–}25)$$

Equating numerators and denominators yields the branch impedances

$$Z_2(s) = \frac{K}{s} \quad \text{and} \quad Z_1(s) = 1 + \frac{\alpha}{s} - Z_2(s) = 1 + \frac{\alpha - K}{s}$$

$$(11-26)$$

In this case we see that $Z_2(s)$ is realizable as a capacitance ($C_2 = 1/K$ F) and $Z_1(s)$ as a resistance ($R_1 = 1\ \Omega$) in series with a capacitance [$C_1 = 1/(\alpha - K)$ F]. The resulting circuit diagram is shown in Figure 11–22(b). For $K = \alpha$ the capacitance C_1 can be replaced by a short circuit because its capacitance is infinite. A gain restriction $K \le \alpha$ is required to keep C_1 from being negative.

As a second design example, consider a voltage-divider realization of the transfer function $Ks/(s + \alpha)$. We can find two voltage-divider realizations by writing the specified transfer function in the following two ways:

$$T(s) = \frac{Ks}{s + \alpha} = \frac{Z_2(s)}{Z_1(s) + Z_2(s)} \tag{11-27a}$$

$$T(s) = \frac{K}{1 + \alpha/s} = \frac{Z_2(s)}{Z_1(s) + Z_2(s)} \tag{11-27b}$$

Equation (11–27a) uses the transfer function as given, while Eq. (11–27b) factors s out of the numerator and denominator. Equating the numerators and denominators in Eqs. (11–27a) and (11–27b) yields two possible impedance assignments:

Using Eq. (11−27a): $Z_2 = Ks \quad \text{and} \quad Z_1 = s + \alpha - Z_2 = (1 - K)s + \alpha$

$$(11-28a)$$

Using Eq. (11−27b): $Z_2 = K \quad \text{and} \quad Z_1 = 1 + \frac{\alpha}{s} - Z_2 = (1 - K) + \frac{\alpha}{s}$

$$(11-28b)$$

The assignment in Eq. (11–28a) yields $Z_2(s)$ as an inductance ($L_2 = K$ H) and $Z_1(s)$ as an inductance [$L_1 = (1 - K)$ H] in series with a resistance ($R_1 = \alpha\ \Omega$). The assignment in Eq. (11–28b) yields $Z_2(s)$ as a resistance ($R_2 = K$) and $Z_1(s)$ as a resistance [$R_1 = (1 - K)\ \Omega$] in series with a capacitance ($C_1 = 1/\alpha$ F). The two realizations are shown in Figure 11–23. Both realizations require $K \le 1$ for the branch impedances to be realizable and both simplify when $K = 1$.

VOLTAGE-DIVIDER AND OP AMP CASCADE CIRCUIT DESIGN

The examples in Figures 11–22 and 11–23 illustrate an important feature of voltage-divider realizations. In general, we can write a transfer function as a quotient of polynomials $T(s) = r(s)/q(s)$. A voltage-divider realization requires the impedances $Z_2(s) = r(s)$ and $Z_1(s) = q(s) - r(s)$ to be physically realizable. A voltage-divider circuit usually places limitations on the gain K. This gain limitation can be overcome by using an OP AMP circuit in cascade with divider circuit.

For example, a voltage-divider realization of the transfer function in Eq. (11–23) requires $K \le \alpha$. When $K > \alpha$, then $T(s)$ is not realizable as a

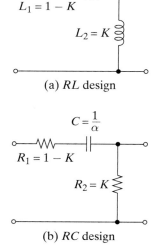

$R = \alpha$

$L_1 = 1 - K$

$L_2 = K$

(a) *RL* design

$C = \frac{1}{\alpha}$

$R_1 = 1 - K$

$R_2 = K$

(b) *RC* design

F I G U R E 1 1 – 2 3 *Circuit realizations of* T(s) = Ks/(s + α) *for* K ≤ 1.

simple voltage divider, since $Z_2(s) = s + \alpha - K$ requires a negative resistance. However, the given transfer function can be written as a two-stage product:

$$T_V(s) = \frac{K}{s + \alpha} = \underbrace{\left[\frac{K}{\alpha}\right]}_{\substack{\text{First} \\ \text{Stage}}} \underbrace{\left[\frac{\alpha}{s + \alpha}\right]}_{\substack{\text{Second} \\ \text{Stage}}} \qquad (11\text{--}29)$$

When $K > \alpha$, the first stage is a positive gain greater than unity and can be realized using the noninverting OP AMP configuration. Using the transfer function from Figure 11–21 for this circuit yields

$$\frac{Z_1(s) + Z_2(s)}{Z_1(s)} = \frac{K}{\alpha} \qquad (11\text{--}30)$$

Arbitrarily choosing $Z_1(s) = R_1 = 1\ \Omega$, we solve for $Z_2(s)$ and find $R_2 = K/(\alpha - 1)$. The second stage in Eq. (11–29) is realizable as a voltage divider because $q(s) - r(s) = s$ is realizable as an inductance.

Alternatively, factoring s out of the second-stage transfer function yields a different RC divider realization:

$$\frac{\alpha/s}{1 + \alpha/s} = \frac{Z_2(s)}{Z_1(s) + Z_2(s)} \qquad (11\text{--}31)$$

Equating numerators and denominators yields $Z_2(s) = \alpha/s$ and $Z_1(s) = 1$. Figure 11–24 shows a cascade connection of a noninverting first stage and the RC divider second stage. The chain rule applies to this circuit, since the first stage has an OP AMP output. The cascade circuit in Figure 11–24 realizes the first-order transfer function $K/(s + \alpha)$ for $K > \alpha$, a gain requirement that cannot be met by the divider circuit alone.

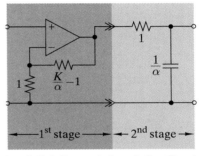

FIGURE   11-24 *Circuit realization of* T(s) = K/(s + α) *for* K > α.

D DESIGN EXAMPLE: 11-15

Design a circuit to realize the following transfer function using only resistors, capacitors, and OP AMPs:

$$T_V(s) = \frac{3000s}{(s + 1000)\,(s + 4000)}$$

SOLUTION:

The given transfer function can be written as a three-stage product.

$$T_V(s) = \underbrace{\left[\frac{K_1}{s + 1000}\right]}_{\substack{\text{First} \\ \text{Stage}}} \underbrace{[K_2]}_{\substack{\text{Second} \\ \text{Stage}}} \underbrace{\left[\frac{K_3 s}{s + 4000}\right]}_{\substack{\text{Third} \\ \text{Stage}}}$$

where the stage gains K_1, K_2, and K_3 have yet to be selected. Factoring s out of the denominator of the first-stage transfer function leads to an RC divider realization:

$$\frac{K_1/s}{1 + 1000/s} = \frac{Z_2(s)}{Z_1(s) + Z_2(s)}$$

Equating numerators and denominators yields

$$Z_2(s) = K_1/s \quad \text{and} \quad Z_1(s) = 1 + (1000 - K_1)/s$$

The first stage $Z_1(s)$ is simpler when we select $K_1 = 1000$. Factoring s out of the denominator of the third-stage transfer function leads to an RC divider realization:

$$\frac{K_3}{1 + 4000/s} = \frac{Z_2(s)}{Z_1(s) + Z_2(s)}$$

Equating numerators and denominators yields

$$Z_2(s) = K_3 \text{ and } Z_1(s) = 1 - K_3 + 4000/s$$

The third stage $Z_1(s)$ is simpler when we select $K_3 = 1$. The stage gains must meet the constraint $K_1 \times K_2 \times K_3 = 3000$ since the overall gain of the given transfer function is 3000. We have selected $K_1 = 1000$ and $K_3 = 1$, which requires $K_2 = 3$. The second stage must have a positive gain greater than 1 and can be realized using a noninverting amplifier with $K_2 = (R_1 + R_2)/R_1 = 3$. Selecting $R_1 = 1 \ \Omega$ requires that $R_2 = 2 \ \Omega$.

Figure 11–25 shows the three stages connected in cascade. The chain rule applies to this cascade connection because the OP AMP in the second stage isolates the *RC* voltage-divider circuits in the first and third stages. The circuit in Figure 11–25 realizes the given transfer function but is not a realistic design because the values of resistance and capacitance are impractical. For this reason we call this circuit a **prototype** design. We will shortly discuss how to scale a prototype to obtain practical element values. ■

FIGURE 11–25

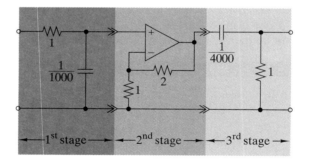

INVERTING OP AMP CIRCUIT DESIGN

The inverting OP AMP circuit is more versatile than the voltage divider because it places fewer restrictions on the form of the transfer function. There are at least four different inverting OP AMP circuit designs for a general first-order transfer function of the form

$$T_V(s) = -K\frac{s + \gamma}{s + \alpha}$$

The first realization results from equating $T_V(s)$ to the inverting OP AMP relationship in Figure 11–21:

$$-\frac{Ks + K\gamma}{s + \alpha} = -\frac{Z_2(s)}{Z_1(s)} \tag{11–32}$$

Equating numerators and denominators yields the impedance $Z_1(s) = s + \alpha$ and $Z_2(s) = Ks + K\gamma$. Both of these impedances are of the form $Ls + R$ and can be realized by an inductance in series with a resistor. These impedance identifications produce the RL circuit in Figure 11–26(a).

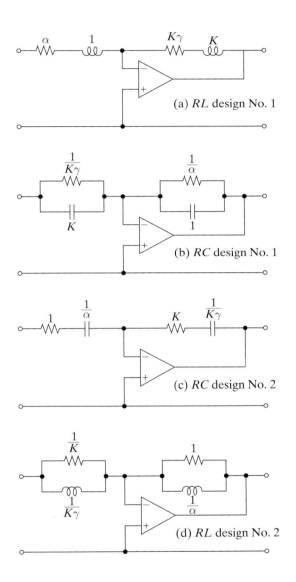

FIGURE   11–26   *Inverting OP AMP circuit realizations of* $T(s) = K(s + \gamma)/(s + \alpha)$.

(a) *RL* design No. 1

(b) *RC* design No. 1

(c) *RC* design No. 2

(d) *RL* design No. 2

A second inverting OP AMP realization is obtained by equating $Z_2(s)$ in Eq. (11–32) to the reciprocal of the denominator and equating $Z_1(s)$ to the reciprocal of the numerator. This assignment yields the impedances $Z_1(s) = 1/(Ks + K\gamma)$ and $Z_2(s) = 1/(s + \alpha)$. Both of these impedances are of the form $1/(Cs + G)$, where Cs is the admittance of a capacitor and G is the admittance of a resistor. Both impedances can be realized by a capacitance in parallel with a resistance. These impedance identifications produce the RC circuit in Figure 11–26(b).

Two more inverting realizations are obtained by factoring s out of the denominator of Eq. (11–32):

$$-\frac{K + K\gamma/s}{1 + \alpha/s} = -\frac{Z_2(s)}{Z_1(s)} \qquad (11\text{–}33)$$

Equating numerators and denominators yields $Z_2(s) = K + K\gamma/s$ and $Z_1(s) = 1 + \alpha/s$. Both of these impedances are realizable as a resistance in series with a capacitance and lead to the RC circuit in Figure 11–26(c). When we equate $Z_2(s)$ in Eq. (11–33) to the reciprocal of the denominator and $Z_1(s)$ to the reciprocal of the numerator, we find

$$Z_2(s) = \frac{1}{1 + \alpha/s} \quad \text{and} \quad Z_1(s) = \frac{1}{K + K\gamma/s} \qquad (11\text{–}34)$$

Both of these impedances are of the form $1/(G + 1/Ls)$, where G is the admittance of a resistor and $1/Ls$ is the admittance of an inductor. As a result, both are realizable by a resistor in parallel with an inductor, as shown by the RL circuit in Figure 11–26(d).

The four circuits in Figure 11–26 show that there are several ways to partition a transfer function into realizable input and feedback impedances for an inverting OP AMP circuit. For a first-order circuit the only limitation is that the parameters K, γ, and α must be positive. These conditions are less restrictive than the gain restrictions for the voltage-divider circuit.

Because it has fewer restrictions, it is often easier to realize transfer functions using the inverting OP AMP circuit. To use inverting circuits, the given transfer function must require an inversion or be realized using an even number of inverting stages. In some cases, the sign in front of the transfer function is immaterial and the required transfer function is specified as $\pm T_V(s)$. Caution: The input impedance of an inverting OP AMP circuit may load the source circuit.

D DESIGN EXAMPLE: 11–16

Design a circuit to realize the transfer function given in Example 11–15 using inverting OP AMP circuits.

SOLUTION:
The given transfer function can be expressed as the product of two inverting transfer functions:

$$T_V(s) = \frac{3000s}{(s + 1000)(s + 4000)} = \underbrace{\left[-\frac{K_1}{s + 1000}\right]}_{\text{First Stage}} \underbrace{\left[-\frac{K_2 s}{s + 4000}\right]}_{\text{Second Stage}}$$

where the stage gains K_1 and K_2 have yet to be selected. The first stage can be realized in an inverting OP AMP circuit since

$$-\frac{K_1}{s + 1000} = -\frac{K_1/1000}{1 + s/1000} = -\frac{Z_2(s)}{Z_1(s)}$$

Equating the $Z_2(s)$ to the reciprocal of the denominator and $Z_1(s)$ to the reciprocal of the numerator yields

$$Z_2 = \frac{1}{1 + s/1000} \quad \text{and} \quad Z_1 = 1000/K_1$$

The impedance $Z_2(s)$ is realizable as a capacitance ($C_2 = 1/1000$ F) in parallel with a resistance ($R_2 = 1\ \Omega$) and $Z_1(s)$ as a resistance ($R_1 = 1000/K_1\ \Omega$). We select $K_1 = 1000$ so that the two resistances in the first stage are equal. Since the overall gain requires $K_1 \times K_2 = 3000$, this means that $K_2 = 3$. The second-stage transfer function can also be produced using an inverting OP AMP circuit:

$$-\frac{3s}{s + 4000} = -\frac{3}{1 + 4000/s} = -\frac{Z_2(s)}{Z_1(s)}$$

Equating numerators and denominators yields $Z_2(s) = R_2 = 3$ and $Z_1(s) = R_1 + 1/C_1 s = 1 + 4000/s$.

Figure 11–27 shows the cascade connection of the two RC OP AMP circuits that realize each stage. The overall transfer function is noninverting because the cascade uses an even number of inverting stages. The chain rule applies here since the first stage has an OP AMP output. The circuit in Figure 11–27 is a prototype design because the values of resistance and capacitance are impractical. ∎

FIGURE   11-27

MAGNITUDE SCALING

The circuits obtained in Examples 11–15 and 11–16 are called prototype designs because the element values are outside of practical ranges. The allowable ranges depend on the fabrication technology used to construct the circuits. For example, monolithic integrated circuit technology limits capacitances to a few hundred picofarads. An OP AMP circuit should have a feedback resistance greater than around 10 kΩ to keep the output current demand within the capabilities of general-purpose OP AMP devices. Other technologies and applications place different constraints on element values.

There are no hard and fast rules here, but, roughly speaking, a circuit is probably realizable by some means if its passive element values fall in the following ranges:

Capacitors: 1 pF to 100 μF

Inductors: 10 μH to 100 mH

Resistors: 10 Ω to 100 MΩ.

The important idea here is that circuit designs like Figure 11–27 are impractical because 1-Ω resistors and 1-mF capacitors are unrealistic values.

It is often possible to scale the magnitude of circuit impedances so that the element values fall into practical ranges. The key is to scale the element values in a way that does not change the transfer function of the circuit. Multiplying the numerator and denominator of the transfer function of a voltage-divider circuit by a scale factor k_m yields

$$T_V(s) = \frac{k_m}{k_m} \frac{Z_2(s)}{Z_1(s) + Z_2(s)} = \frac{k_m Z_2(s)}{k_m Z_1(s) + k_m Z_2(s)} \qquad (11\text{--}35)$$

Clearly, this modification does not change the transfer function, but scales each impedance by a factor of k_m and changes the element values in the following way:

$$R_{after} = k_m R_{before} \quad L_{after} = k_m L_{before} \quad C_{after} = \frac{C_{before}}{k_m} \qquad (11\text{--}36)$$

Equation (11–36) was derived using the transfer function of a voltage-divider circuit. It is easy to show that we would reach the same conclusion if we had used the transfer functions of inverting or noninverting OP AMP circuits.

In general, a circuit is magnitude scaled by multiplying all resistances, multiplying all inductances, and dividing all capacitances by a scale factor k_m. The scale factor must be positive, but can be greater than or less than 1. Different scale factors can be used for each stage of a cascade design, but only one scale factor can be used for each stage. These scaling operations do not change the voltage transfer function realized by the circuit.

Our design strategy is first to create a prototype circuit whose element values may be unrealistically large or small. Applying magnitude scaling to the prototype produces a design with practical element values. Sometimes there may be no scale factor that brings the prototype element values into a practical range. When this happens, we must seek alternative realizations because the scaling process is telling us that the prototype is not a viable candidate.

EXAMPLE 11–17

Magnitude scale the circuit in Figure 11–27 so all resistances are at least 10 kΩ and all capacitances are less than 1 μF.

SOLUTION:
The resistance constraint requires $k_m R \geq 10^4$ Ω. The smallest resistance in the prototype circuit is 1 Ω; therefore, the resistance constraint requires $k_m \geq 10^4$. The capacitance constraint requires $C/k_m \leq 10^{-6}$ F. The largest capacitance in the prototype is 10^{-3} F; therefore, the capacitance constraint requires $k_m \geq 10^3$. The resistance condition on k_m dominates the two constraints. Selecting $k_m = 10^4$ produces the scaled design in Figure 11–28. This circuit realizes the same transfer function as the prototype in Figure 11–27, but uses practical element values. ∎

1st stage   2nd stage

Exercise 11–10

Select a magnitude scale factor for each stage in Figure 11–25 so that both capacitances are 10 nF and all resistances are greater than 10 kΩ.

Answer: $k_m = 10^5$ for the first stage, $k_m = 10^4$ for the second stage, and $k_m = 0.25 \times 10^5$ for the third stage.

SECOND-ORDER CIRCUIT DESIGN

An *RLC* voltage divider can also be used to realize second-order transfer functions. For example, the transfer function

$$T_V(s) = \frac{K}{s^2 + 2\zeta\omega_0 s + \omega_0^2}$$

can be realized by factoring s out of the denominator and equating the result to the voltage-divider input-output relationship:

$$T_V(s) = \frac{K/s}{s + 2\zeta\omega_0 + \omega_0^2/s} = \frac{Z_2(s)}{Z_1(s) + Z_2(s)} \tag{11–37}$$

Equating numerators and denominators yields

$$Z_2(s) = \frac{K}{s} \text{ and } Z_1(s) = s + 2\zeta\omega_0 + \frac{\omega_0^2 - K}{s} \tag{11–38}$$

The impedance $Z_2(s)$ is realizable as a capacitance ($C_2 = 1/K$ F) and $Z_1(s)$ as a series connection of an inductance ($L_1 = 1$ H), resistance ($R_1 = 2\zeta\omega_0$ Ω), and capacitance [$C_1 = 1/(\omega_0^2 - K)$ F]. The resulting voltage-divider circuit is shown in Figure 11–29(a). The impedances in this circuit are physically realizable when $K \leq \omega_0^2$. Note that the resistance controls the damping ratio ζ because it is the element that dissipates energy in the circuit.

When $K > \omega_0^2$, we can partition the transfer function into a two-stage cascade of the form

$$T_V(s) = \left[\frac{K}{\omega_0^2}\right] \underbrace{\left[\frac{\omega_0^2/s}{s + 2\zeta\omega_0 + \omega_0^2/s}\right]}_{} \tag{11–39}$$

$\underbrace{}_{\substack{\text{First}\\\text{Stage}}}$ $\underbrace{}_{\substack{\text{Second}\\\text{Stage}}}$

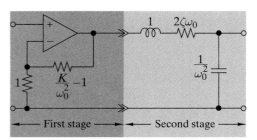

(a) Voltage divider design $K \leq \omega_0^2$

(b) Cascade design $K > \omega_0^2$

(c) Inverting amplifier design

FIGURE 11–29 *Second-order circuit realizations.*

The first stage requires a positive gain greater than unity and can be realized using a noninverting OP AMP circuit. The second stage can be realized as a voltage divider with $Z_2(s) = \omega_0^2/s$ and $Z_1(s) = s + 2\zeta\omega_0$. The resulting cascade circuit is shown in Figure 11–29(b).

An inverting OP AMP design of the second-order function also is possible provided that the given transfer function requires inversion or its sign is not specified. For example, starting with

$$T_V(s) = -\frac{K}{s^2 + 2\zeta\omega_0 s + \omega_0^2}$$

we factor s out of the denominator and equate the result to the inverting OP AMP transfer function:

$$T_V(s) = -\frac{K/s}{s + 2\zeta\omega_0 + \omega_0^2/s} = -\frac{Z_2(s)}{Z_1(s)} \qquad (11\text{–}40)$$

Equating $Z_2(s)$ to the reciprocal of the denominator and $Z_1(s)$ to the reciprocal of the numerator yields

$$Z_1(s) = \frac{s}{K} \quad \text{and} \quad Z_2(s) = \frac{1}{s + 2\zeta\omega_0 + \omega_0^2/s} \qquad (11\text{–}41)$$

The impedance $Z_1(s)$ is realizable as an inductance ($L_1 = 1/K$ H) and the impedance $Z_2(s)$ as the parallel connection of a capacitance ($C_2 = 1$ F), resistance [$R_2 = 1/(2\zeta\omega_0)\ \Omega$], and inductance ($L_2 = 1/\omega_0^2\ H$). The resulting inverting OP AMP realization is shown in Figure 11–29(c).

The examples in Figure 11–29 illustrate a general approach to designing prototype circuits for second-order transfer functions. The approach begins by factoring s out of the denominator. A voltage-divider realization is possible when the resulting denominator minus the numerator is a realizable impedance. When it is not realizable, either a cascade design with a noninverting gain stage or an inverting OP AMP realization can be used. Realizing a second-order transfer function requires resistors, inductors, and capacitors because factoring out an s leaves the denominator of the transfer function in the form

$$s + 2\zeta\omega_0 + \omega_0^2/s = Ls + R + 1/Cs$$

We can use this approach for any value of $\zeta > 0$. However, when $\zeta \geq 1$ the second-order transfer function can be factored into a product of two first-order transfer functions and realized as a cascade connection. Conversely, a transfer function consisting of the product of two first-order functions can be realized as a single second-order stage.

D DESIGN EXAMPLE 11–18

Find a second-order realization of the transfer function given in Example 11–15.

SOLUTION:
The given transfer function can be written as

$$T_V(s) = \frac{3000s}{(s + 1000)(s + 4000)} = \frac{3000s}{s^2 + 5000s + 4 \times 10^6}$$

Factoring s out of the denominator and equating the result to the transfer function of a voltage divider gives

$$\frac{3000}{s + 5000 + 4 \times 10^6/s} = \frac{Z_2(s)}{Z_1(s) + Z_2(s)}$$

Equating the numerators and denominators yields

$$Z_2(s) = 3000 \quad \text{and} \quad Z_1(s) = s + 2000 + 4 \times 10^6/s$$

Both of these impedances are realizable, so a single-stage voltage-divider design is possible. The prototype impedance $Z_1(s)$ requires a 1-H inductor, which is a bit large. A more practical value is obtained using a scale factor of $k_m = 0.1$. The resulting scaled voltage divider circuit is shown in Figure 11–30. ■

FIGURE 11 – 30

DESIGN EVALUATION SUMMARY

Examples 11–15, 11–16, and 11–18 show that there are several ways to realize a given transfer function, as summarized here:

EXAMPLE	FIGURE	DESCRIPTION	R	NUMBER OF		OP AMP
				L	C	
11–15	11–25	RC voltage-divider cascade	4	0	2	1
11–16	11–27	RC inverting cascade	4	0	2	2
11–18	11–30	RLC voltage divider	2	1	1	0

Selecting a final design from among these alternatives involves evaluation of each circuit using additional criteria. For example, inductors are heavy and lossy in low-frequency applications. These applications favor the first two circuits since they are "inductorless" realizations. However, the two *RC* circuits contain OP AMPs, which require a dc power supply. The *RC* voltage-divider cascade uses only one OP AMP, requiring less dc power than the *RC* inverting cascade, which uses two OP AMPs. On the other hand, the inverting circuit has an OP AMP output, so it could drive any reasonably large load impedance. Finally, in some applications, the fact that the passive *RLC* voltage divider does not require a dc power supply could outweigh the disadvantage of the inductor.

A design problem involves more than simply finding a prototype that realizes a given transfer function. In general, the first step in a design problem involves determining an acceptable transfer function, one that meets performance requirements such as the characteristics of the step or frequency response. In other words, we must first design the transfer function and then design several circuits that realize the transfer function. To deal with transfer function design we must understand how performance characteristics are related to transfer functions. The next two chapters provide some background on this issue.

D E ◖ **D E S I G N   A N D   E V A L U A T I O N   E X A M P L E :   1 1 – 1 9** ◗

Given the step response $g(t) = \pm[1 + 4e^{-500t}]u(t)$,

(a) Find the transfer function $T(s)$.
(b) Design two *RC* OP AMP circuits that realize the $T(s)$ found in part (a).
(c) Compare the two designs on the basis of element count, input impedance, output impedance.

SOLUTION:
(a) The transform of the step response is

$$G(s) = \pm\mathcal{L}\{[1 + 4e^{-500t}]u(t)\} = \pm\frac{1}{s} \pm \frac{4}{s + 500} = \pm\frac{5s + 500}{s(s + 500)}$$

and the required transfer function is

$$T(s) = H(s) = sG(s) = \pm\frac{5s + 500}{s + 500}$$

(b) The first design uses an inverting OP AMP configuration. Using the minus sign on the transfer function $T(s)$ and factoring an s out of the numerator and denominator yield

$$-T(s) = -\frac{5 + 500/s}{1 + 500/s} = -\frac{Z_2(s)}{Z_1(s)}$$

Equating numerators and denominators yields $Z_2(s) = 5 + 500/s$ and $Z_1(s) = 1 + 500/s$. The impedance $Z_2(s)$ is realizable as a resistance ($R_2 = 5\ \Omega$) in series with a capacitance ($C_2 = 1/500$ F) and $Z_1(s)$ as a resistance ($R_1 = 1\ \Omega$) in series with a capacitance ($C_1 = 1/500$ F). Using a magnitude scale factor $k_m = 10^5$ produces circuit C1 in Figure 11–31.

The second design uses a noninverting OP AMP configuration. Using the plus sign on the transfer function $T(s)$ and factoring an s out of the numerator and denominator yield

$$T(s) = \frac{5 + 500/s}{1 + 500/s} = \frac{Z_1(s) + Z_2(s)}{Z_1(s)}$$

Equating numerators and denominators yields

$$Z_1(s) = 1 + \frac{500}{s} \quad \text{and} \quad Z_2(s) = 5 + \frac{500}{s} - Z_1(s) = 4$$

The impedance $Z_1(s)$ is realizable as a resistance ($R_1 = 1\ \Omega$) in series with a capacitance ($C_1 = 1/500$ F) and $Z_2(s)$ as a resistance ($R_2 = 4\ \Omega$). Using a scale factor of $k_m = 10^4$ produces circuit C2 in Figure 11–31.

(c) Circuit C1 uses one more capacitor than circuit C2. The OP AMP output on both circuits means that they each have almost zero output impedance. The input impedance to circuit C2 is very large, because its input is the noninverting input of the OP AMP. The input impedance of circuit C1 is $Z_1(s) = k_m(1 + 500/s)$; hence, the scale factor must be selected to avoid loading the source circuit. The final design for circuit C1 in Figure 11–31 uses $k_m = 10^5$, which means that $|Z_1| > 100\ k\Omega$, which should be high enough to avoid loading the source circuit. ∎

C1

C2

FIGURE 11 – 31

D DESIGN EXAMPLE 11–20

Verify that circuit C2 in Figure 11–31 meets its design requirements.

SOLUTION:
One of the important uses of computer-aided analysis is to verify that a proposed design meets the performance specifications. The circuit C2 in Figure 11–31 is designed to produce a specified step response

$$g(t) = [1 + 4e^{-500t}]u(t) \quad \text{V}$$

This response jumps from zero to 5 V at $t = 0$ and then decays exponentially to 1 V at large t. The time constant of the exponential is $1/500 = 2$ ms, which means that the final value is effectively reached after about five time constants, or 10 ms.

F I G U R E   1 1 – 3 2

Figure 11–32 shows the circuit diagram drawn in the Electronics Work-bench circuit window. From the **Analysis** menu we bring up the **Transient Analysis** dialog box where we set the initial conditions to zero (to get a zero-state response), set the start time at zero, the end time at 10 ms (five time constants), and set node 4 (to OP AMP output) as the node for analysis. The step response calculated by Electronics Workbench is shown at the bottom of Figure 11–32. The response jumps to 5 V at $t = 0$ and then exponentially decays to 1 V after about 10 ms. The two cursors are set at $x_1 = 2.0079$ ms (about one time constant) where simulation predicts $y_1 = 2.4657$ and $x_2 = 4.0079$ ms (about two time constants) where simulation predicts $y_2 = 1.5392$. The theoretical values of $g(t)$ at these points are

$$g(t_1) = 1 + 4e^{-1} = 2.4715$$
$$g(t_2) = 1 + 4e^{-2} = 1.5413$$

Thus, theory and simulation agree to three significant figures confirming that circuit C2 meets the design requirement. ∎

SUMMARY

- A network function is defined as the ratio of the zero-state response transform to the input transform. Network functions are either driving-point functions or transfer functions. Network functions are rational functions of s with real coefficients whose complex poles and zeros occur in conjugate pairs.

- Network functions for simple circuits like voltage and current dividers and inverting and noninverting OP AMPs are easy to derive and often useful. Node-voltage or mesh-current methods are used to find the network functions for more complicated circuits. The transfer function of a cascade connection obeys the chain rule when each stage does not load the preceding stage in the cascade.

- The impulse response is the zero-state response of a circuit for a unit impulse input. The transform of the impulse response is equal to the network function. The impulse response contains only natural poles and decays to zero in stable circuits. The impulse response of a linear, time-invariant circuit obeys the proportionality and time-shifting properties. The short pulse approximation is a useful way to simulate the impulse response in practical situations.

- The step response is the zero-state response of a circuit when the input is a unit step function. The transform of the step response transform is equal to the network function times $1/s$. The step response contains natural poles and a forced pole at $s = 0$ that leads to a dc steady-state response in stable circuits. The amplitude of the dc steady-state response can be found by evaluating the network function at $s = 0$. The step response waveform can also be found by integrating the impulse response waveform.

- The sinusoidal steady-state response is the forced response of a stable circuit for a sinusoidal input. With a sinusoidal input the response transform contains natural poles and forced poles at $s = \pm j\omega$ that lead to a sinusoidal steady-state response in stable circuits. The amplitude and phase angle of the sinusoidal steady-state response can be found by evaluating the network function at $s = j\omega$.

- The sinusoidal steady-state response can be found using phasor circuit analysis or directly from the transfer function. Phasor circuit analysis works best when the circuit is driven at only one frequency and several responses are needed. The network function method works best when the circuit is driven at several frequencies and only one response is needed.

- First- and second-order transfer functions can be designed using voltage dividers and inverting or noninverting OP AMP circuits. Higher-order transfer functions can be realized using a cascade connection of first- and second-order circuits. Prototype designs usually require magnitude scaling to obtain practical element values.

PROBLEMS

ERO 11–1 NETWORK FUNCTIONS (SECTS. 11–1, 11–2)

Given a linear circuit in the zero state, find specified network functions and locate their poles and zeros.

FIGURE   P 1 1 – 1

11–1   Connect a voltage source $v_1(t)$ at the input port and an open circuit at the output port of the circuit in Figure P11–1.
   (a) Transform the circuit into the s domain and find the driving-point impedance $Z(s) = V_1(s)/I_1(s)$ and the transfer function $T_V(s) = V_2(s)/V_1(s)$.
   (b) Find the poles and zeros of $Z(s)$ and $T_V(s)$ for $R_1 = R_2 = 2\text{ k}\Omega$, $L = 100$ mH, $C = 250$ nF.

11–2   Connect a current source $i_1(t)$ at the input port and an open circuit at the output port of the circuit in Figure P11–1.
   (a) Transform the circuit into the s domain and find the input impedance $Z(s) = V_1(s)/I_1(s)$ and the transfer function $T_Z(s) = V_2(s)/I_1(s)$.
   (b) Find the poles and zeros of $Z(s)$ and $T_Z(s)$ for $R_1 = R_2 = 1\text{ k}\Omega$, $L = 100$ mH, $C = 50$ nF.

FIGURE   P 1 1 – 3

11–3   Connect a voltage source $v_1(t)$ at the input port and a short circuit at the output port of the circuit in Figure P11–3.
   (a) Transform the circuit into the s domain and find the driving-point impedance $Z(s) = V_1(s)/I_1(s)$ and the transfer function $T_Y(s) = I_2(s)/V_1(s)$.
   (b) Find the poles and zero of $Z(s)$ and $T_Y(s)$ for $R_1 = 100\ \Omega$, $R_2 = 50\ \Omega$, $L_1 = 0.5$ H, and $L_2 = 0.25$ H.

11–4   Connect a current source $i_1(t)$ at the input port and a short circuit at the output port of the circuit in Figure P11–3.
   (a) Transform the circuit into the s-domain and find the driving-point impedance $Z(s) = V_1(s)/I_1(s)$ and the transfer impedance $T_I(s) = I_2(s)/I_1(s)$.
   (b) Find the poles and zero of $Z(s)$ and $T_I(s)$ for $R_1 = 500\ \Omega$, $R_2 = 2000\ \Omega$, and $L_1 = 0.4$ H.

FIGURE   P 1 1 – 5

11–5   Transform the bridged-T circuit in Figure P11–5 into the s domain and show that

$$T_V(s) = \frac{V_2(s)}{V_1(s)} = \frac{R_1C_1R_2C_2s^2 + (R_1C_1 + R_1C_2)s + 1}{R_1C_1R_2C_2s^2 + (R_1C_1 + R_2C_2 + R_1C_2)s + 1}$$

Locate the poles and zeros of $T_V(s)$ for $R_1 = R_2 = 20\text{ k}\Omega$, $C_1 = C_2 = 20$ nF.

11–6   Replace the voltage source in Figure P11–5 by a current source $i_1(t)$, transform this circuit into the s domain, and show that

$$T_Z(s) = \frac{V_2(s)}{I_1(s)} = -\frac{R_1C_1R_2C_2s^2 + (R_1C_1 + R_1C_2)s + 1}{s(R_2C_1C_2s + C_1 + C_2)}$$

Locate the poles and zeros of $T_Z(s)$ for $R_1 = 5\text{ k}\Omega$, $R_2 = 10\text{ k}\Omega$, $C_1 = 0.5\ \mu\text{F}$, and $C_2 = 0.4\ \mu\text{F}$.

11–7 Transform the circuit in Figure P11–7 into the s domain and solve for the transfer function $T_V(s) = V_2(s)/V_1(s)$. Locate the poles and zeros of the transfer function.

11–8 Transform the circuit in Figure P11–8 into the s domain and solve for the transfer function $T_V(s) = V_2(s)/V_1(s)$. Locate the poles and zeros of the transfer function.

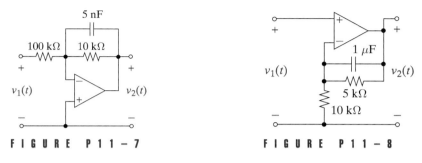

FIGURE P11–7   FIGURE P11–8

11–9 Transform the circuit in Figure P11–9 into the s domain and solve for the transfer function $T_V(s) = V_2(s)/V_1(s)$. Locate the poles and zeros of the transfer function.

FIGURE P11–9

11–10 The input impedance of the circuit in Figure P11–10 is $Z(s) = 10^4(s + 400)/(s + 200)$ Ω. Find the value of C.

$Z(s)$

FIGURE P11–10

ERO 11–2 NETWORK FUNCTIONS AND IMPULSE RESPONSE (SECT. 11–3)

(a) Given a first- or second-order linear circuit, find its impulse response.
(b) Given the impulse responses of a linear circuit, find the network functions.

11–11 Find the transfer function $T_V(s) = V_2(s)/V_1(s)$ of the circuit in Figure P11–11. Find $v_2(t)$ when the input $v_1(t)$ is a unit impulse.

11–12 Find the transfer function $T_V(s) = V_2(s)/V_1(s)$ of the circuit in Figure P11–12. Find $v_2(t)$ when the input $v_1(t)$ is a unit impulse.

FIGURE P11–11   FIGURE P11–12

FIGURE P11-13

11-13 Find the transfer function $T_V(s) = V_2(s)/V_1(s)$ of the circuit in Figure P11-13. Find $v_2(t)$ when the input $v_1(t)$ is a unit impulse.

11-14 Find the transfer function $T_V(s) = V_2(s)/V_1(s)$ of the circuit in Figure P11-14. Find $v_2(t)$ when the input $v_1(t)$ is a unit impulse.

11-15 Find the transfer function $T_V(s) = V_2(s)/V_1(s)$ of the circuit in Figure P11-15. Find $v_2(t)$ when the input $v_1(t)$ is a unit impulse.

FIGURE P11-14   **FIGURE P11-15**

11-16 Find the network functions $Z(s) = V_1(s)/I_1(s)$ and $T_I(s) = I_2(s)/I_1(s)$ for the circuit in Figure P11-16. Find $v_1(t)$ and $i_2(t)$ when the input $i_1(t) = \delta(t)$, $R_1 = 100\ \Omega$, $R_2 = 400\ \Omega$, and $L = 100$ mH.

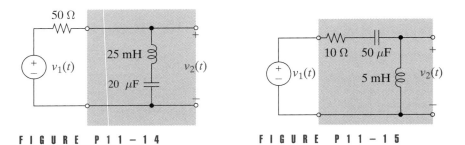

FIGURE P11-16

11-17 Find the network functions $Z(s) = V_1(s)/I_1(s)$ and $T_I(s) = I_2(s)/I_1(s)$ for the circuit in Figure P11-17. Find $v_1(t)$ and $i_2(t)$ when $i_1(t) = \delta(t)$, $R = 1$ kΩ, L = 200 mH, and C = 500 nF.

FIGURE P11-17

FIGURE P11-18

11-18 The input $i_1(t)$ in Figure P11-18 is a unit impulse $\delta(t)$. Find $v_2(t)$.

11-19 A one-port circuit is driven by a unit impulse of current, and the voltage response is found to be $v(t) = \delta(t) + (e^{-100t} + e^{-200t})u(t)$ V. Find the current response when the one-port circuit is driven by a unit impulse of voltage.

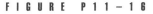

11–20 A one-port circuit is driven by a unit impulse of current, and the voltage response is found to be $v(t) = \delta(t) - (e^{-1000t}\sin(1000t))u(t)$ V. Find the current response when the one-port circuit is driven by a unit impulse of voltage.

ERO 11–3 NETWORK FUNCTIONS AND STEP RESPONSE (SECT. 11–4)

(a) Given a first- or second-order linear circuit, find its step response.
(b) Given the impulse response of a linear circuit, find the step response.
(b) Given the step response of a linear circuit, find the network function.

11–21 In Figure P11–11 the input is $v_1(t) = u(t)$ V. Find $T_V(s) = V_2(s)/V_1(s)$ and $v_2(t)$.

11–22 In Figure P11–12 the input is $v_1(t) = u(t)$ V. Find $T_V(s) = V_2(s)/V_1(s)$ and $v_2(t)$.

11–23 In Figure P11–13 the input is $v_1(t) = u(t)$ V. Find $T_V(s) = V_2(s)/V_1(s)$ and $v_2(t)$.

11–24 The impulse response of a linear circuit is $h(t) = 15(e^{-10t} - e^{-30t})u(t)$. Find the circuit step response $g(t)$.

11–25 In Figure P11–15 the input is $v_1(t) = u(t)$ V. Find $T_V(s) = V_2(s)/V_1(s)$ and $v_2(t)$.

11–26 In Figure P11–16 $R_1 = R_2 = 500$ Ω and $i_1(t) = u(t)$ A. Select the value of inductance L so that $i_2(t) = 0.5(e^{-4000t} - 1)u(t)$ A.

11–27 Find the transfer function and the impulse-response waveform corresponding to each of the following step responses:
(a) $g(t) = [-e^{-2000t}]u(t)$
(b) $g(t) = [1 - e^{-2000t}]u(t)$

11–28 Find the step response waveforms corresponding to each of the following impulse responses.
(a) $h(t) = [-1000e^{-2000t}]u(t)$
(b) $h(t) = \delta(t) - [2000e^{-2000t}]u(t)$

11–29 The step response of a linear circuit is $g(t) = [e^{-2000t}\sin 5000t]u(t)$. Find the poles and zeros of the transfer function.

11–30 A circuit consisting of a resistor and capacitor connected in series is connected across a voltage source $v_S(t) = 10u(t)$. The output voltage is observed to be $v_O(t) = 10[e^{-2000t}]u(t)$. Is the observed output taken across the resistor or the capacitor?

ERO 11–4 NETWORK FUNCTIONS AND SINUSOIDAL STEADY-STATE RESPONSE (SECT. 11–5)

(a) Given a first- or second-order linear circuit with a specified input sinusoid, find the amplitude and phase angle of the sinusoidal steady-state response.
(b) Given one of the responses $H(s)$, $h(t)$, $G(s)$, or $g(t)$, find the sinusoidal steady-state response for a specified input sinusoid.

11–31 Find the amplitude and phase angle of the steady-state output $v_{2SS}(t)$ of the circuit in Figure P11–7 for $v_1(t) = 5 \cos 10^5 t$. Repeat for $v_1(t) = 10 \sin 10^3 t$.

11–32 Find the amplitude and phase angle of the steady-state output $v_{2SS}(t)$ of the circuit in Figure P11–8 for $v_1(t) = 10 \sin 100t$. Repeat for $v_1(t) = 3 \cos 200t$.

11–33 Find the amplitude and phase angle of the steady-state output

$v_{2SS}(t)$ of the circuit in Figure P11–11 for $v_1(t) = 5 \cos 10^4 t$. Repeat for $v_1(t) = 10 \sin 10^3 t$.

11–34 The transfer function of a linear circuit is $T(s) = (s + 200)/(s + 50)$. Select the frequency of the sinusoidal input so that the amplitude of the steady-state output is twice that of the input.

11–35 Find the amplitude and phase angle of the steady-state output $v_{2SS}(t)$ of the circuit in Figure P11–13 for $v_1(t) = 10 \cos 5000t$. Repeat for $v_1(t) = 5 \cos 2000t$.

11–36 In Figure P11–16 $R_1 = R_2 = 1$ kΩ and the input $i_1(t)$ is a 2-MHz sinusoid. Select the value of inductance L so that the amplitude of steady-state response $i_{2SS}(t)$ is 10% of the amplitude of the input sinusoid.

11–37 The step response of a linear circuit is $g(t) = [e^{-1000t}]u(t)$. Find the amplitude and phase angle of the sinusoidal steady-state output for the sinusoidal input $x(t) = 5 \cos 2000t$.

11–38 The circuit in Figure P11–17 is operating in the sinusoidal steady state. Select the values of inductance L and capacitance C so that the sinusoidal input $i_1(t)$ and the steady-state output $i_{2SS}(t)$ have the same amplitude.

11–39 The impulse response of a linear circuit is $h(t) = -\delta(t) + [800e^{-1000t}]u(t)$. Find the amplitude and phase angle of the sinusoidal steady-state output for the sinusoidal input $x(t) = 3 \sin 400t$.

11–40 The impulse response of a linear circuit is $h(t) = 400[e^{-100t} - e^{-5000t}]u(t)$. Find the amplitude and phase angle of the sinusoidal steady-state output for the sinusoidal input $x(t) = 5 \cos 700t$.

ERO 11–5 GENERAL RESPONSES (SECTS. 11–3, 11–4, 11–5)

Given the impulse, step, or ramp response of a linear circuit, find the response due to other specified inputs.

11–41 The impulse response of a linear circuit is $h(t) = 2000[e^{-20000t}]u(t)$. Find the output waveform when the input is $x(t) = 5tu(t)$ V.

11–42 The step response of a linear circuit is $g(t) = 0.5[1 + e^{-200t}]u(t)$. Find the output waveform when the input is $x(t) = [e^{-100t}]u(t)$.

11–43 The step response of a linear circuit is $g(t) = 20[tu(t) - (t - 5)u(t - 5)]$. Find the response waveform when the input is $x(t) = tu(t)$.

11–44 The impulse response of a linear circuit is $H(s) = (s + 2000)/(s + 1000)$. Find the response waveform when the input is $x(t) = 5 e^{-1000t}u(t)$.

11–45 The impulse response of a linear circuit is $h(t) = u(t) - u(t - 10)$. Find the response waveform due to an input $x(t) = u(t)$.

11–46 The response of a linear circuit due to an input $x(t) = tu(t)$ is $y(t) = [e^{-200t}]u(t)$. Find the transform $G(s)$ of the step response of the circuit.

11–47 The transform of the impulse response of a linear circuit is $H(s) = 200/(s + 200)$. Find the response waveform when the input is $x(t) = 10u(t) - 10u(t - 0.1)$.

11–48 The impulse response of a linear circuit is $h(t) = 200u(t) + \delta(t)$. Find the response transform when the input is $x(t) = [e^{-200t}]u(t)$.

11–49 The step response of a linear circuit is $g(t) = 10[e^{-1000t}\sin 2000t]u(t)$. Find the sinusoidal steady-state response for the input $x(t) = 10 \cos 2000t$.

11–50 The impulse response of a linear circuit is $h(t) = \delta(t - 0.002)$. Find the circuit response waveform when the input is $x(t) = 10[e^{-1000t}]u(t)$.

ERO 11–6 NETWORK FUNCTION DESIGN (SECT. 11–6)

Given a transfer function $T(s)$,
(a) Design a prototype circuit that realizes $T(s)$ and meets stated constraints on the available elements.
(b) Magnitude scale the prototype to obtain practical element values.

11–51 **D** Design a circuit to realize the transfer function below using only resistors, capacitors, and OP AMPs. Scale the circuit so that all resistors are greater than 10 kΩ and all capacitors are less than 1 μF.

$$T_V(s) = \pm \frac{10^5}{(s + 200)(s + 2500)}$$

11–52 **D** Design a circuit to realize the transfer function below using only resistors, capacitors, and not more than one OP AMP. Scale the circuit so that all capacitors are exactly 100 pF.

$$T_V(s) = \pm \frac{100(s + 500)}{(s + 200)(s + 2500)}$$

11–53 **D** Design a circuit to realize the transfer function below using only resistors, capacitors, and no more than one OP AMP. Scale the circuit so that its input impedance does not load a source with an internal resistance of 50 Ω.

$$T_V(s) = \pm \frac{1000s}{(s + 50)(s + 2000)}$$

11–54 **D** Design a circuit to realize the transfer function below using only resistors, capacitors, and inductors (no OP AMPs allowed). Scale the circuit so that all inductors are 10 mH or less.

$$T_V(s) = \frac{s^2}{(s + 1000)(s + 5000)}$$

11–55 **D** Design a circuit to realize the transfer function listed below using only resistors, capacitors, and not more than one OP AMP. Scale the circuit so that all resistors are greater than 10 kΩ and all capacitors are less than 1 μF.

$$T_V(s) = \pm \frac{(s + 100)(s + 1000)}{(s + 200)(s + 500)}$$

11–56 **D** Design a circuit to realize the transfer function below using only resistors, capacitors, and OP AMPs. Design the circuit so that the output voltage is across a 10 kΩ resistor.

$$T_V(s) = -\frac{100(s + 600)}{s(s + 200)}$$

11–57 **D** Design a circuit to realize the transfer function below using practical element values:

$$T_V(s) = \pm \frac{10^6}{s^2 + 1000s + 10^6}$$

11–58 **D** Design a circuit to realize the transfer function below using practical element values:

$$T_V(s) = -\frac{s}{s^2 + 2000s + 10^6}$$

11–59 **D** Design a circuit that produces the following step response:

$$g(t) = [11 - 9e^{-50t}]u(t)$$

11–60 **D** Design a circuit that produces the following step response:

$$g(t) = [1 - e^{-50t} - 50te^{-50t}]u(t)$$

INTEGRATING PROBLEMS

11–61 **A** OBSERVING THE CIRCUIT STATE AND INPUT IN OUTPUT RESPONSES

A circuit has a transfer function

$$T(s) = \frac{s}{(s + 2)(s + 4)}$$

(a) A certain input produces a zero-state output $y(t) = [-e^{-4t} + \cos 4t + \sin 4t]u(t)$. What is the input waveform?

(b) A step function input produces $y(t) = [e^{-2t} + e^{-4t}]u(t)$ as the output. Is the circuit in the zero state?

(c) An output $y(t) = [-e^{-2t} + 2e^{-3t}]u(t)$ is observed. Is this a zero-input response? If not, what is the input?

(d) It is claimed that the input $x(t) = 6 \sin(\sqrt{8}\,t)$ produces a steady-state output of $y_{SS} = \sin(\sqrt{8}\,t)$. Is this correct?

11–62 **A D** SECOND-ORDER CIRCUIT STEP RESPONSE

The step response in Figure P11–62 is of the form

$$g(t) = \frac{K}{\alpha^2 + \beta^2}\left[1 - e^{-\alpha t}\cos\beta t - \frac{\alpha}{\beta}e^{-\alpha t}\sin\beta t\right]u(t)$$

(a) **A** Estimate values for the parameters K, α, and β from the plot of $g(t)$.

(b) **A** Find the transfer function $T(s)$ corresponding to $g(t)$.

(c) **D** Design a circuit that realizes the $T(s)$ found in (b).

(d) **D** Use computer-aided circuit analysis to verify your design.

$g(t)$ (V)

FIGURE P11–62

11–63 **A D E** OP AMP RC CIRCUIT DESIGN EVALUATION

This problem uses circuits C1 and C2 in Figure P11–63.

(a) **A** Determine the transfer functions $T_V(s) = V_2(s)/V_1(s)$ for each circuit. Locate the poles and zeros for each function and relate their locations to circuit parameters.

(b) **D** Use a cascade connection of one or more of these circuits to realize the following transfer function:

$$T(s) = \frac{s(s + 300)}{(s + 200)(s + 400)}$$

(c) **E** Show that a cascade connection of these circuits *cannot* realize the following transfer function and explain why in terms of poles and zeros.

$$T(s) = \frac{s(s + 500)}{(s + 200)(s + 400)}$$

(d) **E** Develop a design rule that restricts the locations of the poles and zeros of $T(s)$ to ensure that it can be realized using a cascade connection of these circuits.

11–64 **D** **E** NETWORK FUNCTIONS AND PHASE SHIFT

In the sinusoidal steady state the transfer function $T(j\omega)$ relates the input and output sinusoids. The angle of $T(j\omega)$ is called the phase shift through the circuit and indicates the amount by which the input and output sinusoids are out of phase.

(a) **D** Select the element values of Circuit C1 in Figure 11–64 so that $v_2(t)$ and $v_3(t)$ are 90° out of phase at $\omega = 2$ krad/s.

(b) **E** It is claimed that in the sinusoidal steady state the voltages $v_2(t)$ and $v_3(t)$ in Circuit C2 are always 90° out of phase regardless of frequency. Prove or disprove this claim.

FIGURE   P11–63

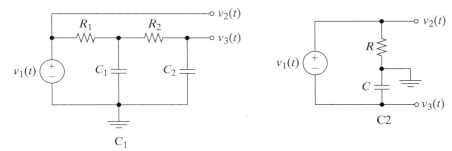

FIGURE   P11–64

11–65 **A** **D** **E** EVALUATING ACTIVE AND PASSIVE CIRCUIT DESIGN OPTIONS

Figure P11–65 shows an RLC voltage divider and an RC OP AMP circuit.

(a) **A** Show that both circuits have a transfer function of the form

$$T_V(s) = \frac{V_2(s)}{V_1(s)} = \frac{s^2}{s^2 + 2\zeta\omega_0 s + \omega_0^2}$$

(b) **D** **E** If possible, select practical element values so that $\omega_0 = 10^6$ rad/s and $\zeta = 0.5$. Which circuit best meets these requirements?

(c) **D** **E** If possible, select element values so that $\omega_0 = 10^2$ rad/s and $\zeta = 2$. Which circuit best meets these requirements?

11–66 **A** SINUSOIDAL STEADY-STATE RESPONSE

The input to the RL circuit in Figure P11–66 is $v_1(t) = V_A \cos \omega t$.

(a) Use the s-domain network function method in Chapter 11 to find the sinusoidal steady-state response $i_{SS}(t)$.

FIGURE   P11–65

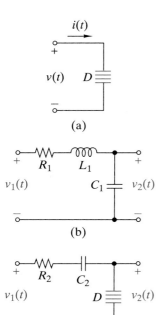

F I G U R E **P 1 1 – 6 6**

(b) Use the phasor circuit analysis method of Chapter 8 to find the sinusoidal current response.

(c) Formulate the circuit differential equation and use the classical methods of Chapter 7 to find the forced component of $i(t)$.

(d) The responses found in (a), (b), and (c) should be identical. Which method do you think is easiest to apply? Discuss.

11–67 **A** **FREQUENCY-DEPENDENT NEGATIVE RESISTOR**

The frequency-dependent negative resistor (FDNR) is an integrated circuit device whose circuit symbol is shown in Figure P11–67(a). Using the passive sign convention, the $i–v$ relationship of the element modeling the device is

$$i(t) = D\frac{d^2v(t)}{dt^2} \quad \text{with} \quad D > 0$$

(a) Find the element impedance $Z(s) = V(s)/I(s)$. Evaluate $Z(j\omega)$ and discuss any reasons you see for the device's name.

(b) In the sinusoidal steady state the voltage across an FDNR is $v(t) = V_A \cos \omega t$. Find the instantaneous power delivered to the element. Is the FDNR active or passive?

(c) FDNR devices are used to produce inductorless circuits with transfer functions like RLC circuits. Show that when $DR_2 = L_1C_1$ and $D/C_2 = R_1C_1$, the RLC circuit in Figure P11–67(b) and the RDC circuit in P11–67(c) have the same transfer function $T(s) = V_2(s)/V_1(s)$.

F I G U R E **P 1 1 – 6 7**

CHAPTER 12

FREQUENCY RESPONSE

The advantage of the straight-line approximation is, of course, that it reduces the complete characteristics to a sum of elementary characteristics.

Hendrik W. Bode, 1945,
American Engineer

12–1 Frequency-response Descriptors

12–2 First-order Circuit Frequency Response

12–3 Second-order Circuit Frequency Response

12–4 The Frequency Response of *RLC* Circuits

12–5 Bode Diagrams

12–6 Bode Diagrams with Complex Critical Frequencies

12–7 Frequency Response and Step Response

Summary

Problems

Integrating Problems

591

In the sinusoidal steady state, both the input and the output of a stable linear circuit are sinusoids of the same frequency. The term *frequency response* refers to the frequency-dependent relation, in both gain and phase, between a sinusoidal input and the resulting sinusoidal steady-state output. The gain and phase relationships can be presented as equations or graphs. When presented as separate graphs of gain and phase versus logarithmic frequency, the result is commonly called a **Bode diagram**.

Hendrick Bode (pronounced Bow Dee) spent most of his distinguished career as a member of the technical staff of the Bell Telephone Laboratories. In the 1930s Bode made major contributions to circuit and feedback amplifier theory as part of the effort to develop the precise frequency-selective amplifiers required by long-distance telephone systems. In describing his work Bode used straight-line approximations as an aid to understanding and simplifying the slide rule calculations required in his day. With the advent of digital computers, the computational benefits of Bode diagrams are no longer important. His straight-line method is still useful today, however, because it allows us to quickly visualize how the poles and zeros of a transfer function affect the frequency response of a circuit. Thus, Bode diagrams are a conceptual tool that simplifies the analysis and design of circuits and systems, just as they did in Bode's day.

The first section of this chapter defines the parameters used to characterize the frequency response. The frequency response of first- and second-order circuits is discussed in the following two sections. The fourth section describes how the frequency-response characteristics of series and parallel *RLC* circuits can be traced to variation in the impedance of specific elements in the circuit. The formal treatment of Bode diagrams in the fifth and sixth sections generalizes the results obtained using basic first- and second-order circuits. The final section discusses the relationship between frequency response and the time-domain step response.

12–1 FREQUENCY-RESPONSE DESCRIPTORS

In Chapter 11, we found that replacing s by $j\omega$ allows us to find the steady-state response of a stable circuit directly from its transfer function. The circuit transfer function influences the sinusoidal steady-state response through the **gain** function $|T(j\omega)|$ and **phase** function $\theta(\omega)$.

$$\text{Output Amplitude} = |T(j\omega)| \times (\text{Input Amplitude})$$

$$\text{Output Phase} = \text{Input Phase} + \theta(\omega)$$

Taken together, the gain and phase functions show how the circuit modifies the input amplitude and phase angle to produce the output sinusoid. These two functions define the **frequency response** of the circuit since they are frequency-dependent functions that relate the sinusoidal steady-state input and output. The gain and phase functions can be expressed mathematically or presented graphically as the frequency response plots in Figure 12–1.

The terminology used to describe the frequency response of circuits and systems is based on the form of the gain plot. For example, at high fre-

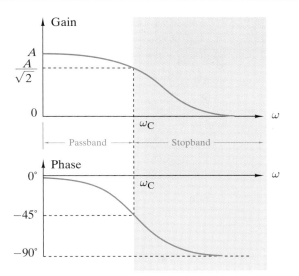

FIGURE   12 – 1 *Frequency   response plots.*

quencies the gain in Figure 12–1 falls off so that output signals in this frequency range are reduced in amplitude. The range of frequencies over which the output is significantly attenuated is called the **stopband**. At low frequencies the gain is essentially constant and there is relatively little attenuation. The frequency range over which there is little attenuation is called a **passband**. The frequency associated with the boundary between a passband and an adjacent stopband is called the **cutoff frequency** ($\omega_C = 2\pi f_C$). In general, the transition from the passband to the stopband is gradual, so the precise location of the cutoff frequency is a matter of definition. The most widely used definition specifies the cutoff frequency to be the frequency at which the gain has decreased by a factor of $1/\sqrt{2} = 0.707$ from its maximum value in the passband.

Again this definition is arbitrary, since there is no sharp boundary between a passband and an adjacent stopband. However, the definition is motivated by the fact that the power delivered to a resistance by a sinusoidal current or voltage waveform is proportional to the square of its amplitude. At a cutoff frequency the gain is reduced by a factor of $1/\sqrt{2}$ and the square of the output amplitude is reduced by a factor of one-half. For this reason the cutoff frequency is also called the **half-power frequency**.

When the gain response is plotted on log-log scales, we obtain the four prototype gain characteristics shown in Figure 12–2. A **low-pass** gain characteristic has a single passband extending from zero frequency (dc) to the cutoff frequency. A **high-pass** gain characteristic has a single passband extending from the cutoff frequency to infinite frequency. A **bandpass** gain has a single passband with two cutoff frequencies, neither of which is zero or infinite. Finally, the **bandstop** gain has a single stopband with two cutoff frequencies, neither of which is zero or infinite.

The **bandwidth** of a gain characteristic is defined as the frequency range spanned by its passband. The bandwidth of a low-pass circuit is equal to its cutoff frequency ($B = \omega_C$). The bandwidth of a high-pass characteristic is infinite since passband extends to infinity. For the bandpass and bandstop

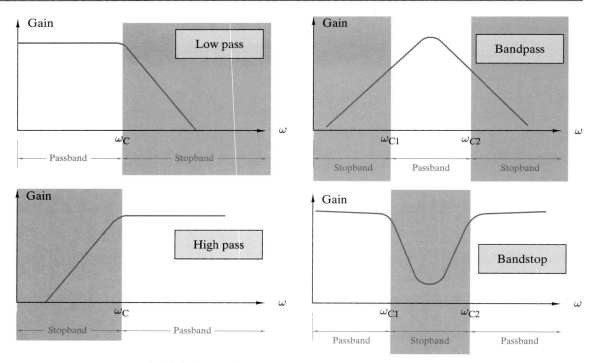

FIGURE 12–2 *Prototype gain responses.*

cases in Figure 12–2, the bandwidth is the difference in the two cutoff frequencies:

$$B = \omega_{C2} - \omega_{C1} \qquad (12-1)$$

For the bandstop case Eq. (12–1) defines the width of the stopband rather than the passbands.

The gain responses in Figure 12–2 have different characteristics at zero and infinite frequency—namely,

	GAIN AT	
	$\omega = 0$	$\omega = \infty$
Low pass	Finite	0
High pass	0	Finite
Bandpass	0	0
Bandstop	Finite	Finite

Since these extreme values form a unique pattern, the type of gain response can be inferred from the values of $|T(0)|$ and $|T(\infty)|$. These end-point values, in turn, are usually determined by the impedance of capacitors and inductors in the circuit. In the sinusoidal steady state the impedances of these elements are

$$Z_C(j\omega) = \frac{1}{j\omega C} \quad \text{and} \quad Z_L(j\omega) = j\omega L \qquad (12\text{--}2)$$

These impedances form a unique pattern at zero and infinite frequency—namely,

	IMPEDANCE AT	
	$\omega = 0$	$\omega = \infty$
Capacitor	Infinite	0
Inductor	0	Infinite

Thus, the capacitor acts like an open circuit at zero frequency and a short circuit at infinite frequency, while the inductor acts like a short circuit at dc and an open circuit at infinite frequency. These observations often allow us to infer the type of gain response directly from the circuit itself without finding the transfer function.

When frequency-response plots use logarithmic scales for the frequency variable, they are normally referred to as **Bode diagrams**. Logarithmic scales are used to compress the data range because the frequency ranges of interest often span several orders of magnitude. The use of a logarithmic frequency scale involves some special terminology. Any frequency range whose end points have a 2:1 ratio is called an **octave**. Any range whose end points have a 10:1 ratio is called a **decade**. For example, the frequency range from 10 Hz to 20 Hz is one octave, as is the range from 20 MHz to 40 MHz. The standard UHF (ultra high frequency) band spans one decade from 0.3 to 3 GHz.

In Bode diagrams the gain $|T(j\omega)|$ is usually expressed in **decibels** (dB), defined as

$$\left|T(j\omega)\right|_{dB} = 20 \log_{10}\left|T(j\omega)\right| \qquad (12\text{--}3)$$

To construct and interpret frequency-response plots, you must have some familiarity with gains expressed in decibels. First note that the gain in dB can be either positive, negative, or zero. A gain of zero dB means that $|T(j\omega)| = 1$ (i.e., the input and output amplitudes are equal). When $|T(j\omega)| > 1$ the output amplitude exceeds the input and $|T(j\omega)|_{dB} > 0$. When $|T(j\omega)| < 1$ the output amplitude is less than the input and $|T(j\omega)|_{dB} < 0$. A cutoff frequency occurs when the gain is reduced from its maximum passband value by a factor $1/\sqrt{2}$. Expressed in dB, this is a gain reduction of

$$20 \log_{10}\left(\frac{1}{\sqrt{2}}|T|_{MAX}\right) = 20 \log_{10}|T_{MAX}| - 20 \log_{10}\sqrt{2}$$
$$\approx |T|_{MAX, dB} - 3 \text{ dB}$$

That is, cutoff occurs when the dB gain is reduced by about 3 dB. For this reason the cutoff is also called the **3-dB down frequency**.

The generalization of this idea is that multiplying $|T(j\omega)|$ by a factor K means $20 \times \log|K|T(j\omega) = |T(j\omega)|_{dB} + 20 \log(K)$. That is, multiplying the gain by K changes the gain in dB by an additive factor K_{dB}. Some multiplicative factors whose dB equivalents are worth remembering are as follows:

- A multiplicative factor $K = 1$ changes $|T(j\omega)|_{dB}$ by 0 dB.
- A multiplicative factor $K = \sqrt{2}$ $(1/\sqrt{2})$ changes $|T(j\omega)|_{dB}$ by about +3 dB (–3 dB).
- A multiplicative factor $K = 2$ (1/2) changes $|T(j\omega)|_{dB}$ by about +6 dB (–6 dB).
- A multiplicative factor $K = 10$ (1/10) changes $|T(j\omega)|_{dB}$ by +20 dB (–20 dB).
- A multiplicative factor $K = 30$ (1/30) changes $|T(j\omega)|_{dB}$ by about +30 dB (–30 dB).
- A multiplicative factor $K = 100$ (1/100) changes $|T(j\omega)|_{dB}$ by +40 dB (–40 dB).
- A multiplicative factor $K = 1000$ (1/1000) changes $|T(j\omega)|_{dB}$ by +60 dB (–60 dB).

In summary, Bode diagrams use logarithmic frequency scales and linear scales for the gain (in dB) and phase (in radians or degrees). Both plots are semilog graphs, although in effect the gain plot is log/log because of the logarithmic definition of $|T(j\omega)|_{dB}$. To create frequency-response plots we must calculate $|T(j\omega)|$ and $\theta(\omega)$ at a sufficient number of frequencies to define adequately the stop and passband characteristics. The most efficient way to generate accurate plots is to use one of the many computer-aided analysis tools. Spreadsheets and math analysis programs like Mathcad can be used to calculate and plot frequency-response curves. Circuit analysis programs such as SPICE have ac analysis options that generate the frequency-response plots. We will illustrate the use of these tools in several examples in this chapter.

As always, some knowledge of the expected response is required to use computer-aided circuit analysis tools effectively. Using computer tools to generate frequency-response plots requires at least some preliminary analysis or a rough sketch of the expected response. To develop the required insight, we treat first- and second-order circuits and then use Bode plots to show how these building blocks combine to produce the frequency response of more complicated circuits.

APPLICATION NOTE: **E X A M P L E   1 2 – 1**

The use of the decibel as a measure of performance pervades the literature and folklore of electrical engineering. The decibel originally came from the definition of power ratios in **bels**.[1]

$$\text{Number of bels} = \log_{10}\frac{P_{OUT}}{P_{IN}}$$

---

[1] The name of the unit honors Alexander Graham Bell (1847–1922), the inventor of the telephone.

The decibel (dB) is more commonly used in practice. The number of decibels is 10 times the number of bels:

$$\text{Number of dB} = 10 \times (\text{Number of bels}) = 10 \log_{10} \frac{P_{\text{OUT}}}{P_{\text{IN}}}$$

When the input and output powers are delivered to equal input and output resistances R, then the power ratio can be expressed in terms of voltages across the resistances.

$$\text{Number of dB} = 10 \log_{10} \frac{v_{\text{OUT}}^2/R}{v_{\text{IN}}^2/R} = 20 \log_{10} \frac{v_{\text{OUT}}}{v_{\text{IN}}}$$

or in terms of currents through the resistances:

$$\text{Number of dB} = 10 \log_{10} \frac{i_{\text{OUT}}^2 \times R}{i_{\text{IN}}^2 \times R} = 20 \log_{10} \frac{i_{\text{OUT}}}{i_{\text{IN}}}$$

The definition of gain in dB in Eq. (12–3) is consistent with these results, since in the sinusoidal steady state the transfer function equals the ratio of output amplitude to input amplitude. The preceding discussion is not a derivation of Eq. (12–3) but simply a summary of its historical origin. In practice Eq. (12–3) is applied when the input and output are not measured across resistances of equal value.

When the chain rule applies to a cascade connection, the overall transfer function is a product

$$T(j\omega) = T_1 \times T_2 \times \cdots T_N$$

where $T_1, T_2, \ldots T_N$ are the transfer functions of the individual stages in the cascade. Expressed in dB, the overall gain is

$$|T(j\omega)|_{\text{dB}} = 20 \log_{10}(|T_1| \times |T_2| \times \cdots |T_N|)$$
$$= 20 \log_{10}|T_1| + 20 \log_{10}|T_2| + \cdots + 20 \log_{10}|T_N|$$
$$= |T_1|_{\text{dB}} + |T_2|_{\text{dB}} + \cdots + |T_N|_{\text{dB}}$$

Because of the logarithmic definition, the overall gain (in dB) is the sum of the gains (in dB) of the individual stages in a cascade connection. The effect of altering a stage or adding an additional stage can be calculated by simply adding or subtracting the change in dB. Since summation is simpler than multiplication, the enduring popularity of the dB comes from its logarithmic definition and not its somewhat tenuous relationship to power ratios.

12–2   FIRST-ORDER CIRCUIT FREQUENCY RESPONSE

FIRST-ORDER LOW-PASS RESPONSE

We begin the study of frequency response with the first-order low-pass transfer function:

$$T(s) = \frac{K}{s + \alpha} \qquad\qquad (12\text{–}4)$$

The constants K and α are real. The constant K can be positive or negative, but α must be positive so that the natural pole at $s = -\alpha$ is in the left half of the s plane to ensure that the circuit is stable. Remember, the concepts of sinusoidal steady state and frequency response do not apply to unstable circuits that have poles in the right half of the s plane or on the j-axis.

To describe the frequency response of the low pass transfer function, we replace s by $j\omega$ in Eq. (12–4)

$$T(j\omega) = \frac{K}{\alpha + j\omega} \tag{12–5}$$

and express the gain and phase functions as

$$|T(j\omega)| = \frac{|K|}{\sqrt{\omega^2 + \alpha^2}}$$

$$\theta(\omega) = \angle K - \tan^{-1}(\omega/\alpha) \tag{12–6}$$

The gain function is a positive number. Since K is real, the angle of K ($\angle K$) is either $0°$ when $K > 0$ or $\pm 180°$ when $K < 0$. An example of a negative K occurs in an inverting OP AMP configuration where $T(s) = -Z_2(s)/Z_1(s)$.

Figure 12–3 shows the gain and phase functions plotted versus normalized frequency ω/α. In the gain plot the left vertical scale is a log scale for $|T(j\omega)|$ and the right scale is the equivalent linear scale for $|T(j\omega)|_{dB}$. The maximum passband gain occurs at $\omega = 0$ where $|T(0)| = |K|/\alpha$. As frequency increases the gain gradually decreases until at $\omega = \alpha$

$$|T(j\alpha)| = \frac{|K|}{\sqrt{\alpha^2 + \alpha^2}} = \frac{|K|/\alpha}{\sqrt{2}} = \frac{|T(0)|}{\sqrt{2}} \tag{12–7}$$

That is, the cutoff frequency of the first-order low-pass transfer function is $\omega_C = \alpha$. The graph of the gain function in Figure 12–3 displays a low-pass characteristic with a finite dc gain and zero infinite frequency gain.

The low- and high-frequency gain asymptotes shown in Figure 12–3 are especially important. The low-frequency asymptote is the horizontal line and the high-frequency asymptote is the sloped line. At low frequencies ($\omega << \alpha$) the gain approaches $|T(j\omega)| \rightarrow |K|/\alpha$. At high frequencies ($\omega >> \alpha$) the gain approaches $|T(j\omega)| \rightarrow |K|/\omega$. The intersection of the two asymptotes occurs when $|K|/\alpha = |K|/\omega$. The intersection forms a "corner" at $\omega = \alpha$, so the cutoff frequency is also called the **corner frequency**.

The high-frequency gain asymptote decreases by a factor of 10 (–20 dB) whenever the frequency increases by a factor of 10 (one decade). As a result, the high-frequency asymptote has a slope of –1 or –20 dB per decade and the low-frequency asymptote has a slope of 0 or 0 dB/decade. These two asymptotes provide a straight-line approximation to the gain response that differs from the true response by a maximum of 3 dB at the corner frequency.

The semilog plot of the phase shift of the first-order low-pass transfer function is shown in Figure 12–3. At $\omega = \alpha$ the phase angle in Eq. (12–12) is

$$\theta(\omega_C) = \angle K - \tan^{-1}\left(\frac{\alpha}{\alpha}\right)$$

$$= \angle K - 45°$$

FIGURE 12–3 *First-order low-pass frequency response plots.*

At low frequency ($\omega \ll \alpha$) the phase angle approaches $\angle K$ and at high frequencies ($\omega \gg \alpha$) the phase approaches $\angle K - 90°$. Almost all of the $-90°$ phase change occurs in the two-decade range from $\omega/\alpha = 0.1$ to $\omega/\alpha = 10$. The straight-line segments in Figure 12–3 provide an approximation of the phase response. The phase approximation below $\omega/\alpha = 0.1$ is $\theta = \angle K$ and above $\omega/\alpha = 10$ is $\theta = \angle K - 90°$. Between these values the phase approximation is a straight line that begins at $\theta = \angle K$, passes through $\theta = \angle K - 45°$ at the cutoff frequency, and reaches $\theta = \angle K - 90°$ at $\omega/\alpha = 10$. The slope of this line segment is $-45°/\text{decade}$ since the total phase change is $-90°$ over a two-decade range. The $-45°/\text{decade}$ slope only applies over a frequency range from one decade below to one decade above the cutoff frequency. Outside of this range the straight-line approximation has zero slope.

To construct the straight-line approximations for a first-order low-pass transfer function, we need two parameters, the value of $T(0)$ and α. The parameter α defines the cutoff frequency and the value of $T(0)$ defines the passband gain $|T(0)|$ and the low-frequency phase $\angle T(0)$. The required quantities $T(0)$ and α can be determined directly from the transfer function $T(s)$ and can often be estimated by inspecting the circuit itself.

Using logarithmic scales in frequency-response plots allows us to make straight-line approximations to both the gain and phase responses. These approximations provide a useful way of visualizing a circuit's frequency response. Often such graphical estimates are adequate for developing analysis and design approaches. For example, the frequency response of the first-order low-pass function can be characterized by calculating the gain and phase over a two-decade band from one decade below to one decade above the cutoff frequency.

FIGURE 12-4

EXAMPLE 12–2

Consider the circuit in Figure 12–4. Find the transfer function $T(s) = V_2(s)/V_1(s)$ and construct the straight-line approximations to the gain and phase responses.

SOLUTION:
Applying voltage division, the voltage transfer function for the circuit is

$$T(s) = \frac{R}{Ls + R} = \frac{R/L}{s + R/L}$$

Comparing this with Eq. (12–4), we see that the circuit has a low-pass gain response with $\alpha = R/L$ and $T(0) = 1$. Therefore, $|T(0)|_{dB} = 0$ dB, $\omega_C = R/L$, and $\angle K = 0°$. Given these quantities, we construct the straight-line approximations shown in Figure 12–5. Note that the frequency scale in Figure 12–5 is normalized by multiplying ω by $L/R = 1/\alpha$.

Circuit Interpretation: The low-pass response in Figure 12–5 can be explained in terms of circuit behavior. At zero frequency the inductor acts like a short circuit that directly connects the input port to the output port to produce a passband gain of 1 (or 0 dB). At infinite frequency the inductor acts like an open circuit, effectively disconnecting the input and output ports and leading to a gain of zero. Between these two extremes the impedance of the inductor gradually increases, causing the circuit gain to decrease. In particular, at the cutoff frequency we have $\omega L = R$, the impedance of the inductor is $j\omega L = jR$, and the transfer function reduces to

$$T(j\omega_C) = \frac{R}{R + jR} = \frac{1}{\sqrt{2}}\angle -45°$$

In other words, at the cutoff frequency the gain is –3 dB and the phase shift –45°. Obviously, the changing impedance of the inductor gives the circuit its low-pass gain features. ■

◖ DESIGN EXAMPLE 12–3

(a) Show that the transfer function $T(s) = V_2(s)/V_1(s)$ in Figure 12–6 has a low-pass gain characteristic.
(b) Select element values so the passband gain is 4 and the cutoff frequency is 100 rad/s.
(c) The design is to be tested using a 2-V sinusoidal input $v_S(t) = 2 \cos(\omega_A t)$ at $\omega_A = 5$ rad/s, 500 rad/s, and 50 krad/s. Use the asymptotic straight-line gain to estimate the amplitude of the steady-state output at each of these frequencies.

FIGURE   12 – 5

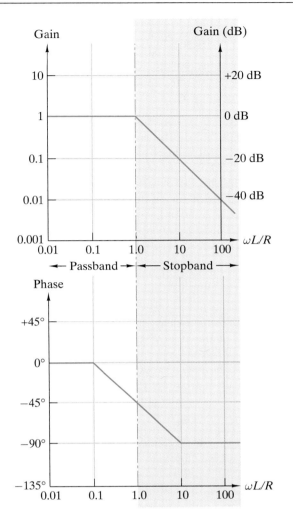

SOLUTION:

(a) The circuit is an inverting amplifier configuration with

$$Z_1(s) = R_1 \quad \text{and} \quad Z_2(s) = \cfrac{1}{C_2 s + \cfrac{1}{R_2}} = \frac{R_2}{R_2 C_2 s + 1}$$

The circuit transfer function is found as

$$T(s) = -\frac{Z_2(s)}{Z_1(s)} = -\frac{R_2}{R_1} \times \frac{1}{R_2 C_2 s + 1}$$

Rearranging the standard low-pass form in Eq. (12–4) as

$$T(s) = \frac{K/\alpha}{s/\alpha + 1}$$

shows that the circuit transfer function has a low-pass form with

FIGURE   12 – 6

$$\omega_C = \alpha = \frac{1}{R_2 C_2} \quad \text{and} \quad T(0) = -\frac{R_2}{R_1}$$

This is an inverting circuit, so the −90° phase swing of the low-pass form runs from $\angle T(0) = -180°$ to $\angle T(\infty) = -270°$, passing through $\angle T(j\omega_C) = -225°$ along the way.

Circuit Interpretation: The low-pass response is easily deduced from known circuit performance. At dc the capacitor acts like an open circuit and the circuit in Figure 12–6 reduces to a resistance inverting amplifier with $K = T(0) = -R_2/R_1$. At infinite frequency the capacitor acts like a short circuit that connects the OP AMP output directly to the inverting input. This connection results in zero output since the node voltage at the inverting input is necessarily zero. In between these two extremes the gain gradually decreases as the decreasing capacitor impedance gradually pulls the OP AMP output down to zero at infinite frequency.

(b) The design constraints require that $\omega_C = 1/R_2 C_2 = 100$ and $|T(0)| = R_2/R_1 = 4$. Selecting $R_1 = 10$ kΩ implies that $R_2 = 40$ kΩ and $C_2 = 250$ nF.

(c) The input frequency $\omega_A = 5$ rad/s falls in the passband below the cutoff frequency at $\omega_C = 100$ rad/s. In passband the asymptotic gain is $K = 4$; hence, the amplitude of the steady-state output is approximately $K \times$ input amplitude $= 4 \times 2 = 8$ V.

The input frequencies at $\omega_A = 500$ rad/s and 50 krad/s fall in the stopband above the cutoff frequency. In the stopband the asymptotic gain is

$$|T(j\omega)| = \underbrace{[K]}_{\substack{\text{Passband}\\\text{Gain}}} \times \underbrace{\left[\frac{\omega_C}{\omega}\right]}_{\substack{\text{Stopband}\\\text{Roll Off}}} = [4] \times \left[\frac{100}{\omega}\right] = \frac{400}{\omega}$$

Hence, at $\omega_A = 500$ rad/s the steady-state output would be approximately $(400/500) \times$ input amplitude $= 1.6$ V. At $\omega_A = 50$ krad/s the steady-state output would be approximately $(400/50000) \times$ input amplitude $= 16$ mV.

APPLICATION NOTE: **EXAMPLE 12–4**

In terms of frequency response, the ideal OP AMP model introduced in Chapter 4 assumes that the device has an infinite gain and an infinite bandwidth. A more realistic model of the device is shown in Figure 12–7(a). The controlled source gain in Figure 12–7(a) is a low-pass transfer function with a dc gain of A and a cutoff frequency ω_C. The straight-line asymptotes of controlled source gain are shown in Figure 12–7(b). The **gain-bandwidth product** $(G = A\omega_C)$ is the basic performance parameter of this model.

With no feedback the OP AMP transfer function is the same as the controlled-source transfer function. The gain-bandwidth product of the open-loop transfer function is

$$G = A\omega_C \quad \text{(Open Loop)}$$

FIGURE 12-7

(a)

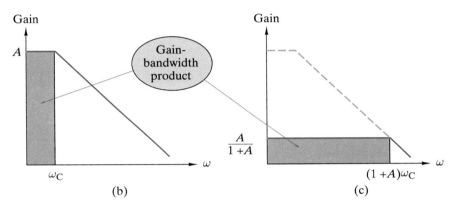

(b)   (c)

The closed-loop transfer function of the circuit in Figure 12–7(a) is found by writing the following device and connection equations:

$$\text{Device equation: } V_O = \frac{A}{s/\omega_C + 1}(V_P - V_N)$$

$$\text{Input connection: } V_P = V_S$$

$$\text{Feedback connection: } V_N = V_O$$

Substituting the connection equations into the OP AMP device equation yields

$$V_O(s) = \frac{A}{s/\omega_C + 1}(V_S(s) - V_O(s))$$

Solving for the closed-loop transfer function produces

$$T(s) = \frac{V_O(s)}{V_S(s)} = \frac{A}{A + 1}\left[\frac{1}{\dfrac{s}{(A + 1)\omega_C} + 1}\right]$$

The straight-line asymptotes of the closed-loop transfer function are shown in Figure 12–7(c).

The closed-loop circuit has a low-pass transfer function with a dc gain of $A/(A + 1)$ and a cutoff frequency of $(A + 1)\omega_C$. The gain-bandwidth product of the closed-loop circuit is

$$G = \left[\frac{A}{A + 1}\right][(A + 1)\omega_C] = A\omega_C \quad \text{(Closed Loop)}$$

which is the same as the open-loop case. In other words, the gain-bandwidth product is invariant and is not changed by feedback. It can be shown that this result is a general one that applies to all linear OP AMP circuits, regardless of the circuit configuration.

Gain-bandwidth product is a fundamental parameter that limits the frequency response of OP AMP circuits. For example, an OP AMP with a gain-bandwidth product of $G = 10^6$ Hz is connected as a noninverting amplifier with a closed–loop gain of 20. The frequency response of the resulting closed-loop circuit has a low pass characteristic with a passband gain of 20 and a cutoff frequency of

$$f_C = \frac{10^6}{20} = 50 \text{ kHz}$$

FIRST-ORDER HIGH-PASS RESPONSE

We next treat the first-order high-pass transfer function

$$T(s) = \frac{Ks}{s + \alpha} \tag{12–8}$$

The high-pass function differs from the low-pass case by the introduction of a zero at $s = 0$. Replacing s by $j\omega$ in $T(s)$ and solving for the gain and phase functions yields

$$|T(j\omega)| = \frac{|K|\omega}{\sqrt{\omega^2 + \alpha^2}}$$

$$\theta(\omega) = \angle K + 90° - \tan^{-1}(\omega/\alpha) \tag{12–9}$$

Figure 12–8 shows the gain and phase functions versus normalized frequency ω/α. The maximum gain occurs at high frequency ($\omega \gg \alpha$) where $|T(j\omega)| \to |K|$. At low frequency ($\omega \ll \alpha$) the gain approaches $|K|\omega/\alpha$. At $\omega = \alpha$ the gain is

$$|T(j\alpha)| = \frac{|K|\alpha}{\sqrt{\alpha^2 + \alpha^2}} = \frac{|K|}{\sqrt{2}} \tag{12–10}$$

which means the cutoff frequency is $\omega_C = \alpha$. The gain response plot in Figure 12–8 displays a high-pass characteristic with a passband extending from $\omega = \alpha$ to infinity and a stopband between zero frequency and $\omega = \alpha$.

The low- and high-frequency gain asymptotes approximate the gain response in Figure 12–8. The high-frequency asymptote ($\omega \gg \alpha$) is the horizontal line whose ordinate is $|K|$ (slope = 0 or 0 dB/decade). The low-frequency asymptote ($\omega \ll \alpha$) is a line of the form $|K|\omega/\alpha$ (slope = +1 or +20 dB/decade). The intersection of these two asymptotes occurs when $|K| = |K|\omega/\alpha$, which defines a corner frequency at $\omega = \alpha$.

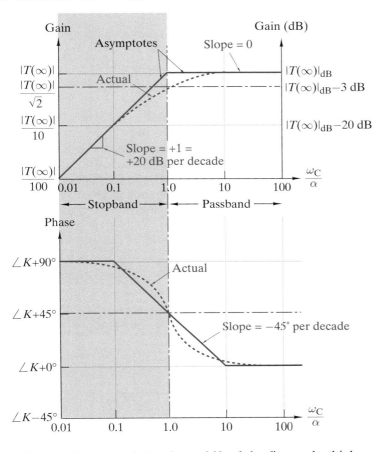

FIGURE 1 2 – 8 *First-order   high-pass frequency response plots.*

The semilog plot of the phase shift of the first-order high-pass function is shown in Figure 12–8. The phase shift approaches $\angle K$ at high frequency, passes through $\angle K + 45°$ at the cutoff frequency, and approaches $\angle K + 90°$ at low frequency. Most of the 90° phase change occurs over the two-decade range centered on the cutoff frequency. The phase shift can be approximated by the straight-line segments shown in the Figure 12–8. As in the low-pass case, $\angle K$ is 0° when K is positive and ±180° when K is negative.

Like the low-pass function, the first-order high-pass frequency response can be approximated by straight-line segments. To construct these lines we need two parameters, $T(\infty)$ and α. The parameter α defines the cutoff frequency, and the quantity $T(\infty)$ gives the passband gain $|T(\infty)|$ and the high-frequency phase angle $\angle T(\infty)$. The quantities $T(\infty)$ and α can be determined directly from the transfer function or estimated directly from the circuit in some cases. The straight line shows that the first-order high-pass response can be characterized by calculating the gain and phase over a two-decade band from one decade below to one decade above the cutoff frequency.

EXAMPLE 12–5

Show that the transfer function $T(s) = V_2(s)/V_1(s)$ in Figure 12–9 has a high-pass gain characteristic. Construct the straight-line approximations to the gain and phase responses of the circuit.

FIGURE 12-9

SOLUTION:
Applying voltage division, the voltage transfer function for the circuit is

$$T(s) = \frac{R}{R + 1/Cs} = \frac{RCs}{RCs + 1}$$

Rearranging Eq. (12–8) as

$$T(s) = \frac{K(s/\alpha)}{s/\alpha + 1}$$

shows that the circuit has a high-pass gain characteristic with $\alpha = RC$ and $T(\infty) = 1$. Therefore, $|T(\infty)|_{dB} = 0$ dB, $\omega_C = 1/RC$, and $\angle T(\infty) = 0°$. Given these quantities, we construct the straight-line gain and phase approximations in Figure 12–10. The frequency scale in Figure 12–10 is normalized by multiplying ω by $RC = 1/\alpha$.

FIGURE 12-10

Circuit Interpretation: The high-pass response in Figure 12–10 can be understood in terms of known circuit behavior. At zero frequency the capacitor acts like an open circuit that effectively disconnects the input signal source, leading to zero gain. At infinite frequency the capacitor acts like a short circuit that directly connects the input to the output, leading to a passband gain of 1 (or 0 dB). Between these two extremes the impedance of the capacitor gradually decreases, causing the gain to increase. In particular, at the cutoff frequency we have $1/\omega C = R$, the impedance of the capacitor is $1/j\omega C = -jR$, and the transfer function is

$$T(j\omega_C) = \frac{R}{R - jR} = \frac{1}{\sqrt{2}} \angle +45°$$

In other words, at the cutoff frequency the gain is –3 dB and the phase shift +45°. Obviously, the decreasing impedance of the capacitor gives the circuit its high-pass gain characteristics. ⬛

⬥ DESIGN EXAMPLE 12–6

(a) Show that the transfer function $T(s) = V_2(s)/V_1(s)$ of the circuit in Figure 12–11 has a high-pass gain characteristic.

FIGURE 12 – 11

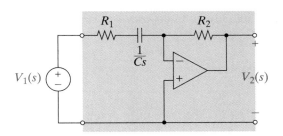

(b) Select the element values to produce a passband gain of 4 and a cutoff frequency of 40 krad/s.

SOLUTION:

(a) The branch impedances of the inverting OP AMP configuration in Figure 12–11 are

$$Z_1(s) = R_1 + \frac{1}{Cs} = \frac{R_1 Cs + 1}{Cs} \quad \text{and} \quad Z_2(s) = R_2$$

and the voltage transfer function is

$$T(s) = -\frac{Z_2(s)}{Z_1(s)} = -\frac{R_2 Cs}{R_1 Cs + 1} = \frac{(-R_2/R_1)s}{s + 1/R_1 C}$$

This results in a high-pass transfer function of the form $Ks/(s + \alpha)$ with $K = -R_2/R_1$ and $\alpha = \omega_C = 1/R_1 C$.

Circuit Interpretation: The high-pass response of this circuit is easily understood in terms of element impedances. At dc the capacitor in Figure 12–11

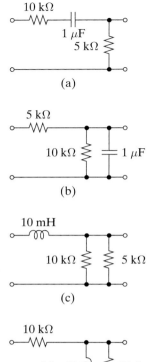

(a)

(b)

(c)

(d)

FIGURE   12 – 12

acts like an open circuit that effectively disconnects the input source, resulting in zero gain. At infinite frequency the capacitor acts like a short circuit that reduces the circuit to an inverting amplifier with $K = T(\infty) = -R_2/R_1$. As the frequency varies from zero to infinity, the gain gradually increases as the capacitor impedance decreases.

(b) The design requirements specify that $1/R_1C = 4 \times 10^4$ and $R_2/R_1 = 4$. Selecting $R_1 = 10 \text{ k}\Omega$ requires $R_2 = 40 \text{ k}\Omega$ and $C = 2.5 \text{ nF}$.   ∎

Exercise 12–1

For each circuit in Figure 12–12 identify whether the gain response has low-pass or high-pass characteristics and find the passband gain and cutoff frequency.

Answers:
(a) High pass, $|T(\infty)| = 1/3$, $\omega_C = 66.7 \text{ rad/s}$
(b) Low pass, $|T(0)| = 2/3$, $\omega_C = 300 \text{ rad/s}$
(c) Low pass, $|T(0)| = 1$, $\omega_C = 333 \text{ krad/s}$
(d) High pass, $|T(\infty)| = 1/3$, $\omega_C = 333 \text{ krad/s}$

Exercise 12–2

A first-order circuit has a step response $g(t) = 5[e^{-2000t}]u(t)$.
(a) Is the circuit frequency response low-pass or high-pass?
(b) Find the passband gain and cutoff frequency.

Answers:
(a) High-pass
(b) $|T(\infty)| = 5$ and $\omega_C = 2000 \text{ rad/s}$

BANDPASS AND BANDSTOP RESPONSES USING FIRST-ORDER CIRCUITS

The first-order high-pass and low-pass circuits can be used in a building block fashion to produce circuits with bandpass or bandstop responses. Figure 12–13 shows a cascade connection of first-order high-pass and low-pass circuits. When the second stage does not load the first, the overall transfer function can be found by the chain rule:

$$T(s) = T_1(s) \times T_2(s) = \underbrace{\left(\frac{K_1 s}{s + \alpha_1}\right)}_{\text{High Pass}} \underbrace{\left(\frac{K_2}{s + \alpha_2}\right)}_{\text{Low Pass}} \qquad (12\text{–}11)$$

Replacing s by $j\omega$ in Eq. (12–11) and solving for the gain response yields

$$|T(j\omega)| = \underbrace{\left(\frac{|K_1|\omega}{\sqrt{\omega^2 + \alpha_1^2}}\right)}_{\text{High Pass}} \underbrace{\left(\frac{|K_2|}{\sqrt{\omega^2 + \alpha_2^2}}\right)}_{\text{Low Pass}} \qquad (12\text{–}12)$$

Note that the gain of the cascade is zero at $\omega = 0$ and at infinite frequency.

FIGURE   12 – 13 *Cascade connection of high-pass and low-pass circuits.*

When $\alpha_1 \ll \alpha_2$ the high-pass cutoff frequency is much lower than the low-pass cutoff frequency, and the overall transfer function has a bandpass characteristic. At low frequencies ($\omega \ll \alpha_1 \ll \alpha_2$) the gain approaches $|T(j\omega)| \to |K_1 K_2| \omega / \alpha_1 \alpha_2$. At mid frequencies ($\alpha_1 \ll \omega \ll \alpha_2$) the gain approaches $|T(j\omega)| \to |K_1 K_2| / \alpha_2$. The low- and mid-frequency asymptotes intersect when $|K_1 K_2| \omega / \alpha_1 \alpha_2 = |K_1 K_2| / \alpha_2$ at $\omega = \alpha_1$ (that is, at the cutoff frequency of the high-pass stage). At high frequencies ($\alpha_1 \ll \alpha_2 \ll \omega$) the gain approaches $|T(j\omega)| \to |K_1 K_2| / \omega$. The high- and mid-frequency asymptotes intersect when $|K_1 K_2| / \omega = |K_1 K_2| / \alpha_2$ at $\omega = \alpha_2$ (that is, at the cutoff frequency of the low-pass stage). The plot of these asymptotes in Figure 12–14 shows that the asymptotic gain exhibits a passband between α_1 and α_2. Input sinusoids whose frequencies are outside of this range fall in one of the two stopbands.

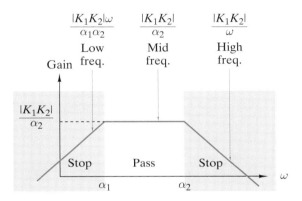

FIGURE 12-14 *Asymptotic gain of a bandpass characteristic.*

In the bandpass cascade connection the input signal must pass both a low- and a high-pass stage to reach the output. In the parallel connection in Figure 12–15 the input can reach the output via either a low- or a high-pass path. The overall transfer function is the sum of the low- and high-pass transfer functions

$$|T(j\omega)| = \underbrace{\left(\frac{|K_1| \omega}{\sqrt{\omega^2 + \alpha_1^2}} \right)}_{\text{High Pass}} + \underbrace{\left(\frac{|K_2|}{\sqrt{\omega^2 + \alpha_2^2}} \right)}_{\text{Low Pass}} \qquad (12\text{–}13)$$

FIGURE 12-15 *Parallel connection of high-pass and low-pass circuits.*

Any sinusoid whose frequency falls in either passband will find its way to the output unscathed. An input sinusoid whose frequency falls in both stopbands will be attenuated.

When $\alpha_1 \gg \alpha_2$ the high-pass cutoff frequency is much higher than the low-pass cutoff frequency, and the overall transfer function has a bandstop gain response as shown in Figure 12–16. At low frequencies ($\omega \ll \alpha_2 \ll \alpha_1$) the high-pass gain is negligible and the over asymptote approaches the passband gain of the low-pass function $|T(j\omega)| \to |K_2|/\alpha_2$. At high frequencies ($\alpha_2 \ll \alpha_1 \ll \omega$) the low-pass gain is negligible and the overall gain approaches the bandpass gain of the high-pass function $|T(j\omega)| \to |K_1|$. With a bandstop function the two passbands normally have the same gain, hence $|K_1| = |K_2|/\alpha_2$. Between these two passbands there is a stopband. For $\omega > \alpha_2$ the low-pass asymptote is $|K_2|/\omega$ and for $\omega < \alpha_1$ the high-pass asymptote is $|K_1|\omega/\alpha_1$. The asymptotes intersect at $\omega^2 = \alpha_1|K_2|/|K_1|$. But equal gains in the two passband frequencies require $|K_1| = |K_2|/\alpha_2$, so the intersection frequency is $\omega = \sqrt{\alpha_1\alpha_2}$. Below this frequency the stopband attenuation is determined by the low-pass function, and above this frequency the attenuation is governed by the high-pass function.

FIGURE 12–16 *Asymptotic gain of a bandstop characteristic.*

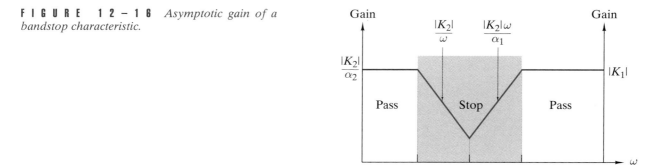

The analysis of the bandpass and bandstop transfer functions in Eqs. (12–12) and (12–13) illustrates that the asymptotic gain plots of first-order functions can help us understand and describe other types of gain response as well. The asymptotic responses in Figures 12–14 and 12–16 are a reasonably good approximation as long as the two first-order cutoff frequencies are widely separated. The straight-line approximation shows us that the frequency range of interest extends from a decade below the lowest cutoff frequency to a decade above the highest. This range could be a very wide indeed, since the two cutoff frequencies may be separated by several decades.

D DESIGN EXAMPLE 12–7

(a) Design a bandpass circuit with a passband gain of 10 and cutoff frequencies at 20 Hz and 20 kHz.
(b) The circuit is to be tested using a 2-V sinusoidal input at $f_A = 6$ Hz, 600 Hz, and 60 kHz. Use the asymptotic straight-line gain to estimate the amplitude of the steady-state output at each of these frequencies.

SOLUTION:
(a) Our design uses a cascade connection of first-order low- and high-pass building blocks. The required transfer function has the form

$$T(s) = \left(\frac{K_1 s}{s + \alpha_1}\right)\left(\frac{K_2}{s + \alpha_2}\right)$$

with the following constraints:

Lower Cutoff Frequency: $\alpha_1 = 2\pi(20) = 40\pi$ rad/s

Upper Cutoff Frequency: $\alpha_2 = 2\pi(20 \times 10^3) = 4\pi \times 10^4$ rad/s

Midband Gain: $\dfrac{|K_1 K_2|}{\alpha_2} = 10$

With numerical values inserted, the required transfer function is

$$T(s) = \underbrace{\frac{s}{s + 40\pi}}_{\text{High Pass}} \times \underbrace{10}_{\text{Gain}} \times \underbrace{\frac{4\pi \times 10^4}{s + 4\pi \times 10^4}}_{\text{Low Pass}}$$

This transfer function can be realized using the high-pass/low-pass cascade circuit in Figure 12–17. The first stage is the RC high-pass circuit from Example 12–5 and the third stage is the RL low-pass circuit from Example 12–2. The noninverting OP AMP second stage serves two purposes: (1) It isolates the first and third stages, so the chain rule applies, and (2) it supplies the midband gain. Using the chain rule, the transfer function of this circuit is

$$T(s) = \underbrace{\left[\frac{s}{s + 1/R_C C}\right]}_{\text{High Pass}} \underbrace{\left[\frac{R_1 + R_2}{R_1}\right]}_{\text{Gain}} \underbrace{\left[\frac{R_L/L}{s + R_L/L}\right]}_{\text{Low Pass}}$$

FIGURE   12 – 1 7

Comparing this to the required transfer function leads to the following design constraints:

High-pass Stage: $R_C C = 1/40\pi$ Let $R_C = 100$ kΩ Then $C = 79.6$ nF

Gain Stage: $(R_1 + R_2)/R_1 = 10$ Let $R_1 = 10$ kΩ Then $R_2 = 90$ kΩ

Low-pass Stage: $R_L/L = 40000\pi$ Let $R_L = 200$ kΩ Then $L = 0.628$ H

(b) The input frequency $f_A = 6$ Hz falls in the lower stopband below the lower cutoff frequency at $f_{C1} = 20$ Hz. In this region the circuit gain is controlled by the high-pass and gain stages. The asymptotic straight-line gain in this region is

$$|T(j\omega)| = \underbrace{\left[\frac{\omega}{2\pi f_{C1}}\right]}_{\substack{\text{High-pass} \\ \text{Stage}}} \times \underbrace{[10]}_{\substack{\text{Gain} \\ \text{Stage}}} \times \underbrace{[1]}_{\substack{\text{Low-pass} \\ \text{Stage}}} = \frac{f}{f_{C1}} \times 10 = \frac{f}{2}$$

and the amplitude of the steady-state output is approximately $(f/2) \times$ input amplitude $= 3 \times 2 = 6$ V.

The input frequency $f_A = 600$ Hz falls in the passband between the lower cutoff frequency at $f_{C1} = 20$ Hz and the upper cutoff frequency at $f_{C2} = 20$ kHz. In this region the circuit gain is controlled by the gain stage and the amplitude of the steady-state output would be approximately $10 \times$ input amplitude $= 20$ V.

The input frequency $f_A = 60$ kHz falls in the upper stopband above the upper cutoff frequency at $f_{C2} = 20$ kHz. In this region the circuit gain is controlled by the low-pass and gain stages. The asymptotic straight-line gain in this region is

$$|T(j\omega)| = \underbrace{[1]}_{\substack{\text{High-pass} \\ \text{Stage}}} \times \underbrace{[10]}_{\substack{\text{Gain} \\ \text{Stage}}} \times \underbrace{\left[\frac{2\pi f_{C2}}{\omega}\right]}_{\substack{\text{Low-pass} \\ \text{Stage}}} = [10] \times \left[\frac{f_{C2}}{f}\right] = \frac{2 \times 10^5}{f}$$

and the amplitude of the steady-state output is approximately $(2 \times 10^5/6 \times 10^4) \times$ input amplitude $= (10/3) \times 2 = 6.67$ V. ∎

EXAMPLE 12–8

Use computer-aided analysis to show that the circuit in Figure 12–18 implements a bandstop gain response.

FIGURE 12–18

SOLUTION:

This circuit implements the block diagram in Figure 12–15. The two first-order circuits have the same input. The upper circuit is the high-pass OP AMP circuit designed in Example 12–6 to have a passband gain of 4 and a cutoff frequency $\alpha_1 = 40$ krad/s. The lower path is the low-pass OP AMP circuit designed in Example 12–3 to have a passband gain of 4 and a cutoff frequency $\alpha_2 = 100$ rad/s. The inverting summer at the far right implements the summing point function in Figure 12–15. Since the cutoff frequency of the low pass circuit is much lower than the cutoff frequency of the high-pass circuit, we expect to see two passbands on either side of a stopband centered at

$$\omega = \sqrt{\alpha_1 \alpha_2} = \sqrt{100 \times 4 \times 10^4} = 2000 \ \text{rad/s}$$

Figure 12–19 shows the circuit as drawn in the Electronics Workbench circuit window. The input voltage source $\mathbf{V}_1$ is set to produce 1-V ac, so the circuit gain is equal to the ac voltage at node 8 (the summer output). Selecting **AC Frequency** from the **Analysis** menu brings up the **AC Frequency Analysis** dialog box shown in the figure. Here we must specify the analysis frequency range. This range should extend from at least a decade below the lowest cutoff frequency ($\alpha_2/2\pi = 16$ Hz) to at least a decade above the highest cutoff frequency ($\alpha_1/2\pi = 6.4$ kHz). To span this range we select a **Start frequency (FSTART)** of 1 Hz and an **End frequency(FSTOP)** of 100 kHz. To span this range we choose a **Decade** sweep and select the summer output (node 8) as the **Node for analysis**. Clicking the simulate button launches an ac frequency analysis that produces the output voltage plot shown at the bottom of Figure 12–19.

As expected, the frequency response of the overall circuit has two passbands surrounding a stopband. In this plot one cursor is set in the lower passband and the other at the stopband minimum. The passband output voltage is $y_1 = 3.9921$ V (gain = 12 dB) at $x_1 = 1.0000 = 1$ Hz, which confirms that the two passbands have gain = 4 as straight-line analysis predicts. The minimum output is $y_2 = 20.1212$ mV (gain = –33.9 dB) at $x_2 = 316.2278$ Hz. This notch frequency corresponds to a radian frequency of $\omega = 2\pi x_2 = 1987$ rad/s, which is very close to the 2000 rad/s predicted by straight-line gain analysis. ■

12–3 S E C O N D - O R D E R   C I R C U I T   F R E Q U E N C Y   R E S P O N S E

We begin our study with the second-order bandpass transfer function:

$$T(s) = \frac{Ks}{s^2 + 2\zeta\omega_0 s + \omega_0^2} \tag{12–14}$$

The second-order polynomial in the denominator is expressed in terms of the damping ratio ζ and undamped natural frequency ω_0 parameters. These parameters play a key role in describing the frequency response of second-order circuits. When the damping ratio is less than 1, the complex poles of $T(s)$ dramatically influence the frequency response.

To gain a qualitative understanding of the effect of complex poles, we write Eq. (12–14) in factored form:

$$T(s) = \frac{Ks}{(s - p_1)(s - p_2)} \tag{12–15}$$

where p_1 and p_2 are the complex poles of $T(s)$. The frequency-response gain function is

$$|T(j\omega)| = \frac{|K|\omega}{|j\omega - p_1||j\omega - p_2|} = \frac{|K|\omega}{M_1 M_2} \tag{12–16}$$

As shown in Figure 12–20, the factors $(j\omega - p_1)$ and $(j\omega - p_2)$ can be interpreted as vectors from the natural poles at $s = p_1$ and $s = p_2$ to the point $s = j\omega$, where ω is the frequency at which the gain is being calculated. The lengths of these vectors are

$$M_1 = |j\omega - p_1|$$

$$M_2 = |j\omega - p_2|$$

The frequency response of a circuit depends on how the circuit responds to sinusoidal inputs of constant amplitude by changing frequencies. The sinusoidal input produces two forced poles on the j-axis whose location depends of the frequency of the input. Changing the input frequency causes the forced pole at $s = j\omega$ to slide up or down the imaginary axis. This, in turn, changes the lengths M_1 and M_2, since the tips of the vectors $j\omega - p_1$ and $j\omega - p_2$ move up or down the j-axis while the tails remain fixed at the location of the natural poles. The length M_1 reaches a minimum when the forced pole at $s = j\omega$ is close to the natural pole at $s = p_1$. Since M_1 appears in the denominator of Eq. (12–16), this minimum tends to produce a maximum or peak in the gain response.

Figure 12–21 shows how different input frequencies change the lengths M_1 and M_2 to produce the gain response curve. The lengths M_1 and M_2 are approximately equal at low frequencies ($\omega << \omega_1$) and again at high frequencies ($\omega >> \omega_3$). At intermediate frequencies ($\omega_1 < \omega < \omega_3$) the lengths differ, especially in the neighborhood of ω_2, where length M_1 reaches a minimum and is very small compared with length M_2. The small value of the length M_1 produces a maximum in the gain response indicative of a phenomenon called resonance. **Resonance** is the increased response observed when the frequency of a periodic driving force like a sinusoid is close to a natural frequency of the circuit. In terms of s-plane geometry, resonance occurs when the point at $s = j\omega$ and the natural pole at $s = p_1$ are close together so the length M_1 is small.

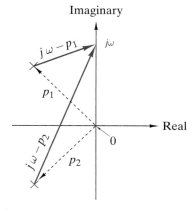

FIGURE 12-20 *s-Plane vectors defining* $|T(j\omega)|$.

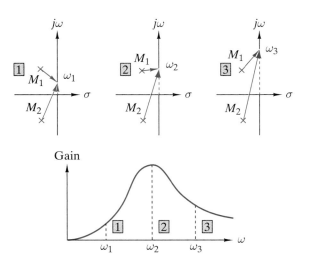

FIGURE 12-21 *Gain response resonance from s-plane geometry.*

The resonant peak decreases when p_1 and p_2 are shifted to the left in the s plane. Replacing s by $s + \alpha$ produces such a shift. The s-domain translation property

$$\mathcal{L}^{-1}\{F(s + \alpha)\} = e^{-\alpha t}f(t)$$

then shows that shifting the poles of $F(s)$ to the left increases the damping in the time domain. Qualitatively, we expect resonant peaks in the frequency response to be less pronounced when circuit damping increases and more pronounced when it decreases.

SECOND-ORDER BANDPASS CIRCUITS

To develop a quantitative understanding of resonance, we replace s by $j\omega$ in Eq. (12–14) to produce

$$T(j\omega) = \frac{Kj\omega}{-\omega^2 + 2\zeta\omega_0 j\omega + \omega_0^2} \tag{12–17}$$

At low frequencies ($\omega \ll \omega_0$) the gain approaches $|T(j\omega)| \to |K|\omega/\omega_0^2$. At high frequencies ($\omega \gg \omega_0$) the gain approaches $|T(j\omega)| \to |K|/\omega$. The low-frequency asymptote is directly proportional to frequency (slope = +1 or +20 dB/decade), while the high-frequency asymptote is inversely proportional to frequency (slope = −1 or −20 dB/decade). As shown in Figure 12–22, the two asymptotes intersect at $\omega = \omega_0$. At this intersection the gain is $|K|/\omega_0$. Note that the frequency scale in Figure 12–16 is logarithmic and normalized by dividing ω by the undamped natural frequency ω_0.

FIGURE 12–22 *Asymptotic gain of a second-order bandpass response.*

The two asymptotes display the characteristics of a bandpass response located around ω_0. For this reason, in a second-order bandpass circuit ω_0 is also called the **center frequency**. To examine the gain response in more detail, we must consider the effect of circuit damping. Factoring $j\omega\omega_0$ out of the denominator of Eq. (12–17) produces

$$T(j\omega) = \frac{K/\omega_0}{2\zeta + j\left(\dfrac{\omega}{\omega_0} - \dfrac{\omega_0}{\omega}\right)} \tag{12–18}$$

Arranging the transfer function in this form shows that only the imaginary part in the denominator varies with frequency. Clearly, the maximum gain occurs when the imaginary part in the denominator of Eq. (12–18) vanishes at $\omega = \omega_0$. The maximum gain is

$$|T(j\omega)|_{\text{MAX}} = \frac{|K|/\omega_0}{2\zeta} \qquad (12\text{–}19)$$

Since the maximum gain is inversely proportional to the damping ratio, the resonant peak will increase as circuit damping decreases, and vice versa. The quantity $|K|/\omega_0$ in the numerator of Eq. (12–19) is the gain at the point in Figure 12–22 where the low- and high-frequency asymptotes intersect.

Figure 12–23 shows plots of the gain of the bandpass function for several values of the damping ratio ζ. For $\zeta < 0.5$ the gain curve lies above the asymptotes and has a narrow resonant peak at ω_0. For $\zeta > 0.5$ the gain curve flattens out and lies below the asymptotes. When $\zeta = 0.5$ the gain curve remains fairly close to the asymptotes.

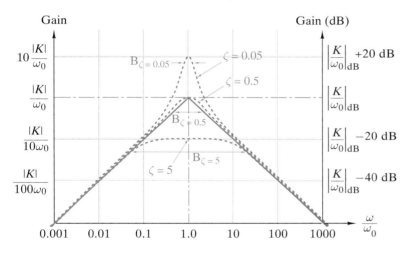

FIGURE 12–23 *Second-order bandpass gain responses.*

The gain response has a bandpass characteristic regardless of the value of the damping ratio. The passband is defined by a cutoff frequency on either side of the center frequency ω_0. To locate the two cutoff frequencies, we must find the values of ω at which the gain is $|T(j\omega)|_{\text{MAX}}/\sqrt{2}$, where the maximum gain is given in Eq. (12–19). We note that when the imaginary part in the denominator of Eq. (12–18) equals $\pm 2\zeta$, the gain is

$$|T(j\omega)| = \frac{|K|/\omega_0}{|2\zeta \pm j2\zeta|} = \frac{\dfrac{|K|/\omega_0}{2\zeta}}{\sqrt{2}} = \frac{|T(j\omega)|_{\text{MAX}}}{\sqrt{2}} \qquad (12\text{–}20)$$

The values of frequency ω that cause the imaginary part to be $\pm j2\zeta$ are the cutoff frequencies because at these points the gain is reduced by a factor of $1/\sqrt{2}$ from its maximum value at $\omega = \omega_0$.

To find the cutoff frequencies, we set the imaginary part in the denominator in Eq. (12–18) equal to $\pm 2\zeta$:

$$\frac{\omega}{\omega_0} - \frac{\omega_0}{\omega} = \pm 2\zeta \qquad (12\text{–}21)$$

which yields the quadratic equation

$$\omega^2 \mp 2\zeta\omega_0\omega - \omega_0 = 0 \tag{12-22}$$

Because of the $\mp$ sign this quadratic has four roots:

$$\omega_{C1}, \ \omega_{C2} = \omega_0(\pm \ \zeta \ \pm \ \sqrt{1 + \zeta^2}) \tag{12-23}$$

Only the two positive roots have physical significance. Since $\sqrt{1 + \zeta^2} > \zeta$, the two positive roots are

$$\omega_{C1} = \omega_0(-\zeta + \sqrt{1 + \zeta^2})$$
$$\omega_{C2} = \omega_0(+\zeta + \sqrt{1 + \zeta^2}) \tag{12-24}$$

Since $\zeta > 0$, these equations show that $\omega_{C1} < \omega_0$ is the lower cutoff frequency, while $\omega_{C2} > \omega_0$ is the upper cutoff frequency. Multiplying ω_{C1} times ω_{C2} produces

$$\omega_0^2 = \omega_{C1}\omega_{C2} \tag{12-25}$$

This result means that the center frequency ω_0 is the geometric mean of the two cutoff frequencies. The **bandwidth (B)** of the passband is found by subtracting ω_{C1} from ω_{C2}:

$$B = \omega_{C2} - \omega_{C1} = 2\zeta\omega_0 \tag{12-26}$$

The bandwidth is proportional to the product of the damping ratio and the natural frequency. When $\zeta < 0.5$ then $B < \omega_0$ and when $\zeta > 0.5$ then $B > \omega_0$.

We can think of $\zeta = 0.5$ as the boundary between two extreme cases. In the **narrowband** case ($\zeta << 0.5$) the complex natural poles are close to the j-axis and the gain response has a resonant peak that produces a bandwidth that is small compared to the circuit's natural frequency. In the **wideband** case ($\zeta >> 0.5$) the gain response is relatively flat because the natural poles are farther from the j-axis because the circuit has more damping. The narrowband response is highly selective, passing only a very restricted range of frequencies, often less than one octave. In contrast, the wideband response usually encompasses several decades within its passband. Both cases are widely used in applications.

The narrow and broadband cases also show up in the phase response. From Eq. (12–18) the phase angle is

$$\theta(\omega) = \angle K - \tan^{-1}\left[\frac{\omega/\omega_0 - \omega_0/\omega}{2\zeta}\right] \tag{12-27}$$

At low frequency ($\omega << \omega_0$) the phase angle approaches $\theta(\omega) \rightarrow \angle K + 90°$. At the center frequency ($\omega = \omega_0$) the phase reduces to $\theta(\omega_0) = \angle K$. Finally, at high frequencies ($\omega >> \omega_0$) the phase approaches $\theta(\omega) \rightarrow \angle K - 90°$. Graphs of the phase response for several values of damping are shown in Figure 12–24. The phase shift at the center frequency is 90° and the total phase change is −180° regardless of the damping ratio. However, the transition from +90° to −90° is very abrupt for the narrowband case ($\zeta = 0.05$) and more gradual for the wideband case ($\zeta = 5$).

Phase

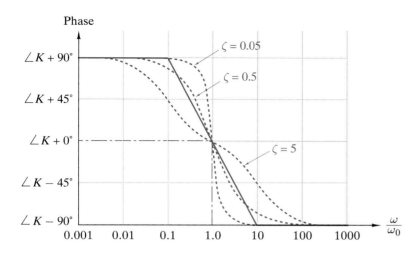

FIGURE   12-24   *Second-order band-pass phase responses.*

EXAMPLE 12-9

(a) Show that the transfer function $T(s) = V_2(s)/V_1(s)$ of the circuit in Figure 12–25 has a bandpass characteristic.

(b) Select the element values to obtain a center frequency of 455 kHz and a bandwidth of 10 kHz.[2]

(c) Calculate the cutoff frequencies for the values selected in (b).

FIGURE   12-25

SOLUTION:

(a) The given circuit is a voltage divider with branch impedances

$$Z_1 = R$$

$$Z_2 = \frac{1}{Cs + \dfrac{1}{Ls}} = \frac{Ls}{LCs^2 + 1}$$

By voltage division, the transfer function is

$$T(s) = \frac{Z_2}{Z_1 + Z_2} = \frac{Ls}{RLCs^2 + Ls + R}$$

$$= \frac{s/RC}{s^2 + s/RC + 1/LC}$$

This transfer function is of the standard bandpass form $Ks/(s^2 + 2\zeta\omega_0 s + \omega_0^2)$.

Circuit Interpretation: The circuit's bandpass response is easily explained by the element impedances at zero and infinite frequency. At zero frequency the inductor acts like a short circuit that forces the output down to zero. At infinite frequency the capacitor is a short circuit again, forcing the

---

2 These values are typical of the bandpass filters in a conventional AM radio receiver.

output to zero. Thus, the circuit has zero gain at dc and zero gain at infinite frequency, which is the characteristic pattern of a bandpass response.

(b) Comparing the circuit's transfer function with the standard form yields the following relationships:

$$\omega_0^2 = \frac{1}{LC}$$

$$\zeta = \frac{\sqrt{L/C}}{2R}$$

The center frequency and bandwidth requirements impose the following numerical values:

$$\omega_0 = 2\pi \times 455 \times 10^3 = 2.86 \times 10^6 \text{ rad/s}$$

$$\zeta = \frac{B}{2\omega_O} = \frac{2\pi \times 10 \times 10^3}{2(2\pi \times 455 \times 10^3)} = 0.0110$$

and the resulting design constraints on the element values are

$$LC = \frac{1}{\omega_0^2} = 1.22 \times 10^{-13} \quad \text{and} \quad \frac{\sqrt{L/C}}{2R} = \zeta = 0.011$$

Selecting $C = 10$ nF yields $L = 12.2$ μH and $R = 1.59$ kΩ.

(c) Using Eq. (12–24), the cutoff frequencies are

$$f_{C1} = \frac{\omega_0(-\zeta + \sqrt{1 + \zeta^2})}{2\pi} = 450.2 \text{ kHz}$$

$$f_{C2} = \frac{\omega_0(+\zeta + \sqrt{1 + \zeta^2})}{2\pi} = 460.2 \text{ kHz}$$

Since the damping ratio is very small, the natural poles are quite close to the j-axis, producing a very narrow resonant peak at 455 kHz. ∎

Exercise 12–3

Find the natural frequency, damping ratio, cutoff frequencies, bandwidth, and maximum gain of the following bandpass transfer functions:

$$\text{(a) } T(s) = \frac{500s}{s^2 + 200s + 10^6}$$

$$\text{(b) } T(s) = \frac{5 \times 10^{-3} s}{2.5 \times 10^{-5}s^2 + 5 \times 10^{-3}s + 1}$$

Answers:

	ω_0(rad/s)	ζ	ω_{C1}(rad/s)	ω_{C2}(rad/s)	B(rad/s)	$|T(j\omega)|_{MAX}$
(a)	1000	0.1	905	1105	200	2.5
(b)	200	0.5	123.6	323.6	200	1

Exercise 12–4

A bandpass circuit has a center frequency of $\omega_0 = 1$ krad/s and a bandwidth of $B = 10$ krad/s.

Find the damping ratio ζ and cutoff frequencies. Is this a narrowband or broad-band circuit?

Answers:
$\zeta = 5$, $\omega_{C1} = 99.02$ rad/s, $\omega_{C2} = 10.099$ krad/s, broadband

Exercise 12-5

A bandpass circuit has cutoff frequencies at $\omega_{C1} = 20$ krad/s and $\omega_{C2} = 50$ krad/s. Find the center frequency ω_0, damping ratio ζ, and bandwidth B.

Answers:
$\omega_0 = 31.6$ krad/s, $\zeta = 0.474$, $B = 30$ krad/s

SECOND-ORDER LOW-PASS CIRCUITS

The transfer function of a second-order low-pass prototype has the form

$$T(s) = \frac{K}{s^2 + 2\zeta\omega_0 s + \omega_0^2} \qquad (12\text{--}28)$$

The frequency response is found by replacing s by $j\omega$.

$$T(j\omega) = \frac{K}{\omega_0^2 - \omega^2 + j2\zeta\omega_0\omega} \qquad (12\text{--}29)$$

At low frequencies ($\omega \ll \omega_0$) the gain approaches $|T(j\omega)| \to |K|/\omega_0^2 = |T(0)|$. At high frequencies ($\omega \gg \omega_0$) the gain approaches $|T(j\omega)| \to |K|/\omega^2$. The low-frequency asymptote is a constant equal to the zero frequency or dc gain. The high-frequency asymptote is inversely proportional to the square of the frequency. As a result, the high-frequency gain asymptote has a slope of -40 dB/decade (slope $= -2$). The two asymptotes intersect to form a corner at $\omega = \omega_0$, as indicated in Figure 12–26. The two asymptotes display a low-pass gain characteristic with a passband below ω_0, and a stopband above ω_0. The passband gain is the dc gain $|T(0)|$ and the slope of the gain asymptote in the stopband is -40 dB/decade.

FIGURE 12-26 *Asymptotic gain of a second-order low-pass response.*

The influence of the damping ratio on the gain response can be illustrated by evaluating the gain at $\omega = \omega_0$:

$$|T(j\omega_0)| = \frac{|K|/\omega_0^2}{2\zeta} = \frac{|T(0)|}{2\zeta} \tag{12–30}$$

This result and the plots in Figure 12–27 show that the actual gain curve lies above the asymptotes when $\zeta < 0.5$, below the asymptotes when $\zeta > 0.5$, and close to the asymptotes when $\zeta = 0.5$. When $\zeta = 1/\sqrt{2}$ then the gain at $\omega = \omega_0$ is

$$|T(j\omega_0)| = \frac{|T(0)|}{\sqrt{2}} \qquad (\zeta = 1/\sqrt{2}) \tag{12–31}$$

F I G U R E   1 2 – 2 7 *Second-order low-pass gain responses.*

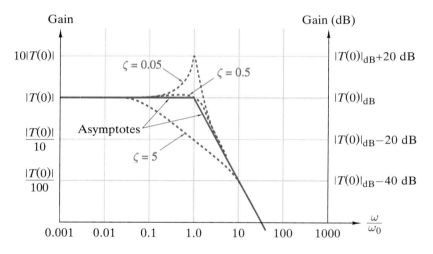

Since the passband gain is $|T(0)|$, this result means that the cutoff frequency is equal to ω_0 when $\zeta = 1/\sqrt{2}$

The distinctive feature of the low-pass case is a narrow resonant peak in the neighborhood of ω_0 for lightly damped circuits. The actual maximum gain of a second-order low-pass response is

$$|T(j\omega)|_{\text{MAX}} = \frac{|T(0)|}{2\zeta\sqrt{1 - \zeta^2}} \tag{12–32}$$

and occurs at

$$\omega_{\text{MAX}} = \omega_0\sqrt{1 - 2\zeta^2} \tag{12–33}$$

Note that the ω_{MAX} is always less than the natural frequency ω_0. Deriving Eqs. (12–32) and (12–33) is straightforward and is left as a problem at the end of this chapter (Problem 12–20).

Figure 12–28 compares the response of the first-order low-pass prototype with the second-order gain response for $\zeta = 1/\sqrt{2}$. The important difference is that the gain slope in the stopband is –2 or –40 dB/decade for the second-order response and only –20 dB/decade for the first-order circuit. The two poles in the second-order case produce a steeper slope because the high-frequency response decreases as $1/\omega^2$, rather than $1/\omega$ as in the first-order case. The generalization is that the high-frequency gain asymptote of an n-pole low-pass function has a slope of $-n$ or $-20n$ dB/decade.

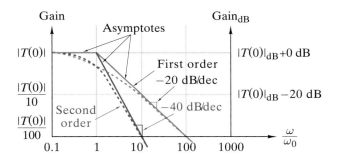

FIGURE 12–28 *Comparison of first- and second-order gain responses.*

EXAMPLE 12–10

(a) Show that the transfer function $T(s) = V_2(s)/V_1(s)$ of the circuit in Figure 12–29 has a low-pass gain characteristic.

FIGURE 12–29

(b) Derive expressions relating the damping ratio ζ and undamped natural frequency ω_0 to the circuit element values. Use these relationships to select the element values to produce $\zeta = 0.1$ and $\omega_0 = 5$ krad/s.

SOLUTION:

(a) The transfer function of this circuit is found in Example 11–6 using node-voltage analysis.

$$T(s) = \frac{V_2(s)}{V_1(s)} = \frac{\mu}{(RCs)^2 + (3 - \mu)RCs + 1}$$

This is a second-order low-pass transfer function of the form $K/(s^2/\omega_0^2 + 2\zeta s/\omega_0 + 1)$.

Circuit Interpretation: At zero frequency both capacitors in Figure 12–29 act like open circuits and the control voltage $V_X(s) = V_1(s)$ since there is no voltage drop across either resistor. As a result, the dc output voltage is $V_2(s) = \mu V_X(s) = \mu V_1$. In other words, the dc gain of the circuit is $T(0) = \mu$. At infinite frequency both capacitors act like short circuits. The shunt capacitor shorts the control node X to ground so that $V_X = 0$, the dependent voltage source μV_X is effectively turned off, and the output voltage is zero. The circuit has a finite gain at dc and zero gain at infinite frequency, which is the characteristic pattern of a low-pass response.

(b) Comparing the circuit transfer function with the standard form yields the following relationships:

$$\omega_0 = \frac{1}{RC} \quad \text{and} \quad \zeta = \frac{1}{2}(3 - \mu)$$

For the specified values the design constraints on the element values are

$$RC = \frac{1}{\omega_0} = 2 \times 10^{-4} \quad \text{Let } R = 10\text{ k}\Omega \quad \text{Then } C = 20\text{ nF}$$

$$\mu = 3 - 2\zeta = 2.8$$

The maximum gain for these conditions is

$$|T(j\omega)|_{MAX} = \frac{|T(0)|}{2\zeta\sqrt{1 - \zeta^2}} = 5.03 \times |T(0)|$$

The peak gain is five times as large as the dc gain. Thus, lightly damped second-order circuits of all types have strong resonant peaks in their gain response. ∎

Exercise 12–6

Find the natural frequency ω_0 damping ratio ζ, dc gain $T(0)$, and maximum gain of the following second-order low-pass transfer functions:

(a) $T(s) = \dfrac{(4000)^2}{(s + 3000)^2 + (4000)^2}$

(b) $T(s) = \dfrac{5}{0.25 \times 10^{-4}s^2 + 10^{-2}s + 1}$

Answers:

(a) $\omega_0 = 5$ krad/s, $\zeta = 0.6$, $T(0) = 0.64$, $T_{MAX} = 0.667$
(b) $\omega_0 = 200$ rad/s, $\zeta = 1.00$, $T(0) = 5$, $T_{MAX} = T(0)$

Exercise 12–7

The impulse response of a second-order low-pass circuit is

$$h(t) = 2000[e^{-1000t} \sin 1000t]u(t)$$

Find the natural frequency ω_0, damping ratio ζ, dc gain $T(0)$, and maximum gain.

Answers:
$\omega_0 = 1414$ rad/s, $\zeta = 0.707$, $T(0) = T_{MAX} = 1$

SECOND-ORDER HIGH-PASS CIRCUITS

To complete the study of second-order circuits, we treat the high-pass case.

$$T(s) = \frac{Ks^2}{s^2 + 2\zeta\omega_0 s + \omega_0^2} \qquad (12\text{--}34)$$

The high-pass frequency response is a mirror image of the low-pass case reflected about $\omega = \omega_0$. At low frequencies ($\omega \ll \omega_0$) the gain approaches $|T(j\omega)| \to |K|\omega^2/\omega_0^2$. At high frequencies ($\omega \gg \omega_0$) the gain approaches $|T(j\omega)| \to |K| = |T(\infty)|$. The high-frequency gain is a constant $|T(\infty)|$ called the infinite frequency gain. The low-frequency asymptote is inversely proportional to the square of the frequency (slope = +2 or +40 dB/decade). The two asymptotes intersect at the natural frequency ω_0, as shown in Figure 12–30. The asymptotes display a high-pass gain response with a passband above ω_0 and a stopband below ω_0. The passband gain is $|T(\infty)|$ and the slope of the gain asymptote in the stopband is +2 or +40 dB/decade.

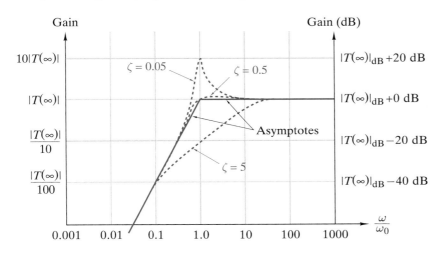

FIGURE 12–30 *Second-order high-pass gain responses.*

Figure 12–30 also shows the effect of the damping ratio on the actual gain curve. Evaluating the gain response at $\omega = \omega_0$ produces

$$\left|T(j\omega_0)\right| = \frac{|K|}{2\zeta} = \frac{|T(\infty)|}{2\zeta} \qquad (12\text{--}35)$$

This result and the plots in Figure 12–30 show that the gain response will be above the asymptotes for $\zeta < 0.5$, below for $\zeta > 0.5$, and close to the asymptotes for $\zeta = 0.5$. When $\zeta = 1/\sqrt{2}$ the gain in Eq. (12–35) is $|T(\infty)|/\sqrt{2}$. That is, ω_0 is the cutoff frequency when $\zeta = 1/\sqrt{2}$.

Again, the distinctive feature of the second-order response is the peak in the neighborhood of ω_0 for lightly damped circuits with $\zeta \ll 0.5$. For the second-order high-pass function the maximum gain is

$$\left|T(j\omega)\right|_{\text{MAX}} = \frac{|T(\infty)|}{2\zeta\sqrt{1-\zeta^2}} \qquad (12\text{--}36)$$

and occurs at

$$\omega_{\text{MAX}} = \frac{\omega_0}{\sqrt{1-2\zeta^2}} \qquad (12\text{--}37)$$

The frequency at which the gain maximum occurs is always above ω_0. Deriving Eqs. (12–36) and (12–37) is straightforward and is similar to the low-pass case.

EXAMPLE 12–11

(a) Show that the transfer function $T(s) = V_2(s)/V_1(s)$ of the circuit in Figure 12–31 has a high-pass characteristic.

FIGURE 12–31

(b) Derive expressions relating the damping ratio ζ and undamped natural frequency ω_0 to the circuit element values.

SOLUTION:

This high-pass circuit is obtained from the low-pass circuit in Figure 12–29 by interchanging the positions of the resistors and capacitors. This interchange is an example of a transformation that converts any low-pass RC circuit into a high-pass RC circuit. The CR-RC transformation is carried out by:

1. Replacing every impedance R by an impedance $1/Cs$, and
2. Replacing every impedance $1/Cs$ by an impedance R.

Applying this transformation at the circuit level converts the low-pass circuit in Figure 12–29 into the high-pass circuit in Figure 12–31. The transformation can also be applied at the transfer function level. The transfer function of the low-pass prototype is given in Example 12–10 as

$$T_{\text{LP}}(s) = \frac{\mu}{(RCs)^2 + (3 - \mu)RCs + 1}$$

$$= \frac{\mu}{(R)^2\left(\frac{Cs}{1}\right)^2 + (3 - \mu)(R)\left(\frac{Cs}{1}\right) + 1}$$

Performing the CR-RC transformation defined previously on the second expression yields

$$T_{\text{HP}}(s) = \frac{\mu}{\left(\frac{1}{Cs}\right)^2\left(\frac{1}{R}\right)^2 + (3 - \mu)\left(\frac{1}{Cs}\right)\left(\frac{1}{R}\right) + 1}$$

$$= \frac{\mu(RCs)^2}{(RCs)^2 + (3 - \mu)RCs + 1}$$

The low-pass transfer function is converted into a high-pass form with an infinite frequency gain of $|T(\infty)| = \mu$. Comparing the denominator of $T_{HP}(s)$ with the standard second-order form yields

$$\omega_0 = \frac{1}{RC} \quad \text{and} \quad \zeta = \frac{3 - \mu}{2}$$

These are the same relationships we obtained for the low-pass prototype in Example 12–10. Note that the *CR-RC* transformation only affects the resistors and capacitors and does not change the dependent source.

Circuit Interpretation: It is not hard to see why the circuit in Figure 12–31 has a high-pass response with zero gain at dc and finite gain at infinite frequency. At zero frequency both capacitors act like open circuits that disconnect the input source, leading to zero output and zero gain. At infinite frequency the capacitors act like short circuits that directly connect the input source to the control voltage node, forcing the condition $V_X(s) = V_1(s)$. As a result, the output is $V_2(s) = \mu V_1(s)$, which means the infinite frequency gain is μ. ∎

Exercise 12–8

The transfer function of a high-pass circuit is $T(s) = 0.2s^2/(s^2 + 10^4 s + 10^6)$. Find the undamped natural frequency ω_0, damping ratio ζ, high-frequency gain $T(\infty)$, and maximum gain of the transfer function.

Answers:
$\omega_0 = 10^3$ rad/s, $\zeta = 5$, $T(\infty) = 0.2$, $T_{MAX} = T(\infty)$

Exercise 12–9

The step response of a second-order high-pass circuit is

$$g(t) = [e^{-1000t} \cos 2000t - 0.5e^{-1000t} \sin 2000t]u(t)$$

Find the natural frequency ω_0, damping ratio ζ, high-frequency gain $T(\infty)$, and maximum gain of the circuit.

Answers:
$\omega_0 = 2236$ rad/s, $\zeta = 0.447$, $T(\infty) = 1$, $T_{MAX} = 1.25$

12–4   THE FREQUENCY RESPONSE OF *RLC* CIRCUITS

The series *RLC* circuit and the parallel *RLC* circuit are canonical examples traditionally used in electrical engineering to illustrate the response of second-order circuits. These circuits give us a special insight because the form of a response is easily related to the circuit elements. We first encountered these circuits in Chapter 7, where they served as vehicles for studying the solution of second-order differential equations. In this section we concentrate on the frequency response of the traditional *RLC* circuits starting with the series circuit in Figure 12–32.

By voltage division, the transfer function of the series *RLC* circuit in Figure 12–32 is

FIGURE   12–32 *Series RLC bandpass circuit.*

$$T(s) = \frac{R}{Ls + 1/Cs + R} = \frac{\frac{R}{L}s}{s^2 + \frac{R}{L}s + \frac{1}{LC}} \quad (12\text{--}38)$$

Comparing this result with the standard bandpass in Eq. (12–14), we conclude that the circuit has a bandpass response with

$$K = \frac{R}{L}$$

$$\omega_0 = \frac{1}{\sqrt{LC}} \quad (12\text{--}39)$$

$$2\zeta\omega_0 = \frac{R}{L}$$

Inserting these conclusions into Eqs. (12–19), (12–24), and (12–26), we find the descriptive parameters of the passband to be

$$|T(j\omega)|_{\text{MAX}} = \frac{K}{2\zeta\omega_0} = 1$$

$$B = 2\zeta\omega_0 = \frac{R}{L} \quad (12\text{--}40)$$

and

$$\omega_{C1} = -\frac{R}{2L} + \sqrt{\left(\frac{R}{2L}\right)^2 + \frac{1}{LC}}$$

$$\omega_{C2} = +\frac{R}{2L} + \sqrt{\left(\frac{R}{2L}\right)^2 + \frac{1}{LC}} \quad (12\text{--}41)$$

For historical reasons, it is traditional to add a fifth descriptive parameter called the **quality factor** Q.[3] We can define the quality factor of any bandpass filter as $Q = \omega_0/B$. In general, this means that for a second-order circuit, $Q = 1/2\zeta$. In the particular case of the series RLC circuit, this requires

$$Q = \frac{1/\sqrt{LC}}{R/L} = \frac{\sqrt{L/C}}{R} \quad (12\text{--}42)$$

The consequences of these results are illustrated graphically in Figure 12–33. The maximum gain is always 1 (or 0 dB) regardless of the bandwidth. The bandwidth is directly proportional to the resistance, so ω_{C1} decreases and ω_{C2} increases as R increases. The boundary between broadband and narrowband occurs when $\zeta = 0.5$, which requires that $Q = 1$. Figure 12–33 shows the response for values of Q one decade above and one decade

---

3 Q was originally defined as a measure of the quality of lossy inductors. In modern usage it is a substitute parameter for the damping ratio of a second-order factor. The relationship between the two parameters is $2\zeta = 1/Q$. Since the two are reciprocally related, their comparative adjectives are reversed; that is, high Q means low damping and low Q means high damping.

below this boundary value. Increasing Q causes the passband to shrink into a high-Q, narrowband response. Conversely, decreasing Q expands the passband into a low-Q, broadband response. But regardless of the value of Q, the maximum gain remains anchored at $\omega = \omega_0$ with a value of 1.

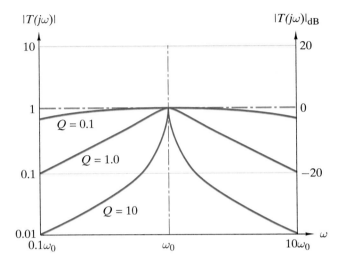

FIGURE 12–33 *Effect of Q on the series RLC bandpass response.*

What is interesting, and worth remembering, about this circuit is the way the impedances of the inductor and capacitor combine to produce the bandpass response. Again, using voltage division in Figure 12–32, we write the transfer function in the form

$$T(j\omega) = \frac{R}{R + Z_{LC}(j\omega)}$$

where

$$Z_{LC}(j\omega) = j\omega L + \frac{1}{j\omega C} = j\left(\omega L - \frac{1}{\omega C}\right)$$

is the impedance of inductor and capacitor in the series leg of the voltage divider. The variation of this impedance affects the gain response as follows:

- At $\omega = 0$ the impedance $Z_{LC}(j\omega)$ is infinite since the capacitor acts like an open circuit. The open circuit disconnects the input source, leading to zero output and $|T(0)| = 0$.

- At $\omega = \infty$ the impedance $Z_{LC}(j\omega)$ is infinite since the inductor acts like an open circuit. The open circuit disconnects the input source, leading to zero output and $|T(\infty)| = 0$.

- At $\omega = \omega_0$ the impedance $Z_{LC}(j\omega)$ is zero since

$$Z_{LC}(j\omega_0) = j\left(\omega_0 L - \frac{1}{\omega_0 C}\right) = j\left(\sqrt{\frac{L}{C}} - \sqrt{\frac{L}{C}}\right) = 0$$

The inductor and capacitor together act like a short circuit that directly connects the input across the output, leading to $|T(j\omega_0)| = 1$.

- At $\omega = \omega_{C1}$ the impedance $Z_{LC}(j\omega) = -jR$ since this value yields

$$T(j\omega_{C1}) = \frac{R}{R - jR} = \frac{1}{\sqrt{2}}\angle + 45°$$

The inductor and capacitor together act like an impedance $-jR$ with the result that the the gain at the lower cutoff frequency is $|T(j\omega_{C1})| = 1/\sqrt{2}$ as required.

- At $\omega = \omega_{C2}$ the impedance $Z_{LC}(j\omega) = +jR$ since this value yields

$$|T(j\omega_{C2})| = \frac{R}{R + jR} = \frac{1}{\sqrt{2}}\angle - 45°$$

The inductor and capacitor together act like an impedance $+jR$ with the result that the the gain at the upper cutoff frequency is $|T(j\omega_{C2})| = 1/\sqrt{2}$ as required.

Figure 12–34 summarizes these observations graphically and shows how the impedance $Z_{LC}(j\omega)$ controls the gain response of the circuit. Knowing

FIGURE 12–34 *Effect of Z_{LC} on the series RLC bandpass response.*

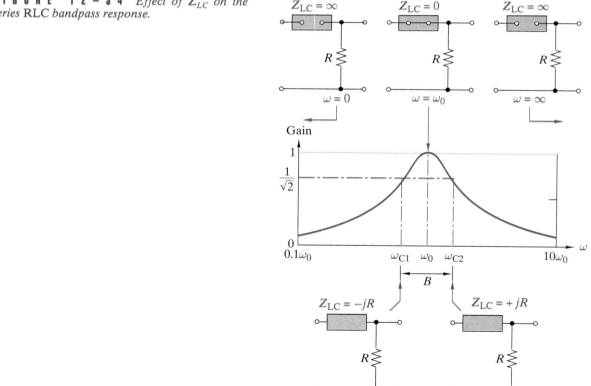

this impedance at five frequencies is enough to describe the bandpass characteristics.

EXAMPLE 12-12

A series *RLC* circuit has a resonant frequency of $\omega_0 = 20$ krad/s, $Q = 5$, and a resistance of 50 Ω. Find the values of L, C, B, ω_{C1}, and ω_{C2}.

SOLUTION:
From the definition of Q, we have $B = \omega_0/Q = 4000$ rad/s. For the series *RLC* circuit, $B = R/L$; hence we can calculate the inductance as $L = R/B = 12.5$ mH and the capacitance as $C = 1/\omega_0^2 L = 0.2$ μF. Inserting these results into Eq. (12–41) yields the cutoff frequencies as

$$\omega_{C1} = -\frac{4000}{2} + \sqrt{\left(\frac{4000}{2}\right)^2 + 20{,}000^2} = 18.1 \text{ krad/s}$$

$$\omega_{C2} = +\frac{4000}{2} + \sqrt{\left(\frac{4000}{2}\right)^2 + 20{,}000^2} = 22.1 \text{ krad/s} \qquad \blacksquare$$

EXAMPLE 12-13

(a) Show that the current transfer function $T(s) = I_2(s)/I_1(s)$ in Figure 12–35 has a bandpass gain response with a center frequency at $\omega_0 = 1/\sqrt{LC}$.
(b) Derive expression relating the descriptive parameters B, Q, ω_{C1}, and ω_{C2} to the circuit parameters R, L, and C.

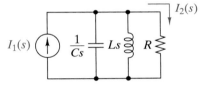

FIGURE   12-35   *Parallel RLC bandpass circuit.*

SOLUTION:
(a) The current transfer function is derived using current division as

$$T(s) = \frac{I_2(s)}{I_1(s)} = \frac{Y_R}{Y_L + Y_C + Y_R} = \frac{1/R}{1/Ls + Cs + 1/R}$$

$$= \frac{\dfrac{s}{RC}}{s^2 + \dfrac{s}{RC} + \dfrac{1}{LC}}$$

Comparing this transfer function to the standard bandpass form in Eq. (12–14), we see that this is a bandpass response with

$$K = \frac{1}{RC} \quad \omega_0 = \frac{1}{\sqrt{LC}} \quad 2\zeta\omega_0 = \frac{1}{RC}$$

(b) In general, the bandwidth of a second-order bandpass response is $B = 2\zeta\omega_0$. Hence for the parallel *RLC* circuit the bandwidth and Q are

$$B = \frac{1}{RC} \quad \text{and} \quad Q = \frac{\omega_0}{B} = \frac{R}{\sqrt{L/C}}$$

Inserting these results into the Eq. (12–24) leads to the following expressions:

$$\omega_{C1} = -\frac{1}{2RC} + \sqrt{\left(\frac{1}{2RC}\right)^2 + \frac{1}{LC}}$$

$$\omega_{C2} = +\frac{1}{2RC} + \sqrt{\left(\frac{1}{2RC}\right)^2 + \frac{1}{LC}}$$　■

Exercise 12–10

A series RLC circuit has cutoff frequencies of $\omega_{C1} = 100$ rad/s and $\omega_{C2} = 10$ krad/s. Find the values of B, ω_0, and Q. Does the circuit have a broadband or narrowband response?

Answers:

$B = 9.9$ krad/s, $\omega_0 = 1$ krad/s, $Q = 0.101$, broadband

Exercise 12–11

A parallel RLC bandpass circuit has a center frequency at $\omega_0 = 200$ krad/s, bandwidth of $B = 10$ krad/s, and a resistance of $R = 10$ kΩ. Find ω_0 and B when the resistance is increased to 40 kΩ.

Answers:

$\omega_0 = 200$ krad/s, $B = 2.5$ krad/s

FIGURE 12–36 *Series RLC bandstop circuit.*

THE SERIES *RLC* BANDSTOP CIRCUIT

The transfer function of the series RLC circuit in Figure 12–36 can be derived by voltage division as

$$T(s) = \frac{Ls + 1/Cs}{Ls + 1/Cs + R} = \frac{s^2 + \frac{1}{LC}}{s^2 + \frac{R}{L}s + \frac{1}{LC}}$$

If we define $\omega_0 = 1/\sqrt{LC}$ then $|T(j\omega_0)| = 0$ because the transfer function has zeros on the j-axis at $s = \pm j\omega_0$. The dc gain is $|T(0)| = 1$ and the infinite frequency gain is $|T(\infty)| = 1$. Thus, this transfer function has finite gain at dc and infinite frequency, and zero gain at an intermediate frequency. These features are indicative of a bandstop response.

The bandstop circuit in Figure 12–36 and the bandpass circuit in Figure 12–32 are both series RLC circuits. The difference is that the bandstop circuit takes its output across the inductor and capacitor in series while the bandpass circuit defines its output across the resistor. In either case, the shape of the frequency response is controlled by the impedance $Z_{LC}(j\omega) = j\omega L + 1/j\omega C$.

Figure 12–37 shows how this impedance controls the shape of the bandstop response. At dc and infinite frequency Z_{LC} acts like an open circuit so the output voltage equals the input voltage and the passband gains are both one. At $\omega = \omega_0$ the Z_{LC} acts like a short circuit so the output voltage and the gain are both zero. At the cutoff frequencies $Z_{LC} = \pm jR$, so the passband gains are reduced to $1/\sqrt{2}$. In sum, the factors defining the shape of the bandstop response are the same as those defining the shape of the bandpass response.

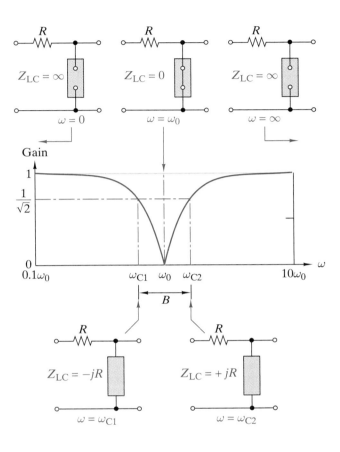

FIGURE   1 2 – 3 7   *Effect of Z_{LC} on the series RLC bandstop response.*

The net result is that the equations relating circuit parameters and the stopband descriptive parameters are the same as the equations for the bandpass case. That is, we can use Eqs. (12–39), (12–40), and (12–41) for both the bandpass and bandstop series *RLC* circuits. The difference is that B, ω_{C1}, and ω_{C2} describe the passband in the former and the stopband in the latter.

D ◖ DESIGN EXAMPLE 12–14 ◗

A series connection of an inductor and capacitor are to be connected across the output of a source whose Thévenin resistance is 50 Ω. The goal is to create a bandstop filter to eliminate a spurious (undesirable) source output at 25 krad/s. Select the values of L and C that eliminate the spurious response and produce a stopband bandwidth less than 1 krad/s.

SOLUTION:

Using Eq. (12–40), we have $B = R/L < 1000$. Since $R = 50\ \Omega$, this constraint requires $L > 50$ mH. Using Eq. (12–39), the bandstop notch must be located at $\omega_0 = 1/\sqrt{LC} = 25{,}000$. Selecting $L = 100$ mH to meet the bandwidth constraint means that $C = 1/(\omega_0^2 L) = 16$ nF.   ∎

Exercise 12–12

A series RLC circuit has a bandstop response with a B of 2 krad/s and a notch at $\omega_0 = 10$ krad/s. Find the circuit gain at $\omega = 9$ krad/s and $\omega = 11$ krad/s.

Answers: 0.726, 0.690

12–5   BODE DIAGRAMS

Bode diagrams are separate graphs of the gain $|T(j\omega)|_{dB}$ and phase $\theta(j\omega)$ versus log-frequency scales. The frequency-response performance of devices, circuits, and systems is often presented in this format. The poles and zeros of the transfer functions can be estimated from Bode plots of experimental frequency-response data. Thus, the Bode plot format is a very useful way to develop an understanding of frequency response of circuits and systems. Any of our computer-aided analysis tools can precisely calculate and plot Bode diagrams. However, often a simple hand-calculated approximation will tell us what we want to know about the frequency response of a circuit.

The purpose of this section is to present a method of quickly drawing straight-line approximations to Bode plot of transfer functions with real poles and zeros. In an analysis situation the straight-line approximations tell us how the poles and zeros of $T(s)$ affect frequency response. In circuit design these straight-line plots serve as a shorthand notation for outlining design approaches or developing design requirements. The straight-line plots help us use computer-aided analysis effectively because they tell us what frequency ranges are important and what features of the response need to be investigated in greater detail.

The straight-line versions of Bode plots are particularly useful when the poles and zeros are located on the real axis in the s plane. As a starting place, consider the transfer function

$$T(s) = \frac{Ks(s + \alpha_1)}{(s + \alpha_2)(s + \alpha_3)} \tag{12–43}$$

where K, α_1, α_2, and α_3 are real. This function has zeros at $s = 0$ and $s = -\alpha_1$, and poles at $s = -\alpha_2$ and $s = -\alpha_3$. All of these critical frequencies lie on the real axis in the s plane. When making Bode plots we put $T(j\omega)$ in a standard format obtained by factoring out α_1, α_2, and α_3:

$$T(j\omega) = \left(\frac{K\alpha_1}{\alpha_2\alpha_3}\right) \frac{j\omega(1 + j\omega/\alpha_1)}{(1 + j\omega/\alpha_2)(1 + j\omega/\alpha_3)} \tag{12–44}$$

Using the following notation

$$\text{Magnitude} = M = |1 + j\omega/\alpha| = \sqrt{1 + (\omega/\alpha)^2}$$

$$\text{Angle} = \theta = \angle(1 + j\omega/\alpha) = \tan^{-1}(\omega/\alpha) \qquad (12\text{–}45)$$

$$\text{Scale Factor} = K_0 = \frac{K\alpha_1}{\alpha_2\alpha_3}$$

we can write the transfer function in Eq. (12–44) in the form

$$T(j\omega) = K_0 \frac{(\omega\, e^{j90°})(M_1\, e^{j\theta_1})}{(M_2\, e^{j\theta_2})(M_3\, e^{j\theta_3})} = \frac{|K_0|\omega M_1}{M_2 M_3}\, e^{j(\angle K_0 + 90° + \theta_1 - \theta_2 - \theta_3)} \qquad (12\text{–}46)$$

The gain (in dB) and phase responses are

$$|T(j\omega)|_{dB} = \underbrace{20\log_{10}|K_0|}_{\text{Scale Factor}} + \underbrace{20\log_{10}\omega}_{\text{Zero}} + \underbrace{20\log_{10} M_1}_{\text{Zero}} - \underbrace{20\log_{10} M_2}_{\text{Pole}} - \underbrace{20\log_{10} M_3}_{\text{Pole}}$$

$$\theta(\omega) = \underbrace{\angle K_0}_{} + \underbrace{90°}_{} + \underbrace{\theta_1}_{} - \underbrace{\theta_2}_{} - \underbrace{\theta_3}_{}$$

$$(12\text{–}47)$$

The terms in Eq. (12–47) caused by zeros have positive signs and increase the gain and phase angle, while the pole terms have negative signs and decrease the gain and phase.

The summations in Eq. (12–47) illustrate a general principle. In a Bode plot, the gain and phase responses are determined by the following types of factors:

1. The scale factor K_0

2. A factor of the form $j\omega$ due to a zero or a pole at the origin

3. Factors of the form $(1 + j\omega/\alpha)$ caused by a zero or pole at $s = -\alpha$.

We can construct Bode plots by considering the contributions of these three factors.

The Scale Factor. The gain and phase contributions of the scale factor are constants that are independent of frequency. The gain contribution $20\log_{10}|K_0|$ is positive when $|K_0| > 1$ and negative when $|K_0| < 1$. The phase contribution $\angle K_0$ is $0°$ when $K_0 > 0$ and $\pm180°$ when $K_0 < 0$.

The Factor $j\omega$. A simple zero or pole at the origin contributes $\pm20\log_{10}\omega$ to the gain and $\pm90°$ to the phase, where the plus sign applies to a zero and the minus to a pole. When $T(s)$ has a factor s^n in the numerator (denominator), it has a zero (pole) of order n at the origin. Multiple zeros or poles at $s = 0$ contribute $\pm20n\log_{10}\omega$ to the gain and $\pm n90°$ to the phase. Figure 12–38 shows that the gain factors contributed by zeros and poles at the origin are straight lines that pass through a gain of 1 (0 dB) at $\omega = 1$.[4]

The Factor $1 + j\omega/a$. The gain contributions of first-order zeros and poles are shown in Figure 12–39. Like the first-order transfer functions studied earlier in this chapter, these factors produce straight-line gain

---

4 Strictly speaking, a circuit with a natural pole at the origin is unstable and does not have a sinusoidal steady-state response. Nevertheless, it is traditional to treat poles at the origin in Bode diagrams because there are practical applications in which such poles are important considerations.

FIGURE 12 – 38 *Gain responses of poles and zeros at s = 0.*

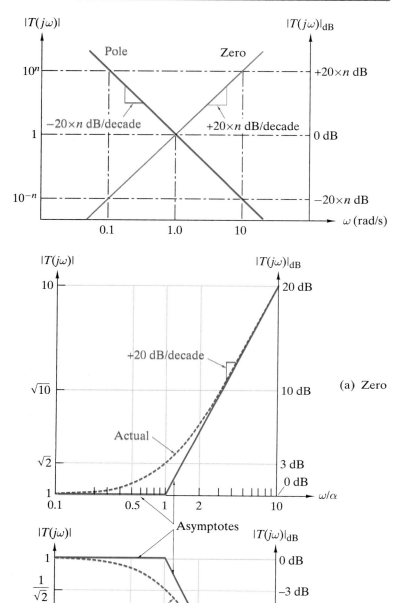

FIGURE 12 – 39 *Gain responses of real poles and zeros at s = −α.*

asymptotes at low and high frequency. In a Bode plot the low-frequency ($\omega \ll \alpha$) asymptotes are horizontal lines at with gain = 1 (0 dB). The high-frequency ($\omega \gg \alpha$) asymptotes are straight lines of the form $\pm\omega/\alpha$ ($\pm 20 \log(\omega/\alpha)$ dB), where the plus sign applies to a zero and the minus to a pole. The high-frequency gain asymptote is proportional to the frequency ω (slope = +1 or +20 dB/decade) for a zero and proportional to $1/\omega$ (slope = −1 or −20 dB/decade) for a pole. In either case, the low- and high-frequency asymptotes intersect at the corner frequency $\omega = \alpha$.

To construct straight-line (SL) gain approximations, we develop a piece-wise linear function $|T(j\omega)|_{SL}$ defined by the asymptotes of each of the factors in $T(j\omega)$. The function $|T(j\omega)|_{SL}$ has a corner frequency at each of the critical frequencies of the transfer function. At frequencies below a corner ($\omega < \alpha$) a first-order factor is represented by a gain of 1. Above the corner frequency ($\omega > \alpha$) the factor is represented by its high-frequency asymptote ω/α. To generate $|T(j\omega)|_{SL}$ we start out below the lowest critical frequency with a low-frequency baseline that accounts for the scale factor K_0 and any poles or zeros at the origin. We increase frequency and change the form of $|T(j\omega)|_{SL}$ whenever we pass a corner frequency. We proceed upward in frequency until we have gone beyond the highest critical frequency, at which point we have a complete expression for $|T(j\omega)|_{SL}$.

The following example illustrates the process.

EXAMPLE 12–15

(a) Construct the Bode plot of the straight-line approximation of the gain of the transfer function

$$T(s) = \frac{12,500\,(s + 10)}{(s + 50)\,(s + 500)}$$

(b) Find the point at which the high-frequency gain falls below the dc gain.

SOLUTION:

(a) Written in the standard form for a Bode plot, the transfer function is

$$T(j\omega) = \frac{5(1 + j\omega/10)}{(1 + j\omega/50)\,(1 + j\omega/500)}$$

The scale factor is $K_0 = 5$ and the corner frequencies are at $\omega_C = 10$ (zero), 50 (pole), and 500 (pole) rad/s. At low frequency ($\omega < 10$ rad/s) all of the first-order factors are represented by their low-frequency asymptotes. As a result $|T(j\omega)| \approx 5(1)/[(1)(1)] = 5$, so the low-frequency baseline is $|T(j\omega)|_{SL} = 5$ for $\omega \leq 10$. At $\omega = 10$ rad/s we encounter the first critical frequency. Beginning at this point the factor $(1 + j\omega/10)$ is represented by its high-frequency asymptote $\omega/10$ and the straight-line gain becomes $|T(j\omega)|_{SL} = 5(\omega/10) = \omega/2$. This expression applies until we pass the critical frequency at $\omega_C = 50$ due to the pole at $s = -50$. After this point the gain contribution due to the pole factor $1/(1 + j\omega/50)$ is represented by its high-frequency asymptote $\omega/50$ and $|T(j\omega)|_{SL} = 5(\omega/10)/(\omega/50) = 25$. This version applies until we pass the final critical frequency at $\omega_C = 500$ beyond which the gain contribution of last pole is approximated by $\omega/500$ and the high-frequency gain rolls off as $|T(j\omega)|_{SL} = 5(\omega/10)/[(\omega/50)(\omega/500)] = 12500/\omega$. In summary, the straight-line approximation to the gain is

$$|T(j\omega)|_{SL} = \begin{vmatrix} 5 & \text{if} & 0 < \omega \leqslant 10 \\ \omega/2 & \text{if} & 10 < \omega \leqslant 50 \\ 25 & \text{if} & 50 < \omega \leqslant 500 \\ 12,500/\omega & \text{if} & 500 < \omega \end{vmatrix}$$

Given this function, we can easily plot the straight-line gain response in Figure 12–40. At low frequency ($\omega < 10$) the gain is flat at value of 5 (14 dB). At $\omega = 10$ the zero causes the gain to increase as ω (slope = +1 or +20 dB/decade). This increasing gain continues until $\omega = 50$, where the first pole cancels the effect of the zero and the gain is flat at a value of 25 (28 dB). The gain remains flat until the final poles causes a corner at $\omega = 500$. Thereafter the gain falls off as $1/\omega$ (slope = –1 or –20 dB/decade).

(b) The dc gain is 5. A quick look at the sketch in Figure 12–40 shows that the high-frequency gain falls below the dc gain in the region above $\omega = 500$, where the straight-line gain is $12,500/\omega$. Hence we estimate the required frequency to be $\omega = 12,500/5 = 2500$ rad/s. ■

FIGURE 12–40

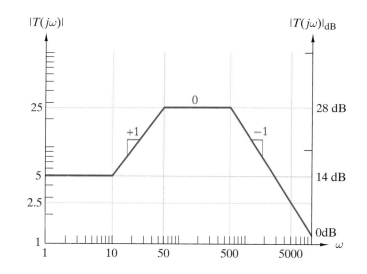

EXAMPLE 12–16

(a) Construct the Bode plot of the straight-line approximation of the gain of the transfer function:

$$T(s) = \frac{10s^2}{(s + 40)(s + 200)}$$

(b) Find the point at which the low-frequency gain falls –40 dB below the passband gain.

SOLUTION:

(a) Writing $T(j\omega)$ in standard form produces

$$T(j\omega) = \frac{1}{800}\left[\frac{(j\omega)^2}{(1 + j\omega/40)(1 + j\omega/200)}\right]$$

In this form $T(j\omega)$ has a scale factor of $K_0 = 1/800$, a double zero ($n = 2$) at the origin, and finite critical frequencies at $\omega = 40$ and 200, both due to poles. At low frequency ($\omega < 40$) the two first-order factors are represented by their low-frequency asymptote. As a result, the low-frequency gain baseline is $|T(j\omega)|_{SL} = \omega^2/800$. This trend continues until we pass the critical frequency at $\omega = 40$. After this point the straight-line gain is

$$|T(j\omega)|_{SL} = (\omega^2/800)/(\omega/40) = \omega/20$$

The gain continues to increase at a reduced rate until we pass the final critical frequency at $\omega = 200$. Beyond this point the straight-line gain is constant at the high frequency of

$$|T(j\omega)|_{SL} = (\omega^2/800)/[(\omega/40)(\omega/200)] = 10$$

In summary, the straight-line approximation to the gain response is

$$|T(j\omega)|_{SL} = \begin{vmatrix} \omega^2/800 & \text{if} & 0 < \omega \leqslant 40 \\ \omega/20 & \text{if} & 40 < \omega \leqslant 200 \\ 10 & \text{if} & 200 < \omega \end{vmatrix}$$

We can easily plot the straight-line gain response shown in Figure 12–41. At low frequency ($\omega < 20$) the gain is increasing as ω^2 (slope = +2 or +40 dB/decade). This upward trend lasts until the critical frequency (a pole) at $\omega = 40$, at which point the gain is $(40)^2/800 = 2$. Thereafter the gain increases as ω (slope = +1 or +20 dB/decade) until the final critical frequency at $\omega = 200$ reduces the slope to zero at a flat gain of 10 (20 dB).

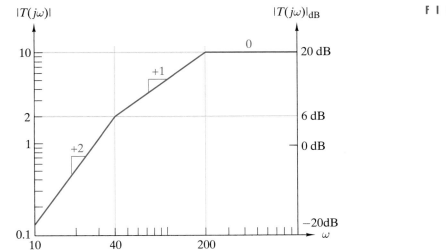

FIGURE 12 – 41

(b) A quick look at the plot in Figure 12–41 shows that the gain response has a high-pass characteristic with a passband above $\omega = 200$, where the passband gain is 20 dB. The low-frequency gain will be −40 dB below this level in the frequency range below $\omega = 40$, where the straight-line gain is $\omega^2/800$. The actual gain must be 20 dB − 40 dB = −20 dB or 0.1. This gain occurs when $\omega^2/800 = 0.1$ at $\omega = 8.9$ rad/s.   ■

Exercise 12–13

(a) Derive an expression for the straight-line approximation to the gain response of the following transfer function:

$$T(s) = \frac{500\,(s + 50)}{(s + 20)(s + 500)}$$

(b) Find the straight-line gains at $\omega = 10$, 30, and 100 rad/s.
(c) Find the frequency at which the high-frequency gain asymptote falls below −20 dB.

Answers:

(a) $|T(j\omega)|_{\text{SL}} = \begin{cases} 2.5 & \text{if} & 0 < \omega \leqslant 2 \\ 50/\omega & \text{if} & 20 < \omega \leqslant 50 \\ 1 & \text{if} & 50 < \omega \leqslant 500 \\ 500/\omega & \text{if} & 500 < \omega \end{cases}$

(b) 8 dB, 4.4 dB, 0 dB
(c) 5 krad/s

Exercise 12–14

Given the following transfer function

$$T(s) = \frac{4000s}{(s + 100)(s + 2000)}$$

(a) Find the straight-line approximation to the gain at $\omega = 50$, 100, 500, 2000, and 4000 rad/s.
(b) Estimate the actual gain at the frequencies in (a).

Answers:
(a) 0 dB, 6 dB, 6 dB, 6 dB, 0 dB
(b) −1 dB, 3 dB, 6 dB, 3 dB, −1 dB

In many situations the straight-line Bode plot tells us all we need to know. When greater accuracy is needed the straight-line gain plot can be refined by adding gain corrections in the neighborhood of the corner frequency. Figure 12–39 shows that the actual gains and the straight-line approximations differ by ±3 dB at the corner frequency. They differ by roughly ±1 dB an octave above or below the corner frequency. When these "corrections" are included, we can sketch the actual response and achieve somewhat greater accuracy. However, making the graphical gain corrections is usually not worth the trouble. First, the gain corrections overlap unless the corner frequencies are separated by more than two octaves. More important, the purpose of a straight-line gain analysis is to provide insight, not to generate accurate frequency-response data. The straight-line plots are useful in preliminary analysis and in the early stages of design. At some point accurate response data will be needed, in which case it is better to use computer-aided analysis rather than trying to "correct the errors" graphically in a straight-line plot.

EXAMPLE 12–17

Use computer-aided analysis to compare the actual and straight-line approximation to the gain of the transfer function in Example 12–15.

SOLUTION:

Figure 12–42 shows a Mathcad worksheet that calculates both the straight-line gain and the actual gain. In Mathcad we can write the transfer function $T(s)$ exactly as it is given in Example 12–15 and then have Mathcad calculate and plot the gain $|T(j\omega)|$. Similarly, we can use an extended conditional statement to write the straight-line gain $|T(j\omega)|_{SL}$ exactly as it appears in Example 12–15 and then have Mathcad calculate and plot this function as well. The frequency range of interest extends from one decade below the lowest critical frequency ($\omega = 10$) to the first decade point above the highest critical frequency ($\omega = 500$). The resulting graph shows that the straight-line gain (the solid line) is a reasonable approximation that captures the major features of the actual gain (the dashed curve). When numerical accuracy is important it is best to have a program like Mathcad calculate values of $|T(j\omega)|$ rather than try graphically to correct the plot of $|T(j\omega)|_{SL}$. ∎

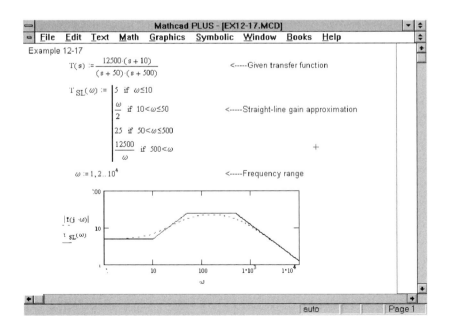

FIGURE   12 – 42

STRAIGHT-LINE PHASE ANGLE PLOTS

Figure 12–43 shows the phase contributions of first-order zeros and poles. The straight-line approximations are similar to the gain asymptotes except that there are two slope changes. The first occurs a decade below the gain corner frequency, and the second occurs a decade above. The total phase changes by 90° over this two-decade range, so the straight-line approximations have slopes of ±45° per decade, where the plus sign applies to a zero and the minus to a pole. Poles and zeros at the origin contribute a constant phase angle of ±n90°, where n is the order of the critical frequency and the plus (minus) sign applies to zeros (poles).

To generate a straight-line phase plot, we begin with the low-frequency phase asymptote. This low-frequency baseline accounts for the effect of the scale factor K_0 and any poles or zeros at the origin. We account for the effect of other critical frequencies by introducing a slope change of ±45°/decade one decade below and one decade above each gain corner frequency. These slope changes generate a straight-line phase plot as we proceed from the low-frequency baseline to a high frequency that is at least a decade above the highest gain corner frequency.

It is important to remember that a decade above the highest corner frequency the phase asymptote is a constant value with zero slope. That is, at high frequency the straight-line phase plot is a horizontal line at $\theta(j\omega) = (m - n)90°$, where m is the number of finite zeros and n is the number of poles.

EXAMPLE 12–18

Find the straight-line approximation to the phase response of the transfer function in Example 12–15.

SOLUTION:

In Example 12–15 the standard from of $T(j\omega)$ is shown to be

$$T(j\omega) = \frac{5(1 + j\omega/10)}{(1 + j\omega/50)(1 + j\omega/500)}$$

The scale factor is $K_0 = 5$ and the corner frequencies are $\omega_C = 10$ (zero), 50 (pole), and 500 (pole) rad/s. At low frequency $T(j\omega) \to K_0 = 5$, so the low-frequency phase asymptote is $\theta(\omega) \to \angle K_0 = 0°$. Proceeding from one decade below the lowest corner frequency (1 rad/s) to one decade above the highest corner frequency (5000 rad/s), we encounter the following slope changes:

FREQUENCY	CAUSED BY	SLOPE CHANGE	NET SLOPE
1	zero at $s = -10$	+45°/decade	+45°/decade
5	pole at $s = -50$	−45°/decade	0°/decade
50	pole at $s = -500$	−45°/decade	−45°/decade
100	zero at s = −10	−45°/decade	−90°/decade
500	pole at $s = -50$	+45°/decade	−45°/decade
5000	pole at $s = -500$	+45°/decade	0°/decade

FIGURE 12–43 *Phase responses of real poles and zeros.*

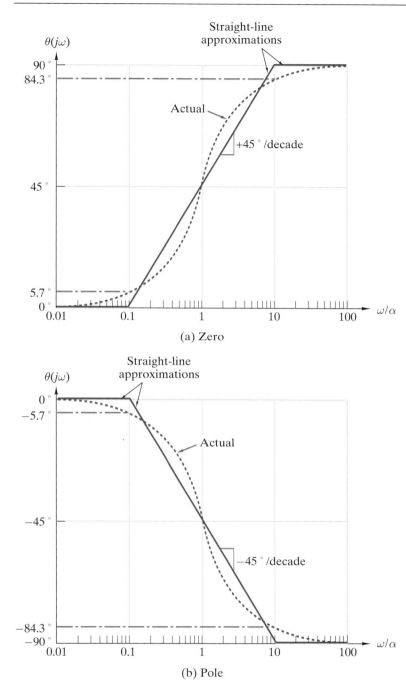

(a) Zero

(b) Pole

FIGURE 12-44

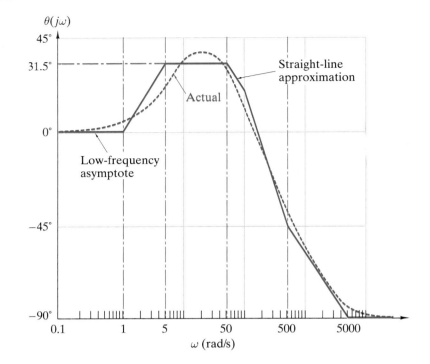

Figure 12–44 shows the straight-line approximation and the actual phase response. ∎

Exercise 12–15

Construct a Bode plot of the straight-line approximation to the phase response of the transfer function in Exercise 12–13. Use the plot to estimate the phase angles at $\omega = 1, 15, 300$, and 10^4 rad/s.

Answers: $0°, -18°, -45°, -90°$

12–6 BODE DIAGRAMS WITH COMPLEX CRITICAL FREQUENCIES

The frequency response of transfer functions with complex poles and zeros can be analyzed using straight-line gain plots. However, complex critical frequencies may produce resonant peaks (or valleys) where the actual gain response departs significantly from the straight-line approximation. Straight-line plots can be used to define a starting place for describing the frequency response of these highly resonant circuits.

Complex poles and zeros occur in conjugate pairs that appear as quadratic factors of the form

$$s^2 + 2\zeta\omega_0 s + \omega_0^2 \qquad (12\text{–}48)$$

where ζ and ω_0 are the damping ratio and undamped natural frequency. In a Bode diagram the appropriate standard form of the quadratic factor is obtained by factoring out ω_0^2 and replacing s by $j\omega$ to obtain

$$1 - (\omega/\omega_0)^2 + j2\zeta(\omega/\omega_0) \qquad\qquad (12\text{–}49)$$

In a Bode diagram this quadratic factor introduces gain and phase terms of the following form:

$$|T(j\omega)|_{dB} = \pm 20 \log_{10}\sqrt{[1 - (\omega/\omega_0)^2]^2 + (2\zeta\omega/\omega_0)^2} \qquad (12\text{–}50a)$$

$$\theta(\omega) = \pm \tan^{-1}\frac{2\zeta\omega/\omega_0}{1 - (\omega/\omega_0)^2} \qquad\qquad (12\text{–}50b)$$

where the plus sign applies to complex zeros of $T(s)$ and the minus sign to complex poles.

Figure 12–45 shows the gain contribution of complex poles and zeros for several values of the damping ratio ζ. The low-frequency ($\omega \ll \omega_0$) asymp-

(a) Zero

(b) Pole

FIGURE 12–45 Gain responses of complex poles and zeros.

totes for these plots is a gain of unity (0 dB). The high-frequency ($\omega \gg \omega_0$) gain asymptotes are of the form $(\omega/\omega_0)^{\pm2}$. Expressed in dB the high-frequency asymptote is $\pm40\,\log_{10}(\omega/\omega_0)$, which in a Bode diagram is a straight line with a slope of ±2 or ±40 dB/decade, where again the plus sign applies to zeros and the minus to poles.

These asymptotes intersect at a corner frequency of $\omega_C = \omega_0$. The gains in the neighborhood of the corner frequency are strong functions of the damping ratio. From our study of second-order low-pass transfer functions we can infer the following. For $\zeta > 1/\sqrt{2}$ the actual gain lies entirely above the asymptotes for complex zeros and entirely below the asymptotes for complex poles. For $\zeta < 1/\sqrt{2}$ the gain is a minimum at $\omega = \omega_0\sqrt{1-2\zeta^2}$ for complex zeros and a maximum for complex poles. These valleys (for zeros) and peaks (for poles) are not particularly conspicuous until $\zeta < 0.5$.

To develop a straight-line gain plot for complex critical frequencies we insert a corner frequency at $\omega = \omega_0$. Below this corner frequency we use the low-frequency asymptote to approximate the gain and above the corner we use the high-frequency asymptote. The actual gain around the corner frequency depends on ζ. But generally speaking, the straight-line gain is within ±3 dB of the actual gain for ζ in the range from about 0.3 to about 0.7. When ζ falls outside this range we can calculate the actual gain at the corner frequency and perhaps a few points on either side of the corner frequency. These gains may give us a better picture of the gain plot in the vicinity of the corner frequency.

However, we should keep in mind that the purpose of straight-line gain analysis is insight into the major features of a circuit's frequency response. If greater accuracy is required, then computer-aided analysis is the best approach. The straight-line gain gives useful results when the resonant peaks and valleys are not too abrupt. When a circuit has lightly damped critical frequencies, the straight-line approach may not be particularly helpful. The following examples illustrate both of these cases.

E X A M P L E   1 2 – 1 9

(a) Construct the straight-line gain plot for the transfer function

$$T(s) = \frac{5000(s+100)}{s^2 + 400s + (500)^2}$$

(b) Use the straight-line plot to estimate the maximum gain and the frequency at which it occurs.

S O L U T I O N :

(a) The transfer function has a real zero at $s = -100$ rad/s and a pair of complex poles with $\zeta = 0.4$ and $\omega_0 = 500$ rad/s. This damping ratio falls in the range (0.3 to 0.7) in which the resonant peak due to the complex poles is not too pronounced. Hence we expect the straight-line gain to give a useful approximation. Written in standard form, $T(j\omega)$ is

$$T(j\omega) = 2\left(\frac{1 + j\omega/100}{1 - (\omega/500)^2 + j400\omega/500}\right)$$

The scale factor is $K_0 = 2$, and there are corner frequencies at $\omega = 100$ rad/s due to the zero and $\omega = 500$ rad/s due to the pair of complex poles. At low frequency $T(j\omega) \to 2$, so the low-frequency ($\omega < 100$) baseline is $|T(j\omega)|_{SL} = 2$. This gain applies until we pass the first critical frequency at $\omega = 100$. Beginning at that point, the zero is represented by its high-frequency asymptote ($\omega/100$) and the straight-line gain becomes

$$|T(j\omega)|_{SL} = 2(\omega/100) = \omega/50$$

This linearly increasing gain applies until we pass the critical frequency at $\omega_0 = 500$. Thereafter the complex poles are represented by their high-frequency asymptote ($500^2/\omega^2$). After this point the gain rolls off as

$$|T(j\omega)|_{SL} = 2(\omega/100)(500^2/\omega^2) = 5000/\omega$$

In summary, the straight-line gain function is

$$|T(j\omega)|_{SL} = \begin{vmatrix} 2 & \text{if} & 0 < \omega \leqslant 100 \\ \omega/50 & \text{if} & 100 < \omega \leqslant 500 \\ 5000/\omega & \text{if} & 500 < \omega \end{vmatrix}$$

Figure 12–46 shows a plot of the straight-line gain. We expect to see a gain peak around $\omega = 500$ rad/s due to the complex poles. The plot in Figure 12–46 shows that the zero at $s = -100$ rad/s causes the gain to bend upward prior to the corner frequency at $\omega = 500$ rad/s. This upward bend enhances the height of the resonant peak caused by the complex poles.

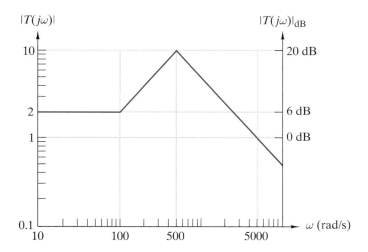

FIGURE 1 2 – 4 6

(b) The straight-line gain predicts a maximum gain of 10 (20 dB) at $\omega = 500$ rad/s. The actual gain at that point is

$$|T(j500)| = 2\left|\frac{1 + j5}{j0.8}\right| = 12.7 \text{ (22 dB)}$$

In this case the straight-line plot gives us a useful approximation of the gain in the vicinity of the resonant peak. ∎

FIGURE 12 – 47

EXAMPLE 12–20

Calculate the straight-line and the actual gain response of the *RLC* circuit shown in Figure 12–47.

SOLUTION:
The analysis of this circuit is in Example 12–9, where its transfer function is shown to be

$$T(s) = \frac{s/RC}{s^2 + s/RC + 1/LC}$$

This is a second-order bandpass transfer function with a center frequency at $\omega_0 = 1/\sqrt{LC}$. The appropriate standard form for straight-line gain analysis is

$$|T(j\omega)| = \left| \frac{j\omega L/R}{1 - LC\omega^2 + j\omega L/R} \right|$$

At low frequency ($\omega \ll \omega_0$) the gain asymptote is $\omega L/R$. At high frequency ($\omega \gg \omega_0$) the gain asymptote is $1/\omega RC$. These asymptotes intersect when $\omega L/R = 1/\omega RC$, which occurs at $\omega^2 = 1/LC$; that is, at the center frequency $\omega_0 = 1/\sqrt{LC}$. The straight-line gain for this circuit is

$$|T(j\omega)|_{\text{SL}} = \left| \begin{matrix} \omega L/R & \text{if} & 0 < \omega \le \omega_0 \\ 1/\omega RC & \text{if} & \omega_0 < \omega \end{matrix} \right.$$

Figure 12–48 shows a Mathcad worksheet that includes the circuit parameters, the circuit transfer function, the straight-line gain function, and a frequency range from a decade below to a decade above the center frequency. The Mathcad worksheet includes a plot of the actual gain (the

FIGURE 12 – 48

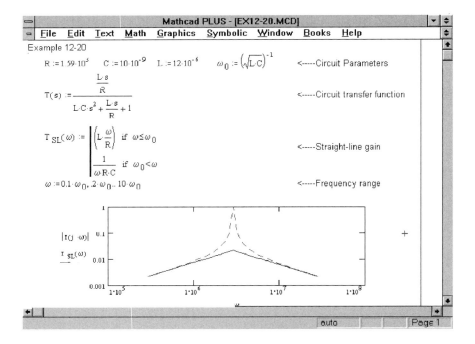

dashed curve) and the straight-line gain (the solid lines). In this case the straight-line gain plot indicates a bandpass response but does not tell us the bandwidth.

In Example 12–9 the circuit element values used are selected to produce a center frequency at 455 kHz and a bandwidth of 10 kHz. The damping ratio of this circuit is

$$\zeta = \frac{B}{2\omega_0} = \frac{10 \times 10^3}{2 \times 455 \times 10^3} = 0.011$$

Since $\zeta \ll 0.5$, the circuit has a very narrow resonant peak at 455 kHz. For highly resonant circuits like this one the straight-line gain is not particularly helpful. ■

Exercise 12–16

(a) Construct the straight-line gain function for the transfer function

$$T(s) = \frac{10^5}{[s^2 + 100s + 10^4](s + 10)}$$

(b) Find the straight-line gain at $\omega = 1, 10, 100, 1000$ rad/s.

Answers:

(a) $|T(j\omega)|_{SL} = \begin{cases} 1 & \text{if} \quad 0 < \omega < 10 \\ 10/\omega & \text{if} \quad 10 < \omega \leqslant 100 \\ 10^5/\omega^3 & \text{if} \quad 100 < \omega \end{cases}$

(b) 0 dB, 0 dB, –20 dB, –80 dB

PHASE PLOTS FOR COMPLEX CRITICAL FREQUENCIES

Figure 12–49 shows the phase contribution from complex poles or zeros for several values of ζ. The low-frequency phase asymptotes are 0° and the high-frequency limits are ±180°. The phase is always ±90° at $\omega = \omega_0$ regardless of the value of the damping ratio. The total phase change is ±180°, and most of this change occurs in a two-decade range from $\omega_0/10$ to $10\omega_0$. As a result, the straight lines in Figure 12–49 offer crude approximations to the phase shift. The shape of the phase curves change radically with the damping ratio, so these straight-line approximations are of little use except at ω_0 and around the end points.

The net result is that phase angle plots for complex critical frequencies are best generated by computer-aided analysis. In practical applications we can often derive useful information from the gain plot alone without generating phase response. However, the converse is not true. When we need phase response we usually need gain as well. The next example shows how to obtain both using a computer-aided analysis tool.

EXAMPLE 12–21

Plot the gain and phase response of the transfer function

$$T(s) = \frac{5000(s + 100)}{s^2 + 400s + (500)^2}$$

F I G U R E   1 2 – 4 9 *Phase responses of complex poles and zeros.*

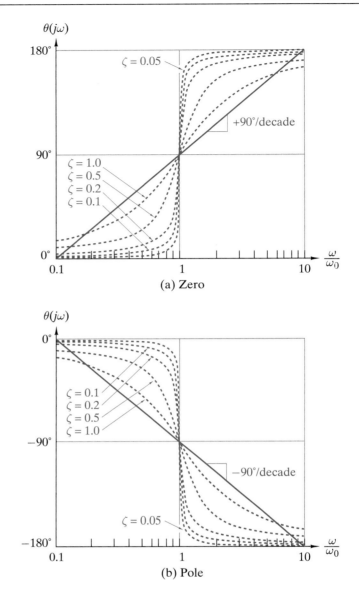

(a) Zero

(b) Pole

SOLUTION:

The MATLAB function `bode` calculates and plots the gain and phase responses of transfer functions of the form $T(s) = n(s)/d(s)$, where $n(s)$ is the numerator polynomial and $d(s)$ is the denominator polynomial. In the MATLAB command window these polynomials are defined by the row matrices num and den that contain the coefficients of the polynomials $n(s)$ and $d(s)$ in descending powers of s. In the present case these matrices are

```
num=[5000,5000*100];
den=[1,400,500^2];
```

The first element in both matrices is the coefficient of the highest power of s followed by the other coefficients in descending powers of s. The order of the polynomial is determined by the total number of entries in the matrix. This means that every polynomial coefficient must be entered in the matrix even if it is zero. The MATLAB statement

```
bode (num,den)
```

produces the Bode plots shown in Figure 12–50.

Bode Diagrams

Example 12-21

At low frequencies the phase angle gradually increases under the influence of the zero at $s = -100$ rad/s. As the input frequency approaches the natural frequency of the complex poles, the phase begins to decrease. As the frequency passes natural frequency at $\omega = 500$ rad/s, the phase rapidly decreases and eventually approaches $-90°$ at high frequency. The net effect of the zero and the two poles is that the phase initially increases and then decreases as the poles come into play. The result is that the phase has a maximum of about $42°$ at around 200 rad/s. ■

12–7 FREQUENCY RESPONSE AND STEP RESPONSE

Frequency and step reponses are often used to describe or specify the performance of devices, circuits, and systems. In Chapter 11 we found that a transfer function contains all of the data needed to construct the step response. In the present chapter we found that a transfer function also gives us the frequency response in the form of gain $|T(j\omega)|$ and phase $\angle T(j\omega)$. This section explores the relationship between the frequency-domain response of a circuit and its time-domain step response.

If we are given a transfer function then obviously we can calculate either the step or the frequency response. If we are given the time-domain step response, we can obtain the transfer function as $T(s) = s\mathcal{L}\{g(t)\}$ and then calculate the gain and phase reponses. In other words, it is a conceptually easy matter to calculate the gain and phase responses when given $g(t)$, or $T(s)$.

But, can we determine the time-domain step response given only the frequency-domain gain response $|T(j\omega)|$? The answer is a qualified yes, provided we can determine the circuit transfer function. It may not always be possible to construct a unique transfer function because gain alone does not necessarily determine a unique phase response. Nonetheless it is true that the transfer function serves as an intermediary that allows us to calculate the response due to any input, including the step response and gain response. It is important to be able to go from one domain to the other because design requirements and device performance are often stated in both the time and frequency domains.

The next series of examples illustrates problems in which we must operate in both domains.

EXAMPLE 12–22

Figure 12–51 shows the straight-line gain response of a low-pass circuit. Find the rise time of the circuit step response.

F I G U R E   1 2 – 5 1

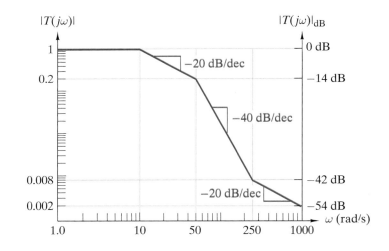

SOLUTION:

The gain plot shows corner frequencies at $\omega = 10$, 50, and 250 rad/s. The following observations determine whether the critical frequencies are poles or zeros:

- The gain slope between $\omega = 10$ and 50 is $m = (-14 - 0)/(\log 50 - \log 10) = -20$ dB/decade, so the critical frequency at $\omega = 10$ is a pole.

- The slope between $\omega = 50$ and 250 is $m = [-42 -(-14)]/(\log 250 - \log 50) = -40$ dB/dec, so the critical frequency at $\omega = 50$ is a pole.

- The gain slope between $\omega = 250$ and 1000 is $m = [-54 - (-42)]/(\log 1000 - \log 250) = -20$ dB/dec, so the critical frequency at $\omega = 250$ is a zero.

Combining these observations, the transfer function must have the form

$$T(s) = \frac{K(s + 250)}{(s + 10)(s + 50)}$$

The dc gain of this transfer function is $T(0) = K/2$. The dc gain in Figure 12–51 is one; hence, $K = 2$ and the required transfer function $T(s)$ is

$$T(s) = \frac{2(s + 250)}{(s + 10)(s + 50)} = \frac{2s + 500}{s^2 + 60s + 500}$$

The MATLAB function step calculates and plots the step response of a transfer function of the form $T(s) = n(s)/d(s)$, where $n(s)$ is the numerator polynomial and $d(s)$ is the denominator polynomial. In the MATLAB command window these polynomials are defined by the row matrices num and den that contain the coefficients of the polynomials $n(s)$ and $d(s)$ in descending powers of s. In the present case these matrices are

```
num=[2,500];
```

```
den=[1,60,500];
```

The first element in both matrices is the coefficient of the highest power of s followed by the other coefficients in descending powers of s. The MATLAB statement

```
step(num,den);
```

produces the step response plot show in Figure 12–52.

Step Response

FIGURE 12 – 52

The plot shows that the final value of the step response $g(\infty) = 1$. The rise time is the time required for the step response to rise from 10% to 90% of its final value. The plot in Figure 12–52 shows that the step response rises to 0.1 (10% rise) at $t = 0.02$ s and to 0.9 (90% rise) at $t = 0.25$ s. Hence, the step response rise time is about $T_R = 0.25 - 0.02 = 0.23$ s.   ■

EXAMPLE 12–23

Characterize the time-domain step response of a second-order bandpass circuit with a maximum gain of 12 dB, a center frequency of 2000 rad/s, and a bandwidth of 100 rad/s.

SOLUTION:
The bandpass circuit transfer function is of the form

$$T(s) = \frac{Ks}{s^2 + 2\zeta\omega_0 s + \omega_0^2}$$

The bandwidth and center frequency values require that $2\zeta\omega_0 = B = 100$ and $\omega_0 = 2000$. The maximum gain occurs at the center frequency. A gain requirement of 12 dB (factor of 4) means that $|T(j\omega_0)| = K/(2\zeta\omega_0) = 4$, or $K = 4B = 400$. Hence the circuit transfer function is

$$T(s) = \frac{400s}{s^2 + 100s + 2000^2}$$

The circuit step response is

$$g(t) = \mathcal{L}^{-1}\left\{\frac{T(s)}{s}\right\} = \mathcal{L}^{-1}\left\{\frac{400}{1999.4} \times \frac{1999.4}{(s + 50)^2 + 1999.4^2}\right\}$$

$$= 0.200\, e^{-50t}\sin(1999.4t) \qquad t > 0$$

The bandpass transfer function has zero at the origin that cancels the forced pole due to the step function input. As a result, the steady-state value of the step response is zero. The standard definition of rise time does not work here because the initial value and the final value are both zero. In this case we use a time-domain parameter called settling time. **Settling time (T_S)** is defined as the time required for the response to enter and remain within specified limits centered about its steady-state value. For step response the limits are often specified as ±2%.

Figure 12–53 shows a graph of the step response. The narrowband frequency response requires a small damping ratio ($\zeta = 0.025$, in this example) so the step response is a slowly decaying damped sine. The decay rate is controlled by an exponential envelope $\pm 0.2e^{-50t}$. This envelope decays to ±2% of its initial value when $e^{-50t} = 0.02$, which means that $T_S = 0.0782$ s. ■

EXAMPLE 12–24

A class of piezoelectric pressure transducers have step reponses of the form

$$v_O(t) = K_V e^{-t/T_C} u(t)$$

FIGURE   12 – 53

where K_V is called the voltage sensitivity of the transducer and T_C is its time constant. In a certain application a transducer is needed to sense sinusoidal pressure waves in the range from 2 Hz to 2000 Hz. The following transducers are available for use in this application.

MODEL NUMBER	VOLTAGE SENSITIVITY (mV/PSI)	TIME CONSTANT (MS)
2612-50	50	600
2612-20	20	240
2612-10	10	120
2612-5	5	60
2612-2	2	24
2612-1	1	12

Select the transducer with the largest sensitivity that meets the frequency response requirement.

SOLUTION:
Because $v_O(t)$ is the step response of the transducer, its transfer function from pressure to output voltage is of the form

$$T(s) = s\mathcal{L}\{K_V e^{-t/T_C}\} = \frac{K_V s}{s + 1/T_C}$$

This transfer function has a first-order high-pass characteristic with a cutoff frequency of $\omega_C = 1/T_C$. The sinusoidal pressure waves range from 2 hz to 2000 Hz. This frequency range must fall in the passband of the transducer. Since the transducer frequency response is high pass, the transducer cutoff frequency must fall below 2 Hz. In equation form this requires that

$$\omega_C = \frac{1}{T_C} < 2\pi \times 2 \quad \text{or} \quad T_C > \frac{1}{4\pi} = 79.6 \text{ ms}$$

The first three transducers in the table meet this requirement. The model with the largest sensitivity is 2612-50 with $K_V = 50$ mV/psi.   ∎

SUMMARY

- The frequency response of a circuit is the variation of the gain $|T(j\omega)|$ and phase $\angle T(j\omega)$ with frequency. The gain function is usually expressed in dB in frequency-response plots. Logarithmic frequency scales are used on frequency-response plots of the gain and phase functions.

- A passband is a range of frequencies over which the steady-state output is essentially constant with very little attenuation. A stopband is a range of frequencies over which the steady-state output is significantly attenuated. The cutoff frequency is the boundary between a passband and the adjacent stopband.

- Circuit gain responses are classified as low pass, high pass, bandpass, and bandstop depending on the number and location of the stop and passbands. The performance of devices and circuits is often specified in terms of frequency-response descriptors such as bandwidth, passband gain, and cutoff frequency.

- The low- and high-frequency gain asymptotes of a first-order circuit intersect at a corner frequency determined by the location of its pole. The total phase change from low to high frequency is ±90°. First-order circuits can be connected to produce bandpass and bandstop responses.

- The low- and high-frequency gain asymptotes of second-order circuits intersect at a corner frequency determined by the natural frequency of the poles. The total phase change from low to high frequency is 180°. Second-order circuits with complex critical frequencies may exhibit narrow resonant peaks and valleys.

- Series and parallel *RLC* circuits provide bandpass and bandstop gain characteristics that are easily related to the circuit parameters.

- Bode plots are graphs of the gain (in dB) and phase angle (in degrees) versus log-frequency scales. Straight-line approximations to the gain and phase can be constructed using the corner frequencies defined by the poles and zeros of $T(s)$. The purpose of the straight-line approximations is to develop a conceptual understanding of frequency response. The straight-line plots do not necessarily provide accurate data at all frequencies, especially for circuits with complex poles.

- Computer-aided circuit analysis programs can accurately generate and plot frequency-response data. The user must have a rough idea of the gain and frequency ranges of interest to use these tools intelligently.

- It is often necessary to relate the time-domain characteristics of a circuit to its frequency response, and vice versa. The transfer function provides a link between the frequency-domain and time-domain responses.

FIGURE   P12-1

PROBLEMS

ERO 12–1 FIRST-ORDER CIRCUIT FREQUENCY RESPONSE (SECTS. 12–1, 12–2)

Given a first-order linear circuit or its transfer function,

(a) Determine frequency-response descriptors and classify the circuit response.
(b) Draw the straight-line approximations of the gain and phase responses.
(c) Use the straight-line approximations to estimate the gain or steady-state output at specified frequencies.
(d) Select circuit parameters to produce specified descriptors.

12–1   (a) Find the transfer function $T_V(s) = V_2(s)/V_1(s)$ for the RL circuit in Figure P12–1.

(b) Determine the dc gain, infinite-frequency gain, and the cutoff frequency ω_C. Is the gain response low-pass, high-pass, or bandpass?

(c) Draw the straight-line approximations of the gain and phase of $T_V(j\omega)$.

(d) Use the straight-line approximations to estimate the gain at $\omega = 0.5\omega_C$, ω_C, and $2\omega_C$. Compare these straight-line gains to the value of $|T_V(j\omega)|$ at the same frequencies.

FIGURE   P12-2

12–2   (a) Find the transfer function $T_V(s) = V_2(s)/V_1(s)$ for the RL circuit in Figure P12–2.

(b) Determine the dc gain, infinite-frequency gain, and the cutoff frequency ω_C. Is the gain response low-pass, high-pass, or bandpass?

(c) Draw the straight-line approximations of the gain and phase of $T_V(j\omega)$.

(d) Use the straight-line gain approximations to estimate the amplitude of the steady-state output for a 5-V sinusoidal input $\omega = 0.5\omega_C$, ω_C, and $2\omega_C$.

FIGURE   P12-3

12–3   (a) Find the transfer function $T_V(s) = V_2(s)/V_1(s)$ for the RC OP AMP circuit in Figure P12–3.

(b) Determine the dc gain, infinite-frequency gain, and the cutoff frequency ω_C. Is the gain response low-pass, high-pass, or bandpass?

(c) Draw the straight-line approximations of the gain and phase of $T_V(j\omega)$.

(d) Use the straight-line approximations to estimate the gain at $\omega = 0.5\omega_C$, ω_C, and $2\omega_C$. Compare these straight-line gains to the value of $|T_V(j\omega)|$ at the same frequencies.

12–4   (a) Find the transfer function $T_V(s) = V_2(s)/V_1(s)$ for the RC OP AMP circuit in Figure P12–4.

(b) Determine the dc gain, infinite-frequency gain, and the cutoff frequency ω_C. Is the gain response low-pass, high-pass, or bandpass?

FIGURE   P12-4

(c) Draw the straight-line approximations of the gain and phase of $T_V(j\omega)$.

(d) Use the straight-line gain approximations to estimate the amplitude of the steady-state output for a 2-V sinusoidal input at $\omega = 0.5\omega_C$, ω_C, and $2\omega_C$.

12–5 **(a)** Find the transfer function $T_V(s) = V_2(s)/V_1(s)$ for the *RC* OP AMP circuit in Figure P12–5.

(b) Determine the dc gain, infinite-frequency gain, and the cutoff frequency ω_C. Is the gain response low-pass, high-pass, or bandpass?

(c) Draw the straight-line approximations of the gain and phase of $T_V(j\omega)$.

(d) Use the straight-line approximations to estimate the gain at $\omega = 0.5\omega_C$, ω_C, and $2\omega_C$. Compare these straight-line gains to the value of $|T_V(j\omega)|$ at the same frequencies.

FIGURE P12–5

12–6 In the *RC* OP AMP circuit in Figure P12–6

(a) Find the transfer function $T_V(s) = V_2(s)/V_1(s)$ for the $R = 10$ kΩ and $C = 100$ nF.

(b) Determine the dc gain, infinite-frequency gain, and the cutoff frequency ω_C. Is the gain response low-pass, high-pass, or bandpass?

(c) Draw the straight-line approximations of the gain and phase of $T_V(j\omega)$.

(d) Use the straight-line gain approximation to estimate the amplitude of the steady-state output for a 5-V sinusoidal input at $\omega = 0.5\omega_C$, ω_C, and $2\omega_C$.

12–7 The circuit in Figure P12–7 is a small-signal model of a transistor circuit. Determine the gain of $T_V(s) = V_2(s)/V_1(s)$ at $\omega = 0$ and $\omega = \infty$. Is this a low-pass, bandpass, or high-pass circuit? What are the passband gain and the cutoff frequency for $R_S = 1$ kΩ, $R_g = 500$ kΩ, $R_O = 10$ kΩ, $R_L = 5$ kΩ, $C_L = 5$ pF, and $\beta = 100$?

FIGURE P12–6

FIGURE P12–7

12–8 The circuit in Figure P12–8 is a small-signal model of a transistor circuit. Determine the gain of $T_V(s) = V_2(s)/V_1(s)$ at $\omega = 0$ and $\omega = \infty$. Is this a low-pass, bandpass, or high-pass circuit? What are the passband gain and the cutoff frequency for $R_S = 0.1$ kΩ, $R_g = 500$ kΩ, $R_D = 10$ kΩ, $R_L = 5$ kΩ, $C_C = 0.05$ μF, and $g = 0.01$ S?

12–9 The circuit in Figure P12–9 produces a bandpass response. Find the passband gain and the cutoff frequencies for $R_1 = R_2 = R_4 = 10$ kΩ, $R_3 = 90$ kΩ, $C_1 = 4$ nF, and $C_2 = 100$ nF.

12–10 The transfer function of a bandstop circuit can be written as

$$T(s) = \frac{20}{s + 40} + \frac{s}{s + 4000}$$

(a) What are the passband gains and cutoff frequencies?
(b) Draw the straight-line approximations of the gain and phase of $T(j\omega)$.
(c) Use the straight-line approximations to estimate the gain at $\omega = 100, 400$, and 2000 rad/s. Compare these straight-line gains to the value of $|T(j\omega)|$ at the same frequencies.

ERO 12–2 SECOND-ORDER CIRCUIT FREQUENCY RESPONSE (SECT. 12–3)

Given a second-order linear circuit or its transfer function,

(a) Determine frequency-response descriptors and classify the circuit response.
(b) Draw the straight-line approximations of the gain and phase responses.
(c) Use the straight-line approximations to estimate the gain or steady-state output at specified frequencies.
(d) Select circuit parameters to produce specified descriptors.

12–11 **(a)** Find the transfer function $T_V(s) = V_2(s)/V_1(s)$ for the *RLC* circuit in Figure P12–11.
(b) Determine the dc gain, infinite-frequency gain, the damping ratio ζ, and the undamped natural frequency ω_0. Is the gain response low-pass, high-pass, or bandpass?

FIGURE P12-12

FIGURE P12-13

FIGURE P12-14

FIGURE P12-15

FIGURE P12-16

(c) Draw the straight-line approximation of the gain of $T_V(j\omega)$.

(d) Use the straight-line approximation to calculate the gain at $\omega = 0.5\omega_0$, ω_0, and $2\omega_0$. Compare these straight-line gains with the values of $|T_V(j\omega)|$ at the same frequencies.

12-12 **(a)** Find the transfer function $T_V(s) = V_2(s)/V_1(s)$ for the *RC* circuit in Figure P12-12.

(b) Determine the dc gain, infinite-frequency gain, the damping ratio ζ, and the undamped natural frequency ω_0. Is the gain response low-pass, high-pass, or bandpass?

(c) Draw the straight-line approximation of the gain of $T_V(j\omega)$.

(d) Use the straight-line approximation to estimate the amplitude of the steady-state output for a 2-V sinusoidal input at $\omega = 0.5\omega_C$, ω_C, and $2\omega_C$.

12-13 **(a)** Find the transfer function $T_V(s) = V_2(s)/V_1(s)$ for the *RLC* circuit in Figure P12-13.

(b) Determine the dc gain, infinite-frequency gain, the damping ratio ζ, and the undamped natural frequency ω_0. Is the gain response low-pass, high-pass, or bandpass?

(c) Draw the straight-line approximation of the gain of $T_V(j\omega)$.

(d) Use the straight-line approximation to estimate the gain at $\omega = 0.5\omega_0$, ω_0, and $2\omega_0$. Compare these straight-line gains with the actual values of $|T_V(j\omega)|$ at the same frequencies.

12-14 **(a)** Find the transfer function $T_V(s) = V_2(s)/V_1(s)$ for the OP AMP circuit in Figure P12-14.

(b) Determine the dc gain, infinite-frequency gain, the damping ratio ζ, and the undamped natural frequency ω_0. Is the gain response low-pass, high-pass, or bandpass?

(c) Draw the straight-line approximation of the gain of $T_V(j\omega)$.

(d) Use the straight-line approximation to estimate the gain at $\omega = 0.5\omega_0$, ω_0, and $2\omega_0$. Compare these straight-line gains with the values of $|T_V(j\omega)|$ at the same frequencies.

12-15 **(a)** Find the transfer function $T_V(s) = V_2(s)/V_1(s)$ for the OP AMP circuit in Figure P12-15.

(b) Determine the dc gain, infinite-frequency gain, the damping ratio ζ, and the undamped natural frequency ω_0. Is the gain response low-pass, high-pass, or bandpass?

(c) Draw the straight-line approximation of the gain of $T_V(j\omega)$.

(d) Use the straight-line approximation to estimate the gain at $\omega = 0.5\omega_0$, ω_0, and $2\omega_0$. Compare these straight-line gains with the values of $|T_V(j\omega)|$ at the same frequencies.

12-16 **(a)** Find the transfer function $T_V(s) = V_2(s)/V_1(s)$ for the OP AMP circuit in Figure P12-16 for $R_1 = 10$ kΩ, $R_2 = 40$ kΩ, $C = 20$ nF.

(b) Determine the dc gain, infinite-frequency gain, the damping ratio ζ, and the undamped natural frequency ω_0. Is the gain response low-pass, high-pass, or bandpass?

(c) Draw the straight-line approximation of the gain of $T_V(j\omega)$.

(d) Use the straight-line approximation to estimate the amplitude of the steady-state output for a 0.5-V sinusoidal input at $\omega = 0.5\omega_0$, ω_0, and $2\omega_0$.

12–17 Find the transfer function $T_V(s) = V_2(s)/V_1(s)$ for the RLC circuit in Figure P12–17. Is this a low-pass, high-pass, bandpass, or bandstop circuit? Derive expressions for the damping ratio ζ and the undamped natural frequency ω_0 in terms of circuit parameters. Select values of the circuit parameters so that $\zeta = 2$ and $\omega_0 = 5$ krad/s. What is the pass-band gain for your choice?

FIGURE P12–17

12–18 Find the transfer function $T_V(s) = V_2(s)/V_1(s)$ for the RC circuit in Figure P12–18. Is this a low-pass, high-pass, bandpass, or bandstop circuit? Derive expressions for the damping ratio ζ and the undamped natural frequency ω_0 in terms of circuit parameters. Select values of the circuit parameters so that $\omega_0 = 2500$ rad/s. What are the passband gains for your choice?

12–19 Find the transfer function $T_V(s) = V_2(s)/V_1(s)$ for the OP AMP circuit in Figure P12–19. Is this a low-pass, high-pass, bandpass, or bandstop circuit? Derive expressions for the damping ratio ζ and the undamped natural frequency ω_0 in terms of circuit parameters. What is the passband gain for the circuit?

FIGURE P12–18

FIGURE P12–19

12–20 The second-order transfer function

$$T(s) = \frac{s^2 + \omega_0^2}{s^2 + 2\zeta\omega_0 s + \omega_0^2}$$

displays a bandstop response with two passbands separated by a stopband. Sketch the gain response of the transfer function. What are the high-frequency and low-frequency gains? What is the gain at $\omega = \omega_0$? Find the cutoff frequencies for the two passbands. *Hint:* They are the same as those for a second-order bandpass response.

ERO 12–3 THE FREQUENCY RESPONSE OF RLC CIRCUITS (SECT. 12–4)

Given a series or parallel RLC circuit connected as a bandpass or bandstop filter,

(a) Find the circuit parameters or frequency-response descriptors.
(b) Select the circuit parameters to achieve specified filter characteristics.
(c) Derive expressions for the frequency-response descriptors.

12–21 A series RLC circuit is designed to have a bandwidth of 8 Mrad/s and an input impedance of 24 Ω at its resonant frequency of 50 Mrad/s. Determine the values L, C, Q, and the upper and lower cutoff frequencies.

12–22 A 20-mH inductor with an internal series resistance of 20 Ω is connected in series with a capacitor and a voltage source with a Thévenin resistance of 60 Ω.

(a) Determine the value of capacitance C required to produce resonance at 5 krad/s.

(b) Calculate the Q and bandwidth of the circuit.

12–23 A series RLC circuit with $R = 100$ Ω, $L = 25$ mH, and $C = 100$ nF is driven by a sinusoidal voltage source with a peak amplitude of 10 V.

(a) Calculate the circuit bandwidth and the upper and lower cutoff frequencies.

(b) Calculate the amplitude of the steady-state voltage across the capacitor and the inductor at the resonant frequency.

12–24 A series RLC circuit has a bandwidth of 10 krad/s and a driving-point impedance of 80 Ω at its resonant frequency of 100 krad/s. Determine the values of L, C, Q, and the upper and lower cutoff frequencies.

12-25 A parallel RLC circuit with $R = 40$ kΩ is to have a resonant frequency of 100 MHz. Calculate the values of L and C required to produce a bandwidth of 100 kHz.

12-26 D A series RLC bandpass filter is required to have a resonance at $f_0 = 100$ kHz. The series-connected L and C are to be driven by a sinusoidal source with a Thévenin resistance of 50 Ω. The following standard capacitors are available in the stock room: 1 μF, 680 nF, 470 nF, 330 nF, 220 nF, and 150 nF. The inductor will be custom designed to match the capacitor used. Select from available capacitors the one that minimizes the circuit bandwidth.

12-27 D A series RLC circuit is to be used as a notch filter to eliminate a bothersome 60-Hz hum in an audio channel. The signal source has a Thévenin resistance of 600 Ω. Select values of L and C so the upper cutoff frequency is below 200 Hz.

12–28 Find the transfer function $T_V(s) = V_2(s)/V_1(s)$ for the bandpass circuit in Figure P12–28. Find the parameters B, Q, ω_{C1}, and ω_{C2} for $R = 50$ Ω, $L = 50$ μH, and $C = 20$ nF.

12–29 The transfer function $T_V(s) = V_2(s)/V_1(s)$ of the circuit in Figure P12–29 has a bandstop filter characteristic. Without solving for the transfer function, use the element impedances to

(a) Explain why the low- and high-frequency passband gains are unity.

(b) Explain why the bandstop notch occurs at $\omega_0 = 1/\sqrt{LC}$.

(c) Derive expressions ω_{C1}, and ω_{C2} in terms of circuit parameters R, L, and C.

12–30 Find the transfer impedance $T_Z(s) = V_2(s)/I_1(s)$ for the bandpass circuit in Figure P12–30. Find the parameters B, Q, ω_{C1}, and ω_{C2} for $R = 5$ Ω, $L = 50$ μH, and $C = 2$ μF.

FIGURE P12–28

FIGURE P12–29

FIGURE P12–30

ERO 12–4 BODE PLOTS (SECTS. 12–5, 12–6)

(a) Construct Bode plots of the asymptotic gain and phase responses of a given circuit or transfer function.
(b) Construct the transfer function corresponding to a given asymptotic gain plot.
(c) Use the asymptotic gain plots to estimate frequency-response descriptors or steady-state output.

12–31 Construct a plot of the straight-line approximation to the gain response of the circuit in Figure P12–31. Use the straight-line gain plots to calculate the amplitude of the steady-state output when the input is $v_1(t) = 10 \sin 30t$ V. Use the circuit transfer function to calculate the amplitude of the steady-state output for the same input and compare the two results.

12–32 Repeat Problem 12–31 using the circuit in Figure P12–32.

12–33 Construct a plot of the straight-line approximation to the gain response of the circuit in Figure P12–33. Is this a low-pass, high-pass, bandpass, or bandstop response? Use the straight-line gain plot to estimate the cutoff frequency and passband gain.

FIGURE P12-31

FIGURE P12-32

FIGURE P12-33

12–34 Construct a plot of the straight-line approximation to the gain response of the circuit in Figure P12–34. Is this a low-pass, high-pass, bandpass, or bandstop response? Use the straight-line gain plot to estimate the cutoff frequency and passband gain.

FIGURE P12-34

12–35 Construct plots of the straight-line approximations to the gain and phase responses of the following transfer functions. Are these low-pass, high-pass, bandpass, or bandstop functions? Estimate the cutoff frequency and passband gain.

(a) $T(s) = \dfrac{(0.02s + 1)}{(0.2s + 1)(2s + 1)}$

(b) $T(s) = \dfrac{(0.2s + 1)}{(0.02s + 1)(2s + 1)}$

12–36 Construct plots of the straight-line approximations to the gain and phase responses of the following transfer functions. Are these low-pass, high-pass, bandpass, or bandstop functions? Estimate the cutoff frequency and passband gain.

(a) $T(s) = \dfrac{s(s + 2)}{(0.04s + 1)(0.1s + 1)}$

(b) $T(s) = \dfrac{s^2}{(0.04s + 1)(0.1s + 1)}$

12–37 Construct a plot of the straight-line approximation to the gain response of each of the following transfer functions. Are these low-pass, high-pass, or bandpass functions? What is the passband gain and the maximum gain?

(a) $T(s) = \dfrac{20}{(0.01s^2 + 0.05s + 1)}$

(b) $T(s) = \dfrac{40s^2}{(4s^2 + 0.2s + 1)}$

12–38 Construct a plot of the straight-line approximation to the gain response of each of the following transfer functions. Are these low-pass, high-pass, or bandpass functions? What is the passband gain and the maximum gain?

(a) $T(s) = \dfrac{5(0.4s + 1)}{(0.5s^2 + 0.5s + 1)}$

(b) $T(s) = \dfrac{5s}{(0.5s^2 + 0.25s + 1)}$

12–39 Find the transfer function corresponding to the straight-line gain plot in Figure P12–39. Calculate the actual gain at each corner frequency and compare it to the straight-line gain shown in the figure.

F I G U R E   P 1 2 – 3 9

12–40 Find the transfer function corresponding to the straight-line gain plot in Figure P12–40. Calculate the actual gain at each corner frequency and compare it to the straight-line gain shown in the figure.

FIGURE P12–40

ERO 12–5 FREQUENCY RESPONSE AND STEP RESPONSE (SECT. 12–7)

(a) Given an asymptotic gain response of a linear circuit, determine its step response.
(b) Given the step response of a linear circuit, construct the asymptotic gain and phase responses.
(c) Construct a transfer function that meets constraint on both the frequency and step responses.

12–41 Figure P12–41 shows the straight-line approximations to the gain response of a linear circuit. Find the time-domain step response $g(t)$.

FIGURE P12–41

12–42 Repeat Problem 12–41 for the gain response in Figure P12–39.
12–43 Repeat Problem 12–41 for the gain response in Figure P12–40.
12–44 The step response of a linear circuit is

$$g(t) = 2 - 4e^{-10t} + 2e^{-20t} \qquad t \geq 0$$

Is the circuit a low-pass, high-pass, bandpass, or bandstop filter? Construct plots of the straight-line approximations to the gain and phase responses of the circuit.

12–45 The step response of a linear circuit is

$$g(t) = 5 - 5e^{-100t} \sin 200\, t \qquad t \geq 0$$

Is the circuit a low-pass, high-pass, bandpass, or bandstop filter? Construct plots of the straight-line approximations to the gain and phase responses of the circuit.

12–46 The step response of a linear circuit is

$$g(t) = 10 - 10e^{-100t} \cos 100\,t \qquad t \geq 0$$

Is the circuit a low-pass, high-pass, bandpass, or bandstop filter? Construct a plot of the straight-line approximation to the gain response of the circuit.

12–47 The step response of a linear circuit is

$$g(t) = -10e^{-100t} + 200e^{-2000t} \qquad t \geq 0$$

Is the circuit a low-pass, high-pass, bandpass, or bandstop filter? Construct plots of the straight-line approximations to the gain and phase responses of the circuit.

12–48 Construct a first-order low-pass transfer function with a dc gain of 10, a bandwidth less than 250 rad/s, and a step response that rises to 50% of its final value in less than 4 ms.

12–49 Construct a first-order high-pass transfer function whose step response decays to 5% of its peak value in less than 10 μs and whose cutoff frequency is less than 100 kHz.

12–50 Construct a second-order bandpass transfer function with a midband gain of 10, a center frequency of 100 kHz, a bandwidth less than 50 kHz, and a step response that decays to less that 20% of its peak value in less than 5 μs.

INTEGRATING PROBLEMS

12–51 **A** DC, AC, AND IMPULSE RESPONSES

A linear circuit is driven by an input $v_1(t) = 4 \cos(1000t)$. The zero-state response for this input is observed to be $v_2(t) = 4 \sin(1000t) + 8 \cos(1000t) - e^{-1000t}[6 \sin(2000t) + 8 \cos(2000t)]$.

(a) Find the phasor representation of the steady-state output when the input is

$$v_1(t) = 4 \cos(2000t).$$

(b) Find the steady-state output when the input is 5 V dc.

(c) Construct the Bode plot of the circuit gain as a function of frequency.

(d) Determine the circuit impulse response.

12–52 **A** BANDPASS FILTER STEP RESPONSES

The step responses of the two second-order bandpass filters are

$$g_1(t) = 0.2e^{-20t} \sin(200t)u(t)$$
$$g_2(t) = (e^{-10t} - e^{-4000t})u(t)$$

(a) Find the center frequency and bandwidth of each filter.

(b) Which filter has a gain less than −20 dB two octaves above and below the center frequency?

(c) Which filter has a gain greater than −3 dB one decade above and below the center frequency?

(d) Which filter has phase shift within ±15° over the frequency range from one octave above to one octave below the center frequency?

12–53 **D** FILTER DESIGN WITH INPUT IMPEDANCE SPECIFIED

Design a circuit whose gain response lies entirely within the unshaded region in Figure P12–53 and whose input impedance is around 50 Ω for most of the frequencies in the passband.

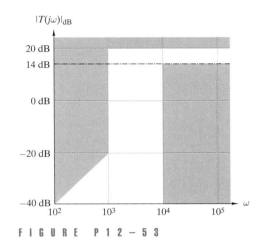

FIGURE P12–53

12–54 **A** HIGH-FREQUENCY MODEL OF A RESISTOR

Figure P12–54 shows a circuit model of a resistor R that includes parasitic capacitance C and parasitic lead inductance L. The purpose of this problem is to investigate the effect of these parasitic elements on the high-frequency characteristics of the resistor device.

(a) Derive an expression for the impedance $Z(s)$ of the circuit model in Figure P12–54 in terms of the circuit parameters R, L, and C.

(b) For $R = 10\ \text{k}\Omega$, $L = 2\ \mu\text{H}$, and $C = 4\ \text{pF}$ find the poles and zeros of $Z(s)$. Construct a plot of the straight-line asymptotes of $|Z(j\omega)|$. Do not express $|Z(j\omega)|$ in decibels. Use a logarithmic scale with the impedance magnitude expressed in ohms. Identify the corner frequencies and slopes (in ohms per decade) in this plot.

(c) Use the straight-line plot in (b) to identify the frequency range (in hertz) over which the resistor device appears to be: (1) a 10-kΩ resistance, (2) a 4-pF capacitance, and (3) a 2-μH inductor.

(d) Over what frequency range can you use this device in a circuit design and treat it as a pure 10-kΩ resistor?

FIGURE P12–54

12–55 **A** **D** COMPENSATOR DESIGN

A modification is required to improve the performance of an existing output transducer. The compensator/transducer combination must meet the following performance specifications: dc gain 0 dB, gain = 0 ±5 dB for ω between 20 rad/s and 200 rad/s, gain slope of −40 dB/dec for ω > 1000 rad/s. The two designs in Figure P12–55 have been proposed

(a) **A** Verify that both designs meet the performance specification.

(b) **E** The compensator is to be built using an *RC* OP AMP circuit

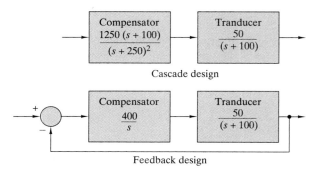

with no more than one OP AMP. Which of the two designs would you recommend and why?

(c) **D** Design an OP AMP circuit to implement your recommendation in (b).

12–56 **A D USING A BODE PLOT**

Figure P12–56 shows the gain response of a second-order filter. The purpose of this problem is to infer things about the performance of the filter with various inputs. The only additional data is that the filter output is an OP AMP with $V_{CC} = \pm 15$ V.

(a) What is the maximum allowable dc input voltage?

(b) Estimate the maximum allowable sinusoidal input voltage at $f = 20$ Hz.

(c) A short rectangular pulse is to be used to evaluate the impulse response of the filter. Estimate the required pulse duration.

(d) What is the final value of the impulse response of the filter? About how long does it take to settle down to this value?

(e) A 0.5-V step function is to be applied at the input. Estimate the final value of the output response and about how long it takes the response to reach 50% of this value?

(f) The signal $v(t) = 0.5 \sin(100t) + 0.25 \sin(1000t)$ V is to be applied at the input. About how long will it take for a steady-state condition to be reached? What will be the ratio of the steady-state output amplitude at $\omega = 100$ rad/s to the amplitude of the output at $\omega = 1000$ rad/s? Express the ratio in dB.

(g) Your boss needs to know the frequency at which the phase shift is $-90°$. What is your estimate?

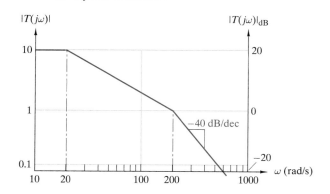

CHAPTER 13

FOURIER SERIES

The series formed of sines or cosines of multiple arcs are therefore adapted to represent between definite limits all possible functions, and the ordinates of lines or surfaces whose form is discontinuous.

Jean Baptiste Joseph Fourier, 1822,
French Mathematician

13–1 Overview of Fourier Analysis

13–2 Fourier Coefficients

13–3 Waveform Symmetries

13–4 Circuit Analysis Using Fourier Series

13–5 RMS Value and Average Power

Summary

Problems

Integrating Problems

The analysis techniques in this chapter have their roots in the works of the French mathematician Joseph Fourier (1768–1830). While studying heat transfer, Fourier found that he could represent discontinuous functions by an infinite series of harmonic sinusoids, or sines of multiple arcs as he called them. Fourier was guided by his physical intuition and presented experimental evidence to support his claim, but he left the question of series convergence unanswered. As a result, his theory was incomplete and his 1807 treatise was initially rejected by the scientific community of his day. Some years later an acceptable proof of the "Fourier theorem" was supplied by the German mathematician P. G. L. Dirichlet (1805–1859). The basic ideas behind Fourier analysis were eventually accepted during his lifetime and are now firmly entrenched in all branches of science and engineering.

The Fourier series resolves a periodic waveform into a sum (albeit infinite) of sinusoidal components. In this sense the series is an extension of the study of composite signals begun in Chapter 5. In the meantime we have learned how to find the sinusoidal steady-state response using phasors (Chapter 8) and using Laplace transforms (Chapter 10). We have also learned how to use Bode plots to describe the steady-state response when the frequency sweeps across a range of values (Chapter 12). In this chapter we use this background together with Fourier analysis to predict steady-state response to a periodic input. This process involves two basic steps: (1) finding the Fourier series of the periodic input waveform and (2) finding the Fourier series of the steady-state response using circuit analysis and the input Fourier series. Step (1) is treated in the first three sections where we find the Fourier series for some common periodic signals. Step (2) is treated in the last two sections where we develop several methods of finding the steady-state response for periodic inputs. The Fourier analysis concepts discussed in this chapter serve as a prelude to the analysis and design of frequency selective circuits called filters.

13–1 OVERVIEW OF FOURIER ANALYSIS

In this chapter we develop a method of finding the steady-state response of circuits to periodic inputs such as the waveforms in Figure 13–1. These periodic waveforms can be written as a Fourier series consisting of an infinite sum of harmonically related sinusoids. More specifically, if $f(t)$ is periodic with period T_0 and is reasonably well behaved, then $f(t)$ can be expressed as a **Fourier series** of the form

$$f(t) = a_0 + a_1 \cos(2\pi f_0 t) + a_2 \cos(2\pi 2 f_0 t) + \ldots + a_n \cos(2\pi n f_0 t) + \ldots$$
$$+ \, b_1 \sin(2\pi f_0 t) + b_2 \sin(2\pi 2 f_0 t) + \ldots + b_n \sin(2\pi n f_0 t) + \ldots \qquad (13–1)$$

or, more compactly,

$$f(t) = \underbrace{a_0}_{\text{dc}} + \underbrace{\sum_{n=1}^{\infty} [a_n \cos(2\pi n f_0 t) + b_n \sin(2\pi n f_0 t)]}_{\text{ac}} \qquad (13–2)$$

The coefficient a_0 is the dc component or average value of $f(t)$. The constants a_n and b_n ($n = 1, 2, 3, \ldots$) are the **Fourier coefficients** of the sinusoids in the ac component. The lowest frequency in the ac component occurs for

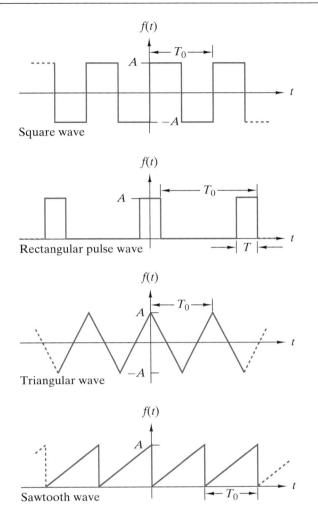

FIGURE 13–1 *Examples of periodic waveforms.*

$n = 1$ and is called the **fundamental frequency** defined as $f_0 = 1/T_0$. The other frequencies are integer multiples of the fundamental called the second harmonic ($2f_0$), third harmonic ($3f_0$) and, in general, the nth harmonic (nf_0).

Since Eq. (13–2) is an infinite series, there is always a question of convergence. We have said that the series converges as long as $f(t)$ is reasonably well behaved. Basically, this means that $f(t)$ is single valued, the integral of $|f(t)|$ over a period is finite, and $f(t)$ has a finite number of discontinuities in any one period. These requirements, called the **Dirichlet conditions**, are sufficient to assure convergence. Every periodic waveform that meets the Dirichlet conditions has a convergent Fourier series. However, there are waveforms that do not meet the Dirichlet conditions that also have convergent Fourier series. That is, while the Dirichlet conditions are sufficient, they are not necessary and sufficient. This limitation does not present a serious problem because the Dirichlet conditions are satisfied by the waveforms generated in physical systems. All of the periodic waveforms in Figure 13–1 meet the Dirichlet requirements.

Before discussing how to construct a Fourier series, it is important to have an overview of why such a series is useful in circuit analysis. The Fourier series resolves a periodic input into a sum of dc and ac components that can be thought of as being generated by separate dc and ac signal sources. For linear circuits we can think of the steady-state response to a periodic input as the sum of the steady-state responses to each of these sources acting alone. Since we know how to find the steady-state response to dc inputs (Chapters 2, 3, and 4) and sinusoidal (ac) inputs (Chapters 8, 11, and 12), we can calculate the steady-state response to a periodic input using superposition. The net result is that both the input and the steady-state output are expressed as Fourier series.

At first glance it may seem awkward to express circuit inputs and outputs as infinite series. However, it turns out that the Fourier series tells us a great deal about signals and circuits. With experience you will see that finding the Fourier series of a waveform is not an end in itself, but the beginning of an analysis process that pervades modern signal processing. The distribution of the amplitudes and phases of the Fourier components is called the **Spectrum** of a periodic signal. The manner in which signal processing alters the input signal spectrum to produce the output signal is one of the key concepts in circuit analysis and design.

13–2 FOURIER COEFFICIENTS

The Fourier coefficients for any periodic waveform $f(t)$ satisfying the Dirichlet conditions can be obtained from the equations

$$a_0 = \frac{1}{T_0}\int_{-T_0/2}^{+T_0/2} f(t)dt$$

$$a_n = \frac{2}{T_0}\int_{-T_0/2}^{+T_0/2} f(t)\cos(2\pi nt/T_0)dt \qquad (13\text{–}3)$$

$$b_n = \frac{2}{T_0}\int_{-T_0/2}^{+T_0/2} f(t)\sin(2\pi nt/T_0)dt$$

The integration limits in these equations extend from $-T_0/2$ to $+T_0/2$. However, the limits can span any convenient interval as long as it is exactly one period. For example, the limits could be from 0 to T_0 or $-T_0/4$ to $3T_0/4$. We will show where Eq. (13–3) comes from in a moment, but first we use these equations to obtain the Fourier coefficients of the sawtooth wave.

EXAMPLE 13–1

Find the Fourier coefficients for the sawtooth wave in Figure 13–1.

SOLUTION:
An expression for a sawtooth wave on the interval $0 \le t \le T_0$ is

$$f(t) = \frac{At}{T_0} \qquad 0 \le t < T_0$$

For this definition of $f(t)$ we use 0 and T_0 as the limits in Eq. (13–3). The first expression in Eq. (13–3) yields a_0 as

$$a_0 = \frac{1}{T_0}\int_0^{T_0}\frac{At}{T_0}dt = \frac{At^2}{2T_0^2}\bigg|_0^{T_0} = \frac{A}{2}$$

This result states that the average or dc value is $A/2$, which is easy to see because the area under one cycle of the sawtooth wave is $AT_0/2$. The second expression in Eq. (13–3) yields a_n as

$$a_n = \frac{2}{T_0}\int_0^{T_0}\frac{At}{T_0}\cos(2\pi nt/T_0)dt$$

$$= \frac{2A}{T_0^2}\left[\frac{\cos(2\pi nt/T_0)}{(2\pi n/T_0)^2} + \frac{t \times \sin(2\pi nt/T_0)}{(2\pi n/T_0)}\right]_0^{T_0}$$

$$= \frac{2A}{T_0^2}\left[\frac{\cos(2\pi n) - \cos(0)}{(2\pi n/T_0)^2}\right] = 0 \quad \text{for all } n$$

Since $a_n = 0$ for all n, there are no cosine terms in the series. The b_n coefficients are found using the third expression in Eq. (13–3):

$$b_n = \frac{2}{T_0}\int_0^{T_0}\frac{At}{T_0}\sin(2\pi nt/T_0)dt$$

$$= \frac{2A}{T_0^2}\left[\frac{\sin(2\pi nt/T_0)}{(2\pi n/T_0)^2} - \frac{t \times \cos(2\pi nt/T_0)}{(2\pi n/T_0)}\right]_0^{T_0}$$

$$= \frac{2A}{T_0^2}\left[\frac{T_0\cos(2\pi n)}{(2\pi n/T_0)}\right] = -\frac{A}{n\pi} \quad \text{for all } n$$

Given the coefficients a_n and b_n found above, the Fourier series for the sawtooth wave is

$$f(t) = \frac{A}{2} + \sum_{n=1}^{\infty}\left[-\frac{A}{n\pi}\right]\sin(2\pi nf_0t) \qquad \blacksquare$$

EXAMPLE 13–2

In this example we use a computer tool to show that a truncated Fourier series approximates a periodic waveform. The waveform is a sawtooth with $A = 10$ and $T_0 = 2$ ms. Calculate the Fourier coefficients of the first 20 harmonics and plot the truncated series representation of the waveform using the first 5 harmonics and the first 10 harmonics.

SOLUTION:
From Example 13–1 the Fourier coefficients for the sawtooth wave are

$$a_0 = A/2 \quad a_n = 0 \quad \text{and} \quad b_n = -\frac{A}{n\pi} \quad \text{for all } n$$

Figure 13–2 shows a Mathcad worksheet that generates truncated Fourier series. The first line defines the waveform amplitude and period. The Fourier coefficients of the first 20 harmonics are calculated in the second line using the index variable $n = 1,2, \ldots 20$. In Mathcad syntax $n = I, J, \ldots$

K means FOR $n = I$ TO K STEP $J - I$ DO. The next line defines a truncated Fourier series $f(k,t)$ consisting of the dc component a_0 plus the sum of the first k harmonics. Specifically, the function $f(5, t)$ is the sum of the dc component plus the first 5 harmonics and $f(10, t)$ is the sum of the dc component plus the first 10 harmonics. The plots of these two functions at the bottom of the worksheet show how function $f(k, t)$ approaches the sawtooth wave as more harmonics are added. Infinitely many harmonics are needed to represent a sawtooth wave exactly. However, the plots suggest that a relatively small number, say 5 or 10, provides a reasonable approximation of the major features of the sawtooth wave.

FIGURE 13 – 2

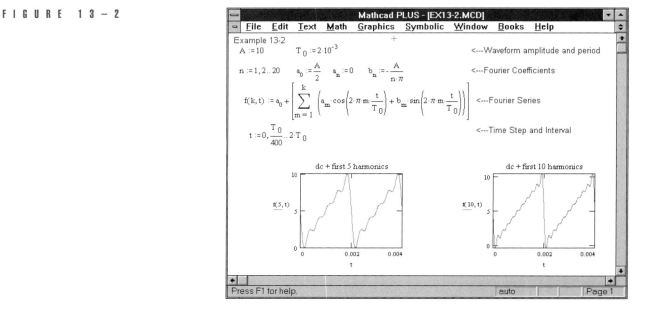

The Mathcad worksheet in Figure 13–2 is a template for generating truncated Fourier series. Changing the amplitude and period in the first line and the Fourier coefficients in the second line generates plots of the truncated Fourier series for other periodic waveforms. The index for n in the second line may need to be changed as well. For example, if only odd harmonics are present in the series, then the index should be $n = 1,3, \ldots 21$. ■

DERIVING EQUATIONS FOR a_n AND b_n

The sawtooth wave example shows how to calculate the Fourier coefficients using Eq. (13–3). We now turn to the derivation of these equations. An equation for a Fourier coefficient is derived by multiplying both sides of Eq. (13–2) by the sinusoid associated with the coefficient and then integrating the result over one period. This multiply and integrate process isolates one coefficient because it turns out that all of the integrations produce zero except one.

The following derivation makes use of the fact that the area under a sine or cosine wave over an integer number of cycles is zero. That is,

$$\int_{-T_0/2}^{+T_0/2} \sin(2\pi k f_0 t)dt = 0 \qquad \text{for all } k$$

$$\int_{-T_0/2}^{+T_0/2} \cos(2\pi k f_0 t)dt = 0 \qquad \text{for } k \neq 0 \qquad (13\text{–}4)$$

$$= T_0 \qquad \text{for } k = 0$$

where k is an integer. These equations state that integrating a sinusoid over $k \neq 0$ cycles produces zero, since the areas under successive half cycles cancel. The single exception occurs when $k = 0$, in which case the cosine function reduces to one and the net area for one period is T_0.

We derive the equation for the amplitude of the dc component a_0 by integrating both sides of Eq. (13–2):

$$\int_{-T_0/2}^{+T_0/2} f(t)dt = \qquad\qquad\qquad\qquad (13\text{–}5)$$

$$a_0 \int_{-T_0/2}^{+T_0/2} dt + \sum_{n=1}^{\infty}\left[a_n \int_{-T_0/2}^{+T_0/2} \cos(2\pi n f_0 t)dt + b_n \int_{-T_0/2}^{+T_0/2} \sin(2\pi n f_0 t)dt \right]$$

$$= a_0 T_0 + \qquad\qquad 0 \qquad + \qquad\qquad 0$$

The integrals of the ac components vanish because of the properties in Eq. (13–4), and the right side of this expression reduces to $a_0 T_0$. Solving for a_0 yields the first expression in Eq. (13–3).

To derive the expression for a_n we multiply Eq. (13–2) by $\cos(2\pi m f_0 t)$ and integrate over the interval from $-T_0/2$ to $+T_0/2$:

$$\int_{-T_0/2}^{+T_0/2} f(t)\cos(2\pi m f_0 t)dt = a_0 \int_{-T_0/2}^{+T_0/2} \cos(2\pi m f_0 t)dt$$

$$+ \sum_{n=1}^{\infty}\left[a_n \int_{-T_0/2}^{+T_0/2} \cos(2\pi m f_0 t)\cos(2\pi n f_0 t)dt + \qquad (13\text{–}6) \right.$$

$$\left. b_n \int_{-T_0/2}^{+T_0/2} \cos(2\pi m f_0 t)\sin(2\pi n f_0 t)dt \right]$$

All of the integrals on the right side of this equation are zero except one. To show this we use identities

$$\cos(x)\cos(y) = \frac{1}{2}\cos(x-y) + \frac{1}{2}\cos(x+y)$$

$$\cos(x)\sin(y) = \frac{1}{2}\sin(x-y) + \frac{1}{2}\sin(x+y)$$

to change Eq. (13–6) into the following form:

$$\int_{-T_0/2}^{+T_0/2} f(t) \cos(2\pi m f_0 t) dt = a_0 \int_{-T_0/2}^{+T_0/2} \cos(2\pi m f_0 t) dt$$

$$+ \sum_{n=1}^{\infty} \left\{ \frac{a_n}{2} \left[\int_{-T_0/2}^{+T_0/2} \cos[2\pi(m-n)f_0 t] dt + \int_{-T_0/2}^{+T_0/2} \cos[2\pi(m+n)f_0 t] dt \right] \right\}$$

(13–7)

$$+ \sum_{n=1}^{\infty} \left\{ \frac{b_n}{2} \left[\int_{-T_0/2}^{+T_0/2} \sin[2\pi(m-n)f_0 t] dt + \int_{-T_0/2}^{+T_0/2} \sin[2\pi(m+n)f_0 t] dt \right] \right\}$$

All of the integrals are now in the form of expressions in Eq. (13–4). Consequently, we see all of the integrals on the right side of Eq. (13–7) vanish, except for one cosine integral when $m = n$. This one survivor corresponds to the $k = 0$ case in Eq. (13–4), and the right side of Eq. (13–7) reduces to

$$\int_{-T_0/2}^{T_0/2} f(t) \cos(2\pi n f_0 t) dt = \frac{a_n}{2} \int_{-T_0/2}^{T_0/2} \cos[2\pi(n-n)f_0 t] dt$$

$$= \frac{a_n}{2} T_0$$

Solving Eq. (13–7) for a_n yields the second expression in Eq. (13–3).

To obtain the expression for b_n we multiply Eq. (13–2) by $\sin(2\pi m f_0 t)$ and integrate over the interval $t = -T_0/2$ to $+T_0/2$. The derivation steps then parallel the approach used to find a_n. The end result is that the dc component integral vanishes and the ac component integrals reduce to $b_n T_0/2$, which yields the expression for b_n in Eq. (13–3).

The derivation of Eq. (13–3) focuses on the problem of finding the Fourier coefficients of a given periodic waveform. Some experience and practice are necessary to understand the implications of this procedure. On the other hand, it is not necessary to go through these mechanics for every newly encountered periodic waveform because tables of Fourier series expansions are available. For our purposes the listing in Figure 13–3 will suffice. For each waveform defined graphically, the figure lists the expressions for a_0, a_n, and b_n as well as restrictions on the integer n.

EXAMPLE 13–3

Verify the Fourier coefficients given for the square wave in Figure 13–3 and write the first three nonzero terms in its Fourier series.

SOLUTION:

An expression for a square wave on the interval $0 < t < $ to T_0 is

$$f(t) = \begin{cases} A & 0 < t < T_0/2 \\ -A & T_0/2 < t < T_0 \end{cases}$$

Using the first expression in Eq. (13–3) to find a_0 yields

Waveform	Fourier Coefficients	Waveform	Fourier Coefficients
Constant (dc) $f(t)$	$a_0 = A$ $a_n = 0$ all n $b_n = 0$ all n	Sawtooth wave $f(t)$	$a_0 = \dfrac{A}{2}$ $a_n = 0$ all n $b_n = -\dfrac{A}{n\pi}$ all n
Cosine wave $f(t)$	$a_0 = 0$ $a_1 = A$ $a_n = 0$ $n \neq 1$ $b_n = 0$ all n	Triangular wave $f(t)$	$a_0 = 0$ $a_n = \dfrac{8A}{(n\pi)^2}$ n odd $a_n = 0$ n even $b_n = 0$ all n
Sine wave $f(t)$	$a_0 = 0$ $a_n = 0$ all n $b_1 = A$ $b_n = 0$ $n \neq 1$	Half-wave rectified sine wave $f(t)$	$a_0 = \dfrac{A}{\pi}$ $a_n = \dfrac{2A/\pi}{1 - n^2}$ n even $a_n = 0$ n odd $b_1 = \dfrac{A}{2}$ $n = 1$ $b_n = 0$ $n \neq 1$
Square wave $f(t)$	$a_0 = 0$ $a_n = 0$ all n $b_n = \dfrac{4A}{n\pi}$ n odd $b_n = 0$ n even	Full-wave rectified sine wave $f(t)$	$a_0 = 2A/\pi$ $a_n = \dfrac{4A/\pi}{1 - n^2}$ n even $a_n = 0$ n odd $b_n = 0$ all n
Rectangular pulse $f(t)$	$a_0 = \dfrac{AT}{T_0}$ $a_n = \dfrac{2A}{n\pi} \sin\left(\dfrac{n\pi T}{T_0}\right)$ $b_n = 0$ all n	Parabolic wave $f(t)$	$a_0 = 0$ $a_n = 0$ all n $b_n = \dfrac{32A}{(n\pi)^3}$ n odd $b_n = 0$ n even

FIGURE 13 – 3   *Fourier coefficients for some periodic waveforms.*

$$a_0 = \frac{1}{T_0}\int_0^{T_0/2} A\,dt + \frac{1}{T_0}\int_{T_0/2}^{T_0}(-A)dt$$

$$= \frac{A}{T_0}\left[\frac{T_0}{2} - 0 - T_0 + \frac{T_0}{2}\right] = 0$$

The result $a_0 = 0$ means that the dc value of the square wave is zero, which is easy to see because the area under a positive half cycle cancels the area under a negative half cycle. Using the second expression in Eq. (13–3) to find a_n produces

$$a_n = \frac{2}{T_0}\int_0^{T_0/2} A\cos(2\pi nt/T_0)dt + \frac{2}{T_0}\int_{T_0/2}^{T_0}(-A)\cos(2\pi nt/T_0)dt$$

$$= \frac{2A}{T_0}\left[\frac{\sin(2\pi nt/T_0)}{2\pi n/T_0}\right]_0^{T_0/2} - \frac{2A}{T_0}\left[\frac{\sin(2\pi nt/T_0)}{2\pi n/T_0}\right]_{T_0/2}^{T_0}$$

$$= \frac{A}{n\pi}[\sin(n\pi) - \sin(0) - \sin(2n\pi) + \sin(n\pi)] = 0$$

Since $a_n = 0$ for all n there are no cosine terms in the series. This makes some intuitive sense because a sinewave with the same fundamental frequency as the square wave fits nicely inside the square wave with zeros crossing at the same points, whereas a cosine with the same frequency does not fit at all. The b_n coefficients for the sine terms are found using the third expression in Eq. (13–3):

$$b_n = \frac{2}{T_0}\int_0^{T_0/2} A\sin(2\pi nt/T_0)dt + \frac{2}{T_0}\int_{T_0/2}^{T_0}(-A)\sin(2\pi nt/T_0)dt$$

$$= \frac{2A}{T_0}\left[-\frac{\cos(2\pi nt/T_0)}{2\pi n/T_0}\right]_0^{T_0/2} - \frac{2A}{T_0}\left[-\frac{\cos(2\pi nt/T_0)}{2\pi n/T_0}\right]_{T_0/2}^{T_0}$$

$$= \frac{A}{n\pi}[-\cos(n\pi) + \cos(0) + \cos(2n\pi) - \cos(n\pi)]$$

$$= \frac{2A}{n\pi}[1 - \cos(n\pi)]$$

The term $[1 - \cos(n\pi)] = 2$ if n is odd and zero if n is even. Hence b_n can be written as

$$b_n = \begin{cases} \dfrac{4A}{n\pi} & n \text{ odd} \\ 0 & n \text{ even} \end{cases}$$

The first three nonzero terms in the Fourier series of the square wave are

$$f(t) = \frac{4A}{\pi}\left[\sin 2\pi f_0 t + \frac{1}{3}\sin 2\pi 3f_0 t + \frac{1}{5}\sin 2\pi 5f_0 t + \dots\right]$$

Note that this series contains only odd harmonic terms.   ■

The triangular wave in Figure 13–3 has a peak amplitude of $A = 10$ and $T_0 = 2$ ms. Calculate the Fourier coefficients of the first nine harmonics.

Answers:
$a_1 = 8.11$, $a_2 = 0$, $a_3 = 0.901$, $a_4 = 0$, $a_5 = 0.324$, $a_6 = 0$, $a_7 = 0.165$, $a_8 = 0$, $a_9 = 0.100$, $b_n = 0$ for all n

ALTERNATIVE FORM OF THE FOURIER SERIES

The series in Eq. (13–1) can be written in several alternative yet equivalent forms. From our study of sinusoids in Chapter 5, we recall that the Fourier coefficients determine the amplitude and phase angle of the general sinusoid. Thus, we can write a general Fourier series in the form

$$f(t) = A_0 + A_1 \cos(2\pi f_0 t + \phi_1) + A_2 \cos(2\pi 2 f_0 t + \phi_2) + \ldots$$

$$+ A_n \cos(2\pi n f_0 t + \phi_n) + \ldots \qquad (13\text{–}8)$$

where

$$A_n = \sqrt{a_n^2 + b_n^2} \quad \text{and} \quad \phi_n = \tan^{-1}\frac{-b_n}{a_n} \qquad (13\text{–}9)$$

The coefficient A_n is the amplitude of the nth harmonic and ϕ_n is its phase angle.[1]

Note that the amplitude A_n and phase angle ϕ_n contain all of the information needed to construct the Fourier series in the form of Eq. (13–8). Figure 13–4 shows how plots of this information are used to display the spectral content of a periodic waveform $f(t)$. The plot of A_n versus nf_0 (or $n\omega_0$) is called the **amplitude spectrum**, while the plot of ϕ_n versus nf_0 (or $n\omega_0$) is called the **phase spectrum**. Both plots are **line spectra** because spectral content can be represented as a line at discrete frequencies.

In theory, a Fourier series includes infinitely many harmonics, although the harmonics tend to decrease in amplitude at high frequency. For example, the summary in Figure 13–3 shows that the amplitudes of the square wave decrease as $1/n$, the triangular wave as $1/n^2$, and the parabolic wave as $1/n^3$. The $1/n^3$ dependence means that the amplitude of the fifth harmonic in a parabolic wave is less than 1% of the amplitude of the fundamental (actually 1/125th of the fundamental). In practical signals the harmonic amplitudes decrease at high frequency so that at some point the higher-order components become negligibly small. This means that we can truncate the series at some finite frequency and still retain the important features of the signal. This is an important consideration in systems with finite bandwidth.

---

1 There is a 180° ambiguity in the value returned by the inverse tangent function in most computational tools. The ambiguity is resolved by the following rule: $b_n < 0$ implies that the angle is in the range 0 to −180°, while $b_n > 0$ implies the 0 to +180° range.

FIGURE **13 – 4** *Amplitude and phase spectra.*

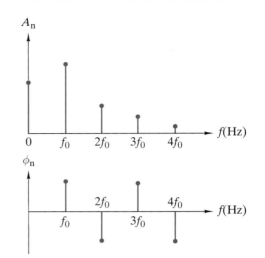

EXAMPLE 13–4

Derive expressions for the amplitude A_n and phase angle ϕ_n of the Fourier series of the sawtooth wave in Figure 13–3. Sketch the amplitude and phase spectra of a sawtooth wave with $A = 5$ and $T_0 = 4$ ms.

SOLUTION:

Figure 13–3 gives the Fourier coefficients of the sawtooth wave as

$$a_0 = \frac{A}{2} \qquad a_n = 0 \qquad b_n = -\frac{A}{n\pi} \qquad \text{for all } n$$

Using Eq. (13–9) yields

$$A_n = \sqrt{a_n^2 + b_n^2} = \begin{cases} \dfrac{A}{2} & n = 0 \\[2mm] \dfrac{A}{n\pi} & n > 0 \end{cases}$$

and

$$\phi_n = \tan^{-1}\frac{-b_n}{a_n} = \begin{cases} \text{undefined} & n = 0 \\ 90° & n > 0 \end{cases}$$

For $A = 5$ and $f_0 = 1/T_0 = 250$ Hz the first four nonzero terms in the series are

$$f(t) = 2.5 + 1.59 \cos(2\pi 250t + 90°)$$

$$+ \; 0.796 \cos(2\pi 500t + 90°) + 0.531 \cos(2\pi 750t + 90°) + \cdots$$

Figure 13–5 shows the amplitude and phase spectra for this signal. Note that the lines in the amplitude spectrum are inversely proportional to frequency. As frequency increases the amplitudes decrease so that the high-frequency components become negligible. ∎

FIGURE 13–5

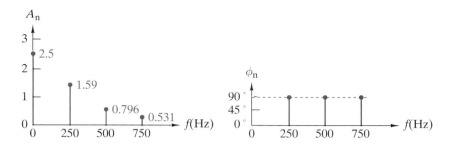

Exercise 13–2

Derive expressions for the amplitude A_n and phase angle ϕ_n for the triangular wave in Figure 13–3 and write an expression for the first three nonzero terms in the Fourier series with $A = \pi^2/8$ and $T_0 = 2\pi/5000$ s.

Answers:

$$A_n = \frac{8A}{(n\pi)^2} \quad \phi_n = 0° \quad n \text{ odd}$$

$$A_n = 0 \quad \phi_n \text{ undefined} \quad n \text{ even}$$

$$f(t) = \cos(5000t) + \frac{1}{9}\cos(15{,}000t) + \frac{1}{25}\cos(25{,}000t) + \cdots$$

13–3 WAVEFORM SYMMETRIES

Many of the Fourier coefficients are zero when a periodic waveform has certain types of symmetries. It is helpful to recognize these symmetries, since they may simplify the calculation of the Fourier coefficients.

The first expression in Eq. (13–3) shows that the amplitude of the dc component a_0 is the average value of the periodic waveform $f(t)$. If the waveform has equal area above and below the time axis, then the integral over one cycle vanishes, the average value is zero, and $a_0 = 0$. The square wave, triangular wave, and parabolic wave in Figure 13–3 are examples of periodic waveforms with zero average value.

A waveform is said to have **even symmetry** if $f(-t) = f(t)$. The cosine wave, rectangular pulse, and triangular wave in Figure 13–3 are examples of waveforms with even symmetry. The Fourier series of an even waveform is made up entirely of cosine terms: that is, all of the b_n coefficients are zero. To show this we write the Fourier series for $f(t)$ in the form

$$f(t) = a_0 + \sum_{n=1}^{\infty} [a_n \cos(2\pi n f_0 t) + b_n \sin(2\pi n f_0 t)] \tag{13–10}$$

Given the Fourier series for $f(t)$, we use the identities $\cos(-x) = \cos(x)$ and $\sin(-x) = -\sin(x)$ to write the Fourier series for $f(-t)$ as follows:

$$f(-t) = a_0 + \sum_{n=1}^{\infty} [a_n \cos(2\pi n f_0 t) - b_n \sin(2\pi n f_0 t)] \tag{13–11}$$

For even symmetry $f(t) = f(-t)$ and the right sides of Eqs. (13–10) and (13–11) must be equal. Comparing the Fourier coefficients term by term, we find that $f(t) = f(-t)$ requires $b_n = -b_n$. The only way this can happen is for $b_n = 0$ for all n.

A waveform is said to have **odd symmetry** if $-f(-t) = f(t)$. The sine wave, square wave, and parabolic wave in Figure 13–3 are examples of waveforms with this type of symmetry. The Fourier series of odd waveforms are made up entirely of sine terms: that is, all of the a_n coefficients are zero. Given the Fourier series for $f(t)$ in Eq. (13–10), we use the identities $\cos(-x) = \cos(x)$ and $\sin(-x) = -\sin(x)$ to write the Fourier series for $-f(-t)$ in the form

$$-f(-t) = -a_0 + \sum_{n=1}^{\infty} [-a_n \cos(2\pi n f_0 t) + b_n \sin(2\pi n f_0 t)] \qquad (13\text{–}12)$$

With odd symmetry $f(t) = -f(-t)$ and the right sides of Eq. (13–10) and (13–12) must be equal. Comparing the Fourier coefficients term by term, we find that odd symmetry requires $a_0 = -a_0$ and $a_n = -a_n$. The only way this can happen is for $a_n = 0$ for all n, including $n = 0$.

A waveform is said to have **half-wave symmetry** if $-f(t - T_0/2) = f(t)$. This requirement states that inverting the waveform $[-f(t)]$ and then time shifting by half a cycle $(T_0/2)$ must produce the same waveform. Basically, this means that successive half cycles have the same waveshape but opposite polarities. In Figure 13–3 the sine wave, cosine wave, square wave, triangular wave, and parabolic wave have half-wave symmetry. The sawtooth wave, half-wave sine, rectangular pulse train, and full-wave sine do not have this symmetry.

With half-wave symmetry the amplitudes of all even harmonics are zero. To show this we use the identities $\cos(x - n\pi) = (-1)^n \cos(x)$ and $\sin(x - n\pi) = (-1)^n \sin(x)$ to write the Fourier series of $-f(t - T_0/2)$ in the form

$$-f(t - T_0/2) = -a_0 + \sum_{n=1}^{\infty} [-(-1)_n a_n \cos(2\pi n f_0 t) -$$

$$(-1)^n b_n \sin(2\pi n f_0 t)] \qquad (13\text{–}13)$$

For half-wave symmetry the right sides of Eqs. (13–10) and (13–13) must be equal. Comparing the coefficients term by term, we find that equality requires $a_0 = -a_0$, $a_n = -(-1)^n a_n$, and $b_n = -(-1)^n b_n$. The only way this can happen is for $a_0 = 0$ and for $a_n = b_n = 0$ when n is even. In other words, the only nonzero Fourier coefficients occur when n is odd.

A waveform may have more than one symmetry. For example, the triangular wave in Figure 13–3 has even symmetry and half-wave symmetry, while the square wave has both odd and half-wave symmetries. The sawtooth wave in Figure 13–3 is an example where an underlying odd symmetry is masked by a dc component. A symmetry that is not apparent until the dc component is removed is sometimes called a **hidden symmetry**.

Finally, whether a waveform has even or odd symmetry (or neither) depends on where we choose to define $t = 0$. For example, the triangular wave in Figure 13–3 has even symmetry because the $t = 0$ vertical axis is located at a local maximum. If the axis is shifted to a zero crossing, the waveform has odd symmetry and the cosine terms in the series are replaced by sine terms. If the vertical axis is shifted to a point between a zero cross and a maximum, then the resulting waveform is neither even nor odd and its Fourier series contains both sine and cosine terms.

EXAMPLE 13-5

Given that $f(t)$ is a square wave of amplitude A and period T_0, use the Fourier coefficients in Figure 13–3 to find the Fourier coefficients of $g(t) = f(t + T_0/4)$.

SOLUTION:

Figure 13–6 compares the square waves $f(t)$ and $g(t) = f(t + T_0/4)$. The square wave $f(t)$ has odd symmetry (sine terms only) and half-wave symmetry (odd harmonics only). Using the coefficients in Figure 13–3, the Fourier series for $f(t)$ is

$$f(t) = \sum \frac{4A}{n\pi} \sin(2\pi nt/T_0) \quad n \text{ odd}$$

The Fourier series for $g(t) = f(t + T_0/4)$ can be written in the form

$$g(t) = f(t + T_0/4) = \sum \frac{4A}{n\pi} \sin[2\pi n(t + T_0/4)/T_0] \quad n \text{ odd}$$

$$= \sum \frac{4A}{n\pi} \sin(2\pi nt/T_0 + n\pi/2)$$

$$= \sum \frac{4A}{n\pi} \cos(2\pi nt/T_0)\sin(n\pi/2)$$

$$= \sum \frac{4A}{n\pi} \cos(2\pi nt/T_0)(-1)^{\frac{n-1}{2}} \quad n \text{ odd}$$

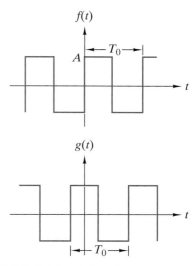

$f(t)$

$g(t)$

FIGURE   1 3 - 6

Figure 13–6 shows that $g(t)$ has even and half-wave symmetry so its Fourier series has only cosine terms and odd harmonics. The Fourier coefficients for $g(t)$ are

$$a_0 = 0 \quad a_n = \begin{cases} 0 & n \text{ even} \\ \left[\dfrac{4A}{n\pi}\right](-1)^{\frac{2n-1}{2}} & n \text{ odd} \end{cases}$$

$$b_n = 0 \quad \text{all } n$$

Shifting the time origin alters the even or odd symmetry properties of a periodic waveform because these symmetries depend on values of $f(t)$ on opposite sides of the vertical axis at $t = 0$. The half-wave symmetry of a waveform is not changed by time shifting because this symmetry only requires successive half cycles to have the same form but opposite polarities. ∎

Exercise 13-3

(a) Identify the symmetries in the waveform $f(t)$ whose Fourier series is

$$f(t) = \frac{2\sqrt{3}A}{\pi}\left[\cos(\omega_0 t) - \frac{1}{5}\cos(5\,\omega_0 t) + \frac{1}{7}\cos(7\,\omega_0 t)\right.$$

$$\left. - \frac{1}{11}\cos(11\,\omega_0 t) + \frac{1}{13}\cos(13\,\omega_0 t) + \dots\right]$$

(b) Write the corresponding terms of the function $g(t) = f(t - T_0/4)$.

Answers:

(a) Even symmetry, half-wave symmetry, zero average value.

(b) $g(t) = \dfrac{2\sqrt{3}A}{\pi}\left[\sin(\omega_0 t) - \dfrac{1}{5}\sin(5\,\omega_0 t) - \dfrac{1}{7}\sin(7\,\omega_0 t)\right.$

$\left. + \dfrac{1}{11}\sin(11\,\omega_0 t) + \dfrac{1}{13}\sin(13\,\omega_0 t) + \dots\right]$

13–4 CIRCUIT ANALYSIS USING FOURIER SERIES

Up to this point we have concentrated on finding the Fourier series description of periodic waveforms. We are now in a position to address circuit analysis problems of the type illustrated in Figure 13–7. This first-order RL circuit is driven by a periodic sawtooth voltage, and the objective is to find the steady-state current $i(t)$.

FIGURE 13–7 *Linear circuit with a periodic input.*

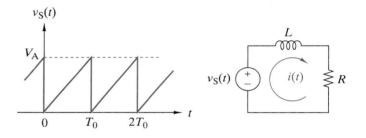

We begin by using the results in Example 13–4 to express the input voltage as a Fourier series in the form

$$v_S(t) = \underbrace{\frac{V_A}{2}}_{\text{dc}} + \underbrace{\sum_{n=1}^{\infty}\frac{V_A}{n\pi}\cos(n\omega_0 t + 90°)}_{\text{ac}} \qquad (13\text{–}14)$$

This result expresses the input driving force as the sum of a dc component plus ac components at harmonic frequencies $n\omega_0 = 2\pi n f_0$, $n = 1, 2, 3, \dots$. Since the circuit is linear, we find the steady-state response caused by each component acting alone and then obtain the total response by superposition.

In the dc steady state the inductor acts like a short circuit, so the steady-state current due to the dc input $V_A/2$ is simply $i_0(t) = V_A/(2R)$. The nth ac component in Eq. 13–14 can be written in phasor form as

$$\mathbf{V}_n = \frac{V_A}{n\pi}\angle 90°$$

The impedance of the series connection at the nth harmonic frequency is

$$Z_n = R + jn\omega_0 L = R\sqrt{1 + (n\omega_0 L/R)^2}\angle\theta_n$$

where $\theta_n = \tan^{-1}(n\omega_0 L/R)$. Hence the phasor representation of the steady-state current due to the nth harmonic of the input is

$$\mathbf{I}_n = \frac{\mathbf{V}_n}{Z_n} = \frac{V_A}{n\pi R} \frac{1}{\sqrt{1 + (n\omega_0 L/R)^2}} \angle(90° - \theta_n)$$

The sinusoidal waveform corresponding to this phasor is

$$i_n(t) = \frac{V_A}{n\pi R} \frac{1}{\sqrt{1 + (n\omega_0 L/R)^2}} \cos(n\omega_0 t + 90° - \theta_n)$$

We have now found the steady-state response of the circuit due to the dc component acting alone and the nth harmonic ac component acting alone. Since the circuit is linear, superposition applies and we find the steady-state response caused the sawtooth input triangular summing the contributions of each of these sources:

$$i(t) = i_0(t) + \sum_{n=1}^{\infty} i_n(t)$$

$$= \frac{V_A}{2R} + \frac{V_A}{R} \sum_{n=1}^{\infty} \frac{1}{n\pi\sqrt{1 + (n\omega_0 L/R)^2}} \cos(n\omega_0 t + 90° - \theta_n) \qquad (13\text{--}15)$$

The Fourier series in Eq. (13–15) represents the steady-state current due to a sawtooth driving force whose Fourier series is in Eq. (13–14).

Given the Fourier series of a periodic input, it is a straightforward procedure to obtain the Fourier series of the steady-state response. But interpreting the analysis result requires some thought since the response is presented as an infinite series. In practice the series converges rather rapidly, so we can calculate specific values or generate plots using computer tools to sum a truncated version of the infinite series. Before doing so let us look closely at Eq. (13–15) to see what we can infer about the response.

First note that if $L = 0$ Eq. (13–15) reduces to

$$i(t) = \frac{V_A/R}{2} + \sum_{n=1}^{\infty} \frac{V_A/R}{n\pi} \cos(n\omega_0 t + 90°)$$

which is the Fourier series of a sawtooth wave of amplitude V_A/R. This makes sense because without the inductor the circuit in Figure 13–7 is a simple resistive circuit in which $i(t) = v_S(t)/R$, so the input and response must have the same waveform.

When $L \neq 0$ the response is not a sawtooth, but we can infer some features of its waveform if we examine the amplitude spectrum. At high frequency $[(n\omega_0 L/R) \gg 1]$ Eq. (13–15) points out that the amplitudes of the ac components are approximately

$$I_n \approx \frac{V_A}{R} \frac{1}{n^2\pi\omega_0 L/R}$$

In the steady-state response the amplitudes of the high-frequency ac components decrease as $1/n^2$, whereas the ac components in the input sawtooth decrease as $1/n$. In other words, the relative amplitudes of the high-frequency components are much smaller in the response than in the input. This makes sense because the inductor's impedance increases with frequency and thereby reduces the amplitudes of the high-frequency ac currents. We would expect the circuit to filter out the high-frequency components in the input and produce a response without the sharp corners and discontinuities in the input sawtooth.

The next example examines this thought for a specific set of parameters.

EXAMPLE 13–6

The parameters of the steady-state waveform in Eq. (13–15) are V_A = 25 V, T_0 = 5 μs, L = 40 μH, and R = 50 Ω. Calculate and plot a truncated Fourier series representation of the steady-state current using the first 5 harmonics and the first 10 harmonics.

SOLUTION:

Figure 13–8 shows a Mathcad worksheet that calculates truncated Fourier series of the steady-state current in Eq. (13–15). The first two lines define the waveform and circuit parameters. The Fourier parameters for the first 20 harmonics are calculated in the third line using the index variable n = 1, 2, . . . 20. The next line defines a truncated Fourier series $i(k, t)$ consisting of the dc component I_0 plus the sum of the first k harmonics. The responses $I(5, t)$ and $I(10, t)$ are plotted at the bottom of the worksheet. There is not much change between the two plots, suggesting that a truncated series converges rather rapidly and gives a reasonable approximation to the steady-state current.

The plots in Figure 13–8 show that the steady-state current is indeed smoother than the input driving force. The reason is that the inductor suppresses the high-frequency components that are important contributors to the discontinuities in the sawtooth waveform. As a result, the response

FIGURE 13–8

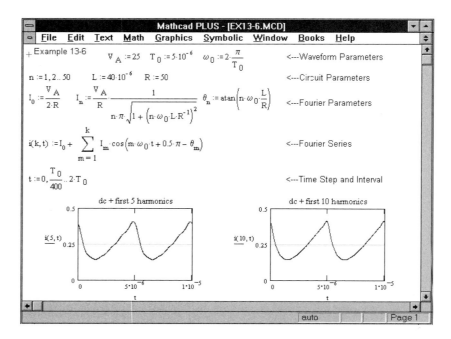

does not have the abrupt changes and sharp corners present in the saw-tooth input.   ■

Exercise 13-4

Derive an expression for the steady-state current of the RL circuit in Figure 13-7 when the input voltage is a triangular wave with amplitude V_A and frequency ω_0.

Answer:

$$i(t) = \sum_{n=1}^{\infty} \frac{8V_A/R}{(n\pi)^2\sqrt{1 + (n\omega_0 L/R)^2}} \cos(n\omega_0 t - \theta_n) \quad n \text{ odd}$$

where

$$\theta_n = \tan^{-1}(n\omega_0 L/R)$$

The transfer function offers another approach to circuit analysis with Fourier series. In Chapter 12 we found that in the sinusoidal steady state the relationship between the input and output sinusoids can be summarized in the following statements:

(Output Amplitude) = (Input Amplitude) × (Magnitude of $T(j\omega)$)
(Output Phase)   =   (Input Phase)   +   (Angle of $T(j\omega)$)

When the driving force is periodic, the input and output signals are represented by their amplitude and phase spectra. When the circuit is linear, the generalization of the preceding result for periodic driving forces is

$$\begin{pmatrix} \text{Output Amplitude} \\ \text{Spectrum} \end{pmatrix} = \begin{pmatrix} \text{Input Amplitude} \\ \text{Spectrum} \end{pmatrix} \times \begin{pmatrix} \text{Magnitude of} \\ T(jn\omega_0) \end{pmatrix}$$

$$\begin{pmatrix} \text{Output Phase} \\ \text{Spectrum} \end{pmatrix} = \begin{pmatrix} \text{Input Phase} \\ \text{Spectrum} \end{pmatrix} + \begin{pmatrix} \text{Angle of} \\ T(jn\omega_0) \end{pmatrix} \qquad (13-16)$$

where the spectrum product and sum operations are carried out at dc ($n = 0$) and each of the harmonic ac frequencies ($n = 1, 2, 3, \ldots$). In some cases the output amplitude spectrum suffices as the solution for a problem.

EXAMPLE 13-7

Figure 13-9 shows a block diagram of a dc power supply. The ac input is a sinusoid that is converted to a full-wave sine by the rectifier. The filter passes the dc component in the rectified sine and suppresses the ac components. The result is an output consisting of a small residual ac ripple riding on top of a much larger dc signal.

Calculate and plot the first 10 harmonics in amplitude spectra of the filter input and output for $V_A = 23.6$ V, $T_0 = 1/60$ s, and a lowpass filter transfer function of

$$T(s) = \frac{(200)^2}{s^2 + 280s + (200)^2}$$

FIGURE 13-9

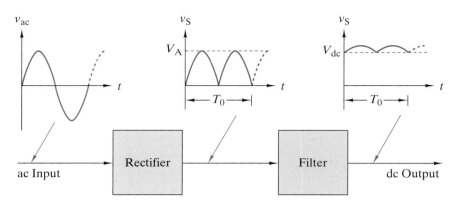

SOLUTION:

The amplitude spectrum of the filter input is obtained using the Fourier coefficients for the full-wave rectified sine in Figure 13–3:

$$V_0 = 2V_A/\pi = 15.02 \text{ V}$$

$$V_n = \begin{cases} 0 & n \text{ odd} \\ \left| \dfrac{4V_A/\pi}{1 - n^2} \right| = \dfrac{30.04}{n^2 - 1} & n \text{ even} \end{cases} \tag{13–17}$$

The magnitude of the transfer function at each of these discrete frequencies is

$$|T(jn\omega_0)| = \frac{(200)^2}{\sqrt{[(200)^2 - (n\omega_0)^2]^2 + (280 \, n\omega_0)^2}} \tag{13–18}$$

To obtain the specified output spectrum, we must generate the product of the input amplitude times the transfer function magnitude for $n = 0, 1, 2, 3, \ldots 10$.

Spreadsheets are ideally suited to making repetitive calculations of this type. Figure 13–10 shows an Excel spreadsheet that implements the required calculations. Column A gives the index n and column B gives the corresponding frequencies in Hz. The input amplitudes in column C are calculated using Eq. (13–17), while the entries in column D are based on Eq. (13–18). Finally, the entries in column E are the product of those in columns C and D. Since the lowpass filter has unity gain at zero frequency, the dc components in the input and output are equal. The first nonzero ac component is the second harmonic, which has an amplitude of 10 V in the input but less than 1 V in the output. By the time we get to the next nonzero harmonic at 240 Hz, the ac amplitudes in the output are entirely negligible. ∎

Exercise 13–5

Derive an expression for the first three nonzero terms in the Fourier series of the steady-state output voltage in Example 13–7.

Answer:
$$v_O(t) = 15.02 + 0.736 \cos(2\pi \, 120t + 21.5°) + 0.0349 \cos(2\pi \, 240t + 10.6°) \text{ V}$$

FIGURE   13 – 10

	A	B	C	D	E	F G H I J
1		Freq	Input	Magnitude	Output	
2	n	Hz	Amplitude	of T(jw)	Amplitude	
3	0	0	15.02	1	15.02	
4	1	60	0	0.39163	0	
5	2	120	10.01	0.075686	0.757868	
6	3	180	0	0.032281	0	
7	4	240	2.00	0.017905	0.035859	
8	5	300	0	0.011386	0	
9	6	360	0.86	0.00788	0.006763	
10	7	420	0	0.005777	0	
11	8	480	0.48	0.004417	0.002106	
12	9	540	0	0.003487	0	
13	10	600	0.30	0.002822	0.000856	

13–5   RMS VALUE AND AVERAGE POWER

In Chapter 5 we introduced the rms value of a periodic waveform as a descriptor of the average power carried by a signal. In this section we relate the rms value of the waveform to the amplitudes of the dc and ac components in its Fourier series. The rms value of a periodic waveform is defined as

$$F_{\text{rms}} = \sqrt{\frac{1}{T_0}\int_0^{T_0}[f(t)]^2\,dt} \qquad (13\text{–}19)$$

The waveform $f(t)$ can be expressed as a Fourier series in the amplitude and phase form as

$$f(t) = A_0 + \sum_{n=1}^{\infty} A_n\cos(n\omega_0 t + \phi_n)$$

Substituting this expression into Eq. (13–19), we can write F^2_{rms} as

$$F^2_{\text{rms}} = \frac{1}{T_0}\int_0^{T_0}\left[A_0 + \sum_{n=1}^{\infty}A_n\cos(n\omega_0 t + \phi_n)\right]^2 dt \qquad (13\text{–}20)$$

Squaring and expanding the integrand on the right side of this equation produces three types of terms. The first is the square of the dc component:

$$\frac{1}{T_0}\int_0^{T_0}[A_0]^2\,dt = A_0^2 \qquad (13\text{–}21)$$

The second is the cross product of the dc and ac components, which takes the form

$$\frac{1}{T_0} \sum_{n=1}^{\infty} 2A_0 \int_0^{T_0} A_n \cos(n\omega_0 t + \phi_n) dt = 0 \tag{13–22}$$

These terms all vanish because they involve integrals of sinusoids over an integer number of cycles. The third and final type of term is the square of the ac components, which can be written as

$$\frac{1}{T_0} \sum_{n=1}^{\infty} \sum_{m=1}^{\infty} \int_0^{T_0} A_n \cos(n\omega_0 t + \phi_n) A_m \cos(m\omega_0 t + \phi_m) dt = \frac{1}{2} \sum_{n=1}^{\infty} A_n^2 \tag{13–23}$$

This rather formidable expression boils down to a simple sum of squares because all of the integrals vanish except when $m = n$.

Combining Eqs. (13–19) through (13–23), we obtain the rms value as

$$F_{\text{rms}} = \sqrt{A_0^2 + \sum_{n=1}^{\infty} \frac{A_n^2}{2}}$$

$$= \sqrt{A_0^2 + \sum_{n=1}^{\infty} \left(\frac{A_n}{\sqrt{2}}\right)^2} \tag{13–24}$$

Since the rms value of a sinusoid of amplitude A is $A/\sqrt{2}$, we conclude that

> *The rms value of a periodic waveform is equal to the square root of the sum of the square of the dc value and the square of the rms value of each of the ac components.*

In Chapter 5 we found that the average power delivered to a resistor is related to its rms voltage or current as

$$P = \frac{V_{\text{rms}}^2}{R} = I_{\text{rms}}^2 R$$

Combining these expressions with the result in Eq. (13–24), we can write the average power delivered by a periodic waveform to the average power delivered by each of its Fourier components:

$$P = \frac{V_0^2}{R} + \sum_{n=1}^{\infty} \frac{V_n^2}{2R} = I_0^2 R + \sum_{n=1}^{\infty} \frac{I_n^2}{2} R$$

$$= P_0 + \sum_{n=1}^{\infty} P_n \tag{13–25}$$

where P_0 is the average power delivered by the dc component and P_n is the average power delivered by the nth ac component. This additive feature is important because it means we can find the total average power by adding the average power carried by the dc plus that carried by each of the ac components.

Caution: In general, we cannot find the total power by adding the power delivered by each component acting alone because the superposition principle does not apply to power. However, the average power carried by harmonic sinusoids is additive because they belong to a special class call orthogonal signals.

EXAMPLE 13-8

Derive an expression for the average power delivered to a resistor by a sawtooth voltage of amplitude V_A and period T_0. Then calculate the fraction of the average power carried by the dc component plus the first three ac components.

SOLUTION:

An equation for the sawtooth voltage is $v(t) = V_A t/T_0$ for the range $0 < t < T_0$. The square of the rms value of a sawtooth is

$$V_{rms}^2 = \frac{1}{T_0} \int_0^{T_0} \left(\frac{V_A t}{T_0} \right)^2 dt = V_A^2 \left[\frac{t^3}{3T_0^3} \right]_0^{T_0} = \frac{V_A^2}{3}$$

The average power delivered to a resistor is

$$P = \frac{V_{rms}^2}{R} = \frac{V_A^2}{3R} = 0.333 \frac{V_A^2}{R}$$

This result is obtained directly from the sawtooth waveform without having to sum an infinite series. The same answer could be obtained by summing an infinite series. The question this example asks is, How much of the average power is carried by the first four components in the Fourier series of the sawtooth wave? The amplitude spectrum of the sawtooth wave (see Example 13–4) is

$$V_n = \begin{cases} \dfrac{V_A}{2} & n = 0 \\ \dfrac{V_A}{n\pi} & n > 0 \end{cases}$$

From Eq. (13–25) the average power in terms of amplitude spectrum is

$$P = \frac{(V_A/2)^2}{R} + \sum_{n=1}^{\infty} \frac{(V_A/n\pi)^2}{2R}$$

which can be arranged in the form

$$P = \frac{V_A^2}{R} \left\{ \underbrace{\frac{1}{(2)^2} + \frac{1}{2(\pi)^2} + \frac{1}{2(2\pi)^2} + \frac{1}{2(3\pi)^2}}_{0.319(96\%)} + \frac{1}{2(4\pi)^2} + \cdots \right\}$$
$$\underbrace{}_{0.333}$$

The infinite series within the braces must sum to 0.333 to match the average power we calculated directly from the waveform itself. The dc component plus the first three ac terms contribute 0.319 to the infinite sum. In other words, these four components alone deliver 96% of the average power carried by the sawtooth wave. ∎

Exercise 13–6

The full-wave sine shown in Figure 13–3 has an rms value of $A/\sqrt{2}$. What fraction of the average power that the waveform delivers to a resistor is carried by the first two nonzero terms in its Fourier series?

A n s w e r :

$$\text{Fraction} = 88/9\pi^2 \text{ or } 99.07\%.$$

EXAMPLE 13–9

Calculate the average power delivered to the 50-Ω resistor in Example 13–6.

SOLUTION:

In Example 13–6 the Fourier series of the current in a series RL circuit is found to be

$$i(t) = I_0 + \sum_{n=1}^{\infty} I_n \cos(n\omega_0 t + \theta_n)$$

where

$$I_0 = \frac{V_A}{R} \qquad I_n = \frac{V_A}{R}\frac{1}{n\pi\sqrt{1 + (n\omega_0 L/R)^2}}$$

We cannot directly calculate the total average power carried by this current because we do not have a closed form expression for the current. In this case we must use a truncated series to estimate the average power. As previously noted, at high frequency ($n\omega_0 L/R \gg 1$) the amplitude I_n decreases as $1/n^2$. According to Eq. (13–25) the average power delivered to the resistor is

$$P = I_0^2 R + \sum_{n=1}^{\infty} \frac{I_n^2}{2} R$$

Because average power depends on the square of the current the power carried by ac components decreases as $1/n^4$ at high frequency. In other words, the series representing the average power converges very rapidly. The numerical values in Example 13–6 are $V_A = 25$ V, $R = 50$ Ω, $L = 40$ μH, $T_0 = 5$ μs, and $\omega_0 = 2\pi/T_0 = 1.26$ Mrad/s. To see how rapidly the series converges we use these values to calculate the first few terms as follows

$$I_0^2 R \quad + \frac{I_1^2}{2}R \quad + \frac{I_2^2}{2}R \quad + \frac{I_3^2}{2}R \quad =$$

$$3.125 + 0.315 + 0.0314 + 0.00697 = 3.478 \text{ W}$$

$$\longleftarrow\!\!-3.44-\!\!\longrightarrow$$

$$\longleftarrow\!\!-\!\!-\!\!-3.471-\!\!-\!\!-\!\!-\!\!\longrightarrow$$

This suggests that $P = 3.48$ W is a reasonable estimate of the average power, an estimate obtained using only four terms in the infinite series. The important point made by this example is that in some situations we may be able to ignore the high-frequency ac components because they are not important contributors to the average power carried by a signal. ■

APPLICATION NOTE: **EXAMPLE 13–10**

A signal processor can be described by the way it operates on the input signal spectrum to produce the output spectrum. Some processors are specifically designed to produce an output spectrum that is different than the input spectrum. For example, the rectifier/filter in Example 13–7 changed an input spectrum containing a single ac component into an output spectrum with a large dc component plus some much smaller ac components. In the next chapter we study filter circuits that are specifically designed to pass the input spectral content in a specified frequency band and to suppress the content at frequencies outside of that band.

In a signal transmission system the goal is to deliver an output signal whose spectrum is an acceptable replica of the input spectrum. These systems are characterized by **transmission bandwidth**, which may be defined as the range of frequencies within which the frequency response is essentially constant. Perfect, distortionless transmission of a periodic signal would require an infinite bandwidth because its Fourier series extends to infinite frequency. This makes distortionless transmission a practical impossibility because the bandwidth of a physical transmission medium is always finite. In practice an acceptable level of distortion is obtained by making the transmission bandwidth large enough to encompass the significant spectral content of the input signal. What do we mean by significant spectral content and acceptable level of distortion?

We can qualitatively illustrate these concepts using the average power carried by a periodic signal. For example, consider a square wave of amplitude V_A and fundamental frequency ω_0. Using the Fourier coefficients from Figure 13–3, the amplitude of the nth harmonic is

$$V_n = \frac{4 V_A}{n\pi} \quad n = 1, 3, 5, \ldots$$

Using Eq. (13–25) the average power the nth harmonic would deliver to a resistance R is

$$P_n = \frac{V_n^2}{2R} = \frac{8V_A^2}{n^2\pi^2 R} \quad n = 1, 3, 5, \ldots$$

Suppose we define significant spectral content to mean those harmonics that deliver an average power of at least 1% of the value delivered by the fundamental. This definition translates into the following constraint:

$$P_n = \frac{8V_A^2}{n^2\pi^2 R} \geq \frac{P_1}{100} = \frac{8V_A^2}{100\,\pi^2 R}$$

which yields the condition $n \leq 10$. Since only odd harmonics are present in a square wave, a transmission bandwidth of $9\omega_0$ would include all of the harmonics that meet our 1% power requirement.

How much distortion of the square wave input would be produced by a transmission system with this bandwidth? Figure 13–11 shows one cycles of the steady-state response of a first-order low-pass system with

FIGURE 13-11

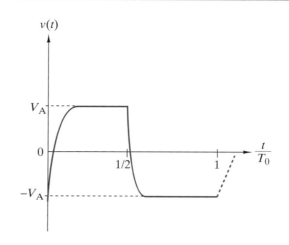

a square wave input and a cutoff frequency of $\omega_C = 9\omega_0$. Clearly the major features of the square wave input are present in the output, but there is some distortion. Whether this level of distortion is acceptable is a question that is impossible to answer without a specific system goal in mind. For example, if the purpose of the system is to detect the presence or absence of a pulse in the output, then this level of distortion is acceptable. On the other hand, if the purpose is to precisely locate the time at which a pulse arrives, then the finite rise and fall times in the output may not be acceptable.

This example is not a definitive discussion of the trade-off between distortion and transmission bandwidth. Rather its purpose is to show how the Fourier series leads to the concept of a signal spectrum and how that concept might be used in the design of signal transmission systems. ∎

SUMMARY

- The Fourier series resolves a periodic waveform into a dc component plus an ac component containing an infinite sum of harmonic sinusoids. The dc component is equal to the average value of the waveform. The amplitudes of the sine and cosine terms in the ac component are called Fourier coefficients.

- The fundamental frequency of the ac component is determined by the period T_0 of the waveform ($f_0 = 1/T_0$). The harmonic frequencies in the ac component are integer multiples of the fundamental frequency.

- Waveform symmetries cause the amplitudes of some terms in a Fourier series to be zero. Even symmetry causes all of the sine terms in the ac component to be zero. Odd symmetry causes all of the cosine terms to be zero. Half-wave symmetry causes all of the even harmonics to be zero.

- An alternative form of the Fourier series represents each harmonic in the ac component by its amplitude and phase angle. A plot of amplitudes versus frequency is called the amplitude spectrum. A plot of phase angles versus frequency is called the phase spectrum. A periodic waveform has spectral components at the discrete frequencies present in its Fourier series.

- The steady-state response of a linear circuit for a periodic driving force can be found by first finding the steady-state response due each term in the Fourier series of the input. The Fourier series of the steady-state response is then found by adding (superposing) responses due to each term acting alone. The individual responses can be found using either phasor or s-domain analysis.

- The rms value of a periodic waveform is equal to the square root of the sum of the square of the dc value and the square of the rms value of each of the ac components. The average power delivered by a periodic waveform is equal to the average power delivered by the dc component plus the sum of average powers delivered by each of the ac components.

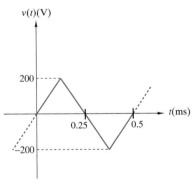

FIGURE P13-1

PROBLEMS

ERO 13-1 THE FOURIER SERIES (SECTS. 13-1, 13-2, 13-3)

(a) Given an equation or graph of a periodic waveform, find its Fourier coefficients.
(b) Given an equation or graph describing the Fourier series of periodic waveform, determine properties of the waveform and identify the waveform.

13-1 Find the Fourier series for the periodic square wave in Figure P13-1.

13-2 Find the Fourier series for the periodic triangular wave in Figure P13-2.

13-3 Find the Fourier series for the periodic full-wave rectified sine wave in Figure P13-3.

13-4 A half-wave rectified sine wave of the type shown in Figure 13-3 has an amplitude of 50 V and a fundamental frequency of 100 Hz. Write an expression for the first four nonzero terms in the Fourier series and plot the amplitude spectrum of the signal.

13-5 A composite waveform is formed by adding 5-V dc to a 1-kHz square wave with a peak-to-peak amplitude of 5 V. Write an expression for the first four nonzero terms in the Fourier series and plot the amplitude spectrum of the signal.

13-6 A parabolic wave of the type shown in Figure 13-3 has a peak-to-peak amplitude of 5 V and a fundamental frequency of 1 kHz. Write an expression for the first four nonzero terms in the Fourier series and plot the amplitude spectrum of the signal.

13-7 The equation for the first cycle ($0 \le t \le T_0$) of a periodic waveform is $v(t) = V_A(t/T_0 - 1)$.
  (a) Sketch the first two cycles of the waveform.
  (b) Derive expressions for the Fourier coefficients a_n and b_n.

13-8 The equation for the first cycle ($0 \le t \le T_0$) of a periodic pulse train is

$$v(t) = V_A[u(3t - T_0) - u(3t - 3T_0)]$$

FIGURE P13-2

FIGURE P13-3

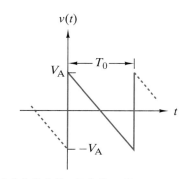

$v(t)$

V_A — T_0 —|

$-V_A$

(a) Sketch the first two cycles of the waveform.

(b) Derive expressions for the Fourier coefficients a_n and b_n.

13–9 Find the Fourier series for the periodic waveform in Figure P13–9.

13–10 The equation for the first cycle ($0 \le t \le T_0$) of a periodic waveform is

$$v(t) = V_A[u(3t - T_0) - u(3t - 2T_0)]$$

(a) Sketch the first two cycles of the waveform.

(b) Derive expressions for the Fourier coefficients a_n and b_n.

13–11 The Fourier coefficients of the triangular wave $f(t)$ are given in Figure 13–3. Use these to find the Fourier coefficients when the waveform is shifted by a quarter of a cycle; that is, find the Fourier coefficients of the periodic waveform $g(t) = f(t - T_0/4)$.

13–12 The four terms in the Fourier series of a periodic signal are

$$v(t) = 25 \left[\sin(100\pi t) - \frac{1}{9} \sin(300\pi t) + \frac{1}{25} \sin(500\pi t) - \right.$$

$$\left. \frac{1}{49} \sin(700\pi t) + \cdots \right]$$

(a) Find the period and fundamental frequency in rad/s and Hz.

(b) Does the waveform have even or odd symmetry?

(c) Plot the waveform using the truncated series to confirm your answer in (b). Can you identify the waveform from this plot?

13–13 The four terms in the Fourier series of a periodic signal are

$$v(t) = 30 + 20 \left[\cos(400\,t) - \frac{1}{5} \cos(800\,t) + \frac{1}{7} \cos(1200\,t) - \right.$$

$$\left. \frac{1}{21} \cos(1600\,t) + \cdots \right]$$

(a) Find the period and fundamental frequency in rad/s and Hz.

(b) Does the waveform have even or odd symmetry?

(c) Plot the waveform using the truncated series above to confirm your answer in (b). Can you identify the waveform from this plot?

13–14 The Fourier coefficients a_n and b_n of the half-wave rectified sine wave are given in Figure 13–3. Use these coefficients to derive expressions for the amplitude A_n and phase angle ϕ_n of the Fourier series and sketch the amplitude and phase spectra of the waveform for $A = 26$ and $f_0 = 60$ Hz.

13–15 The Fourier coefficients a_n and b_n of the rectangular pulse waveform are given in Figure 13–3. Use these coefficients to derive expressions for the amplitude A_n and phase angle ϕ_n of the Fourier series and sketch the amplitude and phase spectra of the waveform for $A = 5$, $f_0 = 100$ kHz, and $T = T_0/10$.

ERO 13–2 FOURIER SERIES AND CIRCUIT ANALYSIS (SECT. 13–4)

Given a linear circuit with a periodic input waveform, find the Fourier series representation of the steady-state response.

13–16 The periodic waveform in Figure P13–16 is applied to the RL circuit shown in the figure.

FIGURE P13–16

(a) Derive an expression for the Fourier series of the steady-state output voltage $v_O(t)$.

(b) Write the first four nonzero terms in the Fourier series for $v_O(t)$ when $V_A = 125$ V, $T_0 = \pi/2$ ms, $T = T_0/2$, $R = 200$ Ω, and $L = 100$ mH.

13–17 The periodic triangular wave in Figure P13–17 is applied to the RC circuit shown in the figure. The Fourier coefficients of the input are

$$a_0 = a_n = 0 \quad \text{all } n$$

$$b_n = 0 \quad n \text{ even}$$

$$b_n = \frac{8V_A}{(n\pi)^2} \sin\left(n\frac{\pi}{2}\right) \quad n \text{ odd}$$

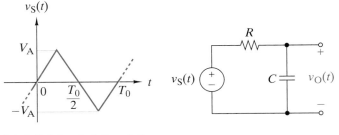

FIGURE P13–17

(a) Derive an expression for the Fourier series of the steady-state output voltage $v_O(t)$.

(b) Write the first four nonzero terms in the Fourier series for $v_O(t)$ when $V_A = 15$ V, $T_0 = 20\pi$ μs, $R = 10$ kΩ, and $C = 500$ pF.

13–18 The periodic sawtooth wave in Figure P13–18 drives the OP AMP circuit shown in the figure.

FIGURE P 1 3 – 1 8

(a) Derive an expression for the Fourier series of the steady-state output voltage $v_O(t)$.

(b) Write the first four nonzero terms in the Fourier series for $v_O(t)$ when $V_A = 5$ V, $T_0 = 400\pi$ μs, $R_1 = 10$ kΩ, $R_2 = 50$ kΩ, and $C = 10$ nF.

13–19 The periodic sawtooth wave in Figure P13–18 drives the OP AMP circuit shown in the figure.

(a) Derive an expression for the Fourier series of the steady-state input current $i(t)$.

(b) Write the first four nonzero terms in the Fourier series for $i(t)$ when $V_A = 4$ V, $T_0 = 800\pi$ μs, $R_1 = 10$ kΩ, $R_2 = 50$ kΩ, and $C = 400$ nF.

13–20 The periodic triangular wave in Figure P13–20 is applied to the *RLC* circuit shown in the figure. The Fourier coefficients of the input are

$$a_0 = a_n = 0 \quad \text{all } n$$

$$b_n = 0 \quad n \text{ even}$$

$$b_n = \frac{8V_A}{(n\pi)^2} \sin\left(n\frac{\pi}{2}\right) \quad n \text{ odd}$$

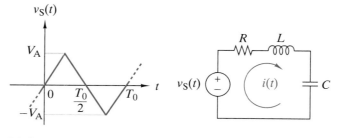

FIGURE P 1 3 – 2 0

(a) Derive an expression for the Fourier series of the steady-state current $i(t)$.

(b) Write the first four nonzero terms in the Fourier series for $v_O(t)$ when $V_A = 1$ V, $T_0 = 240\pi$ μs, $R = 1$ Ω, $L = 2$ mH, and $C = 800$ nF.

(c) What term in the Fourier series tends to dominate the response? Explain.

13–21 A square wave with $V_A = 10$ V and $T_0 = 20\pi$ ms drives a circuit with a transfer function $T(s) = s/(s + 300)$. Find the amplitude of the first four nonzero terms in the Fourier series of the steady-state output. What term in the Fourier series tends to dominate the response? Explain.

13–22 A differentiator is a linear signal processor whose output is proportional to the time derivative of its input. The input-output relationship for a certain differentiator is

$$v_O = \frac{1}{1000} \frac{dv_{IN}(t)}{dt}$$

The input is a triangular wave with $V_A = 12$ V and $T_0 = 10\pi$ ms. Construct plots of the amplitude spectrum of the input and the output. Discuss the spectral changes caused by the differentiator.

13–23 An ideal low-pass filter has a gain of one for sinusoidal inputs with $\omega \leq \omega_C$, and a gain of zero for sinusoidal inputs with $\omega > \omega_C$. A square wave with $V_A = 8$ V and $T_0 = 2\pi$ ms is the input to an ideal filter with $\omega_C = 4$ krad/s. Construct plots of the amplitude spectrum of the input and the output. Discuss the spectral changes caused by the low pass filter.

13–24 An ideal time delay is a signl processor whose output at any time t is equal to the input at an earlier time $t - T_D$, that is, $v_O(t) = v_{IN}(t - T_D)$. For a time delay of $T_D = 1$ ms and an input

$$v_{IN}(t) = 10 + 10\cos(2\pi500t) + 3\cos(2\pi1000t) + 2\cos(2\pi4000t)$$

construct plots of the amplitude spectrum and the phase spectrum of the input and output. Discuss the spectral changes caused by the time delay.

13–25 A square wave with $V_A = 5\pi$ V and $T_0 = 20\pi$ ms drives a first-order filter with unity passband gain. The amplitudes of the first four ac components in the steady-state output voltage are observed to be 5.49 V at ω_0, 4.34 V at $3\omega_0$, 3.28 V at $5\omega_0$, and 2.56 V at $7\omega_0$. Is the filter high pass or low pass? Estimate the cutoff frequency of the filter.

ERO 13–3 RMS VALUE AND AVERAGE POWER (SECT. 13–5)

Given a linear circuit with a periodic input waveform, find the fraction of the average power carried by specified components and average power delivered to a specified load.

13–26 Derive a general expression for the rms value of the rectangular pulse waveform shown in Figure 13–3 and use this expression to find the total average power the waveform delivers to a resistor R. Use the Fourier series given in Figure 13–3 to determine the fraction of the total average power carried by the fundamental frequency.

13–27 Derive a general expression for the rms value of the triangular wave shown in Figure 13–3 and use this expression to find the total average power the waveform delivers to a resistor R. Use the Fourier series given in Figure 13–3 to determine the fraction of the total average power carried by the first three nonzero harmonics.

13–28 Derive a general expression for the rms value of the half-wave rectified sinewave shown in Figure 13–3 and use this expression to find the total average power the waveform delivers to a resistor R. Use the Fourier series given in Figure 13–3 to determine the fraction of the total average power carried by the dc component.

13–29   Find the rms value of the periodic waveform in Figure P13–29 and write an expression for the total average power the waveform delivers to a resistor R. Find the Fourier coefficients of the waveform. What fraction of the total average power is carried by dc component plus the first three nonzero ac components in the Fourier series of the waveform?

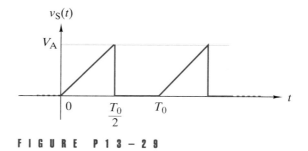

FIGURE P 1 3 – 2 9

13–30   Find the rms value of the periodic waveform in Figure P13–30 and write an expression for the total average power the waveform delivers to a resistor R. Find the Fourier coefficients of the waveform. What fraction of the total average power is carried by the dc component? What fraction is carried by the ac components?

FIGURE P 1 3 – 3 0

13–31   A first-order low-pass filter has a cutoff frequency of 100 rad/s and a passband gain of 6 dB. The input to the filter is $v_1(t) = 30 \sin 200t + 10 \cos 600t$ V. Find the rms value of the steady-state output.

13–32   The Fourier series of the voltage across a resistor is

$$v(t) = \frac{2V_A}{\pi}\left[\sum_{n=1}^{\infty} \frac{(-1)^n}{n} \sin n\omega_0 t\right]$$

Estimate the fraction of the total average power carried by the fundamental component. Repeat for the third harmonic.

13–33   The Fourier series of the voltage across a resistor is

$$v(t) = V_A \left[\frac{\pi^2}{12} - \cos \omega_0 t + \frac{1}{4} \cos 2\omega_0 t - \frac{1}{9} \cos 3\omega_0 t + \dots \right]$$

Estimate the fraction of the average power carried by the dc component plus the fundamental component.

13–34   The input to the circuit in Figure P13–34 is the voltage given in Problem 13–32 with $V_A = 5$ V and $T_0 = 50\pi$ μs. Calculate the average power delivered to the 50-Ω resistor.

13–35   The input to the circuit in Figure P13–34 is the voltage given in Problem 13–33 with $V_A = 4$ V and $T_0 = 50\pi$ μs. Calculate the average power delivered to the 50-Ω resistor.

FIGURE P13-34

INTEGRATING PROBLEMS

13–36   **A**   FOURIER SERIES FROM A BODE PLOT

Figure P13–36 shows the straight-line Bode plots of the magnitude and phase angle of a transfer function. The circuit is driven by a square wave with $V_A = 10$ V and $T_0 = 20\pi$ ms.

(a) Write the first five terms in the Fourier series of the input.

(b) Use the straight-line Bode Plot and the input Fourier series in (a) to obtain an approximation to the first five terms in the Fourier series of the steady-state output.

(c) Find the function $T(s)$ that matches the Bode plots shown in the figure. Use $T(j\omega)$ and the Fourier series of the input found in (a) to calculate the exact values of the first five terms of the Fourier series of the steady-state output.

(d) Compare your results in (b) and (c) and discuss the differences.

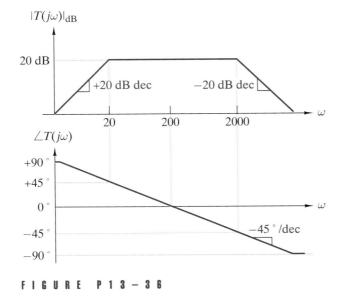

FIGURE P13-36

📖 13–37 Ⓐ STEADY-STATE RESPONSE FOR A PERIODIC IMPULSE TRAIN

A periodic impulse train can approximate a pulse train when the individual pulse durations are very short compared with the circuit response time. This example explores the response of a first-order circuit to a periodic impulse train. A linear circuit whose impulse response is

$$h(t) = \frac{1}{T_0} e^{-t/T_0} u(t)$$

is driven by a periodic impulse train

$$x(t) = T_0 \sum_{n=-\infty}^{\infty} \delta(t - nT_0)$$

(a) Use Eq. (13–3) to find the Fourier coefficients of $x(t)$. Write a general expression for the Fourier series of the input $x(t)$.

(b) Drive expressions for the Fourier coefficients of the steady-state output $y(t)$ when the input is the periodic impulse train $x(t)$.

(c) Write a general expression for the Fourier series of the steady-state output $y(t)$.

(d) Use a computer tool to generate a plot of a truncated Fourier series that gives a reasonable approximation of the steady-state response found in (c).

13–38 Ⓐ PHASOR REPRESENTATION OF THE FOURIER SERIES

Show that a periodic waveform with an average value of zero can be written as

$$f(t) = Re\left\{ \sum_{n=1}^{\infty} C_n e^{jn\omega_0 t} \right\}$$

where C_n is the phasor representation of the nth harmonic.

📖 13–39 Ⓓ FILTER DESIGN REQUIREMENTS

A periodic signal with odd symmetry is to be applied to inputs of a bank of second-order bandpass filters. Each filter must pass one specified harmonic with a gain of 0 dB and have a gain less than –20 dB at the two adjacent harmonic frequencies. What quality factor Q is required in each of the second-order filters?

13–40 Ⓔ SPECTRUM ANALYZER CALIBRATION

A certain spectrum analyzer measures the average power delivered to a calibrated resistor by the individual harmonics of periodic waveforms. The calibration of the analyzer has been checked by applying a 1-MHz square wave and the following results reported:

f(MHz)	1	3	5	7	9	11
P(dBm)	12.1	2.56	–1.88	–4.80	–6.98	–8.73

The reported power in dBm is $P = 10 \, Log(P_n)$, where P_n is the average power delivered by the nth harmonic in mW. Is the spectrum analyzer correctly calibrated?

CHAPTER 14

ANALOG FILTER DESIGN

In its usual form the electric wave-filter transmits currents of all frequencies lying within one or more specified ranges, and excludes currents of all other frequencies.

George A. Campbell, 1922,
American Engineer

14–1 Frequency-domain Signal Processing

14–2 Design with First-order Circuits

14–3 Design with Second-order Circuits

14–4 Low-pass Filter Design

14–5 High-pass Filter Design

14–6 Bandpass and Bandstop Filter Design

Summary

Problems

Integrating Problems

The electric filter was independently invented during World War I by George Campbell in the United States and by K. W. Wagner in Germany. Electric filters and vacuum tube amplifiers were key technologies that triggered the growth of telephone and radio communication systems in the 1920s and 1930s. The emergence of semiconductor electronics in the 1960s, especially the integrated circuit OP AMP, allowed the functions of filtering and amplification to be combined into what are now called active filters.

This chapter treats an important type of electric wave filter made possible by operational amplifiers. These circuits are often called active filters, although the term *analog filter* is also common. Modern integrated circuit technology allows signal-processing systems to use both analog and digital filters, sometimes on the same chip. Although analog and digital filters have some common attributes, the subject of digital filtering is covered in courses in discrete-time signal processing.

Some signal-processing functions performed by analog filters are described in the first section of this chapter. The second section introduces filter design techniques using a cascade connection of first-order circuits. The third section treats the design of second-order low-pass, high-pass, and bandpass circuits. The fourth section develops cascade realizations of low-pass transfer functions using the first- and second-order building blocks. The fifth section shows how the low-pass filter serves as a prototype for designing high-pass filters. The final section illustrates that low-pass and high-pass filters are the building blocks for bandpass and bandstop filters.

14–1 FREQUENCY-DOMAIN SIGNAL PROCESSING

The spectrum of a signal is a description of the amplitude and phase angle of its sinusoidal components. When viewed as a spectrum, we can think of a signal as the sum of a large number of voltage sources, each producing a sinusoid with a specified frequency, amplitude, and phase angle. In frequency-domain signal processing the spectrum of the output signal is obtained by selectively altering the amplitudes and phase angles of the frequencies contained in the input signal. The idea that signals can be described and processed in terms of their frequency components is one of the fundamental concepts in electrical engineering.

For example, the touch-tone telephone shown in Figure 14–1 transmits two frequencies when the button for a digit is pressed. One frequency comes from the low group (697, 770, 852, and 941 Hz) and the other from the high group (1209, 1336, and 1477 Hz). When the no. 3 button is pressed the instrument transmits a sum of sinusoids at 697 Hz and 1477 Hz. When the no. 5 button is pressed the dial signal is a combination of 770 Hz and 1336 Hz. The details of the time-domain waveform produced are not important. The important point is that the touch-tone telephone produces signals whose frequency spectrum lies in the range from 697 Hz to 1477 Hz. The telephone-switching equipment decodes a dial signal by detecting the frequencies it contains.

A **filter** is a signal processor that transmits signals in some frequency bands and rejects or attenuates signals in other bands. Figure 14–2 shows how an input spectrum containing three frequencies is modified by the four basic filter responses. The low-pass filter passes frequencies below its cutoff frequency ω_C and attenuates the high-frequency components in the stopband above the cutoff frequency. The high-pass filter passes frequencies above its cutoff frequency and attenuates the low-frequency components below cutoff. The bandpass filter passes the intermediate frequency

FIGURE 14–1 *Touch-tone dialing frequencies.*

and attenuates the low- and high-frequency components on either side. Finally, the bandstop filter is the inverse of the bandpass case, attenuating the intermediate frequency and passing the low- and high-frequency components.

This chapter introduces methods of designing filter circuits with prescribed frequency-domain signal-processing characteristics. A filter design problem is defined by specifying a frequency-response gain function $|T(j\omega)|$. The objective is to design one or more circuits that produce the desired gain function. In general, a given gain response can be realized by many different circuits, including passive circuits containing only resistors, capacitors, and inductors. We limit our study to **active RC filter** realizations involving resistors, capacitors, and OP AMPs. These filters offer the following advantages:

1. They combine amplifier gain with the frequency-response characteristics of passive RLC filters.

2. The transfer function can be divided into stages that can be designed independently and then connected in cascade to realize the required gain function.

3. They are often smaller and less expensive than RLC filters because they do not require inductors, which can be quite large in low-frequency applications.

The design approaches described produce circuits that approximate the specified filter gain characteristics. The filter phase response is not directly controlled, since the design process concentrates on the gain response. Designing circuits with specified phase characteristics involves similar ideas but will not be treated in detail.

14–2 DESIGN WITH FIRST-ORDER CIRCUITS

In this section we begin the study of filter design problems in which characteristics of the gain response $|T(j\omega)|$ are specified. Our objective is to obtain a circuit whose transfer function has the specified gain response and whose element values are in practical ranges. The design process begins by constructing a transfer function whose gain response meets the specifica-

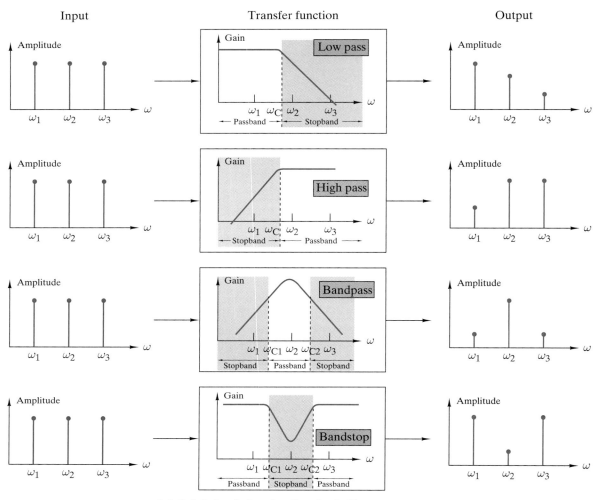

FIGURE 14 – 2 *Four basic filter responses.*

tion. Except in the case of simple first- and second-order filters, we partition the transfer function into a product of the form

$$T(s) = T_1(s) \times T_2(s) \times T_3(s) \times \dots T_n(s) \qquad (14–1)$$

where each $T_k(s)$ in this equation is a first-order transfer function with one pole and no more than one finite zero. We then design prototype first-order circuits using the voltage divider, noninverting amplifier, and the inverting amplifier building blocks in Figure 14–3. The circuit element values in the prototype may be unrealistically large or small, so we use a magnitude scale factor to bring the element values into practical ranges.

The required transfer function $T(s)$ is realized by connecting the scaled circuit for each stage in cascade, as shown in Figure 14–4. The cascade connection produces the required $T(s)$ because the building blocks in Figure 14–3 have OP AMP outputs. As a result, loading does not change the stage transfer functions and the chain rule in Eq. (14–1) applies. The nonin-

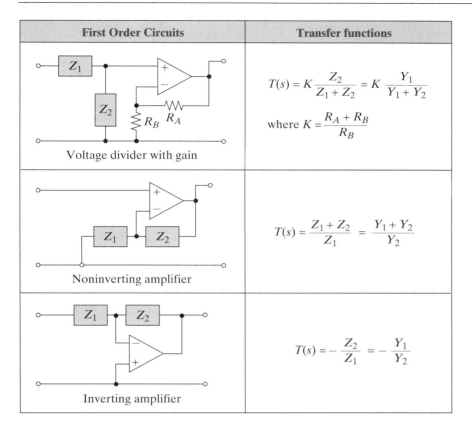

FIGURE   14—3   *First-order circuit building blocks.*

First Order Circuits	Transfer functions
Voltage divider with gain	$$T(s) = K\frac{Z_2}{Z_1 + Z_2} = K\frac{Y_1}{Y_1 + Y_2}$$ where $K = \dfrac{R_A + R_B}{R_B}$
Noninverting amplifier	$$T(s) = \frac{Z_1 + Z_2}{Z_1} = \frac{Y_1 + Y_2}{Y_2}$$
Inverting amplifier	$$T(s) = -\frac{Z_2}{Z_1} = -\frac{Y_1}{Y_2}$$

verting and inverting amplifier circuits in Figure 14–3 inherently have low-impedance outputs. Adding a noninverting amplifier to the output of the basic voltage-divider circuit in Figure 14–3 produces a circuit that can be connected in cascade without changing the stage transfer function.

The end result is a cascade circuit whose gain response has the attributes required by the specification. The aforementioned procedure does not produce a unique solution for the following reasons:

1. Several transfer functions may exist that adequately approximate the required gain characteristics.

2. Partitioning the selected transfer function into a product of first-order transfer functions can be carried out in several different ways.

3. Any of the three building blocks in Figure 14–3 can be used to realize the first-order transfer functions in the selected partitioning.

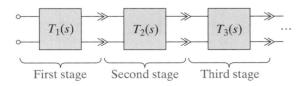

First stage   Second stage   Third stage

FIGURE 14 — 4 *A cascade connection.*

4. There usually are several ways to assign numerical values to circuit elements in the selected building block for each stage.

5. Each stage normally requires a different magnitude scale factor to bring the assigned numerical values into practical ranges.

In summary, since there are many choices, there are no unique answers to the circuit design problem. The following examples illustrate the design of active *RC* circuits.

D **DESIGN EXAMPLE 14–1**

Design two active *RC* circuits that realize a first-order low-pass filter with a cutoff frequency of 2 krad/s and a passband gain of 5.

SOLUTION:
In Chapter 12 we found that a first-order low-pass transfer function has a pole at $s = -\alpha$, cutoff frequency of $\omega_C = \alpha$, and passband gain of $T(0)$. For design puposes it is convenient to write the first-order lowpass transfer function in the Bode plot standard format

$$T(s) = \frac{K}{s/\alpha + 1}$$

Written in this form, we see that the design specification requires that $\omega_C = \alpha = 2000$ and that passband gain $T(0) = K = 5$. Thus, the required transfer function is

$$T(s) = \frac{\pm 5}{s/2000 + 1}$$

where the $\pm$ sign indicates that the phase response is not specified.

First Design: Using the voltage divider circuit in Figure 14–3, we use the plus sign and partition $T(s)$ as

$$T(s) = [K]\left[\frac{Y_1}{Y_1 + Y_2}\right] = [+5]\left[\frac{1}{s/2000 + 1}\right]$$

from which we identify $K = 5$, $R_A = (5 - 1)R_B$, $Y_1 = 1$, and $Y_2 = s/2000$. Selecting $R_B = 1\ \Omega$, we obtain $R_A = 4\ \Omega$. The admittance Y_1 can be realized by a resistor $R_1 = 1\ \Omega$, and admittance Y_2 by a capacitor $C_2 = 1/2000$ F. These results lead to the prototype design in Figure 14–5(a). To obtain practical element values, we apply a magnitude scale factor of $k_m = 10^4$ to produce the final design shown in Figure 14–5(b).

Second Design: Using the inverting amplifier circuit in Figure 14–3, we use the minus sign and write $T(s)$ as

$$T(s) = \left[-\frac{Y_1}{Y_2}\right] = \left[-\frac{5}{s/2000 + 1}\right]$$

from which we identify $Y_1 = 5$ and $Y_2 = s/2000 + 1$. The admittance Y_1 can be realized by a resistor $R_1 = 1/5\ \Omega$. The admittance Y_2 is of the form $C_2 s +$

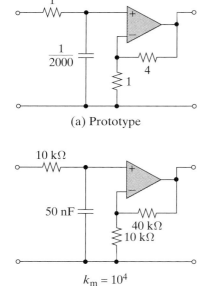

(a) Prototype

10 kΩ

50 nF

40 kΩ
10 kΩ

$k_m = 10^4$

(b) Final design

G_2 and hence can be realized by capacitor $C_2 = 1/2000$ F connected in parallel with a resistor $R_2 = 1\ \Omega$. These results lead to the prototype design in Figure 14–6(a). To obtain practical element values, we apply a magnitude scale factor of $k_m = 5 \times 10^4$ to produce the final design shown in Figure 14–6(b). ∎

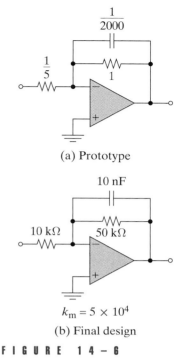

(a) Prototype

(b) Final design

$k_m = 5 \times 10^4$

FIGURE 14–6

D ◖ DESIGN EXAMPLE 14–2 ◗

(a) Construct a transfer function $T(s)$ with the straight-line gain response shown in Figure 14–7.

(b) Design an active RC circuit to realize the $T(s)$ found in (a).

FIGURE 14–7

SOLUTION:

(a) The gain response shows slope changes at $\omega = 20$ rad/s, 100 rad/s, and 1000 rad/s. Since the net slope changes at $\omega = 20$ rad/s and 1000 rad/s are both −20 dB/decade, the denominator of $T(s)$ must have poles at $s = -20$ rad/s and $s = -1000$ rad/s. Since the net slope change at $\omega = 100$ rad/s is +20 dB/decade, the numerator of $T(s)$ must have a zero at $s = -100$ rad/s. These critical frequencies account for all of the slope changes, so the required transfer function has the form

$$T(s) = K\frac{s/100 + 1}{(s/20 + 1)(s/1000 + 1)}$$

where the gain K is yet to be determined. The dc gain of the proposed transfer function is $T(0) = K$ and the specified dc gain in Figure 14–7 is +20 dB; hence $K = 10$.

(b) To design first-order circuits, we partition $T(s)$ into a product of first-order transfer functions:

$$T(s) = T_1(s)T_2(s) = \left(\frac{s/100 + 1}{s/20 + 1}\right)\left(\frac{10}{s/1000 + 1}\right)$$

This partitioning is not unique because the overall gain $K = K_1 \times K_2 = 10$ can be distributed between the two functions in any number of ways (for example, $K_1 = 2$ and $K_2 = 5$). We could move the zero from the first function to the second or reverse the order in which the functions are assigned to stages in the cascade design. Selecting the inverting amplifier building block in Figure 14–3 for both stages leads to the design sequence shown in Figure 14–8.

The transfer function of the inverting amplifier is $-Z_2(s)/Z_1(s)$, so the selected building block places constraints on Z_1 and Z_2, as indicated in the second row in Figure 14–8. Equating $Z_1(s)$ to the reciprocal of the transfer function numerator and $Z_2(s)$ to the reciprocal of the denominator yields impedances of the form $1/(Cs + G)$. An impedance of this form is a resistor in parallel with a capacitor, as shown in the third and fourth rows in Figure 14–8. The element values of the prototype designs shown in the fifth row are not within practical ranges. To make all resistors 10 kΩ or greater, we use a magnitude scale factor of $k_m = 10^4$ in the first stage and $k_m = 10^5$ in the second stage. Applying these scale factors to the prototype designs produces the final designs shown in the last row of Figure 14–8. ■

D DESIGN EXAMPLE 14–3

(a) Construct a transfer function $T(s)$ with the straight-line gain response shown in Figure 14–9.
(b) Design an active RC circuit to realize the $T(s)$ found in (a).

SOLUTION:
(a) The straight-line gain response shows slope changes at $\omega = 10$ rad/s, 100 rad/s, 500 rad/s, and 5000 rad/s. The net slope changes at $\omega = 10$ rad/s and 5000 rad/s are -20 dB/decade. The net slope changes at $\omega = 100$ rad/s and 500 rad/s are $+20$ dB/decade. Therefore, $T(s)$ must have poles at $s = -10$ rad/s and $s = -5000$ rad/s and zeros at $s = -100$ rad/s and $s = -500$ rad/s. The required transfer function has the form

$$T(s) = K\frac{(s/100 + 1)(s/500 + 1)}{(s/10 + 1)(s/5000 + 1)}$$

where the gain K is yet to be determined. The gain plot in Figure 14–9 indicates that the required dc gain is 0 dB. The dc gain of the proposed transfer function is $T(0) = K$, so we conclude that $K = 1$.

Item	First Stage	Second Stage
Transfer function	$T_1(s) = -\dfrac{(s/100) + 1}{(s/20) + 1}$	$T_2(s) = -\dfrac{10}{(s/1000) + 1}$
Design constraints	Inverting Amplifier $\dfrac{Z_2}{Z_1} = \dfrac{(s/100) + 1}{(s/20) + 1}$	Inverting Amplifier $\dfrac{Z_2}{Z_1} = \dfrac{10}{(s/1000) + 1}$
Z_1	$\dfrac{1}{(s/100) + 1}$	$\dfrac{1}{10}$
Z_2	$\dfrac{1}{(s/20) + 1}$	$\dfrac{1}{(s/1000) + 1}$
Prototype designs		
Final designs		

FIGURE 14 – 8 *Design sequence for Example 14–2.*

FIGURE 14-9

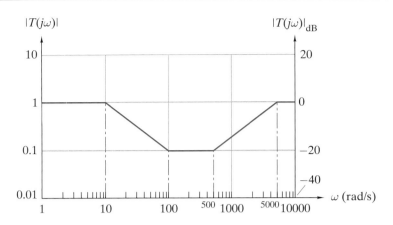

(b) To begin the circuit design, we partition $T(s)$ into the following product of first-order transfer functions:

$$T(s) = T_1(s)T_2(s) = \left(\frac{s/100 + 1}{s/10 + 1}\right)\left(\frac{s/500 + 1}{s/5000 + 1}\right)$$

This partitioning of $T(s)$ is not unique, since we could, for example, interchange the zeros or reverse the order in which the stages are listed. The design sequence shown in Figure 14–10 uses a voltage-divider circuit for the first stage and a noninverting amplifier for the second stage.

Factoring an s out of the numerator and denominator of the stage transfer functions produces the design constraints in the second row of Figure 14–10. Equating numerators and denominators of the design constraint equations leads to the RC circuits shown in the fourth and fifth rows of Figure 14–10. The gain of the voltage-divider stage is unity ($K = 1$) so the OP AMP in the first stage is a voltage follower, as indicated in the prototype design. However, the output of the first stage drives the noninverting input of the second-stage OP AMP. Since OP AMP inputs draw negligible current, no loading occurs when the voltage follower in the first stage is eliminated. The final design in the last row of Figure 14–10 eliminates the first-stage OP AMP and uses magnitude scale factors of $k_m = 10^6$ and $k_m = 5 \times 10^7$ to make all resistors greater than 10 kΩ.   ■

D DESIGN EXERCISE: 14–1

Design an active RC circuit that realizes a first-order high-pass filter with a cutoff frequency of 500 rad/s and a passband gain of 10.

Answers:

There are no unique answers. Two possible unscaled prototypes are shown in Figure 14–11.

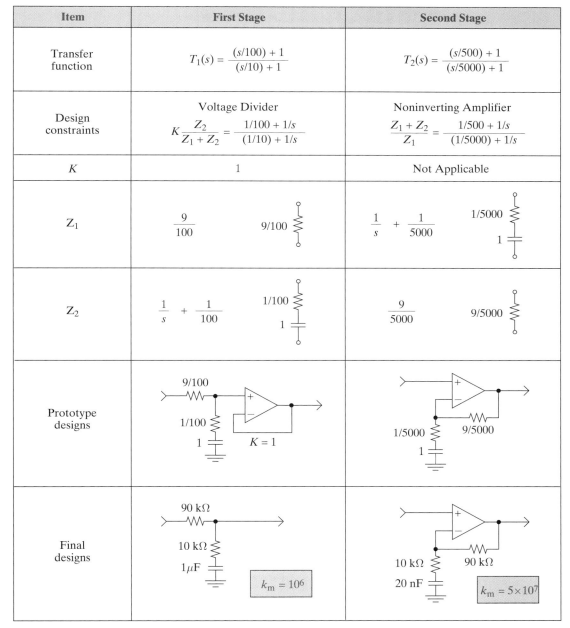

Item	First Stage	Second Stage
Transfer function	$T_1(s) = \dfrac{(s/100) + 1}{(s/10) + 1}$	$T_2(s) = \dfrac{(s/500) + 1}{(s/5000) + 1}$
Design constraints	Voltage Divider $K\dfrac{Z_2}{Z_1 + Z_2} = \dfrac{1/100 + 1/s}{(1/10) + 1/s}$	Noninverting Amplifier $\dfrac{Z_1 + Z_2}{Z_1} = \dfrac{1/500 + 1/s}{(1/5000) + 1/s}$
K	1	Not Applicable
Z_1	$\dfrac{9}{100}$	$\dfrac{1}{s} + \dfrac{1}{5000}$
Z_2	$\dfrac{1}{s} + \dfrac{1}{100}$	$\dfrac{9}{5000}$
Prototype designs	$K = 1$	
Final designs	90 kΩ, 10 kΩ, 1 μF, $k_\mathrm{m} = 10^6$	10 kΩ, 90 kΩ, 20 nF, $k_\mathrm{m} = 5 \times 10^7$

FIGURE   14 – 10   *Design sequence for Example 14–3.*

FIGURE   14 – 11

DESIGN EXERCISE: 14–2

(a) Construct a transfer function $T(s)$ with the asymptotic gain response shown in Figure 14–12.

(b) Design an active RC circuit to realize the $T(s)$ found in (a).

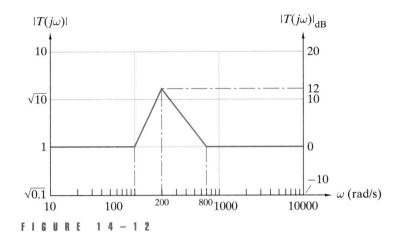

FIGURE 14 – 12

Answers:

(a) $T(s) = \dfrac{(s/100 + 1)^2\,(s/800 + 1)}{(s/200 + 1)^3}$

(b) There is no unique answer. One possible unscaled prototype design is shown in Figure 14–13.

FIGURE 14 – 13

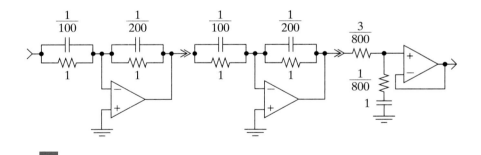

14–3 DESIGN WITH SECOND-ORDER CIRCUITS

Cascade connections of first-order circuit can only produce transfer functions with poles on the negative real axis in the s plane. The complex poles available in second-order circuits produce highly selective filter characteristics that cannot be realized using only real poles. While passive RLC circuits can produce complex poles, our interest centers on active RC circuits. In this regard we need second-order building block circuits analogous to the first-order circuits in Figure 14–3.

SECOND-ORDER LOW-PASS CIRCUITS

The active RC circuit in Figure 14–14(a) is analyzed in Chapter 11 (Example 11–6), where its voltage transfer function is shown to be

$$T(s) = \frac{V_2(s)}{V_1(s)} = \frac{\mu}{R_1R_2C_1C_2s^2 + (R_1C_1 + R_2C_2 + R_1C_2 - \mu R_1C_1)s + 1} \quad (14\text{–}2)$$

In circuit applications the dependent source in Figure 14–14(a) is replaced by the noninverting OP AMP circuit in Figure 14–14(b). The OP AMP circuit simulates the dependent source because it draws negligible input current and has very low output resistance. To serve as a replacement for the dependent source, the gain of the noninverting OP AMP circuit must be

$$\frac{R_A + R_B}{R_B} = \mu \quad (14\text{–}3)$$

where μ is the gain of the dependent source in Figure 14–14(a).

In second-order circuit design we select element values to produce specified values of the natural frequency ω_0 and damping ratio ζ. Comparing the denominator of Eq. (14–2) with the standard form $(s/\omega_0)^2 + 2\zeta(s/\omega_0) + 1$ gives

$$\omega_0 = \frac{1}{\sqrt{R_1R_2C_1C_2}} \text{ and } 2\zeta = \sqrt{\frac{R_2C_2}{R_1C_1}} + \sqrt{\frac{R_1C_2}{R_2C_1}} + (1 - \mu)\sqrt{\frac{R_1C_1}{R_2C_2}}$$

$$(14\text{–}4)$$

Since this yields two design constraints and five unknown element values, the designer has more options than are actually needed. Two common methods of selecting element values are discussed next.

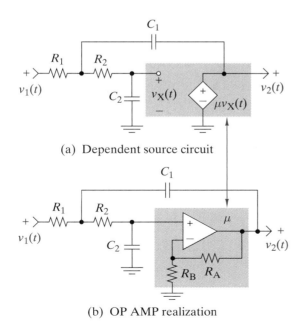

(a) Dependent source circuit

(b) OP AMP realization

FIGURE 14–14 *A second-order low-pass circuit.*

The equal R and equal C method requires that $R_1 = R_2 = R$ and $C_1 = C_2 = C$. Using the equal element constraints together with Eqs. (14–3) and (14–4) yields the conditions

$$RC = \frac{1}{\omega_0} \text{ and } \frac{R_A}{R_B} = 2(1 - \zeta) \qquad (14-5)$$

The unity gain method requires that $R_1 = R_2 = R$ and $\mu = 1$. Using these constraints together with Eqs. (14–3) and (14–4) yields the conditions

$$R_A = 0, \ R\sqrt{C_1 C_2} = \frac{1}{\omega_0}, \text{ and } \frac{C_2}{C_1} = \zeta^2 \qquad (14-6)$$

Since $R_A = 0$, the noninverting OP AMP circuit in Figure 14–14(b) can be a voltage follower, so neither feedback resistor in the OP AMP circuit is needed.

Finally, combining Eqs. (14–2) and (14–3), we find that the dc gain is $T(0) = (R_A + R_B)/R_B$. That is, the dc gain of the transfer function equals the gain of the OP AMP circuit.

Note: Under the equal element or unity gain approaches, the designer does not control the dc gain since the OP AMP circuit gain is determined by the damping ratio or is set to unity. In some cases the design conditions may not specify the dc gain because what really matters in a filter is the relative gains in passband and stopband. An additional gain stage can be added if the dc gain is specified.

D ▶ DESIGN EXAMPLE 14–4

(a) Construct a second-order low-pass transfer function with a natural frequency of $\omega_0 = 1000$ rad/s and with $|T(0)| = |T(j\omega_0)|$.
(b) Design second-order circuits that realize the transfer function found in (a).

SOLUTION:
(a) The transfer function has the form

$$T(s) = \frac{K}{(s/1000)^2 + 2\zeta(s/1000) + 1}$$

where K and ζ are to be determined. The dc gain of this function is $|T(0)| = |K|$. The gain at the natural frequency is $|T(j1000)| = |K|/2\zeta$. The condition $|T(0)| = |T(j\omega_0)|$ requires $\zeta = 0.5$. Hence a transfer function meeting the design statement is

$$T(s) = \frac{K}{(s/1000)^2 + (s/1000) + 1}$$

Note that the design conditions do not specify a value for the dc gain $|T(0)| = K$.

(b) We use both methods of assigning component values to obtain alternative designs.

Equal R and Equal C Design: Inserting the design constraints $\omega_0 = 1000$ rad/s and $\zeta = 0.5$ into Eq. (14–5) yields $RC = 10^{-3}$ s and $R_A = R_B$. Selecting $R = R_A = 10$ kΩ requires $C = 100$ nF and $R_B = 10$ kΩ and produces the circuit in Figure 14–15(a). The transfer function realized by the equal element value design has a dc gain of $K = (R_A + R_B)/R_B = 2$.

FIGURE 14-15

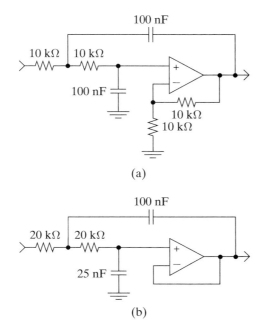

(a)

(b)

Unity Gain Design: Inserting the design constraints $\omega_0 = 1000$ rad/s and $\zeta = 0.5$ into Eq. (14–6) yields $R\sqrt{C_1 C_2} = 10^{-3}$ rad/s and $C_2 = 0.25C_1$. Selecting $C_1 = 100$ nF dictates that $C_2 = 25$ nF and $R = 20$ kΩ and yields the circuit in Figure 14–15(b). The unity gain design uses two fewer resistors since the OP AMP is connected as a voltage follower. The transfer function realized by the method has a dc gain $K = 1$, which is less than our equal R and equal C design. ∎

SECOND-ORDER HIGH-PASS CIRCUITS

The active RC circuit in Figure 14–16(a) has a second-order high-pass voltage transfer function of the form

$$T(s) = \frac{V_2(s)}{V_1(s)} = \frac{\mu R_1 R_2 C_1 C_2 s^2}{R_1 R_2 C_1 C_2 s^2 + (R_2 C_2 + R_1 C_1 + R_1 C_2 - \mu R_2 C_2)s + 1} \quad (14\text{–}7)$$

This high-pass circuit is obtained from the low-pass circuit in Figure 14–14(a) by interchanging the locations of the resistors and capacitors.[1]

---

1 Both circuits belong to a family of circuits originally proposed by R. P. Sallen and E. L. Key in "A Practical Method of Designing RC Active Filters," *IRE Transactions on Circuit Theory*, Vol. CT-2, pp. 74–85, 1955. In 1955 the controlled source was derived using vacuum tubes.

FIGURE 14-16 *A second-order high-pass circuit.*

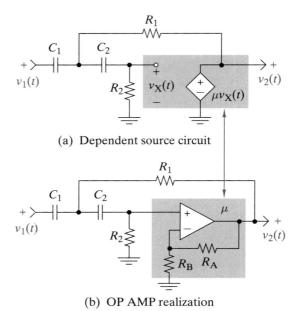

(a) Dependent source circuit

(b) OP AMP realization

Derivation of the transfer function in Eq. (14–7) is left as an exercise (see Problem 14–26).

In design applications the dependent source in Figure 14–16(a) is replaced by the noninverting OP AMP circuit in Figure 14–16(b). To serve as a replacement for the dependent source, the gain of the noninverting OP AMP circuit must be

$$\frac{R_A + R_B}{R_B} = \mu \qquad (14\text{–}8)$$

where μ is the gain of the noninverting OP AMP circuit that simulates the dependent source in Figure 14–16(a). Comparing the denominator of Eq. (14–7) with the standard form $(s/\omega_0)^2 + 2\zeta\,(s/\omega_0) + 1$ yields the following results:

$$\omega_0 = \frac{1}{\sqrt{R_1 R_2 C_1 C_2}} \quad \text{and} \quad 2\zeta = \sqrt{\frac{R_1 C_1}{R_2 C_2}} + \sqrt{\frac{R_1 C_2}{R_2 C_1}} + (1 - \mu)\sqrt{\frac{R_2 C_2}{R_1 C_1}} \qquad (14\text{–}9)$$

As is the case with the low-pass circuit, we have two design constraints and five unknown element values. Two often used methods of defining element values are discussed next.

The equal R and equal C method requires that $R_1 = R_2 = R$ and $C_1 = C_2 = C$. Inserting these constraints into Eqs. (14–8) and (14–9) yields the conditions

$$RC = \frac{1}{\omega_0} \quad \text{and} \quad \frac{R_A}{R_B} = 2(1 - \zeta) \qquad (14\text{–}10)$$

which is the same as the result in Eq. (14–5) for the low-pass circuit. The unity gain method requires that $C_1 = C_2 = C$ and $\mu = 1$. Inserting these constraints into Eqs. (14–8) and (14–9) yields the conditions

$$R_A = 0, \; C\sqrt{R_1 R_2} = \frac{1}{\omega_0}, \; \text{and} \; \frac{R_1}{R_2} = \zeta^2 \qquad (14\text{--}11)$$

Since $R_A = 0$, the noninverting OP AMP circuit in Figure 14–16(b) can be a voltage follower, so neither feedback resistor is needed.

Combining Eqs. (14–7) and (14–8) shows that the passband gain of the high-pass transfer function is $|T(\infty)| = (R_A + R_B)/R_B$. As is the case for the low-pass circuit, the passband gain equals the gain of the OP AMP circuit.

Note: Under the equal element or unity gain approaches, the designer does not control the passband gain since the OP AMP circuit gain is determined by the damping ratio or is set to unity. An additional gain stage can be added if a different value of gain is specified.

The next example illustrates the problem of achieving both a specified damping ratio and a passband gain.

D DESIGN   EXAMPLE   14–5

(a) Design a second-order high-pass transfer function with a passband gain of 0 dB and a cutoff frequency of $\omega_C = \omega_0 = 20$ krad/s.
(b) Design active RC circuits that realize the transfer function.

SOLUTION:
(a) A high-pass transfer function with a natural frequency of $\omega_0 = 20$ krad/s has the form

$$T(s) = \frac{K(s/20{,}000)^2}{(s/20{,}000)^2 + 2\zeta(s/20{,}000) + 1}$$

where K and ζ are to be determined. The passband gain of this transfer function is $|T(\infty)| = |K|$. The specification calls for a passband gain of 0 dB so $|K| = 1$. For the natural frequency also to be the cutoff frequency requires that $|T(j\omega_0)| = |K|/\sqrt{2}$. This, in turn, means that $|T(j20{,}000)| = |K|/\sqrt{2}\zeta = |K|/\sqrt{2}$, or $2\zeta = \sqrt{2}$. A transfer function that meets the specification is

$$T(s) = \frac{(s/20{,}000)^2}{(s/20{,}000)^2 + \sqrt{2}(s/20{,}000) + 1}$$

(b) We use both methods of assigning component values to obtain alternative designs.

Equal R and Equal C Design: Substituting the conditions $\omega_0 = 20$ krad/s and $\zeta = 1/\sqrt{2}$ into Eq. (14–10) yields $RC = 5 \times 10^{-5}$ s and $R_A = (2 - \sqrt{2})R_B = 0.586R_B$. Selecting $C = 5$ nF and $R_B = 50$ kΩ requires $R = 10$ kΩ and $R_A = 29.3$ kΩ. The passband gain of the resulting second-order circuit is $|T(\infty)| = K = (R_A + R_B)/R_B = 1.586$ and not unity (0 dB), as given in the specification. To overcome this problem we partition the transfer function as

$$T(s) = \underbrace{\left(\frac{1}{1.586}\right)}_{\text{First Stage}} \underbrace{\left(\frac{1.586(s/20{,}000)^2}{(s/20{,}000)^2 + \sqrt{2}(s/20{,}000) + 1}\right)}_{\text{Second Stage}}$$

FIGURE 14 – 17

(a)

(b)

The second stage has a passband gain of $K = 1.586$ and can be realized using the second-order high-pass circuit designed previously. The first stage is a voltage divider providing an attenuation of 1/1.586 to bring the overall passband gain down to the 0 dB level specified. Figure 14–17(a) shows the resulting two-stage cascade design.

Unity Gain Design: Substituting the conditions $\omega_0 = 20$ krad/s and $\zeta = 1/\sqrt{2}$ into Eq. (14–11) yields $R_A = 0$, $C\sqrt{R_1 R_2} = 5 \times 10^{-5}$ rad/s, and $R_1 = 0.5R_2$. Selecting $R_2 = 10$ kΩ dictates that $R_1 = 5$ kΩ and $C = 7.071$ nF. The transfer function realized by the method has a passband gain $|T(\infty)| = (R_A + R_B)/R_B$ $= 1$, which matches the 0 dB condition in the specification. Since no gain correction is needed, the single stage design in Figure 14–17(b) suffices. ■

SECOND-ORDER BANDPASS CIRCUITS

The active RC circuit in Figure 14–18 has a second-order bandpass transfer function of the form

$$T(s) = \frac{V_2(s)}{V_1(s)} = \frac{-R_2 C_2 s}{R_1 R_2 C_1 C_2 s^2 + (R_1 C_1 + R_1 C_2)s + 1} \tag{14–12}$$

Derivation of the transfer function in Eq. (14–12) is left as an exercise (see Problem 14–27). This bandpass circuit has two negative feedback paths, one provided by the resistor R_2 and the other by the capacitor C_2. This arrangement identifies this circuit as a member of a class of active RC circuits called multiple feedback circuits.[2]

---

2 For an extensive discussion of this family of circuits, see Wai-Kai Chen, Ed., *The Circuits and Filters Handbook*, Boca Raton, Fla., CRC Press, 1995, Chapter 76, p. 2372ff. The bandpass circuit in Figure 14–18 is sometimes called the Delyannis-Friend circuit. See M. E. Van Valkenburg, *Analog Filter Design*, New York, Holt, Rinehart and Winston, 1982, p. 203.

FIGURE 14-18 *A second-order band-pass circuit.*

The key descriptors of a bandpass filter are its center frequency ω_0 and its bandwidth $2\zeta\omega_0$. Comparing the denominator of Eq. (14–12) with the standard form $(s/\omega_0)^2 + 2\zeta\,(s/\omega_0) + 1$ yields the following results:

$$\omega_0 = \frac{1}{\sqrt{R_1 R_2 C_1 C_2}} \quad \text{and} \quad 2\zeta = \sqrt{\frac{R_1 C_1}{R_2 C_2}} + \sqrt{\frac{R_1 C_2}{R_2 C_1}} \qquad (14\text{–}13)$$

In the multiple feedback bandpass case we have two design constraints and four unknown element values. One way to assign element values is to use equal capacitors ($C_1 = C_2 = C$), in which case Eq. (14–13) yields the conditions

$$C\sqrt{R_1 R_2} = \frac{1}{\omega_0} \quad \text{and} \quad \frac{R_1}{R_2} = \zeta^2 \qquad (14\text{–}14)$$

For this choice of element values the gain at the center frequency is $|T(j\omega_0)| = R_2/2R_1 = 1/2\zeta$. As is the case for the low-pass and high-pass circuits, we cannot independently choose the damping ratio and the passband gain.

Bandpass filters are traditionally described in terms of center frequency ω_0 and a quality factor $Q = 1/2\zeta$. For second-order functions the filter bandwidth is

$$B = 2\zeta\omega_0 = \omega_0/Q$$

When $Q > 1$ ($\zeta < 0.5$) the filter is said to be narrowband because the bandwidth is less than the center frequency. Conversely, $Q < 1$ ($\zeta > 0.5$) describes a broadband filter. The active RC circuit shown in Figure 14–18 is best suited to narrowband applications. The design of broadband filters is discussed in Sect. 14–6.

DESIGN EXAMPLE 14–6

Design an active RC bandpass circuit that has a center frequency of 10 kHz and a bandwidth of 2 kHz.

SOLUTION:
From the definitions of center frequency and bandwidth we have

$$\omega_0 = 2\pi \times 10^4 = 6.283 \times 10^4 \text{ rad/s} \quad \text{and} \quad Q = \frac{\omega_0}{B} = 5$$

Selecting $R_2 = 100$ kΩ and using $\zeta = 1/2Q = 0.1$ in Eq. (14–14) yields $R_1 = 1$ kΩ and $C_1 = C_2 = 1.592$ nF. The passband gain produced by this design is $|T(j\omega_0)| = 1/2 \zeta = 5$. ∎

F I G U R E   1 4 – 1 9

D   DESIGN EXERCISE: 14–3

Design an active *RC* circuit that has the same transfer function $T(s) = V_2(s)/V_1(s)$ as the *RLC* circuit in Figure 14–19.

Answer: One possible solution is shown in Figure 14–20.

F I G U R E   1 4 – 2 0

DISCUSSION: *At first glance the* RLC *filter in Figure 14–19 seems simpler than the active* RC *circuit in Figure 14–20. However, active* RC *filters offer the advantages discussed in the first section of this chapter. Briefly, the advantages are that they produce both gain and frequency selectivity, they can be connected in cascade to produce high-order filters, and they are often smaller and less expensive in low-frequency applications.*

D   DESIGN EXERCISE: 14–4

Design a second-order active *RC* high-pass circuit with $\omega_0 = 500$ rad/s and $Q = 0.5$ using only 10-kΩ resistors.

Answers: Two possible solutions are shown in Figure 14–21.

F I G U R E   1 4 – 2 1

14–4   LOW-PASS FILTER DESIGN

The gain of the ideal low-pass filter response in Figure 14–22 is unity (0 dB) in the passband and zero ($-\infty$ dB) in the stopband. The transfer functions of physically realizable (real) low-pass filters are rational functions that can only approximate the ideal response. These rational functions have asymptotic high-frequency responses that roll off as ω^{-n} ($-20n$ dB/decade), where n is the number of poles in $T(s)$. Increasing the number of poles improves the approximation, as indicated in Figure 14–22. On the other hand, increasing the number of poles makes the circuit realizing $T(s)$ more complicated. Since cost and circuit complexity are usually related, there is an important trade-off between circuit complexity and how closely different filters approximate the ideal response.

FIGURE 14—22 *Ideal and real low-pass filter asymptotic responses.*

Since we can only approximate the ideal response filter, design requirements specify how closely we must approach the ideal. A commonly used way to specify this is to require that the gain response fall within the unshaded region in Figure 14–23. Many different transfer functions can meet this restriction, as illustrated by two responses shown in the figure. The allowable region in Figure 14–23 is defined by four parameters. The parameter T_{MAX} is the maximum gain in the passband, called simply the **passband gain**. Within the passband the gain $|T(j\omega)|$ must remain in the range

$$\frac{T_{MAX}}{\sqrt{2}} \leq |T(j\omega)| \leq T_{MAX}$$

and must equal $T_{MAX}/\sqrt{2}$ at the **cutoff frequency** ω_C. In the stopband the gain must decrease to and remain below T_{MIN} for all $\omega \geq \omega_{MIN}$. In sum, the parameters T_{MAX} and ω_C specify the passband response and the parameters T_{MIN} and ω_{MIN} specify how rapidly the stopband response must decrease.

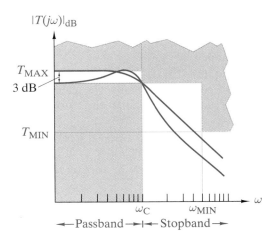

FIGURE 14—23 *Gain responses meeting a low-pass filter specification.*

To design a low-pass filter we must construct a transfer function whose gain $|T(j\omega)|$ approximates the ideal filter response within the tolerances allowed by the four parameters T_{MAX}, T_{MIN}, ω_C, and ω_{MIN}. Such a low-pass transfer function has the form

$$T_n(s) = \frac{K}{q_n(s)}$$

where $q_n(s)$ is an nth-order polynomial defining the poles. Thus, filter design involves two major tasks: (1) Construct a transfer function $T_n(s)$ whose gain response meets the filter performance specification and (2) devise one or more circuits that produce the transfer function $T_n(s)$.

We will consider three approaches to the first task: first-order cascade responses, Butterworth responses, and Chebychev responses. For the second task we partition the transfer function into a product of first- and second-order functions each of which can be realized using the building block circuits described in the preceding sections. A circuit realizing the transfer function is obtained by a cascade connection of these first- and second-order stages.

FIRST-ORDER CASCADE RESPONSES

A simple way to produce an nth-order low-pass filter is to connect n identical first-order low-pass filters in cascade. When we connect n identical filters in cascade the overall transfer function is

$$T_n(s) = \underbrace{\left[\frac{K}{s/\alpha + 1}\right] \times \left[\frac{K}{s/\alpha + 1}\right] \times \cdots \left[\frac{K}{s/\alpha + 1}\right]}_{n \text{ Stages}} = \frac{K^n}{(s/\alpha + 1)^n}$$

$$\text{(14–15)}$$

In a design problem we must select K, α, and n such that the gain response of this transfer function meets the performance requirements defined by T_{MAX}, T_{MIN}, ω_C, and ω_{MIN}.

The transfer function in Eq. (14–15) produces a gain response of

$$\left|T_n(j\omega)\right| = \frac{|K|^n}{\left[\sqrt{1 + (\omega/\alpha)^2}\right]^n} \qquad \text{(14–16)}$$

Note that the low-frequency gain asymptote is $|K|^n$ and the high-frequency asymptote is $(|K|\alpha/\omega)^n$. These asymptotes intersect (i.e., are equal) at a corner frequency located at $\omega = \alpha$. However, when $n > 1$ this corner frequency is *not* the filter cutoff frequency.

The maximum gain in Eq. (14–16) occurs at $\omega = 0$, where gain is $|T(0)| = |K|^n$. To meet the passband gain requirement, we select $|K|^n = T_{MAX}$. The passband specification also requires the gain at the cutoff frequency to be

$$\left|T_n(j\omega_C)\right| = \frac{|K|^n}{\left[\sqrt{1 + (\omega_C/\alpha)^2}\right]^n} = \frac{T_{MAX}}{\sqrt{2}}$$

With $|K|^n = T_{MAX}$ we can equate the denominators on the right side of this expression and solve for α as

$$\alpha = \frac{\omega_C}{\sqrt{2^{1/n} - 1}} \qquad \text{(14–17)}$$

Each first-order function in the cascade has a corner frequency at $\omega = \alpha$ whose value depends on ω_C and the number of poles n. Since $2^{1/n} - 1 \le 1$ it follows that $\alpha \ge \omega_C$; that is, the corner frequency is higher than the cutoff frequency for all $n > 1$. Said another way, increasing the order of a cascade filter requires that we increase the corner frequency to maintain the same bandwidth.

Figure 14–24 shows normalized gain responses of first-order low-pass cascades for $n = 1$ to $n = 8$. All of these responses meet the passband requirements. As expected, the stopband attenuation increases as the number of poles n increases since the high-frequency asymptote is proportional to $(\alpha/\omega)^n$. We can estimate the number of poles needed to meet the stopband requirements from the graphs in Figure 14–24. For example, suppose the stopband requirement is $T_{MIN}/T_{MAX} = 0.01$ (–40 dB) at $\omega_{MIN} = 10\omega_C$. In Figure 14–24 we see that at $\omega/\omega_C = 10$ the normalized gain $|T(j\omega)|/T_{MAX}$ is –32 dB for $n = 2$ and –42 dB for $n = 3$. Hence $n = 3$ is the smallest number of poles in a first-order cascade response that meets these requirements.

In summary, to construct a first-order cascade transfer function, we proceed as follows. Given values of T_{MAX}, T_{MIN}, ω_C, and ω_{MIN}, we set $|K|^n = T_{MAX}$ and use Figure 14–24 (or trial and error) to find the smallest integer n that meets the stopband requirements. Given n and ω_C, we calculate α using Eq. (14–17) and obtain the required transfer function $T_n(s)$ from Eq. (14–15). The transfer function is partitioned into a product of identical first-order functions, each of which can be obtained using a first-order circuit. The design process is relatively simple and yields identical stages, which may reduce manufacturing and logistical support costs. The next example illustrates these features.

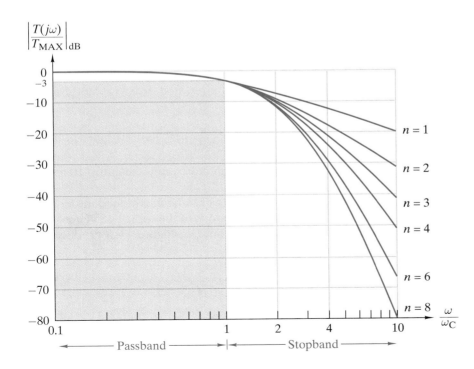

FIGURE   14 – 24   First-order low-pass cascade filter responses.

D **DESIGN EXAMPLE 14-7**

(a) Construct a first-order cascade transfer function that meets the following requirements: $T_{MAX} = 10$ dB, $\omega_C = 200$ rad/s, $T_{MIN} = -20$ dB, and $\omega_{MIN} = 800$ rad/s.

(b) Design a cascade of active RC circuits that realizes the transfer function developed in (a).

SOLUTION:

(a) The stopband requirement is that the normalized gain be less than -30 dB at $\omega/\omega_C = 800/200 = 4$. Figure 14–24 suggests that $n = 7$ might work, but we cannot be sure since the curve for $n = 7$ is not shown. To resolve this question, we use trial and error. Using Eq. (14–17), we calculate corner frequencies for $n = 7$ and $n = 8$.

$$\text{For } n = 7, \quad \alpha = \frac{200}{\sqrt{2^{1/7} - 1}} = 620 \quad \text{rad/s}$$

$$\text{For } n = 8, \quad \alpha = \frac{200}{\sqrt{2^{1/8} - 1}} = 665 \quad \text{rad/s}$$

Next we use Eq. (14–16) to calculate the corresponding normalized stopband gains:

$$\text{For } n = 7, \quad \left| \frac{T_7(j\omega_{MIN})}{T_{MAX}} \right| = 20 \log \left(\frac{1}{\left[\sqrt{1 + (800/620)^2} \right]^7} \right) = -29.8 \text{ dB}$$

$$\text{For } n = 8, \quad \left| \frac{T_8(j\omega_{MIN})}{T_{MAX}} \right| = 20 \log \left(\frac{1}{\left[\sqrt{1 + (800/665)^2} \right]^8} \right) = -31.1 \text{ dB}$$

Strictly speaking, $n = 8$ is the smallest integer for which the stopband gain meets the design requirement, although $n = 7$ comes within 0.2 dB (roughly 2.3%). Using $n = 8$, we solve for K. Since $T_{MAX} = 10$ dB (factor of $10^{1/2}$), we can write the passband requirement as $|K|^8 = 10^{1/2}$; hence $K = 1.155$. So, finally, the required first-order cascade transfer function is

$$T_8(s) = \left(\frac{1.155}{s/665 + 1} \right)^8$$

Note that the corner frequency is 665 rad/s while the cutoff frequency is 200 rad/s.

(b) The transfer function developed in (a) can be partitioned into a product of eight identical first-order functions of the form $1.155/(s/665 + 1)$. Using the voltage-divider building block in Figure 14–3 produces the following design constraints:

$$K \frac{Z_2}{Z_1 + Z_2} = 1.155 \times \frac{1/s}{1/665 + 1/s}$$

These conditions yield $K = (R_A + R_B)/R_B = 1.155$, $Z_2(s) = 1/s$, and $Z_1(s) = 1/665$. Selecting $R_B = 100$ kΩ leads to $R_A = 15.5$ kΩ. Using the magnitude scale factor of $k_m = 10^7$ on the RC voltage divider circuit produces the first-order low-pass circuit in Figure 14–25. A cascade connection of eight such circuits produces a low-pass filter that meets the design requirements. The gain of each stage is the 8th root of the overall passband gain since the overall gain is the product of the gain in each of the identical eight stages. The OP AMP at the output of each stage also provides isolation so that the cascade connection of eight such stages does not cause loading effects that change the frequency response of the individual stages. ∎

FIGURE   14 − 25

BUTTERWORTH LOW-PASS RESPONSES

All Butterworth low-pass filters produce a gain response of the form

$$|T_n(j\omega)| = \frac{|K|}{\sqrt{1 + (\omega/\omega_C)^{2n}}} \qquad (14\text{--}18)$$

where ω_C is the cutoff frequency and n is the number of poles in $T_n(s)$. By inspection, the maximum passband gain occurs at $\omega = 0$, where the gain is $|T(0)| = |K| = T_{MAX}$. At the cutoff frequency the gain is $|T_n(j\omega_C)| = |K|/\sqrt{2}$ for all n. Thus, selecting $|K| = T_{MAX}$ ensures that the Butterworth gain response in Eq. (14–18) satisfies the passband requirements for all values of n.

The low-frequency gain asymptote in Eq. (14–18) is $|K|$, while the high-frequency asymptote is $|K| (\omega_C/\omega)^n$. These two asymptotes intersect at a corner frequency of $\omega = \omega_C$. In a Butterworth response the corner frequency is always equal to the cutoff frequency, which means that it is independent of the order of the filter. This feature is different from the first-order cascade response, where the corner frequency increases with filter order.

Figure 14–26 compares normalized first-order cascade and Butterworth gain responses for $n = 4$. Both responses meet the passband requirements,

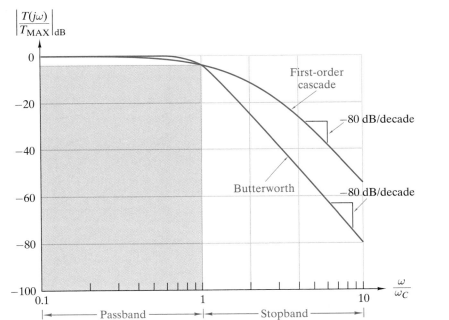

FIGURE   14 − 26   First-order cascade and Butterworth low-pass filter responses for n = 4.

and both have high-frequency asymptotic slopes of $-20n = -80$ dB/decade. However, the Butterworth response provides greater attenuation in the stopband because its corner frequency is lower and hence it approaches its high-frequency asymptote at lower frequencies than does the first-order cascade response.

With the Butterworth response we can analytically solve for the required number of poles. Since $|K| = T_{MAX}$, we insert the stopband requirements in Eq. (14–18) to obtain

$$|T(j\omega_{MIN})| = \frac{T_{MAX}}{\sqrt{1 + (\omega_{MIN}/\omega_C)^{2n}}} \leq T_{MIN} \qquad (14\text{--}19)$$

Solving this equation for n yields the constraint

$$n \geq \frac{1}{2} \frac{\ln[(T_{MAX}/T_{MIN})^2 - 1]}{\ln[\omega_{MIN}/\omega_C]} \qquad (14\text{--}20)$$

For example, the stopband requirement of $T_{MAX}/T_{MIN} = 10^{3/2}$ (30 dB) at $\omega_{MIN} = 4\omega_C$ yields

$$n \geq \frac{1}{2} \frac{\ln[(10^{3/2})^2 - 1]}{\ln[4]} = 2.49$$

The smallest integer meeting this constraint is $n = 3$.

Figure 14–27 shows normalized plots of the Butterworth gain response in Eq. (14–18) for $n = 1$ to $n = 6$. These graphs can be used to estimate the number of poles required to meet the stopband requirements. For example, consider the requirements used previously, namely, $|T_{MIN}/T_{MAX}|_{dB} = -30$ dB at $\omega_{MIN} = 4\omega_C$. In Figure 14–27 we see that a $\omega/\omega_C = 4$, $|T(j\omega)/T_{MAX}|_{dB}$ is -24 dB for $n = 2$, and is -36 dB for $n = 3$. Hence $n = 3$ is the smallest num-

FIGURE   14–27 *Butterworth low-pass filter responses.*

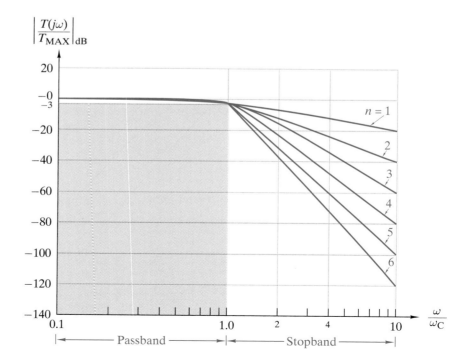

ber of poles in a Butterworth response that meets these stopband requirements. Thus, either Eq. (14–20) or Figure 14–27 can be used to determine n.

To design low-pass filters we need polynomials $q_n(s)$ such that the transfer function

$$T_n(s) = \frac{K}{q_n(s)}$$

produces the Butterworth gain response in Eq. (14–18). We begin with a
first-order polynomial of the form $q_1(s) = s/\omega_C + 1$, which produces the gain
response

$$|T_1(j\omega)| = \frac{|K|}{\sqrt{1 + (\omega/\omega_C)^2}} \qquad (14\text{–}21)$$

This response has the Butterworth form for $n = 1$. Next consider a second-
order polynomial of the form $q_2(s) = (s/\omega_C)^2 + 2\zeta(s/\omega_C) + 1$, which produces
the gain response

$$|T_2(j\omega)| = \frac{|K|}{\sqrt{[1 - (\omega/\omega_C)^2]^2 + (2\zeta\omega/\omega_C)^2}}$$

$$= \frac{|K|}{\sqrt{1 + (4\zeta^2 - 2)(\omega/\omega_C)^2 + (\omega/\omega_C)^4}} \qquad (14\text{–}22)$$

When $4\zeta^2 = 2$ this response has the Butterworth form for $n = 2$. As a final
example, consider a third-order polynomial of the form $q_3(s) = (s/\omega_C + 1)[(s/\omega_C)^2 + 2\zeta(s/\omega_C) + 1]$, which produces a gain response

$$|T_3(j\omega)| = \frac{|K|}{\sqrt{1 + (\omega/\omega_C)^2}\sqrt{[1 - (\omega/\omega_C)^2]^2 + 2\zeta(\omega/\omega_C)^2}}$$

$$= \frac{|K|}{\sqrt{1 + (4\zeta^2 - 1)(\omega/\omega_C)^2 + (4\zeta^2 - 1)(\omega/\omega_C)^4 + (\omega/\omega_C)^6}} \qquad (14\text{–}23)$$

When $4\zeta^2 = 1$ this response has the Butterworth form of Eq. (14–18) for
$n = 3$.

The preceding analysis shows that the polynomials

$$q_1(s) = s/\omega_C + 1$$
$$q_2(s) = (s/\omega_C)^2 + \sqrt{2}(s/\omega_C) + 1$$
$$q_3(s) = (s/\omega_C + 1)[(s/\omega_C)^2 + (s/\omega_C) + 1]$$

produce the Butterworth gain response in Eq. (14–18) for $n = 1, 2$, and 3.
Proceeding in this fashion, we can generate the normalized ($\omega_C = 1$) polynomials in Table 14–1 for values of n up to 6.

A general Butterworth low-pass transfer function with a cutoff frequency of ω_C is

$$T_n(s) = \frac{K}{q_n(s/\omega_C)} \qquad (14\text{–}24)$$

where $q_n(s)$ is the nth-order normalized polynomial in Table 14–1. With
$|K| = T_{MAX}$ this transfer function meets the passband conditions because
the normalized polynomials produce Butterworth gain responses. The
order of the polynomial, and hence the number of poles, is determined
from the stopband requirement using Figure 14–27 or Eq. (14–20).

TABLE 14–1 **NORMALIZED POLYNOMIALS THAT PRODUCE BUTTERWORTH RESPONSES**

ORDER	NORMALIZED DENOMINATOR POLYNOMIALS
1	$(s + 1)$
2	$(s^2 + 1.414s + 1)$
3	$(s + 1)(s^2 + s + 1)$
4	$(s^2 + 0.7654s + 1)(s^2 + 1.848s + 1)$
5	$(s + 1)(s^2 + 0.6180s + 1)(s^2 + 1.618s + 1)$
6	$(s^2 + 0.5176s + 1)(s^2 + 1.414s + 1)(s^2 + 1.932s + 1)$

Once we have constructed the required $T_n(s)$, we partition it into a product of first- and second-order functions and realize each function using the building blocks developed in the previous sections. The next example illustrates the design procedure for a Butterworth low-pass filter.

◗ **DESIGN EXAMPLE 14–8**

(a) Construct a Butterworth low-pass transfer function such that T_{MAX} = 20 dB, ω_C = 1000 rad/s, and T_{MIN} = –20 dB at ω_{MIN} = 4000 rad/s.
(b) Design a cascade of active RC circuits that realizes the transfer function found in part (a).

SOLUTION:
(a) The passband gain condition requires that K = 10 in Eq. (14–24). The stopband requirement specifies that the gain at ω_{MIN}/ω_C = 4 must be 40 dB (factor of 100) less than the passband gain. Using Eq. (14–20) yields

$$n \geq \frac{1}{2} \frac{\ln[(100)^2 - 1]}{\ln[4]} = 3.32$$

from which we see that n = 4 is the lowest-order polynomial that meets the stopband requirements. Using the fourth-order polynomial from Table 14–1, the required Butterworth low-pass transfer function is

$$T(s) = \frac{K}{q_4(s/\omega_C)}$$

$$= \frac{10}{\left[\left(\frac{s}{1000}\right)^2 + 0.7654\left(\frac{s}{1000}\right) + 1\right]\left[\left(\frac{s}{1000}\right)^2 + 1.848\left(\frac{s}{1000}\right) + 1\right]}$$

(b) The transfer function developed in (a) can be partitioned as follows:

$$\frac{T(s)}{10} = \left[\frac{1}{\left(\frac{s}{1000}\right)^2 + 0.7654\left(\frac{s}{1000}\right) + 1}\right]\left[\frac{1}{\left(\frac{s}{1000}\right)^2 + 1.848\left(\frac{s}{1000}\right) + 1}\right]$$

Figure 14–28 shows a design sequence for these second-order transfer functions using the Sallen-Key low-pass circuit shown in Figure 14–14. The first and second rows in the figure give the transfer functions for each stage and the stage parameters. Using the equal R and equal C design method [see Eq. (14–5)] leads to the design constraints in the fourth row. The element values selected in the fifth row produce the final design shown in the last row of Figure 14–28.

Item	First Stage	Second Stage
Prototype transfer function	$\dfrac{1}{(s/1000)^2 + 0.7654(s/1000) + 1}$	$\dfrac{1}{(s/1000)^2 + 1.848(s/1000) + 1}$
Stage parameters	$\omega_0 = 1000 \quad \zeta = 0.7654/2 = 0.3827$	$\omega_0 = 1000 \quad \zeta = 1.848/2 = 0.924$
Stage prototype		
Design constraints	$RC = \dfrac{1}{\omega_0} = 0.001$ $\dfrac{R_A}{R_B} = 2(1 - \zeta) = 1.23$	$RC = \dfrac{1}{\omega_0} = 0.001$ $\dfrac{R_A}{R_B} = 2(1 - \zeta) = 0.152$
Element values	Let $R = 100$ kΩ, then $C = 10$ nF Let $R_B = 100$ kΩ, then $R_A = 123$ kΩ	Let $R = 100$ kΩ, then $C = 10$ nF Let $R_B = 100$ kΩ, then $R_A = 15.2$ kΩ
Final designs		

FIGURE 14 − 2 8   *Design sequence for Example 14–8.*

With the Sallen-Key circuit the gain of each stage is determined by the stage damping ratio. The final designs in Figure 14–28 produce an overall passband gain of

$$K = K_1 K_2 = 2.23 \times 1.152 = 2.57$$

which is less than the requirement of $K = 10$. To meet the gain requirement, we add a third stage with a gain of $K_3 = 10/2.57 = 3.89$. Figure 14–29 shows a final three-stage cascade design that meets all of the design requirements. ■

First stage   Second stage   Third stage

F I G U R E   1 4 – 2 9

CHEBYCHEV LOW-PASS RESPONSES

All Chebychev low-pass filters produce a gain response of the form

$$|T_n(j\omega)| = \frac{|K|}{\sqrt{1 + C_n^2(\omega/\omega_C)}} \tag{14–25}$$

where $C_n(x)$ is an nth-order Chebychev polynomial defined by

$$C_n(x) = \cos[n \times \cos^{-1}(x)] \quad \text{for} \quad x \leq 1 \tag{14–26a}$$

and

$$C_n(x) = \cosh[n \times \cosh^{-1}(x)] \quad \text{for} \quad x > 1 \tag{14–26b}$$

In the passband ($x = \omega/\omega_C \leq 1$) the function $C_n(x)$ in Eq. (14–26a) is a cosine function that varies between -1 and $+1$. Hence $C_n^2(x)$ varies between 0 and 1 with the result that the passband gain of the Chebychev response in Eq. (14–25) varies over the range

$$\frac{|K|}{\sqrt{2}} \leq |T_n(j\omega)| \leq |K| \quad \text{for} \quad \omega \leq \omega_C$$

and, in particular at $\omega = \omega_C$,

$$|T_n(j\omega_C)| = \frac{|K|}{\sqrt{1 + C_n^2(1)}} = \frac{|K|}{\sqrt{2}}$$

for all values of n. By selecting $|K| = T_{MAX}$ we ensure that the Chebychev response in Eq. (14–25) meets the passband gain requirements for all n.

Like Butterworth responses, the Chebychev order is determined by the stopband requirements. Inserting these requirements into Eq. (14–25) produces

$$|T_n(j\omega_{\text{MIN}})| = \frac{|T_{\text{MAX}}|}{\sqrt{1 + C_n^2(\omega_{\text{MIN}}/\omega_C)}} \leq T_{\text{MIN}} \qquad (14\text{–}27)$$

Solving this constraint for $C_n(\omega_{\text{MIN}}/\omega_C)$ yields

$$C_n(\omega_{\text{MIN}}/\omega_C) \geq \sqrt{\left(\frac{T_{\text{MAX}}}{T_{\text{MIN}}}\right)^2 - 1}$$

In the stopband ($x = \omega/\omega_C \geq 1$) the function $C_n(x)$ is defined by the hyperbolic cosine function in Eq. (14–26b). Inserting this definition into the preceding constraint and then solving for n yields

$$n \geq \frac{\cosh^{-1}(\sqrt{(T_{\text{MAX}}/T_{\text{MIN}})^2 - 1})}{\cosh^{-1}(\omega_{\text{MIN}}/\omega_C)} \qquad (14\text{–}28)$$

In Example 14–8 the Butterworth response requires $n = 4$ to meet the stopband conditions $T_{\text{MAX}}/T_{\text{MIN}} = 100$ (40 dB) at $\omega_{\text{MIN}} = 4\omega_C$. Inserting these conditions into Eq. (14–28) yields

$$n \geq \frac{\cosh^{-1}(\sqrt{(100)^2 - 1})}{\cosh^{-1}(4)} = 2.57$$

The Chebychev response can meet the same stopband conditions using $n = 3$.

Figure 14–30 compares the Butterworth and Chebychev gain responses for $n = 4$. Both responses have stopband asymptotic slopes of –80 dB/decade. The Butterworth response is relatively flat in the passband and has a smooth transition to its high-frequency asymptote. In contrast, the Chebychev passband gain displays a succession of resonant peaks and valleys with a more abrupt transition to the stopband asymptote. Because

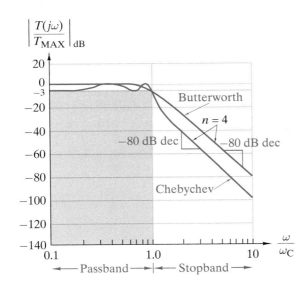

FIGURE   14 – 3 0 *Butterworth and Chebychev low-pass filter responses for* n = 4.

TABLE **14–2** **NORMALIZED POLYNOMIALS THAT PRODUCE CHEBYCHEV RESPONSES**

ORDER	NORMALIZED DENOMINATOR POLYNOMIALS
1	$(s + 1)$
2	$[(s/0.8409)^2 + 0.7654(s/0.8409) + 1]$
3	$[(s/0.2980) + 1][(s/0.9159)^2 + 0.3254(s/0.9159) + 1]$
4	$[(s/0.9502)^2 + 0.1789(s/0.9502) + 1)][(s/0.4425)^2 + 0.9276(s/0.4425) + 1]$
5	$[(s/0.1772) + 1][(s/0.9674)^2 + 0.1132(s/0.9674) + 1][(s/0.6139)^2 + 0.4670(s/0.6139) + 1]$
6	$[(s/0.9771)^2 + 0.0781(s/0.9771) + 1][(s/0.7223)^2 + 0.2886(s/0.7223) + 1][(s/0.2978)^2 + 0.9562(s/0.2978) + 1]$

$C_n(\omega)$ varies as cosine function in the passband, the resonant peaks in the Chebychev gain characteristic are all equal to T_{MAX} and the valleys all equal to $T_{MAX}/\sqrt{2}$. For this reason the Chebychev gain response in Eq. (14–25) is called the **equal-ripple response**. The net result of these resonances is that the Chebychev response transitions to its high-frequency asymptote rather abruptly.

To design a low-pass filter we need polynomials $q_n(s)$ such that the transfer function

$$T_n(s) = \frac{K}{q_n(s)}$$

produces a Chebychev gain response. Derivation of these polynomials involves complex variable theory beyond the scope of our study.[3] The analysis carried out in the reference allows us to write the normalized ($\omega_C = 1$) polynomials in Table 14–2. The transfer function of an nth-order Chebychev low-pass filter with a cutoff frequency of ω_C is then written as follows

$$T_n(s) = \frac{K}{q_n(s/\omega_C)} \qquad n \text{ odd} \qquad (14\text{–}29a)$$

or

$$T_n(s) = \frac{K/\sqrt{2}}{q_n(s/\omega_C)} \qquad n \text{ even} \qquad (14\text{–}29b)$$

where $q_n(s)$ is the nth-order normalized polynomial in Table 14–2. When we select $|K| = T_{MAX}$ the scale factor adjustment of $1/\sqrt{2}$ in Eq. (14–29b) is needed to ensure that $|T_n(j\omega)|$ meets the passband requirements when n is even. As with the Butterworth response, the order of the denominator polynomial can be found analytically from Eq. (14–28) or graphically using the normalized gain plots in Figure 14–31.

---

3 For example, see M. E. Van Valkenburg, *Analog Filter Design*, Chicago: Holt, Rinehart, and Winston, 1982, pp. 233–241.

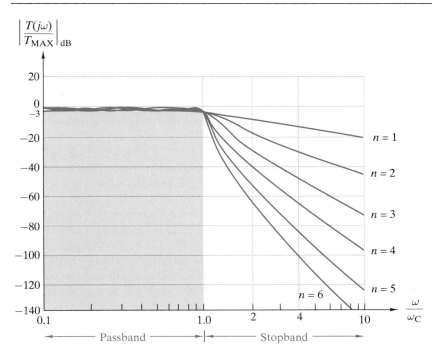

FIGURE   14-31   Chebychev low-pass filter responses.

DESIGN EXAMPLE 14-9

(a) Construct a Chebychev low-pass transfer function that meets a passband requirement of T_{MAX} = 20 dB and ω_C = 10 rad/s, and a stopband requirement of T_{MIN} = −30 dB at ω_{MIN} = 50 rad/s.

(b) Design a cascade of active RC circuits that produces the transfer function found in part (a).

SOLUTION:

(a) The passband gain condition requires K = 10. The stopband requirement specifies that the stopband gain at ω_{MIN}/ω_C = 5 must be 50 dB (factor of $10^{5/2}$) less than the passband gain. Using Eq. (14–28) yields

$$n \geq \frac{\cosh^{-1}(\sqrt{(10^{2.5})^2 - 1})}{\cosh^{-1}(5)} = 2.81$$

Hence n = 3 is the smallest integer meeting the stopband requirement. Using the third-order polynomial from Table 14–2, we construct the required Chebychev low-pass transfer function:

$$T(s) = \frac{K}{q_3(s/\omega_C)}$$

$$= \frac{10}{\left[\left(\dfrac{s}{0.298 \times 10}\right) + 1\right]\left[\left(\dfrac{s}{0.9159 \times 10}\right)^2 + 0.3254\left(\dfrac{s}{0.9159 \times 10}\right) + 1\right]}$$

(b) The Chebychev transfer function developed in (a) can be partitioned as follows:

$$\frac{T(s)}{10} = \left[\frac{1}{\left(\dfrac{s}{0.298 \times 10}\right) + 1} \right] \left[\frac{1}{\left(\dfrac{s}{0.9159 \times 10}\right)^2 + 0.3254 \left(\dfrac{s}{0.9159 \times 10}\right) + 1} \right]$$

realizing this partition requires a first-order stage in cascade with a second-order stage.

The design sequence in Figure 14–32 begins with the required transfer functions in the first row. The next three rows show the stage parameters, stage prototype, and design constraints for the selected prototypes. The constraints are then used to select element values that produce the final designs shown in the last row of Figure 14–32. The second stage is a Sallen-Key circuit, so its gain ($K_2 = 2.675$) is determined by the second-stage damping ratio. The first stage is a voltage divider circuit whose gain can be adjusted without changing its pole location. Thus, we can adjust the first-stage gain at $K_1 = 10/K_2$ so that the overall passband gain is $K_1 \times K_2 = K = 10$. A cascade connection of the final designs in the last two of Figure 14–32 meets all design requirements without introducing an additional gain stage. ∎

Exercise 14–5

What is the minimum order of the first-order cascade, Butterworth, and Chebychev transfer functions that meet the following stopband conditions?

(a) $T_{MIN}/T_{MAX} = -20$ dB at $4\omega_C$
(b) $T_{MIN}/T_{MAX} = -30$ dB at $5\omega_C$
(c) $T_{MIN}/T_{MAX} = -40$ dB at $6\omega_C$
(d) $T_{MIN}/T_{MAX} = -60$ dB at $8\omega_C$

Answers:

	First Order	Butterworth	Chebychev
(a)	$n = 3$	$n = 2$	$n = 2$
(b)	$n = 4$	$n = 3$	$n = 2$
(c)	$n = 5$	$n = 3$	$n = 3$
(d)	$n = 7$	$n = 4$	$n = 3$

COMPARISON OF LOW-PASS FILTER RESPONSES

We have described low-pass filter design methods for three responses, the first-order cascade, the Butterworth, and the Chebychev. At this point we want to compare the methods and discuss how we might choose between them. In filter applications the gain response is obviously important. Figure 14–33 shows the three straight-line asymptotes (solid lines) and the actual gain responses (dashed curves) for $n = 4$. All three responses meet the same passband requirements, have the same cutoff frequency, and have high-frequency asymptotes with slopes of $-20n = -80$ dB/decade. However, the three responses have different corner frequencies. At one extreme the corner frequency of the first-order cascade response is above the cutoff frequency, so its actual gain response approaches its asymptote very gradually. At the other extreme the Chebychev corner frequency lies below the

Item	First Stage	Second Stage
Prototype transfer function	$\dfrac{1}{(s/2.98) + 1}$	$\dfrac{1}{(s/9.159)^2 + 0.3254(s/9.159) + 1}$
Stage parameters	$\omega_0 = 2.980$	$\omega_0 = 9.159 \quad \zeta = 0.3254/2 = 0.1627$
Stage prototype		
Design constraints	$RC = \dfrac{1}{\omega_0} = 0.3357$ $\dfrac{R_A}{R_B} = K_1 - 1 = 2.74$	$RC = \dfrac{1}{\omega_0} = 0.1092$ $\dfrac{R_A}{R_B} = 2(1 - \zeta) = 1.67$
Element values	Let $R = 100$ kΩ, then $C = 3.36\ \mu$F Let $R_B = 100$ kΩ, then $R_A = 274$ kΩ	Let $R = 100$ kΩ, then $C = 1.092\ \mu$F Let $R_B = 100$ kΩ, then $R_A = 167$ kΩ
Final designs		

FIGURE 14-32   *Design sequence for Example 14–9.*

cutoff frequency and the actual response has a resonant peak that fills the gap between the corner frequency and the cutoff frequency. This resonance causes the Chebychev response to decrease rapidly in the neighborhood of the cutoff frequency. The Butterworth response has its corner frequency at the cutoff frequency, so its gain response falls between these two extremes.

FIGURE 14–33 *First-order cascade, Butterworth, and Chebychev low-pass responses for* n = 4.

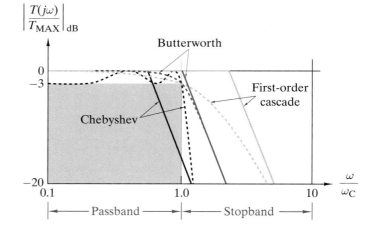

The differences in gain response can be understood by examining the pole-zero diagram in Figure 14–34. The Butterworth poles are evenly distributed on a circle of radius ω_C. The Chebychev poles lie on an ellipse whose minor axis is much smaller than ω_C. As a result, the Chebychev poles are closer to the j-axis, have lower damping ratios, and produce a gain response with pronounced resonant peaks. These resonant peaks lead to the equal-ripple response in the passband and the steep gain slope in the neighborhood of the cutoff frequency. At the other extreme the first-order cascade response has a fourth-order pole (quadruple pole) located on the negative real axis. The distance from the j-axis to the first-order cascade poles is much larger than ω_C, which explains the rather leisurely way its

FIGURE 14–34 *First-order cascade, Butterworth, and Chebychev pole locations for* n = 4.

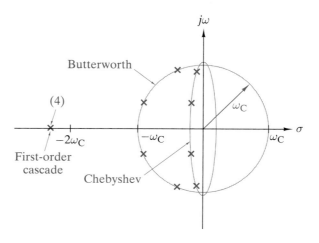

gain response transitions from the passband to stopband asymptote. As might be expected, the Butterworth poles fall between these two extremes.

This discussion illustrates the following principle. For any given value of n, the Chebychev response produces more stopband attenuation than the Butterworth response, which, in turn, produces more than the first-order cascade response. If stopband performance is the only consideration, then we should choose the Chebychev response. However, the Chebychev response comes at a price.

Figure 14–35 shows the step response of these three low-pass filters for $n = 4$. The step response of the Chebychev filter has lightly damped oscillations that produce a large overshoot and a long settling time. These undesirable features of the Chebychev step response are a direct result of the low-damping-ratio complex poles that produce the desirable features of its gain response. At the other extreme the step response of the first-order cascade filter rises rapidly to its final value without overshooting. This result should not be surprising since the remote poles of the first-order cascade produce exponential waveforms that have relatively short durations. In other words, the desirable features of the first-order cascade step response are a direct result of the remote real poles that produce the undesirable features of its gain response. Not surprisingly, the step response of the Butterworth filter lies between these two extremes.

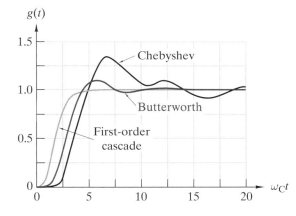

FIGURE 14–35 *First-order cascade, Butterworth, and Chebychev step responses for n = 4.*

Finally, consider the element values in the circuit realizations of these filters. Examination of Table 14–2 reveals that each pair of complex poles in a Chebychev filter has a different ω_0 and a different ζ. These parameters define the constraints on the element values for each stage in the filter. As a result, each stage in a cascade realization of a Chebychev filter has a different set of element values. In contrast, the stages in a first-order cascade filter can be exactly the same. From a manufacturing point of view, it may be better to produce and stock identical circuits rather than uniquely different circuits.

The essential point is that stopband attenuation is important, but it does not tell the whole story. Filter design, and indeed all design, involves trade-offs between conflicting requirements. The choice of a design approach is driven by the weight assigned to conflicting requirements.

14–5 HIGH-PASS FILTER DESIGN

High-pass transfer functions can be derived from low-pass prototypes using a transformation of the complex frequency variable s. If a low-pass transfer function $T_{LP}(s)$ has a passband gain of T_{MAX} and cutoff frequency of ω_C, then the transfer function $T_{HP}(s)$ defined as

$$T_{HP}(s) = T_{LP}(\omega_C^2/s) \qquad (14\text{–}30)$$

has high-pass characteristics with the same cutoff frequency and passband gain. On a logarithmic frequency scale this transformation amounts to a horizontal reflection of the gain response about ω_C. In other words, replacing s by ω_C^2/s changes a low-pass transfer function into a high-pass function without changing the parameters defining passband and stopband performances.

The first-order low-pass transfer function provides a simple example of the low-pass to high-pass transformation. Given a first-order low-pass function

$$T_{LP}(s) = \frac{T_{MAX}}{\dfrac{s}{\omega_C} + 1}$$

replacing s by ω_C^2/s yields a first-order high-pass transfer function

$$T_{HP}(s) = \frac{T_{MAX}}{\dfrac{\omega_C^2/s}{\omega_C} + 1} = \frac{T_{MAX}s}{s + \omega_C}$$

Note that the cutoff frequency and passband gain of the high-pass function are the same as those of the low-pass function from which it was derived.

Figure 14–36 shows the high-pass responses derived from the fourth-order Butterworth and Chebychev low-pass responses in Figure 14–30. From these plots we see that the transformation interchanges the pass and stopbands. The cutoff frequency is unchanged because it lies at the boundary between the two bands. Moreover, the same maximally flat or equal-ripple features appear in the passband of the high-pass response. Finally, the high-pass stopband gain is a mirror image about $\omega/\omega_C = 1$ of the low-pass stopband response. For example, if a low-pass function has −40 dB gain one octave above cutoff, then the corresponding high-pass function will have −40 dB gain one octave below cutoff.

A high-pass filter design problem is defined by specifying four parameters: T_{MAX}, ω_C, T_{MIN}, and ω_{MIN}. To obtain a high-pass transfer function that meets these requirements, we first construct a low-pass prototype $T_{LP}(s)$. The passband parameters for the low-pass prototype are the same as T_{MAX} and ω_C for the high pass. The stopband requirement for the low-pass prototype is T_{MIN} at $\omega_C^2/s\omega_{MIN}$. Given these design requirements, we construct a low-pass prototype $T_{LP}(s)$ using the methods described in Sect. 14–4. Replacing s by ω_C^2/s transforms the low-pass prototype into a high-pass transfer function $T_{HP}(s)$. The high-pass function is then partitioned into first- and second-order factors and realized using high-pass building blocks discussed in previous sections.

The next examples illustrate the method of designing high-pass filters using a low-pass prototype.

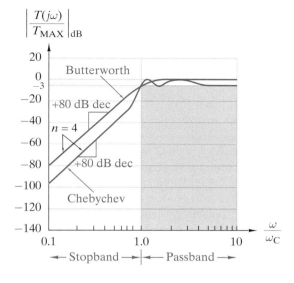

$$\left|\frac{T(j\omega)}{T_{\text{MAX}}}\right|_{\text{dB}}$$

← Stopband →|← Passband →

D **DESIGN EXAMPLE 14−10**

(a) Construct a Butterworth high-pass transfer function with $T_{\text{MAX}} = 20$ dB, $\omega_C = 10$ rad/s, and a stopband requirement of $T_{\text{MIN}} = -10$ dB at $\omega_{\text{MIN}} = 3$ rad/s.

(b) Design a cascade of active *RC* circuits that realizes the transfer function found in part (a).

SOLUTION:

(a) The passband requirements for the low-pass prototype are $K = 10$ and $\omega_C = 10$ rad/s. The stopband requirement for the low-pass prototype is $T_{\text{MIN}} = -10$ dB at $\omega_{\text{MIN}} = \omega_C^2/3 = 33.3$. The total range from T_{MIN} to T_{MAX} is 30 dB; hence $T_{\text{MAX}}/T_{\text{MIN}} = 10^{3/2}$. Using Eq. (14−20) yields

$$n \geq \frac{1}{2}\frac{\ln[(10^{3/2})^2 - 1]}{\ln(33.3/10)} = 2.87$$

We find that $n = 3$ is the lowest-order Butterworth response that meets the low-pass design requirements. Using the third-order polynomial from Table 14−1, the low-pass prototype is

$$T_{\text{LP}}(s) = \frac{K}{q_3(s/10)} = \frac{10}{\left[\left(\dfrac{s}{10}\right)^2 + \left(\dfrac{s}{10}\right) + 1\right]\left[\left(\dfrac{s}{10}\right) + 1\right]}$$

The high-pass transfer function is obtained by replacing s with $\omega_C^2/s = 100/s$:

$$T_{\text{HP}}(s) = \frac{10}{\left[\left(\dfrac{100/s}{10}\right)^2 + \left(\dfrac{100/s}{10}\right) + 1\right]\left[\left(\dfrac{100/s}{10}\right) + 1\right]}$$

$$= \left[\frac{(s/10)^2}{(s/10)^2 + (s/10) + 1}\right]\left[\frac{10(s/10)}{(s/10) + 1}\right]$$

The function $T_{HP}(s)$ is a high-pass transfer function that meets the filter design requirements.

(b) Figure 14–37 shows a design sequence for realizing $T_{HP}(s)$ using the first- and second-order high-pass active *RC* building blocks. The stage transfer functions given in the first row yield the stage parameters in the second row. The stage prototypes are a second-order high-pass section with unity gain and a first-order high-pass section with an adjustable gain. The stage prototypes together with the stage parameters yield the design constraints in the fourth row. Selecting element values within these constraints produces the final stage designs shown in the last row of Figure 14–37. The required passband gain is included in the first-order circuit, so no gain correction stage is required in this example. In other words, connecting the two stages in Figure 14–37 in cascade meets all design requirements, including the passband gain of 20 dB.   ■

EXAMPLE 14–11

Use Electronics Workbench to verify the high-pass design developed in Example 14–10.

SOLUTION:

One of the most important uses of computer-aided analysis is to simulate a proposed circuit design and confirm that it meets the performance requirements. To verify the design in Example 14–10, we must confirm that the frequency response of the two-stage circuit in the last row of Figure 14–37 has a high-pass characteristic with the following features:

1. A passband gain of 20 dB (gain = 10).
2. A cutoff frequency of 10 rad/s (gain = $10/\sqrt{2}$ = 7.07 at f = 10/2π = 1.59 Hz).
3. Stopband gain less than –10 dB (gain < 0.316) below 3 rad/s (0.477 Hz).

Figure 14–38 shows the two-stage circuit as constructed in the Electronics Workbench circuit window. The input source V_1 is set to produce a 1 V ac input, so the ac voltage at node 9 (the second OP AMP output) is the overall filter circuit gain response. Using the **AC Analysis** command from the **Analysis** menu, we sweep the frequency of the source V_1 from 0.3 Hz to 4 Hz, and select node 9 as the node for analysis. The **Simulate** command produces an ac analysis run leading to the gain response shown in Figure 14–38. The gain plot shows that

1. The circuit has a high-pass response with a passband of 10.
2. The x2 cursor is set at x2 = 1.59 Hz where the gain is found to be y2 = 7.07 confirming that the cutoff frequency is 10 rad/s.
3. The x1 cursor is set at x1 = 0.477 Hz where the gain is found to be y1 = 0.269 (–11.4 dB) confirming that the stopband gain is below –10 dB at 3 rad/sec.

The plot confirms that the circuit meets the design requirements.   ■

Item	First Stage	Second Stage
Prototype transfer function	$\dfrac{(s/10)^2}{(s/10)^2 + (s/10) + 1}$	$\dfrac{10\,(s/10)}{(s/10) + 1}$
Stage parameters	$\omega_0 = 10 \quad \zeta = 0.5 \quad K_1 = 1$	$\omega_0 = 10 \quad K_2 = 10$
Stage prototype		
Design constraints	$\sqrt{R_1 R_2}\,C = 1/\omega_0 = 0.1$ $R_1/R_2 = \zeta^2 = 0.25$	$RC = 1/\omega_0 = 0.1$ $R_A/R_B = K_2 - 1 = 9$
Element values	Let $R_2 = 100$ kΩ, then $R_1 = 25$ kΩ and $C = 2\ \mu$F	Let $R = 50$ kΩ, then $C = 2\ \mu$F Let $R_B = 10$ kΩ, then $R_A = 90$ kΩ
Final designs		

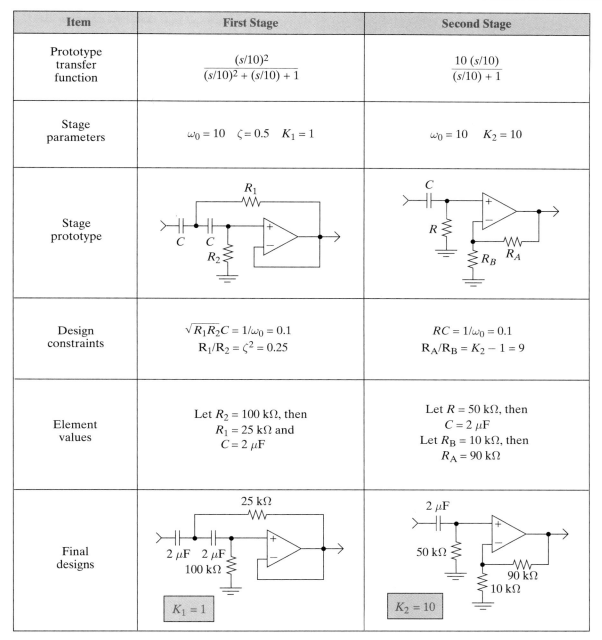

FIGURE 14−37 *Design sequence for Example 14–10.*

FIGURE 14–38

Exercise 14–6

Construct first-order cascade and Butterworth high-pass transfer functions that have passband gains of 0 dB, cutoff frequencies of 50 rad/s, and $T_{MIN} = -40$ dB at $\omega_{MIN} = 10$ rad/s.

Answers:

$$T_{FO}(s) = \frac{s^8}{(s + 15)^8}, \; T_{BU}(s) = \frac{s^3}{(s + 50)(s^2 + 50s + 2500)}$$

14–6 BANDPASS AND BANDSTOP FILTER DESIGN

In Chapter 12 we found that the cascade connection in Figure 14–39 can produce a bandpass filter. When the cutoff frequency of the low-pass filter (ω_{CLP}) is higher than the cutoff frequency of the high-pass filter (ω_{CHP}), the interval between the two frequencies is a passband separating two stopbands. The low-pass filter provides the high-frequency stopband and the high-pass filter the low-frequency stopband. Frequencies between the two cutoffs fall in the passband of both filters and are transmitted through the cascade connection, producing the passband of the resulting bandpass filter.

FIGURE 14–39 *Cascade connection of high-pass and low-pass filters.*

When $\omega_{CLP} \gg \omega_{CHP}$, two bandpass cutoff frequencies are approximately $\omega_{C1} \approx \omega_{CHP}$ and $\omega_{C2} \approx \omega_{CLP}$. Under these conditions the center frequency and bandwidth of the bandpass filter are

$$\omega_0 = \sqrt{\omega_{CHP}\omega_{CLP}} \text{ and } B = \omega_{CLP} - \omega_{CHP}$$

and the ratio of the center frequency over the bandwidth is approximately

$$\frac{\omega_0}{B} = Q \approx \sqrt{\frac{\omega_{CHP}}{\omega_{CLP}}} \ll 1$$

Since the quality factor is less than 1, this method of bandpass filter design produces a broadband filter, as contrasted with the narrowband response ($Q > 1$) produced by the active RC bandpass circuit studied earlier in Sect. 14–3.

◀D▶ DESIGN EXAMPLE 14–12

Construct a Butterworth bandpass filter with a passband gain of 0 dB and cutoff frequencies at $\omega_{C1} = 10$ rad/s and $\omega_{C2} = 50$ rad/s. The stopband gains must be less than −20 db at 2 rad/s and 250 rad/s.

SOLUTION:

In the upper stopband the −20-dB gain is specified at 250 rad/s $= 5\omega_{C2}$. Using either Eq. (14–20) or Figure 14–27 shows that this requirement can be met by a second-order Butterworth polynomial. Hence the required low-pass function is

$$T_{LP}(s) = \frac{1}{(s/50)^2 + \sqrt{2}(s/50) + 1}$$

In the lower stopband the −20-dB gain is specified at 2 rad/s = $\omega_{C1}/5$. Hence a second-order Butterworth polynomial will meet this requirement as well and the required high-pass function is

$$T_{HP}(s) = \frac{(s/10)^2}{(s/10)^2 + \sqrt{2}(s/10) + 1}$$

When circuits realizing these two transfer functions are connected in cascade the overall transfer function is

$$T_{HP}(s) \times T_{LP}(s) = \left[\frac{(s/10)^2}{(s/10)^2 + \sqrt{2}(s/10) + 1} \right] \times \left[\frac{1}{(s/50)^2 + \sqrt{2}(s/50) + 1} \right]$$

$$= \frac{2500\,s^2}{s^4 + 60\sqrt{2}s^3 + 3600s^2 + 30000\,\sqrt{2}s + 250000}$$

With the transfer function expressed as a quotient of polynomials, we use MATLAB to generate Bode plots of the frequency response. In the MATLAB command window, we define the numerator and denominator polynomials using the row matrices num and den.

```
num=[2500 0 0];

den=[1 60*sqrt(2) 3600 30000*sqrt(2) 250000];
```

These matrices list the polynomial coefficients in descending order. The MATLAB command

```
bode(num,den)
```

produces the Bode plots shown in Figure 14–40. The gain response displays a bandpass characteristic with a lower cutoff frequency at 10 rad/s and an upper cutoff frequency at 50 rad/s. These two cutoff frequencies come from the high-pass and low-pass functions, respectively. The phase shift swings from +180° at low frequency to −180° at high frequency passing through zero at the passband center frequency of $\sqrt{10 \times 50}$ = 22.36 rad/s. ∎

FIGURE 14–40

Example 14-12
Bandpass Response

Frequency (rad/sec)

Figure 14–41 shows the dual situation in which a high-pass and a low-pass filter are connected in parallel to produce a bandstop filter. When $\omega_{CLP} \ll \omega_{CHP}$, the region between the two cutoff frequencies is a stopband separating two passbands. The low-pass filter provides the low-frequency passband via the lower path and the high-pass filter the high-frequency passband via the upper path. Frequencies between the two cutoffs fall in the stopband of both filters and are not transmitted through either path in the parallel connection. As a result, the two filters produce the stopband of the resulting bandstop filter. When $\omega_{CHP} \gg \omega_{CLP}$, two cutoff frequencies are approximately $\omega_{C1} \approx \omega_{CLP}$ and $\omega_{C2} \approx \omega_{CHP}$.

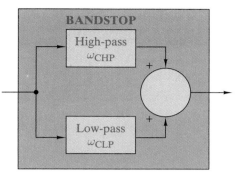

D **DESIGN EXAMPLE 14−13**

Construct a fourth order Butterworth bandstop filter with a passband gain of 0 dB and cutoff frequencies at $\omega_{C1} = 10$ rad/s and $\omega_{C2} = 50$ rad/s.

SOLUTION:

In Example 14–12 we obtained a Butterworth bandpass response using a cascade connection of a second-order high-pass function and a low-pass function with $\omega_{CLP} = 50 >> \omega_{CHP} = 10$. To obtain a bandstop response, we interchange the cutoff frequencies and write the two functions as

$$T_{LP}(s) = \frac{1}{(s/10)^2 + \sqrt{2}(s/10) + 1}$$

$$T_{HP}(s) = \frac{(s/50)^2}{(s/50)^2 + \sqrt{2}(s/50) + 1}$$

We now have $\omega_{CLP} = 10 << \omega_{CHP} = 50$ which leads to a bandstop response. For the parallel connection the overall transfer function is the sum of the two transfer functions.

$$T_{LP}(s) + T_{HP}(s) = \frac{1}{(s/10)^2 + \sqrt{2}(s/10) + 1} + \frac{(s/50)^2}{(s/50)^2 + \sqrt{2}(s/50) + 1}$$

$$= \frac{s^4 + 10\sqrt{2}s^3 + 200s^2 + 5000\sqrt{2}s + 250000}{s^4 + 60\sqrt{2}s^3 + 3600s^2 + 30000\sqrt{2}s + 250000}$$

With the transfer function expressed as a quotient of polynomials, we use MATLAB to generate Bode plots of the frequency response. In the MATLAB command window, we define the numerator and denominator polynomials using the row matrices num and den.

```
num=[1 10*sqrt(2) 200 5000*sqrt(2) 250000];
```

```
den=[1 60*sqrt(2) 3600 30000*sqrt(2) 250000];
```

These matrices list the polynomial coefficients in descending order. The MATLAB command

```
bode(num,den)
```

produces the Bode plots shown in Figure 14–42. The gain displays a bandstop response with a lower cutoff frequency at 10 rad/s and an upper cutoff frequency at 50 rad/s. These two cutoff frequencies come from the low-

FIGURE 14-42

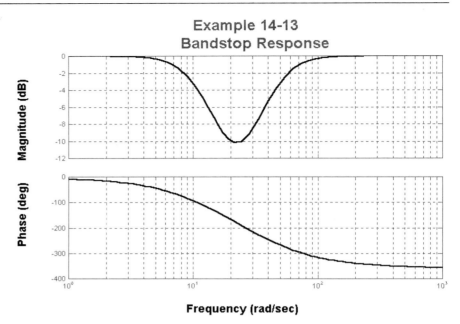

Example 14-13
Bandstop Response

pass and high-pass functions, respectively. The phase shift changes from +0° at low frequency to −360° at high frequency passing through −180° at the stopband center frequency of $\sqrt{10 \times 50} = 22.36$ rad/s. ∎

When the two cutoff frequencies are widely separated, we can realize broadband bandpass and bandstop filters using a cascade or parallel connection of low-pass and high-pass filters. The design problem reduces to designing separate low-pass and high-pass filters and then connecting them in cascade or parallel to obtain the required overall response. In the bandpass case the active *RC* circuits building blocks can be connected in cascade without violating the chain rule. The bandstop case requires a summation, which can be implemented using an OP AMP summer.

Exercise 14–7

Construct Butterworth low-pass and high-pass transfer functions whose cascade connection produces a bandpass function with cutoff frequencies at 20 rad/s and 500 rad/s, a passband gain of 0 dB, and a stopband gain less than −20 dB at 5 rad/s and 2000 rad/s.

Answer:

$$T(s) = \left[\frac{500^2}{s^2 + 707s + 500^2}\right]\left[\frac{s^2}{s^2 + 28.3s + 400}\right]$$

Exercise 14–8

Develop Butterworth low-pass and high-pass transfer functions whose parallel connection produces a bandstop filter with cutoff frequencies at 2 rad/s and 800 rad/s, passband gains of 20 dB, and stopband gains less than −30 dB at 20 rad/s and 80 rad/s.

Answer:

$$T_{LP}(s) = \frac{80}{(s^2 + 2s + 4)(s + 2)}$$

$$T_{HP}(s) = \frac{10s^3}{(s^2 + 800s + 800^2)(s + 800)}$$

SUMMARY

- A filter design problem is defined by specifying attributes of the gain response such as a straight-line gain plot, cutoff frequency, passband gain, and stopband attenuation. The first step in the design process is to construct a transfer function $T(s)$ whose gain response meets the specification requirements.

- In the cascade design approach, the required transfer function is partitioned into a product of first- and second-order transfer functions, which can be independently realized using basic active RC building blocks.

- Transfer functions with real poles and zeros can be realized using the voltage divider, noninverting amplifier, or inverting amplifier building blocks. Transfer functions with complex poles can be realized using second-order active RC circuits.

- Transfer functions meeting low-pass filter specifications can be constructed using first-order cascade, Butterworth, or Chebychev poles. First-order cascade filters are easy to design but have poor stopband performance. Butterworth responses produce maximally flat passband responses and more stopband attenuation than a first-order cascade with the same number of poles. The Chebychev responses produce equal-ripple passband responses and more stopband attenuation than the Butterworth response with the same number of poles.

- A high-pass transfer function can be constructed from a low-pass prototype by replacing s with ω_C^2/s. Bandpass (bandstop) filters can be constructed using a cascade (parallel) connection of a low-pass and a high-pass filter.

PROBLEMS

ERO 14–1 DESIGN WITH FIRST-ORDER CIRCUITS (SECT. 14–2)

Given a filter specification or a Bode plot of a straight-line gain,

(a) Construct a transfer function $T(s)$ that has the specified characteristics.
(b) Design a cascade of first-order circuits that realizes $T(s)$.

14–1   **D** Design a first-order low-pass filter with a cutoff frequency of 500 Hz and a passband gain of 14 dB. Scale the circuit so that the element values are in practical ranges.

14–2   **D** Design a first-order high-pass filter with a passband gain of 10 dB, a gain of at least 9 dB at 3 kHz, and a gain of less than −5 dB at

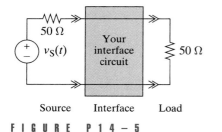

FIGURE P14-4

FIGURE P14-5

300 Hz. Scale the circuit so that the element values are in practical ranges.

14-3 ⬤ Design a cascade connection of first-order low-pass and high-pass circuits to obtain a bandpass filter with cutoff frequencies at 200 rad/s and 3000 rad/s. The passband gain of the bandpass circuit must be at least 30 dB. Scale the circuit so that the element values are in practical ranges.

14-4 ⬤ The open-circuit voltage of the source in Figure P14-4 is

$$v_S(t) = 5 + 0.02 \sin(4000\,\pi\,t) \quad V$$

The purpose of the bypass capacitor is to reduce the ac voltage delivered to the 20-Ω load below 10 mV (peak). Select a value of C.

14-5 ⬤ The open-circuit voltage of the signal source in Figure P14-5 is

$$v_S(t) = 5 + 0.02 \sin(50000\,\pi\,t) \quad V$$

Design an interface circuit so that no dc voltage appears across the 50-Ω load and the ac voltage to the load has a peak amplitude of at least 2 V.

14-6 ⬤ The open-circuit voltage of the signal generator in Figure P14-5 is a 5-kHz sinusoid with a 2-V peak amplitude plus a spurious (undesirable) sinusoid at 100 kHz with a 0.5-V peak amplitude. Design an interface circuit filter so that the voltage delivered to the 50-Ω load is a 5-kHz sinusoid with a peak amplitude of at least 4 V and a 100-kHz sinusoid with a peak amplitude of less than 200 mV.

14-7 ⬤ The straight-line gain plot in Figure P14-7 emphasizes the frequencies below 50 rad/s and deemphasizes the frequencies above 250 rad/s.

(a) Construct a transfer function that has this gain response using only real poles and zeros.

(b) Design a cascade of first-order active RC circuits to realize the $T(s)$ found in (a). Scale the circuits so that the element values are in practical ranges.

FIGURE P14-7

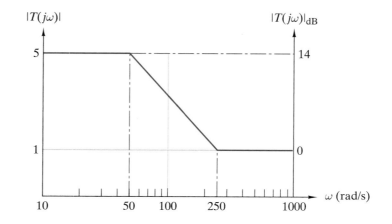

14–8   **D** The straight-line plot in Figure P14–8 has a low-pass characteristic.
   (a) Construct a transfer function $T(s)$ that has this gain response
   using only real poles and zeros.
   (b) Design a cascade of first-order active RC circuits to realize the
   transfer function found in (a). Scale the circuits so that the element values are in practical ranges.

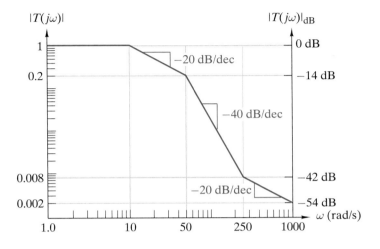

14–9   **D** The straight-line gain response in Figure P14–9 emphasizes
the frequencies between 200 rad/s and 1000 rad/s and deemphasizes
the frequencies below 20 rad/s and above 10 krad/s.
   (a) Construct a transfer function $T(s)$ that has this gain response
   using only real poles and zeros.
   (b) Design a cascade of first-order active RC circuits to realize the
   $T(s)$ found in (a). Scale the circuits so that the element values are
   in practical ranges.

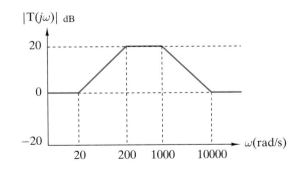

14–10   **D** The gain response in Figure P14–10 has a passband between
100 rad/s and 8 krad/s and –20 dB/decade roll off in the stopbands.
   (a) Construct a transfer function $T(s)$ that has this gain response
   using only real poles and zeros.

FIGURE P14−10

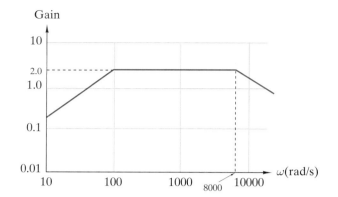

(b) Design a cascade of first-order active *RC* circuits to realize the transfer function found in (a). Scale the circuits so that the element values are in practical ranges.

14–11 **D** The high-pass straight-line gain response in Figure P14–11 provides a two-pole stopband roll off at low frequencies.
  (a) Construct a transfer function $T(s)$ that has this gain response using only real poles and zeros.
  (b) Design an active *RC* circuit to realize the $T(s)$ found in (a) using only one OP AMP. Scale the circuit so that the element values are in practical ranges.

FIGURE P14−11

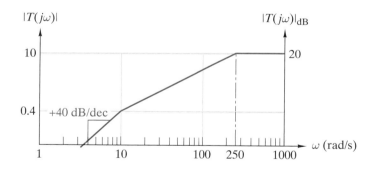

14–12 **D** The straight-line gain response in Figure P14–12 is the design specification for an audio preamplifier. Note that frequencies are specified in Hz.
  (a) Construct a transfer function $T(s)$ that has this gain response using only real poles and zeros.
  (b) Design an active *RC* circuit to realize the $T(s)$ found in (a) using only one OP AMP. Scale the circuit so that the element values are in practical ranges.

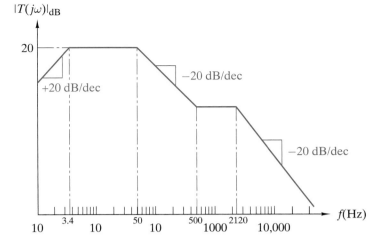

14–13 **D** A first-order low-pass power amplifier has a dc gain of 5 dB and a cutoff frequency of 2 kHz. Design an active RC pre-amp to connect in cascade with the power amplifier so that the overall circuit has a first-order low-pass response with a dc gain of 20 dB and a cutoff frequency of 6 kHz. Scale the circuit so that the element values are in practical ranges.

14–14 **D** A first-order low-pass amplifier has a dc gain of 5 dB and a cutoff frequency of 10 kHz. Design an active RC pre-amp circuit to connect in cascade with the amplifier so that the resulting combination has a bandpass characteristic with cutoff frequencies at 50 Hz and 10 kHz and a passband gain of 20 dB. Scale the circuit so that the element values are in practical ranges

14–15 **D** Your boss asks you to design a cost-effective RC circuit that realizes the transfer function given below. You are cautioned that cost effective means using as few devices as possible and using standard values wherever possible. Also, your boss would really like you to use some of those 10-nF capacitors left over from your research project that he cancelled.

$$T(s) = \pm \frac{Ks}{(s + 10^2)(s + 10^5)}$$

ERO 14–2 DESIGN WITH SECOND-ORDER CIRCUITS (SECT. 14–3)

A Given a circuit that realizes a second-order transfer function, develop a method of selecting circuit parameters to achieve specified filter characteristics.

D Given a filter specification, develop a second-order transfer function that has the required characteristics and design an active RC circuit that realizes the transfer function.

Design second-order active RC circuits to meet the following requirements:

Problem	Type	ω_0(rad/s)	ζ	Constraints
14–16 **D**	Low pass	500	0.5	Use 20-nF capacitors.
14–17 **D**	Low pass	2000	0.4	Gain of 10 dB at dc.

14–18   **D** Low pass   1000   ?   Cutoff frequency at 1 krad/s.

14–19   **D** High pass   2000   ?   20-dB gain at corner frequency.

14–20   **D** High pass   200   0.5   Use 20-kΩ resistors.

14–21   **D** High pass   1000   0.75   High-frequency gain of 40 dB.

14–22   **D** Bandpass   500   ? $Q = 2$.

14–23   **D** Bandpass   2000   ?   Bandwidth of 300 rad/s.

14–24   **D**   A second-order bandpass filter is required with cutoff frequencies at 97 kHz and 103 kHz. What are the values of ω_0, Q, and ζ? Design an active RC circuit that meets the design requirements.

14–25   **D** Design an active RC bandpass filter with a center frequency at 5 kHz and a bandwith of 2 kHz. For your design, what is the minimum amplitude of a sinusoidal input at the center frequency that will saturate the OP AMP? Assume that $V_{CC} = 15$ V.

14–26   Show that the Sallen-Key high-pass circuit in Figure 14–16 has the transfer function given in Eq. (14–7).

14–27   Show that the active RC bandpass circuit in Figure 14–18 has the transfer function given in Eq. (14–12).

14–28   The Sallen-Key low-pass circuit in Figure 14–14 can be designed using the conditions $R_A = R_B = R_1 = R_2 = R$. Develop a method of selecting R, C_1, and C_2 to achieve specified values of ω_0 and ζ.

14–29   Figure P14–29 shows an active RC low-pass circuit with two voltage followers. Show that the circuit has a transfer function

$$T(s) = \frac{V_2(s)}{V_1(s)} = \frac{1}{RC_1RC_2s^2 + RC_2s + 1}$$

Develop a method of selecting R, C_1, and C_2 to achieve specified values of ω_0 and ζ.

FIGURE   P14–29

14–30   Figure P14–30 shows an active RC high-pass circuit with two voltage followers. Show that the circuit has a transfer function

$$T(s) = \frac{V_2(s)}{V_1(s)} = \frac{RC_1RC_2s^2}{RC_1RC_2s^2 + RC_1s + 1}$$

Develop a method of selecting R, C_1, and C_2 to achieve specified values of ω_0 and ζ.

ERO 14-3 LOW-PASS FILTER DESIGN (SECT. 14-4)

Given a low-pass filter specification,

(a) Construct a transfer function $T(s)$ that meets the specification using first-order cascade, Butterworth, or Chebychev poles.
(b) Design a cascade of first- and second-order circuits that realizes $T(s)$.

14-31 Construct a transfer function with a first-order cascade low-pass gain response for $n = 5$, $\omega_C = 400$ rad/s, and a gain of 7 dB at its cutoff frequency. Find the gain (in dB) at $2\omega_C$, $5\omega_C$, and $10\omega_C$. Plot the straight-line gain response on the range $0.1\omega_C \leq \omega \leq 10\omega_C$ and sketch the actual response.

14-32 Construct a transfer function with a third-order Butterworth low-pass response with $\omega_C = 2500$ rad/s and a passband gain of 10 dB. Find the gain (in dB) at $2\omega_C$, $5\omega_C$, and $10\omega_C$. Plot the straight-line gain response on the range $0.1\omega_C \leq \omega \leq 10\omega_C$ and sketch the actual response.

14-33 Construct a transfer function with a fourth-order Chebychev low-pass response with $\omega_C = 500$ rad/s and a passband gain of 0 dB. Find the gain (in dB) at $2\omega_C$, $5\omega_C$, and $10\omega_C$. Plot the straight-line gain response on the range $0.1\omega_C \leq \omega \leq 10\omega_C$ and sketch the actual response.

14-34 **D** A low-pass filter specification requires $\omega_C = 2$ krad/s, a passband gain of 0 dB, and a stopband gain less than -30 dB at 10 krad/s.
(a) Find the lowest-order (n) of a first-order cascade response that meets these requirements.
(b) Repeat (a) using a Butterworth response and a Chebychev response.

14-35 **D** Design a fourth-order low-pass filter with $\omega_C = 500$ rad/s and a passband gain of 0 dB using first-order cascade poles.

14-36 **D** Design a third-order low-pass filter with $\omega_C = 300$ rad/s and a passband gain of 10 dB using Butterworth poles.

14-37 **D** Design a third-order low-pass filter with $\omega_C = 5$ krad/s and a passband gain of 5 using Chebychev poles.

14-38 **D** A low-pass filter specification requires $\omega_C = 300$ rad/s, a passband gain of 20 dB, and a stopband gain less than -15 dB at 2 krad/s.
(a) Construct a transfer function $T(s)$ that meets the specification using Butterworth poles.
(b) Design an active RC circuit that realizes the transfer function in (a) using only 100-nF capacitors.

14-39 **D** Design a low-pass filter with 0-dB passband gain, a cutoff frequency of 2.2 kHz, and stopband gains less than -20 dB at 7.4 kHz and -30 dB at 16 kHz. Calculate the gains realized by your design at 2.2 kHz, 7.4 kHz, and 16 kHz.

⬚ 14–40 **D** A pesky signal at 200 kHz is interfering with a desired signal at 25 kHz. A careful analysis suggests that reducing the interfering signal by 30 dB will eliminate the problem, provided the desired signal is not reduced by more than 3 dB. Design an active *RC* filter that meets these requirements.

ERO 14–4 HIGH-PASS, BANDPASS, AND BANDSTOP FILTER DESIGN (SECTS. 14–5, 14–6)

Given a high-pass, broadband bandpass, or bandstop filter specification,

(a) Construct a transfer function $T(s)$ that meets the specification using first-order cascade, Butterworth, or Chebychev poles.
(b) Design a cascade or parallel connection of first- and second-order circuits that realizes $T(s)$.

14–41 **D** A high-pass filter specification requires $\omega_C = 20$ krad/s, a passband gain of 20 dB, and a stopband gain less than –20 dB at 5 krad/s.
   (a) Construct a high-pass transfer function $T(s)$ that meets the specification using first-order cascade poles.
   (b) Repeat (a) using Butterworth poles.
   (c) Which of these responses would you use and why?

14–42 **D** A high-pass filter specification requires $\omega_C = 25$ krad/s, a passband gain of 0 dB, a stopband gain less than –20 dB at 10 krad/s, and a stopband gain less than –60 dB at 5 krad/s.
   (a) Construct a high-pass transfer function $T(s)$ that meets the specification using Chebychev poles.
   (b) Calculate the actual stopband gain of $T(s)$ at $\omega = 5$ krad/s and 10 krad/s. Which stopband requirement determines the number of poles in $T(s)$?

14–43 **D** A high-pass filter specification requires $\omega_C = 8$ krad/s, a passband gain of 10 dB, and a stopband gain less than –10 dB at 1 krad/s.
   (a) Construct a transfer function $T(s)$ that meets the specification using first-order cascade poles.
   (b) Design an active *RC* circuit that realizes the transfer function in (a) using only 100-kΩ resistors.

14–44 **D** Design a third-order high-pass filter with $\omega_C = 200$ rad/s and a passband gain of 14 dB using Butterworth poles.

14–45 **D** Design a third-order high-pass filter with $\omega_C = 300$ rad/s and a passband gain of 15 dB using Chebychev poles.

⬚ 14–46 **D** A bandpass filter specification requires cutoff frequencies at 100 rad/s and 2 krad/s, a passband gain of 0 dB, and stopband gains less than –20 dB at 25 rad/s and 8 krad/s.
   (a) Construct a bandpass transfer function $T(s)$ that meets the specification using Butterworth poles.
   (b) Repeat (a) using Chebychev poles.
   (c) Calculate the stopband gain at 25 rad/s and 8 krad/s for the transfer functions found in (a) and (b).
   (d) Which one of these responses would you choose and why?

⬚ 14–47 **D** A bandstop filter specification requires cutoff frequencies at 3 krad/s and 40 krad/s, passband gains of 0 dB, a stopband gain less than –15 dB at 10 krad/s.

(a) Construct a transfer function $T(s)$ that meets the specification.

(b) Calculate the actual stopband gain of $T(s)$ at $\omega = 3$, 10, and 40 krad/s.

14–48 **D** A bandpass filter specification requires cutoff frequencies at 100 rad/s and 3 krad/s, a passband gain of 0 dB, and stopband gains less than –40 dB at 5 rad/s and at 60 krad/s.

(a) Construct a transfer function $T(s)$ that meets the specification.

(b) Calculate the actual stopband gain of $T(s)$ at $\omega = 5$ rad/s and 60 krad/s.

14–49 **D** Design a Butterworth bandpass filter that passes sinusoidal tones at 1500 Hz and 3300 Hz with a gain of at least –3 dB and has gains no greater than –40 dB for sinusoidal tones at 60 Hz and 40 kHz.

14–50 **D** Design a bandpass filter that has cutoff frequencies at 100 rad/s and 2000 rad/s and has stopband gains at least 30 dB below the passband gain at 10 rad/s and 20 krad/s.

INTEGRATING PROBLEMS

14–51 **E** UNIVERSAL FILTER CIRCUIT

The circuit in Figure P14–51 consists of two OP AMP integrators and an inverting OP AMP summer. In this circuit the capacitors all have the same value C and the resistors all have the same value R, except the resistor labelled R_X. It is claimed that this circuit produces second-order low-pass, high-pass, and bandpass responses. Specifically it is claimed that $T_1(s) = V_1(s)/V_S(s)$ is high pass, $T_2(s) = V_2(s)/V_S(s)$ is bandpass, and $T_3(s) = V_3(s)/V_S(s)$ is low pass.

(a) Prove or disprove this claim.

(b) If the claim is true, develop a method of selecting R, C, and R_X to achieve specified values of ω_0 and ζ.

(c) If the claim is false, suggest a simple modification that would make it true.

FIGURE P14-51

14–52 **D** MODIFYING AN EXISTING DESIGN

An existing digital data channel uses a second-order Butterworth low-pass filter with a cutoff frequency of 10 kHz. Field experience reveals that noise is causing the equipment performance to fall below advertised levels. Analysis by the engineering department suggests that decreasing the filter gain by 20 dB at 30 kHz will solve the problem, provided the cutoff frequency does not change. The manufacturing department reports that adding anything more than one second-order stage will cause a major redesign and an unacceptable slip in scheduled deliveries. Design a second-order low-pass filter to connect in cascade with the existing second-order Butterworth filter. The frequency response of the circuit consisting of your second-order circuit in cascade with the original second-order Butterworth circuit must have a cutoff frequency of 10 kHz and reduce the gain at 30 kHz by at least 20 dB.

14–53 A E SECOND-ORDER CASCADE RESPONSE

The transfer function of a second-order Butterworth low-pass filter is

$$T_2(s) = \frac{K}{\left(\dfrac{s}{\omega_0}\right)^2 + \sqrt{2}\left(\dfrac{s}{\omega_0}\right) + 1}$$

where ω_0 is the cutoff frequency of the second-order response. A second-order cascade response is obtained by a cascade connection of n stages each with the transfer function given above. The overall transfer function of the n-stage cascade is $T_2(s) \times T_2(s) \times \ldots T_2(s) = [T_2(s)]^n$.

(a) Derive an expression relating the cutoff frequency (ω_C) of an n-stage second-order cascade to the cutoff frequency (ω_0) of the individual stages. *Hint:* Review the derivation of the first-order cascade response.

(b) Derive an expression relating the corner frequency (ω_{CORNER}) to the cutoff frequency (ω_C) of an n-stage second-order cascade.

(c) Plot the normalized gain response of a two-stage second-order cascade response on the range $0.1\omega_C \leq \omega \leq 10\omega_C$. Plot the normalized gain response of a first-order cascade response with the same number of poles. Which response produces the most stopband attenuation and why?

(d) Discuss the advantages of the second-order cascade over other low-pass responses such as Butterworth or Chebychev.

14–54 A FILTER INSERTION LOSS

Filter characteristics are sometimes specified in terms of insertion loss, defined as the ratio of average power delivered to a load when the filter is not present over the average power with the filter present.

(a) Circuit C1 in Figure P14–54 shows a sinusoidal source connected directly to a resistive load. Use phasor circuit analysis to derive an expression for the average power delivered to the load in the sinusoidal steady state in terms of V_A, R_0.

(b) Circuit C2 shows the same source and load with an inductor inserted between them. With the inductor inserted use phasor circuit analysis to derive an expression for the average power delivered to the load in the sinusoidal steady state.

(c) Insertion loss is the ratio of power found in (a) and the power found in (b) above. Derive an expression for the insertion loss of the inductor in terms of R_0, L, and ω.

(d) Show that the insertion loss is unity at zero frequency and explain this result in terms of the inductor impedance. Show that the insertion loss is infinite at infinite frequency and explain this result physically. Find the frequency at which the insertion loss is +3 dB.

14–55   **A D E** INVERSE SYSTEM

During a missile flight test a telemetry system monitors a quantity $x(t)$ and records an output signal $y(t)$. The transfer function relating $x(t)$ and $y(t)$ has been calibrated and shown to have a bandpass characteristic of the form

$$\frac{Y(s)}{X(s)} = \frac{Ks}{s^2 + 2\zeta\omega_0 s + \omega_0^2}$$

where K, ζ, and ω_0 are known parameters. During post-test analysis it is necessary to estimate the quantity $x(t)$ from the recorded output $y(t)$.

(a) **A** For the given transfer function show that $x(t)$ can be recovered from $y(t)$ as follows:

$$x(t) = \frac{1}{K}\frac{dy(t)}{dt} + \frac{2\zeta\omega_0}{K}y(t) + \frac{\omega_0^2}{K}\int_0^t y(x)dx$$

(b) **D** Design an RC OP AMP circuit that implements this relationship.

(c) **E** Suppose that small imperfections cause the signal recorded by the telemetry system to be the desired signal plus a small dc offset. That is, the recorded signal is $y(t) + V_{dc}$, where V_{dc} is a small (but unknown and unpredictable) dc value. Describe the way this small offset will affect the output signal produced by the circuit you designed in (b) above.

14–56   **A D E** THIRD-ORDER BUTTERWORTH CIRCUIT

(a) **A** Show that the circuit in Figure P14–56 produces a third-order Butterworth low-pass filter with a cutoff frequency of $\omega_C = 1/RC$ and a passband gain of $K = 4$.

(b) **D** Use the Butterworth design method in Section 14–4 to produce a two-stage circuit with the same third-order low-pass filter response.

(c) **E** Rank order the two circuits using each of the following criteria: (1) smallest number of components required, (2) smallest OP AMP gain bandwidth required, (3) smallest number of different component values required, and (4) ease of fine-tuning the cutoff frequency in production.

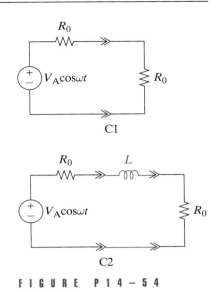

FIGURE   P 1 4 – 5 4

FIGURE   P 1 4 – 5 6

CHAPTER 15

MUTUAL INDUCTANCE

The results which I had by this time obtained with magnets led me to believe that the battery current through one wire, did, in reality, induce a similar current through the other wire, but that it continued for an instance only . . .

Michael Faraday, 1831
British Physicist

15–1 Coupled Inductors

15–2 The Dot Convention

15–3 Energy Analysis

15–4 The Ideal Transformer

15–5 Transformers in the Sinusoidal Steady State

15–6 Transformer Equivalent Circuits

Summary

Problems

Integrating Problems

Ampere's discovery that a force existed between two adjacent current-carrying wires led Micheal Faraday to investigate whether the current in one wire would affect the current in the other. He wound two separate insulted wires around an iron ring. One winding was connected to a battery by a switch and the other to a galvanometer. He observed that current in one wire did indeed induce a current in the other, but only when the switch was opened or closed. He correctly deduced that it was the change in the magnetic field that created the inductive effect, and he thereby discovered a basic principle that has come to be known as Faraday's law.

Faraday's iron ring with its two separate windings was actually a crude forerunner of an electrical device called a transformer. The development of practical transformers in 1882 by the British engineers Lucien Gauland and Josiah Gibbs gave the ac electrical power system one of its major advantages. Thomas Edison had pioneered the development of electrical power generation and distribution with his 110-V dc system. The low system voltage severely limited the distances over which dc electrical power could be transferred without encountering unacceptable losses. The transformer allows ac voltages to be increased to high levels for improved transmission efficiency and then reduced to safer levels at the load. Early ac systems used 1000-V lines while current-day systems use much higher levels. The development of the transformer was one of the key technology advances that lead to widespread electrical power systems of today.

The first section in this chapter develops the *i–v* relationships for a pair of windings called coupled inductors. The central feature of these relationships is that the time rate of change of current in one winding induces a voltage in the other, and vice versa. Because it is necessary to keep track of the polarity of these induced voltages, the second section introduces the dot convention—a special set of reference marks associated with coupled inductors. Energy calculations are used in the third section to define a parameter that indicates the degree of coupling between windings. The concept of perfect coupling is then used in the fourth section to develop an ideal model of the transformer device. The final two sections deal with phasor analysis of transformer circuits operating in the sinusoidal steady state.

15-1   COUPLED INDUCTORS

The *i–v* characteristic of the inductor results from the magnetic field produced by current through a coil of wire. A constant current produces a constant magnetic field that forms closed loops of magnetic flux lines in the vicinity of the inductor. A changing current causes these closed loops to expand or contract, thereby cutting the turns in the winding that makes up the inductor. Faraday's law states that voltage across the inductor is equal to the time rate of change of the total flux linkage. In earlier chapters we expressed this relationship between a time-varying current and an induced voltage in terms of an circuit parameter called inductance L.

Now suppose that a second inductor is brought close to the first so that the flux from the first inductor links with the turns of the second inductor. If the current in the first inductor is changing, then this flux linkage will generate a voltage in the second inductor. The magnetic coupling between the changing current in one inductor and the voltage generated in a second inductor produces **mutual inductance**.

I–V CHARACTERISTICS

The *i–v* characteristics of coupled inductors unavoidably involve describing the effects observed in one inductor due to causes occurring in the other. We will use a double subscript notation because it clearly identifies the various cause-and-effect relationships. The first subscript indicates the inductor in which the effect takes place, and the second identifies the inductor in which the cause occurs. For example, $v_{12}(t)$ is the voltage across inductor no. 1 due to causes occurring in inductor no. 2, whereas $v_{11}(t)$ is the voltage across inductor no. 1 due to causes occurring in inductor no. 1 itself.

We begin by assuming that the inductors are far apart as shown in Figure 15–1(a). Under these circumstances there is no magnetic coupling between the two. A current $i_1(t)$ passes through the N_1 turns of the first inductor and $i_2(t)$ through N_2 turns in the second. Each inductor produces a flux

$$\text{Inductor 1} \qquad \text{Inductor 2}$$
$$\phi_1(t) = k_1 N_1 i_1(t) \qquad \phi_2(t) = k_2 N_2 i_2(t) \qquad (15\text{–}1)$$

where k_1 and k_2 are proportionality constants. The flux linkage in each inductor is proportional to the number of turns:

$$\text{Inductor 1} \qquad \text{Inductor 2}$$
$$\lambda_{11}(t) = N_1 \phi_1(t) \qquad \lambda_{22}(t) = N_2 \phi_2(t) \qquad (15\text{–}2)$$

By Faraday's law the voltage across an inductor is equal to the time rate of change of the flux linkage. Using Eqs. (15–1) and (15–2) together with relationship between voltage and time rate of change of flux linkage gives

$$\text{Inductor 1:} \quad v_{11}(t) = \frac{d\lambda_{11}(t)}{dt} = N_1 \frac{d\phi_1(t)}{dt} = [k_1 N_1^2]\frac{di_1(t)}{dt}$$
$$\text{Inductor 2:} \quad v_{22}(t) = \frac{d\lambda_{22}(t)}{dt} = N_2 \frac{d\phi_2(t)}{dt} = [k_2 N_2^2]\frac{di_2(t)}{dt} \qquad (15\text{–}3)$$

Equations (15–3) provide the *i–v* relationships for the inductors when there is no mutual coupling. These results are the same as previously found in Chapter 6.

Now suppose the inductors are brought close together so that part of the flux produced by each inductor intercepts the other, as indicated in Figure 15–1(b). That is, part (but not necessarily all) of the fluxes $\phi_1(t)$ and $\phi_2(t)$ in Eq. (15–1) intercept the opposite inductor. We describe the cross coupling using the double subscript notation:

$$\text{Inductor 1} \qquad \text{Inductor 2}$$
$$\phi_{12}(t) = k_{12} N_2 i_2(t) \qquad \phi_{21}(t) = k_{21} N_1 i_1(t) \qquad (15\text{–}4)$$

The quantity $\phi_{12}(t)$ is the flux intercepting inductor 1 due to the current in inductor 2, and $\phi_{21}(t)$ is the flux intercepting inductor 2 due to the current in inductor 1. The total flux linkage in each inductor is proportional to the number of turns:

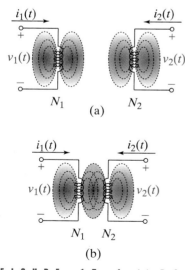

FIGURE 15–1 *(a) Inductors separated, only self-inductance present. (b) Inductors coupled, both self- and mutual inductance present.*

Inductor 1   Inductor 2
$$\lambda_{12}(t) = N_1\phi_{12}(t) \qquad \lambda_{21}(t) = N_2\phi_{21}(t)$$ (15–5)

By Faraday's law, the voltage across a winding is equal to the time rate of change of the flux linkage. Using Eqs. (15–4) and (15–5) together with derivative relationship, the time rate of change of flux linkages and voltages gives

Inductor 1: $v_{12}(t) = \dfrac{d\lambda_{12}(t)}{dt} = N_1\dfrac{d\phi_{12}(t)}{dt} = [k_{12}N_1N_2]\dfrac{di_2(t)}{dt}$

(15–6)

Inductor 2: $v_{21}(t) = \dfrac{d\lambda_{21}(t)}{dt} = N_2\dfrac{d\phi_{21}(t)}{dt} = [k_{21}N_1N_2]\dfrac{di_1(t)}{dt}$

The expressions in Eq. (15–6) are the *i–v* relationships describing the cross coupling between inductors when there is mutual coupling.

When the magnetic medium supporting the fluxes is linear, the superposition principle applies, and the total voltage across the inductors is the sum of the results in Eqs. (15–3) and (15–6):

Inductor 1: $v_1(t) = v_{11}(t) + v_{12}(t)$

$$= [k_1N_1^2]\dfrac{di_1(t)}{dt} + [k_{12}N_1N_2]\dfrac{di_2(t)}{dt}$$

Inductor 2: $v_2(t) = v_{21}(t) + v_{22}(t)$ (15–7)

$$= [k_{21}N_1N_2]\dfrac{di_1(t)}{dt} + [k_2N_2^2]\dfrac{di_2(t)}{dt}$$

We can identify four inductance parameters in these equations:

$$L_1 = k_1N_1^2 \qquad L_2 = k_2N_2^2$$ (15–8)

and

$$M_{12} = k_{12}N_1N_2 \qquad M_{21} = k_{21}N_2N_1$$ (15–9)

The two inductance parameters in Eq. (15–8) are the **self-inductance** of the inductors. The two parameters in Eq. (15–9) are the **mutual inductances** between the two inductors. In a linear magnetic medium, $k_{12} = k_{21} = k_M$. As a result, we can define a single mutual inductance parameter M as

$$M = M_{12} = M_{21} = k_M N_1 N_2$$ (15–10)

Using the definitions in Eqs. (15–8) and (15–10), the *i–v* characteristics of two coupled inductors are

Inductor 1: $v_1(t) = L_1\dfrac{di_1(t)}{dt} + M\dfrac{di_2(t)}{dt}$

(15–11)

Inductor 2: $v_2(t) = M\dfrac{di_1(t)}{dt} + L_2\dfrac{di_2(t)}{dt}$

Coupled inductors involve three inductance parameters, the two self-inductances L_1 and L_2 and the mutual inductance M.

The preceding development assumes that the cross coupling is additive. Additive coupling means that a positive rate of change of current in induc-

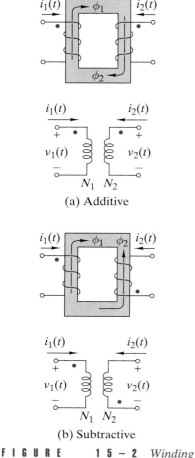

(a) Additive

(b) Subtractive

FIGURE 15-2 *Winding orientations and corresponding reference dots. (a) Additive. (b) Subtractive.*

tor 2 induces a positive voltage in inductor 1, and vice versa. The additive assumption produces the positive sign on the mutual inductance terms in Eq. (15–11). Unhappily it is possible for a positive rate of change of current in one inductor to induce a negative voltage in the other. To account for additive and subtractive coupling, the general form of the coupled inductor *i–v* characteristics includes a ± sign on the mutual inductance terms:

$$\text{Inductor 1:} \quad v_1(t) = L_1 \frac{di_1(t)}{dt} \pm M \frac{di_2(t)}{dt}$$

$$\text{Inductor 2:} \quad v_2(t) = \pm M \frac{di_1(t)}{dt} + L_2 \frac{di_2(t)}{dt}$$

(15–12)

When applying these element equations, it is necessary to know when to use a plus sign and when to use a minus sign.

15–2 THE DOT CONVENTION

The parameter *M* is positive, so the question is, what sign should be placed in front of this positive parameter in the *i–v* relationships in Eq. (15–12)? The correct sign depends on two things: (1) the spatial orientation of the two windings, and (2) the reference marks given to the currents and voltages.

Figure 15–2 shows the additive and subtractive spatial orientation of two coupled inductors. In either case, the direction of the flux produced by a current is found using the right-hand rule treated in physics courses. In the additive case, currents i_1 and i_2 both produce clockwise fluxes ϕ_1 and ϕ_2. In the subtractive case, the currents produce opposing fluxes because ϕ_1 is clockwise and ϕ_2 is counterclockwise. The sign for the mutual inductance term is positive for the additive orientation and negative for the subtractive case.

In general, it is awkward to show the spatial features of the windings in circuit diagrams. The dots shown near one terminal of each winding in Figure 15–2 are special reference marks indicating the relative orientation of the windings. The reference directions for the currents and voltages are arbitrary. They can be changed as long as we follow the passive sign convention. However, the dots indicate physical attributes of the windings that make up the coupled inductors. They are not arbitrary. They cannot be changed.

The correct sign for the mutual inductance term hinges on how the reference marks for currents and voltages are assigned relative to the dots. For a given winding orientation, Figure 15–3 shows all four possible current and voltage reference assignments under the passive sign convention. In cases A and B the fluxes are additive, so the mutual inductance term is positive. In cases C and D the fluxes are subtractive and the mutual inductance term is negative. From these results we derive the following rule:

> *Mutual inductance is additive when both current reference directions point toward or both point away from dotted terminals; otherwise, it is subtractive.*

Because we always use the passive sign convention, the rule can be stated in terms of voltages as follows:

Mutual inductance is additive when the voltage reference marks are both positive or both negative at the dotted terminals; otherwise, it is subtractive.

Because the current reference directions or voltage polarity marks can be changed, a corollary of this rule is that we can always assign reference directions so that the positive sign applies to the mutual inductance. This corollary is important because a positive sign is built into the mutual inductance models in circuit analysis programs like SPICE.

It may seem that all of this discussion about signs and dots is much ado about nothing. Not so. First, selecting the wrong sign can have nontrivial consequences because the polarity of the output signal is reversed. If the signal is a command to your car's autopilot, then you really need to know whether stepping on the brake pedal will slow the car down or speed it up. Another problem is that the two coupled inductors may appear in different parts of a circuit diagram and may be assigned voltage and current reference marks for other reasons. In such circumstances it is important to understand the underlying principle to select the correct sign for the mutual inductance term.

The following examples and exercises illustrate selecting the correct sign and applying the *i–v* characteristics in Eq. (15–12)

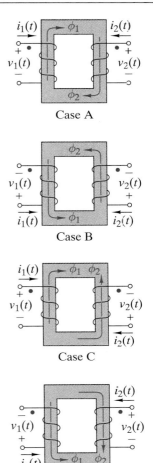

F I G U R E   1 5 – 3   *All possible current and voltage reference marks for a fixed winding orientation.*

EXAMPLE 15–1

The source voltage in Figure 15–4 is $v_S(t) = 10 \cos 100t$ V. Find the output voltage $v_2(t)$.

SOLUTION:

The sign of the mutual inductance term is positive because both current reference directions are toward the dots. Since the load connected across inductor 2 is an open circuit, $i_2(t) = 0$ and the *i–v* equations of the coupled inductors reduce to

$$\text{Inductor 1:} \quad 10 \cos 100t = 0.01 \frac{di_1(t)}{dt} + 0$$

$$\text{Inductor 2:} \quad v_2(t) = 0.002 \frac{di_1(t)}{dt} + 0$$

Solving the first equation for di_1/dt yields

$$\frac{di_1(t)}{dt} = 1000 \cos 100t$$

Substituting this equation for $i_1(t)$ into the second equation yields

$$v_2(t) = 2 \cos 100t \text{ (correct) V}$$

If we had incorrectly chosen a minus sign, the mutual inductance would have produced a voltage of

$$v_2(t) = -2 \cos 100t \text{ (incorrect)}$$

which differs from the correct answer by a signal inversion.   ∎

$L_1 = L_2 = 10$ mH
$M = 2$ mH

F I G U R E   1 5 – 4

FIGURE 15 – 5

EXAMPLE 15–2

Solve for $v_x(t)$ in terms of $i_1(t)$ for the coupled inductors in Figure 15–5.

SOLUTION:

In this case the signs of mutual inductance terms are negative because the reference direction for $i_1(t)$ points toward the dot for inductor 1 and the reference direction for $i_2(t)$ points away from the dot for inductor 2. The coupled inductor i–v equations are

$$\text{Inductor 1:} \quad v_1(t) = L_1 \frac{di_1(t)}{dt} - M \frac{di_2(t)}{dt}$$

$$\text{Inductor 2:} \quad v_2(t) = -M \frac{di_1(t)}{dt} + L_2 \frac{di_2(t)}{dt}$$

A KCL constraint at node A requires that $i_1(t) = -i_2(t)$. Therefore, these i–v equations can be written in the form

$$\text{Inductor 1:} \quad v_1(t) = L_1 \frac{di_1(t)}{dt} - M \frac{d[-i_1(t)]}{dt}$$

$$\text{Inductor 2:} \quad v_2(t) = -M \frac{di_1(t)}{dt} + L_2 \frac{d[-i_1(t)]}{dt}$$

A KVL constraint around the loop requires that $v_x(t) = v_1(t) - v_2(t)$. Subtracting the second equation from the first yields

$$v_x(t) = L_1 \frac{di_1(t)}{dt} + M \frac{di_1(t)}{dt} + M \frac{di_1(t)}{dt} + L_2 \frac{di_1(t)}{dt}$$

$$= (L_1 + L_2 + 2M) \frac{di_1(t)}{dt}$$

In this case the two mutual inductance terms add to produce an equivalent inductance of

$$L_{EQ} = L_1 + L_2 + 2M.$$

If the plus sign in the coupled inductor i–v relationships was appropriate, the mutual inductance terms would subtract to produce an equivalent inductance of $L_{EQ} = L_1 + L_2 - 2M$. Clearly it is important to have the right sign in the i–v relationships. ∎

Exercise 15–1

Find $v_1(t)$ and $v_2(t)$ for the circuit in Figure 15–6.

FIGURE 15 – 6

$$L_1 = 0.2 \text{ mH}, \ L_2 = 0.5 \text{ mH}$$
$$M = 0.3 \text{ mH}$$

Answers:
$v_1(t) = -10 \sin 10^4 t - 3 \cos 5 \times 10^3 t$ V
$v_2(t) = -15 \sin 10^4 t - 5 \cos 5 \times 10^3 t$ V

15-3   ENERGY ANALYSIS

Calculating the total energy stored in a pair of coupled inductors reveals a fundamental limitation on allowable values of the self- and mutual inductances. To uncover this limitation, we first calculate the total power absorbed. Multiplying the first equation in Eq. (15–12) by $i_1(t)$ and the second equation by $i_2(t)$ produces

$$p_1(t) = v_1(t)i_1(t) = L_1 i_1(t) \frac{di_1(t)}{dt} \pm M i_1(t) \frac{di_2(t)}{dt}$$

$$p_2(t) = v_2(t)i_2(t) = \pm M i_2(t) \frac{di_1(t)}{dt} + L_2 i_2(t) \frac{di_2(t)}{dt}$$

(15–13)

The quantities $p_1(t)$ and $p_2(t)$ are the powers absorbed with inductors 1 and 2. The total power is the sum of the individual inductor powers:

$$p(t) = p_1(t) + p_2(t) \tag{15–14}$$

$$= L_1 \left[i_1(t) \frac{di_1(t)}{dt} \right] \pm M \left[i_1(t) \frac{di_2(t)}{dt} + i_2(t) \frac{di_1(t)}{dt} \right] + L_2 \left[i_2(t) \frac{di_2(t)}{dt} \right]$$

Each of the bracketed terms in Eq. (15–14) is a perfect derivative. Specifically,

$$i_1(t) \frac{di_1(t)}{dt} = \frac{1}{2} \frac{di_1^2(t)}{dt}$$

$$i_2(t) \frac{di_2(t)}{dt} = \frac{1}{2} \frac{di_2^2(t)}{dt} \tag{15–15}$$

$$i_1(t) \frac{di_2(t)}{dt} + i_2(t) \frac{di_1(t)}{dt} = \frac{di_1(t)i_2(t)}{dt}$$

Therefore, the total power in Eq. (15–14) is

$$p(t) = \frac{d}{dt} \left[\frac{1}{2} L_1 i_1^2(t) \pm M\, i_1(t)i_2(t) + \frac{1}{2} L_2\, i_2^2(t) \right] \tag{15–16}$$

Because power is the time rate of change of energy, the quantity inside the brackets in Eq. (15–16) is the total energy stored in the two inductors. That is,

$$w(t) = \frac{1}{2} L_1 i_1^2(t) \pm M\, i_1(t)i_2(t) + \frac{1}{2} L_2 i_2^2(t) \tag{15–17}$$

In Eq. (15–17) the self-inductance terms are always positive. However, the mutual-inductance term can be either positive or negative. At first glance it appears that the total energy could be negative. But the total energy must be positive: Otherwise the coupled inductors could deliver net energy to the rest of the circuit.

The condition $w(t) \geq 0$ places a constraint on the values of the self- and mutual inductances. First, if $i_2(t) = 0$, then $w(t) \geq 0$ in Eq. (15–17) requires $L_1 > 0$. Next, if $i_1(t) = 0$, then $w(t) \geq 0$ in Eq. (15–17) requires $L_2 > 0$. Finally, if $i_1(t) \neq 0$ and $i_2(t) \neq 0$, then we divide Eq. (15–17) by $[i_2(t)]^2$ and defined a variable $x = i_1/i_2$. With these changes, the energy constraint $w(t) > 0$ becomes

$$\frac{w(t)}{i_2^2(t)} = f(x) = \frac{1}{2} L_1 x^2 \pm Mx + \frac{1}{2} L_2 \geq 0 \qquad (15\text{–}18)$$

The minimum value of $f(x)$ occurs when

$$\frac{df(x)}{dx} = L_1 x \pm M = 0 \quad \text{hence} \quad x_{\min} = \mp \frac{M}{L_1} \qquad (15\text{–}19)$$

The value $x_{\min}$ yields the minimum of $f(x)$ because the second derivative of $f(x)$ is positive. Substituting $x_{\min}$ back into Eq. (15–18) yields the condition

$$f(x_{\min}) = \frac{1}{2} L_1 \frac{M^2}{L_1^2} - \frac{M^2}{L_1} + \frac{1}{2} L_2 = \frac{1}{2} \left[-\frac{M^2}{L_1} + L_2 \right] \geq 0 \qquad (15\text{–}20)$$

The constraint in Eq. (15–20) means that the stored energy in a pair of coupled inductors is positive if

$$L_1 L_2 \geq M^2 \qquad (15\text{–}21)$$

Energy considerations dictate that in any pair of coupled inductors, the product of the self-inductances must exceed the square of the mutual inductance.

The constraint in Eq. (15–21) is usually written in terms of a new parameter called the **coupling coefficient** k:

$$k = \frac{M}{\sqrt{L_1 L_2}} \leq 1 \qquad (15\text{–}22)$$

The parameter k ranges from 0 to 1. If $M = 0$ then $k = 0$ and the coupling between the inductors is zero. The condition $k = 1$ requires **perfect coupling** in which all of the flux produced by one inductor links the other. Perfect coupling is physically impossible, although careful design can produce coupling coefficients of 0.99 and higher.

The next section discusses a transformer model that assumes perfect coupling ($k = 1$). Computer-aided circuit analysis programs like SPICE specify the parameters of a pair of coupled inductors in terms of the self-inductances L_1 and L_2 and the coupling coefficient k.

Exercise 15–2

What is the coupling coefficient of the coupled inductors in Figure 15–6.

Answer:
$k = 0.949$

15–4 THE IDEAL TRANSFORMER

A **transformer** is an electrical device that utilizes magnetic coupling between two inductors. Transformers find application in virtually every type of electrical system, but especially in power supplies and commercial power grids. Some example devices from these applications are shown in Figure 15–7.

In Figure 15–8 the transformer is shown as an interface device between a source and a load. The winding connected to the source is called the **primary winding**, and the one connected to the load is called the **secondary winding**. In most applications the transformer is a coupling device that transfers signals (especially power) from the source to the load. The basic purpose of the device is to change voltage and current levels so that the conditions at the source and load are compatible.

Transformer design involves two primary goals: (1) to maximize the magnetic coupling between the two windings, and (2) to minimize the power loss in the windings. The first goal produces nearly perfect coupling ($k \approx 1$) so that almost all of the flux in one winding links the other. The second goal produces nearly zero power loss so that almost all of the power delivered to the primary winding transfers to the load. The **ideal transformer** is a circuit element in which coupled inductors are assumed to have perfect coupling and zero power loss. Using these two idealizations, we can derive the i–v characteristics of an ideal transformer.

PERFECT COUPLING

Perfect coupling means that all of the flux in the first winding links the second, and vice versa. Equation (15–1) defines the total flux in each winding as

$$
\begin{array}{cc}
\text{Winding 1} & \text{Winding 2} \\
\phi_1(t) = k_1 N_1 i_1(t) & \phi_2(t) = k_2 N_2 i_2(t)
\end{array}
\qquad (15\text{–}23)
$$

where k_1 and k_2 are proportionality constants. Equation (15–4) defines the cross coupling using the double subscript notation

$$
\begin{array}{cc}
\text{Winding 1} & \text{Winding 2} \\
\phi_{12}(t) = k_{12} N_2 i_2(t) & \phi_{21}(t) = k_{21} N_1 i_1(t)
\end{array}
\qquad (15\text{–}24)
$$

In this equation $\phi_{12}(t)$ is the flux intercepting winding 1 due to the current in winding 2, and $\phi_{21}(t)$ is the flux intercepting winding 2 due to the current in winding 1. Perfect coupling means that

$$
\phi_{21}(t) = \phi_1(t) \quad \text{and} \quad \phi_{12}(t) = \phi_2(t)
\qquad (15\text{–}25)
$$

Comparing Eqs. (15–23) and (15–24) shows that perfect coupling requires $k_1 = k_{21}$ and $k_2 = k_{12}$. But in a linear magnetic medium $k_{12} = k_{21} = k_M$, so perfect coupling implies

$$
k_1 = k_2 = k_{12} = k_{21} = k_M
\qquad (15\text{–}26)
$$

Substituting the perfect coupling conditions in Eq. (15–26) into the i–v characteristics in Eq. (15–7) gives

Iron core

Air core

Toroid

Powerline

FIGURE 15–7 *Examples of transformer devices.*

FIGURE 15–8 *Transformer connected at a source/load interface.*

$$v_1(t) = [k_M N_1^2] \frac{di_1(t)}{dt} \pm [k_M N_1 N_2] \frac{di_2(t)}{dt}$$

$$v_2(t) = \pm [k_M N_1 N_2] \frac{di_1(t)}{dt} + [k_M N_2^2] \frac{di_2(t)}{dt}$$

(15–27)

Factoring N_1 out of the first equation and $\pm N_2$ out of the second produces

$$v_1(t) = N_1 \left([k_M N_1] \frac{di_1(t)}{dt} \pm [k_M N_2] \frac{di_2(t)}{dt} \right)$$

$$v_2(t) = \pm N_2 \left([k_M N_1] \frac{di_1(t)}{dt} \pm [k_M N_2] \frac{di_2(t)}{dt} \right)$$

(15–28)

Dividing the second equation by the first shows that perfect coupling implies

$$\frac{v_2(t)}{v_1(t)} = \pm \frac{N_2}{N_1} = \pm n$$

(15–29)

where the parameter n is called the **turns ratio**.

With perfect coupling the secondary voltage is proportional to the primary voltage, so they have the same waveshape. For example, when the primary voltage is $v_1(t) = V_A \sin \omega t$, the secondary voltage is $v_2(t) = \pm n V_A \sin \omega t$. When the turns ratio $n > 1$, the secondary voltage amplitude is larger than the primary and the device is called a **step-up transformer**. Conversely, when $n < 1$, the secondary voltage is smaller than the primary and the device is called a **step-down transformer**. The ability to increase or decrease ac voltages is a basic feature of transformers. Commercial power systems use transmission voltages of several hundred kilovolts. For residential applications the transmission voltage is reduced to safer levels (typically 220/110 V_{rms}) using step-down transformers.

The $\pm$ sign in Eq. (15–29) reminds us that mutual inductance can be additive or subtractive. Selecting the correct sign is important because of signal inversion. The sign depends on the reference marks given the primary and secondary currents relative to the dots indicating the relative winding orientations. The rule for the ideal transformer is a corollary of the rule for selecting the sign of the mutual inductance term for coupled inductors.

> *The sign in Eq. (15–29) is positive when the reference directions for both currents point toward or both point away from a dotted terminal; otherwise it is negative.*

Exercise 15–3

The transformer in Figure 15–9 has perfect coupling and a turns ratio of $n = 0.1$. The input voltage is $v_S(t) = 120 \sin 377t$ V.

(a) What is the secondary voltage?
(b) What is the secondary current for a 50-Ω load?
(c) Is this a step-up or step-down transformer?

Answers:
(a) $v_2(t) = +12 \sin 377t$ V
(b) $i_2(t) = -0.24 \sin 377t$ A
(c) Step down

F I G U R E   1 5 – 9

ZERO POWER LOSS

The ideal transformer model also assumes that there is no power loss in the transformer. With the passive sign convention, the quantity $v_1(t)i_1(t)$ is the power in the primary winding and $v_2(t)i_2(t)$ is the power in the secondary winding. Zero power loss requires

$$v_1(t)i_1(t) + v_2(t)i_2(t) = 0 \qquad (15\text{–}30)$$

This equation states that whatever power enters the transformer on one winding immediately leaves via the other winding. This not only means that there is no energy lost in the ideal transformer, but also that there is no energy stored within the element. The zero-power loss constraint can be rearranged as

$$\frac{i_2(t)}{i_1(t)} = -\frac{v_1(t)}{v_2(t)} \qquad (15\text{–}31)$$

But under the perfect coupling assumption $v_2(t)/v_1(t) = \pm n$. With zero power loss and perfect coupling the primary and secondary currents are related as

$$\frac{i_2(t)}{i_1(t)} = \mp\frac{1}{n} \qquad (15\text{–}32)$$

The correct sign in this equation depends on the orientation of the current reference directions relative to the dots describing the transformer structure.

With both perfect coupling and zero power loss, the secondary current is inversely proportional to the turns ratio. A step-up transformer ($n > 1$) increases the voltage and decreases the current, which improves transmission line efficiency because the i^2R losses in the conductors are smaller.

i–v CHARACTERISTICS

Equations (15–29) and (15–32) define the i–v characteristics of the ideal transformer circuit element.

$$v_2(t) = \pm n\, v_1(t)$$
$$i_2(t) = \mp\frac{1}{n} i_1(t) \qquad (15\text{–}33)$$

where $n = N_2/N_1$ is the turns ratio. The correct sign in these equations depends on the assigned reference directions and transformer dots, as previously discussed.

Because the turns ratio is a key transformer parameter, it is worth relating it to the inductance parameters. Looking back at Eq. (15–27), we see that in a linear magnetic medium with perfect coupling the inductance parameters are $L_1 = k_M N_1^2$, $L_2 = k_M N_2^2$, and $M = k_M N_1 N_2$. As a result, the turns ratio can be expressed in terms of the inductance parameters as

$$n = \frac{N_2}{N_1} = \sqrt{\frac{L_2}{L_1}} = \frac{M}{L_1} = \frac{L_2}{M} \qquad (15\text{–}34)$$

Although actual transformer devices are not ideal, they do approach perfect coupling. Hence, these relations are useful approximations with transformers that are tightly coupled ($k \approx 1$).

Using the ideal transformer model requires some caution. The relationships in Eq. (15–33) state that the voltages are independent of the currents. For example, if the secondary winding is connected to an open circuit then $i_2(t) = 0$. The ideal transformer characteristics require that $i_1(t) = 0$ regardless of the value of $v_1(t)$. In effect this means that the inductance parameters are all infinite, but infinite in such a way that their ratios in Eq. (15–34) are still equal to the turns ratio. Equally important is the fact that the element equations appear to apply to all signals, including constant (dc) signals. In theory, a transformer with perfect coupling and zero power loss would pass dc signals. In practice, some transformers approach these conditions but no real transformer actually achieves this state of perfection. Remember that the element equations of the ideal transformer are an idealization of mutual coupling between inductors, and that time-varying signals are necessary; otherwise, the coupling terms $M di_1/dt$ and $M di_2/dt$ are zero.

EXAMPLE 15–3

The input to the primary of an ideal transformer with a turns ratio of $n = 10$ is $v_1(t) = 120 \sin 377t$ V. The load connected to the secondary winding is a 50-Ω resistor. Assuming additive coupling, find the secondary voltage, the secondary current, the secondary power, the primary current, and the primary power.

SOLUTION:
With additive coupling the secondary voltage is

$$v_2(t) = +nv_1(t) = 10 \times 120 \sin 377\,t = 1200 \sin 377t \ \text{V}$$

From Ohm's law the secondary current is

$$i_2(t) = -\frac{v_2(t)}{R_L} = -\frac{1200}{50} \sin 377\,t = -24 \sin 377t \ \text{A}$$

The minus sign in this expression comes about because the reference marks for $v_2(t)$ and $i_2(t)$ follow the passive sign convention with respect to the secondary winding and not the load resistance. The power in the secondary winding is

$$p_2(t) = v_2(t)i_2(t) = -28.8 \sin^2 377t \ \text{kW}$$

The minus sign here means that the secondary delivers power to the load resistor. With additive coupling the primary current is

$$i_1 = -ni_2(t) = -10 \times (-24 \sin 377\,t) = 240 \sin 377t \ \text{A}$$

and the primary power is

$$p_1(t) = v_1(t)i_1(t) = 28.8 \sin^2 377t \ \text{kW}$$

The plus sign here means that the primary winding is absorbing power from the source. Note that $p_1(t) + p_2(t) = 0$ as required by the zero power loss assumption. ∎

EQUIVALENT INPUT RESISTANCE

Because a transformer changes the voltage and current levels, it effectively changes the load resistance seen by a source in the primary circuit. To derive the equivalent input resistance, we write the device equations for the ideal transformer shown in Figure 15–10.

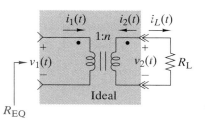

FIGURE   15 – 10 *Equivalent resistance seen on the primary winding.*

$$\text{Resistor:} \qquad v_2(t) = R_L i_L(t)$$

$$\text{Transformer:} \qquad v_2(t) = n\, v_1(t) \quad \text{and} \quad i_2(t) = -\frac{1}{n} i_1(t)$$

Dividing the first transformer equation by the second and inserting the load resistance constraint yields

$$\frac{v_2(t)}{i_2(t)} = \frac{i_L(t)R_L}{i_2(t)} = -n^2 \frac{v_1(t)}{i_1(t)}$$

Applying KCL at the output interface tells us that $i_L(t) = -i_2(t)$. Therefore, the equivalent resistance seen on the primary side is

$$R_{EQ} = \frac{v_1(t)}{i_1(t)} = \frac{1}{n^2} R_L \qquad\qquad (15\text{–}35)$$

The equivalent load resistance seen on the primary side depends on the turns ratio and the load resistance. Adjusting the turns ratio can make R_{EQ} equal to the source resistance. Transformer coupling can produce the conditions for maximum power transfer when the source and load resistances are not equal.

The derivation leading to Eq. (15–35) used the ideal transformer with the dot markings and reference directions in Figure 15–10. However, the final result does not depend on the location of the dot marks relative to voltage and current reference directions. In other words, Eq. (15–35) yields the input resistance for any ideal transformer with a turns ratio of n and a load R_L.

EXAMPLE 15–4

A stereo amplifier has an output resistance of 600 Ω. The input resistance of the speaker (the load) is 8 Ω. Select the turns ratio of a transformer to obtain maximum power transfer.

SOLUTION:

The maximum power transfer theorem in Chapter 3 states that the source and load resistance must be matched (equal) to achieve maximum power. Directly connecting the amplifier (600 Ω) to the speaker (8 Ω) produces a mismatch. If a transformer is inserted as shown in Figure 15–11, then the equivalent load resistance seen by the amplifier is

$$R_{EQ} = \frac{1}{n^2} R_L = \frac{1}{n^2} 8$$

To produce a resistance match we need $R_{EQ} = 600 = 8/n^2$ or a turns ratio of $n = 1/8.66$. ∎

F I G U R E   1 5 – 1 1

Exercise 15–4

The load resistor in Figure 15–10 is 50 Ω, and the number of turns on the primary and secondary windings are $N_1 = 540$ and $N_2 = 108$. Find the equivalent resistance on the primary side.

Answer:
$R_{EQ} = 1250 \ \Omega$

F I G U R E   1 5 – 1 2 *The autotransformer connection.*

APPLICATION NOTE: **E X A M P L E   1 5 – 5**

In a transformer the primary and secondary windings are magnetically coupled but are usually electrically isolated. Transformer performance in some applications can be improved by electrically connecting the two magnetically coupled windings in a configuration called an **autotransformer**. Figure 15–12 shows a two-winding ideal transformer connected in an autotransformer step-up configuration.

The rating of the ideal transformer when connected in the usual electrically isolated two-winding configuration is $P_{\text{rated}} = v_1 i_1 = -v_2 i_2$. Connected in the autotransformer configuration in Figure 15–12, the power delivered to the load is

$$P_{\text{load}} = (v_1 + v_2)(-i_2)$$

For an ideal transformer $v_2 = nv_1$ and $i_2 = -i_1/n$, where $n = N_2/N_1$. Hence the power delivered to the load is

$$P_{\text{load}} = (v_1 + nv_1)(i_1/n) = \left(1 + \frac{1}{n}\right) v_1 i_1$$

$$= \left(1 + \frac{1}{n}\right) P_{\text{rated}}$$

The autotransformer configuration delivers more power to the load than the power rating of the two-winding transformer. Put differently, the autotransformer can supply a specified load power using a transformer with a lower power rating. Autotransformers are normally used when the turns ratio is less than 3:1, so this advantage can be significant. A disadvantage is that the electrical isolation provided by the usual transformer configuration is lost.

15–5 T R A N S F O R M E R S   I N   T H E   S I N U S O I D A L   S T E A D Y   S T A T E

By far the most common application of transformers occurs in electric power systems where they operate in the sinusoidal steady state. In this context we describe transformers in terms of phasors and impedances and deal with average power transfer. The ac analysis of a transformer begins with the time-domain element equations for a pair of coupled inductors

$$v_1(t) = L_1 \frac{di_1(t)}{dt} \pm M \frac{di_2(t)}{dt}$$

$$v_2(t) = \pm M \frac{di_1(t)}{dt} + L_2 \frac{di_2(t)}{dt}$$

Transforming these equations into the phasor domain involves replacing waveforms by phasors using the derivative property to obtain the phasors for di_1/dt and di_2/dt:

$$\mathbf{V}_1 = j\omega L_1 \mathbf{I}_1 \pm j\omega M \mathbf{I}_2$$
$$\mathbf{V}_2 = \pm j\omega M \mathbf{I}_1 + j\omega L_2 \mathbf{I}_2$$

(15–36)

where

1. $\mathbf{V}_1$ and $\mathbf{I}_1$ are the phasors representing ac voltage and current of the first winding, and $j\omega L_1$ is the impedance of the self-inductance of first winding.

2. $\mathbf{V}_2$ and $\mathbf{I}_2$ are the phasors representing ac voltage and current of the second winding, and $j\omega L_2$ is the impedance of the self-inductance of second winding.

3. $j\omega M$ is the impedance of the mutual inductance between the two windings.

The self-inductance impedances relate phasor voltage and current at the same pair of terminals, while the mutual inductance impedance relates the phasor voltage at one pair of terminals to the phasor current at the other pair. The phasor domain equations can also be written in terms of reactances as

$$\mathbf{V}_1 = jX_1 \mathbf{I}_1 \pm jX_M \mathbf{I}_2$$
$$\mathbf{V}_2 = \pm jX_M \mathbf{I}_1 + jX_2 \mathbf{I}_2$$

where the three reactances are $X_1 = \omega L_1$, $X_2 = \omega L_2$, and $X_M = \omega M$. The degree of coupling between windings is indicated by the coupling coefficient, which can be written in terms of reactances as

$$k = \frac{M}{\sqrt{L_1 L_2}} = \frac{X_M}{\sqrt{X_1 X_2}}$$

In either form, energy considerations dictate that $0 \le k \le 1$.

Figure 15–13 shows the phasor-domain version of transformer coupling between a source and a load. Following our previous notation, the winding connected to the source is called the **primary,** and the winding connected to the load is called the **secondary.** Although the transformer is a bilateral

FIGURE 15–13 *Phasor circuit model of the two-winding transformer.*

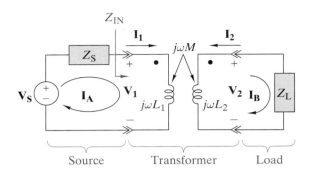

device, we normally think of signal and power transfer as passing from the primary to the secondary winding.

Our immediate objective is to write circuit equations for the transformer using the mesh currents $\mathbf{I}_A$ and $\mathbf{I}_B$ in Figure 15–13. Applying KVL around the primary circuit (mesh A) and secondary circuit (mesh B), we obtain the following equations:

$$
\begin{aligned}
\text{Mesh A:} &\quad Z_S\mathbf{I}_A + \mathbf{V}_1 = \mathbf{V}_S \\
\text{Mesh B:} &\quad -\mathbf{V}_2 + Z_L\mathbf{I}_B = 0
\end{aligned}
\tag{15–37}
$$

The reference directions for the inductor currents in Figure 15–13 are both directed in at dotted terminals, so the mutual inductance coupling is additive and the plus signs in Eq. (15–36) apply. Using KCL we see that the reference directions for the currents lead to the relations $\mathbf{I}_1 = \mathbf{I}_A$ and $\mathbf{I}_2 = -\mathbf{I}_B$. The **I–V** relationships of the coupled inductors in terms of the mesh currents are

$$
\begin{aligned}
\mathbf{V}_1 &= +j\omega L_1\mathbf{I}_A + j\omega M(-\mathbf{I}_B) \\
\mathbf{V}_2 &= +j\omega M\mathbf{I}_A + j\omega L_2(-\mathbf{I}_B)
\end{aligned}
\tag{15–38}
$$

Substituting the inductor voltages from Eq. (15–38) into the KVL equations in Eq. (15–37) yields

$$
\begin{aligned}
\text{Mesh A:} &\quad (Z_S + j\omega L_1)\mathbf{I}_A - j\omega M\mathbf{I}_B = \mathbf{V}_S \\
\text{Mesh B:} &\quad -j\omega M\mathbf{I}_A + (Z_L + j\omega L_2)\mathbf{I}_B = 0
\end{aligned}
\tag{15–39}
$$

This set of mesh equations provides a complete description of the circuit ac response. Once we solve for the mesh currents, we can calculate every phasor voltage and current using Kirchhoff's laws and element equations.

EXAMPLE 15–6

The source circuit in Figure 15–13 has $Z_S = 0 + j20\ \Omega$ and $\mathbf{V}_S = 2500\angle 0°$ V at $\omega = 377$ rad/s. The transformer has $L_1 = 2$ H, $L_2 = 0.2$ H, and $M = 0.6$ H. The load impedance is $Z_L = 25 + j15\ \Omega$. Find $\mathbf{I}_A$, $\mathbf{I}_B$, $\mathbf{V}_1$, $\mathbf{V}_2$, Z_{IN}, and the average power delivered by the source at input interface.

SOLUTION:
The transformer impedances are

$$
j\omega L_1 = j377 \times 2 = j754\ \Omega
$$

$$
j\omega L_2 = j377 \times 0.2 = j75.4\ \Omega
$$

$$
j\omega M = j377 \times 0.6 = j226\ \Omega
$$

Using these impedances in Eq. (15–39) yields the following mesh equations:

$$\text{Mesh A:} \quad (j20 + j754)\mathbf{I}_A - j226\mathbf{I}_B = 2500 \angle 0°$$
$$\text{Mesh B:} \quad -j226\mathbf{I}_A + (25 + j15 + j75.4)\mathbf{I}_B = 0$$

Solving these equations for the mesh currents yields

$$\mathbf{I}_A = 4.36 - j7.49 = 8.67 \angle -59.8° \quad \text{A}$$
$$\mathbf{I}_B = 15.0 - j14.6 = 20.9 \angle -44.2° \text{ A}$$

The winding voltages are found to be

$$\mathbf{V}_1 = \mathbf{V}_S - Z_S\mathbf{I}_A = 2350 - j87.2 = 2350 \angle -2.1° \text{ V}$$
$$\mathbf{V}_2 = Z_L\mathbf{I}_B = 594 - j140 = 610 \angle -13.2° \text{ V}$$

The input impedance seen by the source circuit is

$$Z_{IN} = \frac{\mathbf{V}_1}{\mathbf{I}_A} = R_{IN} + jX_{IN} = 145 + j229 \ \Omega$$

The average power delivered by the source at the input interface is

$$P_{IN} = \frac{|\mathbf{I}_A|^2}{2} R_{IN} = \frac{(8.67)^2}{2} 145 = 5.45 \text{ kW}$$

Exercise 15–5

Repeat Example 15–6 when the dot on the secondary winding in Figure 15–13 is moved to the bottom terminal and all other reference marks stay the same.

Answers:
$\mathbf{I}_A = 4.36 - j7.49$ A; $\mathbf{I}_B = -15.0 + j14.6$ A
$\mathbf{V}_1 = 2350 - j87.2$ V; $\mathbf{V}_2 = -594 + j140$ V; $Z_{IN} = 145 + j\,229\ \Omega$; $P_{IN} = 5.45$ kW

The method used in the preceding example illustrates a general approach to the analysis of transformer circuits. The steps in the method are as follows:

STEP 1. Write KVL equations around the primary and secondary circuits using assigned mesh currents, source voltages, and inductor voltages.

STEP 2. Write the I–V characteristics of the coupled inductors in terms of the mesh currents using the dot convention to determine whether the coupling is additive or subtractive.

STEP 3. Use the I–V relationships from step 2 to eliminate the inductor voltages from the KVL equations obtained in step 1 to obtain mesh-current equations.

The next two examples illustrate this method of formulating mesh equations.

EXAMPLE 15–7

Find $\mathbf{I}_A$, $\mathbf{I}_B$, $\mathbf{V}_1$, $\mathbf{V}_2$, and the impedance seen by the voltage source in Figure 15–14.

FIGURE 15–14

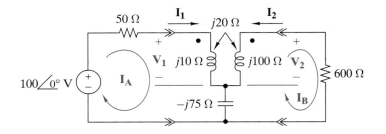

SOLUTION:

STEP 1. The KVL equations around meshes A and B are

Mesh A: $(50 - j75)\mathbf{I}_A - (-j75)\mathbf{I}_B + \mathbf{V}_1 = 100 \angle 0°$
Mesh B: $- (-j75)\mathbf{I}_A + (600 - j75)\mathbf{I}_B - \mathbf{V}_2 = 0$

STEP 2. For the assigned reference directions, the coupling is additive. By KCL we have $\mathbf{I}_1 = \mathbf{I}_A$ and $\mathbf{I}_2 = -\mathbf{I}_B$. Hence, the element equations for the coupled inductors in terms of the mesh currents are

$$\mathbf{V}_1 = j10\mathbf{I}_A + j20(-\mathbf{I}_B)$$
$$\mathbf{V}_2 = j20\mathbf{I}_A + j100(-\mathbf{I}_B)$$

STEP 3. Using these equations to eliminate the inductor voltages from the KVL equations found in step 1 yields the following mesh equations:

Mesh A: $(50 - j75 + j10)\mathbf{I}_A + (j75 - j20)\mathbf{I}_B = 100 \angle 0°$
Mesh B: $(j75 - j20)\mathbf{I}_A + (600 - j75 + j100)\mathbf{I}_B = 0$

Solving these equations for the mesh currents produces

$$\mathbf{I}_A = 0.756 + j0.896 = 1.17 \angle 49.8 \text{ A}$$
$$\mathbf{I}_B = 0.0791 - j0.0726 = 0.107 \angle -42.5° \text{ A}$$

Given the mesh currents, we find the inductor voltages from the **I–V** relations:

$$\mathbf{V}_1 = j10\mathbf{I}_A - j20\mathbf{I}_B = -10.4 + j5.98 = 12.0 \angle 150° \text{ V}$$
$$\mathbf{V}_2 = j20\mathbf{I}_A - j100\mathbf{I}_B = -25.2 + j7.21 = 26.2 \angle 164° \text{ V}$$

Finally, the impedance seen by the input voltage source is

$$Z_{IN} = \frac{\mathbf{V}_S}{\mathbf{I}_A} = 55.0 - j65.2 \ \Omega$$

∎

EXAMPLE 15–8

Figure 15–15 shows an ideal transformer connected as an autotransformer. Find the voltage and average power delivered to the load for $\mathbf{V}_S$ = 500∠0° V, $Z_S = j10$ Ω, $Z_L = 50 + j0$ Ω, $N_1 = 200$, and $N_1 = 280$.

SOLUTION:

The three-step method of writing mesh equations can be used here with a modification to the second step:

STEP 1. The KVL equations around meshes A and B are

$$\text{Mesh A:} \quad Z_S\mathbf{I}_A + \mathbf{V}_1 = \mathbf{V}_S$$
$$\text{Mesh B:} \quad -\mathbf{V}_1 - \mathbf{V}_2 + Z_L\mathbf{I}_B = 0$$

STEP 2. For the assigned reference directions, the coupling is additive so the ideal transformer voltages and currents are related as

$$\mathbf{V}_2 = n\mathbf{V}_1 \quad \text{and} \quad \mathbf{I}_2 = -n\mathbf{I}_1$$

By KCL we have $\mathbf{I}_1 = \mathbf{I}_A - \mathbf{I}_B$ and $\mathbf{I}_2 = -\mathbf{I}_B$. Hence, the ideal transformer constrains the two mesh currents as

$$\mathbf{I}_A = (n + 1)\mathbf{I}_B$$

STEP 3. Using these results to eliminate $\mathbf{I}_A$ and $\mathbf{V}_2$ from the KVL equation in step 1 yields

$$\text{Mesh A:} \quad (n + 1)Z_S\mathbf{I}_B + \mathbf{V}_1 = \mathbf{V}_S$$
$$\text{Mesh B:} \quad -(n + 1)\mathbf{V}_1 + Z_L\mathbf{I}_B = 0$$

Notice that we cannot eliminate both $\mathbf{V}_1$ and $\mathbf{V}_2$ from the KVL equations since voltage and current are independent in an ideal transformer. Substituting the numerical values produces

$$\text{Mesh A:} \quad j24\mathbf{I}_B + \mathbf{V}_1 = 500 \angle 0°$$
$$\text{Mesh B:} \quad -2.4\mathbf{V}_1 + 50\mathbf{I}_B = 0$$

Solving these equations for $\mathbf{I}_B$ and $\mathbf{V}_1$ yields

$$\mathbf{I}_B = 10.3 - j11.9 = 15.7 \angle -49.1° \text{ A}$$
$$\mathbf{V}_1 = 215 - j247 = 328 \angle -49.0° \text{ A}$$

Given the $\mathbf{I}_B$, we find the output quantities as

$$\mathbf{V}_L = Z_L\mathbf{I}_B = 515 - j595 = 787 \angle -49.1° \text{ A}$$
$$P_L = \frac{|\mathbf{I}_B|^2}{2} R_L = \frac{15.7^2}{2} 50 = 6.16 \text{ kW}$$

Exercise 15–6

Find $\mathbf{V}_L$, $\mathbf{V}_1$, $\mathbf{V}_2$, and Z_{IN} in the circuit shown in Figure 15–16.

FIGURE 15 – 16

Answers: $\mathbf{V}_L = 31.6 \angle{-108°}$ V, $\mathbf{V}_1 = 23.7 \angle{71.6°}$ V, $\mathbf{V}_2 = 23.7 \angle{71.6°}$ V, $Z_{IN} = 0 + j20 \ \Omega$.

15–6 TRANSFORMER EQUIVALENT CIRCUITS

Replacing the two-winding model by an equivalent circuit without mutual inductance provides insight into transformer characteristics. Two circuits are equivalent if they have the same **I–V** characteristics at specified terminal pairs. Since coupled inductors involve two terminal pairs, equivalence requires identical characteristics at both pairs. Although there are many equivalent circuits for coupled inductors, we only cover two examples. The first works best for tight coupling and leads to transformer models used in power systems. The second works best in the loosely coupled tuned circuits in communication systems.

In the sinusoidal steady state the element equations for a transformer in the two-winding configuration with additive coupling are

$$
\begin{aligned}
\text{Primary Circuit:} \quad & \mathbf{V}_1 = +j\omega L_1 \mathbf{I}_1 + j\omega M \mathbf{I}_2 \\
\text{Secondary Circuit:} \quad & \mathbf{V}_2 = +j\omega M \mathbf{I}_1 + j\omega L_2 \mathbf{I}_2
\end{aligned}
\tag{15–40}
$$

Solving the primary circuit equation for $\mathbf{I}_1$ yields

$$
\mathbf{I}_1 = \frac{\mathbf{V}_1}{j\omega L_1} - \frac{M}{L_1} \mathbf{I}_2
\tag{15–41}
$$

Using this result to eliminate $\mathbf{I}_1$ from the secondary circuit equation in Eq. (15–40) produces

$$
\mathbf{V}_2 = \frac{M}{L_1} \mathbf{V}_1 + j\omega L_2 \left(1 - \frac{M^2}{L_1 L_2}\right) \mathbf{I}_2
\tag{15–42}
$$

Defining a real parameter $a = M/L_1$ and recognizing that the term $M^2/L_1 L_2$ is the square of the coupling coefficient (k^2) allow us to write the **I–V** characteristics in Eq. (15–40) as

$$
\begin{aligned}
\text{Primary Circuit:} \quad & \mathbf{I}_1 = \frac{\mathbf{V}_1}{j\omega L_1} - a\mathbf{I}_2 \\
\text{Secondary Circuit:} \quad & \mathbf{V}_2 = a\mathbf{V}_1 + j\omega L_2(1 - k^2)\mathbf{I}_2
\end{aligned}
\tag{15–43}
$$

The first equation is a KCL constraint in the primary circuit, and the second is a KVL constraint in the secondary circuit. Taken together, these two equations describe the equivalent circuit shown in Figure 15–17, where the ideal transformer is governed by the element equations

$$\mathbf{V}'_2 = a\mathbf{V}'_1 \quad \text{and} \quad \mathbf{I}'_1 = -a\mathbf{I}'_2 \qquad (15\text{--}44)$$

However, applying KVL in the primary circuit yields $\mathbf{V}_1 = \mathbf{V}'_1$ and applying KCL in the secondary circuit yields $\mathbf{I}_2 = \mathbf{I}'_2$. Hence, the element equations of the ideal transformer can also be expressed as

$$\mathbf{V}'_2 = a\mathbf{V}_1 \quad \text{and} \quad \mathbf{I}'_1 = -a\mathbf{I}_2 \qquad (15\text{--}45)$$

which is the form required to relate the two equations in Eq. (15–43) to the equivalent circuit in Figure 15–17.

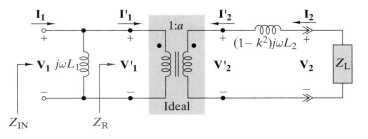

FIGURE   15 – 17   *Two-winding transformer equivalent circuit with an ideal transformer.*

The fact that we derived this equivalent circuit using additive coupling is not critical to the form of the equivalent circuit. If we had used subtractive coupling the only change would be that one of the dots on the ideal transformer would move to a lower terminal.

The equivalent circuit in Figure 15–17 points out several important features of transformers with perfect coupling ($k = 1$):

(1) The turns ratio of the ideal transformer in the equivalent circuit is $a = M/L_1$. As we have previously discussed, with perfect coupling Eq. (15–34) points out that

$$n = \frac{N_2}{N_1} = \frac{M}{L_1} = a \qquad (15\text{--}46)$$

In other words, with perfect coupling the physical transformer and the ideal transformer in the equivalent circuit have the same turns ratio.

(2) The self-inductance of the secondary winding produces an impedance $j\omega L_2(1 - k^2)$ connected in series with the load impedance. With perfect coupling ($k = 1$) this impedance is zero and the voltage across the load becomes $\mathbf{V}_2 = a\mathbf{V}_1 = n\mathbf{V}_1$. That is, the secondary voltage is equal to the primary voltage times the transformer turns ratio.

(3) In Section 15–4 we showed that the resistance seen on the primary side of an ideal transformer is equal to the resistance on the secondary side divided by the square of the turns ratio. The same concept applies to impedances in the phasor domain. The total impedance connected to the secondary of the ideal transformer is $j\omega L_2(1 - k^2) + Z_L$. The **reflected impedance** Z_R seen on the primary side of the ideal transformer is

$$Z_R = \frac{j\omega L_2(1 - k^2)}{a^2} + \frac{Z_L}{a^2} \qquad (15\text{--}47)$$

The first term on the right side of this result is called the **leakage reactance**. With perfect coupling ($k = 1$ and $a = n$) the leakage reactance is zero and the reflected impedance reduces to Z_L/n^2.

(4) With perfect coupling the primary current consists of two components:

$$\mathbf{I_1} = \frac{\mathbf{V_1}}{j\omega L_1} + \frac{\mathbf{V_1}}{Z_L/n^2}$$

$$= \mathbf{I_M} + \mathbf{I_L}$$

(15–48)

The load component $\mathbf{I_L}$ depends on the load impedance and the transformer turns ratio. The other component (designated $\mathbf{I_M}$) is called the **magnetizing** current. This component is always present even under noload ($Z_L = \infty$) conditions and represents the primary current required to initially establish the magnetic field that couples the primary and secondary windings.

The conclusions reached in preceding discussions require perfect coupling, a condition not achieved by real transformers. However, transformers in electrical power systems have tight coupling ($k \approx 1$) so that these conclusions serve as the basis for developing the detailed models used in power system analysis.

EXAMPLE 15–9

A power transformer is operating with $\omega = 377$ rad/s and $Z_L = 125 + j90$ Ω. The self-inductances of the primary and secondary windings are $L_1 = 0.4$ H and $L_2 = 10$ H. When $\mathbf{V_1} = 400 \angle 0°$ V, find the primary current for $k = 1$ and $k = 0.999$.

SOLUTION:
For $k = 1$ the unspecified transformer parameters are

$$M = k\sqrt{L_1 L_2} = 2 \quad \text{and} \quad a = \frac{M}{L_1} = 5$$

and the reflected impedance in Eq. (15–47) is

$$Z_R = \frac{j\omega L_2(1 - k^2)}{a^2} + \frac{Z_L}{a^2} = 0 + \frac{125 + j90}{5^2}$$

$$= 5 + j3.6 \ \Omega$$

The reflected impedance is connected in parallel with the self-impedance of the primary winding, namely $j\omega L_1 = j377 \times 0.4 = j150.8 \ \Omega$. Hence, the primary current is

$$\mathbf{I_1} = \frac{400 \angle 0°}{j150.8} + \frac{400 \angle 0°}{5 + j3.6}$$

$$= -j2.65 + 52.69 - j37.94$$

$$= 66.5 \angle -37.6° \ \text{A}$$

For $k = 0.999$ the unspecified transformer parameters become

$$M = k\sqrt{L_1 L_2} = 1.998 \quad \text{and} \quad a = \frac{M}{L_1} = 4.995$$

and the reflected impedance in Eq. (15–47) is

$$Z_R = \frac{j\omega L_2(1 - k^2)}{a^2} + \frac{Z_L}{a^2} = j0.302 + \frac{125 + j90}{4.995^2}$$

$$= 5.01 + j3.607 \ \Omega$$

Hence the primary current is

$$I_1 = \frac{400 \ \angle 0°}{j150.8} + \frac{400 \ \angle 0°}{5.01 + j3.607}$$

$$= -j2.65 + 49.63 - j38.72$$

$$= 64.6 \ \angle 39.8° \ A$$

which is essentially the same as the result obtained for perfect coupling. Note that the magnetizing component of the primary current is small compared to the load component. ∎

Find the reflected impedance and the total impedance seen on the primary side of a transformer with $L_1 = 20$ mH, $L_2 = 100$ mH, $k = 0.99$, $\omega = 2500$ rad/s, and $Z_L = 15 + j0 \ \Omega$.

Answer: $Z_R = 3.06 + j1.015 \ \Omega$, $Z_T = 2.93 + j1.17 \ \Omega$.

Figure 15–18 shows three uncoupled inductors connected in a T-circuit configuration. This circuit is equivalent to a pair of coupled inductors and is particularly useful when the inductors are loosely coupled ($k \ll 1$). To demonstrate the equivalence we write mesh-current equations of the T-circuit using the mesh currents I_A and I_B:

$$V_1 = [j\omega(L_1 \mp M) \pm j\omega M]I_A - (\pm j\omega M)I_B$$
$$-V_2 = -(\pm j\omega M)I_A + [j\omega(L_2 \mp M) \pm j\omega M]I_B$$

But the phasor currents I_1 and I_2 in Figure 15–18 are related to the mesh currents as $I_1 = I_A$ and $I_2 = -I_B$, and so these mesh equations reduce to

$$V_1 = j\omega L_1 I_1 \pm j\omega M I_2$$
$$V_2 = \pm j\omega M I_1 + j\omega L_2 I_2$$

These equations are the same as the **I–V** relations for coupled inductors in Eqs. (15–36). Hence, the T-circuit in Figure 15–18 is equivalent to a pair of coupled inductors with additive (plus sign) or subtractive (minus sign) coupling.

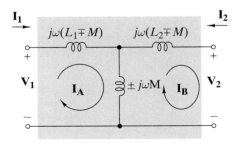

FIGURE 15 − 18 *T-equivalent circuit of a two-winding transformer.*

For additive coupling, the equivalent T-circuit involves three uncoupled inductances whose values are $L_1 - M$, $L_2 - M$, and M. For subtractive coupling the three inductances are $L_1 + M$, $L_2 + M$, and $-M$. In either case, one of the inductances in the equivalent T-circuit is negative. Negative inductances can be used in circuit analysis even though they are not physically realizable using passive elements.

EXAMPLE 15–10

Use the equivalent T-circuit in Figure 15–18 to find the input impedance of the transformer in Example 15–6.

SOLUTION:
The transformer in Example 15–6 had additive coupling with $L_1 = 2$ H, $L_2 = 0.2$ H, $M = 0.6$ H, $\omega = 377$ rad/s, and $Z_L = 25 + j15$ Ω. Consequently we have

$$j\omega(L_1 - M) = j528 \ \Omega$$

$$j\omega(L_2 - M) = -j151 \ \Omega$$

$$j\omega M = j226 \ \Omega$$

Note that the inductance $L_2 - M = -0.4$ H is negative. Figure 15–19 shows the equivalent T-circuit together with the load. The input impedance to this circuit is

$$Z_{IN} = j528 + j226 \,\|(-j151 + Z_L)$$

$$= j528 + \cfrac{1}{\cfrac{1}{j226} + \cfrac{1}{25 - j136}}$$

$$= 145 + j228 \ \Omega$$

This result agrees with the answer in Example 15–6, where the input impedance was found using mesh-current analysis.   ∎

FIGURE   15 – 19

Z_{IN} Equivalent T-circuit

Exercise 15–8

Use an equivalent T-circuit to find the input impedance of a transformer with subtractive coupling and $L_1 = 20$ mH, $L_2 = 250$ mH, $M = 70$ mH, $\omega = 2000$ rad/s, and $Z_L = 500 + j100$ Ω.

Answer:
$Z_{IN} = 16.1 + j20.7$ Ω

SUMMARY

- Mutual inductance describes magnetic coupling between two inductors. The mutual inductance parameter relates the voltage induced in one inductor to the rate of change of current in the other inductor. The induced voltage can be either positive (additive coupling) or negative (subtractive coupling).

- The dot convention describes the physical orientation of the two magnetically coupled inductors. Mutual inductance coupling is additive when both current reference arrows point toward or away from dotted terminals; otherwise, it is subtractive. The reference directions for the currents can always be selected so that the coupling is additive.

- The degree of coupling is indicated by the coupling coefficient. Energy analysis shows that the coupling coefficient must lie between zero and one. A coupling coefficient of unity is called perfect coupling and means that all of the flux produced by first winding links with a second winding, and vice versa. A coupling coefficient of zero indicates no magnetic linkage between the windings.

- A transformer is an electrical device based on the mutual inductance coupling between two windings. Transformers find applications in almost all electrical systems, especially in power supplies and the electrical power grid. The transformer winding connected to the power source is called the primary winding and the winding connected to the load is called the secondary winding.

- The ideal transformer is a circuit element in which the primary and secondary windings are assumed to be perfectly coupled and to have no power loss. In an ideal transformer the voltages and currents in the primary and secondary windings are related by the turns ratio, which is the ratio of the number of turns in the secondary winding to that in the primary winding.

- In the sinusoidal steady state, transformers can be analyzed using phasors to determine steady-state currents, voltages, impedances, and average power transfer. The phasor analysis of transformers is carried out using a modified mesh-current approach.

- Transformers have many different equivalent circuits. These equivalent circuits provide insight into the performance of transformers and may simplify certain types of calculations.

PROBLEMS

ERO 15–1 MUTUAL INDUCTANCE (SECT. 15–1, 15–2, 15–3)

(a) Given the current through or voltage across two coupled inductors, find the unspecified currents or voltages.
(b) Find characteristics of coupled inductors when given some device parameters and circuit connections.

15–1 The input to the coupled inductors in Figure P15–1 is a voltage source $v_S(t) = 10 \sin 2000t$ V.

$L_1 = 10$ mH, $L_2 = 5$ mH
$M = 7$ mH

FIGURE P15-1

$L_1 = L_2 = 3$ mH
$M = 2$ mH

FIGURE P15-4

FIGURE P15-7

L_{EQ}

FIGURE P15-8

(a) Write the i–v relationships for the coupled inductors using the reference marks in the figure.

(b) Solve for $v_2(t)$ when the output terminals are open circuited ($i_2 = 0$).

15–2   The input to the coupled inductors in Figure P15–1 is a voltage source $v_S(t) = 25 \sin 1000t$ V.

(a) Write the i–v relationships for the coupled inductors using the reference marks in the figure.

(b) Solve for $i_1(t)$ and $i_2(t)$ when the output terminals are short circuited ($v_2 = 0$).

15–3   In Figure P15–1 the outputs are $i_2(t) = 0$ and $v_2(t) = 10(e^{-1000t})u(t)$ V.

(a) Write the i–v relationships for the coupled inductors using the reference marks given.

(b) Solve for the source voltage $v_S(t)$.

15–4   The input to the coupled inductors in Figure P15–4 is a current source $i_S(t) = 0.5 \sin 1000t$ A. Solve for $v_1(t)$ and $v_2(t)$ when the output terminals are open circuited ($i_2 = 0$).

15–5   The input to the coupled inductors in Figure P15–4 is a current source $i_S(t) = 0.5 \sin 1000t$ A. Solve for $v_1(t)$ and $i_2(t)$ when the output terminals are short circuited ($v_2 = 0$).

15–6   A pair of coupled inductors have $L_1 = 1.2$ H and $L_2 = 2.4$ H. With the output open circuited ($i_2(t) = 0$), the coil voltages are observed to be $v_1(t) = 3 \sin 1000t$ V and $v_2(t) = 4 \sin 1000t$ V. Find the mutual inductance and the coupling coefficient. Is the coupling additive or subtractive?

15–7   In Figure P15–7 $L_1 = L_2 = 10$ mH, $M = 9$ mH, and $i_1(t) = 2 \cos(1000t)$ A. Find the input voltage $v_x(t)$.

15–8   In Figure P15–8 $L_1 = 200$ mH, $L_2 = 400$ mH and $k = 0.98$. Find the equivalent inductance L_{EQ}.

15–9   A pair of perfectly coupled inductors have $L_1 = 40$ mH and $L_2 = 60$ mH. What is the mutual inductance?

15–10   The self-inductances of two coupled inductors are found to be 110 mH and 26.6 mH. When the two coupled inductors are connected in series the total inductance is found to be 237 mH. What is the coupling coefficient.

ERO 15–2 THE IDEAL TRANSFORMER (SECT. 15–4)

Given a circuit containing ideal transformers,

(a) Find specified voltages, currents, powers, and equivalent circuits.
(b) Select the turns ratio to meet prescribed conditions.

15–11   The primary voltage in an ideal transformer is a 120-V, 60-Hz sinusoid. The secondary voltage is a 24-V, 60-Hz sinusoid. The secondary terminals are connected to an 8-Ω resistive load.

(a) Find the transformer turns ratio.

(b) Write expressions for the primary current and voltage.

15–12   The number of turns in the primary and secondary of an ideal transformer are $N_1 = 50$ and $N_2 = 600$. The secondary current is observed to be $i_2(t) = 3 \sin 377t$ A. Write an expression for the primary current. Assume that the magnetic coupling is additive.

15–13 The turns ratio of the second ideal transformer in Figure P15–13 is $n = 5$. Find the equivalent resistance indicated in the figure.

FIGURE P15-13

15–14 Select the turns ratio of the second ideal transformer in Figure P15–13 so that the same power is delivered to each resistor.

15–15 Figure P15–15 shows an ideal transformer connected as an auto-transformer. Find $i_L(t)$ and $i_S(t)$ when $v_S(t) = 100 \sin 400t$ V and $R_L = 100\ \Omega$.

15–16 Two subsystems are to be connected in cascade using a trans-former. The output resistance of the first is 50 Ω, and the input resis-tance to the second is 4.8 kΩ. Select the transformer turns ratio so that maximum power is transferred.

15–17 Select the turns ratio of an ideal transformer in the interface cir-cuit shown in Figure P15–17 so that the maximum available power is delivered to the 1-kΩ output resistor.

FIGURE P15-15

FIGURE P15-17

15–18 A voltage source with Thévenin parameters $v_T = 5 \sin 1000t$ V and $R_T = 50\ \Omega$ drives the primary winding of an ideal transformer with $n = 5$. Find the power delivered to a 300-Ω load connected across the sec-ondary winding.

15–19 A voltage source with $v_S = 2200 \sin 377t$ V drives the primary wind-ings of the two ideal transformers as shown in Figure P15–19. The

FIGURE P15-19

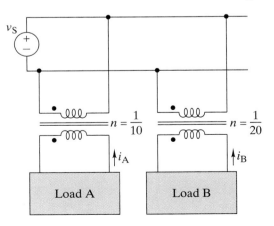

load currents in the figure are $i_A = 10 \sin 377t$ A and $i_B = 20 \sin 377t$ A. Derive an expression for the power delivered by the voltage source.

15–20 A power supply module requires an input sinusoid with an amplitude of 30 V peak to peak. The available ac source supplies a sinusoid with an amplitude of 120 V rms. Find the turns ratio of an ideal transformer to interface the source to the power supply.

ERO 15–3 TRANSFORMERS IN THE SINUSOIDAL STEADY STATE (SECT. 15–5, 15–6)

Given a linear transformer in a linear circuit operating in the sinusoidal steady state, find phasor voltages and currents, average powers, and equivalent impedances.

15–21 The input voltage to the transformer in Figure P15–21 is a sinusoid $v_1(t) = 10 \cos 2000t$ V. With the circuit operating in the sinusoidal steady state, use mesh-current analysis to find the phasor output voltage $\mathbf{V}_2$ and the equivalent impedance seen by the input source.

FIGURE P15–21

15–22 A transformer operating in the sinusoidal steady state has inductances $L_1 = 200$ mH, $L_2 = 450$ mH, and $k = 0.95$. The source has a peak amplitude of 90 V and a frequency of 377 rad/s . The load connected to the secondary of the transformer is a 75-Ω resistor. Assume additive coupling.
(a) Find the phasor responses $\mathbf{I}_1$, $\mathbf{I}_2$, $\mathbf{V}_1$, and $\mathbf{V}_2$.
(b) Find the input impedance seen by the source.
(c) Find the average power delivered by the source.

15–23 The circuit in Figure P15–23 is operating in the sinusoidal steady state with $v_S(t) = 20 \cos 20000t$ V. Find the steady-state response $i(t)$.

15–24 Repeat Problem 15–23 when the input is $v_S(t) = 20 \cos 10000t$ V.

15–25 The circuit in Figure P15–25 is operating in the sinusoidal steady state with $v_S(t) = 150 \cos 4000t$ V. The resistors in series with the coupled inductors simulate winding resistances. Find input impedance Z_{IN}, and the phasor response $\mathbf{V}_1$ and $\mathbf{V}_L$.

FIGURE P15–23

FIGURE P15–25

15–26   The frequency of the voltage source in Figure P15–25 is adjustable. Find the frequency at which the input impedance Z_{IN} is purely real.

15–27   The circuit in Figure P15–27 is operating in the sinusoidal steady state with $v_S(t) = 10 \sin 2000t$ V. Find the impedance seen by the voltage source, input current $i(t)$, and the average power supplied by the voltage source.

FIGURE P15–27

15–28   Replace the coupled inductors in Figure P15–27 by their T-equivalent circuit. Use circuit reduction to find the equivalent impedance seen by the voltage source $v_S(t) = 10 \sin 4000t$ V and the input current $i(t)$.

15–29   A transformer operating in the sinusoidal steady state has inductances $L_1 = 800$ mH, $L_2 = 320$ mH, and $M = 500$ mH. The load connected to the secondary is a 60-Ω resistor. The source connected to the primary side has a peak amplitude of 150 V and a frequency of $\omega = 377$ rad/s. Find the impedance seen by the voltage source, the average power delivered by the source, and the average power delivered to the load.

15–30   A transformer operating in the sinusoidal steady state has inductances $L_1 = 510$ mH, $L_2 = 2$ H, and $M = 1$ H. The load connected to the secondary is $Z_L = 200 + j150$ Ω. The source connected to the primary side has a peak amplitude of 2500 V and a frequency of $\omega = 377$ rad/s. Find the impedance seen by the voltage source, the average power delivered by the source, and the average power delivered to the load.

15–31   The magnitudes of the steady-state ac voltages and currents in a transformer are measured with the secondary winding open circuited and short circuited. When the secondary is open circuited the measurements are $|V_1| = 100$ V, $|I_1| = 120$ mA, and $|V_2| = 220$ V. When the secondary is short circuited the measurements are $|I_1| = 10$ A and $|I_2| = 2.2$ A. All measurements were made at $f = 400$ Hz. Find L_1, L_2, and M.

15–32   The secondary winding of an ideal transformer with a turns ratio of $n = 10$ is connected to a load $Z_L = 500 + j200$ Ω. The primary is connected to a voltage source with a peak amplitude of 500 V and an internal impedance of $Z_S = j2$ Ω. Find the average power delivered to the load.

15–33   A certain resistive load requires 25 kW at a voltage of 480 V rms. The available supply is 2400 V rms. A transformer is to be used to reduce the supply voltage to 480 V, and a circuit breaker is to be installed on the primary side to protect the equipment. The available circuit breakers have tripping currents of 5 A, 10 A, 15 A, and 20 A. Which breaker would you choose in this application?

15–34 A 60-Hz, 4400-V(rms) to 220-V (rms) step-down transformer delivers 25 kW to a resistive load. Treat the transformer as ideal and estimate the rms primary and secondary currents.

15–35 A transformer that can be treated as ideal has 480 turns in the primary winding and 120 turns in the secondary winding. The primary is connected to a 60-Hz source with a peak amplitude of 480 V. The secondary delivers 5 kW to a resistive load. Find the primary and secondary currents and the impedance seen by the source.

INTEGRATING PROBLEMS

15–36 Ⓐ DIFFERENTIAL EQUATIONS FOR COUPLED INDUCTORS

The purpose of this problem is to derive a differential equation describing the dynamic response of a pair of coupled inductors. Figure P15–36 shows the circuit with assigned reference directions for the inductor currents and voltages.

(a) Write device and connection equations for the circuit.

(b) Reduce the equations in (a) to a single second-order differential equation in the current $i_2(t)$.

(c) Show that with perfect coupling ($L_1L_2 = M^2$) reduces the second-order equation in (b) to a first-order equation with a time constant of $L_1/R_1 + L_2/R_2$.

FIGURE P15–36

15–37 Ⓐ TRANSFER FUNCTION FOR COUPLED INDUCTORS

Figure P15–36 shows two coupled inductors in the time domain. The purpose of this problem is to solve for the s-domain transfer function to obtain the frequency response and step response of a transformer.

(a) Write the device and connection equations for the circuit in the time domain.

(b) Assuming zero initial conditions, transform the equations in (a) into the s domain and solve for the transfer function $V_2(s)/V_S(s)$.

(c) Show that the transfer function has a second-order bandpass characteristic.

(d) Find the center frequency, upper and lower cutoff frequencies, and the bandwidth when $L_1 = 0.25$ H, $L_2 = 1$ H, $M = 0.499$ H, $R_1 = 50$ Ω, and $R_L = 200$ Ω. Does the transformer have a wideband or narrowband bandpass frequency response?

(e) Find the step response of the transformer using the element values in (d).

15–38 ⬛ THREE-WINDING IDEAL TRANSFORMER

Figure P15–38 shows a three-winding transformer that can be treated as ideal.

(a) Perfect coupling between all winding implies that

$$v_2(t) = \frac{N_2}{N_1} v_1(t) \quad \text{and} \quad v_3(t) = \frac{N_3}{N_1} v_1(t)$$

Show that zero power loss implies that

$$N_1 i_1(t) + N_2 i_2(t) + N_3 i_3(t) = 0$$

(b) When a resistor R_2 is connected across winding no. 2 and a resistor R_3 is connected across winding no. 3, show that the equivalent resistance seen at winding no. 1 is

$$\frac{1}{R_{EQ}} = \left(\frac{N_2}{N_1}\right)^2 \frac{1}{R_2} + \left(\frac{N_3}{N_1}\right)^2 \frac{1}{R_3}$$

Ideal

FIGURE P15–38

CHAPTER 16

POWER IN THE SINUSOIDAL STEADY STATE

George Westinghouse was, in my opinion, the only man on the globe who could take my alternating current (power) system under the circumstances then existing and win the battle against prejudice and money power.

Nikola Tesla, 1932,
American Engineer

16–1 Average and Reactive Power

16–2 Complex Power

16–3 AC Power Analysis

16–4 Load-flow Analysis

16–5 Three-phase Circuits

16–6 Three-phase AC Power Analysis

Summary

Problems

Integrating Problems

In the last decade of the nineteenth century there were two competing electrical power systems in this country. The direct current approach was developed by Thomas Edison, who installed the first of his dc power systems at the famous Pearl Street Station in New York City in 1882. By 1884 there were more than 20 such power stations operating in the United States. The competing alternating current technology was hampered by the lack of practical motors for the single-phase systems initially produced. The three-phase ac induction motor that met this need was the product of the mind of the brilliant Serbian immigrant Nikola Tesla (1856–1943). Tesla was educated in Europe and in 1884 emigrated to the United States, where he initially worked for Edison. In 1887 Tesla founded his own company to develop his inventions, eventually producing some 40 patents on three-phase equipment and controls. The importance of Tesla's work was recognized by George Westinghouse, a hard-driving electrical pioneer who purchased the rights to Tesla's patents. In the early 1890s the competition between the older dc system championed by Edison and the newer ac technology sponsored by Westinghouse developed into a heated controversy called the "war of the currents." The showdown came over the equipment to be installed in a large electric power station at Niagara Falls, New York. The choice of an ac system for Niagara Falls established a trend that led to large interconnected ac systems that form the backbone of modern industry today.

In this chapter we study the flow of electrical power in the sinusoidal steady state using phasors. The first section shows that the instantaneous power at an interface can be divided into two components called real and reactive power. The unidirectional real component produces a net transfer of energy from the source to the load. The oscillatory reactive power represents an interchange between the source and load with no net transfer of energy. The concept of complex power developed in the second section combines these two components into a single complex entity relating power flow to the interface voltage and current phasors. The third and fourth sections develop the basic tools for using complex power to analyze power flow in ac circuits. The final two sections discuss the three-phase circuits and systems that deal with the large blocks of electrical power required in a modern industrial society.

16-1   AVERAGE AND REACTIVE POWER

We begin our study of electric power circuits with the two-terminal interface in Figure 16–1. In power applications we normally think of one circuit as the source and the other as the load. Our objective is to describe the flow of power across the interface when the circuit is operating in the sinusoidal steady state. To this end, we write the interface voltage and current in the time domain as sinusoids of the form

$$v(t) = V_A \cos(\omega t + \theta)$$

$$i(t) = I_A \cos \omega t \qquad (16\text{--}1)$$

In Eq. (16–1) V_A and I_A are real, positive numbers representing the peak amplitudes of the voltage and current, respectively.

In Eq. (16–1) we have selected the $t = 0$ reference at the positive maximum of the current $i(t)$ and assigned a phase angle to $v(t)$ to account for

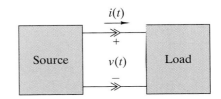

F I G U R E   1 6 – 1   *A two-terminal interface.*

the fact that the voltage maximum may not occur at the same time. In the phasor domain the angle $\theta = \phi_V - \phi_I$ is the angle between the phasors $\mathbf{V} = V_A\angle\phi_V$ and $\mathbf{I} = I_A\angle\phi_I$. In effect, choosing $t = 0$ at the current maximum shifts the phase reference by an amount $-\phi_I$ so that the voltage and current phasors become $\mathbf{V} = V_A\angle\theta$ and $\mathbf{I} = I_A\angle 0$.

A method of relating power to phasor voltage and current will be presented in the next section, but at the moment we write the instantaneous power in the time domain.

$$
\begin{aligned}
p(t) &= v(t) \times i(t) \\
&= V_A I_A \cos(\omega t + \theta)\cos \omega t
\end{aligned}
\tag{16–2}
$$

This expression for instantaneous power contains dc and ac components. To separate the components, we first use the identity $\cos(x + y) = \cos x \cos y - \sin x \sin y$ to write $p(t)$ in the form

$$
\begin{aligned}
p(t) &= V_A I_A[\cos \omega t \cos \theta - \sin \omega t \sin \theta] \cos \omega t \\
&= [V_A I_A \cos \theta]\cos^2 \omega t - [V_A I_A \sin \theta]\cos \omega t \sin \omega t
\end{aligned}
\tag{16–3}
$$

Using the identities $\cos^2 x = \frac{1}{2}(1 + \cos 2x)$ and $\cos x \sin x = \frac{1}{2} \sin 2x$, we write $p(t)$ in the form

$$
p(t) = \underbrace{\left[\frac{V_A I_A}{2} \cos \theta\right]}_{\text{dc Component}} + \underbrace{\left[\frac{V_A I_A}{2} \cos \theta\right] \cos 2\omega t - \left[\frac{V_A I_A}{2} \sin \theta\right] \sin 2\omega t}_{\text{ac Component}}
\tag{16–4}
$$

Written this way, we see that the instantaneous power is the sum of a dc component and a double-frequency ac component. That is, the instantaneous power is the sum of a constant plus a sinusoid whose frequency is 2ω, which is twice the angular frequency of the voltage and current in Eq. (16–1).

Note that instantaneous power in Eq. (16–4) is periodic. In Chapter 5 we defined the average value of a periodic waveform as

$$
P = \frac{1}{T}\int_0^T p(t)dt
$$

where $T = 2\pi/2\omega$ is the period of $p(t)$. In Chapter 5 we also showed that the average value of a sinusoid is zero, since the area under the waveform during a positive half cycle is canceled by the area under a subsequent negative half cycle. Therefore, the **average value** of $p(t)$, denoted as P, is equal to the constant or dc term in Eq. (16–4):

$$
P = \frac{V_A I_A}{2} \cos \theta
\tag{16–5}
$$

The amplitude of the $\sin 2\omega t$ term in Eq. (16–4) has a form much like the average power in Eq. (16–5), except it involves $\sin \theta$ rather than $\cos \theta$. This amplitude factor is called the **reactive power** of $p(t)$, where reactive power Q is defined as

$$
Q = \frac{V_A I_A}{2} \sin \theta
\tag{16–6}
$$

Substituting Eqs. (16–5) and (16–6) into Eq. (16–4) yields the instantaneous power in terms of the average power and reactive power:

$$p(t) = \underbrace{P(1 + \cos 2\omega t)}_{\text{Unipolar}} - \underbrace{Q \sin 2\omega t}_{\text{Bipolar}} \qquad (16\text{--}7)$$

The first term in Eq. (16–7) is said to be unipolar because the factor $1 + \cos 2\omega t$ never changes sign. As a result, the first term is either always positive or always negative depending on the sign of P. The second term is said to be bipolar because the factor $\sin 2\omega t$ alternates signs every half cycle.

The energy transferred across the interface during one cycle $T = 2\pi/2\omega$ of $p(t)$ is

$$W = \int_0^T p(t)dt$$

$$= P\underbrace{\int_0^T (1 + \cos 2\omega t)dt}_{\text{Net Energy}} - Q\underbrace{\int_0^T \sin 2\omega t \, dt}_{\text{No Net Energy}} \qquad (16\text{--}8)$$

$$= \qquad P \times T \qquad - \qquad 0$$

Only the unipolar term in Eq. (16–7) provides any net energy transfer, and that energy is proportional to the average power P. With the passive sign convention the energy flows from source to load when $W > 0$. Equation (16–8) shows that the net energy will be positive if the average power $P > 0$. Equation (16–5) points out that the average power P is positive when $\cos \theta > 0$, which in turn means $|\theta| < 90°$. We conclude that

> *The net energy flow in Figure 16–1 is from source to load when the angle between the interface voltage and current is bounded by −90° < θ < 90°; otherwise the net energy flow is from load to source.*

The bipolar term in Eq. (16–7) is a power oscillation that transfers no net energy across the interface. In the sinusoidal steady state the load in Figure 16–1 borrows energy from the source circuit during part of a cycle and temporarily stores it in the load inductance or capacitance. In another part of the cycle the borrowed energy is returned to the source unscathed. The amplitude of the power oscillation is called reactive power because it involves periodic energy storage and retrieval from the reactive elements of the load. The reactive power can be either positive or negative depending on the sign of $\sin \theta$. However, the sign of Q tells us nothing about the net energy transfer, which is controlled by the sign of P.

We are obviously interested in average power, since this component carries net energy from source to load. For most power system customers the basic cost of electrical service is proportional to the net energy delivered to the load. Large industrial users may also pay a service charge for their reactive power. This may seem unfair, since reactive power transfers no net energy. However, the electric energy borrowed and returned by the load is generated within a power system that has losses. From a power company's viewpoint, the reactive power is not free because there are losses in the system connecting the generators in the power plant to source/load interface at which the lossless interchange of energy occurs.

In ac power circuit analysis, it is necessary to keep track of both the average power and reactive power. These two components of power have the same dimensions, but because they represent quite different effects they traditionally are given different units. The average power is expressed in watts (W) while reactive power is expressed in VARs, which is an acronym for "volt-amperes reactive."

Exercise 16–1

Using the reference marks in Figure 16–1, calculate the average and reactive power for the following voltages and currents. State whether the load is absorbing or delivering net energy.

(a) $v(t) = 168 \cos(377t + 45°)$ V, $i(t) = 0.88 \cos 377t$ A
(b) $v(t) = 285 \cos(2500t - 68°)$ V, $i(t) = 0.66 \cos 2500t$ A
(c) $v(t) = 168 \cos(377t + 45°)$ V, $i(t) = 0.88 \cos(377t - 60°)$ A
(d) $v(t) = 285 \cos(2500t - 68°)$ V, $i(t) = 0.66 \sin 2500t$ A

  A n s w e r s :
(a) $P = +52.3$ W, $Q = +52.3$ VAR,   absorbing
(b) $P = +35.2$ W, $Q = -87.2$ VAR,   absorbing
(c) $P = -19.1$ W, $Q = +71.4$ VAR,   delivering
(d) $P = +87.2$ W, $Q = +35.2$ VAR,   absorbing

16–2 COMPLEX POWER

It is important to relate average and reactive power to phasor quantities because steady-state analysis is conveniently carried out using phasors. In our previous work the magnitude of a phasor represented the peak amplitude of a sinusoid. However, in power circuit analysis it is convenient to express phasor magnitudes in rms (root mean square) values. In this chapter phasor voltages and currents are expressed as

$$\mathbf{V} = V_{rms}e^{j\phi_V} \quad \text{and} \quad \mathbf{I} = I_{rms}e^{j\phi_I} \qquad (16\text{–}9)$$

Notice that the phasor magnitudes are the rms amplitude of the corresponding sinusoid.

Equations (16–5) and (16–6) express average and reactive power in terms of peak amplitudes V_A and I_A. In Chapter 5 we showed that the peak and rms values of a sinusoid are related by $V_{rms} = V_A/\sqrt{2}$. The expression for average power easily can be converted to rms amplitudes since we can write Eq. (16–5) as

$$P = \frac{V_A I_A}{2} \cos \theta = \frac{V_A}{\sqrt{2}} \frac{I_A}{\sqrt{2}} \cos \theta$$

$$= V_{rms} I_{rms} \cos \theta \qquad (16\text{–}10)$$

where $\theta = \phi_V - \phi_I$ is the angle between the voltage and current phasors. By similar reasoning, Eq. (16–6) becomes

$$Q = V_{rms} I_{rms} \sin \theta \qquad (16\text{–}11)$$

Using rms phasors the **complex power** (S) at a two-terminal interface is defined as follows:

$$S = \mathbf{V}\mathbf{I}^* \tag{16–12}$$

That is, the complex power at an interface is the product of the voltage phasor times the conjugate of the current phasor. Substituting Eq. (16–9) into this definition yields

$$S = \mathbf{V}\mathbf{I}^* = V_{rms}e^{j\phi_V}I_{rms}e^{-j\phi_I}$$
$$= [V_{rms}\,I_{rms}]e^{j(\phi_V - \phi_I)} \tag{16–13}$$

Using Euler's relationship and the fact that the angle is $\theta = \phi_V - \phi_I$, we can write complex power as

$$S = [V_{rms}I_{rms}]e^{j\theta}$$
$$= [V_{rms}I_{rms}]\cos\theta + j[V_{rms}I_{rms}]\sin\theta \tag{16–14}$$
$$= P + jQ$$

The real part of the complex power S is the average power, and the imaginary part is the reactive power. Although S is a complex number, it is not a phasor. However, it is a convenient variable for keeping track of the two components of power when the voltage and currents are expressed as phasors.

The power triangles in Figure 16–2 provide a convenient way to remember complex power relationships and terminology. We confine our study to cases in which net energy is transferred from source to load. In such cases $P > 0$ and the power triangles fall in the first or fourth quadrant, as indicated in Figure 16–2.

The magnitude $|S| = V_{rms}I_{rms}$ is called **apparent power** and is expressed using the unit volt-ampere (VA). The ratio of the average power to the apparent power is called the **power factor** (pf). Using Eq. (16–10), we see that the power factor is

$$pf = \frac{P}{|S|} = \frac{V_{rms}I_{rms}\cos\theta}{V_{rms}I_{rms}} = \cos\theta \tag{16–15}$$

Since pf = cos θ, the angle θ is called the **power factor angle**.

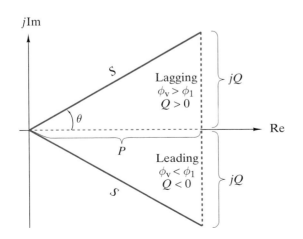

When the power factor is unity the phasors **V** and **I** are in phase ($\theta = 0°$) and the reactive power is zero since $\sin\theta = 0$. When the power factor is less than unity, the reactive power is not zero and its sign is indicated by the modifiers lagging or leading. The term *lagging power factor* means the current phasor lags the voltage phasor so that $\theta = \phi_V - \phi_I > 0$. For a lagging power factor S falls in the first quadrant in Figure 16–2 and the reactive power is positive since $\sin\theta > 0$. The term *leading power factor* means the current phasor leads the voltage phasor so that $\theta = \phi_V - \phi_I < 0$. In this case S falls in the fourth quadrant in Figure 16–2 and the reactive power is negative since $\sin\theta < 0$. Most industrial and residential loads have lagging power factors.

The apparent power rating of electrical power equipment is an important design consideration. The ratings of generators, transfomers, and transmission lines are normally stated in kVA. The ratings of most loads are stated in kW and power factor. The wiring must be large enough to carry the required current and insulated well enough to withstand the rated voltage. However, only the average power is potentially available as useful output, since the reactive power represents a lossless interchange between the source and device. Because reactive power increases the apparent power rating without increasing the net energy output, it is desirable for electrical devices to operate with zero reative power or, equivalently, at unity power factor.

Exercise 16–2

Determine the average power, reactive power, and apparent power for the following voltage and current phasors. State whether the power factor is lagging or leading.

(a) $\mathbf{V} = 208\angle{-90°}$ V (rms), $\mathbf{I} = 1.75\angle{-75°}$ A (rms)
(b) $\mathbf{V} = 277\angle{+90°}$ V (rms), $\mathbf{I} = 11.3\angle{0°}$ A (rms)
(c) $\mathbf{V} = 120\angle{-30°}$ V (rms), $\mathbf{I} = 0.30\angle{-90°}$ A (rms)
(d) $\mathbf{V} = 480\angle{+75°}$ V (rms), $\mathbf{I} = 8.75\angle{+105°}$ A (rms)

Answers:
(a) $P = 352$ W, $Q = -94.2$ VAR, $|S| = 364$ VA, leading
(b) $P = 0$ W, $Q = +3.13$ kVAR, $|S| = 3.13$ kVA, lagging
(c) $P = 18$ W, $Q = +31.2$ VAR, $|S| = 36$ VA, lagging
(d) $P = 3.64$ kW, $Q = -2.1$ kVAR, $|S| = 4.20$ kVA, leading

---

COMPLEX POWER AND LOAD IMPEDANCE

In many cases power circuit loads are described in terms of their power ratings at a specified voltage or current level. To find voltages and current elsewhere in the circuit, it is necessary to know the load impedance. For this reason we need to relate complex power and load impedance.

Figure 16–3 shows the general case for a two-terminal load. For the assigned reference directions the load produces the element constraint $\mathbf{V} = \mathbf{ZI}$. Using this constraint in Eq. (16–12), we write the complex power of the load as

$$S = \mathbf{V} \times \mathbf{I^*} = \mathbf{ZI} \times \mathbf{I^*} = Z|\mathbf{I}|^2$$

$$= (R + jX)\, I_{rms}^2$$

F I G U R E   1 6 – 3 *A two-terminal impedance.*

where R and X are the resistance and reactance of the load, respectively. Since $S = P + jQ$, we conclude that

$$R = \frac{P}{I_{rms}^2} \quad \text{and} \quad X = \frac{Q}{I_{rms}^2} \tag{16-16}$$

The load resistance and reactance are proportional to the average and reactive power of the load, respectively.

The first condition in Eq. (16–16) demonstrates that resistance cannot be negative, since P cannot be negative for a passive circuit. That is, in the sinusoidal steady state a passive circuit cannot produce average power; otherwise, perpetual motion would be possible and the energy crisis would be a small footnote in the great sweep of human history. The second condition in Eq. (16–16) points out that when the reactive power is positive the load is inductive, since $X_L = \omega L$ is positive. Conversely, when the reactive power is negative the load is capacitive, since $X_C = -1/\omega C$ is negative. The terms *inductive load, lagging power factor,* and *positive reactive power* are synonymous, as are the terms *capacitive load, leading power factor,* and *negative reactive power.*

EXAMPLE 16-1

At 440 V (rms) a two-terminal load draws 3 kVA of apparent power at a lagging power factor of 0.9. Find

(a) I_{rms}
(b) P
(c) Q
(d) the load impedance.

Draw the power triangle for the load.

SOLUTION:

(a) $I_{rms} = |S|/V_{rms} = 3000/440 = 6.82$ A (rms)
(b) $P = V_{rms}I_{rms} \cos \theta = 3000 \times 0.9 = 2.7$ kW
(c) For $\cos \theta = 0.9$ lagging, $\sin \theta = 0.436$ and $Q = V_{rms}I_{rms} \sin \theta = 1.31$ kVAR
(d) $Z = (P + j Q)/(I_{rms})^2 = (2700 + j1310)/46.5 = 58.0 + j28.2 \ \Omega$.

Figure 16–4 shows the power triangle for this load. ■

FIGURE 16-4

Find the impedance of a two-terminal load under the following conditions.
(a) $\mathbf{V} = 120\angle30°$ V (rms) and $\mathbf{I} = 20\angle75°$ A (rms)
(b) $|S| = 3.3$ kVA, $Q = -1.8$ kVAR, and $I_{rms} = 7.5$ A
(c) $P = 3$ kW, $Q = 4$ kVAR, and $V_{rms} = 880$ V
(d) $V_{rms} = 208$ V, $I_{rms} = 17.8$ A, and $P = 3$ kW

Answers:
(a) $Z = 4.24 - j4.24$ Ω
(b) $Z = 49.2 - j32$ Ω
(c) $Z = 92.9 + j124$ Ω
(d) $Z = 9.47 \pm j6.85$ Ω

16–3 AC POWER ANALYSIS

The nature of ac power analysis can be modeled in terms of the phasor voltage, phasor current, and complex power at the source/load interface in Figure 16–1. Two different types of problems are treated using this model. In the **direct analysis** problem the source and load circuit are given and we are required to find the steady-state responses at one or more interfaces. This type of problem is essentially the same as the phasor circuit analysis problems in Chapter 8, except that we calculate complex powers as well as phasor responses. In the **load-flow** problem we are required to adjust the source so that a prescribed complex power is delivered to the load at a specified interface voltage magnitude. This type of problem arises in electrical power systems where the objective is to supply changing energy demands at a fixed voltage level.

The following examples are direct analysis problems that illustrate the computational tools needed to deal with the load-flow problems discussed in the next section. One of the useful tools is the principle of the **conservation of complex power**, which can be stated as follows:

In a linear circuit operating in the sinusoidal steady state, the sum of the complex powers produced by each independent source is equal to the sum of the complex power absorbed by all other two-terminal elements in the circuit.

To apply this principle, it is important to distinguish between the complex power "produced by" and "absorbed by" a two-terminal element. To do so, we modify our practice of using the passive sign convention. We continue to use the passive convention (current reference directed in at the + voltage mark) for loads, but use the active convention (current reference directed out at the + voltage mark) for sources. In either case, the complex power is always calculated as $S = \mathbf{VI}^* = P + jQ$. The net result is that the average power produced by a source is positive and the average power absorbed by a load is positive.

EXAMPLE 16–2

(a) Calculate the complex power absorbed by each parallel branch in Figure 16–5.

FIGURE 16–5

(b) Calculate the complex power produced by the source and the power factor of the load seen by the source.

SOLUTION:

(a) The voltage across each branch is $15\angle0°$ V and the branch impedances are $Z_1 = 100$ and $Z_2 = 60 - j200$. Therefore, the branch currents are

$$\mathbf{I}_1 = \frac{15\angle0°}{100} = 0.15\angle0° \text{ A}$$

$$\mathbf{I}_2 = \frac{15\angle0°}{60 - j200} = 0.0718\angle73.3° \text{ A}$$

The reference marks for the voltage across and currents through these loads follow the passive sign convention. Hence the complex powers absorbed by the loads are

$$S_1 = (15\angle0°)\mathbf{I}_1^* = (15\angle0°)(0.15\angle-0°) = 2.25\angle0° \text{ VA}$$

$$S_2 = (15\angle0°)\mathbf{I}_2^* = (15\angle0°)(0.0718\angle-73.3°) = 1.08\angle-73.3° \text{ VA}$$

(b) Using KCL, the source current $\mathbf{I}$ is

$$\mathbf{I} = (\mathbf{I}_1 + \mathbf{I}_2) = (0.15\angle0° + 0.0718\angle73.3°)$$

$$= 0.171 + j0.0688 = 0.184\angle21.9° \text{ A}$$

The reference marks for the source current $\mathbf{I}$ and source voltage $15\angle0°$ conform to the active sign convention, and the complex power produced by the source is

$$S = (15\angle0°)\mathbf{I}^* = (15\angle0°)(0.184\angle-21.9°)$$

$$= 2.76\angle-21.9° \text{ VA}$$

The power factor is $\cos(-21.9°) = 0.928$ leading. Alternatively, we can use the conservation of complex power to obtain the complex power produced by the source as the sum of the complex powers delivered to the passive elements:

$$S_1 + S_2 = 2.25\angle0° + 1.08\angle-73.3°$$

$$= 2.25 + j0 + 0.310 - j1.03$$

$$= 2.56 - j1.03 = 2.76\angle-21.9° \text{ VA}$$

This result is the same as the answer obtained previously.   ■

EXAMPLE 16–3

Find the complex power produced by each source in Figure 16–6 when $\mathbf{V}_{S1} = 440\angle0°$ V (rms) and $\mathbf{V}_{S2} = 500\angle0°$ V (rms).

FIGURE 16-6

SOLUTION:
Since both voltage sources are connected to ground, the voltage at node A is the only unknown node voltage in the circuit. By inspection, the node equation at node A is

$$\left[\frac{1}{75} + \frac{1}{j25} + \frac{1}{j30}\right]\mathbf{V}_A = \frac{\mathbf{V}_{S1}}{j25} + \frac{\mathbf{V}_{S2}}{j30} = -j34.27$$

which yields $\mathbf{V}_A = 452 - j82.2$ V. The current supplied by source no. 1 is

$$\mathbf{I}_{S1} = \frac{\mathbf{V}_{S1} - \mathbf{V}_A}{j25} + \frac{\mathbf{V}_{S1} - \mathbf{V}_{S2}}{j45} = 3.29 + j1.81 \text{ A (rms)}$$

The complex power supplied by source no. 1 is

$$S_{S1} = \mathbf{V}_{S1}\mathbf{I}_{S1}^* = 440 \times (3.29 - j1.83)$$
$$= 1450 - j796 \quad \text{VA}$$

The current supplied by source no. 2 is

$$\mathbf{I}_{S2} = \frac{\mathbf{V}_{S2} - \mathbf{V}_A}{j30} + \frac{\mathbf{V}_{S2} - \mathbf{V}_{S1}}{j45} = 2.74 - j2.93 \quad \text{A (rms)}$$

and the complex power supplied by source no. 2 is

$$S_{S2} = \mathbf{V}_{S2}\mathbf{I}_{S2}^* = 500 \times (2.74 + j2.93)$$
$$= 1370 + j1460 \quad \text{VA} \qquad \blacksquare$$

EXAMPLE 16-4

Given the transformer circuit in Figure 16–7, find the complex power produced by the input voltage source and the power absorbed by the 100-Ω load resistor.

FIGURE 16-7

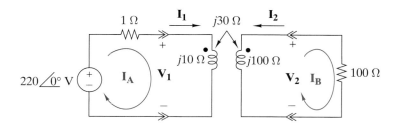

SOLUTION:
Using the voltages and mesh currents in Figure 16–7, the KVL equations around the primary and secondary circuits are

$$1 \times \mathbf{I}_A + \mathbf{V}_1 = 220\angle 0°$$
$$100\,\mathbf{I}_B - \mathbf{V}_2 = 0$$

The coil currents are related to the mesh currents as $\mathbf{I}_1 = \mathbf{I}_A$ and $\mathbf{I}_2 = -\mathbf{I}_B$. Both coil currents are directed inward at the winding dots, so the mutual inductive coupling is additive and the i–v characteristics of the transformer in terms of the mesh currents are

$$\mathbf{V}_1 = j10\mathbf{I}_A + j30(-\mathbf{I}_B)$$

$$\mathbf{V}_2 = j30\mathbf{I}_A + j100(-\mathbf{I}_B)$$

Substituting these i–v characteristics into the KVL equation produces the two mesh current equations for the transformer:

$$(1 + j10)\mathbf{I}_A - j30\mathbf{I}_B = 220$$

$$-j30\mathbf{I}_A + (100 + j100)\mathbf{I}_B = 0$$

Solving for the two mesh current produces

$$\mathbf{I}_A = 20 - j20 \text{ A (rms)}$$

$$\mathbf{I}_B = 6 + j0 \text{ A (rms)}$$

Using mesh current $\mathbf{I}_A$, the complex power produced by the source is

$$S_{IN} = (220\angle 0°)\mathbf{I}_A^* = 4400 + j4400 \quad \text{VA}$$

and the power absorbed by the 100-Ω output resistor is

$$S_{OUT} = (100\mathbf{I}_B)\mathbf{I}_B^* = 3600 + j0 \text{ VA} \qquad ■$$

Exercise 16–4

Calculate the complex power delivered by each source in Figure 16–8.

FIGURE 16 - 8

Answers: $S_1 = 0.4 + j0.8$ VA, $S_2 = 1.6 + j1.2$ VA

16–4 LOAD-FLOW ANALYSIS

The analysis of ac electrical power systems is one of the major applications of phasor circuit analysis. Although the loads on power systems change throughout the day, these variations are extremely slow compared with the period of the 50/60-Hz sinusoid involved.[1] Consequently, electrical power system analysis can be carried out using steady-state concepts and phasors. In fact, it was the study of the steady-state performance of ac power equipment that led Charles Steinmetz to advocate phasors in the first place.

In this section we treat ac power analysis using the simple model of an electrical power system in Figure 16–9. This model is a series circuit with

---

1 In the United States commercial ac power systems operate at 60 Hz. In most of the rest of the world the standard operating frequency is 50 Hz.

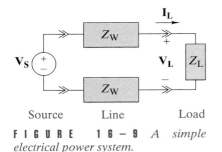

FIGURE 16-9 *A simple electrical power system.*

an ac source connected to a load via power lines whose wire impedances are Z_W. In a **load-flow problem** the complex power delivered to the load is specified and we are asked to find either the source voltage for a given load voltage or the load voltage for a given source voltage. The analysis approach is similar to the unit output method. That is, we begin with conditions at the load and work backward through the circuit to establish the required source voltage. The load-flow problem is different from the maximum power transfer problem studied in Chapter 8. In a maximum power transfer problem the source is fixed and the load is adjusted to achieve a conjugate match. Conjugate matching does not apply to electrical power systems because the load power is fixed and the source is adjusted to meet customer demands.

Large industrial customers are charged for their reactive power, so in some cases it is desirable to reduce the load reactance. Since power system loads are normally inductive, the net reactive power of the load can be reduced by adding a capacitor in parallel with the load. The amount of the negative reactive power taken by the capacitor is selected to cancel some or all of the positive reactance power drawn by the inductive load. Physically, this means that the oscillatory interchange of energy represented by reactive power takes place between the capacitor and inductance in the load, rather than between the load inductance and the lossy power system.

Adding parallel capacitance is called **power factor correction**, since the net power factor of the composite load is increased. If the power factor is increased to unity, then the net reactance is zero and the load is in resonance. Power factor correction reduces the reactive power drain on the power system but does not change the average power delivered to the load.

The following examples illustrate ac power analysis problems, including load flow and power factor correction.

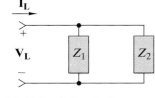

FIGURE 16-10

EXAMPLE 16-5

In this problem the two parallel connected impedances in Figure 16–10 are the load in the power system model in Figure 16–9. With $V_L = 480$ V (rms), load Z_1 draws an average power of 10 kW at a lagging power factor of 0.8 and load Z_2 draws 12 kW at a lagging power factor of 0.75. The line impedances shown in Figure 16–9 are $Z_W = 0.35 + j1.5 \ \Omega$.

(a) Find the total complex power delivered to the composite load.
(b) Find the apparent power delivered by the source and the source voltage phasor $\mathbf{V}_S$.
(c) Calculate the transmission efficiency of the system.

SOLUTION:
(a) In this example we are given the power factor (pf) and the average power (*P*) for each load. Because the power factor angle is $\theta = \cos^{-1}(\text{pf})$ and $Q = P \tan \theta$, we can find the complex power for each load as $S = P + P \tan(\cos^{-1} \text{pf})$, hence

$$S_{L1} = 10 + j10 \tan(\cos^{-1} 0.8) = 10 + j7.5 \text{ kVA}$$

$$S_{L2} = 12 + j12 \tan(\cos^{-1} 0.75) = 12 + j10.6 \text{ kVA}$$

The total complex power delivered to the composite load is

$$S_L = S_{L1} + S_{L2} = 22 + j18.1 = 28.5\angle 39.4° \text{ kVA}$$

(b) Using the load voltage as the phase reference, we find the load current to be

$$\mathbf{I}_L^* = \frac{S_L}{\mathbf{V}_L} = \frac{28500\angle 39.4°}{480\angle 0°} = 59.4\angle 39.4° \quad \text{A (rms)}$$

or $\mathbf{I}_L = 59.4\angle -39.4°$ A (rms). To find the source power, we need to find the complex power lost in the transmission to the line:

$$S_W = 2|\mathbf{I}_L|^2(R_W + jX_W) = 2 \times (59.4)^2 (0.35 + j1.5)$$

$$= 2.47 + j10.6 \text{ kVA}$$

Using the conservation of complex power, we find that the source must produce

$$S_S = S_L + S_W = 24.5 + j28.7 = 37.7\angle 49.5° \quad \text{kVA}$$

For the series model in Figure 16-9 the source and load currents are equal so the source power is $S_S = \mathbf{V}_S(\mathbf{I}_L)^*$. Given the source power and load current, we find that the required source voltage is

$$\mathbf{V}_S = \frac{S_S}{\mathbf{I}_L^*} = \frac{37700\angle 49.5°}{59.4\angle 39.4°} = 635\angle 10.1° \quad \text{V (rms)}$$

(b) The transmission efficiency is defined in terms of source and load average power as

$$\eta = \frac{P_L}{P_S} \times 100\% = \frac{22}{24.5} \times 100 = 89.8\% \qquad \blacksquare$$

EXAMPLE 16-6

A power system modeled by the circuit in Figure 16-9 delivers a load power of $S_L = 25 + j10$ kVA when the source voltage is $\mathbf{V}_S = 600\angle 0°$. The line impedances are $Z_W = 0.35 + j1.5$ Ω. Find the load current, load voltage, and transmission efficiency.

SOLUTION:
In the preceding example the known quantities are the load power and load voltage. In this example the load power and the source voltage are given. As a result both the load current and the load voltage are unknowns. We can write two constraints on these unknowns. First, the specified complex power delivered to the load requires that

$$S_L = 25000 + j10000 = \mathbf{V}_L\mathbf{I}_L^*$$

Next, writing a KVL equation around the loop in Figure 16-9 yields $\mathbf{V}_S - \mathbf{V}_L - 2Z_W\mathbf{I}_L = 0$. Inserting the known quantities and solving for $\mathbf{V}_L$ produces

$$\mathbf{V}_L = 600 - (0.7 + j3)\mathbf{I}_L$$

Substituting this result into the complex power constraint yields

$$25000 + j1000 - 600\,\mathbf{I}_L^* + (0.7 + j3)\mathbf{I}_L\mathbf{I}_L^* = 0$$

We have a single complex constraint in which the only unknown is the load current phasor $\mathbf{I}_L$. This condition requires both the real part and the imaginary part of the constraint to vanish. Hence, the constraint can be written in terms of two real value functions. Writing the load current phasor as $\mathbf{I}_L = I_1 + jI_2$, hence $\mathbf{I}_L\mathbf{I}_L^* = I_1^2 + I_2^2$, the constraints on the real part and the imaginary part are

$$\text{Real Part:} \quad 25000 - 600\,I_1 + 0.7(I_1^2 + I_2^2) = 0$$
$$\text{Imag. Part:} \quad 10000 + 600\,I_2 + 3(I_1^2 + I_2^2) = 0$$

We have now set up the equations so that MATLAB can be used to solve the complex constraints. In the MATLAB command window we write the following statements:

```
syms I1 I2
FR=25000-600*I1+0.7*(I1^2+I^2);
FI=10000+600*I2+3*(I1^2+I2^2);
```

The first statement declares I1 and I2 to be symbolic variables. The next two statements define the real part constraint FR and the imaginary part constraint FI as functions of the form $f(x, y) = 0$. The MATLAB solve command is set up to solve equations of this type. Accordingly, we write the command

```
[I1,I2]=solve(FR,FI)
```

and MATLAB responds with the values of I1 and I2 that satisfy the conditions FR=0 and FI=0, namely,

```
I1=

[55000/949-700/2847*2669^(1/2)]
[55000/949+700/2847*2669^(1/2)]

I2=

[-246200/2847+1000/949*2669^(1/2)]
[-246200/2847-1000/949*2669^(1/2)]
```

MATLAB returns two possible solutions for the load current. The first solution is

$$\mathbf{I}_L = I_1 + jI_2$$
$$= \left(\frac{55000}{949} - \frac{700}{2847}\sqrt{2669}\right) + j\left(\frac{-246200}{2847} + \frac{1000}{949}\sqrt{2669}\right)$$
$$= 45.253 - j32.083 \text{ A}$$

which produces a load voltage phasor of

$$\mathbf{V}_L = 600 - (0.7 + j3)\mathbf{I}_L = 472.208 - j113.333 \text{ V}$$

The apparent power delivered by the source is

$$S_S = \mathbf{V}_S\mathbf{I}_L{}^* = (600 + j0)(45.253 + j32.038)$$
$$= 27.2 + j19.2 \ \text{kVA}$$

and the transmission efficiency for the first solution is

$$\eta = \frac{P_L}{P_S} \times 100\% = \frac{25}{27.2} \times 100 = 91.9\%$$

The second MATLAB solution is

$$\mathbf{I}_L = I_1 + jI_2$$
$$= \left(\frac{55000}{949} + \frac{700}{2847}\sqrt{2669}\right) + j\left(\frac{-246200}{2847} - \frac{1000}{949}\sqrt{2669}\right)$$
$$= 70.658 - j140.916 \ \text{A}$$

which produces a load voltage phasor of

$$\mathbf{V}_L = 600 - (0.7 + j3)\mathbf{I}_L = 127.790 - j113.333 \ \text{V}$$

The apparent power delivered by the source is

$$S_S = \mathbf{V}_S\mathbf{I}_L{}^* = (600 + j0)(70.658 + j140.915)$$
$$= 42.39 + j84.55 \ \text{kVA}$$

and the transmission efficiency for the second solution is

$$\eta = \frac{P_L}{P_S} \times 100\% = \frac{25}{42.395} \times 100 = 59.0\%$$

Several comments are in order. This type of load flow problem may have several solutions, two in this case. On the basis of transmission efficiency we would choose the first solution over the second. This simple example illustrates some features of the general load flow problem in large power systems. Such systems have a number of generating stations with different efficiencies and many geographically separated load centers. The load flow problem involves controlling the output of the different generating stations to meet the total load requirements in a way that maximizes the system efficiency. ■

EXAMPLE 16–7

Using the loads and line impedances defined in Example 16–5,

(a) Find the parallel capacitance needed so that the load power factor is at least 0.95. The power system frequency is 60 Hz.
(b) Find the transmission efficiency with the capacitor connected.

SOLUTION:
(a) To determine the capacitance, it is necessary to relate the reactive power of a capacitor to the load voltage. Since the capacitor is in parallel with the load, the current through it is $\mathbf{I}_C = j\omega C\mathbf{V}_L$. Therefore, the reactive power of the capacitor can be written as

$$Q_C = |\mathbf{I}_C|^2 X_C = |j\omega C\mathbf{V}_L|^2 \left(\frac{-1}{\omega C}\right)$$

$$= |\mathbf{V}_L|^2(-\omega C)$$

Note that Q_C is negative. In Example 16–5 the complex power of the two parallel loads is found to be

$$S_L = P_L + jQ_L = 22 + j18.1 \quad \text{kVA}$$

When the capacitor is connected the complex power delivered to the composite load becomes

$$S_L = P_L + j(Q_L + Q_C)$$

Since the capacitor is a reactive element, it only changes the reactive power and not the average power. To correct the power factor to at least 0.95 requires that

$$\cos\theta = \frac{P_L}{\sqrt{P_L^2 + (Q_L + Q_C)^2}} \geq 0.95$$

Solving for $-Q_C$ yields

$$-Q_C \geq Q_L - P_L\sqrt{1/(0.95)^2 - 1} = 10.9 \quad \text{kVAR}$$

which means that a lower bound on the capacitance is

$$C = \frac{-Q_C}{V_L^2\omega} \geq \frac{10,900}{(480)^2(2\pi60)} = 125 \ \mu\text{F}$$

This yields a lower bound on the required capacitance. An upper bound is found by increasing the load-power factor to unity ($\cos\theta = 1$). The required capacitive reactance is

$$-Q_C = Q_L = 18.1 \quad \text{kVAR}$$

which yields an upper bound on the capacitance of

$$C = \frac{-Q_C}{V_L^2\omega} \leq \frac{18,100}{(480)^2(2\pi60)} = 208 \ \mu\text{F}$$

Thus, the design requirement can be met by any capacitance in the range from 125 μF to 208 μF.

(b) With the minimum capacitance of $C = 125$ μF connected in parallel with the load, the apparent power delivered to the composite load is $S_L = 22 + j7.2$ kVA. Using the load voltage for the phase reference, we find that the load current is

$$\mathbf{I}_L^* = \frac{S_L}{\mathbf{V}_L} = \frac{22,000 + j7200}{480\angle0°}$$

$$= 48.2\angle18.1° \quad \text{A (rms)}$$

The apparent power lost in the line is

$$S_W = |\mathbf{I}_L|^2 \, 2(R_W + jX_W) = 1.62 + j6.97 \quad \text{kVA}$$

Hence with power factor correction the source produces

$$S_S = S_L + S_W = 23.6 + j14.2 \ \text{kVA}$$

and the transmission efficiency is

$$\eta = \frac{P_L}{P_S} \times 100\% = \frac{22}{23.6} \times 100 = 93.2\%$$

In Example 16–5 we found the transmission efficiency without power factor correction to be 89.8%. With power factor correction the source delivers the same average power to the load with an increase in efficiency. Reactive power is a burden to a power system even though it represents a lossless interchange of energy at the terminals of the load. ∎

Exercise 16–5

Find the source voltage and apparent power required to deliver 2400 V (rms) to a load that draws 25 kVA at a 0.85 lagging power factor from a line with a total line impedance of $2Z_W = 4 + j20 \ \Omega$.

   Answers: 2.55 kV (rms) and 26.6 kVA at a lagging power factor of 0.82

APPLICATION NOTE: EXAMPLE 16−8

The electrical power for most residential customers in the United States is supplied by the 60-Hz, 110/220-V (rms) single-phase, three-wire system modeled in Figure 16–11. The term *single phase* means that the phasors representing the two source voltages are in phase. The three lines connecting the sources and loads are labeled A, B, and N (for neutral). The impedances Z_W and Z_N are small compared with the load impedances, so the load voltages differ from the source voltages by only a few percent. The impedances Z_1 and Z_2 connected from Lines A or B to neutral represent small appliance and lighting loads which require 110 V (rms) service. The impedances Z_3 connected between lines A and B are heavier loads that require 220 V (rms) service, such as water heaters or clothes dryers.

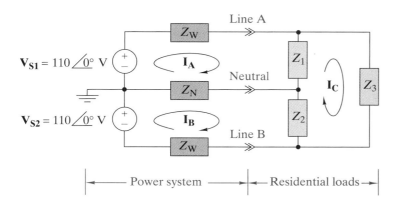

FIGURE 16−11 *Residential power distribution circuit.*

When the two source voltages are exactly equal and $Z_1 = Z_2$ the system is said to be balanced. Under balanced conditions the current in the neutral wire is zero. To show why the neutral current is zero, we write two mesh-current equations:

$$\text{Mesh A: } (Z_W + Z_1 + Z_N)\mathbf{I}_A - Z_N\mathbf{I}_B - Z_1\mathbf{I}_C = \mathbf{V}_{S1}$$

$$\text{Mesh B: } -Z_N\mathbf{I}_A + (Z_W + Z_2 + Z_N)\mathbf{I}_B - Z_2\mathbf{I}_C = \mathbf{V}_{S2}$$

For balanced conditions $\mathbf{V}_{S1} = \mathbf{V}_{S2}$ and $Z_1 = Z_2 = Z_L$. Subtracting the mesh B equation from the mesh A equation yields the condition

$$(Z_W + Z_L + 2Z_N)(\mathbf{I}_A - \mathbf{I}_B) = 0$$

This condition requires $\mathbf{I}_A - \mathbf{I}_B = 0$, since the impedance sum cannot be zero for all loads. Therefore, the net current in the neutral line is zero and theoretically the neutral wire can be disconnected. In practice, the balance is never perfect and the neutral line is included for safety reasons. But even so, the current in the neutral is usually less than the line currents, so losses in the feeder lines are reduced.

16–5 THREE-PHASE CIRCUITS

The three-phase system shown in Figure 16–12 is the predominant method of generating and distributing ac electrical power. The system uses four lines (A, B, C, N) to transmit power from the source to the loads. The symbols stand for the three phases A, B, and C, and a neutral line labeled N. The three-phase generator in Figure 16–12 is modeled as three independent sources, although the physical hardware is a single unit with three separate windings. Similarly, the loads are modeled as three separate impedances, although the actual equipment may be housed within a single container.

F I G U R E 1 6 – 1 2 *A three-phase ac electrical power system.*

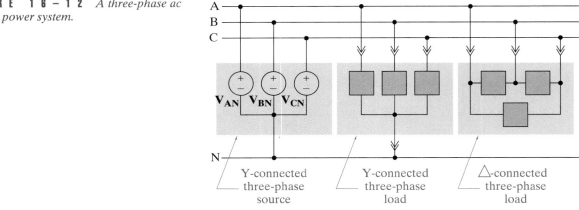

A three-phase system involves three voltages and three currents. In a balanced three-phase system the generator produces phasor voltages that are equal in magnitude and symmetically displaced in phase at 120° intervals. When the three-phase load is balanced (equal impedances), the resulting phasor currents have equal magnitude and are symmetically displaced in phase at 120° intervals. Thus, a **balanced three-phase** system is one in which the phasor currents and voltages have equal magnitudes and phase differences of 120°.

The terminology Y-connected and Δ-connected refers to the two ways the source and loads can be electrically connected. Figure 16–13 shows the same electrical arrangement as Figure 16–12 with the elements rearranged to show the Y and Δ nature of the connections (the Δ is upside down in the figure). The circuit diagrams in the two figures are electrically equivalent, but we will use the form in Figure 16–12 because it highlights the purpose of the system. You need only remember that in a Y-connection the three elements are connected from line to neutral, while in the Δ-connection they are connected from line to line. In most systems the source is Y-connected while the loads can be either Y or Δ, although the latter is more common.

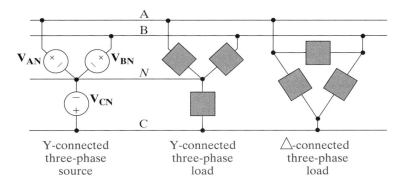

Y-connected
three-phase
source

Y-connected
three-phase
load

Δ-connected
three-phase
load

FIGURE 16–13 *A three-phase power system with the loads rearranged.*

Three-phase sources usually are Y-connected because the Δ-connection involves a loop of voltage sources. Large currents may circulate in this loop if the three voltages do not exactly sum to zero. In analysis situations, a Δ-connection of ideal voltage sources is awkward because it is impossible to determine the current in each source.

We use a double subscript notation to identify voltages in the system. The reason is that there are at least six voltages to deal with: three line-to-line voltages and three line-to-neutral voltages. If we use the usual plus and minus reference marks to define all of these voltages, our circuit diagram would be hopelessly cluttered and confusing. Hence we use two subscripts to define the points across which a voltage is defined. For example, $\mathbf{V}_{XY}$ means the voltage between points X and Y, with an implied plus reference mark at the first subscript (X) and an implied minus at the second subscript (Y).

The three line-to-neutral voltages are called the **phase voltages** and are written in double subscript notation as $\mathbf{V}_{AN}$, $\mathbf{V}_{BN}$, and $\mathbf{V}_{CN}$. Similarly, the three line-to-line voltages, called simply the **line voltages**, are identified as $\mathbf{V}_{AB}$, $\mathbf{V}_{BC}$, and $\mathbf{V}_{CA}$. From the definition of the double subscript notation it follows that $\mathbf{V}_{XY} = -\mathbf{V}_{YX}$. Using this result and KVL, we derive the relationships between the line voltages and phase voltages:

$$\mathbf{V}_{AB} = \mathbf{V}_{AN} + \mathbf{V}_{NB} = \mathbf{V}_{AN} - \mathbf{V}_{BN}$$

$$\mathbf{V}_{BC} = \mathbf{V}_{BN} + \mathbf{V}_{NC} = \mathbf{V}_{BN} - \mathbf{V}_{CN} \qquad (16\text{–}17)$$

$$\mathbf{V}_{CA} = \mathbf{V}_{CN} + \mathbf{V}_{NA} = \mathbf{V}_{CN} - \mathbf{V}_{AN}$$

A balanced three-phase source produces phase voltages that obey the following two constraints:

$$|\mathbf{V}_{AN}| = |\mathbf{V}_{BN}| = |\mathbf{V}_{CN}| = V_P$$

$$\mathbf{V}_{AN} + \mathbf{V}_{BN} + \mathbf{V}_{CN} = 0 + j0$$

That is, the phase voltages have equal amplitudes (V_P) and sum to zero. There are two ways to satisfy these constraints:

Positive Phase Sequence	Negative Phase Sequence	
$\mathbf{V}_{AN} = V_P\angle 0°$	$\mathbf{V}_{AN} = V_P\angle 0°$	
$\mathbf{V}_{BN} = V_P\angle{-}120°$	$\mathbf{V}_{BN} = V_P\angle{-}240°$	(16–18)
$\mathbf{V}_{CN} = V_P\angle{-}240°$	$\mathbf{V}_{CN} = V_P\angle{-}120°$	

Figure 16–14 shows the phasor diagrams for the positive and negative phase sequences. It is apparent that both sequences involve three equal-length phasors that are separated by angles of 120°. As a result, the sum of any two phasors cancels the third. In the positive sequence the phase B voltage lags the phase A voltage by 120°. In the negative sequence phase B lags by 240°. It also is apparent that we can convert one phase sequence into the other by simply interchanging the labels on lines B and C. From a circuit analysis viewpoint, there is no conceptual difference between the two sequences. Consequently, in analysis problems we will use the positive phase sequence unless otherwise stated.

However, the phrase *no conceptual difference* does not mean that phase sequence is unimportant. It turns out that three-phase motors run in one direction when the positive sequence is applied, and in the opposite direction for the negative sequence. This could be a matter of some importance if the motor is driving a conveyor belt at a sewage treatment facility. In practice, it is essential that there be no confusion about which is line A, B, and C and whether the source phase sequence is positive or negative.

A simple relationship between the line and phase voltages is obtained by substituting the positive phase sequence voltages from Eq. (16–18) into the phasor sums in Eq. (16–17). For the first sum

$$\mathbf{V}_{AB} = \mathbf{V}_{AN} - \mathbf{V}_{BN}$$

$$= V_P\angle 0° - V_P\angle{-}120°$$

$$= V_P(1 + j0) - V_P(-1/2 - j\sqrt{3}/2) \qquad (16\text{–}19)$$

$$= V_P(3/2 + j\sqrt{3}/2)$$

$$= \sqrt{3}V_P\angle 30°$$

Using the other sums, we find the other two positive sequence line voltages as

FIGURE 16–14 *Two possible phase sequences: (a) Positive. (b) Negative.*

$$\mathbf{V}_{BC} = \sqrt{3}V_P\angle -90°$$

$$\mathbf{V}_{CA} = \sqrt{3}V_P\angle -210° \qquad (16-20)$$

Figure 16–15 shows the phasor diagram of these results. The line voltage phasors have the same amplitude and are displaced from each other by 120°. Hence they obey equal-amplitude and zero-sum constraints like the phase voltages.

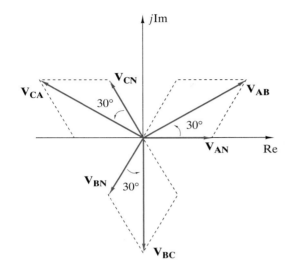

FIGURE 16-15 *Phasor diagram showing phase and line voltages for the positive phase sequence.*

If we denote the amplitude of the line voltages as V_L, then

$$V_L = \sqrt{3}V_P \qquad (16-21)$$

In a balanced three-phase system the line voltage amplitude is $\sqrt{3}$ times the phase voltage amplitude. This ratio appears in equipment descriptions such as 277/480 V three phase, where 277 is the phase voltage and 480 the line voltage.

It is necessary to choose one of the phasors as the zero-phase reference when defining three-phase voltages and currents. Usually the reference is the line A phase voltage (i.e., $\mathbf{V}_{AN} = V_P\angle 0°$), as illustrated in Figures 16–14 and 16–15. Unless otherwise stated, $\mathbf{V}_{AN}$ will be used as the phase reference in this chapter.

Exercise 16-6

A balanced Y-connected three-phase source produces positive sequence phase voltages with 2400-V (rms) amplitudes. Write expressions for the phase and line voltage phasors.

Answers:

$\mathbf{V}_{AN} = 2400\angle 0°$	$\mathbf{V}_{BN} = 2400\angle -120°$	$\mathbf{V}_{CN} = 2400\angle -240°$
$\mathbf{V}_{AB} = 4160\angle +30°$	$\mathbf{V}_{BC} = 4160\angle -90°$	$\mathbf{V}_{CA} = 4160\angle -210°$

Given that $\mathbf{V}_{BC} = 480\angle+135°$ in a balanced, positive sequence, three-phase system, write expressions for the three phase-voltage phasors.

A n s w e r :

$$\mathbf{V}_{AN} = 277\angle-135° \qquad\qquad \mathbf{V}_{BN} = 277\angle+105° \qquad\qquad \mathbf{V}_{CN} = 277\angle-15°$$

16–6 T H R E E - P H A S E A C P O W E R A N A L Y S I S

This section treats the analysis of balanced three-phase circuits. We first treat the direct analysis problem beginning with the Y-connected source and load shown in Figure 16–16. In a direct analysis problem we are given the source phase voltages $\mathbf{V}_{AN}$, $\mathbf{V}_{BN}$, and $\mathbf{V}_{CN}$ and the load impedances Z. Our objective is to determine the three line currents $\mathbf{I}_A$, $\mathbf{I}_B$, and $\mathbf{I}_C$ and the total complex power delivered to the load.

F I G U R E 1 6 – 1 6 *A balanced three-phase system with a Y-connected source and load.*

Y-CONNECTED SOURCE AND Y-CONNECTED LOAD

The load in Figure 16–16 is balanced because the phase impedances in the legs of the Y are equal. With the neutral point at the load connected, the voltage across each phase impedance is a phase voltage. Using $\mathbf{V}_{AN}$ as the phase reference, we find that the line currents are

$$\mathbf{I}_A = \frac{\mathbf{V}_{AN}}{Z} = \frac{V_P\angle 0°}{|Z|\angle\theta} = \frac{V_P}{|Z|}\angle-\theta$$

$$\mathbf{I}_B = \frac{\mathbf{V}_{BN}}{Z} = \frac{V_P\angle-120°}{|Z|\angle\theta} = \frac{V_P}{|Z|}\angle-120°-\theta \qquad (16\text{–}22)$$

$$\mathbf{I}_C = \frac{\mathbf{V}_{CN}}{Z} = \frac{V_P\angle-240°}{|Z|\angle\theta} = \frac{V_P}{|Z|}\angle-240°-\theta$$

Figure 16–17 shows the phasor diagram of the line currents and phase voltages.

The line current phasors in Eq. (16–22) and Figure 16–17 have the same amplitude I_L, where

$$I_L = \frac{V_P}{|Z|} \text{ (Y-connected Load)} \qquad (16\text{–}23)$$

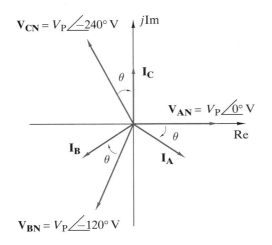

FIGURE 16-17 *Line currents and phase voltages in a balanced three-phase system.*

The line currents have equal amplitudes and are symmetrically disposed at 120° intervals, so they obey the zero-sum condition $\mathbf{I}_A + \mathbf{I}_B + \mathbf{I}_C = 0$. Applying KCL at the neutral point of the load in Figure 16–16, we find that $\mathbf{I}_N = \mathbf{I}_A + \mathbf{I}_B + \mathbf{I}_C = 0$.

Thus, in a balanced Y-Y circuit there is no current in the neutral line. The neutral connection could be replaced by any impedance whatsoever, including infinity, without affecting the power delivered to the load. In other words, the neutral wire can be disconnected without changing the circuit response. Real systems may or may not have a neutral wire, but in solving three-phase problems it is helpful to draw the neutral line because it serves as a reference point for the phase voltages.

The total complex power delivered to the load is

$$
\begin{aligned}
S_{\mathrm{L}} &= \mathbf{V}_{\mathrm{AN}}\mathbf{I}_A^* + \mathbf{V}_{\mathrm{BN}}\mathbf{I}_B^* + \mathbf{V}_{\mathrm{CN}}\mathbf{I}_C^* \\
&= (V_{\mathrm{P}}\angle 0)(I_{\mathrm{L}}\angle\theta) + (V_{\mathrm{P}}\angle -120°)(I_{\mathrm{L}}\angle 120° + \theta) + \\
&\quad (V_{\mathrm{P}}\angle -240°)(I_{\mathrm{L}}\angle 240° + \theta) \\
&= 3V_{\mathrm{P}}I_{\mathrm{L}}\angle\theta \qquad\qquad\qquad\qquad\qquad\qquad (16\text{–}24)
\end{aligned}
$$

Since $V_{\mathrm{P}} = V_{\mathrm{L}}/\sqrt{3}$, the expression for complex power can also be written using the line voltage:

$$
S_{\mathrm{L}} = \sqrt{3}V_{\mathrm{L}}I_{\mathrm{L}}\angle\theta \qquad\qquad (16\text{–}25)
$$

In either Eq. (16–24) or (16–25) the power factor angle θ is the angle of the per-phase impedance of the Y-connected load.

EXAMPLE 16-9

When the line voltage is 480 V (rms), a balanced Y-connected load draws an apparent power of 40 kVA at a lagging power factor of 0.9.

(a) Find the per-phase impedance of the load.
(b) Find the three line current phasors using $\mathbf{V}_{\mathrm{AN}}$ as the phase reference.

SOLUTION:

(a) For the given conditions the phase voltage, line current, and power factor angle are

$$V_P = \frac{V_L}{\sqrt{3}} = \frac{480}{\sqrt{3}} = 277 \quad \text{V (rms)}$$

$$I_L = \frac{|S_L|}{\sqrt{3}V_L} = \frac{4 \times 10^4}{\sqrt{3} \times 480} = 48.1 \text{ A (rms)}$$

$$\theta = \cos^{-1}(0.9) = 25.8°$$

and the per-phase impedance of the Y-connected load is

$$Z_L = \frac{V_P}{I_L}\angle\theta = \frac{277}{48.1}\angle 25.8°$$
$$= 5.18 + j2.51 \quad \Omega$$

(b) With $\mathbf{V}_{AN} = 277\angle 0°$ as the phase reference, we calculate the line current $\mathbf{I}_A$ as

$$\mathbf{I}_A = \frac{\mathbf{V}_{AN}}{Z_L} = I_L\angle -\theta = 48.1\angle -25.8° \text{ A (rms)}$$

Hence for a balanced load the other two line current phasors are

$$\mathbf{I}_B = I_L\angle -120° - \theta = 48.1\angle -145.8° \text{ A (rms)}$$
$$\mathbf{I}_C = I_L\angle -240° - \theta = 48.1\angle -265.8° \text{ A (rms)} \quad ■$$

Exercise 16–8

A Y-connected load with $Z = 10 + j4$ Ω/phase is driven by a balanced, positive sequence three-phase generator with $V_L = 4.16$ kV (rms). Using $\mathbf{V}_{AN}$ as the phase reference,

(a) Find the line currents.
(b) Find the average and reactive power delivered to the load.

Answers:
(a) $\mathbf{I}_A = 223\angle -21.8°$ A $\mathbf{I}_B = 223\angle -141.8°$ A $\mathbf{I}_C = 223\angle -261.8°$ A
(b) $P_L = 1.49$ MW $Q_L = 0.597$ MVAR

Y-CONNECTED SOURCE AND △-CONNECTED LOAD

We now turn to the balanced △-connected load shown in Figure 16–18. When the objective is to determine the line currents and total complex power, it is convenient to replace the △-connected load by an equivalent Y-connected load. Using Figure 16–19, we can easily determine the required transformation. Looking between any two terminals in the △-connected load we see an impedance $Z\|2Z$. Similarly, looking between any two terminals in the Y-connected load we see an impedance $2Z_Y$. For these two circuits to be equivalent we must see the same equivalent impedance between any two terminals. Hence, equivalence requires

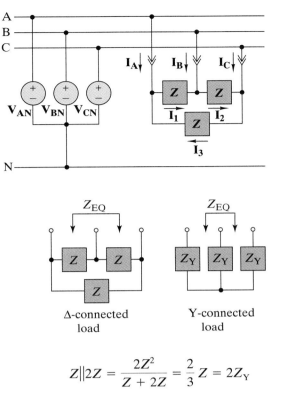

FIGURE   16-19 *Equivalent Y-connected and Δ-connected loads.*

Δ-connected
load

Y-connected
load

$$Z \| 2Z = \frac{2Z^2}{Z + 2Z} = \frac{2}{3} Z = 2Z_Y$$

or

$$Z_Y = \frac{Z}{3} \qquad (16\text{–}26)$$

Using the Δ-to-Y transformation, we reduce the problem to a circuit in which the source and load are Y-connected. The resulting Y-Y configuration can then be analyzed to determine line currents and total power, as discussed previously.

However, when we need to know the current or power delivered to each leg of the delta load, we must determine the phase currents $\mathbf{I}_1$, $\mathbf{I}_2$, and $\mathbf{I}_3$ shown in Figure 16–18. The phase currents can be expressed in terms of the phase impedances and the line voltage. Assuming the positive phase sequence and using $\mathbf{V}_{AN}$ as the phase reference, these expressions are

$$\mathbf{I}_1 = \frac{\mathbf{V}_{AB}}{Z} = \frac{V_L \angle 30°}{|Z| \angle \theta} = \frac{V_L}{|Z|} \angle 30° - \theta$$

$$\mathbf{I}_2 = \frac{\mathbf{V}_{BC}}{Z} = \frac{V_L \angle -90°}{|Z| \angle \theta} = \frac{V_L}{|Z|} \angle -90° - \theta \qquad (16\text{–}27)$$

$$\mathbf{I}_3 = \frac{\mathbf{V}_{CA}}{Z} = \frac{V_L \angle -210°}{|Z| \angle \theta} = \frac{V_L}{|Z|} \angle -210° - \theta$$

The phase currents have the same amplitude I_P defined as

$$I_P = \frac{V_L}{|Z|} \quad (\Delta\text{-connected Load}) \qquad (16\text{–}28)$$

Although of no immediate physical importance, note that the phase currents sum to zero because they have equal amplitudes and are symmetrically disposed at 120° intervals.

Using the results in Eq. (16–27), the complex power delivered to each leg of the delta load is

$$S_1 = \mathbf{V}_{AB} \times \mathbf{I}_1^* = (V_L \angle 30°)(I_P \angle \theta - 30°) = V_L I_P \angle \theta$$
$$S_2 = \mathbf{V}_{BC} \times \mathbf{I}_2^* = (V_L \angle -90°)(I_P \angle \theta + 90°) = V_L I_P \angle \theta \qquad (16\text{–}29)$$
$$S_3 = \mathbf{V}_{CA} \times \mathbf{I}_3^* = (V_L \angle -210°)(I_P \angle \theta + 210°) = V_L I_P \angle \theta$$

The total complex power delivered to the Δ-connected load is

$$S_L = S_1 + S_2 + S_3 = 3V_L I_P \angle \theta \qquad (16\text{–}30)$$

where the power factor angle θ is the angle of the per-phase impedance of the Δ-connected load.

The result in Eq. (16–30) can be put in the same form as Eq. (16–25) by replacing the phase current I_P by the line current I_L. As noted previously, the line currents for a balanced Δ-connected load can be calculated using a Δ-to-Y transformation. In the transformed circuit the line current amplitude is $I_L = V_P/|Z_Y|$, where Z_Y is the impedance in each leg of the equivalent Y-connected load. But in a balanced system $V_P = V_L/\sqrt{3}$, and according to Eq. (16–26) $Z_Y = Z/3$. Therefore, the amplitudes of the line and phase currents in a Δ-connected load are related as follows:

$$I_L = \frac{V_L/\sqrt{3}}{|Z/3|} = \sqrt{3}\frac{V_L}{|Z|} = \sqrt{3}I_P \qquad (16\text{–}31)$$

When $I_P = I_L/\sqrt{3}$ is substituted into Eq. (16–30), we obtain

$$S_L = \sqrt{3}V_L I_L \angle \theta \qquad (16\text{–}32)$$

Equations (16–25) and (16–32) are identical, which means that the relationship applies to balanced three-phase loads whether Y- or Δ-connected. In either case the power factor angle θ is the angle of the per-phase impedance of the load because the transformation $Z_Y = Z/3$ does not alter the phase angle of the phase impedance.

E X A M P L E   1 6 – 1 0

A Δ-connected load with $Z = 40 + j30\ \Omega$/phase is driven by a balanced, positive sequence three-phase generator with $V_L = 2400$ V (rms). Using $\mathbf{V}_{AN}$ as the phase reference,

(a) Find the phase currents.
(b) Find the line currents.
(c) Find the average and reactive power delivered to the load.

S O L U T I O N :
(a) The first phase current is

$$\mathbf{I}_1 = \frac{\mathbf{V}_{AB}}{Z} = \frac{2400 \angle 30°}{40 + j30} = 48.0 \angle -6.87° \text{ A (rms)}$$

Since the circuit is balanced, the other phase currents all have amplitudes of $I_P = 48.0$ A and are displaced at $120°$ intervals.

$$\mathbf{I}_2 = 48.0\angle-126.87° \text{ A (rms)}$$

$$\mathbf{I}_3 = 48.0\angle-246.87° \text{ A (rms)}$$

(b) Using $V_P = 2400/\sqrt{3}$ and Δ-Y transformation with $Z_Y = (40 + j30)/3$,
we find that the phase A line current is

$$\mathbf{I}_A = \frac{\mathbf{V}_{AN}}{Z_Y} = \frac{2400/\sqrt{3}}{(40 + j30)/3} = 83.1\angle-36.9° \text{ A (rms)}$$

Since the circuit is balanced, the other line currents have amplitudes of
$I_L = 83.1$ A and are displaced at $120°$ intervals.

$$\mathbf{I}_B = 83.1\angle-156.9° \text{ A (rms)}$$

$$\mathbf{I}_C = 83.1\angle-276.9° \text{ A (rms)}$$

(c) The complex power delivered to the load is

$$S_L = \sqrt{3}V_L I_L\angle\theta = \sqrt{3} \times 2400 \times 83.1\angle36.9°$$
$$= 345\angle36.9° \text{ kVA}$$

Therefore, $P_L = 276$ kW and $Q_L = 207$ kVAR. ∎

Exercise 16–9

Exercise 16–8 involved a balanced Y-connected load with $Z = 10 + j4$ Ω/phase
and a line voltage of $V_L = 4.16$ kV (rms). In this exercise these same parameters
apply to a Δ-connected load. Using $\mathbf{V}_{AN}$ as the phase reference,

(a) Find phase currents.
(b) Find the line currents.
(c) Find the average and reactive power delivered to the load.

Answers:
(a) $\mathbf{I}_1 = 386\angle+8.20°$ A, $\mathbf{I}_2 = 386\angle-111.8°$ A, $\mathbf{I}_3 = 386\angle-231.8°$ A
(b) $\mathbf{I}_A = 669\angle-21.8°$ A, $\mathbf{I}_B = 669\angle-141.8°$ A, $\mathbf{I}_C = 669\angle-261.8°$ A
(c) $P_L = 4.47$ MW, $Q_L = 1.79$ MVAR

THREE-PHASE LOAD-FLOW ANALYSIS

The analysis and examples given thus far treat direct analysis problems in
which the source parameters are given and the power flow is unknown. In
a load-flow problem we are required to find the source or load voltages
when the power flow is given.

A simple model of the three-phase circuit for the load-flow problem is
shown in Figure 16–20. The impedance Z_W represents the wire impedances
of the power lines connecting the source and load. It is clear even in this
very simple case that including all three phases in a circuit diagram is un-
wieldy. In more complicated situations, including all three phases in circuit
diagrams tends to obscure the working of the system. In a balanced three-
phase system we have seen that once we find one of the line currents or
voltages, the others are easily derived by shifting the known response at

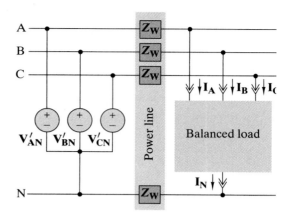

120° intervals. Thus, in effect, we really do not need all three phases in the circuit diagram to analyze balanced three-phase systems.

A simpler representation of a balanced three-phase system is obtained by omitting the neutral and using one line to represent all three phases. Figure 16–21 is a single-line representation of the circuit in Figure 16–20. In power system terminology a **bus** is a group of conductors that serve as a common connection for two or more circuits. In the single-line diagram in Figure 16–21 the buses are represented by short horizontal lines. Bus no. 1 is a generator bus connecting a three-phase source represented by the circle and a transmission line represented by Z_W. Bus no. 2 is a load bus connecting the transmission line to a balanced three-phase load represented by an arrow indicating the delivery of complex power. The line voltages and complex power flow are written beside the buses using a subscript that identifies the bus.

F I G U R E 1 6 – 2 1 *Single-line representation of a three-phase system.*

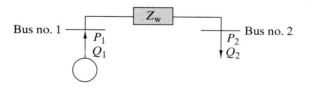

Using single-line diagrams makes three-phase load-flow analysis similar to the single-phase two-wire load-flow problems in Sect. 16–5. The only complication here is that we must account for the power in all three phases and we must distinguish between the line and phase voltages. In a three-phase load-flow problem the bus voltages and line currents are represented by their phasor magnitudes without reference to their phase angles. A load-flow problem does not require that we find these phase angles since the required phase information is carried by the specified complex power.

EXAMPLE 16-11

The single-line diagram in Figure 16-22 shows a power system with a generator bus connected to a load bus via power lines with $Z_W = 4 + j16 \ \Omega$. The line voltage at the load bus is $V_{L2} = 4800$ V (rms) and the load draws $P_2 = 100$ kW at 0.8 lagging power factor. Find the complex power produced by the source, the line voltage at the source bus, and the transmission efficiency.

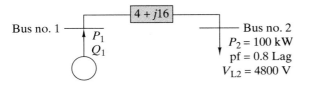

Bus no. 1 ⎯⎯⎯⎯ $4 + j16$ ⎯⎯⎯⎯ Bus no. 2

P_1
Q_1

$P_2 = 100$ kW
pf = 0.8 Lag
$V_{L2} = 4800$ V

FIGURE 16-22

SOLUTION:

The total complex power delivered to the three-phase load connected to bus no. 2 is

$$S_2 = 100 + j100 \tan(\cos^{-1} 0.8) = 100 + j75$$

$$= 125\angle 36.9° \text{ kVA}$$

Using Eq. (16-32) to calculate the line current at the load bus yields

$$I_L = \frac{|S_2|}{\sqrt{3} \ V_{L2}} = \frac{125000}{\sqrt{3} \times 4800} = 15.04 \text{ A (rms)}$$

The total complex power absorbed by the power line is

$$S_W = 3I_L^2(R_W + jX_W) = 3(15.04)^2(4 + j16)$$

$$= 2.71 + j10.8 \text{ kVA}$$

The preceding computation includes the three lines for phases A, B, and C but does not include the neutral wire since the circuit is balanced and there is no current in the neutral line. Using the conservation of complex power yields the source power as

$$S_1 = S_2 + S_W = 102.7 + j85.8$$

$$= 133.8\angle 39.9° \text{ kVA}$$

We now use Eq. (16-32) again to find the line voltage at the source bus as

$$V_{L1} = \frac{|S_1|}{\sqrt{3}I_L} = \frac{133800}{\sqrt{3} \times 15.04}$$

$$= 5.14 \text{ kV (rms)}$$

The transmission efficiency is

$$\eta = \frac{P_2}{P_1} \times 100\% = \frac{100}{102.7} \times 100 = 97.4\% \qquad ■$$

EXAMPLE 16–12

The transmission line in Figure 16–23 has a maximum rated capacity of 250 kVA at line voltage of 4160 V (rms). When operating at rated capacity, the resistive and reactive voltage drops in the line are 2.5% and 6% of the rated voltage, respectively.

FIGURE 16–23

(a) Find the wire impedance Z_W.
(b) The system is operating with $P_1 = 61.1$ kW and $S_2 = 75$ kVA at pf = 0.8 lagging. Find the line voltage at the load bus.

SOLUTION:

(a) When the power line is operating at maximum rated capacity, the line current is

$$I_L = \frac{|S_L|}{\sqrt{3}\, V_L} = \frac{250{,}000}{\sqrt{3} \times 4160} = 34.7 \text{ A (rms)}$$

The magnitudes of the voltage across line resistance and reactance are

$$I_L R_W = 4160 \times 0.025 = 104 \text{ V}$$

$$I_L X_W = 4160 \times 0.060 = 250 \text{ V}$$

Hence the wire impedance of the line is

$$Z_W = \frac{104 + j250}{I_L} = 3 + j7.2 \ \Omega$$

(b) When the system operates with $P_1 = 61.1$ kW and $S_2 = 75$ kVA at pf = 0.8 lagging, the average power lost in the line is $P_W = P_1 - S_2 \times 0.8 = 1.1$ kW. The line current is

$$I_L = \sqrt{\frac{P_W}{3\, R_W}} = \sqrt{\frac{1100}{3 \times 3}} = 11.06 \text{ A (rms)}$$

and the line voltage at load bus is

$$V_{L2} = \frac{|S_2|}{\sqrt{3} I_L} = 3.92 \text{ kV (rms)} \qquad \blacksquare$$

Exercise 16–10

A balanced three-phase load draws 8 kVA at lagging power factor of 0.9 when the phase voltage is 277 V (rms). The load is fed by a balanced three-phase source via lines with $Z_W = 0.3 + j5 \ \Omega$.

(a) Find the line current.
(b) Find the line voltage at the source.

Answers:
(a) $I_L = 9.63$ A
(b) $V_L = 526$ V

APPLICATION NOTE: EXAMPLE 16-13

Electrical power is generated and transmitted in three-phase form even though most residential and small commercial customers are single-phase loads. The primary reason is that three-phase devices and transmission lines are smaller and more efficient than the corresponding single-phase equipment handling the same power. Part of the advantage enjoyed by three phase is that the instantaneous power in a balanced three-phase circuit or device is constant.

To show that $p_T(t)$ is constant, we write the instantaneous power in each phase of the balanced three-phase circuit:

$$p_A(t) = v_{AN}(t) \times i_A(t) = [\sqrt{2}V_P \cos(\omega t)] \times [\sqrt{2}I_L \cos(\omega t - \theta)]$$

$$p_B(t) = v_{BN}(t) \times i_B(t) = \sqrt{2}V_P \cos(\omega t - 120°)] \times$$
$$[\sqrt{2}I_L \cos(\omega t - 120° - \theta)]$$

$$p_C(t) = v_{CN}(t) \times i_C(t) = [\sqrt{2}V_P \cos(\omega t - 240°)] \times$$
$$[\sqrt{2}I_L \cos(\omega t - 240° - \theta)]$$

where $\sqrt{2}V_P$ is the peak amplitude of each line-to-neutral voltage and $\sqrt{2}I_L$ is the peak amplitude of each line current. Using the trigonometric identity

$$[\cos x] \times [\cos y] = \frac{1}{2}\cos(x - y) + \frac{1}{2}\cos(x + y)$$

the phase powers can be put into the form

$$p_A(t) = V_P I_L \cos\theta + V_P I_L \cos(2\omega t - \theta)$$

$$p_B(t) = V_P I_L \cos\theta + V_P I_L \cos(2\omega t - 240° - \theta)$$

$$p_C(t) = V_P I_L \cos\theta + V_P I_L \cos(2\omega t - 480° - \theta)$$

Each phase power is a constant or dc term $V_P I_L \cos\theta$ plus a double-frequency ac term. The double-frequency terms all have the same amplitude $V_P I_L$ and are symmetrically disposed at 120° intervals because −480° is the same as −120°. Therefore, the double-frequency sinusoidal terms sum to zero and the total instantaneous power is

$$p_T(t) = p_A(t) + p_B(t) + p_C(t) = 3V_P I_L \cos\theta$$

The fact that the total instantaneous power is constant means that operating three-phase generators and motors involves constant mechanical torque. As a result, there is smoother operation with less vibration at the electromechanical interfaces of the system.

SUMMARY

- In the sinusoidal steady state the instantaneous power at a two-terminal interface contains average and reactive power components. The average power represents a unidirectional transfer of energy from source to load. The reactive power represents a lossless interchange of energy between the source and load.

- In ac power analysis problems amplitudes of voltage and current phasors are expressed in rms values. Complex power is defined as the product of the voltage phasor and the conjugate of the current phasor. The real part of the complex power is the average power in watts (W), the imaginary part is the reactive power in volt-amperes reactive (VAR), and the magnitude is the apparent power in volt-amperes (VA).

- In a direct analysis problem the load impedance, line impedances, and source voltage are given, and the load voltage, current, and power are the unknowns. In a load-flow problem the source and load powers are given, and the unknowns are line voltages and currents at different points.

- Three-phase systems transmit power from source to load on four lines labeled A, B, C, and N. The three line-to-neutral voltages $\mathbf{V}_{AN}$, $\mathbf{V}_{BN}$, and $\mathbf{V}_{CN}$ are called the phase voltages. The three line-to-line voltages $\mathbf{V}_{AB}$, $\mathbf{V}_{BC}$, and $\mathbf{V}_{CA}$ are called line voltages.

- In a balanced three-phase system, (1) the neutral wire carries no current, (2) the line-voltage amplitude V_L is related to the phase voltage amplitude V_P by $V_L = \sqrt{3}\, V_P$, (3) the three line currents $\mathbf{I}_A$, $\mathbf{I}_B$ and $\mathbf{I}_C$ have the same amplitude I_L, and (4) the total complex power delivered to a Y- or a Δ-connected load is $\sqrt{3}V_L I_L \angle \theta$, where θ is the angle of the phase impedance.

PROBLEMS

ERO 16–1 COMPLEX POWER (SECTS. 16–1, 16–2)

Given a linear circuit operating in the sinusoidal steady state,

(a) Find the complex power delivered at a specified interface.
(b) Find the phasor voltages or currents required to deliver a specified complex power at an interface.
(c) Find the load impedance required to draw a specified complex power at a given voltage level.

16–1   Calculate average power, reactive power, and power factor for the circuit in Figure P16–1 using voltages and currents below. State whether the circuit is absorbing or delivering net energy.
   (a) $v(t) = 170 \cos(\omega t - 45°)$ V, $i(t) = 12 \cos(\omega t - 30°)$ A
   (b) $v(t) = 280 \cos(\omega t + 90°)$ V, $i(t) = 1.5 \cos(\omega t - 30°)$ A

16–2   Calculate average power, reactive power, and power factor for the circuit in Figure P16–1 using voltages and currents listed below. State whether the circuit is absorbing or delivering net energy.
   (a) $v(t) = 135 \sin(\omega t)$ V, $i(t) = 1.5 \cos(\omega t + 30°)$ A
   (b) $v(t) = 370 \cos(\omega t)$ V, $i(t) = 11.5 \cos(\omega t + 130°)$ A

FIGURE P16–1

16–3 Calculate average power, reactive power, and power factor for the circuit in Figure P16–1 using the conditions given below.
  (a) $\mathbf{V} = 22.5\angle 0°$ V (rms), $\mathbf{I} = 250\angle -105°$ mA (rms)
  (b) $\mathbf{V} = 120\angle 35°$ V (rms), $\mathbf{I} = 12.5\angle 115°$ A (rms)

16–4 Calculate average power, reactive power, and power factor for the circuit in Figure P16–1 using the conditions given below.
  (a) $\mathbf{V} = 2.4\angle 45°$ kV (rms), $Z_L = 250\angle -10.5°$ Ω
  (b) $Z_L = 300 - j400$ Ω, $\mathbf{I} = 120\angle 25°$ mA (rms)

16–5 Find the power factor of the circuit in Figure P16–1 under the following conditions. State whether the power factor is lagging or leading.
  (a) $S = 1000 + j250$ kVA
  (b) $|S| = 12$ kVA, $Q = -9$ kVA, $\cos \theta > 0$.

16–6 The instantaneous power absorbed by the load in Figure P16–1 is
$p(t) = 2400 + 2400 \cos 2\omega t - 800 \sin 2\omega t$ VA.
  (a) Find P, Q, and the power factor.
  (b) Find the load impedance when the voltage level is 440 V (peak).

16–7 At 2400 V (rms) the load in Figure P16–1 absorbs an apparent power of 20 kVA at a power factor of 0.75 lagging.
  (a) Find the average power, reactive power, and the rms current delivered to the load.
  (b) Find the load impedance.

16–8 The load in Figure P16–1 has a lagging power factor of 0.8 and draws 12.1 A (rms) from a 2400-V (rms) line.
  (a) Find the apparent power delivered to the load and the power factor.
  (b) Find the load impedance.

16–9 The load in Figure P16–1 draws 4.8 kW and 2.4 kVARS from a 440 V (rms) 60-Hz source. Find the load power factor and impedance.

16–10 An electrical load is rated at 220 V (rms), 20 A (rms), and 4 kW. Find the load impedance.

ERO 16–2 AC POWER ANALYSIS (SECT. 16–3)

Given a linear circuit operating in the sinusoidal steady state,

(a) Find the complex power associated with any element.
(b) Find the load impedance required to draw a given complex power from a source.

16–11 A load consisting of a 100-Ω resistor in series with a 150-mH inductor is connected across a 60-Hz voltage source that delivers 240 V (rms). Find the complex power delivered to the load.

16–12 A load consisting of a 50-Ω resistor in series with a 10-μF capacitor is connected across a 133-Hz voltage source that delivers 110 V (rms). Find the complex power delivered to the load.

16–13 A load consisting of a resistor and capacitor connected in parallel absorbs a complex power $S = 10 - j126$ VA when connected to a 440-V (rms) 60-Hz line. Find the values of R and C.

16–14 The load in Figure P16–14 consists of a 400-Ω resistor in series with an inductor whose reactance is 200 Ω. The load draws 1.2 A (rms) at 60 Hz from the source, and the wire impedance of the line is $Z_W = 1 + j10$ Ω.

FIGURE P16–14

(a) Find the rms source voltage.

(b) Find the complex power absorbed by the load and line.

(c) Calculate the transmission efficiency (η).

16–15 The load in Figure P16–14 consists of a 60-Ω resistor in series with an inductor whose reactance is 40 Ω. The voltage source produces 220 V (rms) at 60 Hz, and the wire impedance of the line is $Z_W = 2 + j5\ \Omega$.

(a) Find the rms current in the line.

(b) Find the complex power absorbed by the load and line.

(c) Calculate the transmission efficiency (η).

16–16 The transmission line in Figure P16–14 has a maximum rated capacity of 50 kVA at 2400 V (rms). When operated at rated capacity, the resistive voltage drop in the line is 5% and the reactive drop is 12% of the rated voltage. Find the line impedance. Find the transmission efficiency when the load is a 500-Ω resistor and the source voltage is 2400 V (rms).

16–17 The power line in Figure 16–14 has a transmission efficiency of $\eta = 90\%$ when it delivers 25 kW to the load. To improve efficiency, a second identical transmission line is added in parallel with this line. Find the transmission efficiency when the two lines in parallel deliver 25 kW to the load.

16–18 The three load impedances in Figure P16–18 are: $Z_1 = 25 + j8\ \Omega$, $Z_2 = 16 + j8\ \Omega$, and $Z_3 = 10 + j5\ \Omega$.

(a) Find the current in lines A, B, and N.

(b) Find the complex power produced by each source.

F I G U R E   P 1 6 – 1 8

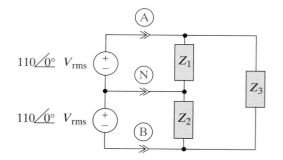

16–19 The three loads in Figure P16–18 are absorbing the following complex powers:

$$S_1 = 1250 + j500\ \text{VA},\ S_2 = 800 + j0\ \text{VA},\ S_3 = 2000 + j400\ \text{VA}.$$

(a) Find the current in lines A, B, and N.

(b) Find the complex power produced by each source.

16–20 A load is rated at 440 V (rms), 30 A (rms), and 10 kW. The load is supplied by 440-V source via a two-wire line with wire impedances of $0.3 + j2\ \Omega$. Find the voltage and apparent power delivered to the load.

ERO 16–3 LOAD-FLOW ANALYSIS (SECT. 16–4)

Given a specified power flow in a linear circuit, find the unknown voltages and power levels.

16–21 The load in Figure P16–14 draws 30 kW at 2400 V (rms) with a power factor of 0.75 lagging. The wire impedance is 2.3 + $j8$ Ω. Find the complex power produced by the source and the transmission efficiency (η).

16–22 The apparent power delivered to the load in Figure P16–14 is 20 kVA at a power factor of 0.8 lagging when the load voltage is 2400 V (rms). The wire impedance is 3 + $j10$ Ω. Find the source voltage and power factor.

16–23 The source in Figure P16–14 delivers 37 kW when the apparent power delivered to the load is 35 + $j20$ kVA. The wire impedance is 2.1 + $j12$ Ω. Find the load and source voltages.

16–24 The two loads in Figure P16–24 absorb complex powers of $S_1 = 12 + j6$ kVA and $|S_2| = 25$ kVA at 0.9 lagging power factor. The load voltage is $|V_L| = 2$ kV (rms) and the line impedances are $Z_W = 2 + j8$ Ω.
(a) Find the line current and source voltage.
(b) Find the complex power produced by the source.
(c) Calculate the transmission efficiency (η).

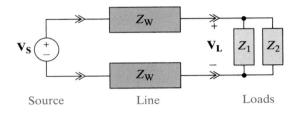

Source   Line   Loads

FIGURE P16–24

16–25 The two loads in Figure P16–24 absorb complex powers of $|S_1| = 16$ kVA at 0.75 lagging power factor and $|S_2| = 25$ kVA at unity power factor. The load voltage is $|V_L| = 277$ V (rms) and the line impedances are $Z_W = 0.1 + j0.5$ Ω.
(a) Find the line current and source voltage.
(b) Find the complex power absorbed by the line and the source power factor.

16–26 When the voltage across a load is 440 (rms) it absorbs 5 kW at 0.75 lagging power factor. A source is connected to this load via a two-wire line with wire impedances of 0.5 + $j4$ Ω. Find the transmission efficiency.

16–27 The load in Figure P16–27 operates at 440 V (rms), 60 Hz, and draws 33 kVA at 0.75 power factor lagging.
(a) Calculate the value of C required to produce an overall power factor of at least 0.95.
(b) Calculate the transmission efficiency with and without the capacitor found in (a).

Source   Line   Loads

FIGURE P16–27

16–28   A load draws an average power of 25 kVA at a lagging power factor 0.7 from a 60-Hz voltage source whose output is 208 V (rms). Find the capacitance to be connected in parallel with the load so that the combined load has a power factor of unity.

16–29   The load in Figure P16–29 operates at 60 Hz. With $\mathbf{V}_L = 12\angle 0°$ kV (rms) the load draws 1.2 MVA at 0.8 power factor lagging. The first source voltage is $\mathbf{V}_{S1} = 12.7 + j0.96$ kV (rms). Find the complex power supplied by each of the sources.

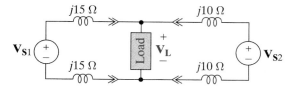

16–30   A two-wire power line connecting points A and B has wire impedances of $Z_W = 2 + j4\ \Omega$. The voltage between wires at point A is $200 + j200$ V (rms) and at point B is $200 + j100$ V (rms). What is the direction of power flow in the line?

ERO 16–4 THREE-PHASE POWER (SECTS. 16–5, 16–6)

Given a balanced three-phase circuit operating in the sinusoidal steady state,

(a) Find the line (phase) voltages or currents when given the phase sequence and the phase (line) voltages or currents.
(b) Find the line current and total complex power (load impedance) when given the line or phase voltage and the load impedance (total complex power).
(c) Find unknown voltages required to produce a specified power flow.

16–31   In a balanced Y-connected three-phase circuit, the magnitude of the line voltage is 208 V (rms) and the phase sequence is positive.
 (a) Write the line and phase-voltage phasors in polar form using $\mathbf{V}_{AN}$ as the reference phasor.
 (b) Draw a phasor diagram of the line and phase voltages.

16–32   In a balanced Δ-connected three-phase circuit, the magnitude of the line voltage is 2400 V (rms) and the phase sequence is positive.
 (a) Write the line and phase voltage phasors in polar form using $\mathbf{V}_{AB}$ as the reference phasor.
 (b) Draw a phasor diagram of the line and phase voltages.

16–33   In a balanced Δ-connected three-phase circuit, the magnitude of the phase current is $I_P = 23.1$ A (rms) and the phase sequence is negative.
 (a) Write the line and phase current phasors in polar form using $\mathbf{I}_A$ as the reference phasor.
 (b) Draw a phasor diagram of the line and phase currents.

16–34   A balanced Y-connected three-phase load with a per-phase impedance of $15 + j5\ \Omega$ operates with a line voltage magnitude of 480 V (rms) using a positive phase sequence. Using $\mathbf{V}_{AN}$ as the reference phasor:
 (a) Find the line current phasors in rectangular form.
 (b) Calculate the total complex power delivered to the load.
 (c) Draw a phasor diagram showing the line currents.

16–35 A balanced Δ-connected three-phase load with a per-phase impedance of $5 + j2$ Ω operates with a line voltage magnitude of 480 V (rms) using a negative phase sequence. Using $\mathbf{V}_{AB}$ as the reference phasor:
 (a) Find the line current phasors in rectangular form.
 (b) Calculate the total complex power delivered to the load.
 (c) Draw a phasor diagram showing the line currents.

16–36 A balanced Δ-connected three-phase load with a per-phase impedance of $50 + j15$ Ω operates with a line voltage magnitude of 480 V (rms) using a positive phase sequence. Using $\mathbf{I}_A$ as the reference phasor,
 (a) Find the line current phasors in rectangular form.
 (b) Calculate the total complex power delivered to the load.

16–37 A balanced Δ-connected three-phase load with a per-phase impedance of $6 - j8$ Ω is connected in parallel with a balanced Y-connected three-phase load with a per-phase impedance of $12\angle30°$ Ω. The line voltage is $V_L = 480$ V (rms). Find the magnitude of the line current and the power factor of the combined loads.

16–38 In a balanced Δ-connected three-phase load, the phase A line current is $\mathbf{I}_A = 50\angle-20°$ A (rms) with a line voltage of $\mathbf{V}_{AB} = 480\angle30°$ V (rms). Find the phase impedance of the load assuming a positive phase sequence.

16–39 A balanced Y-connected three-phase load absorbs 30 kVA at a power factor of 0.75 lagging when the line voltage magnitude is 480 V (rms).
 (a) Find the magnitude of the line current.
 (b) Calculate the resistance and reactance of the per-phase impedance.

16–40 A balanced Y-connected three-phase load absorbs 40 kW when the line current magnitude is 50 A (rms) and the line voltage magnitude is 480 V (rms). Find the resistance and reactance of the per-phase impedance assuming a lagging power factor.

16–41 A balanced Δ-connected three-phase load absorbs 30 kVA at a power factor of 0.75 lagging when the line voltage magnitude is 7200 V (rms).
 (a) Find the magnitude of the line current.
 (b) Calculate the resistance and reactance of the per-phase impedance.

16–42 Two three-phase loads are connected in parallel. The first is a balanced Δ-connected circuit absorbing 30 kVA at a power factor of 0.85 lagging. The second is a balanced Y-connected load with a per-phase impedance of $20 + j15$ Ω. The magnitude of the line voltage at the loads is 480 V (rms). Find the magnitude of the total line current and the total complex power delivered to the loads.

16–43 The wire impedances in the single-line diagram of Figure P16–43 are $Z_W = 0.6 + j4$ Ω per phase. The balanced load connected to Bus

No. 2 is rated at $S_2 = 16 + j12$ kVA when the line voltage is 440 V (rms). The generator connected to Bus No. 1 produces a line voltage of 420 V (rms). Find the line voltage at Bus No. 2 and the complex power produced by the source.

16–44   The source at Bus No. 1 in Figure P16–43 is a Y-connected generator with an internal impedance of $j2$ Ω/phase and a Thévenin voltage of 4160 V/phase. The load connected to Bus No. 2 is a Y-connected three-phase load with a phase impedance of $80 + j60$ Ω/phase. The power lines connecting the two buses have wire impedances of $2 + j5$ Ω/phase.
(a) Find the line current.
(b) Find the complex power delivered to the load and phase voltage at Bus No. 2.
(c) Find the total complex power produced by the source.
(d) Find the transmission efficiency (η).

16–45   The three-phase line in Figure P16–43 has wire impedances of $Z_w = 0 + j8$ Ω per phase. The line voltage at Bus No. 2 is 7200 V (rms), and the load absorbs 800 kVA at 0.8 power factor lagging. Find the line current, the line voltage at Bus No. 1, and the complex power produced by the source.

16–46   The three-phase line in Figure P16–43 has wire impedances of $Z_w = 3 + j6$ Ω per phase. The line voltage at Bus No. 1 is 12.5 kV (rms), and the load at Bus No. 2 absorbs 600 kW at 0.9 power factor lagging. Find the line current and the line voltage at Bus No. 2.

16–47   The three-phase line in Figure P16–43 has wire impedances of $Z_w = 2.5 + j8$ Ω per phase. The generator at Bus No. 1 produces an average power of $P_1 = 400$ kW. The load at Bus No. 2 absorbs an average power of $P_2 = 390$ kW at 0.9 power factor lagging. Find the line current and the line voltages at Bus No. 1 and Bus No. 2.

16–48   The three-phase load connected to Bus No. 2 in Figure P16–48 draws an average power of $P_2 = 100$ MW at a power factor of 0.8 lagging. At Bus No. 2 the phase A voltage phasor is $90\angle 0°$ kV (rms). At Bus No. 3 the phase A voltage phasor is $90\angle 12°$ kV (rms). Find the complex power produced by each source.

F I G U R E   P 1 6 – 4 8

16–49   The balanced three-phase load draws 60 kW at a lagging power factor of 0.8. The line voltage at the load is 2400 V (rms). A second unit drawing 60 kW at a lagging power factor of 0.9 is to be connected in parallel with the first. The line connecting the loads to the source is rated at 30 A (rms).
(a) Can both units be operated at the same time without exceeding the line rating?

(b) Connecting a capacitor in parallel with the two loads could correct the power factor to unity. Would power factor correction allow the two loads to operate at the same time without exceeding the rating of the line?

16–50 A balanced Δ-connected three-phase load has per-phase impedances of $60 + j30 \; \Omega$. The magnitude of the line voltage at the load is 208 V (rms).

(a) The line connecting the load to the source has wire impedances of $Z_W = 5 + j15 \; \Omega$ per phase. Find the transmission efficiency (η).

(b) A second line in parallel with the first has been proposed as a way to improve efficiency. The wire impedances of the proposed new line are $Z_W = 2 + j15 \; \Omega$ per phase. Find the transmission efficiency (η) with the two lines connected in parallel.

INTEGRATING PROBLEMS

16–51 A UNBALANCED THREE-PHASE LOAD

A balanced three-phase source with $V_P = 208$ V (rms) drives a Y-connected three-phase load with phase impedances $Z_A = 100 \; \Omega$, $Z_B = 100 \; \Omega$, and $Z_C = 50 + j100 \; \Omega$.

(a) Calculate the line currents, line voltages, and total complex power delivered to the load.

(b) Repeat part (a) when a zero-resistance neutral wire is connected between the source and load.

16–52 A D THREE-PHASE POWER FACTOR CORRECTIONS

The line voltage in Figure P16–52 is 480 V (rms). The balanced three-phase load operates 8 hours/day and draws 75 kVA at a lagging power factor of 0.75. The electric power supplier charges 7¢/kW-hr when the load power factor is greater than 0.95 and 9¢/kW-hr when the power factor is less than 0.95. The following three-phase capacitor banks are commercially available:

FIGURE P16–52

480-V THREE-PHASE CAPACITOR EQUIPMENT					
KVAR	PART NUMBER	UNIT PRICE	KVAR	PART NUMBER	UNIT PRICE
10	1N0240A05	$500	40	1N0240A17	$900
20	1N0240A09	$600	50	1N0240A19	$1100
30	1N0240A11	$700	60	1N0240A23	$1500

If one or more capacitor banks are purchased to increase the power factor above 0.95, about how long will it take for the accumulated savings in operating costs to equal the equipment capital investment?

16–53 A THREE-PHASE TRANSFORMER BANK

Figure P16–53 shows a three-phase transformer bank modeled as three identical ideal transformers. The primary windings are Δ connected, and the secondary windings are Y connected. The line current

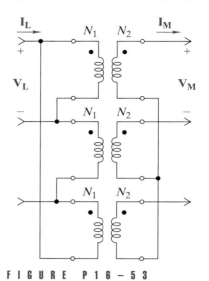

and line voltage phasors on the primary side are denoted as $\mathbf{V}_L$ and $\mathbf{I}_L$, respectively. Derive an expression relating the line voltage and line current phasors on the secondary side of the transformers to $\mathbf{V}_L$ and $\mathbf{I}_L$.

16–54 **E** COMPARISON OF THREE-PHASE AND SINGLE-PHASE SYSTEMS

The three-phase and single-phase power systems in Figure P16–54 operate under the following conditions:
(a) The two systems deliver the same total complex power to the load.
(b) The two systems have the same line-to-line voltage.
(c) The two systems have the same transmission efficiency.
(d) The distance from source to load is the same.
(e) The resistance of a line is proportional to its length divided by the cross-sectional area of the wire.

Show that the transmission line in the three-phase system requires 25% less copper than the single-phase system.

FIGURE P16-54

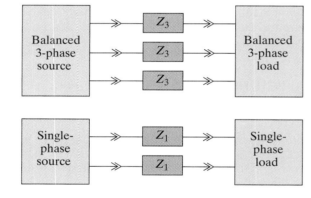

APPENDIX A

STANDARD VALUES

To reduce inventory costs the electrical industry has agreed on standard values and tolerances for commonly used discrete components such as resistors, capacitors, and zener diodes. The standard values for resistors and capacitors with 5%, 10%, and 20% tolerances are shown in Table A–1. Standard resistances are obtained by multiplying the values in the table by different powers of 10. For example, multiplying the values for ±20% tolerance by 10^4 yields a decade range of standard resistances of 100 kΩ, 150 kΩ, 220 kΩ, 330 kΩ, 470 kΩ, and 680 kΩ. Standard values for tolerances down to ±0.1% are defined, although the resulting proliferation of values tends to defeat the purpose of standardization.

T A B L E A–1   STANDARD VALUES FOR RESISTORS AND CAPACITORS

VALUE	TOLERANCES	VALUE	TOLERANCES	VALUE	TOLERANCES
10	±5%, ±10%, ±20%	22	±5%, ±10%, ±20%	47	±5%, ±10%, ±20%
11	±5%	24	±5%	51	±5%
12	±5%, ±10%	27	±5%, ±10%	56	±5%, ±10%
13	±5%	30	±5%	62	±5%
15	±5%, ±10%, ±20%	33	±5%, ±10%, ±20%	68	±5%, ±10%, ±20%
16	±5%	36	±5%	75	±5%
18	±5%, ±10%	39	±5%, ±10%	82	±5%, ±10%
20	±5%	43	±5%	91	±5%

APPENDIX B

SOLUTION OF
LINEAR EQUATIONS

The purpose of this appendix is to review the methods of solving systems of linear algebraic equations. Circuit analysis often requires solving linear algebraic equations of the type

$$5x_1 - 2x_2 - 3x_3 = 4$$
$$-5x_1 + 7x_2 - 2x_3 = -10 \qquad \text{(B–1)}$$
$$-3x_1 - 3x_2 + 8x_3 = 6$$

where x_1, x_2, and x_3 are unknown voltages or currents. Often some of the unknowns may be missing from one or more of the equation. For example, the equations

$$5x_1 - 2x_2 = 5$$
$$-4x_1 + 7x_2 = 0$$
$$-3x_2 + 8x_3 = 0$$

involve three unknowns with one variable missing in each equation. Such equations can always be put in the standard square form by inserting the missing unknowns with a coefficient of zero.

$$5x_1 - 2x_2 - 0x_3 = 5$$
$$-4x_1 + 7x_2 - 0x_3 = 0 \qquad \text{(B–2)}$$
$$0x_1 - 3x_2 + 8x_3 = 0$$

Equations (B–1) and (B–2) will be used to illustrate the different methods of solving linear equations.

CRAMER'S RULE

Cramer's rule states that the solution of a system of linear equations for any unknown x_k is found as the ratio of two determinants

$$x_k = \frac{\Delta_k}{\Delta} \tag{B-3}$$

where Δ and Δ_k are determinants derived from the given set of equations. A **determinant** is a square array of numbers or symbols called **elements**. The elements are arranged in horizontal rows and vertical columns and are bordered by two vertical straight lines. In general, a determinant contains n^2 elements arranged in n rows and n columns. The value of the determinant is a function of the value and position of its n^2 elements.

The **system determinant** Δ in Eq. (B–3) is made up of the coefficients of the unknowns in the given system of equations. For example, the system determinant for Eq. (B–1) is

$$\Delta = \begin{vmatrix} 5 & -2 & -3 \\ -5 & 7 & -2 \\ -3 & -3 & 8 \end{vmatrix}$$

and for Eq. (B–2) is

$$\Delta = \begin{vmatrix} 5 & -2 & 0 \\ -4 & 7 & 0 \\ 0 & -3 & 8 \end{vmatrix}$$

These two equations are examples of the general 3×3 determinant

$$\Delta = \begin{vmatrix} a_{11} & a_{12} & a_{13} \\ a_{21} & a_{22} & a_{23} \\ a_{31} & a_{32} & a_{33} \end{vmatrix} \tag{B-4}$$

where a_{ij} is the element in the ith row and jth column.

The determinant Δ_k in Eq. (B–3) is derived from the system determinant by replacing the kth column by the numbers on the right side of the system of equations. For example, Δ_1 for Eq. (B–1) is

$$\Delta_1 = \begin{vmatrix} 4 & -2 & -3 \\ -10 & 7 & -2 \\ 6 & -3 & 8 \end{vmatrix}$$

and Δ_3 for Eq. (B–2) is

$$\Delta_3 = \begin{vmatrix} 5 & -2 & 5 \\ -4 & 7 & 0 \\ 0 & -3 & 0 \end{vmatrix}$$

These examples are 3×3 determinants because the system determinants from which they are derived are 3×3.

In summary, using Cramer's rule to solve linear equations boils down to evaluating the determinants formed using the coefficients of the unknowns and the right side of the system of equations.

EVALUATING DETERMINANTS

The **diagonal rule** gives the value of a 2×2 determinant as the difference in the product of the elements on the main diagonal $(a_{11}a_{22})$ and the product of the elements on the off diagonal $(a_{21}a_{12})$. That is, for a 2×2 determinant

$$\Delta = \begin{vmatrix} a_{11} & a_{12} \\ a_{21} & a_{22} \end{vmatrix} = a_{11}a_{22} - a_{21}a_{12} \tag{B-5}$$

The value of 3×3 and higher-order determinants can be found using the method of minors. Every element a_{ij} has a **minor** M_{ij}, which is formed by deleting the row and column containing a_{ij}. For example, the minor M_{21} of the general 3×3 determinant in Eq. (B–4) is

$$M_{21} = \begin{vmatrix} a_{12} & a_{13} \\ a_{32} & a_{33} \end{vmatrix} = a_{12}a_{33} - a_{32}a_{13}$$

The **cofactor** C_{ij} of the element a_{ij} is its minor M_{ij} multiplied by $(-1)^{i+j}$.

$$C_{ij} = (-1)^{i+j}M_{ij}$$

The signs of the cofactors alternate along any row or column. The appropriate sign for cofactor C_{ij} is found by starting in position a_{11} and counting plus, minus, plus, minus . . . along any combination of rows or columns leading to the position a_{ij}.

To use the **method of minors** we select one (and only one) row or column. The determinant is the sum of the products of the elements in the selected row or column and their cofactors. For example, selecting the first column in Eq. (B–4), we obtain Δ as follows:

$$\Delta = a_{11}C_{11} + a_{21}C_{21} + a_{31}C_{31}$$

$$= a_{11}(-1)^2 \begin{vmatrix} a_{22} & a_{23} \\ a_{32} & a_{33} \end{vmatrix} + a_{21}(-1)^3 \begin{vmatrix} a_{12} & a_{13} \\ a_{32} & a_{33} \end{vmatrix} + a_{31}(-1)^4 \begin{vmatrix} a_{12} & a_{13} \\ a_{22} & a_{23} \end{vmatrix}$$

$$= a_{11}(a_{22}a_{33} - a_{32}a_{23}) - a_{21}(a_{12}a_{33} - a_{32}a_{13}) + a_{31}(a_{12}a_{23} - a_{22}a_{13})$$

An identical expression for Δ is obtained using any other row or column. For determinants greater than 3×3 the minors themselves can be evaluated using this approach. However, a system of equations leading to determinants larger than 3×3 is probably better handled using computer tools.

EXAMPLE B–1

Solve for the three unknowns in Eq. (B–1) using Cramer's rule.

SOLUTION:
Expanding the system determinant about the first column yields

$$\Delta = \begin{vmatrix} 5 & -2 & -3 \\ -5 & 7 & -2 \\ -3 & -3 & 8 \end{vmatrix} = 5 \begin{vmatrix} 7 & -2 \\ -3 & 8 \end{vmatrix} - (-5) \begin{vmatrix} -2 & -3 \\ -3 & 8 \end{vmatrix} + (-3) \begin{vmatrix} -2 & -3 \\ 7 & -2 \end{vmatrix}$$

$$= 5[7 \times 8 - (-2)(-3)] - (-5)[(-2) \times 8 - (-3)(-3)] + (-3)[(-2)(-2) - (7)(-3)]$$

$$= 250 - 125 - 75 = 50$$

Expanding Δ_1 about the first column yields

$$\Delta_1 = \begin{vmatrix} 4 & -2 & -3 \\ -10 & 7 & -2 \\ 6 & -3 & 8 \end{vmatrix} = 4\begin{vmatrix} 7 & -2 \\ -3 & 8 \end{vmatrix} - (-10)\begin{vmatrix} -2 & -3 \\ -3 & 8 \end{vmatrix} + (6)\begin{vmatrix} -2 & -3 \\ 7 & -2 \end{vmatrix}$$

$$= 200 - 250 + 150 = 100$$

Expanding Δ_2 about the first column yields

$$\Delta_2 = \begin{vmatrix} 5 & 4 & -3 \\ -5 & -10 & -2 \\ -3 & 6 & 8 \end{vmatrix} = 5\begin{vmatrix} -10 & -2 \\ 6 & 8 \end{vmatrix} - (-5)\begin{vmatrix} 4 & -3 \\ 6 & 8 \end{vmatrix} + (-3)\begin{vmatrix} 4 & -3 \\ -10 & -2 \end{vmatrix}$$

$$= -340 + 250 + 114 = 24$$

Expanding Δ_3 about the first column yields

$$\Delta_3 = \begin{vmatrix} 5 & -2 & 4 \\ -5 & 7 & -10 \\ -3 & -3 & 6 \end{vmatrix} = 5\begin{vmatrix} 7 & -10 \\ -3 & 6 \end{vmatrix} - (-5)\begin{vmatrix} -2 & 4 \\ -3 & 6 \end{vmatrix} + (-3)\begin{vmatrix} -2 & 4 \\ 7 & -10 \end{vmatrix}$$

$$= 60 - 0 + 24 = 84$$

Now, applying Cramer's rule, we solve for the three unknowns.

$$x_1 = \frac{\Delta_1}{\Delta} = \frac{100}{50} = 2$$

$$x_2 = \frac{\Delta_2}{\Delta} = \frac{24}{50} = 0.48$$

$$x_3 = \frac{\Delta_3}{\Delta} = \frac{84}{50} = 1.68 \qquad \blacksquare$$

Exercise B–1

Evaluate Δ, Δ_1, Δ_2, and Δ_3 for Eq. (B–2).

Answer: 216, 280, 160, and 60

MATRICES AND LINEAR EQUATIONS

Circuit equations can be formulated and solved in matrix format. By definition, a **matrix** is a rectangular array written as

$$\mathbf{A} = \begin{bmatrix} a_{11} & a_{12} & a_{13} & \cdots & a_{1n} \\ a_{21} & a_{22} & a_{23} & \cdots & a_{2n} \\ \cdots & \cdots & \cdots & \cdots & \cdots \\ a_{m1} & a_{m2} & a_{m3} & \cdots & a_{mn} \end{bmatrix} \qquad (B\text{–}6)$$

The matrix $\mathbf{A}$ in Eq. (B–6) contains m rows and n columns and is said to be of order m by n (or $m \times n$). The matrix notation in Eq. (B–6) can be abbreviated as follows:

$$\mathbf{A} = [a_{ij}]_{mn} \qquad (B-7)$$

where a_{ij} is the element in the ith row and jth column.

SOME DEFINITIONS

Different types of matrices have special names. A **row matrix** has only one row ($m = 1$) and any number of columns. A **column matrix** has only one column ($n = 1$) and any number of rows. A **square matrix** has the same number of rows as columns ($m = n$). A **diagonal matrix** is a square matrix in which all elements not on the main diagonal are zero ($a_{ij} = 0$ for $i \neq j$). An **identity matrix** is a diagonal matrix for which the main diagonal elements are all unity ($a_{ii} = 1$).

For example, given

$$\mathbf{A} = [1 \quad -2 \quad 0 \quad 4] \quad \mathbf{B} = \begin{bmatrix} 3 \\ -2 \\ 6 \\ 0 \end{bmatrix} \quad \mathbf{C} = \begin{bmatrix} 1 & 0 & -7 \\ -3 & 12 & 0 \\ 0 & 0 & -4 \end{bmatrix} \quad \mathbf{U} = \begin{bmatrix} 1 & 0 & 0 & 0 \\ 0 & 1 & 0 & 0 \\ 0 & 0 & 1 & 0 \\ 0 & 0 & 0 & 1 \end{bmatrix}$$

we say that $\mathbf{A}$ is a 1×4 row matrix, $\mathbf{B}$ is a 4×1 column matrix, $\mathbf{C}$ is a 3×3 square matrix, and $\mathbf{U}$ is a 4×4 identity matrix.

The **determinant** of a square matrix $\mathbf{A}$ (denoted det $\mathbf{A}$) has the same elements as the matrix itself. For example, given

$$\mathbf{A} = \begin{bmatrix} 4 & -6 \\ 1 & -2 \end{bmatrix} \text{ then det } \mathbf{A} = \begin{vmatrix} 4 & -6 \\ 1 & -2 \end{vmatrix} = -8 + 6 = -2$$

The **transpose** of a matrix $\mathbf{A}$ (denoted $\mathbf{A}^T$) is formed by interchanging the rows and columns. For example, given

$$\mathbf{A} = \begin{bmatrix} 1 & 2 & 0 & 8 \\ 4 & 7 & -1 & -3 \end{bmatrix} \text{ then } \mathbf{A}^T = \begin{bmatrix} 1 & 4 \\ 2 & 7 \\ 0 & -1 \\ 8 & -3 \end{bmatrix}$$

The **adjoint** of a square matrix $\mathbf{A}$ (denoted adj $\mathbf{A}$) is formed by replacing each element a_{ij} by its cofactor C_{ij} and then transposing.

$$\text{adj } \mathbf{A} = [C_{ij}]^T \qquad (B-8)$$

For example, if

$$\mathbf{A} = \begin{bmatrix} -3 & 2 \\ 0 & 5 \end{bmatrix} \text{ then } C_{11} = 5 \quad C_{12} = 0 \quad C_{21} = -2 \quad C_{22} = -3$$

and therefore

$$\text{adj } \mathbf{A} = \begin{bmatrix} 5 & 0 \\ -2 & -3 \end{bmatrix}^T = \begin{bmatrix} 5 & -2 \\ 0 & -3 \end{bmatrix}$$

MATRIX ALGEBRA

The matrices $\mathbf{A}$ and $\mathbf{B}$ are equal if and only if they have the same number of rows and columns, and $a_{ij} = b_{ij}$ for all i and j. Matrix addition is only possible when two matrices have the same number of rows and columns. When two matrices are of the same order, their sum is obtained by adding the corresponding elements: that is,

$$\text{If } \mathbf{C} = \mathbf{A} + \mathbf{B} \text{ then } c_{ij} = a_{ij} + b_{ij} \tag{B-9}$$

For example, given

$$\mathbf{A} = \begin{bmatrix} -1 & 4 \\ -3 & -2 \end{bmatrix} \text{ and } \mathbf{B} = \begin{bmatrix} 3 & 0 \\ 2 & -4 \end{bmatrix} \text{ then } \mathbf{C} = \mathbf{A} + \mathbf{B} = \begin{bmatrix} 2 & 4 \\ -1 & -6 \end{bmatrix}$$

Multiplying a matrix $\mathbf{A}$ by a scalar constant k is accomplished by multiplying every element by k; that is, $k\mathbf{A} = [ka_{ij}]$. In particular, if $k = -1$ then $-\mathbf{B} = [-b_{ij}]$, and applying the matrix addition rule yields matrix **subtraction**.

$$\text{If } \mathbf{C} = \mathbf{A} - \mathbf{B} \text{ then } c_{ij} = a_{ij} - b_{ij} \tag{B-10}$$

Multiplication of two matrices $\mathbf{AB}$ is defined only if the number of columns in $\mathbf{A}$ equals the number of rows in $\mathbf{B}$. In general, if $\mathbf{A}$ is of order $m \times n$ and $\mathbf{B}$ is of order $n \times r$, then the product $\mathbf{C} = \mathbf{AB}$ is a matrix of order $m \times r$. The element c_{ij} is found by summing the products of the elements in the ith row of $\mathbf{A}$ and the jth column of $\mathbf{B}$.

$$c_{ij} = [a_{i1}\ a_{i2}\ \ldots\ a_{in}] \begin{bmatrix} b_{1j} \\ b_{2j} \\ .. \\ .. \\ .. \\ b_{nj} \end{bmatrix} = a_{i1}b_{1j} + a_{i2}b_{2j} + \cdots a_{in}b_{nj}$$

$$= \sum_{k=1}^{n} a_{ik}b_{kj} \tag{B-11}$$

In other words, matrix multiplication is a row by column operation.

Matrix multiplication is not commutative so usually $\mathbf{AB} \neq \mathbf{BA}$. Two important exceptions are (1) the product of a square matrix $\mathbf{A}$ and an identity matrix $\mathbf{U}$ for which $\mathbf{UA} = \mathbf{AU} = \mathbf{A}$, and (2) the product of a square matrix $\mathbf{A}$ and its **inverse** (denoted $\mathbf{A}^{-1}$) for which $\mathbf{A}^{-1}\mathbf{A} = \mathbf{AA}^{-1} = \mathbf{U}$. A closed-form formula for the inverse of a square matrix is

$$\mathbf{A}^{-1} = \frac{\text{adj } \mathbf{A}}{\det \mathbf{A}} \tag{B-12}$$

That is, the inverse can be found by multiplying the adjoint matrix of $\mathbf{A}$ by the scalar $1/\det \mathbf{A}$. If $\det \mathbf{A} = 0$ then $\mathbf{A}$ is said to be **singular** and $\mathbf{A}^{-1}$ does not exist. Equation (B-12) is useful for deriving properties of the inverse of a matrix. It is not, however, a very efficient way to calculate the inverse of a matrix of order greater than 3×3.

Exercise B-2

Given:

$$\mathbf{A} = \begin{bmatrix} -5 & 7 \\ 7 & 11 \end{bmatrix} \text{ and } \mathbf{B} = \begin{bmatrix} 3 & -1 \\ 6 & -2 \end{bmatrix}$$

Calculate AB, BA, $\mathbf{A}^{-1}$, and $\mathbf{B}^{-1}$.

Answers:

$$\mathbf{AB} = \begin{bmatrix} 27 & -9 \\ 87 & -29 \end{bmatrix} \quad \mathbf{BA} = \begin{bmatrix} -22 & 10 \\ -44 & 20 \end{bmatrix}$$

$$\mathbf{A}^{-1} = \frac{1}{104}\begin{bmatrix} -11 & 7 \\ 7 & 5 \end{bmatrix} \quad \mathbf{B}^{-1} \text{ does not exist}$$

MATRIX SOLUTION OF LINEAR EQUATIONS

The three linear equations in Eq. (B–1) are

$$\begin{aligned} 5x_1 - 2x_2 - 3x_3 &= 4 \\ -5x_1 + 7x_2 - 2x_3 &= -10 \\ -3x_1 - 3x_2 + 8x_3 &= 6 \end{aligned}$$

These equations are expressed in matrix form as follows:

$$\begin{bmatrix} 5 & -2 & -3 \\ -5 & 7 & -2 \\ -3 & -3 & 8 \end{bmatrix}\begin{bmatrix} x_1 \\ x_2 \\ x_3 \end{bmatrix} = \begin{bmatrix} 4 \\ -10 \\ 6 \end{bmatrix} \tag{B–13}$$

The left side of Eq. (B–13) is the product of a 3×3 square matrix and a 3×1 column matrix of unknowns. The elements in the square matrix are the coefficients of the unknown in the given equations. The matrix product on the left side in Eq. (B–13) produces a 3×1 matrix, which equals the 3×1 column matrix on the right side. The elements of the 3×1 on the right side are the constants on the right sides of the given equations.

In symbolic form we write the matrix equation in Eq. (B–13) as

$$\mathbf{AX} = \mathbf{B} \tag{B–14}$$

where

$$\mathbf{A} = \begin{bmatrix} 5 & -2 & -3 \\ -5 & 7 & -2 \\ -3 & -3 & 8 \end{bmatrix}, \quad \mathbf{X} = \begin{bmatrix} x_1 \\ x_2 \\ x_3 \end{bmatrix} \text{ and } \mathbf{B} = \begin{bmatrix} 4 \\ -10 \\ 6 \end{bmatrix}$$

Left multiplying Eq. (B–14) by $\mathbf{A}^{-1}$ yields

$$\mathbf{A}^{-1}\mathbf{AX} = \mathbf{A}^{-1}\mathbf{B}$$

But by definition $\mathbf{A}^{-1}\mathbf{A} = \mathbf{U}$ and $\mathbf{UX} = \mathbf{X}$; therefore

$$\mathbf{X} = \mathbf{A}^{-1}\mathbf{B} \tag{B–15}$$

To solve linear equations by matrix methods we calculate the product $\mathbf{A}^{-1}\mathbf{B}$.

To implement the matrix approach we must first find $\mathbf{A}^{-1}$ using Eq. (B–12). The determinant of the coefficient matrix is

$$\det \mathbf{A} = \begin{vmatrix} 5 & -2 & -3 \\ -5 & 7 & -2 \\ -3 & -3 & 8 \end{vmatrix} = 50$$

The cofactors of the first row of the coefficient matrix are

$$C_{11} = - \begin{vmatrix} +7 & -2 \\ -3 & 8 \end{vmatrix} = 50 \quad C_{12} = \begin{vmatrix} -5 & -2 \\ -3 & 8 \end{vmatrix} = 46$$

$$C_{13} = - \begin{vmatrix} -5 & +7 \\ -3 & -3 \end{vmatrix} = 36$$

The cofactors for the second and third rows are

$$C_{21} = 25 \quad C_{22} = 31 \quad C_{23} = 21$$

$$C_{31} = 25 \quad C_{32} = 25 \quad C_{33} = 25$$

Now, using Eq. (B–12), we obtain $\mathbf{A}^{-1}$ as

$$\mathbf{A}^{-1} = \frac{\text{adj } \mathbf{A}}{\det \mathbf{A}} = \frac{1}{50} \begin{bmatrix} 50 & 46 & 36 \\ 25 & 31 & 21 \\ 25 & 25 & 25 \end{bmatrix}^{T} = \frac{1}{50} \begin{bmatrix} 50 & 25 & 25 \\ 46 & 31 & 25 \\ 36 & 21 & 25 \end{bmatrix}$$

Using Eq. (B–15), we solve for the column matrix of unknowns as

$$\begin{bmatrix} x_1 \\ x_2 \\ x_3 \end{bmatrix} = \mathbf{X} = \mathbf{A}^{-1}\mathbf{B} = \frac{1}{50} \begin{bmatrix} 50 & 25 & 25 \\ 46 & 31 & 25 \\ 36 & 21 & 25 \end{bmatrix} \begin{bmatrix} 4 \\ -10 \\ 6 \end{bmatrix} = \frac{1}{50} \begin{bmatrix} 100 \\ 24 \\ 84 \end{bmatrix}$$

which yields $x_1 = 2$, $x_2 = 24/50$, and $x_3 = 84/50$. These are, of course, the same results previously obtained using Cramer's rule.

USING COMPUTER TOOLS

Computer tools for solving linear equations range from inexpensive hand-held calculators to sophisticated software packages capable of solving hundreds of equations. At the intermediate level are math analysis software packages such as MATLAB and Mathcad. Under what circumstances should you consider using these computer tools in linear circuit analysis?

There are no hard and fast rules here. Somewhere around three or four equations the burden of hand calculations becomes mildly excruciating. Sets of equations with N up to 20 or 30 are routinely solved using computer tools, except when the equations are ill conditioned (several equations are almost linearly dependent). Well-conditioned systems of equations with $N = 50$ or more can be solved using sophisticated numerical methods. On the other hand, these sophisticated computer tools probably don't buy you very much in linear circuit applications. If you encounter a problem that requires solving (say) 20 or more linear equations, you should redefine the problem so that it can be partitioned into smaller pieces.

MATLAB is a software package for matrix-based computations and data analysis. The original MATLAB was a main frame resident software written in the 1970s to provide a "matrix laboratory" for linear algebra and matrix theory courses. Since then personal computer versions have been developed that fit on modest computer platforms. In its present form MATLAB is an interactive system and programming language whose capabilities extend far beyond the original matrix laboratory application.

To solve the matrix equation in Eq. (B–13) using MATLAB, we first enter the **A** and **B** matrices in the MATLAB command window. In this window the command line prompt is either the character string << or EDU<<, depending on the version in use. This prompt indicates that MATLAB is ready to accept data and commands entered via the keyboard. For example, a matrix can be entered by typing its elements one row at a time. Elements in the same row are separated by spaces. The end of a row is indicated by a semicolon or an ↵enter keystroke. The row-by-row entry is enclosed by a left bracket [at the beginning and a right bracket] at the end of the last row.

For example, the **A** matrix in Eq. (B–13) can be entered by typing

```
A=[5  -2  -3
   -5   7  -2
   -3  -3   8]
```

After the last ↵enter keystroke, MATLAB responds by listing the elements of **A.**

```
A=

     5  -2  -3
    -5   7  -2
    -3  -3   8
```

This echo check allows you to verify that the elements of **A** have been entered correctly. The **A** can also be entered by typing

```
A=[5 -2 -3;-5 7 -2;-3 -3 8];
```

The semicolon following the closing right bracket suppresses the Matlab echo check if you have supreme confidence that you have entered the matrix elements without error. An individual matrix element can be referenced by enclosing its subscripts in parenthesis. For instance, the command line query

```
A(3,1)
```

asks $a_{31} = ?$ and produces the MATLAB response

```
A(3,1) =
      -3
```

The **B** matrix for Eq. (B–13) can be entered as

```
B = [4; -10; 6];
```

The matrix equations in Eq. (B–13) can be solved for the unknown column matrix **X** by forming the matrix product $\mathbf{A}^{-1}\mathbf{B}.$ In the MATLAB command window we enter the statement

```
X=inv(A)*B
```

to which MATLAB responds with

```
X=
     2.0000
     0.4800
     1.6800
```

We can verify this result by checking to see that the matrix product **AX** is equal to the matrix **B**. In the command window we type

```
A*X
```

to which MATLAB responds with

```
ans=
      4
    -10
      6
```

which is the matrix **B** on the right side of Eq. (B–13).

Figure B–1 shows a Mathcad worksheet that performs the same matrix operations. The format and appearance of the mathematical operations in the worksheet are virtually self-defining. Again we have checked the answer by forming the matrix product **AX** in final step.

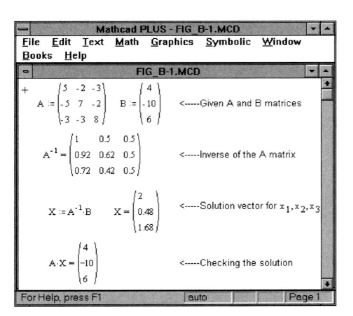

FIGURE   B − 1   A   Mathcad worksheet.

APPENDIX C

COMPLEX NUMBERS

Using complex numbers to represent signals and circuits is a fundamental tool in electrical engineering. This appendix reviews complex-number representations and arithmetic operations. These procedures, though rudimentary, must be second nature to all who aspire to be electrical engineers. Exercises are provided to confirm your mastery of these basic skills.

COMPLEX-NUMBER REPRESENTATIONS

A complex number z can be written in rectangular form as

$$z = x + jy \qquad (C-1)$$

where j represents $\sqrt{-1}$. Mathematicians customarily use i to represent $\sqrt{-1}$, but i represents current in electrical engineering so we use the symbol j instead.

The quantity z is a two-dimensional number represented as a point in the complex plane, as shown in Figure C–1. The x component is called the **real part** and y (not jy) the **imaginary part** of z. A special notation is sometimes used to indicate these two components:

$$x = \mathrm{Re}\{z\} \quad \text{and} \quad y = \mathrm{Im}\{z\} \qquad (C-2)$$

where $\mathrm{Re}\{z\}$ means the real part and $\mathrm{Im}\{z\}$ the imaginary part of z.

Figure C–1 also shows the polar representation of the complex number z. In polar form a complex number is written

$$z = M \angle \theta \qquad (C-3)$$

where M is called the **magnitude** and θ the **angle** of z. A special notation is also used to indicate these two components.

$$|z| = M \quad \text{and} \quad \angle z = \theta \qquad \text{(C-4)}$$

where $|z|$ means the magnitude and $\angle z$ the angle of z.

The real and imaginary parts and magnitude and angle of z are all shown geometrically in Figure C–1. The relationships between the rectangular and polar forms are easily derived from the geometry in Figure C–1:

$$\text{Rectangular to Polar} \quad M = \sqrt{x^2 + y^2} \quad \theta = \tan^{-1}\frac{y}{x}$$

$$\text{Polar to Rectangular} \quad x = M \cos \theta \quad y = M \sin \theta \qquad \text{(C-5)}$$

The inverse tangent relation for θ involves an ambiguity that can be resolved by identifying the correct quadrant in the z-plane using the signs of the two rectangular components. [See Exercise C–1(b) and (c).]

Another version of the polar form is obtained using Euler's relationship:

$$e^{j\theta} = \cos \theta + j \sin \theta \qquad \text{(C-6)}$$

We can write the polar form as

$$z = Me^{j\theta} = M \cos \theta + jM \sin \theta \qquad \text{(C-7)}$$

This polar form is equivalent to Eq. (C–3), since the right side yields the same polar-to-rectangular relationships as Eq. (C–5). Thus, a complex number can be represented in three ways:

$$z = x + jy \quad z = M\angle\theta \quad z = Me^{j\theta} \qquad \text{(C-8)}$$

The relationships between these forms are given in Eq. (C–5).

The quantity $z*$ is called the conjugate of the complex number z. The asterisk indicates the **conjugate** of a complex number formed by reversing the sign of the imaginary component. In rectangular form the conjugate of $z = x + jy$ is written as $z* = x - jy$. In polar form the conjugate is obtained by reversing the sign of the angle of z, $z* = Me^{-j\theta}$. The geometric interpretation in Figure C–2 shows that conjugation simply reflects a complex number across the real axis in the complex plane.

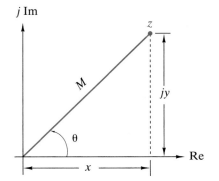

F I G U R E   C – 1 *Graphical representation of complex numbers.*

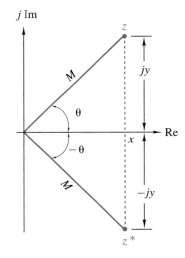

F I G U R E   C – 2 *Graphical representation of conjugate complex numbers.*

Exercise C–1

Convert the following complex numbers to polar form.
(a) $1 + j\sqrt{3}$ (b) $-10 + j20$ (c) $-2000 - j8000$ (d) $60 - j80$

A n s w e r s : (a) $2e^{j60°}$ (b) $22.4e^{j117°}$ (c) $8246e^{j256°}$ (d) $100e^{j307°}$

Exercise C–2

Convert the following complex numbers to rectangular form.
(a) $12e^{j90°}$ (b) $3e^{j45°}$ (c) $400\angle\pi$ (d) $8e^{-j60°}$ (e) $15e^{j\pi/6}$

A n s w e r s : (a) $0 + j12$ (b) $2.12 + j2.12$ (c) $-400 + j0$ (d) $4 - j6.93$ (e) $13 + j7.5$

Exercise C–3

Evaluate the following expressions.

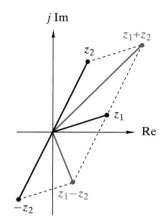

j Im

z_1+z_2

z_2

z_1

Re

z_1-z_2

$-z_2$

FIGURE   C – 3 *Graphical representation of addition and subtraction of complex numbers.*

(a) $\text{Re}(12e^{j\pi})$ (b) $\text{Im}(100\angle 60°)$ (c) $\angle(-2 + j6)$ (d) $\text{Im}[(4e^{j\frac{\pi}{4}})^*]$

Answers:
(a) -12 (b) 86.6 (c) $108.4°$ (d) -2.83

ARITHMETIC OPERATIONS: ADDITION AND SUBTRACTION

Addition and subtraction are defined in terms of complex numbers in rectangular form. Two complex numbers

$$z_1 = x_1 + jy_1 \quad \text{and} \quad z_2 = x_2 + jy_2 \tag{C–9}$$

are added by separately adding the real parts and imaginary parts. The sum $z_1 + z_2$ is defined as

$$z_1 + z_2 = (x_1 + x_2) + j(y_1 + y_2) \tag{C–10}$$

Subtraction follows the same pattern except that the components are subtracted:

$$z_1 - z_2 = (x_1 - x_2) + j(y_1 - y_2) \tag{C–11}$$

Figure C–3 shows a geometric interpretation of addition and subtraction. In particular, note that $z + z^* = 2x$ and $z - z^* = j2y$.

MULTIPLICATION AND DIVISION

Multiplication and division of complex numbers can be accomplished with the numbers in either rectangular or polar form. For complex numbers in rectangular form the multiplication operation yields

$$z_1 z_2 = (x_1 + jy_1)(x_2 + jy_2)$$
$$= (x_1 x_2 + j^2 y_1 y_2) + j(x_1 y_2 + x_2 y_1) \tag{C–12}$$
$$= (x_1 x_2 - y_1 y_2) + j(x_1 y_2 + x_2 y_1)$$

For numbers in polar form the product is

$$z_1 z_2 = (M_1 e^{j\theta_1})\,(M_2 e^{j\theta_2})$$
$$= (M_1 M_2)e^{j(\theta_1 + \theta_2)} \tag{C–13}$$

Multiplication is somewhat easier to carry out with the numbers in polar form, although both methods should be understood. In particular, the product of a complex number z and it conjugate z^* is the square of its magnitude, which is always positive.

$$zz^* = (Me^{j\theta})(Me^{-j\theta}) = M^2 \tag{C–14}$$

For complex numbers in polar form the division operation yields

$$\frac{z_1}{z_2} = \frac{Me^{j\theta_1}}{M_2 e^{j\theta_2}}$$
$$= \left(\frac{M_1}{M_2}\right)e^{j(\theta_1 - \theta_2)} \tag{C–15}$$

When the numbers are in rectangular form the numerator and denominator of the quotient are multiplied by the conjugate of the denominator.

$$\frac{z_1 \, z_2^*}{z_2 \, z_2^*} = \frac{(x_1 + jy_1)(x_2 - jy_2)}{(x_2 + jy_2)(x_2 - jy_2)}$$

Applying the multiplication rule from Eq. (C–12) to the numerator and denominator yields

$$\frac{z_1}{z_2} = \frac{(x_1x_2 + y_1y_2) + j(x_2y_1 - x_1y_2)}{x_2^2 + y_2^2} \qquad (C-16)$$

Complex division is easier to carry out with the numbers in polar form, although both methods should be understood.

Exercise C–4

Evaluate the following expressions using $z_1 = 3 + j4$, $z_2 = 5 - j7$, $z_3 = -2 + j3$, and $z_4 = 5\angle-30°$:

(a) z_1z_2 (b) $z_3 + z_4$ (c) z_2z_3/z_4 (d) $z_1{}^* + z_3z_1$ (e) $z_2 + (z_1z_4)^*$

Answers:
(a) $43 - j$ (b) $2.33 + j0.5$ (c) $-0.995 + j6.12$ (d) $-15 - j3$ (e) $28 - j16.8$

Exercise C–5

Given $z = x + jy = Me^{j\theta}$, evaluate the following statements:

(a) $z + z^*$ (b) $z - z^*$ (c) z/z^* (d) z^2 (e) $(z^*)^2$ (f) zz^*

Answers:
(a) $2x$ (b) $j2y$ (c) $e^{j2\theta}$ (d) $x^2 - y^2 + j2xy$ (e) $x^2 - y^2 - j2xy$ (f) $x^2 + y^2$

Exercise C–6

Given $z_1 = 1$, $z_2 = -1$, $z_3 = j$, and $z_4 = -j$, evaluate (a) z_1/z_3 (b) z_1/z_4 (c) z_3z_4 (d) z_3z_3 (e) z_4z_4 (f) $z_2z_3^*$.

Answers:
(a) $-j$ (b) $+j$ (c) 1 (d) -1 (e) -1 (f) j

Exercise C–7

Evaluate the expression $T(\omega) = j\omega/(j\omega + 10)$ at $\omega = 5, 10, 20, 50, 100$.

Answers: $0.447\angle63.4°, 0.707\angle45°, 0.894\angle26.6°, 0.981\angle11.3°, 0.995\angle5.71°$

APPENDIX D

REFERENCES

CONTEMPORARY BOOKS ON CIRCUIT ANALYSIS

Balabanian, N., *Electric Circuits*, New York, NY: McGraw-Hill, 1994.

Dorf, R. C. and Svoboda, J. A., *Introduction to Electric Circuits*, 4th ed., New York, NY: John Wiley & Sons, 1999.

Hayt, W. H., Jr., and J. E. Kemmerly, *Engineering Circuit Analysis*, 5th ed., New York, NY: McGraw-Hill, 1993.

Huelsman, L. P., *Basic Circuit Theory*, 3rd ed., Englewood Cliffs, NJ: Prentice Hall, 1991.

Irwin, J. D. and Wu, Chwan-Hwa, *Basic Engineering Circuit Analysis*, 6th ed., Englewood Cliffs, NJ: Prentice Hall, 1999.

Johnson, D. E., J. R. Johnson, J. L. Hilburn, and P. D. Scott, *Electric Circuit Analysis*, 3rd ed., Englewood Cliffs, NJ: Prentice Hall, 1997.

Nilsson, J. W. and S. A. Riedel, *Electric Circuits*, 6th ed., Reading, MA: Addison-Wesley, 1999.

CLASSICAL BOOKS ON CIRCUIT ANALYSIS

Bode, H. W., *Network Analysis and Feedback Amplifier Design*, Princeton, NJ: D. Van Nostrand, 1945.

Gardner, M. F., and J. L. Barnes, *Transients in Linear Systems*, New York, NY: John Wiley & Sons, 1942.

Guillemin, E. A., *Introductory Circuit Theory*, New York, NY: John Wiley & Sons, 1953.

Skilling, H. H., *Transient Electric Currents*, New York, NY: McGraw-Hill, 1937.

Van Valkenburg, M. E., *Network Analysis*, Englewood Cliffs, NJ: Prentice Hall, 1st ed., 1955, 2nd ed., 1964, 3rd ed., 1974.

CIRCUIT DESIGN

Franca, J. E., and Y. Tsividis, Editors, *Design of Analog-Digital VLSI Circuits for Telecommunications and Signal Processing*, 2nd ed., Englewood Cliffs, NJ: Prentice Hall, 1994.

Huelsman, L. P., *Active and Passive Analog Filter Design*, New York, NY: McGraw-Hill, 1993.

McConnell, R. L., W. L. Cooley, and N. T. Middleton, *Electrical Engineering Design Compendium*, Reading, MA: Addison-Wesley Pub. Co., 1993.

Van Valkenburg, M. E., *Analog Filter Design*, Chicago, IL: Holt, Rinehart, & Winston, 1982.

Williams, J., *Analog Circuit Design*, Boston, MA: Butterworth-Heinemann, 1991.

CIRCUIT ANALYSIS AND ELECTRONICS

Horenstein, M. N., *Microelectronic Circuits and Devices*, 2nd ed., Englewood Cliffs, NJ: Prentice Hall, 1996.

Savant, C. J., Jr., M. S. Roden, and G. L. Carpenter, *Electronic Design: Circuits and Systems,* 2nd ed., Redwood City, CA: Bengamin/Cummings, 1991.

Senturia, S. D., and B. D. Wedlock, *Electronic Circuits and Applications*, New York, NY: Kreiger, 1993.

Wait, J. V., L. P. Huelsman, and G. A. Korn, *Introduction to Operational Amplifier Theory and Applications*, 2nd ed., New York, NY: McGraw-Hill, 1993.

CIRCUIT ANALYSIS AND ELECTRIC POWER SYSTEMS

del Toro, V., *Electric Power Systems*, Englewood Cliffs, NJ: Prentice Hall, 1992.

Elgerd, O. I., *Electrical Energy Systems Theory: An Introduction*, New York, NY: McGraw-Hill, 1982.

El-Hawary, M. E., *Electrical Power Systems*, Piscataway, NJ: IEEE Press, 1995.

SIGNAL AND SYSTEM ANALYSIS

Cooper, G. R. and C. D. MacGillem, *Continuous and Discrete Signals and Systems Analysis*, 3rd ed., Philadelphia, PA: Saunders, 1991.

Dorf, R. C., *Modern Control Systems*, 6th ed., Reading, MA: Addison-Wesley, 1992.

Kuo, B. C., *Automatic Control Systems*, 6th ed., Englewood Cliffs, NJ: Prentice Hall, 1991.

Oppenheim, A. V., A. S. Willsky, and I. T. Young, *Signals and Systems*, 2nd ed., Englewood Cliffs, NJ: Prentice Hall, 1991.

COMPUTER-AIDED CIRCUIT ANALYSIS

Etter, D. M., *Engineering Problem Solving with MATLAB*, 2nd ed., Englewood Cliffs, NJ: Prentice Hall, 1999.

Goody, R. W., *PSpice for Windows-A Circuit Simulation Primer*, Englewood Cliffs, NJ: Prentice Hall, 1995.

Herniter, M. E., *Schematic Capture with PSpice,* New York, NY: Merrill/Macmillan, 1994.

Rashid, M. H., *SPICE for Circuits and Electronics Using PSpice*, Englewood Cliffs, NJ: Prentice Hall, 1990.

Tuinenga, P., *SPICE: A Guide to Circuit Simulation & Analysis Using PSpice,* 2nd ed., Englewood Cliffs, NJ: Prentice Hall, 1992.

HANDBOOKS

Chen, Wai-Kai, editor-in-chief, *The Circuits and Filter Handbook*, Boca Raton, FL: CRC Press, 1995.

Dorf, R. C., *The Electrical Engineering Handbook*, Boca Raton, FL: CRC Press, 1993.

Harper, C. A., editor-in-chief, *Handbook of Components for Electronics*, New York, NY: McGraw-Hill, 1977.

ANSWERS
TO SELECTED PROBLEMS

CHAPTER ONE

1–3 (a) 2.2 MV;   (b) 2.3 μF;
     (c) 6.2 kΩ;   (d) 0.752 MJ;
     (e) 0.235 mH;

1–5 (a) 10^6;   (b) 10^{-3};
     (c) 10^3;   (d) 10^{-6}

1–7 (a) 1.08×10^7 C;   (b) 250 A

1–9 $i(0) = 0$ A; $i(1) = 6$ A; $i(3) = 18$ A

1–10 $q(0) = 0$; $q(5) = 0.625$ C

1–13 (a) 4.167 A;   (b) 48 hrs

1–15   2.16 MJ

1–19   33.3 mA

1–20   10 mW

1–26   49.05 dB

CHAPTER TWO

2–1 (a) $i = 1.2$ mA; $p = 14.4$ mW;
     (b) $v = -2$ V;
     (c) $p = -120$ mW;
     (d) $v = 0$; $p = 0$

2–3 (a) $i = 2$ mA; $v = 2$ V; $p = 4$ mW;
     (b) $i = 5$ A; $v = 0$; $p = 0$;
     (c) $i = 5$ A; $v = 1$ kV; $p = 5$ kW

2–11 (a) Nodes A, B, C; Loops 1,2; 2,3,4; 1,3,4

(b) Series, 3 & 4; Parallel, 1 & 2

(c) KCL: $i_1 + i_2 + i_3 = 0$;

$-i_3 + i_4 = 0$;

$-i_1 - i_2 - i_4 = 0$

KVL: $-v_1 + v_2 = 0$;

$-v_2 + v_3 + v_4 = 0$;

$-v_1 + v_3 + v_4 = 0$

(d) $i_3 = i_4 = -2$ mA

2–13 (a) Nodes A, B, C, D; Loops 1,3,2; 2,4,5; 3,6,4; 1,5,6; 2,3,5,6

(b) Series, none; Parallel, none

(c) KCL: $i_2 + i_3 + i_4 = 0$;

$i_1 - i_3 + i_6 = 0$;

$-i_1 - i_2 - i_5 = 0$;

$-i_4 + i_5 - i_6 = 0$

KVL: $-v_1 + v_2 - v_3 = 0$;

$-v_2 + v_4 + v_5 = 0$;

$v_3 + v_6 - v_4 = 0$

(d) $v_2 = 1$ V; $v_1 = 9$ V; $v_6 = 0$ V

2–15 (a) Nodes A, B, C, D; Loops 1,3,2; 2,4,5; 3,6,4; 1,5,6

(b) KCL: $i_2 + i_3 + i_4 = 0$;

$i_1 - i_3 + i_6 = 0$;

$-i_1 - i_2 - i_5 = 0$;

$-i_4 + i_5 - i_6 = 0$

(c) $i_2 = -3$ mA; $i_1 = -2$mA; $i_6 = 17$ mA

2–20 C1: $i_x = -1$ A; $v_x = 5$ V; C2: $i_x = 0.4$ A; $v_x = 12$ V

2–21 C3: $i_x = 1.41$ mA; $v_x = 7.89$ V; C4: $i_x = 1.89$ mA; $v_x = 17.2$ V

2–22 $v_4 = -5$ V; $v_5 = 5$ V; $v_6 = 0$ V;

$i_1 = -50$ mA; $i_2 = 100$ mA;

$i_3 = -50$ mA; $i_4 = -50$ mA;

$i_5 = 50$ mA; $i_6 = 0$

2–24 (a) $i_x = -1$ mA; $v_x = -10$ V;

(b) $-i_1 + 0.01 - 0.005 + i_x = 0$

2–30 $R_{AB} = 60$ Ω; $R_{BC} = 0$ Ω; $R_{AC} = 60$ Ω; $R_{CD} = 50$ Ω; $R_{BD} = 50$ Ω; $R_{AD} = 110$ Ω

2–31 $R_{AB} = 90$ Ω; $R_{AC} = 60$ Ω; $R_{AD} = 40$ Ω; $R_{BC} = 130$ Ω; $R_{BD} = 110$ Ω; $R_{CD} = 40$ Ω

2–33 (a) $v_S = 8$ V; $R = 40$ Ω;

(b) $i_S = 150.4$ mA; $R = 133$ Ω;

(c) $i_S = 350.3$ mA; $R = 30.75$ Ω

2–40 $R = 81$ Ω

2–42 C1: $v_x = 0.1v_S$; C2: $v_x = 1.898$ V

2–43 C1: $i_x = 2.5$ A; C2: $v_x = 0.666$ V

2–44 C3: $i_x = 1.67$ A; $v_x = 2.5$ V; C4: $i_x = 0.6$ A; $v_x = 12$ V

2–50 $x = 61.8$ %

2–51 $i_x = -1.25$ A; $v_x = -12.5$ V

2–52 $i_x = 62.5$ mA

2–53 $i_x = 0.833$ A; $v_x = 1.67$ V

2–54 $i_x = -0.1$ A; $v_x = -0.75$ V

Chapter Three

3–5 (b) $i_x = -0.5$ m A; $v_x = 0$

3–6 (b) $i_x = -5.36$ A; $v_x = 10.7$ V

3–7 (b) $i_x = 73.7$ mA; $v_x = 21.05$ V;
   (c) $p_{total} = -7.37$ W

3–8 (b) $i_x = 0.420$ mA; $v_x = 104$ mV;
   (c) $p_{R2} = 0.411$ mW

3–9 (b) $i_x = -0.50$ mA; $v_x = 10$ V;
   (c) $p_{R1} = 2.50$ mW

3–11 (b) $i_x = -167$ mA; $v_x = 0$;
   (c) $p_{vs} = -10$ W

3–13 (b) $i_x = 3.35$ mA; $v_x = -3$ V;
   (c) $p_{S1} = -40.2$ mW

3–21 $K = \dfrac{R_1 R_3}{R_1 + R_2 + R_3}$

3–23 $K = \dfrac{R_2 R_3 - R_1 R_4}{R_1 + R_2 + R_3 + R_4}$

3–25 $i_O = 63.5$ mA

3–29 $i_O = 0.923$ mA

3–31 $v_O = R(0.333 i_1 - i_2)$

3–34 $k_1 = 0.8333$; $k_2 = 0.08333$; $k_3 = 0.05556$

3–36 (a) $v_T = R_1 i_S$; $i_N = R_1 i_S/(R_1 + R_2)$; $R_T = R_1 + R_2$;
   (b) and (c) $i_L = R_1 i_S/(R_1 + R_2 + R_L)$

3–38 (a) $v_T = 40$ V; $R_T = 40$ kΩ;
   (b) If $R_L = 10$ kΩ; $p_L = 6.4$ mW;
   (c) If $R_L = 20$ kΩ; $p_L = 8.89$ mW; If $R_L = 40$ kΩ; $p_L = 10$ mW;
      If $R_L = 80$ kΩ; $p_L = 8.89$ mW

3–40 (a) $v_T = 21$ V; $R_T = 18$ Ω; $i_N = 1.167$ A;
   (b) $p_L = 5.63$ W;
   (c) $p_L = 4.44$ W

3–42 (a) $v_T = 8$ V; $R_T = 1$ kΩ;
   (b) If $R_L = 500$ Ω; $p_L = 14.2$ mW; If $R_L = 1$ kΩ; $p_L = 16$ mW;
      If $R_L = 2$ kΩ; $p_L = 14.2$ mW

3–44 (b) $v_T = 15$ V; $R_T = 5$ kΩ

3–51 (a) $R_L = 13.3$ Ω; $p_{max} = 33.3$ mW;
   (b) $R_L = \infty$; $v_{max} = 1.33$ V;
   (c) $R_L = 0$; $i_{max} = 100$ mA

3–52 (a) $R_L = 600\ \Omega$; $p_{max} = 38.4$ mW;

(b) $R_L = \infty$; $v_{max} = 9.6$ V;

(c) $R_L = 0$; $i_{max} = 16$ mA

3–54 $v_L = 20$ V; $p_L = 100$ mW; $R = 12$ kΩ

3–61 For a series resistor $R = 25\ \Omega$; many other designs are possible

3–62 For a series resistor $R = 6230\ \Omega$; many other designs are possible

3–63 For a series resistor $R = 400\ \Omega$ (nominal);
$R = 100 + 470\ \|\ 680 = 378 \pm 5\%\ \Omega$

Chapter Four

4–2 (a) $-200/3$; (b) $-5/3$;

(c) $p_{IN} = 0.2$ mW; $p_{OUT} = 22.2$ mW

4–3 $v_3 = 2.27$ V; $K_I = 3.33 \times 10^5$

4–4 (a) $v_2 = -0.398$ V; $K_I = -49.8$;

(b) $R_{IN} = 249\ \Omega$

4–5 (a) $i_2 = -0.461$ mA; $K_V = 3.98$;

(b) $R_{IN} = 62.8$ kΩ

4–10 10.58 kΩ

4–14 (a) $0.6 < v_S < 8.6$ V;

(b) $0.6 < v_S < 4.6$ V

4–15 (a) $0.7 < v_S < 6.02$ V;

(b) $0.7 < v_S < 3.89$ V

4–16 (a) $R_B = 113$ kΩ;

(b) $R_B = 226$ kΩ

4–23 (a) $v_O = 2v_S$;

(b) $p_S = 0$; $p_O = 1.8$ mW

4–24 (a) $v_O = -2v_{S1} + 0.6v_{S2}$;

(c) $8.33 < v_{S2} < 58.3$ V

(d) $v_{S2} = 16.667$ V

4–25 (a) $R_1 = 54$ kΩ; $R_2 = 18$ kΩ; $R_3 = 9$ kΩ;

(b) $-10.5 < v_1 < 19.5$ V

4–26 $v_O = K(v_S - V_{BB})$; $K = (R_1 + R_2)/R_1$

4–27 (a) $v_{max} = 5$ V; $i_{max} = 0.417$ mA; $p_{max} = 0.521$ mW;

(b) $v_O = -8.70$ V; $i_O = -1.74$ mA; $p_O = 15.2$ mW

4–46 (a) $V_{OH} = 15$ V; $V_{OL} = 0$ V;

(b) $v_O = V_{OH}$ if $v_S > 7.5$ V;

$v_O = V_{OL}$ if $v_S < 7.5$ V

4–47 (a) $V_{OH} = 15$ V; $V_{OL} = 0$ V;

(b) $v_O = V_{OH}$ if $v_S > -2$ V;

$v_O = V_{OL}$ if $v_S < -2$ V

4–55 (a) $K = (R_F/R)(2 + R/R_F)^{-1}$;
　　　(b) $V_{REF} = 15$ V; $R_F = 100.05$ kΩ

CHAPTER FIVE

5–4　　$5 \times [u(0.003 - t) + u(t - 0.005)]$

5–6　(a) $V_A = 10$ V; $T_C = 0.5$ s;
　　　(b) $V_A = 10$ V; $T_C = 2$ s;
　　　(c) $V_A = -10$ V; $T_C = 50$ ms;
　　　(d) $V_A = -10$ V; $T_C = 20$ s

5–8　　$V_A = 13.9$ V; $T_C = 3.92$ ms

5–12　$T_0 = 1$ ms; $\phi = 2.68$ rad; $T_S = -0.426$ ms; $V_A = 22.36$ V

5–14　$a = 10$ V; $b = 10$ V;
　　　$\phi = -0.785$ rad $= -45°$

5–19 (a) $u(t) - u(t - 2)$;

　　　(b) $u(t) - 2u(t - 1) + u(t - 2)$

5–20　$V_A = 10$ V; $V_B = 5$ V; $\alpha = 1.386 \times 10^5$ rad/s

5–21　$V_A = 9.951$ V; $\alpha = 994.7$ rad/s; $\beta = 1257$ rad/s

5–24　$V_A = 10$ V; $V_B = 12$ V; $\beta = 2.513 \times 10^5$ rad/s

5–29　$V_A = 24.5$ V; $\alpha = 235$ rad/s; $\beta = 1257$ rad/s

5–34　$V_p = 3$ V; $V_{pp} = 3$ V; $V_{avg} = 1.5$ V; $V_{rms} = 1.87$ V

5–35　$V_p = 2$ mV; $V_{pp} = 3$ mV; $V_{avg} = 0.25$ mV; $V_{rms} = 1.12$ mV

5–36　$V_A = 5$ V; $V_B = 10$ V

CHAPTER SIX

6–2　　$\beta = 5$ krad/s; $i_C(t) = -15 \sin(\beta t)$ mA; $p_C(t) = -0.225 \sin(2\beta t)$ W;
　　　$w_C(t) = 22.5 [1 + \cos(2\beta t)]$ μJ; both

6–3　　$\alpha = 5$ krad/s; $v_C(t) = 80 [1 - \exp(-\alpha t)]$ V; $p_C(t) = 4[\exp(-\alpha t) - \exp(-2\alpha t)]$ W; $w_C(t) = 400[1 - \exp(-\alpha t)]^2$ μJ; absorbing

6–7　　50 V; 100 V; 100 V

6–8　　$\alpha = 1$ krad/s; $v_L(t) = 2 \exp(-\alpha t)$ V; $p_L(t) = 2[\exp(-\alpha t) - \exp(-2\alpha t)]$ W; $w_L(t) = [1 - \exp(-\alpha t)]^2$ mJ; absorbing

6–11　$\alpha = 1$ krad/s; $i_L(t) = 20[1 - \exp(-\alpha t)]$ mA; $p_L(t) = 20[\exp(-\alpha t) - \exp(-2\alpha t)]$ mW; $w_L(t) = 10[1 - \exp(-\alpha t)]^2$ μJ; absorbing

6–13　0.125 A; 0.5 A; 1.5 A; absorbing

6–14　$\alpha = 5$ krad/s; $i_C(t) = -1.25\exp(-\alpha t)$ A; $p_C(t) = -31.25 \exp(-2\alpha t)$ W; delivering

6–20　$C = 1$ μF

6–24　$v_O(t) = 10 - 2500t$ V; $t = 10$ ms

6–25　No

6–26　$v_O(t) = 0.25 \times V_A \sin(2000t)$ V; $V_A < 60$ V

6–28 $v_O(t) = -50 \times R \cos(10^6 t)$ µV; $R < 300$ kΩ

6–36   C1: $C_{EQ} = 3.5$ µF; C2: $L_{EQ} = 1.065$ mH

6–39 $C_{EQ} = 5.339$ µF

6–40 $C_{AB} = 4.128$ µF

6–42   Four strings in parallel each consisting of four 20 µF in series

6–46 (a) 1.03 mF;
(b) 11.9 kJ;
(c) 11.5 MW;
(d) 11.9 kW

CHAPTER SEVEN

7–1 (a) $v(t) = -15e^{-1500t}$ V;
(b) $i(t) = 10 - 30\,e^{-1000t}$ mA

7–3   C1: 0.367 ms;
C2: 0.4 ms

7–4   C1: 0.831 ms;
C2: 99.9 ms

7–7 (c) C1: $v_C(t) = V_A\left[-1 + 2\exp\left(-\dfrac{t}{RC}\right)\right]$;

C2: $i_L(t) = \dfrac{V_A}{R}\left[-1 + 2\exp\left(-\dfrac{Rt}{L}\right)\right]$

7–11 $v_C(t) = 4 - e^{-150t}$ V

7–13 $v_C(t) = 11.5(1 - e^{-10400t})$ V

7–21 (a) $s + 1000 = 0$; $T_C = 1$ ms; $v_C(0) = -5$ V;
(b) $v_T(t) = 5$ V for t > 0

7–23 (a) $s + 1000 = 0$; $T_C = 1$ ms; $v_C(0) = 5$ V;
(b) $v_T(t) = 20\cos(1000t)$ V for t > 0

7–25 (a) $s + 1000 = 0$; $T_C = 1$ ms;
(b) $R = 600$ Ω; $C = 1.67$ µF

7–32 $v(t) = 2 + e^{-5t}[3\cos(10t) + 4\sin(10t)]$ for $t > 0$

7–33   C1: $\omega_0 = 50$ krad/s; $\zeta = 0.005$; underdamped; C2: $\omega_0 = 31.62$ Mrad/s; $\zeta = 0.158$; underdamped

7–34 (c) $v_C(t) = 13.3\,e^{-10000t} - 3.33\,e^{-40000t}$ V;
$i_L(t) = -3.33\,e^{-10000t} + 3.33\,e^{-40000t}$ mA

7–36 (c) $v_C(t) = 12.5\,e^{-4000t} - 2.5\,e^{-20000t}$ V;
$i_L(t) = -1.25\,e^{-4000t} + 1.25\,e^{-20000t}$ mA

7–38 (c) $v_C(t) = -1.69\,e^{-1667t}\sin(9860t)$ V; $i_L(t) = 1.67\,e^{-1667t}[\cos(9860t) + 0.169\sin(9860t)]$ mA

7–46 $v_O(t) = -7.73\,e^{-5359t} + 107.73\,e^{-74640t}$ V

7–51 (a) $\omega_0 = 4.243$ krad/s; $\zeta = 1.061$;
(b) $v_C(0) = 0$; $v_C(\infty) = 10$ V; $i_L(0) = 6$ mA; $i_L(\infty) = 0$; $v_T = 10$ V; (c) $C = 133$ nF

7–52 (a) $\omega_0 = 141.4$ rad/s; $\zeta = 0.707$;
(b) $v_C(0) = 10$ V; $v_C(\infty) = 10$ V; $i_L(0) = -0.5$ mA; $i_L(\infty) = 0$; $v_T = 10$ V; (c) $C = 1$ µF

7–56 $v_T(t) = 4.375\ e^{-500t}$ V

7–63 (a) $\alpha = 190$ krad/s; $\beta = 866$ krad/s;
(b) $s^2 + 3.8 \times 10^5 s + 7.872 \times 10^{11} = 0$;
(c) $L = 5.79$ mH; $C = 219$ pF

Chapter Eight

8–1 $\mathbf{V}_1 = 70.7 + j70.7$ V; $\mathbf{V}_2 = 150 + j200$ V; $v_1(t) + v_2(t) = 349\cos(\omega t + 50.8°)$ V

8–3 (a) $v_1(t) = 10\cos(10^4 t - 30°)$ V;
(b) $v_2(t) = 60\cos(10^4 t - 220°)$ V;
(c) $i_1(t) = 5\cos(10^4 t + 90°)$ A;
(d) $i_2(t) = 2\cos(10^4 t + 270°)$ A

8–8 $v_3(t) = 69.95\cos(\omega t + 120.4°)$ V

8–10 $v_2(t) = 10\cos(\omega t - 143.1°)$ V

8–11 (a) $Z_{EQ} = 50 + j40\ \Omega$;
(b) $Z_{EQ} = 10 - j20\ \Omega$;
(c) $Z_{EQ} = 16.5 - j15.5\ \Omega$;
(d) $Z_{EQ} = 4.5 - j12.5\ \Omega$

8–12 $Z_{EQ} = 57.99 + j74.68 = 94.55\angle 52.2°\ \Omega$

8–14 $Z_{EQ} = 0.333 + j124.5 = 124.5\angle 89.8°\ \Omega$

8–19 $L = 0.181$ mH; $C = 6.03$ nF

8–21 $i(t) = 70.7\cos(2500t - 45°)$ mA

8–23 $i_R(t) = 141\cos(377t - 62.05°)$ mA; $i_C(t) = 265\cos(377t + 27.95°)$ mA

8–25 $i_R(t) = 10\cos(1000t + 45°)$ mA; $i_C(t) = 10\cos(1000t + 135°)$ mA

8–27 $Z = 162.2 - j27.03\ \Omega$; $v_x(t) = 116\cos(1000t + 35.5°)$ V

8–29 $\mathbf{I}_x = 0.125 - j1.13 = 1.14\angle -83.7°$ A

8–33 $\mathbf{V}_x = 34.11 - j25.14 = 42.37\angle -36.38°$ V

8–36 $Z_T = 30 + j40$ kΩ; $\mathbf{V}_T = 300 + j0$ V

8–41 $v_x(t) = 121.1\cos(4000t - 84.09°)$ V

8–43 $K = j1.031$; $Z_{IN} = 9897 + j101\ \Omega$

8–45 $\mathbf{I}_x = -2.37 + j0.237$ mA; $\mathbf{V}_y = 1.83 + j12.62$ V

8–47 $K = 0.847 - j0.028$; $Z_{IN} = 21.32 - j9.71$ kΩ

8–51 $\mathbf{V}_L = 35 + j0$ V; $\mathbf{I}_L = 0.7 - j0.7$ A; $P = 12.25$ W

8–53 $R = 1280\ \Omega$; $C = 1.06$ µF

8–55 (a) $P_L = 144$ mW;
(b) $P_{MAX} = 450$ mW;
(c) $Z_L = 1 - j2$ kΩ

8–61 (a) $R = 125\ \Omega$; $L = 60.47$ mH;

(b) $\mathbf{V}_1 = 2.6 - j3.04$ V; $\mathbf{V}_2 = 7.4 + j3.04$ V

CHAPTER NINE

9–1 (a) $F(s) = \dfrac{A(s - \alpha)}{s(s + \alpha)}$

(b) $F(s) = \dfrac{A(s - \alpha)}{s(s + \alpha)^2}$

9–3 $F(s) = \dfrac{A(\beta - s)^2}{s(s^2 + \beta^2)}$

9–5 $F(s) = \dfrac{A(s + \alpha)^2}{(s + \alpha)^2 + \beta^2}$

9–7 (a) $F(s) = \dfrac{-5s}{(s + 10)(s + 20)}$

(b) $F(s) = \dfrac{450000\ s}{(s^2 + 100^2)(s^2 + 200^2)}$

9–9 $F(s) = \dfrac{(s + 36.5)(s + 68.5)}{(s + 50)^2}$

9–13 (a) $f(t) = \dfrac{A}{T}\,tu(t) - \dfrac{A}{T}\,(t - T)u(t - T) - Au(t - T)$

(b) $F(s) = \dfrac{A[1 - (1 + Ts)e^{-Ts}]}{Ts^2}$

9–21 (a) $f_1(t) = 36(e^{-20t} - e^{-45t})u(t)$
(b) $f_2(t) = 900te^{-30t}\ u(t)$
(c) $f_3(t) = 37.5e^{-18t}\sin(24t)u(t)$

9–23 (a) $f_1(t) = \{2.67 - 1.5e^{-2t} - 0.167e^{-6t}\}u(t)$
(b) $f_2(t) = \{0.333 + 0.238\cos(3\sqrt{5}t) + 0.429\cos(4\sqrt{5}t)\}u(t)$

9–25 (a) $f(t) = \{t + 0.25 + 0.289e^{-3t}\cos(3\sqrt{3}t + 150°)\}u(t)$
(b) $f(t) = \{23.4 - 13.4e^{-10t} - 24te^{-10t}\}u(t)$

9–27 (a) $\gamma = 20$; $K = 2/3$;
(b) $\gamma = 30$; $K = 1$;
(c) $\gamma = 40$; $K = 4/3$

9–31 (a) $y(t) = 5e^{-10t}u(t)$
(b) $y(t) = \{10 - 20e^{-10000t}\}u(t)$

9–32 $y(t) = \{-4.95e^{-100t} + 4.95\cos(10t) + 0.495\sin(10t)\}u(t)$

9–38 $y(t) = \{10 - 11.1e^{-t} + 1.11e^{-10t}\}u(t)$

9–39 (b) $i_{\text{L}}(t) = \{e^{-1000t}\,[5\cos(3000t) + 1.67\sin(3000t)]\}u(t)$ mA

9–46 (a) Initial value does not apply; FV = 0;
(b) IV = 1/3; Final value does not apply

9–48 (a) IV = 0; Final value does not apply;
(b) IV = 10; FV = 2

CHAPTER TEN

10–1 $Z(s) = \dfrac{R(RCs + 1)}{2RCs + 1}$

$R = 10 \text{ k}\Omega; C = 50 \text{ nF}$

10–3 $Z(s) = \dfrac{(s + j4472)(s - j4472)}{(s + 2764)(s + 7236)}$

10–5 $Z(s) = \dfrac{1000\,[(s + 2000)^2 + 2000^2)}{s(s + 2000)}$

10–7 $Z(s) = \dfrac{(RCs + 0.382)(RCs + 2.618)}{Cs(RCs + 2)}$

10–13 $V_C(s) = \dfrac{V_A(RCs - 1)}{s(RCs + 1)}$

10–15 $I_L(s) = 2.5 \times 10^{-3}\,\dfrac{(s + 160000)}{s(s + 100000)}$

10–17 $I_L(s) = \dfrac{-30}{s^2 + 1000s + 2000000}$

10–19 (a) $I_L(s) = \dfrac{(V_A - V_B)C}{LCs^2 + R_2Cs + 1}$;

$V_C(s) = \dfrac{V_A}{s} - \dfrac{(V_A - V_B)}{s(LCs^2 + R_2Cs + 1)}$

(b) $i_L(t) = 5te^{-1000t}u(t) \text{ A};$

$v_C(t) = \{5 + 5(1 + 1000\,t)e^{-1000t}\}\,u(t) \text{ V}$

10–21 (a) $V_C(s) = \left(\dfrac{Ls}{RLCs^2 + Ls + R}\right)V_S(s)$

10–27 $C = 50 \text{ pF}$

10–31 (b) $I_2(s) = \dfrac{R_1CsV_1(s)}{(R_1 + R_2)LCs^2 + (L + R_1R_2C)s + R_1}$

(c) $i_2(t) = 8.03e^{-10040t}\sin(10200\,t)u(t) \text{ mA}$

10–33 (b) $V_2(s) = \dfrac{R_1R_2C_1C_2s^2V_1(s)}{R_1R_2C_1C_2s^2 + (R_1C_1 + R_2C_2 + R_1C_2)s + 1}$

(c) $v_2(t) = [-2.493\,e^{-50.21t} + 12.49e^{-251.6t}]u(t) \text{ V}$

10–39 (a) $\Delta(s) = [(RCs)^2 + (3 - \mu)RCs + 1]/R^2$;

(b) $\mu = 3; RC = 1/5000$

10–41 (b) $v_2(t) = [-0.826e^{-90.9t}]u(t) \text{ V}$

10–45 $V_O(s) = \dfrac{4000}{s^2 + 3000s + 800000}$

10–50 $R_1C_1 = 1/100; R_2C_2 = 1/500$

10–52 $i(t) = \{0.1 - 200te^{-1000t}\}u(t) \text{ A}$

CHAPTER ELEVEN

11–1 (a) $Z(s) = \dfrac{(Ls + R_2)(R_1 Cs + 1)}{LCs^2 + (R_1 + R_2)Cs + 1}$;

$T_V(s) = \dfrac{R_2}{Ls + R_2}$

(b) $Z(s) = 2000 \dfrac{(s + 20000)(s + 2000)}{(s + 1026)(s + 38970)}$;

$T_V(s) = \dfrac{20000}{s + 20000}$

11–3 (a) $Z(s) = \dfrac{(R_1 + R_2)L_1 s + R_1 R_2}{L_1 s + R_2}$

$T_Y(s) = \dfrac{L_1 s}{(R_1 + R_2)L_1 s + R_1 R_2}$

(b) $Z(s) = 150 \left(\dfrac{s + 66.7}{s + 100} \right)$;

$T_Y(s) = \dfrac{1}{150} \left(\dfrac{s}{s + 66.7} \right)$

11–5 $T_V(s) = \dfrac{(s + 2500)^2}{(s + 955)(s + 6545)}$

11–7 $T_V(s) = -\dfrac{2000}{s + 20000}$

11–11 $h(t) = 2000 e^{-20000t} u(t)$

11–13 $h(t) = 5000 e^{-5000t} \sin(5000\, t) u(t)$

11–15 $h(t) = \delta(t) - \{2309\, e^{-1000t} \cos(1732\, t - 60°)\} u(t)$

11–19 $i(t) = \delta(t) - \{0.99\, e^{-101t} + 1.01\, e^{-201t}\} u(t)$

11–21 $g(t) = 0.1\{1 - e^{-20000t}\} u(t)$

11–23 $g(t) = \{0.5 + 0.707\, e^{-5000t} \cos(5000\, t + 270°)\} u(t)$

11–25 $g(t) = \{1.155\, e^{-1000t} \cos(1732\, t + 30°)\} u(t)$

11–27 (a) $T(s) = \dfrac{-s}{s + 2000}$;

$h(t) = -\delta(t) + 2000\, e^{-2000t} u(t)$

(b) $T(s) = \dfrac{2000}{s + 2000}$;

$h(t) = 2000\, e^{-2000t} u(t)$

11–31 $v_{2SS}(t) = 0.098 \cos(10^5 t + 101.3°)$ V

$v_{2SS}(t) = 0.999 \sin(10^3 t + 177.1°)$ V

11–33 $v_{2SS}(t) = 0.447 \cos(10^4 t - 26.56°)$ V;

$v_{2SS}(t) = 0.999 \sin(10^3 t - 2.86°)$ V

11–35 $v_{2SS}(t) = 4.47 \cos(5000\,t - 63.4°)$ V;

   $v_{2SS}(t) = 2.49 \cos(2000\,t - 23.56°)$ V

11–37 $y_{SS}(t) = 2.24 \cos(1000\,t + 63.4°)$ V

11–41 $y(t) = 0.25 \times 10^{-4}\{-1 + 20000\,t + e^{-20000t}\}u(t)$

11–43 $y(t) = 10t^2$ for $0 < t < 5$; $y(t) = 100t - 250$ for $5 < t$

11–47 $y(t) = 10(1 - e^{-200t})$ for $0 < t < 0.1$; $y(t) = 10(e^{-200(t-0.1)} - e^{-200t})$ for $0.1 < t$

11–62   (a) $\alpha = 4.77 \times 10^5$ rad/s; $\beta = 2.24 \times 10^6$ rad/s; $K = 3.16 \times 10^{13}$

Chapter Twelve

12–1   (a) $T_V(s) = \dfrac{500}{s + 1000}$;

   (b) $|T_V(0)| = 0.5$; $|T_V(j\infty)| = 0$; $\omega_C = 1$ krad/s; low pass

12–3   (a) $T_V(s) = \dfrac{800}{s + 200}$;

   (b) $|T_V(0)| = 4$; $|T_V(j\infty)| = 0$; $\omega_C = 200$ rad/s; low pass

12–5   (a) $T_V(s) = \dfrac{8s}{3s + 2000}$;

   (b) $|T_V(0)| = 0$; $|T_V(j\infty)| = 2.67$; $\omega_C = 667$ rad/s; high pass

12–7   low pass; $K = 0.665$; $\omega_C = 60$ Mrad/s

12–11   (a) $T_V(s) = \dfrac{2000000}{s^2 + 3000\,s + 4 \times 10^6}$

   (b) $\omega_0 = 2000$ rad/s; $\zeta = 0.75$; low pass

12–13   (a) $T_V(s) = \dfrac{4000\,s}{s^2 + 6000\,s + 16 \times 10^6}$;

   (b) $\omega_0 = 4000$ rad/s; $\zeta = 0.75$; bandpass

12–15   (a) $T_V(s) = \dfrac{-RC_1 s}{R^2 C_1 C_2 s^2 + 2RC_2 s + 1}$

   (b) $\omega_0 = 5000$ rad/s; $\zeta = 0.25$; bandpass

12–17   (a) $T_V(s) = \dfrac{RLCs^2}{RLC\,s^2 + Ls + R}$;

   (b) $\omega_0 = 5000$ rad/s; $\zeta = 2$; high pass

12–21 $L = 3$ μH; $C = 133$ pF; $Q = 6.25$; $\omega_{C2} = 54.2$ Mrad/s; $\omega_{C1} = 46.2$ Mrad/s

12–23   (a) $B = 4$ krad/s; $\omega_{C1} = 18.1$ krad/s; $\omega_{C2} = 22.1$ krad/s;

   (b) $\mathbf{V}_L = j50$ V; $\mathbf{V}_C = -j50$ V

12–25 $L = 63.7$ nH; $C = 39.8$ pF

12–27 $L = 555$ mH; $C = 12.7$ μF; other solutions are possible.

12–31 $\quad |T(j\omega)|_{SL} = \begin{vmatrix} 1/11 & \text{if } 0 < \omega \leqslant 10; \\ \omega/110 & \text{if } 10 < \omega \leqslant 110; \\ 1 & \text{if } 110 < \omega \end{vmatrix}$

12–33 $\quad |T(j\omega)|_{SL} = \begin{vmatrix} 1 & \text{if } 0 < \omega \leqslant 100; \\ 100^2/\omega^2 & \text{if } 100 < \omega \end{vmatrix}$

12–35 (a) $|T(j\omega)|_{SL} = \begin{vmatrix} 1 & \text{if } 0 < \omega \leqslant 0.5; \\ 0.5/\omega & \text{if } 0.5 < \omega \leqslant 5; \\ 2.5/\omega^2 & \text{if } 5 < \omega \leqslant 50; \\ 0.05/\omega & \text{if } 50 < \omega \end{vmatrix}$

$\quad\quad$ (b) $|T(j\omega)|_{SL} = \begin{vmatrix} 1 & \text{if } 0 < \omega \leqslant 0.5; \\ 0.5/\omega & \text{if } 0.5 < \omega \leqslant 5; \\ 0.1 & \text{if } 5 < \omega \leqslant 50; \\ 5/\omega & \text{if } 50 < \omega \end{vmatrix}$

12–37 (a) $|T(j\omega)|_{SL} = \begin{vmatrix} 1 & \text{if } 0 < \omega \leqslant 100; \\ 100/\omega^2 & \text{if } 10 < \omega \end{vmatrix}$

$\quad\quad$ (b) $|T(j\omega)|_{SL} = \begin{vmatrix} 40\omega^2 & \text{if } 0 < \omega \leqslant 0.5; \\ 10 & \text{if } 0.5 < \omega \end{vmatrix}$

12–41 $\quad g(t) = \{2.025\,(e^{-100t} - e^{-8000t})\}u(t)$

12–43 $\quad g(t) = \{1 + 11.03\,(e^{-200t} - e^{-1000t})\}u(t)$

12–45 $\quad T(s) = \dfrac{5\,(s^2 + 50000)}{s^2 + 200s + 50000}$; bandstop

Chapter Thirteen

13–5 $\quad f(t) = 5 + 3.18\ \sin(2000\pi t) + 1.06\ \sin(6000\pi t) + 0.637\ \sin(10000\pi t)$

13–7 $\quad a_0 = -V_A/2;\ a_n = 0;\ b_n = -V_A/n\pi$

13–9 $\quad a_0 = 0;\ a_n = 0;\ b_n = 2V_A/n\pi$

13–11 $\quad a_0 = 0;\ a_n = 0;\ b_n = 8A/(n\pi)^2$ for n odd; $b_n = 0$ for n even

13–16 (b) $v_O(t) = 62.55 + 35.6\ \cos(4000t - 63.4°) - 4.36\ \cos(12000t - 80.5°) + 1.58\ \cos(20000t - 84.3°)$ V

13–17 (b) $v_O(t) = 10.87\ \sin(10^5 t - 26.6°) - 0.749\ \sin(3 \times 10^5 t - 56.3°) + 0.181\ \sin(5 \times 10^5 t - 68.2°) - 0.068\ \sin(7 \times 10^5 t - 74.1°)$ V

13–19 (b) $i(t) = -127\ \sin(2500t - 5.71°) - 63.6\ \sin(5000t - 2.86°) - 42.4\ \sin(7500t - 1.91°) - 31.8\ \sin(10000t - 1.43°)$ μA

13–25 $\quad$ Highpass with $\omega_C = 350$ rad/s

13–27 $\quad V_{rms} = \dfrac{V_A}{\sqrt{3}}$; 99.9%

13–29 $\quad V_{rms} = \dfrac{V_A}{\sqrt{6}}$; 87.8%

13–31 $\quad$ 19.1 V

13–33 $\quad$ 78.4%

CHAPTER FOURTEEN

14–1 $T(s) = \dfrac{5.012}{s/3142 + 1}$

14–3 $T(s) = \left(\dfrac{s}{s + 200}\right)(31.6)\left(\dfrac{1}{s/3000 + 1}\right)$

14–5 $T(s) = \dfrac{150\,s}{s + 40000}$; many other answers are possible

14–7 $T(s) = \dfrac{s + 250}{s + 50}$

14–24 $\omega_0 = 628$ krad/s; $Q = 16.66$; $\zeta = 0.0300$

14–31 $T(s) = \dfrac{\sqrt{10}}{(s/1037 + 1)^5}$

14–32 $T(s) = \dfrac{\sqrt{10}}{(s/2500 + 1)[(s/2500)^2 + (s/2500) + 1]}$

14–33 $T(s) = \dfrac{0.707}{[q_1(s)][q_2(s)]}$

           $q_1(s) = (s/475)^2 + 0.1789(s/475) + 1$

           $q_2(s) = (s/221)^2 + 0.9276(s/221) + 1$

14–34 (a) First order cascade $n = 4$;
         (b) Butterworth $n = 3$; Chebychev $n = 2$

14–41 (a) First order cascade $n = 24$;
         (b) Butterworth $n = 4$; Chebychev $n = 3$

14–43 (a) $T(s) = \left(\dfrac{1.778\,s}{s + 5149}\right)^2$

CHAPTER FIFTEEN

15–1   (b) $v_2(t) = 7\sin(2000t)$ V

15–2   (b) $i_1(t) = -125\cos(1000t)$ A;
         $i_2(t) = 175\cos(1000t)$ A

15–4   (b) $v_1(t) = 1.5\cos(1000t)$ V;
         $v_2(t) = -\cos(1000t)$ V

15–11 (a) $n = 1/5$;
         (b) $v_1(t) = 120\cos(120\pi t)$ V;
         $i_1(t) = 0.6\cos(120\pi t)$ A

15–12 $i_1(t) = -36\sin(377t)$ A

15–17 $n = 2.29$

15–21 $Z_{IN} = 64.37 + j54.25\ \Omega$;
         $v_2(t) = 11.03\cos(2000t + 28.07°)$

15–23 $i(t) = 20.1\cos(20000t - 161.2°)$ mA

15–26   Two solutions possible:
$\omega = 7.27, 7.94$ krad/s

15–28 $Z_{IN} = 2.5 + j1.5$ kΩ;

$i(t) = 3.43 \cos(4000t - 121°)$ mA

Chapter Sixteen

16–1   (a) $P = 985$ W; $Q = -264$ VAR; pf $= 0.966$; absorbing;
(b) $P = -105$ W; $Q = 182$ VAR; pf $= -0.5$; delivering

16–3   (a) $P = -1.46$ W; $Q = 5.43$ VAR; pf $= -0.259$; delivering;
(b) $P = 260$ W; $Q = -1477$ VAR; pf $= 0.174$; absorbing

16–5   (a) pf $= 0.97$: lagging; (b) pf $= 0.661$; leading

16–7   (a) $P = 15$ kW; $Q = 13.2$ kVAR; $|\mathbf{I}| = 8.33$ A
(b) $Z_L = 216 + j190$ Ω

16–11 $S = 436 + j247$ VA

16–13 $R = 19.4$ kΩ; $C = 1.73$ μF

16–15   (a) $|\mathbf{I}| = 2.71$ A;
(b) $S_L = 440 + j294$ VA; $S_W = 29.4 + j73.4$ VA; $\eta = 93.8$ %

16–17 $\eta = 94.7$ %

16–21 $S_S = 31.3 + j30.9$ kVA; $\eta = 95.9$ %

16–23 $\mathbf{V}_L = 1847$ V; $\mathbf{V}_S = 2190 + j409$ V

16–25   (a) $\mathbf{I}_L = 134 - j38.2$ A; $\mathbf{V}_S = 342 + j126$ V;
(b) $S_W = 3.86 + j19.3$ kVA; pf $= 0.807$

16–31   (a) $\mathbf{V}_{AN} = 120\angle 0°$ V; $\mathbf{V}_{BN} = 120\angle -120°$ V; $\mathbf{V}_{CN} = 120\angle 120°$ V;
$\mathbf{V}_{AB} = 208\angle 30°$ V; $\mathbf{V}_{BC} = 208\angle -90°$ V; $\mathbf{V}_{CA} = 208\angle 150°$ V

16–33   (a) $\mathbf{I}_A = 40\angle 0°$ A; $\mathbf{I}_B = 40\angle 120°$ A; $\mathbf{I}_C = 40\angle -120°$ A;
$\mathbf{I}_1 = 23.1\angle -30°$ A; $\mathbf{I}_2 = 23.1\angle 90°$ A; $\mathbf{I}_3 = 23.1\angle -150°$ A

16–37 $|\mathbf{I}_L| = 88.9$ A; pf $= 0.786$

16–39   (a) $|\mathbf{I}_L| = 36.1$ A; (b) $Z = 5.76 - j5.08$ Ω

16–43 $|\mathbf{V}_{L2}| = 315.7$ V; $S_1 = 8.88 + j10.4$ kVA

INDEX

ac response, 326, 371
Active circuit, 149
Active device, 149
  OP AMP, 169
  transistor, 164
Active *RC* filter, 705
Admittance, 387, 487
  parallel connection, 388, 492
  transfer, 544
Ammeter, 48
  d'Arsonval meter, 48
Ampere (unit), 5
Amplifier:
  differential, 183
  instrumentation, 214
  inverting, 183
  noninverting, 172
  operational, 169
  summing, 180, 193
  voltage follower, 177
Amplitude, 227
Amplitude spectrum, 679
Analog-to-Digital Converter, 194
Apparent power, 797
Autotransformer, 774
Available power, 119, 417
Average power, 414, 794
  of a periodic signal, 251, 690
Average value, 250

Balanced three-phase, 811
  Δ-connected load, 816
  Y-connected load, 814
Bandpass filter, 593
  design, 744
  with first-order circuits, 608
  narrowband, 618
  second order, 616, 720
  wideband, 618
Bandstop filter, 593
  design, 744
  with first order circuits, 609
Bandwidth, 593, 618, 693
Bilateral element, 18
Bode diagrams, 595, 634
  complex poles and zeros, 644
  real poles and zeros, 635
Bridge circuits, 99
  impedance, 386
  Maxwell, 386
  Wheatstone, 199
Bus, 820
Butterworth poles, 730

Capacitance, 266
Capacitive reactance, 384
Capacitor, 265
  average power, 414
  energy, 268

Capacitor (*cont.*)
  impedance, 380, 486
  i–v relationship, 266
  parallel connection, 285
  series connection, 286
  state variable, 268
Cascade connection, 551
Causal waveform, 249
Center frequency, 616
Chain rule, 551
Characteristic equation:
  first order, 302
  second order, 333
Charge, 4
Chebychev poles, 734
Circuit, 2, 23
Circuit analysis, 3
Circuit design, 3. *See also* filter design
  phasor domain, 408
  s domain, 523, 527
  of interface circuits, 121
  of network functions, 566
  with OP AMPs, 191
Circuit determinant, 511
Circuit reduction, 48
Circuit simulation, 54
Circuit theorems:
  maximum power transfer, 119, 417
  Norton's, 105, 399, 501
  proportionality, 98, 393, 495
  superposition, 102, 395, 496
  Thévenin's, 105, 399, 501
Comparator, 203
Complex frequency, 338
Complex numbers, A-12
  Arithmetic operations, A-14
  conjugate, A-13
  exponential form, A-13
  imaginary part, A-12
  real part, A-12
Complex power, 796
  and load impedance, 798
  conservation of, 800
Computer tools, 54
  Electronics Workbench, 151, 189, 349, 355, 579,
    612, 742
  Excel, 232, 327, 514, 688
  MATLAB, 129, 253, 410, 446, 512, 524, 560, 653,
    745, 747, 805
  Mathcad, 93, 276, 457, 641, 648, 673, 686
  PSpice, 55, 130, 151, 162, 306, 397

Conductance, 18, 387
Connection constraints, 22
  phasor domain, 378
  s domain, 482
Convolution, 559
Corner frequency, 598
Coulomb (unit), 4
Coupled inductors, 762
  energy, 767
  i–v relationships, 764
Coupling coefficient, 768
Cramer's rule, A-3
Critically damped response, 39
Current, 4
  mesh, 88
  short circuit, 107
Current division, 45, 388, 493
Current gain, 150
Current source, 21
  current controlled, 150
  Norton equivalent, 106, 399, 501
  voltage controlled, 150
Current transfer function, 544
Cutoff frequency, 593, 723

Δ-to-Y transformation, 817

Damped ramp, 245
Damped sine, 245
Damping, 339
Damping ratio, 346
Digital-to-Analog converter, 205
Direct current (dc), 225
Decade, 595
Decibel, 595
Delay time, 263, 559
Dependent sources, 149, 484
Determinant, A-3
Device, 17
Differential amplifier, 183
Differential equation:
  first order, 301
  second order, 331
  solution by Laplace transforms, 459
Differentiator, 281
Dirichlet conditions, 671
Distortion, 693
Dot convention, 764
Double exponential, 247
Driving point impedance, 542
Duality, 36, 278

Element, 17
Energy, 4
  capacitor, 267
  coupled inductors, 767
  inductor, 274
En route objectives, 3
Equivalent circuits, 35
  admittances, 388, 492
  capacitors, 285
  dc, 288
  impedances, 384, 387, 491
  inductors, 286
  transformer, 780
  Norton, 105, 399, 501
  resistors, 35
  sources, 38
  summary of, 41
  Thévenin, 105, 399, 501
Euler's relationship, A-13
Evaluation, 3
Even symmetry, 681
Exponential waveform, 231
  complex, 373
  properties of, 232

Farad (unit), 266
Filter, 275, 398, 704
Filter design, 705
  bandpass, 744
  bandstop, 744
  high pass, 740
  low pass, 722
  with first-order circuits, 705
  with second-order circuits, 714
Final-value theorem, 468
First-order circuits, 301
  design with, 705
  differential equation, 300
  frequency response, 325
  RC and *RL* circuits, 299
  sinusoidal response, 323
  step response, 309
  zero-input response, 301
  zero-state response, 315
Forced pole, 512
Forced response, 309
Fourier series, 670
  alternative form, 679
  coefficients, 670
  in circuit analysis, 684
  table of, 677

Frequency:
  angular, 238
  center, 616
  complex, 338
  corner, 598
  critical, 444
  cutoff, 593, 618, 723
  fundamental, 247, 671
  half power, 593
  harmonic, 247, 671
  natural, 338
  neper, 338
  radian, 338
  resonant, 391
Frequency response, 592
  and step response, 651
  Bode diagrams, 592, 634
  Butterworth, 727
  Chebychev, 732
  descriptors, 593
  first order, 597
  first-order cascade, 724
  RLC circuits, 627
  second order, 613
Fundamental frequency, 247, 671

Gain-bandwidth product, 602
Gain function, 592
Ground, 9

Half-power frequency, 593
Half-wave symmetry, 682
Harmonic frequency, 247, 671
Henry (unit), 273
Hertz (unit), 238
High-pass filter, 593
  design, 740
  first order, 604
  second order, 625, 717

Ideal models:
  current source, 21
  low-pass filter, 722
  OP AMP, 171
  switch, 19
  transformer, 771
  voltage source, 20
Impedance, 380, 486
  and complex power, 798
  driving point, 542
  equivalent, 384, 491

Impedance (*cont.*)
  input, 543
  magnitude scaling, 573
  parallel connection, 387
  series connection, 383, 491
  transfer, 544
Impulse, 228
Impulse response, 554
  from transfer function, 554
  from step response, 557
Inductance, 273
  mutual, 763
  self, 763
Inductive reactance, 384
Inductor, 272
  average power, 414
  energy, 274
  impedance, 380, 486
  i–v relationship, 273
  parallel connection, 287
  series connection, 286
  state variable, 275
Initial-value theorem, 468
Input, 97
Input impedance, 543
Instantaneous power, 794
Instrumentation amplifier, 214
Instrumentation systems, 197
Integrator, 279
Interface, 4
Interface circuit, 121
Inverting amplifier, 183

Joule (unit), 5

Kirchhoff's laws, 22
  current (KCL), 23
  phasor domain, 378
  s domain, 482
  voltage (KVL), 25

Ladder circuit, 48
Lagging power factor, 798
Laplace transformation, 433
  differentiation property, 438
  final-value property, 468
  initial-value property, 468
  integration property, 437
  inverse, 435
  linearity property, 435

  table of pairs, 442
  table of properties, 441
  translation properties, 439, 440
  uniqueness property, 435
Leading power factor, 798
Linear circuit, 97
Linear element, 18
Line current, 814
Line spectrum, 679
Line voltage, 811
Load flow, 800, 803
  three phase, 819
Loading, 551
Lookback resistance, 113
Loop, 23
Low-pass filter, 593
  design, 722
  first order, 597
  ideal, 722
  second order, 715

Magnitude scaling, 573
MATLAB, A-10
Matrix, A-5
  and linear equations, A-8
  inverse, A-7
Matrix algebra, A-6
Maximum signal transfer, 118
  current, 118
  power, 119, 416
  voltage, 118
Mesh, 88
Mesh current, 88
Mesh-current analysis, 88
  fundamental property, 89
  phasor domain, 404
  s domain, 519
  summary of, 97
  with current sources, 93
  with dependent sources, 158
Minor, A-4
Mutual inductance, 763

Natural frequency, 338
Natural pole, 512
Natural response, 309
Neper frequency, 338
Network function, 518, 539
  design of, 566
  determining, 544

Network function (*cont.*)
  driving-point impedance, 542
  transfer functions, 543
Node, 23
Node voltage, 74
Node-voltage analysis, 74
  fundamental property, 74
  phasor domain, 403
  s domain, 506
  summary of, 87
  with dependent sources, 154
  with OP AMPs, 187
  with voltage sources, 82
Noninverting amplifier, 172
Nonlinear element, 18, 115
Norton equivalent circuit, 105
  phasor domain, 399
  s domain, 501

Octave, 595
Odd symmetry, 682
Ohm (unit), 18
Ohm's law, 18
OP AMP, 169
  dependent source model, 171
  effect of finite gain, 174
  gain-bandwidth product, 602
  ideal model, 171
  in the *s* domain, 484
  notation, 170
  operating modes, 170
OP AMP circuits:
  bandpass filter, 720
  differential amplifier, 183
  differentiator, 281
  high-pass filter, 607, 717
  instrumentation amplifier, 214
  integrator, 279
  inverting amplifier, 178
  low-pass filter, 601, 715
  noninverting amplifier, 172
  noninverting summer, 193
  subtractor, 183
  summary of, 280
  summing amplifier, 180
  voltage follower, 177
Open circuit, 19
Open-circuit voltage, 107
Operational amplifier.
    see OP AMP

Output, 97
Overdamped response, 339
Overshoot, 559

Parallel connection, 28
  admittances, 388, 492
  capacitors, 285
  current division in, 45, 388, 493
  inductors, 287
  resistors, 36
Partial fraction expansion, 448
  of improper rational functions, 453
  with complex poles, 450
  with multiple poles, 455
Passband, 593
Passband gain, 723
Passive sign convention, 9
Peak to peak value, 249
Peak value, 249
Periodic waveform, 240, 248
  rms value of, 251, 690
Phase angle, 236
Phase current, 817
Phase function, 592
Phase sequence, 812
Phase spectrum, 679
Phase voltage, 811
Phasor, 371
  additive property, 373
  derivative property, 373
  diagram, 372
  domain, 382
  rotating, 373
Phasor circuit analysis, 377
  current division, 388
  mesh current, 404
  node voltage, 403
  parallel equivalence, 387
  proportionality, 393
  series equivalence, 383
  superposition, 395
  Thévenin/Norton, 399
  transformer, 775
  voltage division, 384
Planar circuit, 88
Pole, 444
  Butterworth, 730
  Chebychev, 734
  first-order cascade, 725
  forced, 512

Pole (*cont.*)
  multiple, 455
  observable, 526
  natural, 512
  number of, 527
  simple, 448
Pole-zero diagram, 444
Port, 121
Potentiometer, 44
Power, 6
  apparent, 797
  available, 119, 417
  average, 414, 794
  complex, 796
  maximum transfer, 119, 417
  reactive, 794
Power factor, 797
Power factor angle, 797
Power factor correction, 804
Primary winding, 769
Proportionality, 98, 393, 495

Quality factor (Q), 628

Radian frequency, 238
Ramp, 229
Rational function, 448
RC and RL circuits, 299
Reactance, 384
Reactive power, 794
Residue, 448
  sum of, 452
Resistance, 18, 384
  lookback, 113
  standard values, A-1
Resistor, 17
  average power, 415
  impedance, 380, 486
  i–v relationship, 18
  power, 18
  tolerance, 201
Resonance, 391, 615
Resonant frequency, 391
Response:
  ac response, 326, 371
  dc steady state, 557
  critically damped, 339
  forced, 309, 512
  frequency, 592
  impulse, 554

natural, 309, 512
overdamped, 339
sinusoidal steady state, 326, 371, 561
steady state, 326
step, 309, 345, 556
underdamped, 339
with periodic inputs, 684
zero input, 301, 497
zero state, 315, 497
Rise time, 263, 559
RLC circuits, 330
  bandpass, 628
  bandstop, 632
  critically damped, 347
  overdamped, 347
  step response, 345
  underdamped, 347
  zero-input response, 332, 341
Root-mean-square (rms) value, 251
Rotating phasor, 373

s domain, 432
s plane, 445
Secondary winding, 769
Second-order circuits:
  bandpass, 616, 720
  frequency response, 613
  high-pass, 625, 717
  low-pass, 621, 715
  parallel RLC, 340
  series RLC, 330
  step response, 345
  zero-input response, 332
Self inductance, 763
Series connection, 28
  capacitors, 286
  impedances, 385, 491
  inductors, 286
  resistors, 35
  voltage division in, 42 , 384, 492
Settling time, 263, 654
Short circuit, 19
Short-circuit current, 107
Siemens (unit), 18
Signal, 4
  conditioning, 197
  dc, 225
  composite, 243
  exponential, 231
  impulse, 228

Signal (*cont.*)
    ramp, 229
    sinusoidal, 235
    step function, 227
    test, 542
Signum function, 243
Single phase, 809
Sinusoidal steady state, 326, 371, 571
Sinusoidal waveform, 235
    damped, 245
    period, 236
    properties, 240
    phasor representation, 372
Source:
    current, 21
    dependent, 149
    equivalent, 38
    Norton equivalent, 106, 399, 501
    practical, 22
    Thévenin equivalent, 106, 399, 501
    three phase, 810
    voltage, 20
Source transformation, 39
    phasor domain, 399
    s domain, 501
Spectrum, 672, 679
Stable, 526, 555
State variable, 268, 275, 300, 301
Steady state, 326
Step function, 227
Step response:
    first-order circuit, 309
    from impulse response, 557
    from transfer function, 556
    second-order circuit, 345
Stopband, 593
Supermesh, 94
Supernode, 83
Superposition, 102, 395, 496
Susceptance, 387
Switch, 19
    analog, 20
    ideal, 19

Test signals, 542
Thévenin equivalent circuit, 105
    phasor domain, 399
    s domain, 501
    with dependent sources, 163
    with nonlinear loads, 115

Three phase power analysis, 814
    load flow, 819
    single-line diagram, 820
    Y-Y connection, 814
    Y-Δ connection, 816
Time constant, 231
    RC circuit, 302
    RL circuit, 303
Tolerance, 201
    worst-case gain, 202
Transmission bandwidth, 693
Transconductance, 150
Transducer, 197
Transfer function, 543
    current divider, 544
    inverting amplifier, 545
    noninverting amplifier, 545
    voltage divider, 544
Transformer, 769
    ac analysis, 775
    autotransformer, 774
    equivalent circuits, 780
    ideal, 769
    step-down, 770
    step-up, 770
    turns ratio, 770
Transform pair, 432
Transistor, 164
Transresistance, 150
Turns ratio, 770

Undamped natural frequency, 346
Underdamped response, 329
Unit output method, 99, 394
Unit impulse, 228
Unit ramp, 229
Unit step function, 227
Units, table of, 5

VA (unit), 797
VARs (unit), 796
Volt (unit), 6
Voltage, 5
    line, 811
    node, 74
    open circuit, 107
    phase, 811
Voltage division, 42, 384, 492
Voltage follower, 177
Voltage gain, 150

Voltage source, 20
    current controlled, 150
    Thévenin equivalent, 106, 399, 501
    voltage controlled, 150
Voltage transfer function, 544

Watt (unit), 6
Waveform, 225

Waveform symmetries, 681
Weber (unit), 272
Wheatstone bridge, 199
Worst case analysis, 201

Zero, 444
Zero-input response, 301, 497
Zero-state response, 315, 497